Neglected Tropical Diseases

This book series publishes contributions from individuals engaged in any of the fields related to neglected tropical diseases clearly connected with their exceptional status as neglected. Specific focus of each title lies on the discussion of relevant issues in particular geographic areas of the world. Each volume of the series deals with the unique situation of one region and is built up by a team of authors lead by an expert for the geographic area discussed. This series provides a forum for wealthy discussion on the topic of neglected infectious diseases with a clear focus on basic scientific topics, clinically relevant issues as well as policy issues associated with the area. Topics to be covered: detection, diagnosis, monitoring, vaccine and drug development, new treatments, translational research (link basic research and health system research), clinical aspects, epidemiology, development of new surveillance and control strategies, public health/health policy issues.

John O. Gyapong • Boakye A. Boatin
Editors

Neglected Tropical Diseases - Sub-Saharan Africa

Second Edition

Springer

Editors
John O. Gyapong
Centre for Neglected Diseases Research
University of Health and Allied Sciences
Ho, Ghana

Boakye A. Boatin
Lymphatic Filariasis Support Centre
University of Ghana
Accra, Ghana

ISSN 2194-8275 ISSN 2194-8283 (electronic)
Neglected Tropical Diseases
ISBN 978-3-031-53903-9 ISBN 978-3-031-53901-5 (eBook)
https://doi.org/10.1007/978-3-031-53901-5

This Springer imprint is published by the registered company Springer Nature Switzerland AG
The registered company address is: Gewerbestrasse 11, 6330 Cham, Switzerland

This book on Neglected Tropical Disease in Sub-Saharan Africa is dedicated to the memory of:
Dr Likezo Mubila for her dedication to the control of NTDs
Dr Dominque Kyelem for his steadfast drive to research in lymphatic filariasis
Dr Ricardo Thompson for his leadership and commitment as a research scientist in NTDs.
Dr Mwele Malecela for her vision and leadership in the efforts toward the research, control, elimination, and eradication of the neglected tropical diseases

Foreword to the Second Edition

I congratulate Professors Gyapong and Boatin on their second edition of *Neglected Tropical Diseases—Sub-Saharan Africa*. Africa remains the epicenter of the world's neglected tropical diseases (NTDs), generally considered the most common afflictions of people who live in extreme poverty. In parallel, there also remains no shortage of poverty on the African continent. According to the World Bank, Africa's population accounts for roughly two-thirds of our global citizens living in destitution. Poverty and disease go hand in glove, especially for the NTDs—the NTDs overwhelmingly thrive among the poor, and in turn the NTDs promote poverty because of their long-term or chronic deleterious effects including disfigurement, blindness, growth failures, cognitive delays, and adverse reproductive outcomes.

In specific regard to the NTDs, what does it mean that approximately 460 million Africans live at or below World Bank extreme poverty levels? From my perspective, this translates to a stark reality that probably all these individuals likely suffer from at least one NTD, including hookworm infection or other soil-transmitted helminthiases, schistosomiasis, lymphatic filariasis, onchocerciasis, scabies, yaws, and many others. Thus, while great progress has been made in terms of mass drug administration and preventive treatments to reduce the prevalence or transmission of Africa's major NTDs, these conditions remain highly prevalent. But poverty is not the sole driver—Africa is undergoing sweeping changes due to other social and physical determinants. Among the former is progressive urbanization with more and more people living in large African cities, which are gradually becoming megacities of more than 10 million people. We are also starting to see shifts with some soil-transmitted helminth infections and schistosomiasis now occurring in urban and peri-urban locations. Political instability is also a driver especially in some of Africa's fragile nation states where it is difficult to implement control and elimination programs for the NTDs. And then, climate change and its associated higher temperatures and higher rainfall patterns appear to be shifting the geographic ranges of several NTDs including schistosomiasis and hookworm infection, among others.

Therefore, it is important to recognize that the NTDs are not static and as Africa's population begins coalescing into megacities such as Kinsasha, Lagos, or Dar, we should expect shifts in the epidemiology and biodiversity of their NTD pathogens. Solving Africa's NTD problem will require a multi-pronged approach including social scientists to design culturally sensitive interventions. We must expand mass treatments, especially for diseases such as schistosomiasis, a disease for which only about 10–15% of children have access to praziquantel, together with case detection and treatment for NTDs such as HAT and leishmaniasis. Mass treatment needs to be accompanied by intersectoral approaches that embrace sanitation, clean water, and other environmental measures. In parallel, vector control will be needed for leishmaniasis, HAT, and other NTDs. The good news is that through these approaches we are beginning to see declines in the number of cases of lymphatic filariasis, onchocerciasis, and HAT, so it is now possible to realistically discuss the potential for eliminating these NTDs. For others, we need to enlarge programs of research and development, which must include capacity enhancement for African scientists.

I am excited that Professors Gyapong and Boatin have embarked on a new edition. Regular updates will be needed to keep up with these dramatic Anthropocene shifts. The fight to combat NTDs on the African continent continues.

Houston, TX, USA Peter Hotez

Preface

The second edition of this book has been necessitated by some advances in the field of neglected tropical diseases (NTDs) since 2016 when the first edition was published. Inasmuch as NTDs occur in neglected populations across the world, we focus on the major challenges of sub-Saharan Africa (SSA). In the first edition, 12 of the major NTDs in Africa were presented. They included Buruli ulcer, Guinea worm, human African trypanosomiasis, leishmaniasis, leprosy, Loa loa and lymphatic filariasis, onchocerciasis, podoconiosis, schistosomiasis, soil-transmitted helminths, trachoma, and yaws. In this edition, all these have been revised at varying detail based on new knowledge that has become available in the last 5–6 years. Leishmaniasis is presented as two separate sections within the chapter. Scabies has been added. Additionally, we have dedicated a chapter to the elimination of onchocerciasis in view of the new knowledge and dynamics supporting a paradigm change from control to elimination. The areas which transcend all the NTDs in sub-Saharan Africa such as health systems and their role in NTDs, the social and economic impact, and the role of vector control in NTDs have also been revised and best read together with the specific diseases. The issue of female genital schistosomiasis (FGS) is given prominence in the chapter on "Social and economic impact of NTDs," and a new chapter on "Drug donations" has been included.

The structure and the common theme for all the chapters were retained. However, important variations on the theme given new available knowledge are evident in the sections within the chapters. The new chapter on "Drug donations" provides additional detailed information and cross-references to the donated medicines that are used in the management of virtually all the diseases presented in the book. Discussion on diagnostics—new and or improved—especially for those diseases that are targeted for elimination is a common thread in the chapters. Future control tools including drugs as well as critical research needed to help overcome the challenges that have been identified for each disease are highlighted again.

Cross-cutting challenges that continue to persist for the control/elimination of the diseases have been reemphasized to maintain the need to try and overcome them. Additionally, new specific challenges, for example, the presence of Guinea worm in dogs just as one approached the "end game" in its eradication, are presented. With these in mind, the section on the expected situation for each disease in the next decade where appropriate is highlighted for each of the chapters.

We have been fortunate to be able to maintain most of the original authors—almost entirely all African research scientists—to undertake the revision of their chapters. Several new authors have also been brought in. As before, they also have experience in the ministries of health in the control of the diseases, have research experience in tropical medicine, and have had practical experiences in managing NTDs in their countries. It is regretted that some of the authors of the first edition were unable to take part in preparing this second edition, but we are grateful for their support and advice.

The revision for this edition of the book took far longer than one expected, the challenge being mostly in obtaining fresh and current data as well as relevant information through time. Despite this, the authors stayed the course as a team and in close collaboration with each other. Their contributions, as well as that of the many people who are not listed in the book but helped in diverse ways, are greatly acknowledged.

The hope remains the same: that this second edition of the book will provide some basic information on the specific NTDs and the special ways that NTDs present in SSA and will be a useful resource read for all who are interested in doing something about NTDs in SSA. It is also our wish that the wide range of references on NTDs will be useful to readers.

Ho, Ghana John O. Gyapong
Accra, Ghana Boakye A. Boatin

Acknowledgments

We are grateful to Mrs Grace Martey, Office of Research, Innovation and Development of the University of Ghana, our copy editor for her ferocious eye for detail to every chapter and for her enormous help and suggestions.

We acknowledge our colleagues who were part of the first edition but for various reasons were unable to participate in the second edition. They include Dr Dieudonne P. Sankara, Dr Andrew S. Kokor, Dr Junerlyn Agua-Agum, and Dr Daniel Argaw. We thank them for their original concept and contribution.

The first edition of this work was partly supported by a grant (Grant Number: G1001337) from the UK Medical Research Council (MRC) and the UK Department for International Development (DFID) under the MRC/DFID Concordat agreement, through the Liverpool School of Tropical Medicine Council and managed by the School of Public Health, University of Ghana.

This second edition was done while Professor John Gyapong was on sabbatical leave at the Global Health Program of the Graduate School of Arts and Sciences, Georgetown University, Washington DC, USA.

About the Book

This book provides an overview of the major neglected tropical diseases (NTDs) occurring in sub-Saharan Africa, such as leishmaniasis, Buruli ulcer, and schistosomiasis. In well-structured chapters, the epidemiology and biology of these parasitic diseases will be discussed in detail. Further, diagnostics and therapeutic approaches as well as prevention strategies will be reviewed. The book will be of interest to basic researchers and clinicians engaged in infectious disease, tropical medicine, and parasitology and a must-have for scientists specialized in the characteristics of the sub-Saharan region.

Contents

About the Editors

John O. Gyapong is Public Health Physician and Epidemiologist. He studied Medicine at the Kwame Nkrumah University of Science and Technology, Kumasi, Ghana, and later studied Public Health (MSc) and Epidemiology (PhD) at the London School of Hygiene and Tropical Medicine of the University of London. His main area of research is infectious diseases epidemiology, especially lymphatic filariasis and other neglected tropical diseases and malaria. He established and managed the Ghana Filariasis Elimination Programme for eight (8) years where he was also in charge of the Onchocerciasis Control. For twelve (12) years, he was Director for Research and Development of the Ghana Health Service where he was responsible for health systems and implementation research.

He served as Vice-Chancellor of the University of Health and Allied Sciences, Ho, from 2016 to 2022. From 2011 to 2016, Professor Gyapong was the Pro-Vice-Chancellor of the University of Ghana responsible for Research, Innovation and Development. He was the Vice-Dean and Professor of Epidemiology and Disease Control at the School of Public Health of the University of Ghana and Adjunct Professor of International Health at the Georgetown University in Washington DC, USA. He serves on several international research review committees and boards and has over 150 publications in peer-reviewed journals.

Boakye A. Boatin is Public Health Physician and Epidemiologist with degrees in medicine, international public health, and epidemiology. He studied medicine in Ghana and undertook postgraduate studies in the Liverpool School of Tropical Medicine (Community Health), University of Liverpool, and epidemiology at the London School of Tropical Medicine and Hygiene (LSTMH) of the University of London. His research focuses on infectious diseases epidemiology, particularly onchocerciasis, schistosomiasis, and human African trypanosomiasis. Until his retirement from the World Health Organization (WHO), he led research in integrated community-based interventions and lymphatic filariasis at the Special Programme for Tropical Diseases (TDR) and had a short stint as the head of the Aids Medicines and Diagnostics (AMDS) Unit at the World Health Organization.

He worked at the Onchocerciasis Control Programme in West Africa (WHO) for 15 years, first as Chief Epidemiologist and then as Head of Planning Evaluation and Transfer and later as Director of the Programme until its closure. He is currently Adjunct Professor in Tropical Health at the Institute of Parasitology, McGill University, Canada, and Senior Scientific Advisor to the Lymphatic Filariasis Support Center for Africa, Ghana. He served for a long period on the Mectizan Expert Committee and several review committees. He has authored chapters on onchocerciasis in several books, has many publications in peer review journals, and has co-edited a book on *Neglected Tropical Diseases in Sub-Saharan Africa*.

An Overview of Neglected Tropical Diseases in Sub-Saharan Africa

John O. Gyapong

Contents

Abstract

Neglected tropical diseases (NTDs) affect many neglected and marginalized populations worldwide, but the burden in sub-Saharan Africa is rather overwhelming. Many of the endemic communities are of very low socioeconomic status with very limited access to health services. Investing to overcome the global impact of NTDs will yield a very high economic rate of return and impact significantly on the quality of life of these populations. To scale up interventions to achieve control, elimination, or eradication of NTDs, programs must be integrated into the regular health system of endemic countries. Efforts to expand

J. O. Gyapong (✉)
Centre for Neglected Tropical Diseases, Institute of Health Research, University of Health and Allied Sciences, Ho, Ghana
e-mail: jgyapong@uhas.edu.gh

J. O. Gyapong, B. A. Boatin (eds.), *Neglected Tropical Diseases - Sub-Saharan Africa*, Neglected Tropical Diseases, https://doi.org/10.1007/978-3-031-53901-5_1

global coverage and targeting of NTDs must therefore involve national and international harmonization with coordination of the activities of partnerships devoted to control of these diseases. The continued support of major donors beyond the initial commitments announced during the London Declaration meetings remains crucial to funding the implementation of programs. The positioning of NTDs within the Universal Health Coverage (UHC) agenda and the Strategic Development Goals (SDGs) could galvanize stakeholder support for these NTDs. Ultimately, we need to address the social structures in which NTDs flourish and invest in research and development for new diagnostics and drugs.

Introduction

Neglected tropical diseases (NTDs) are a diverse group of diseases that prevail in tropical and subtropical conditions in developing countries. They affect more than one billion people and cost developing economies billions of dollars every year. Populations living in poverty, with inadequate sanitation, and in close contact with infectious vectors and domestic animals and livestock are those worst affected. Thus, they thrive mainly among the poorest and deprived populations. Most NTDs are found primarily in low- and middle-income countries of Africa, Asia, and Latin America. Within these countries, the affected populations are in themselves very neglected in many ways and are usually in the lowest socioeconomic status. Populations where people have little access to clean water, have improper ways of disposing of human waste, and therefore live in unsanitary environments tend to have a high burden of NTDs (Hotez et al. 2006; WHO 2010, 2013).

The World Health Organization has prioritized 20 of these NTDs in 149 endemic countries for focused global attention (Table 1). This list is by no means exhaustive since there are some other diseases in neglected populations that are not on this list; however, this list represents the biggest disease burden they face (WHO 2020).

NTDs cost developing communities billions of dollars each year in direct health costs, loss of productivity, and reduced socioeconomic and educational attainment

Table 1 The 20 WHO prioritized neglected tropical diseases for action

Helminth	*Protozoa*
Cysticercosis/taeniasis	Chagas disease
Dracunculiasis (Guinea-worm disease)	Human African trypanosomiasis (sleeping sickness)
Echinococcosis	Leishmaniases
Foodborne trematodiases	*Bacteria*
Lymphatic filariasis	Buruli ulcer
Onchocerciasis (river blindness)	Leprosy (Hansen disease)
Schistosomiasis	Trachoma
Soil-transmitted helminthiases	Yaws
Virus	*Others*
Dengue and chikungunya	Mycetoma
Rabies	Scabies
	Snakebite envenoming

(WHO 2012a). The burden of these diseases is extremely high in sub-Saharan Africa (SSA). For example, approximately 40% of the global burden of lymphatic filariasis (LF) is found in SSA while all the remaining cases of guinea worm disease (GWD) are also found in the same region (WHO 2010, 2020). In addition, these diseases place considerable financial constrain on patients and their families. Human African trypanosomiasis in the Democratic Republic of the Congo costs affected households in a typical rural community more than 40% of their annual household income (Lutumba et al. 2007), and up to 75% of households affected by visceral leishmaniasis in Bangladesh (Anoopa et al. 2006, Ozaki et al. 2011), India (Sundar. 2010), Nepal (Uranw et al. 2013), and Sudan (Meheus et al. 2013) experience some financial catastrophe in obtaining diagnosis and treatment, even when tests and medicines are free of charge.

The German government under GTZ (now GIZ) played an important role in the NTD movement by co-sponsoring with the World Health Organization (WHO) two key meetings of leading stakeholders in 2003 and 2005 in Berlin. These meetings achieved two important outcomes: (1) unified support for an integrated approach in addressing NTD control and elimination efforts, and (2) the brand "neglected tropical diseases" was coined and has since become part of the global health nomenclature (WHO 2004, 2006). This progress reflects the continuity of initiatives and long-standing support and dedication of the global NTD community, from the first meeting of NTD global partners convened by WHO in 2007 to bring together various disease initiatives under the umbrella of the NTD "brand," to the pledges made in the 2012 London Declaration on Neglected Tropical Diseases and the 2017 meeting of global partners. They demonstrate the immense potential that can be unlocked by working in partnership to ensure that NTDs have a prominent position on the global health agenda (WHO 2020).

Prior to the Berlin meetings, several global directives in the form of World Health Assembly resolutions had been passed to mobilize political and social capital to address these diseases individually and many more have been passed since then (Table 2; WHO 2015).

These resolutions were all very comprehensive; for example, the 50th World Health Assembly held in Geneva in May 1997 called on Member States to take advantage of recent advances in the understanding of lymphatic filariasis and the new opportunities for its elimination by developing national plans leading to its elimination, as well as for the monitoring and evaluation of program activities; to strengthen local programs and their integration with the control of other diseases, particularly at the community level, in order to implement simple, affordable, acceptable, and sustainable activities based on community-wide treatment strategies, but supplemented where feasible by vector control and improved sanitation; to strengthen capabilities for training, research, laboratory diagnosis disease management, and data management in order to improve clinical, epidemiological, and operational activities directed toward eliminating lymphatic filariasis as a public health problem; and to mobilize support of all relevant sectors, affected communities, and nongovernmental organizations for the elimination of the disease (WHA50.29).

Table 2 Selected resolutions of the World Health Assembly concerning neglected tropical diseases

Subject area	Resolution	Title	Year
Neglected tropical diseases	WHA66.12	Neglected tropical diseases	2013
Schistosomiasis	WHA65.21	Elimination of schistosomiasis	2012
Chagas disease	WHA63.20	Chagas disease: Control and elimination	2010
Leishmaniases	WHA60.13	Control of leishmaniasis	2007
Buruli ulcer	WHA57.1	Surveillance and control of *Mycobacterium ulcerans* disease (Buruli ulcer)	2004
Dracunculiasis	WHA57.9	Eradication of dracunculiasis	2004
Human African trypanosomiasis	WHA56.7	Pan African tsetse and trypanosomiasis eradication campaign	2003
Dengue and dengue haemorrhagic fever	WHA55.17	Prevention and control of dengue fever and dengue haemorrhagic feve	2002
Schistosomiasis and soil-transmitted helminthiases	WHA54.19	Schistosomiasis and soil-transmitted helminth infections	2001
Trachoma	WHA51.11	Global elimination of blinding trachoma	1998
Chagas disease	WHA51.14	Elimination of transmission of chagas disease	1998
Leprosy	WHA51.15	Elimination of leprosy as a public health problem	1998
Lymphatic filariasis	WHA50.29	Elimination of lymphatic filariasis as a public health problem	1997
Human African trypanosomiasis	WHA50.36	African trypanosomiasis	1997
Onchocerciasis	WHA47.32	Onchocerciasis control through ivermectin distribution	1994
Dengue and dengue haemorrhagic fever	WHA46.31	Dengue prevention and control	1993
Endemic treponematoses	WHA31.58	Control of endemic treponematoses	1978
Leprosy	WHA30.36	Leprosy control	1977

The Assembly also invited other specialized agencies of the United Nations system, bilateral development agencies, nongovernmental organizations, and other groups concerned to increase cooperation in the elimination of lymphatic filariasis through the support of national and international programs relevant to the prevention and elimination of lymphatic filariasis. Finally, they urged the Director-General of the WHO to inform various specialized agencies, United Nations organizations, bilateral development agencies, nongovernmental organizations, and other concerned groups about the imperative for enhanced cooperation in eradicating lymphatic filariasis as a public health problem. Their aim is to garner support for both global and national elimination activities (WHA50.29).

This is probably the most comprehensive global commitment one could get for a disease elimination program. However, by the year 2005, only 8 out of the 38 endemic countries in Africa had active lymphatic filariasis elimination programs, and of these, only Burkina Faso, Ghana, Togo, and Zanzibar were treating their entire national populations at risk. Clearly, targeting individual diseases was not the most efficient way of dealing with the huge burden of NTDs; hence in 2013, a more encompassing resolution WHA66.12 called for monitoring progress in achieving the targets for NTDs set in WHO's roadmap for accelerating work to overcome the global impact of NTDs and intensified, integrated measures and planned investments to improve the health and social well-being of the affected populations (WHA66.12).

In the last decade, there has been significant progress in fighting these diseases, with new interventions and tools, increased donor support, and increased country commitment. The establishment of public–private partnerships has vastly facilitated progress toward the elimination and control of NTDs: pharmaceutical companies have donated nearly three billion tablets of safe, quality-assured medicines annually to support the control and elimination of NTDs in countries where they are endemic (WHO 2020).

More recent data show that the burden of disease is going down though not significantly in Africa as a result of many interventions, strategies, and buy-in from many stakeholders. In 2021, 1.65 billion people were reported to require mass or individual treatment and care for NTDs, down from 2.19 billion in 2010, a reduction of 25%; however, over the same period the reduction in Africa was only 5% (Fig. 1; WER 2021; WHO 2023). The majority of these individuals continued to require mass treatment for diseases amenable to preventive chemotherapy, while the numbers requiring other NTD services such as individual disease management or care were much smaller, approximating a few million.

Global progress in reducing the number of people requiring interventions against NTDs, which is one of the road map's overarching indicators, as well as Sustainable Development Goal indicator 3.3.5, was driven by three main factors: first, a number of countries have eliminated at least one NTD; second, although not yet certified, verified, or validated, several countries have reached the post-intervention surveillance phase for at least one NTD; third, remapping for a number of NTDs has led to the reclassification from endemic to nonendemic status of several geographical areas within a few countries. Improved livelihoods and living conditions may have also contributed, although in uneven and varying ways, as several countries endemic for NTDs are affected by social instability entailing destruction and population displacement (Bangert et al. 2017; WHO 2020, 2023).

In spite of the above, at the end of 2022, WHO's South-East Asia Region still has the largest number of people requiring interventions against NTDs at 857 million (51.8% of the global total), followed by the African Region (584 million, 35.3%). The remaining four regions account for 212 million people requiring interventions, or 12.9% of the global total.

There are many factors that have contributed to the belief that something can really be done about these diseases that have hitherto been perceived to have no remedies. As a result, these World Health Assembly resolutions have managed to draw attention by mobilizing resources and political capital to support these initiatives. These factors include, among others:

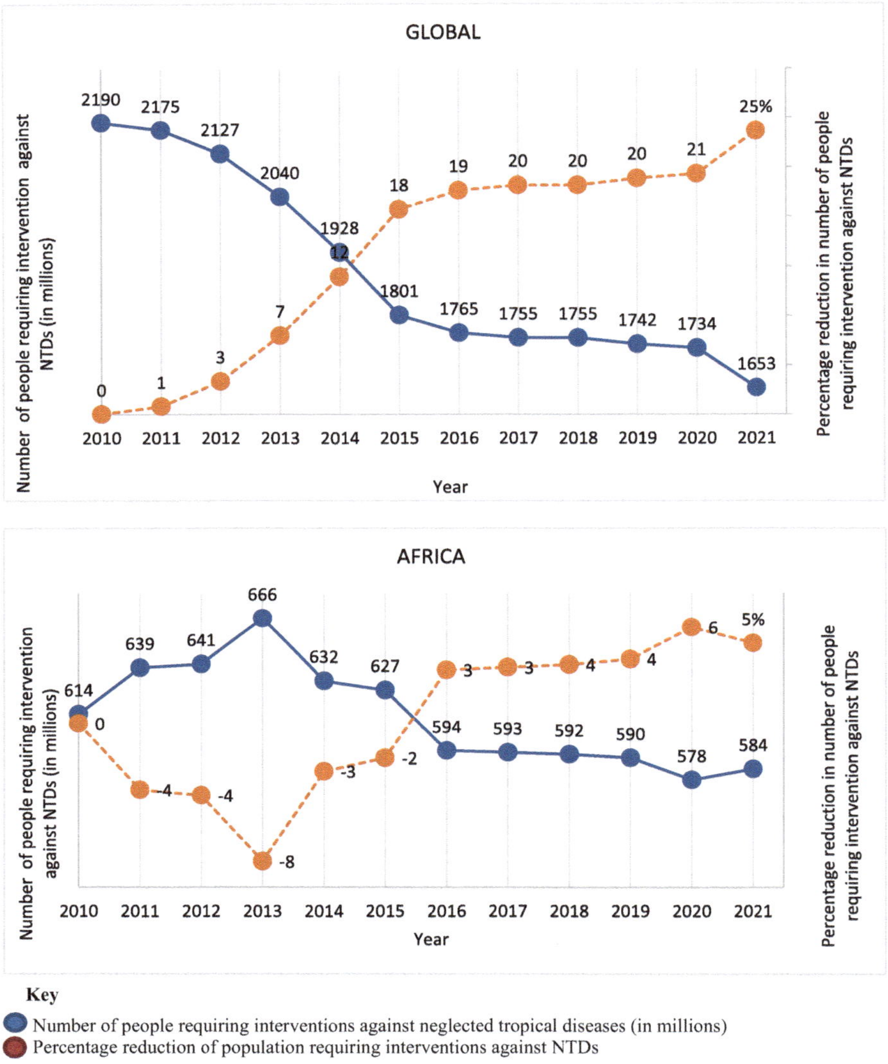

Fig. 1 Number of people requiring interventions against NTDs and associated percentage reduction globally and regionally, 2010–2021 (Adapted from WHO 2023)

- Better understanding of the diseases.
- Improved delivery mechanisms including opportunities for integration.
- Investment case including cost–benefit evaluation, funding, and economic considerations.
- Moral, political, and social determinants.
- Drug donations.

Better Understanding the Diseases

Research over many years has played a critical role in providing the necessary information to understand these diseases and the subsequent development of interventions and program creation. As a result, we have a better understanding of the

- Epidemiology of these diseases.
- Transmission dynamics and vector biology.
- Socioeconomic factors.

These have also led to

- Development of drugs and diagnostics.
- Testing of new interventions.
- Operational delivery of interventions.
- Development of monitoring and evaluation tools to assess the impact of these interventions.
- Development of better surveillance tools.

Thus, knowledge generated in recent years has demonstrated aptly that the biological potential of controlling, eliminating, and even eradicating these NTDs do exist, and ample diagnostic tools and strategies are available to assess the burden and distribution of these diseases and that it is possible to create and execute control programs effectively if we can mobilize the necessary economic, social, and political capital (Fig. 2).

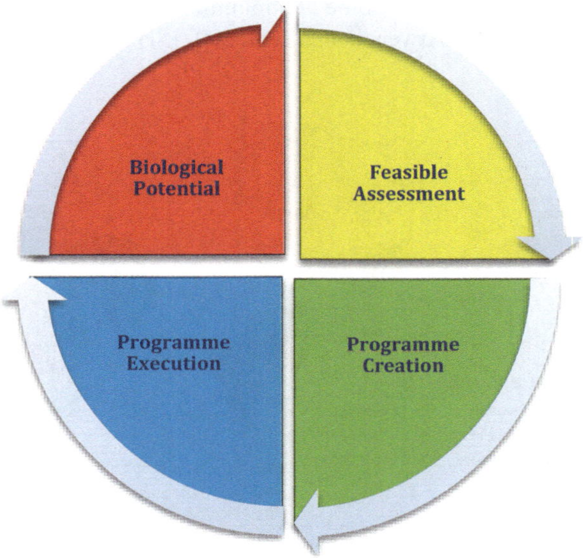

Fig. 2 Control program planning process

Improved Delivery Mechanisms and Opportunities for Integration

Various mapping studies and other available anecdotal information suggest that most countries in SSA have more than three of these diseases (Fig. 3). Strategies, tools, and interventions available for combating these diseases can be used for more than one disease. Therefore, the integration and co-implementation of these strategies, tools, and interventions are key to the control/elimination of the diseases (Gyapong et al. 2010; Molyneux et al. 2005).

Figure 4 illustrates how improved water and sanitation, drugs (praziquantel, albendazole, and ivermectin), and vector control could be delivered as an intervention package to fight several NTDs and malaria. As a result, the World Health Organization (WHO) recommends two main strategies for NTD control: Preventive Chemotherapy and Transmission Control (PCT) and Innovative Intensified Disease Management (IDM) (WHO 2010, 2015).

PCT focuses on diseases for which a strategy exists as well as on tools and the availability of safe and effective drugs that make it feasible to implement large-scale preventive chemotherapy. The diseases amenable to the PCT strategy include cysticercosis, dracunculiasis (guinea-worm disease), foodborne trematode infections, lymphatic filariasis, onchocerciasis, schistosomiasis, and soil-transmitted helminthiasis. Blinding trachoma control through the SAFE strategy—combining drug treatment with hygiene and environmental management—can be linked to helminth control interventions to improve the overall health of affected communities (WHO 2010, 2015).

One of the issues of concern with the PCT strategy is the potential for drug resistance to appear with long-term use of these medicines in the communities. Even though some cases of non-response have been reported for ivermectin treatment for onchocerciasis in some communities in Ghana, there is no evidence of drug resistance (Osei-Atweneboana et al. 2007). There are some who think it is only a matter

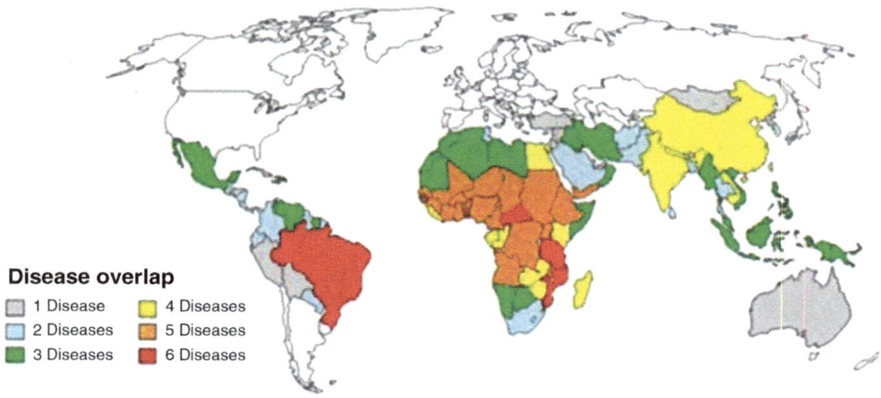

Fig. 3 Extent of disease overlap

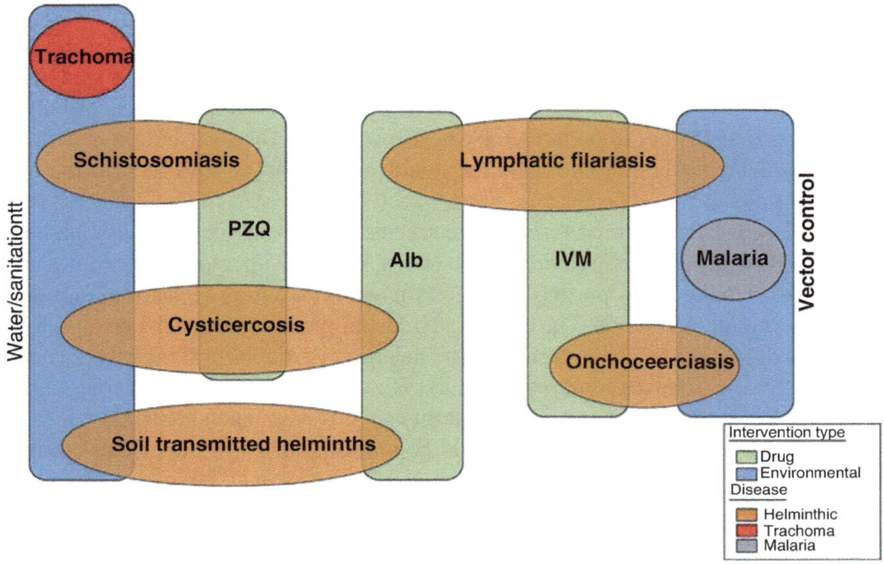

Fig. 4 Some strategies, tools, and interventions for NTD control

of time for drug resistance to occur, as has been reported with many parasites and chemotherapeutic agents. Research for new drug discoveries, repurposing of drugs, and multiple drug therapy is therefore encouraged (King et al. 2020; Jacobson 2022).

IDM focuses on diseases for which cost-effective control tools do not exist and where large-scale use of existing tools is limited. The diseases include Buruli ulcer, Chagas disease, human African trypanosomiasis, and leishmaniasis and share the following characteristics:

- Difficult and costly to manage in terms of diagnosis, treatment, and follow-up.
- Burden is poorly understood.
- Lack of appropriate control tools.
- Relatively lower investment in research and development.
- People affected often live in remote rural areas with limited access to diagnosis and treatment (WHO 2010, 2015).

Having prioritized NTDs for control, having mapped their distribution and armed with interventions that work, the biggest challenge has been to deliver these interventions through a health system in the midst of severe human resource constraints and other health system challenges (Gyapong et al. 2010). The health worker–population ratio is extremely high in Africa with some areas not served at all. To improve access, there is the need to engage other cadres of staff for the delivery of these interventions and explore other delivery mechanisms such as school-based and community-based distribution. The community-directed treatment approach, which provides opportunities for health services to work closely with the community to deliver interventions, has been shown to be highly feasible (Amazigo et al. 2007; Gyapong et al. 2000; WHO 2008). These studies have found community volunteers to be capable, motivated, and reliable; however, they need to be provided with incentives. With this approach, the community decides the timing of the distribution and selects distributors to be trained by health workers. The distribution is done at the convenience of community, and the health worker helps with monitoring and supervision.

Such large-scale community-based treatments could be associated with inadvertent exposure of some populations even when standard operating procedures are adhered to. This requires putting in place efficient community education, monitoring, and evaluation systems. The challenge of dealing with serious adverse events in any mass drug distribution exercise can be daunting particularly if an adverse reaction like Steven Johnson syndrome occurs. How strong is our pharmacovigilance infrastructure to pick up these occasional mishaps? (Gyapong et al. 2003).

Integrated approaches to NTDs should be mainstreamed within various components of national health systems; for example, planning should be incorporated into overall national health planning and budgeting, data management should be included in health management information systems at all levels, and delivery of medicines should be coordinated through national medicines supply and logistics systems. Diligent monitoring for the safe administration of treatment for NTDs and reporting and responding to adverse events align with the objectives of national pharmacovigilance programs and demonstrate a core element of universal health coverage and high-quality people-centered care. Integrated NTD interventions, from prevention to diagnosis, treatment, care, and rehabilitation, can and should be delivered through community or primary- or secondary-care facilities in the national health system. Efforts to expand global coverage and targeting of NTDs must involve national and international harmonization. We need coordination of the activities of partnerships devoted to the control or elimination of these diseases. Programs with similar delivery strategies and interventions, such as those for lymphatic filariasis, onchocerciasis and soil-transmitted helminthiasis, could be managed on the same platform and together. To scale up neglected tropical disease (NTD) interventions to achieve complete eradication, programs must be integrated into the regular health system of countries with the principles of the following:

- Where things fit well; do them together.
- Where things don't, do them separately.
- Look for ways of coordinating efforts to deliver in a more cost-effective way.
- Make integration an "attitude," not a strategy.

Investment Case

Investment case (IC) for the control, elimination, or eradication of these diseases has been done in various forms, including the traditional cost–benefit evaluation of proposed interventions. The "critical elements" of an IC include the proposed investment, the rationale for the investment, the management, and governance. The final product is practical in nature, going beyond a description of what to do, by describing how to do it with respect to some core methodological issues. A reasonable projected cost based on an Investment Case has garnered political and social support, especially when there is an indication that these interventions will not jeopardize existing health systems but rather offer opportunities for synergies with health system activities (Molyneux 2008; Seddoh et al. 2013).

Although the resources for NTDs are often not commensurate with the vast need, NTD interventions are one of the best buys in global public health. The end of NTDs is expected to result in an estimated net benefit to affected individuals of about US$ 25 for every US$ 1 invested in preventive chemotherapy, representing a 30% annualized rate of return, and to contribute significantly toward universal health coverage and social protection for the least well-off (Fitzpatrick et al. 2017). In 2015, it was estimated that investing to overcome the global impact of neglected tropical diseases makes the case that the elimination and control of NTDs will be a "litmus test" for universal health coverage. Endemic countries could contribute by increasing domestic investments and scaling up interventions. Large middle-income economies can also play an important role in developing new diagnostics and medicines and in influencing market dynamics. The report of the Uniting to Combat NTDs coalition estimates cash and in-kind aid at about US$ 300 million in 2014, excluding donated medicines. Investing to overcome the global impact of neglected tropical diseases sets investment targets for universal coverage against NTDs that are more than double current levels of foreign aid—as much as ten times when including investments in vector control. It is unlikely that an increase in aid of this magnitude can be achieved in the current global health-financing climate. NTD control must become an integral part of national health plans and budgets if it is to achieve the scale of universal coverage (WHO 2015). Thus, with the concept of Universal Health Coverage (UHC) high on the agenda, tackling NTDs supports UHC, which means that all individuals and communities receive the health services they need without suffering financial hardship (WHO 2019). Actions against NTDs and their monitoring and evaluation reinforce each other. NTD interventions reach some of the world's most remote communities and can thus improve the potential for equitable access to health care services for these populations. Investment in NTDs can have important benefits for both health and economies (Fitzpatrick and Engels 2016; Seddoh et al. 2013).

NTDs and the Sustainable Development Goals (SDGs)

Sustainable Development Goal 3 (SDG3) aims to ensure healthy lives and ensure well-being for all at all ages, with the third sub-objective focusing on ending epidemics and NTDs by 2030. Ambitious as this may be, it galvanizes stakeholder actions to achieve these objectives. Thus, the SDGs can be achieved only if the NTD goals are met. Successful interventions against NTDs contribute to meeting other SDGs, such as alleviating poverty (SDG1) and hunger (SDG2), enabling people to pursue an education (SDG4) and lead productive working lives (SDG8) and promoting equality, for example, regarding gender (SDG5 & SDG10). Progress toward other Goals can accelerate the achievement of NTD goals. For example, wider provision of clean water, sanitation, and hygiene (WASH) (SDG6) is believed to help eliminate or control NTDs; the availability of resilient infrastructure (SDG9) should facilitate delivery of medicines and outreach to remote communities; the goals of sustainable cities (SDG11) and climate action (SDG13) can support the environmental management necessary for control of disease vectors. Attaining all SDGs and NTD goals is founded on strong global partnerships (SDG17). The interlinkages with the 2030 Agenda for Sustainable development are expected to encourage the NTD community to think differently about the impact of interventions and to work proactively across sectors and disciplines to ensure progress toward sustainable development. Ending the epidemic of NTDs could therefore have an impact on and improve prospects for attaining the SDGs (Bangert et al. 2017; WHO 2020).

Moral, Political, and Social Determinants

The "duty to rescue" is a moral one! The potential of getting rid of a disease forever or controlling it to insignificant levels will protect future generations from its scourge, but beyond all, the notion of disease control/eradication as a public good is one that cannot be overlooked. First, an ethical analysis presents a dimension of the investment case for control/eradication. This can be called the "moral investment" that is seldom discussed in the literature. This is an important dimension to emphasize since it has long been recognized that social and political commitment is essential for the successful control/eradication of a disease. Social and political commitment involves moral motivation or the ethical reasons to act. It is thus important to understand those reasons and how they are relevant in decisions involving large-scale public health interventions. Second, as moral beings, members of the global community have a fundamental interest in identifying what ethical obligations they have to one another. In the context of disease control/eradication, such obligations can impact the lives of millions of people and reflect choices about the kind of world in which we want to live: one where all are free from the burden of disease or one where inequity exists and only some have that luxury of good health (Hotez et al. 2006).

Political and societal support is therefore crucial for initiating and delivering these programs and must be mobilized at all costs. Broad social perception of the importance of the disease is essential; without it, there is no program! Polio, guinea worm disease, and lymphatic filariasis elimination/eradication were launched with the high-level political and technical consensus inherent in World Health Assembly resolutions. Polio eradication from the onset had tremendous societal and political support because of the awareness of the disease in developed and endemic countries. Resource mobilization by Rotary International, their network of volunteers, and the overwhelming support of civil society groups has been the lifeblood of the program. Chinese President Zemin, South African President Mandela, and US

President Clinton have heightened the program's visibility and through that raised lots of resources for the program. The guinea worm eradication initiative has also relied heavily on political advocacy, benefiting tremendously from the support of former heads of state such as US President Carter.

Advocacy and funding provide countries with the necessary support for delivering NTD interventions. Considerable progress has been made both globally and domestically. For example, Brazil, India, and Indonesia contribute significant funding for leprosy and other NTD programs. In some countries, there have been some increases in overall funding available for integrated NTD programs. In view of the growing commitment of the global community to attaining the SDGs and universal health coverage, particularly in the decade of action for the SDGs, this roadmap builds on the experiences and lessons learned and the momentum of the past decade, resulting in increased geographical and population coverage (WHO 2015).

Neglected Tropical Diseases (NTDs) as a brand on the other hand had not had that much visibility with heads of states until President Bush committed USD 350 million to their control. The NTD forum in London and the follow-up in Paris convened by the Gates Foundation has raised the profile of NTD elimination even more. Total commitments at this forum were way more than USD 800 million. The challenge is to maintain the commitment of central-level authorities for a campaign that targets a very small proportion of the national morbidity burden in the poorest communities. Sustaining societal-level support is complicated by the logistic difficulties of routinely supplying, supervising, and ensuring surveillance in remote rural areas and the fatigue of multiple years of national immunization days. A lot more advocacy is therefore required, especially at the local level, to maintain the required steam (Hotez et al. 2006; Molyneux et al. 2005).

Drug Donations and Supply Chain

In this edition, there is a new chapter on drug donations for the NTD program; what is presented here, however, is a brief overview of this issue. Investing in drug/vaccine development for interventions targeted for elimination is a critical part of the equation. This is where big pharmaceutical industry comes in. Once these products are developed, they need to be tested for their applicability in the field. Once they are proven to be efficacious, access to these medicines at reasonable costs becomes an issue. Given that these diseases occur mainly in the poorest communities, most affected individuals would be unable to afford these essential drugs. The decision by many of the pharmaceutical companies to donate or supply at production cost to the programs is therefore highly commendable and a good example of appropriate corporate social responsibility. In the case of LF elimination, GlaxoSmithKline and Merck decided to donate these medicines to the global program for as long as they are needed as part of corporate social responsibility. Diethylcarbamazine (DEC), for instance, is not a donated product in Southeast Asia, so it must be procured at production cost. Table 3 shows a list of pharmaceutical companies that contribute to the NTD portfolio of medicines.

Table 3 Medicines for controlling neglected tropical diseases donated by the pharmaceutical industry

Company	Medicine	Quantity donated	Disease	Commitment	Coordinator
Bayer	Nifurtimox	7,750,000 tablets total	Chagas disease	2021–2025	WHO
	Nifurtimox (120 mg)	300,000 tablets annually	Human African trypanosomiasis	2021–2025	WHO
	Nifurtimox (30 mg)	20,000 tablets annually	Human African trypanosomiasis	2021–2025	WHO
	Suramin	10,000 vials annually	Human African trypanosomiasis	2021–2025	WHO
	Niclosamide (400 mg)	2,800,000 tablets total	Taeniasis/cysticercosis	2020–2024	WHO
	Praziquantel (600 mg)	1,339,000 tablets total	Taeniasis/cysticercosis	2020–2024	WHO
Chemo Ibérica S.A.	Benznidazole (12.5 mg)	3000 tablets total	Chagas disease	2020–2022	WHO
(Fundación Mundo Sano)	Benznidazole (100 mg)	105,000 tablets total	Chagas disease	2020–2022	WHO
Eisai	Diethylcarbamazine citrate	2,200,000,000 tablets total	Lymphatic filariasis	Until elimination	WHO
Gilead Sciences	Liposomal amphotericin B	380,000 vials total	Visceral leishmaniasis	2016–2021	WHO
Sanofi	Eflornithine	Unlimited	Human African trypanosomiasis	Until 2025	WHO
	Melarsoprol	Unlimited	Human African trypanosomiasis	Until 2025	WHO
	Pentamidine	Unlimited	Human African trypanosomiasis	Until 2025	WHO
	Fexinidazole	Unlimited	Human African trypanosomiasis	Until 2025	WHO
Novartis	Multidrug therapy[1]	Unlimited	Leprosy	2021–2025	WHO
	Clofazimine	Unlimited	Severe erythema nodosum leprosum reactions	2021–2025	WHO
	Triclabendazole	600,000 tablets total	Fascioliasis	2016–2022	WHO
EMS	Azithromycin	Up to 153,000,000 tablets	Yaws	2021–2025	WHO
Pfizer	Azithromycin	Unlimited	Trachoma	1998–2025	International Trachoma Initiative
Johnson & Johnson	Mebendazole	200,000,000 tablets annually	Soil-transmitted helminthiases (SAC)[2]	Until 2025	WHO

GlaxoSmithKline	Albendazole	600,000 000 tablets annually	Lymphatic filariasis	Until elimination	WHO
		400,000,000 tablets annually	Soil-transmitted helminthiases (SAC)[2]	Until elimination	WHO
Merck & co.	Praziquantel	250,000,000 tablets annually	Schistosomiasis (SAC)[2]	Unlimited	WHO
MSD	Ivermectin	Unlimited	Onchocerciasis	Until elimination	Mectizan donation program
		Unlimited	Lymphatic filariasis in co-endemic countries	Until elimination[3]	Mectizan donation program
		100,000,000 treatments annually	Lymphatic filariasis for triple-therapy MDA	Until 2025	Mectizan donation program

Source: WHO (2020)

A strong, responsive supply chain is thus necessary to ensure access to high-quality, affordable medicines and health products that are accessible to the target populations. The pharmaceutical industry donated nearly 1.35 billion treatments in 2013, representing a 35% increase since 2011. A set of forms is available to facilitate application, review, and reporting and to improve program coordination and integration. This puts the Ministry of Health in control by centralizing all country requests for medicines and providing oversight for progressive ownership of control programs (WHO 2015). Now it is of the order of at least 1.5 billion treatments every year. Forecasting, securing donations of medicines, coordinating delivery, reverse logistics, education, and training can be particularly challenging. These processes involve ensuring that medicines manufactured at various locations worldwide are accessible to patients and communities living in some of the places that are most difficult to access. Efficient management will optimize the allocation of valuable donated or procured medicines and ensure that they are available at the right place and time, while minimizing wastage. National systems should invest specific resources in the control of NTDs (WHO 2020).

Beyond the drug donation, some of the pharmaceutical companies such as GlaxoSmithKline and Merck have in addition made substantial investments in support for program delivery at the global and country level, operational research, capacity building in endemic areas, and in many other areas. These investments have been critical and timely for the success of the program.

Control, Eliminate, or Eradicate?

Disease control, elimination, and subsequent eradication are desirable goals that many health systems hope to achieve. It represents the ultimate in global equity and the definitive outcome of good public health practice. The eradication of smallpox about four decades ago demonstrated that disease eradication could bring lasting benefits to society. As a result, today, several diseases have been identified as potentially eradicable with massive investments by several stakeholders. In 1997, the Dalhem conference on the eradication of infectious diseases provided some working definitions for disease elimination and eradication. These definitions have since been revised based on a better understanding of epidemiology and control of infectious diseases. The Ernst Strüngmann Forum helped to build consensus around these working definitions (Cochi and Dowdle 2011; Dowdle 1999).

Control refers to "reduction of disease incidence, prevalence, morbidity, and/or mortality to a locally acceptable level as a result of deliberate efforts; continued intervention measures are required to maintain the reduction."

Elimination of transmission (interruption of transmission) means "reduction to zero of the incidence of infection caused by a specific pathogen in a defined geographical area, with minimal risk of reintroduction, as a result of deliberate efforts; continued actions to prevent re-establishment of transmission may be required." The process of documenting elimination of transmission is called "verification."

Table 4 World Health Organization NTD targets

Control	Elimination	Elimination as a public health problem	Eradication
Buruli ulcer Cysticercosis/teniasis Dengue Echinococcus Foodborne trematodes Leishmaniasis (cutaneous) Mycetoma Scabies Snake bite envenoming	Human African trypanosomiasis (gambiense) Leprosy Onchocerciasis	Chagas disease Human African trypanosomiasis (rhodesiense) Leishmaniasis (visceral) Lymphatic filariasis Rabies Schistosomiasis Soil transmitted helminthiasis Trachoma	Dracunculiasis Yaws

Eradication means "permanent reduction to zero of a specific pathogen, as a result of deliberate efforts, with no more risk of reintroduction." The process of documenting eradication is called "certification."

Elimination as a public health problem is a term related to both infection and disease, defined by the achievement of measurable targets set by WHO in relation to a specific disease. When reached, continued action is required to maintain the targets and/or to advance interruption of transmission. Documentation of elimination as a public health problem is called validation.

Extinction: Eradication of a specific pathogen so that it no longer exists in nature or in the laboratory, which may occur with or without deliberate work.

Current global eradication initiatives are poliomyelitis and guinea worm. In the short-term, measles and rubella could get onto the list. Potentially, cysticercosis and lymphatic filariasis may be included in the list in the long term. Onchocerciasis and yaws are not on the list of the International Task Force for Disease Eradication but have received considerable attention in recent times. Since we did the first edition, there has been ample evidence to suggest that control of onchocerciasis could be moved to the elimination stage. A second chapter of onchocerciasis, which focuses on elimination, is therefore presented in this edition. The targets for the 20 NTDs as defined by WHO are presented in Table 4 (WHO 2020).

The Role of Research

Research has been very key to this whole process. Basic research to understand the biology and transmission dynamics of these diseases continues to open opportunities for the development of diagnostic tools and drugs. Several clinical trials are ongoing to help optimize these tools for field use. The importance of research in identifying solutions and options for overcoming implementation obstacles in health systems and programs has been pursued vigorously. Such implementation research has helped to resolve many operational challenges in a multi-disciplinary approach. This form of research addresses implementation bottlenecks, identifies optimal approaches for a particular setting, and promotes the uptake of research findings; ultimately, it leads to improved health care and its delivery. There are however many disease-specific research questions that need answers, which are discussed in subsequent chapters. Research, development, and innovation are therefore crucial to finding appropriate solutions to these disease-specific questions. Basic, operational, and implementation research would be needed to answer various questions and for establishing a baseline for the prevalence of an NTD and determining when to stop mass drug administration (MDA).

The research and development of new interventions, diagnostics, tools, and treatment approaches must therefore be supported, in collaboration with other stakeholders, by means that include product development partnerships. Research is needed into the behavioral and social aspects of communities' needs and perceptions in enhancing treatment compliance and healthy behaviors in the context of NTDs. The UNICEF/UNDP/World Bank/WHO Special Programme for Research and Training in Tropical Diseases (TDR) and the Coalition for Operational Research on Neglected Tropical Diseases (COR-NTD) provide leadership and direction on research priorities and support. In addition, many research funding agencies continue to provide research grants to answer specific questions that address basic understanding of these diseases, intervention trials to find new products, use of molecular epidemiology, mathematical modeling, and new technologies such as "big data," artificial intelligence, digital health, satellite imagery, and drones. They include the US National Institutes of Health (NIH), the Wellcome Trust, the European and Developing Countries Clinical Trials Partnership (EDCTP), and many other foundations. Through these initiatives, local capacity is also built for country programs.

The WHO 2030 NTD Roadmap

WHO's first roadmap for NTD in 2012 became a rallying point for the NTD community to mobilize logistics, medicines, and funding for endemic countries (WHO 2012b). One component was the establishment of the NTD Supply Chain Forum, which is a public–private partnership between WHO, pharmaceutical companies, nongovernmental organizations, logisticians, donors, and countries where NTDs are endemic. The 2021 roadmap outlines specific, measurable targets for 2030 with interim milestones for 2023 and 2025 for the eradication, elimination, and control of each of the 20 diseases and disease groups, as well as crosscutting targets aligned with WHO's Thirteenth General Programme of Work, 2019–2023 and the SDGs. In view of the growing commitment of the global community to attaining the SDGs and universal health coverage, particularly in the decade of action for the SDGs, this roadmap builds on the experiences and lessons learned and the momentum of the past decade. It includes the strategies and approaches for achieving these targets, with crosscutting themes for several diseases. The roadmap and the 2030 targets were based on extensive consultation with the NTD community. It also reflects input from more than 100 bilateral interviews with disease experts and modelers, donors, and other partners, as well as more than 300 responses gathered from an online consultation. It was therefore shaped by the perspectives of Member States and a wide range of stakeholders. It is therefore expected that there will be buy-in from all stakeholders to facilitate the realization of the goals for the benefit of endemic communities (WHO 2020).

NTDs and Climate Change

Climate change poses a substantial risk to NTD resurgence, as well as other infectious diseases like malaria, which could see an additional 60,000 deaths annually by 2050. Despite these risks, the connection between climate and health, especially regarding NTDs, remains underexplored, and efforts to address health and climate crises have often been separate. The African continent is highly susceptible to climate change impacts, including rising temperatures and sea levels, failed rainy seasons, extended periods of droughts, and reduced surface water. These events can, lead to scarcities of food and water, increase political instability, and worsen existing tensions, creating a multiplier effect for those at risk of conflict, natural disasters, and NTDs. The challenge arises from a combination of physical and social determinants that act synergistically. As a result, the African region's health systems have been strained, resulting in limited public health improvements in NTD control. Many studies have focused on the association between NTDs and climate variables like precipitation and temperature, which play a critical role in NTD transmission. While the discussions on climate change and NTDs are increasing, research on extreme weather events such as droughts and floods and their influence on NTDs is limited in Africa (Bryson et al. 2020; Tidman et al. 2021; Kelly-Hope et al. 2023).

Vector-borne NTDs may expand to previously non-endemic areas due to changing climate, while drought can worsen NTD burdens in resource-limited settings. These are the same low- and middle-income countries vulnerable to climate change that take a toll on NTDs with surgical indications. Interrupted NTD control programs during global events like the COVID-19 pandemic highlight the need for integrated strategies that consider surgical care and support calls for increased research on the climate associations of NTDs and the incorporation of surgical services into control programs (Shirley et al. 2022). An ecological study on the impact of complex challenges like political instability, climate change, and population displacement on NTD programs in Africa has established the relationship between NTD endemicity and the risk of conflict-related and climate-related displacements. The analysis reveals high-risk areas with large populations co-endemic for multiple NTDs and facing conflict and disaster-related displacement. The study calls for a coordinated response from national and international stakeholders to address the challenges posed by conflict, climate change, and displacement and meet NTD targets (Tidman et al. 2021).

To better understand the complex interactions between conflict, natural disasters that may occur as a result of climate change, and NTDs, advanced subnational analyses are needed. These analyses should consider various factors like health infrastructure, community accessibility, and NTD programs. Identifying spatial-temporal patterns of displacement due to conflict and natural disasters will help pinpoint vulnerable areas and guide long-term interventions. There is a need for more research and policy development in this area. Prioritizing research and exploring uncharted aspects of climate's impact on NTDs and infectious diseases can help protect progress and work toward elimination in Africa.

Conclusion

Despite the significant progress that has been made in the control of NTDs, the burden remains high for the endemic populations, who are some of the most vulnerable and marginalized people in the world. The control of NTDs is a good public health investment with potentially high economic return. The political platform provided by World Health Assembly resolutions, coupled with drug donations by pharmaceutical companies and financial support from the international community, sets the scene for success. The remaining chapters in this book address some of these NTDs that are prevalent in Africa, focusing on the following:

1. Epidemiology of the disease: Distribution, burden, and public health impact of disease in sub-Saharan Africa.
2. Basic biology: Life cycle, mode of transmission, disease presentation, diagnosis of the infection/disease.
3. Diagnostic tools: What is available? Are they adequate? Is their use supported by evidence? How cost-effective is their application? Operationally feasible?
4. Treatment, control, and elimination strategies: What are the available options? Are currently available strategies effective? Are they evidence-based? Is their application sustainable/cost-effective?
5. Chemotherapy-based strategies: Is the choice of first-line treatment evidence-based? What are the gaps in knowledge, for example, dose and regimen, geographic/regional variation in effect, and their implications on treatment and control, etc.? Any hosts' variation in efficacy, for example, between children and adults and implications for control? Opportunities for combination therapies and would they improve therapeutic efficacy and/or slow resistance development? Reliability of data used over the years for policy recommendations? Any signs of non-response or resistance to any of the drugs?
6. Challenges of program implementation
7. Further research for policy and control
8. Outlook for the next decade

References

Amazigo U, Okeibunor J, Matovu V, Zouré H, Bump J, Seketeli A (2007) Performance of predictors: evaluating sustainability in community-directed treatment projects of the African programme for onchocerciasis control. Soc Sci Med 64(10):2070–2082

Anoopa SD, Bern C, Varghese B, Chowdhury R, Haque R, Ali M et al (2006) The economic impact of visceral leishmaniasis on households in Bangladesh. Trop Med Int Health 11:757–764

Bangert M, Molyneux DH, Lindsay SW, Fitzpatrick C, Engels D (2017) The cross-cutting contribution of the end of neglected tropical diseases to the sustainable development goals. Infect Dis Poverty 6:e73

Bryson JM, Bishop-Williams KE, Berrang-Ford L, Nunez EC, Lwasa S, Namanya DB (2020) Indigenous health adaptation to climate change research team, Harper SL. Neglected tropical diseases in the context of climate change in East Africa: a systematic scoping review. Am J Trop Med Hyg. 102(6):1443–1454. https://doi.org/10.4269/ajtmh.19-0380. PMID: 32228798; PMCID: PMC7253121

Cochi SL, Dowdle WR (2011) Disease eradication in the 21st century. The MIT Press, Cambridge. isbn:978-0-262-01673-5

Dowdle WR (1999) The principle of disease eradication and eradication. MMWR

Fitzpatrick C, Engels D (2016) Leaving no one behind: a neglected tropical disease indicator and tracers for the sustainable development goals. Int Health 8(Suppl1):i15–i18

Fitzpatrick C, Nwankwo U, Lenk E, de Vlas SJ, Bundy DA (2017) An investment case for ending neglected tropical diseases (chapter 17). In: Holmes KK, Bertozzi S, Bloom BR, Jha P (eds) Major infectious diseases, 3rd edition. 6. Disease control priorities. International Bank for Reconstruction and Development; World Bank, Washington, DC, pp 411–431. http://dcp-3.org/sites/default/files/chapters/DCP3%20Major%20Infectious%20Diseases_Ch17.pdf

Gyapong M, Gyapong JO, Owusu-Banahene G (2000) Community-directed treatment: the way forward to eliminating lymphatic filariasis as a public health problem in Ghana. Ann Tropical Med Parasitol 95(1):77–86

Gyapong JO, Chinbuah MA, Gyapong M (2003) *Tropical medicine and international health 2003* inadvertent exposure of pregnant women to ivermectin and albendazole during mass drug administration (MDA) for lymphatic filariasis. Trop Med Int Health 8(12):1093–1101

Gyapong JO, Gyapong M, Yellu N, Anakwah K, Amofah G, Bockarie M, Adjei S (2010) Integration of control of neglected tropical diseases into health-care systems: challenges and opportunities. Lancet 375(9709):160–165

Hotez PJ, Ottesen E, Fenwick A, Molyneux D (2006) The neglected tropical diseases: the ancient afflictions of stigma and poverty and the prospects for their control and elimination. Adv Exp Biol Med 582:22–33

Jacobson J (2022) Compressing the timelines for development and delivery: accelerating access to triple-drug therapy to eliminate lymphatic filariasis. Am J Trop Med Hyg 106(5_Suppl):4–12

Kelly-Hope LA, Harding-Esch EM, Willems J et al (2023) Conflict-climate-displacement: a cross-sectional ecological study determining the burden, risk and need for strategies for neglected tropical disease programmes in Africa. BMJ Open 13:e071557. https://doi.org/10.1136/bmjopen-2023-071557

King CL, Weil GJ, Kazura JW (2020) Single-dose triple-drug therapy for *Wuchereria bancrofti* - 5-year follow-up. N Engl J Med 382(20):1956–1957

Lutumba P, Makieya E, Shaw A, Meheus F, Boelaert M (2007) Human African trypanosomiasis in a rural community, Democratic Republic of Congo. Emerg Infect Dis 13:248–254

Meheus F, Abuzaid AA, Baltussen R, Younis BM, Balasegaram M, Khalil EA et al (2013) The economic burden of visceral leishmaniasis in Sudan: an assessment of provider and household costs. Am J Trop Med Hyg 89:1146–1153

Molyneux DH (2008) Combating the "other diseases" of MDG 6: changing the paradigm to achieve equity and poverty reduction? Trans R Soc Trop Med Hyg 102(6):509–519

Molyneux DH, Hotez PJ, Fenwick A (2005) "Rapid-impact interventions": how a policy of integrated control for Africa's neglected tropical diseases could benefit the poor. PLoS Med 2:e336DOI

Osei-Atweneboana MY, Eng JK, Boakye DA, Gyapong JO, Prichard RK (2007) Prevalence and intensity of Onchocerca volvulus infection and efficacy of ivermectin in endemic communities in Ghana: a two-phase epidemiological study. Lancet 369(9578):2021–2029. https://doi.org/10.1016/S0140-6736(07)60942-8

Ozaki M, Islam S, Rahman KM, Rahman A, Luby SP, Bern C (2011) Economic consequences of post-kala- azar dermal leishmaniasis in a rural Bangladeshi community. Am J Trop Med Hyg 85:528–534

Seddoh A, Onyeze A, Gyapong JO, Holt J, Bundy D (2013) Towards an investment case for neglected tropical diseases. Lancet 9908:1898–1955

Shirley H, Grifferty G, Yates EF, Raykar N, Wamai R, McClain CD (2022) The connection between climate change, surgical care and neglected tropical diseases. Annals of. Glob Health 88(1):68. https://doi.org/10.5334/aogh.3766

Sundar S (2010) Household cost-of-illness of visceral leishmaniasis in Bihar, India. Trop Med Int Health 15(Suppl 2):50–54

Tidman R, Abela-Ridder B, Ruiz de Castañeda R (2021) The impact of climate change on neglected tropical diseases: a systematic review. Trans R Soc Trop Med Hyg 115(2):147–168. https://doi.org/10.1093/trstmh/traa192

Uranw S, Meheus F, Baltussen R, Rijal S, Boelaert M (2013) The household costs of visceral leishmaniasis care in South-Eastern Nepal. PLoS Negl Trop Dis 7:e2062

Weekly Epidemiological Record (2021) Neglected tropical diseases: impact of COVID-19 and WHO's response: 2021 update, vol 96(38). WHO, Geneva, pp 461–468. https://apps.who.int/iris/handle/10665/345383, Accessed 14 May 2023

World Health Organisation (2004) Intensified control of the neglected diseases. Berlin, 10–12 December 2003. Report of an international workshop. WHO/CDS/CPE/2004.45

World Health Organisation (2006) Strategic and technical meeting on intensified control of neglected tropical diseases, Berlin, 18–20 April 2005. A renewed effort to combat entrenched communicable diseases of the poor - Report of an international workshop. WHO/CDS/NTD/2006.1

World Health Organisation (2008) Community directed interventions for major health problems in Africa. multi-country study final report. https://doi.org/10.2174/TDR.07. Geneva, Switzerland ISBN 978 92 4 159660 2

World Health Organisation (2010). Working to overcome the global impact of neglected tropical diseases. First WHO report on Neglected Tropical Diseases. WHO/HTM/NTD/2010.1

World Health Organisation (2013) Sustaining the drive to overcome the global impact of neglected tropical diseases. Second WHO report on neglected tropical diseases. WHO/HTM/NTD/2013.1

World Health Organisation (2015) Investing to overcome the global impact of neglected tropical diseases. Third WHO report on neglected tropical diseases WHO/HTM/NTD/2015.1

World Health Organization (2012a). Economic costs of selected neglected tropical diseases (Table 3.2.1.1). In: Working to overcome the global impact of neglected tropical diseases: first WHO report on neglected tropical diseases. Geneva: https://apps.who.int/iris/handle/10665/44440

World Health Organization (2012b) Accelerating work to overcome the global impact of neglected tropical diseases: a roadmap for implementation. Geneva: https://apps.who.int/iris/handle/10665/70809

World Health Organization (2019) Universal health coverage. WHO fact sheet. Geneva. https://www.who.int/news-room/fact-sheets/detail/universal-health- coverage-(uhc)

World Health Organization (2020) Ending the neglect to attain the sustainable development goals: a road map for neglected tropical diseases 2021–2030. WHO, Geneva

World Health Organization (2023) Global report on neglected tropical diseases 2023. Geneva Licence: CC BY-NC-SA 3.0 IGO

Buruli Ulcer in Sub-Saharan Africa

Ghislain Sopoh, Yves Thierry Barogui, and Kingsley Asiedu

Contents

G. Sopoh (✉)
Regional Institute of Public Health, University of Abomey-Calavi, Ouidah, Bénin

Y. T. Barogui
Team, Neglected Tropical Diseases Programme, Tropical and Vector-borne Diseases, World Health Organization, African Region, Libreville, Gabon

K. Asiedu
Department of Control of Neglected Tropical Diseases, World Health Organization, Geneva, Switzerland

J. O. Gyapong, B. A. Boatin (eds.), *Neglected Tropical Diseases - Sub-Saharan Africa*, Neglected Tropical Diseases, https://doi.org/10.1007/978-3-031-53901-5_2

Abstract

Buruli ulcer is a chronic skin infection caused by an environmental *Mycobacterium ulcerans*. The disease has been documented in over 33 countries, but the bulk of the disease burden occurs in Sub-Saharan Africa. The majority of people affected are children below 15 years of age. The mode of transmission is still not known. However, there have been significant advances in the pathogenesis, diagnosis, and treatment in the past decade. The public health strategy aimed at reducing morbidity and disability based on early detection of cases and antibiotic treatment. This requires intensive health education in the affected communities to encourage the population to report early to the health facilities. Research to develop rapid point-of-care diagnostic tests and to simplify antibiotic treatment are in progress and expected to be in practical application in the field within the next decade.

Introduction

Buruli ulcer (BU) is one of the 20 neglected tropical diseases caused by *Mycobacterium ulcerans*. It has been reported in 33 countries; however, the highest disease burden is in Sub-Saharan Africa. The disease was first described by Sir Albert Cook in Uganda in 1897. However, it was not until 1938 that a group of researchers in Australia were able to culture and characterize the causative organism (MacCallum et al. 1948). The organism grows at a lower temperature of 29–33° and produces a unique toxin called *mycolactone*, which causes the destruction of the tissue (George et al. 1999). In 1998, WHO established the Global Buruli Ulcer Initiative to address the growing public health problem of the disease, especially in West Africa. In the same year, it organized the first international conference on Buruli ulcer control and research in Yamoussoukro, Côte d'Ivoire, which led to the Yamoussoukro declaration. This declaration contributed to the mobilization of national and international support to advance the Buruli ulcer control and research. The WHO strategy for Buruli ulcer control is based on early detection and antibiotic treatment. The disease presents as a nodule, plaque, and edema. With time, all ulcerate with the typical undermining edges and white-yellowish necrotic base. The diagnosis is based on clinical criteria but requires training and experience. A number of other skin conditions can present like Buruli ulcer. In the laboratory, Buruli ulcer can be confirmed by four tests (direct microscopy, histopathology, culture, and polymerase chain reaction (PCR)) (WHO 2014). However, because of sensitivity and specificity as well as the faster turnaround time, PCR analysis targeting the insertion sequence IS *2404* has become the standard laboratory diagnosis of BU. Many endemic countries have research laboratories that are able to perform this test, but it requires the transport of samples from rural endemic communities to these laboratories, which are based in the cities. The current recommended antibiotic regimen is a combination of rifampicin and clarithromycin given for 8 weeks (WHO 2012). In some cases, surgery (skin grafting) and physiotherapy may be necessary depending on the stage, location, and extent of the disease. The development of point-of-care diagnostic tests is a research priority now. In November 2013, WHO and FIND (the Foundation for Innovative New Diagnostics) convened a meeting of experts to review and prioritize diagnostic technologies under development (WHO 2014). Among the various tests, two are well advanced for field evaluations. These are direct detection of the toxin—*mycolactone*—by thin layer chromatography and antigen capture test. While efforts are being made to understand the mode of transmission, it is essential that national control programs intensify efforts to improve early detection of cases.

Definition

Buruli ulcer (BU) is an infectious disease caused by *Mycobacterium ulcerans*. The disease is characterized by extensive necrosis of the subcutaneous tissue that, if left untreated, results in disabling sequelae (WHO 2000).

Epidemiology

Geographical Distribution

BU is prevalent in countries with tropical and subtropical climates. It has been reported in over 33 countries with variable incidence. Although the foci are geo-graphically circumscribed, they are always around aquatic ecosystems (rivers, lakes, artificial or natural wetlands, irrigation systems, etc.) (WHO 2000). Table 1 presents a summary of the countries in which BU has been reported. Figure 1 shows the global distribution of BU in 2012. In 2012, cases were reported from 14 countries, most of which are in Africa, where efforts to control the disease have been focused during the past decade.

Burden of the Disease in Sub-Saharan Africa

The true prevalence of BU in the various endemic countries is not known, and underreporting is suspected in some countries, such as Nigeria (Chukwuekezie et al. 2007). The annual number of reported new cases of BU, which had increased

Table 1 Countries in which BU has been reported

Regions	Countries
Africa	Angola, Benin, Burkina Faso, Cameroon, Congo, Côte d'Ivoire, Democratic Republic of Congo, Gabon, Ghana, Guinea, Equatorial Guinea, Kenya, Liberia, Nigeria, Uganda, Sudan, Togo, and Central African Republic (Smith 1970; Bär et al. 1998; Meyers et al. 1996; Ouoba et al. 1998; WHO 2003; Janssens et al. 2005; Walsh et al. 2008, 2009, 2010, 2011; Minime-Lingoupou et al. 2010)
Latin America	French Guiana (De Gentile et al. 1992), Mexico (Walsh et al. 2011), Peru (Guerra et al. 2008), Suriname (WHO 2008b; Walsh et al. 2011)
Asia	China (Faber et al. 2000), Japan (Kondo et al. 2009; Watanabe et al. 2010)
South pacific	Australia (Goutzamanis and Gilbert 1995; Johnson et al. 1996), Papua New Guinea (Lavender et al. 2007)

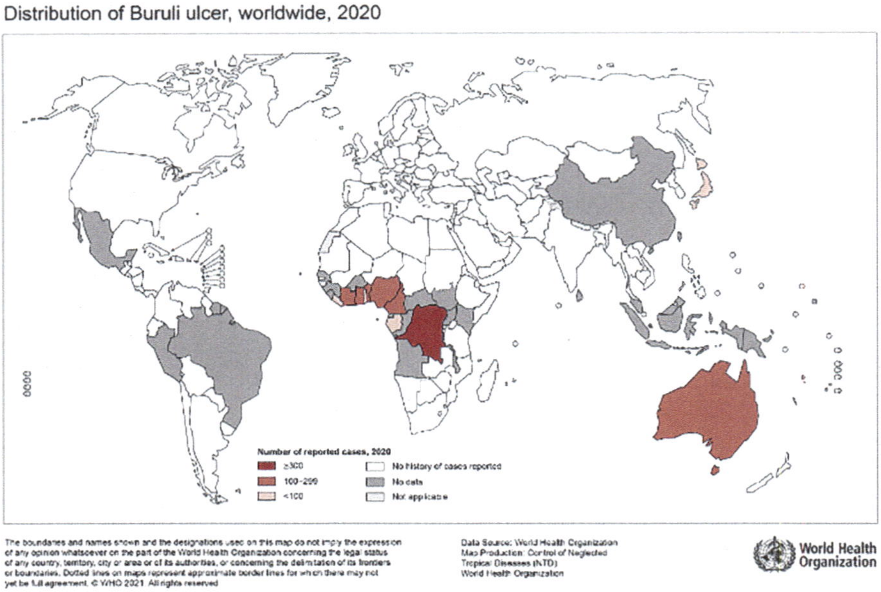

Fig. 1 Distribution of BU by country as of 2020 (Source: WHO). Relative endemicity is denoted as high (*red*), moderate (*pink*), or low (*light pink*)

in the early and mid-1990s to about 10,000, has more recently stabilized to about 5000.

In 2020, 1458 new cases of Buruli ulcer were reported globally, 1233 of them from the African Region. The data available are limited for three reasons: (1) within endemic countries where cases are being reported regularly, control activities are limited in geographical scope, and the data may not therefore reflect the burden at the national level; (2) there are large areas where limited or no activities are being carried out, and the extent of the disease is therefore relatively unknown; and (3) limited knowledge of the disease, its focal distribution, and the fact that it affects mainly poor rural communities all contribute to low reporting of cases. Only Côte d'Ivoire and Ghana have performed national surveys (Kanga and Kacou 2001; Amofah et al. 2002) from which Côte d'Ivoire reported 32 cases per 100,000 inhabitants and Ghana a prevalence of 20.7 cases per 100,000 inhabitants. However, not all of these cases have been confirmed by laboratory tests.

Public Health Impact in Sub-Saharan Africa

The psychologic, social, and economic impacts of the disease on affected families and communities are important (Asiedu and Etuaful 1998; Aujoulat et al. 2003; Ackumey et al. 2012). These include (1) stigma of affected individuals perceived to be attributable in some affected communities' sorcery or witchcraft, (2) anxiety about the progression of the disease, (3) the debilitating consequences of pain and disabilities, and (4) the disruption of school attendance. Others include the high cost of treatment when this is not subsidized, high indirect cost for affected individuals and relatives in terms of lost productivity or depletion of income, and transport cost to centers for daily treatment.

Reservoir and Transmission Hypotheses

Epidemiological studies show that the disease appears in endemic foci often localized around aquatic niches (Merritt et al. 2005; Sizaire et al. 2006; Duker et al. 2006). In Uganda, Lunn et al. (1965) and Barker (1972) described cases primarily from the Nile Valley and bordering marshes. Bayley (1971) described the cases of three BU patients living around a river in Ghana. Ravisse (1977) described cases in Cameroon originating from the Nyong River and the surrounding swamps. Oluwasanmi et al. (1976) described cases in Nigeria located in an area near an artificial lake bordering the University of Ibadan. Johnson et al. (2005) showed that there was an inverse relationship between an area's prevalence of BU and its distance from the Couffo River in Benin. Kibadi et al. (2008) reported three patients originating from villages near the Cuango/Kwango River in Angola and the Democratic Republic of Congo.

Cultures of mycobacteria have been obtained from different grasses (*Hyparrhenia rufa*, *Imperata cylindrica*, *Panicum maximum*, and *Pennisetum catabasis*) (Barker et al. 1970). There is only one reported case of cultivation of *M. ulcerans* from an environmental niche—an aquatic insect (*Gerris* sp.) (Portaels et al. 2008b). However, the DNA of *M. ulcerans* has been detected, albeit not cultivated, in many aquatic insect and animal species (Portaels et al. 2001; Marsollier et al. 2002, 2004; Eddyani et al. 2004) as well as in mosquitoes (Johnson et al. 2007; Wallace et al. 2010). Therefore, a possible role of aquatic insects in the transmission of the disease from the environment to humans has been suggested (Portaels 1989, 2004; Portaels et al. 1999; Silva et al. 2007; Marsollier et al. 2007). Indeed, Portaels hypothesized that *M. ulcerans*, being concentrated by filtering water bodies, such as aquatic larvae, mollusks, and small fishes, could be eaten by and then infect predatory insects such as water bugs. These bugs may transmit the bacteria to humans through a bite. Humans could also be contaminated directly from the environment through trauma. A study by Marsollier and colleagues subsequently demonstrated that aquatic bugs, such as *Naucoris cimicoides*, can transmit the disease to mice in a laboratory setting (Portaels et al. 1999; Marsollier et al. 2007).

M. ulcerans infection has been described in other mammals (koalas, alpacas, possums, horses, and cats) in Australia (Portaels et al. 2001; Elsner et al. 2008; van Zyl et al. 2010; Fyfe et al. 2010). However, the search for the bacteria in small terrestrial mammals in Benin has yielded no results (Durnez et al. 2010).

The transmission of the disease from man to man has never been demonstrated. However, the development of typical BU after a human bite has been reported (Debacker et al. 2002, 2003). The hypothesized explanation for this observation is that the introduction of *M. ulcerans*, previously present on the skin of the affected individual, was facilitated by the bite (Debacker et al. 2003).

Pathogenesis

The incubation period for BU is variable and relies heavily on the host's immune response (Andersen 1965). Some authors have estimated the incubation period to last between 1 and 3 months (Andersen 1965; Barker 1973). Others think that the incubation period may range from 2 weeks to 3 years (Meyers et al. 1974). In experimentally infected animals, the incubation period varied from 7 to 150 days and depended on the size of the inoculum (Fenner 1950; Marsollier et al. 2002).

The course of the infection after the introduction of the bacteria into the skin is also poorly documented. Silva et al. (2009) developed a model based on the microbiological and immunological aspects of the disease, as shown in Fig. 2. This model implies that *M. ulcerans* behaves as an intracellular parasite that induces an inflammatory cellular response. During the active phase of the infection, *M. ulcerans* parasitizes macrophages and multiplies within them, continuously colonizing incoming monocytes or macrophages and progressively invading healthy tissues. The persistent influx of leukocytes to the site of active infection provides immune cells that interact with *M. ulcerans*, triggering leukocyte chemotaxis and cell-mediated immunity (CMI). The ability of this CMI to halt the progress of the infection,

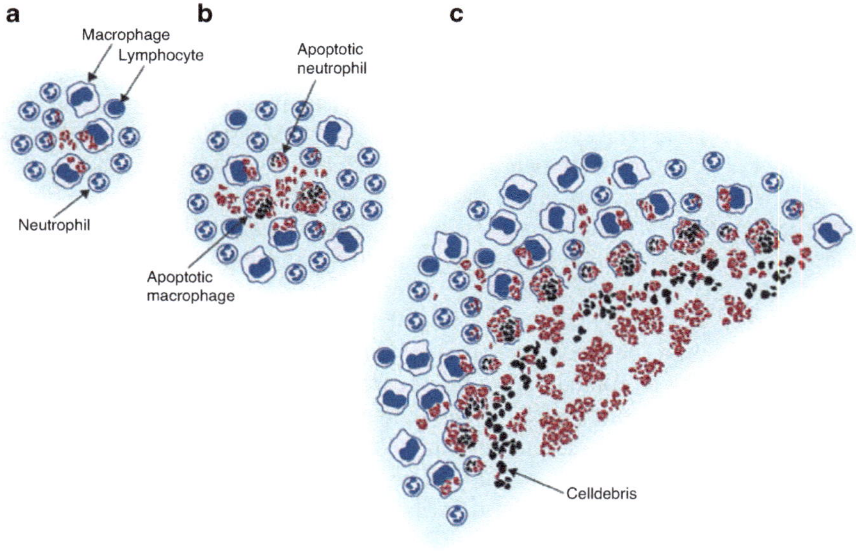

Fig. 2 Schematic representation of the initiation and establishment of a BU lesion (Source: Silva et al. 2009). (**a**) Early stage of infection with *M. ulcerans* (red bacilli) phagocytosed by neutrophils and macrophages in the acute inflammatory infiltrate. (**b**) A more advanced stage characterized by the presence of an area with an inflammatory cellular infiltrate with intraneutrophil and intramacrophage bacilli and apoptotic neutrophils and macrophages. (**c**) The advanced stage of the lesion, with extensive necrotic acellular areas containing abundant clumps of extracellular bacilli, cellular debris, neutrophils, and macrophages with intracellular bacilli at the edges of the necrotic areas

leading to mediated immunopathology or self-cure, depends upon factors that include the dose of infection, the virulence of the bacteria, and the immuno-competence of the host. The band of cellular infiltrate with *M. ulcerans* multiplying within macrophages therefore represents the front where crucial events in the development of the disease occur. As the front advances and invades healthy tissues, leukocyte lysis at the trailing edge of the infiltrate creates a continuously enlarging cellular necrotic area with freed bacilli that probably multiply extracellularly. In advanced disease, the necrotic lesion further expands, with extensive coagulation necrosis of the subcutaneous tissue, dermis, and epidermis, extensive vasculitis, and, in some cases, ulceration.

Over time, immunity becomes protective with the production of a granulomatous tissue reaction and healing.

According to this view of BU pathogenesis, inflammatory infiltrates in advanced lesions occupy a smaller area than the necrotic zones do because of the destruction of the continuously attracted infiltrates of immune cells by mycolactone rather than because of mycolactone-induced inhibition of cellular responses. That is, in advanced lesions of BU, the inflammatory cellular infiltrates are minor in spatial but not in immunological terms.

Another theory is based on the action of the mycolactone. *M. ulcerans* is unique in that the mycolactone it produces is necessary for its virulence (George et al. 1999). Indeed, after penetration into the dermis or subcutaneous tissue, the bacteria are latent for a variable duration and then proliferate and produce mycolactone. This cytotoxin has an affinity for fat cells (Dobos et al. 2001). Once secreted, mycolactone therefore causes the apoptosis of adipocytes and necrosis (George et al. 2000). This necrosis is a good environment for the proliferation of *M. ulcerans*. At this stage, the cutaneous manifestations appear as nodules, plaques, or ulcerations as the necrosis extends toward the surface (van der Werf et al. 2003). The mycolactone also induces the destruction of continuously produced inflammatory infiltrates, resulting in a depressed or absent local immune response from the host (Adusumilli et al. 2005; Oliveira et al. 2005; Torrado et al. 2007a, b, 2010; Simmonds et al. 2009; Fraga et al. 2011). Gradually, following treatment, the host will develop a CMI response (Peduzzi et al. 2007; Silva et al. 2009). At this stage, a specific inflammatory reaction appears with the formation of granulomas that will isolate the bacteria, circumscribe the evolution of the disease, and lead to progressive scarring. The disease may then disappear, leaving a scar behind.

Clinical Characteristics

BU develops in three stages: the non-ulcerated stage, with four clinical forms (papule, nodule, edema, or plaque), the ulcerated stage, and the healing stage. In Africa, nonulcerative forms represent 26% of the clinical presentation of the disease, when ulcerative forms are 74% (World Health Organization 2012).

The *papule* is defined as a painless, raised skin lesion less than 1 cm in diameter. The surrounding skin is reddened. This form is commonly seen in Australia.

The *nodule* is a lesion that extends from the skin into the subcutaneous tissue. It is 1–2 cm in diameter. It is usually painless but may be pruritic and the surrounding skin may be discolored compared to adjacent areas. This form is commonly seen in Africa (Fig. 3).

The *plaque* is a firm, painless, elevated, well-demarcated lesion more than 2 cm in diameter with irregular edges. The skin over the lesion is often reddened or otherwise discolored (Figs. 4 and 5).

The *edema* is a diffuse, extensive, usually nonpitting swelling. The affected area has ill-defined margins, is firm and painless, and involves part or all of a limb or other parts of the body. There may be color changes over the affected region (Fig. 6).

The *ulcer* has undermined edges and is indurated peripherally. The floor of the ulcer may have a white cotton wool-like appearance from the necrotic slough. The ulcer is usually painless unless there is a secondary bacterial infection. When there

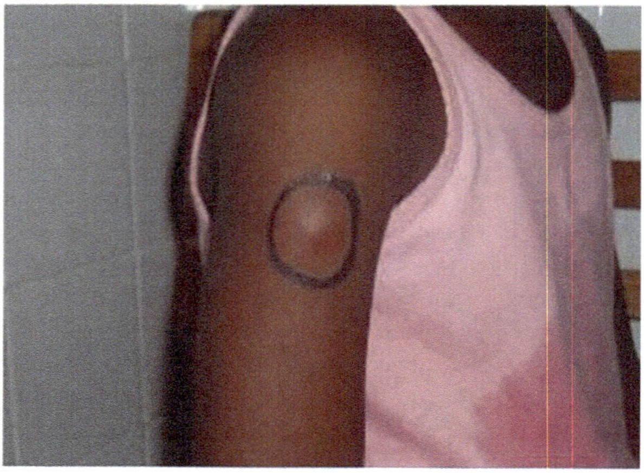

Fig. 3 Nodule on the upper limb

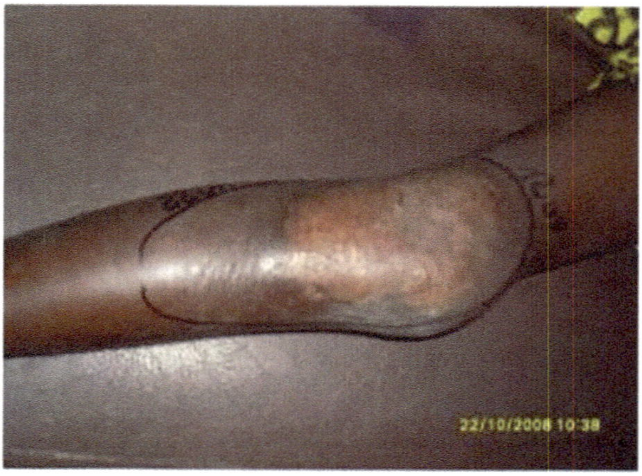

Fig. 4 Plaque on the lower limb

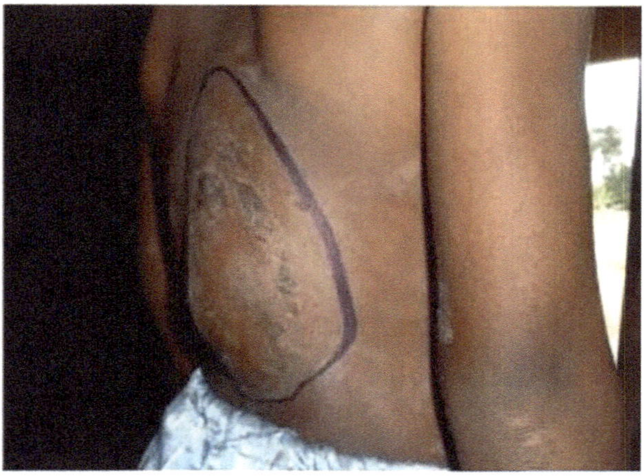

Fig. 5 Plaque on the back

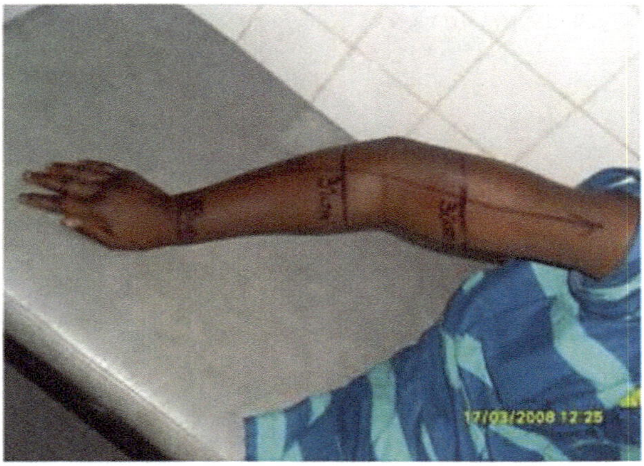

Fig. 6 Edema on the upper limb

is more than one ulcer and the ulcers are close together, they often communicate beneath intact skin (World Health Organization 2001a, b, 2004; Figs. 7, 8, 9, and 10).

The *healing stages* are inactive forms characterized by a particular type of scar, with or without sequelae, that is hypopigmented, starry, retractile, and fibrous (Fig. 11).

Other clinical forms have been described:

- *Bone involvement* can be true osteomyelitis, focal or multifocal. The overlying skin is often intact with no obvious lesion. Osteomyelitis may occur as a primary condition or as a metastatic condition, sometimes at a distance from a cutaneous lesion or after a cutaneous lesion has healed. *M. ulcerans* osteomyelitis is initially painless but subsequently frankly painful and well localized. There is usually an identifiable area of increased warmth. A swelling then appears that may progress to a fistula that discharges necrotic material. Incision of the swelling

Fig. 7 Multiple lesions

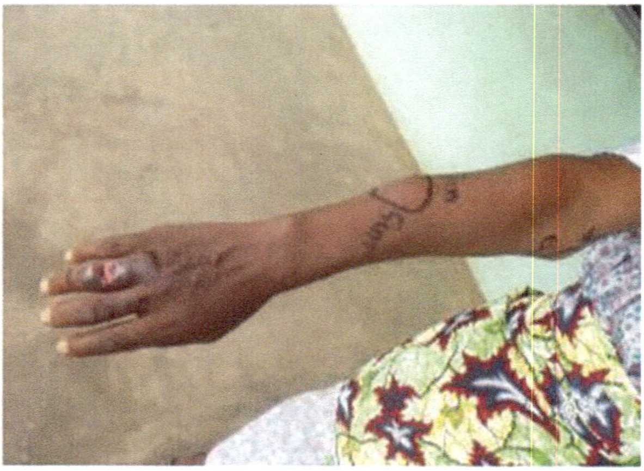

Fig. 8 Typical BU ulcer

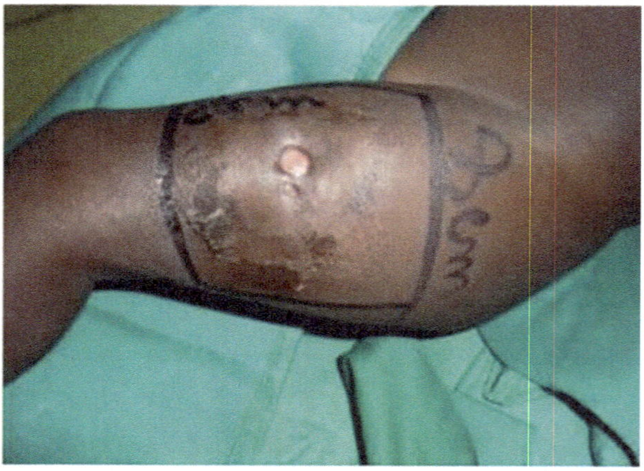

Fig. 9 Extensive ulcers

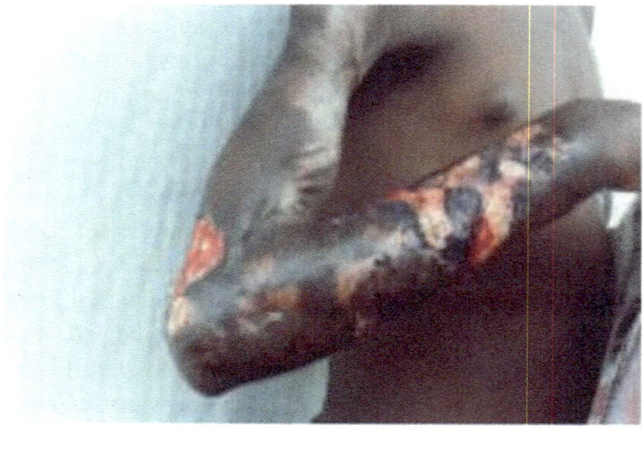

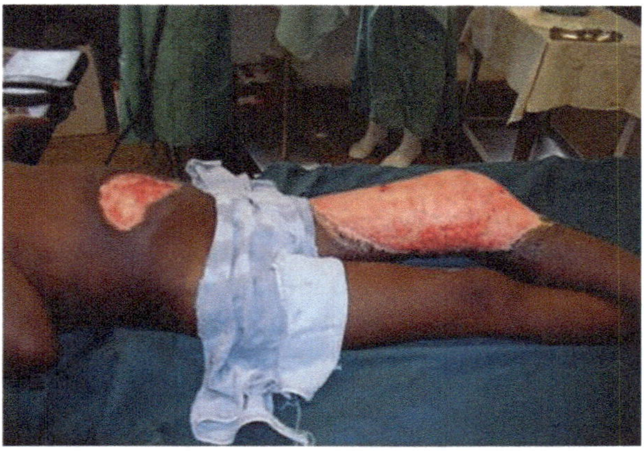

Fig. 10 Extensive ulcers

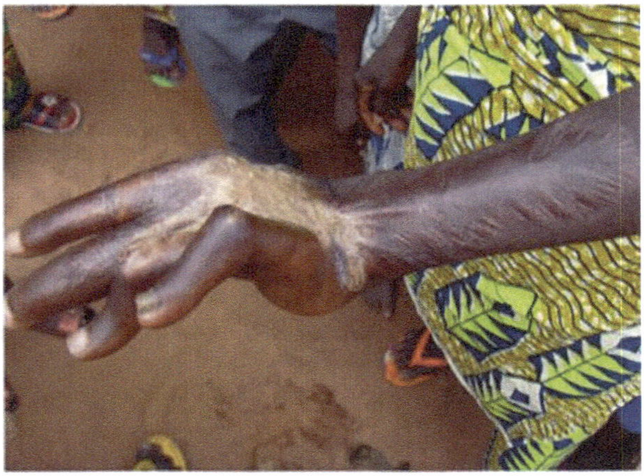

Fig. 11 Complication of healing

reveals gelatinous tissue, and, beneath this, the bone has a moth-eaten appearance. Unlike open (contiguous) osteitis, the bone is the site of necrosis to a variable extent, similar to that seen in tuberculous osteomyelitis. Reactive (contiguous) osteitis occurs as a consequence of deep destruction of overlying soft tissues. Occasionally, the bone is exposed to the point of devascularization, necrosis of cortical bone, sequestration, and osteomyelitis. The macroscopic appearance is then that of white dead bone of almost normal appearance and texture (World Health Organization 2001a, b). Portaels et al. first demonstrated the invasion of *M. ulcerans* into the bone (Portaels et al. 2008a).

- *Disseminated and mixed forms* are described as the simultaneous presence of different forms of the disease, including bone and joint involvement, in the same patient (World Health Organisation 2004).
- *HIV/AIDS coinfections*: Buruli ulcer and HIV coinfection is an emerging area and complicates clinical management. The first case of Buruli and HIV coin-

fection was reported in the Democratic Republic of the Congo in 1992 (World Health Organization 1997). A study conducted in Benin from 2002 to 2003 found that HIV prevalence among patients with Buruli ulcer was higher (2.6%, 11/426) than among controls (0.3%, 2/613) (Johnson et al. 2008). In Benin, 6 (3.6%) out of 156 patients treated at Pobè Buruli Ulcer Treatment Center in 2006 were positive for HIV, and in 2010, 2 (1.5%) out of 135 patients were HIV positive. In Cameroon, an HIV prevalence of 33% has been reported among adult Buruli ulcer patients (compared to 5% in the general population) (World Health Organization 2010b). HIV weakens the immune system and tends to make Buruli ulcer progress more aggressive and could possibly affect the response to antibiotic treatment. Coinfected patients are often adults (>15 years old) who present with multifocal lesions and osteomyelitis (Johnson et al. 2002; Toll et al. 2005).

- *A recurrent case* is defined as a patient with previous surgical (and/or antibiotic) treatment for *M. ulcerans* who presents with another lesion or lesions at the same or a different site within 1 year of the completion of the last treatment (World Health Organisation 2004).
- *Paradoxical reaction* is a recently recognized phenomenon, and some cases previously classified as recurrent lesions may have been due to paradoxical reactions (Barogui et al. 2016; Ruf et al. 2011). Such reactions occur during or long after antibiotic treatment, with new inflammatory disease (presenting as a nodule/swelling, plaque, or edema) leading to the extension of the existing ulcer or a new lesion on a different part of the body, usually with pus formation and pain. These are sometimes seen on parts of the body where there was no evidence of disease before antibiotic treatment, perhaps as a result of subclinical infection. Cultures of tissue or pus are usually sterile, although acid-fast bacilli can still be seen and polymerase chain reaction (PCR) for *M. ulcerans* IS2404 remains positive. Histopathological examination of the lesion demonstrates an intense immunological reaction within and around the lesion (World Health Organization 2012).

In addition to the standard classification of the disease into non-ulcerated and ulcerated forms, an additional classification based on lesion size was introduced for two reasons: (1) small lesions are more likely to heal with antibiotic treatment, and (2) small lesions reflect the impact of health education efforts promoting early diagnosis and therefore help monitor the progress of different countries.

There are three *categories* of lesions:

- Category 1: single lesion less than 5 cm in diameter.
- Category 2: single lesion between 5 and 15 cm in diameter.
- Category 3: single lesion greater than 15 cm in diameter, multiple lesions, lesion(s) at critical sites (eye, breast, genitalia), and osteomyelitis (World Health Organization 2012).

Complications and Sequelae

Contractures may result from scarring caused by lesions over or close to joints. Ankyloses may follow.

There may be continuous minor bleeding or a sudden major *hemorrhage*.

Secondary bacterial infection may be caused by organisms such as staphylococci, streptococci, *Pseudomonas* sp., and *Corynebacterium* sp. Secondary infection may progress to cellulitis and septicemia. Infection may extend beneath the deep fascia to involve tendon sheaths, muscle, blood vessels, nerves, bone, and joints or may destroy periorbital tissue with loss of the eye.

Hypertrophic scars and keloids may develop at infection and surgical sites, including skin graft donor sites.

Squamous cell carcinoma (Marjolin's ulcer) may appear in an unstable scar or a persistent ulcer many years after the initial infection with *M. ulcerans* (World Health Organization 2001a, b).

Clinical Diagnosis

The diagnosis of BU is based on epidemiological, clinical, and biological criteria. The following clinico-epidemiological features are important diagnostic clues:

- Most patients live in or have traveled to a known endemic area.
- Most patients are children under 15 years of age.
- About 85% of lesions are on the limbs.
- Lower-limb lesions are twice as common as upper-limb lesions.
- Non-ulcerated lesions are almost painless or minimally painful (although ulcers may be painful in the presence of a secondary bacterial infection).
- In the absence of a secondary bacterial infection or other coinfections in ulcerated lesions, there are often no constitutional symptoms (such as fever).
- Enlarged lymph nodes are not a feature of *M. ulcerans* disease.

Biology

The biological diagnosis is based on a positive result of one or several of the following tests performed on samples collected from suspected lesions:

- Direct smear examination (DSE) showing acid-fast bacilli (AFB).
- Polymerase chain reaction (PCR).
- Culture of *M. ulcerans* on Löwenstein–Jensen medium.
- Histopathology showing lesions consistent with BU and AFB on Ziehl–Neelsen stain (World Health Organisation 2004).
- PCR is currently the standard laboratory diagnosis of BU.

The advantages and limitations of these tests are summarized in Table 2.

Swabs, fine-needle aspiration, and biopsy (punch or surgical) are used to collect specimens. Fine-needle aspiration has recently been proposed as a minimally invasive technique for the diagnosis of non-ulcerated forms of BU. Its effectiveness has been reported by several authors (Phillips et al. 2009; Eddyani et al. 2009; Cassisa et al. 2010; Herbinger et al. 2010), and a tutorial has been prepared on how to perform this technique (http://www.stopburuli.org/fna-e-tutorial/index_en.html).

Swabs and fine-needle aspiration are noninvasive procedures that can be undertaken at any level (community, health centers, hospitals) during routine management or for case finding in communities.

Specimens obtained by *swabs* should be taken from the undermined edges of a clinically diagnosed BU lesion. Physicians or experienced health workers can perform this technique.

Fine-needle aspiration (FNA) is mainly used to obtain samples from clinically diagnosed non-ulcerated lesions (nodules, plaques, and edema). This technique is necessary for up to 30% of patients (depending on the setting) and is simple enough to be applied more widely in the field. FNA may also be used in some ulcerated lesions in which it is difficult to take swabs because of healing edges. Only physicians

Table 2 Summary of the advantages and disadvantages of confirmatory testing for BU

Method	Advantages	Disadvantages
Direct smear examination	• Easy to perform at a local level • Does not require expensive materials and equipment • Rapid results • Uses swabs, fine-needle aspiration, and biopsy samples	• Low sensitivity (<60%) • Needs trained personnel • Needs external quality assurance
Polymerase chain reaction (PCR)	• Results are fairly rapid • Uses swabs, fine-needle aspiration, and biopsy samples • High sensitivity (>95%)	• Requires a sophisticated laboratory • Expensive to perform • Needs trained personnel • Requires strict quality control
Culture of *M. Ulcerans*	• Uses swabs, fine-needle aspiration, and biopsy samples • Proves that the AFB are alive (viable)	• Requires a sophisticated laboratory • Needs trained personnel • Results take longer than 8 weeks • Low sensitivity (20–60%) • Not immediately useful for patient management
Histopathology	• Sensitivity is about 90% • Results are fairly rapid • Useful in establishing a differential diagnosis and monitoring response to treatment	Requires a sophisticated laboratory Expensive to perform Needs trained personnel Requires an invasive procedure (i.e., biopsy)

Source World Health Organization (2010a)

or experienced health workers should perform this technique; ongoing training and regular supervision should be provided to health workers to improve their skills. Extreme care should be taken when performing fine-needle aspiration around the head and neck area (especially around the eyes) and the genitalia. When necessary, an expert clinician should perform this technique to minimize any unintended damage to important organs or structures.

The WHO recommends that a maximum of two swabs or two fine-needle aspirations be taken from each lesion. Repeat sampling may be indicated if the results of the PCR of the initial samples are negative despite a strong clinical diagnosis. Samples obtained from swabs and fine-needle aspirations are sufficient in most cases.

Punch or surgical *biopsy* may be used when the diagnosis is in the direct interests of the patient (e.g., when swabs and fine-needle aspiration have been unsuccessfully attempted or abandoned). Surgical biopsy may be preferable when larger diagnostic sample specimens are required for histopathological analyses (World Health Organization 2010a).

All samples collected for immediate analysis can be placed in a sterile container with no additives. Samples to be transported should be kept cool (ideally at 4 °C), such as in an insulated container with a frozen cooling block, if the analysis will be performed within 24 h. If the analysis will be performed after 24 h or more, the recommendation is to either keep the sample at 4 °C (not frozen) when refrigeration facilities are available or place it in a transport medium (Liquid Middlebrook 7H9 broth supplemented with polymyxin B, amphotericin B, nalidixic acid, trimethoprim, and azlocillin [PANTA]) when refrigeration facilities are not available. Supplementation with 0.5% agar achieves a semisolid medium (World Health Organization 2001a). Specimens kept in semisolid transport medium may still be culture positive after as long as 26 weeks (Eddyani et al. 2008).

The *direct smear examination* is generally performed after Ziehl–Neelsen staining. The bacteria appeared on a blue background. They are small acid-fast bacilli (AFB) of 2–3 I.Lm in length and 0.3–0.5 I.Lm in width, straight or curved, isolated or grouped.

The *culture* of *M. ulcerans* is performed on a standard Löwenstein–Jensen medium at a low temperature (between 30 °C and 32 °C). However, *M. ulcerans* can survive at 37 °C for up to 13 days (Eddyani and Portaels 2007). The growth of the bacterium is very slow, and the results are often available only after 6–8 weeks. The sensitivity of culture is low (<60%) (Herbinger et al. 2009; Mensah-Quainoo et al. 2008).

PCR consists of amplifying small amounts of DNA to obtain easily detectable quantities in the laboratory. It targets an insertion sequence, *IS2404*, specific for *M. ulcerans* and presents at least 220 copies per genome. The use of the PCR technique for the detection of *M. ulcerans* was first suggested by Portaels et al. (1997). PCR targeting of *IS2404* was later developed in Australia (Ross et al. 1997; Stinear et al. 1999). The sensitivity of PCR remains high, even on punch biopsies (Stienstra et al. 2003; Rondini et al. 2003; Phillips et al. 2005) or FNA (Phillips et al. 2009; Eddyani et al. 2009; Cassisa et al. 2010; Herbinger et al. 2010). However, PCR is not suitable for the monitoring of treatment success, and currently, cultures are considered the only valid confirmatory test for the detection of viable bacilli (Beissner et al. 2010).

A *histopathological examination* is performed on biopsies fixed in 10% formalin and after hematoxylin eosin, Gram, Ziehl–Neelsen, and Grocott staining. The main histopathological lesion characteristic of BU is that of coagulative necrosis. The specific histopathological features have been described previously (Hayman and McQueen 1985; Hayman 1993; Guarner et al. 2003).

The histological examination of early lesions shows necrosis of dermal collagen, described as "fibrillary," resulting in an eosinophilic coagulum. In addition to involving the dermis, this necrosis may extend into the widened septa of the subcutaneous fat (Hayman and McQueen 1985). The collagen necrosis may be described as caseous. There is little cellular reaction despite the presence of large numbers of organisms. Although initially fibrillary, later there is a complete loss of all cellular and structural detail, resulting in an amorphous eosinophilic coagulum in the affected dermis, histopathologically indistinguishable from that seen in tuberculous infection. Together with fat and collagen, there is also necrosis and degeneration of elastic fibers. Fragments of elastic fibers may also be seen within the cytoplasm of giant cells in the granulomas adjacent to the necrotic areas. In addition to elastolysis, there is also elastosis; an increase in elastic tissue occurs with the fibrosis in the widened septa prior to any degenerative changes. Duplication or irregular thickening of the internal elastic lamina also occurs in the arteries in the septa in association with the intimal proliferation described previously. Slight calcification, seen histopathologically, is noted in many cases in association with fat necrosis. The calcification appears to be confined to the proteinaceous areas between the lipid collections within the necrotic fat, developing around the foci of mycobacterial proliferation (Hayman 1993).

Recurrent or persistent infection produces a granulomatous reaction with epithelioid macrophages, variable numbers of giant cells of the Langerhans type, and relatively few organisms. This type of reaction is associated with more successful treatment of the disease and appears analogous to the tuberculoid form of leprosy (Hayman 1993).

Although the histopathological lesions usually seen in BU disease are well described, defining the histopathological criteria for a diagnosis of BU disease remains of clinical and public health importance as it would allow earlier treatment, leading to less deforming sequelae.

Other Useful Exploration Methods

Radiography shows evidence of bone damage, which can occur even below apparently healthy skin or apparently early lesions, such as nodules (Portaels et al. 2008a). The images are not specific for BU. N'Zi et al. (1998) and Hans-Moévi et al. (2003) presented radiographic images of osteitis, demineralization, periosteal reaction, bone condensation, osteolysis, cortical thickening, and geodes.

Ultrasonography is an attractive, simple, and noninvasive method useful to assess the dermoepidermal damage caused by *M. ulcerans* and to follow the evolution of these lesions during treatment (Leigheb et al. 2008). Although the images are not specific, these can help distinguish, for example, early-stage nodules (shown as hyperechoic imbibitions of the adipose tissue and hypoechoic areas due to adiponecrosis) from late-stage ones (shown as hypoechoic images with round or ragged forms, due to the presence of secondary granulomas).

Other techniques based on the detection of the toxin in tissue samples are currently under investigation. The World Health Organization, in collaboration with FIND (The Foundation for Innovative New Diagnostics), has started a thorough evaluation of an innovative diagnostic test developed by researchers at Harvard University that can lead to rapid confirmation of Buruli ulcer in a patient. Initial results from recent field trials in Benin and Ghana show that the test can detect mycolactone, a toxin produced by the bacteria that causes tissue damage and leads to Buruli ulcer (Converse et al. 2014). The most important aspect of this new test, which is more sensitive than microscopy, is that it can be carried out by technicians with minimal training in district hospital laboratories.

Differential Diagnosis

Many conditions can present like BU. Table 3 presents a summary of the diseases that can present like BU according to the clinical form.

Table 3 Summary of the diseases that can present like BU according to the clinical form

BU clinical form	Differential diagnosis	
Nodule	Painful nodule	Insect bite, furuncle, anthrax, abscess, infected sebaceous cyst, furunculoid myiasis, erythema nodosum, hidradenitis, chondrodermatitis nodularis
	Mildly or not painful nodule	Lipome, onchocercal nodule, nodular leishmaniasis
	Isolated noninflammatory nodules	Foreign-body granuloma, gummas, dermatofibroma
Edema	Inflammatory edema	Infectious cutaneous cellulitis, staphylococcal or salmonella osteomyelitis, necrotizing fasciitis, phlegmon, injuries
	Edema	Renal/cardiac insufficiency, anemia, protein-energy malnutrition, erysipelas
	Noninflammatory edema	Kaposi's sarcoma, elephantiasis, actinomycosis, onchocercoma
Plaque	Hematoma, psoriasis, mycosis, leprosy, phlegmon, erysipelas, lupus vulgaris, eczema, urticaria, sarcoidosis	
Ulcer	Tropical phagedenic ulcer, ulcer of venous or arterial origin, ulcerated cutaneous tuberculosis, necrotizing fasciitis, ulcerated carcinoma or melanoma, leishmaniasis, sickle cell disease ulcer, injection abscess, African histoplasmosis, sporotrichosis, tuberculosis, lymphadenitis	
Scar with or without sequelae Bone involvement	Dystrophic burn scar, ankylosis secondary to skin or bone tuberculosis All other banal (staphylococcus, streptococcus, etc.) or specific (*Mycobacterium tuberculosis*) germs causing osteitis or osteomyelitis	

Sources: Guédénon and Portaels (2000), World Health Organization (2000, 2001a)

Control Tools and Strategies

The mode of transmission and incubation period of BU are not fully known. Thus, primary prevention remains a challenge. However, prevention may rely on avoidance of contact with the causative agent. This, obviously, is not easy to implement, particularly for farmers, fishermen, and others who are in contact with the contaminated environment during their daily tasks. However, wearing protective clothing when farming (Pouillot et al. 2007) and the use of soap for immediate cleansing of any skin injuries (Nackers et al. 2007) may reduce rates of infection. The use of protected sources of water for domestic purposes may reduce exposure to water contaminated by *M. ulcerans* and, consequently, may reduce the prevalence of BU (Debacker et al. 2006). Primary prevention activities should thus encourage health education on personal and environmental hygiene, the use of protective clothes when possible, the proper care of all injuries, and the avoidance of contact with natural water bodies.

Immunity can also be boosted by vaccination if a vaccine is available. A vaccine against *M. ulcerans* will be useful to protect children in hyperendemic foci, but it may also play a role in therapy by shortening the duration of antibiotic treatment and preventing recurrences and severe forms of the disease. Currently, there is no specific vaccine against *M. ulcerans*, but there is some evidence for cross-reactive protection against BU by the *M. bovis* BCG vaccine used against tuberculosis. However, the BCG vaccine produces only short-term protection against BU disease. There is considerable evidence supporting the notion that the development of a vaccine is feasible. Studies have indicated that some household contacts of BU disease patients have been exposed to *M. ulcerans* without developing the disease.

This may be because they developed a protective immune response or because of different issues, such as an inadequate infective dose of the organism or the wrong conditions for bacterial proliferation in exposed skin. In addition, there is much anecdotal evidence that some people develop pathological lesions after infection with *M. ulcerans* that heal in a short time without treatment (Huygen et al. 2009).

Currently, health education focuses its efforts on identifying, reporting, and treating early forms of the disease.

The surveillance system in BU is based on the involvement of village volunteers and the use of the WHO's BU02 form containing information on registered BU cases (World Health Organization 2000). The aim of the WHO's control strategy is to minimize the morbidity and disability associated with BU. To achieve this objective, the WHO recently made the following recommendations to endemic countries:

(1) Assess the actual magnitude of BU within each country.
(2) Strengthen the case confirmation capacity of national laboratories in line with the WHO recommendations.
(3) Intensify, at all levels, education on BU, especially in the affected communities, with a view to promoting early case detection.
(4) Ensure that cases are detected at an early stage to reduce the frequency of disabilities.

(5) Provide access to specific antibiotic treatment and surgical and rehabilitation services free of charge or at reduced cost for people affected by BU.
(6) Improve BU mapping and surveillance in affected countries and promote cross-border exchange of information.
(7) Support research through active international cooperation on epidemiology; social and economic determinants and the impact, prevention, and development of new diagnostic tools; and simplification of treatment with orally administered medicines.
(8) Mobilize additional resources for BU control.
(9) Promote effective collaboration with other sectors to control the disease.
(10) Promote the social and economic rehabilitation of people negatively affected by the disease.
(11) Strengthen further the primary healthcare system in the affected areas to improve the integration and implementation of control and disability prevention activities (World Health organization 2009).

Chemotherapy-Based Strategies

The WHO recommends combination therapy for BU, using rifampin (10 mg/kg per day) and clarithromycin (7.5 mg/kg body weight by mouth twice a day for a total of 8 weeks). The extended-release formulation of clarithromycin may be used at 15 mg/kg body weight once daily. The aim of the treatment is to cure *M. ulcerans* infection and to achieve a disability-free outcome with minimal scarring (World Health Organization 2012). Thus, it is expected that most category 1 and some category 2 lesions will completely heal with antibiotic treatment alone. However, most category 3 lesions will still require surgery (excision, skin grafting, or amputation in severe cases) in addition to antibiotics. Multiple small lesions and lesions located at critical sites may heal with antibiotics alone (World Health Organization 2008b).

There is growing evidence of the efficacy of some rifampicin-based oral therapies.

When BU osteomyelitis is suspected, the first line of treatment should be with rifampicin and streptomycin for 8 weeks. Surgery is usually required to remove nonviable bone and to hasten healing. The response to treatment can be monitored radiologically.

Recurrence is rare after a full 8-week course of antibiotics has been completed. If antibiotics were given for fewer than 8 weeks in the first course or poor adherence is suspected, further antibiotic treatment or surgery should be considered. The duration of a second course of antibiotics depends on the situation, but care must be taken not to give streptomycin for more than 90 days in total.

Paradoxical reactions usually resolve without further antibiotic treatment. Clinicians should not therefore rush to restart or extend antibiotic treatment. Paradoxical reactions can be managed conservatively with drainage of pus and routine dressings. Samples should be taken for culture and histopathology. Underlying osteomyelitis should be eliminated. If the lesion due to a paradoxical reaction does not appear to be improving within 6 weeks of conservative treatment, surgery should be considered, and samples should be taken for microbiological and histopathological analyses to look for progressive active infection as an alternative diagnosis. If this reaction occurs during the course of antibiotic treatment, however, the full course of 8 weeks should be completed.

Although further studies are required to improve our understanding of BU/HIV coinfection, the WHO recommends that the management of such cases follow the guidelines for managing TB/HIV coinfection. First, HIV counseling and testing should be offered for all patients presenting with BU. Second, coinfected patients should be screened for tuberculosis. Third, as for TB, coinfected patients may receive early antiretroviral treatment to ensure a better response to treatment irrespective of CD4 count. Due to the interaction between rifampicin and some ARVs, the NNRTI component of the ART regimen should be changed from NVP to EFV, and if protease inhibitors are being used, the TB/HIV guidelines should be consulted for suggested management. Combination antibiotic treatment for BU should be commenced before initiating ART for HIV. Full guiding principles of management of BU/HIV co-infected patients have been recently published by World Health Organization (2015).

Other Treatment

Surgery

Until 2004, surgical intervention was the only therapeutic management technique recommended for BU. Surgical intervention typically requires large excisions of necrotic tissues with a rim of healthy tissue wider than 2 cm. Excision is usually followed by a thin-skin autograft (World Health Organization 2001b), and in some instances, multiple surgical interventions are required. Unfortunately, surgical intervention frequently results in serious physical sequelae (Barogui et al. 2009) and may have a negative psychological impact on the patient (Aujoulat et al. 2003).

Currently, it has been demonstrated that all forms of Buruli ulcer—papules, nodules, plaques, edema, and ulcers—however extensive, respond well to antibiotic treatment. Wide surgical removal of infected tissue is therefore no longer necessary to achieve microbiological cure. Conservative surgery (debridement, skin grafting, and scar revision) may still be required only to speed healing of large lesions and to minimize scarring that might limit movement (World Health Organization 2012).

The best time to undertake debridement and skin grafting, when indicated, is yet to be determined. Grafting should be late enough to allow all the *M. ulcerans* organisms to be killed by antibiotics but early enough to promote rapid recovery and a return to normal activities. Skin grafts should be applied to healthy vascular surfaces. It provides healthy new skin that is stronger and more flexible and is thus better able to withstand minor trauma. A sound conservative approach is to allow 8 weeks of antibiotic treatment before surgical intervention (World Health Organization 2012).

Wound and Scar Management

Wound care is an important component of treatment but has received little attention so far. Various antiseptics and consumables are used for dressing in sub-Saharan Africa endemic areas. They are nonspecific and have no action on the mycobacteria. They are used to prevent secondary infection and necrosis and to prepare the wound for surgery (skin grafting). The most widely used antiseptics are povidone-iodine, Dakin, Mercryl, eosin, and hydrogen peroxide.

Since 2012, the WHO recommendations for wound management are as follows (World Health Organization 2012):

- Manage systemic conditions appropriately.
- Protect the wound from trauma.
- Maintain a clean wound base and control infection.
- Maintain a moist wound environment.
- Control peri-wound lymphedema and edema.

Scar tissue resulting from slow normal healing may cause adhesions to and between underlying structures, which limit movement and are painful. Thick or tight scars resist injuries poorly, may limit movement, and detract from appearance and function. Their surface is usually dry, may crack or ulcerate, and is easily damaged by the sun. Such scars are easily injured during work or play. Note that scars that split frequently or ulcerate may, over many years, develop into squamous-cell carcinomas. Good scar management lessens complications. Management may involve moisturizing, massage, limb elevation, correct positioning of joints, and application of a pressure bandage for up to 1–2 years after natural healing or grafting (World Health Organization 2012).

Prevention of Disabilities

This component of treatment has also received less attention since the beginning of the Global BU initiative (GBUI). However, within the past 4 years, the prevention of disability (POD) has gained greater attention as an important part of BU management. The WHO has published two important documents (World Health Organization 2006, 2008a) to allow for the easy integration of POD activities into the primary healthcare system (World Health Organization 2008a).

Challenges of Program Implementation

According to the new road map for neglected tropical diseases 2021–2030 (WHO 2020), Buruli ulcer is the target for control and three programmatic objectives were set to be targeted by endemic control programs by the end of 2030 as follows:

1. Proportion of cases in category III (late stage) at diagnosis: <10%.
2. Proportion of laboratory-confirmed cases: >95%.
3. Proportion of confirmed cases who have completed a full course of antibiotic treatment: >98%.

These targets reflect the improvements in the accuracy of clinical diagnosis, correct sampling techniques, and the impact of early detection efforts (village education and active surveillance). To achieve the above-mentioned 2030 targets, the WHO recommends to national BU programs:

- To develop or revise their integrated national strategic plans, taking into consideration recent developments in Buruli ulcer control.
- To promote integrated skin NTD surveillance, including Buruli ulcer, and document all cases, using electronic and paper of the BU01 and BU02 new forms.
- To strengthen the healthcare system at all levels through capacity building of health workers on clinical diagnosis, sampling, laboratory confirmation, treatment guidelines, and prevention of disabilities.
- To enhance political commitment among endemic countries and partners to mobilize funds and human resources.
- To build community engagement and mobilization to support program implementation.

In order to improve the accuracy of clinical diagnosis of Buruli ulcer, a facility will be set up for centers in which Internet access is available to upload clinical data and photographs of lesions for secondary remote confirmation. Laboratories should be strengthened for PCR confirmation and should participate in external quality assurance programs and internal quality control programs for microscopy in collaboration with national tuberculosis control programs.

Further Research for Policy and Control

With regard to the epidemiological aspects, and in order to improve the surveillance system, the distribution and dynamics of the evolution of the disease over time should be better documented.

Regarding clinical aspects, better insights are needed on paradoxical reactions and BU/HIV coinfections.

The efficacy of oral treatment should be assessed.

New promising medicines should be evaluated to provide new treatment options including reduction in duration of treatment as well as the best time for surgery when indicated.

Wound dressing and prevention of disability implementation impacts on the limitation of social consequences of the disease should be documented.

Outlook for the Next Decade

While studies to elucidate the mode of transmission continue so that possible preventive strategies can be developed, the short- to medium-term priorities are to improve the diagnosis and treatment of patients (WHO 2020).

For diagnosis, there are two potential rapid tests that are likely to be in field use in the coming years. The first is the direct detection of mycolactone using simplified thin-layer chromatography. Preliminary results from human samples are promising, and a larger field trial was planned for 2015–2016 in Benin, DRC, and Ghana. The second is the antigen capture test converted into lateral flow format (Dreyer et al. 2015).

Improve detection of viable *M. ulcerans* in wound samples to distinguish treatment failures and paradoxical reactions through methods such as mycolactone detection and 16S rRNA.

References

Ackumey MM, Gyapong M, Pappoe M, Kwakye-Maclean C, Weiss MG (2012) Illness meanings and experiences for pre-ulcer and ulcer conditions of Buruli ulcer in the Ga-west and ga-south municipalities of Ghana. BMC Public Health 12:264

Adusumilli S, Mve-Obiang A, Sparer T, Meyers W, Hayman J, Small PL (2005) *Mycobacterium ulcerans* toxic macrolide, mycolactone modulates the host immune response and cellular location of *M. ulcerans* in vitro and in vivo. Cell Microbiol 7:1295–1304

Amofah G, Bonsu F, Tetteh C, Okrah J, Asamoa K, Asiedu K, Addy J (2002) Buruli ulcer in Ghana: results of a national case search. Emerg Infect Dis 8:167–170

Andersen FO (1965) Mycobacterial skin ulcers-clinical experience. Cent Afr J Med 11:131–135

Asiedu K, Etuaful S (1998) Socioeconomic implications of Buruli ulcer in Ghana: a three-year review. Am J Trop Med Hyg 59(6):1015–1022

Aujoulat I, Johnson C, Zinsou C, Guedenon A, Portaels F (2003) Psychosocial aspects of health seeking behaviors of patients with Buruli ulcer in southern Benin. Trop Med Int Health 8:750–759

Bär W, Rüsch-Gerdes S, Richter E, de Bär MG, Dittmer C, Papsdorf H, Stosiek P, de Rijk PB, Meyers WM, Portaels F (1998) Mycobacterium ulcerans infection in a child from Angola: diagnosis by direct detection and culture. Trop Med Int Health 3:189–196

Barker DJ (1972) The distribution of Buruli disease in Uganda. Trans R Soc Trop Med Hyg 66:867–874

Barker DJ (1973) Epidemiology of *Mycobacterium ulcerans* infection. Trans R Soc Trop Med Hyg 67:43–50

Barker DJP, Clancey JK, Morrow RH, Rao S (1970) Transmission of Buruli disease. Br Med J 4:558

Barogui Y, Johnson RC, van der Werf TS, Sopoh G, Dossou A, Dijkstra PU, Stienstra Y (2009) Functional limitations after surgical or antibiotic treatment for Buruli ulcer in Benin. Am J Trop Med Hyg 81:82–87

Barogui YT, Klis SA, Johnson RC, Phillips RO, van der Veer E, van Diemen C, van der Werf TS, Stienstra Y (2016) Genetic susceptibility and predictors of paradoxical reactions in Buruli ulcer. PLoS Negl Trop Dis 10(4):e0004594. https://doi.org/10.1371/journal.pntd.0004594

Bayley AC (1971) Buruli ulcer in Ghana. Br Med J 2:401–402

Beissner M, Herbinger KH, Bretzel G (2010) Laboratory diagnosis of Buruli ulcer disease. Future Microbiol 5:363–370

Cassisa V, Chauty A, Marion E, Ardant MF, Eyangoh S, Cottin J, Aubry J, Koussemou H, Lelièvre B, Férec S, Tekaia F, Johnson C, Marsollier L (2010) Use of fine-needle aspiration for diagnosis of *mycobacterium ulcerans* infection. J Clin Microbiol 48:2263–2264

Chukwuekezie O, Ampadu E, Sopoh G, Dossou A, Tiendrebeogo A, Sadiq L, Portaels F, Asiedu K (2007) Buruli ulcer, Nigeria. Emerg Infect Dis 13:782–783

Converse PJ, Xing Y, Kim KH, Tyagi S, Li S-Y et al (2014) Accelerated detection of mycolactone production and response to antibiotic treatment in a mouse model of mycobacterium ulcerans disease. PLoS Negl Trop Dis 8(1):e2618

De Gentile PL, Mahaza C, Rolland F, Carbonnelle B, Verret JL, Chabasse D (1992) Cutaneous ulcer from *mycobacterium ulcerans*. Apropos of 1 case in French Guiana. Bull Soc Pathol Exot 85:212–214

Debacker M, Zinsou C, Aguiar J, Meyers W, Portaels F (2002) *Mycobacterium ulcerans* disease (Buruli ulcer) following human bite. Lancet 360:1830

Debacker M, Zinsou C, Aguiar J, Meyers WM, Portaels F (2003) First case of *mycobacterium ulcerans* disease (Buruli ulcer) following a human bite. Clin Infect Dis 36:e67–e68

Debacker M, Portaels F, Aguiar J, Steunou C, Zinsou C, Meyers W, Dramaix M (2006) Risk factors for Buruli ulcer, Benin. Emerg Infect Dis 12:1325–1331

Dobos KM, Small PL, Deslauriers M, Quinn FD, King CH (2001) *Mycobacterium ulcerans* cytotoxicity in an adipose cell model. Infect Immun 69:7182–7186

Dreyer A, Röltgen K, Dangy JP, Ruf MT, Scherr N, Bolz M, Tobias N, Charles Moes C, Vettiger A, Stinear T, Pluschke G (2015) Identification of the mycobacterium ulcerans protein MUL_3720 as a promising target for the development of a diagnostic test for Buruli ulcer. PLoS Negl Trop Dis 9(2):e0003477. https://doi.org/10.1371/journal.pntd.0003477

Duker AA, Portaels F, Hale M (2006) Pathways of *mycobacterium ulcerans* infection: a review. Environ Int 32:567–573

Durnez L, Suykerbuyk P, Nicolas V, Barriere P, Verheyen E, Johnson CR, Leirs H, Portaels F (2010) Terrestrial small mammals as reservoirs of *mycobacterium ulcerans* in Benin. Appl Environ Microbiol 76:4574–4577

Eddyani M, Portaels F (2007) Survival of *mycobacterium ulcerans* at 37 degrees C. Clin Microbiol Infect 13:1033–1035

Eddyani M, Ofori-Adjei D, Teugels G, De Weirdt D, Boakye D, Meyers WM, Portaels F (2004) Potential role for fish in transmission of *mycobacterium ulcerans* disease (Buruli ulcer): an environmental study. Appl Environ Microbiol 70:5679–5681

Eddyani M, Debacker M, Martin A, Aguiar J, Johnson CR, Uwizeye C, Fissette K, Portaels F (2008) Primary culture of *mycobacterium ulcerans* from human tissue specimens after storage in semisolid transport medium. J Clin Microbiol 46:69–72

Eddyani M, Fraga AG, Schmitt F, Uwizeye C, Fissette K, Johnson C, Aguiar J, Sopoh G, Barogui Y, Meyers WM, Pedrosa J, Portaels F (2009) Fine-needle aspiration, an effi cient sampling technique for bacteriological diagnosis of nonulcerative Buruli ulcer. J Clin Microbiol 47:1700–1704

Elsner L, Wayne J, O'Brien CR, McCowan C, Malik R, Hayman JA, Globan M, Lavender CJ, Fyfe JA (2008) Localised *mycobacterium ulcerans* infection in a cat in Australia. J Feline Med Surg 10:407–412

Faber WR, Arias-Bouda LM, Zeegelaar JE, Kolk AH, Fonteyne PA, Toonstra J, Portaels F (2000) First reported case of *Mycobacterium ulcerans* infection in a patient from China. Trans R Soc Trop Med Hyg 94:277–279

Fenner F (1950) The signifi cance of the incubation period in infectious diseases. Med J Aust 2:813–818

Fraga AG, Cruz A, Martins TG, Torrado E, Saraiva M, Pereira DR, Meyers WM, Portaels F, Silva MT, Castro AG, Pedrosa J (2011) *Mycobacterium ulcerans* triggers T cell immunity followed by local and regional but not systemic immunosuppression. Infect Immun 79(1):421–430

Fyfe JA, Lavender CJ, Handasyde KA, Legione AR, O'Brien CR, Stinear TP, Pidot SJ, Seemann T, Benbow ME, Wallace JR, McCowan C, Johnson PD (2010) A major role for mammals in the ecology of *Mycobacterium ulcerans*. PLoS Negl Trop Dis 4:e791

George KM, Chatterjee D, Gunawardana G, Welty D, Hayman J, Lee R, Small PL (1999) Mycolactone: a polyketide toxin from *mycobacterium ulcerans* required for virulence. Science 283:854–857

George KM, Pascopella L, Welty DM, Small PL (2000) A *mycobacterium ulcerans* toxin, myco-lactone, causes apoptosis in Guinea pig ulcers and tissue culture cells. Infect Immun 68:877–883

Goutzamanis JJ, Gilbert GL (1995) *Mycobacterium ulcerans* infection in Australian children: report of eight cases and review. Clin Infect Dis 21:1186–1192

Guarner J, Bartlett J, Whitney EA, Raghunathan PL, Stienstra Y, Asamoa K, Etuaful S, Klutse E, Quarshie E, van der Werf TS, van der Graaf WT, King CH, Ashford DA (2003) Histopathologic features of *Mycobacterium ulcerans* infection. Emerg Infect Dis 9:651–656

Guédénon A, Portaels F (2000) Ulcere de Buruli: guide diagnostic à l'usage des professionnels de la santé. DGIS/DGCI, Belgique

Guerra H, Palomino JC, Falconí E, Bravo F, Donaires N, Van Marck E, Portaels F (2008) *Mycobacterium ulcerans* disease, Peru. Emerg Infect Dis 14:373–377

Hans-Moévi A, Sossou R, Boco V, Goudjo KPW, Johnson RC, Odulami H (2003) Aspets radiologiques des atteintes osseuses de la maladie de Buruli. J Afr Image 3:25–30

Hayman J (1993) Out of Africa: observations on the histopathology of *Mycobacterium ulcerans* infection. J Clin Pathol 46:5–9

Hayman J, McQueen A (1985) The pathology of *Mycobacterium ulcerans* infection. Pathology 17:594–600

Herbinger KH, Adjei O, Awua-Boateng NY, Nienhuis WA, Kunaa L, Siegmund V, Nitschke J, Thompson W, Klutse E, Agbenorku P, Schipf A, Reu S, Racz P, Fleischer B, Beissner M, Fleischmann E, Helfrich K, van der Werf TS, Löscher T, Bretzel G (2009) Comparative study of the sensitivity of different diagnostic methods for the laboratory diagnosis of Buruli ulcer disease. Clin Infect Dis 48:1055–1064

Herbinger KH, Beissner M, Huber K, Awua-Boateng NY, Nitschke J, Thompson W, Klutse E, Agbenorku P, Assiobo A, Piten E, Wiedemann F, Fleischmann E, Helfrich K, Adjei O, Löscher T, Bretzel G (2010) Efficiency of fine-needle aspiration compared with other sampling techniques for laboratory diagnosis of Buruli ulcer disease. J Clin Microbiol 48:3732–3734

Huygen K, Adjei O, Affolabi D, Bretzel G, Demangel C, Fleischer B, Johnson RC, Pedrosa J, Phanzu DM, Phillips RO, Pluschke G, Siegmund V, Singh M, van der Werf TS, Wansbrough-Jones M, Portaels F (2009) Buruli ulcer disease: prospects for a vaccine. Med Microbiol Immunol 198:69–77

Janssens PG, Pattyn SR, Meyers WM, Portaels F (2005) Buruli ulcer: an historical overview with updating to 2005. Bull Séances Acad R Sci Outre-Mer 51:265–299

Johnson PD, Veitch MG, Leslie DE, Flood PE, Hayman JA (1996) The emergence of *Mycobacterium ulcerans* infection near Melbourne. Med J Aust 164:76–78

Johnson RC, Ifebe D, Hans-Moevi A, Kestens L, Houessou R, Guédénon A, Meyers WM, Portaels F (2002) Disseminated *mycobacterium ulcerans* disease in an HIV-positive patient: a case study. AIDS 16:1704–1705

Johnson RC, Makoutodé M, Sopoh GE, Elsen P, Gbovi J, Pouteau LH, Meyers WM, Boko M, Portaels F (2005) Buruli ulcer distribution in Benin. Emerg Infect Dis 11:500–501

Johnson PD, Azuolas J, Lavender CJ, Wishart E, Stinear TP, Hayman JA, Brown L, Jenkin GA, Fyfe JA (2007) *Mycobacterium ulcerans* in mosquitoes captured during outbreak of Buruli ulcer, southeastern Australia. Emerg Infect Dis 13:1653–1660

Johnson RC, Nackers F, Glynn JR, de Biurrun BE, Zinsou C, Aguiar J, Tonglet R, Portaels F (2008) Association of HIV infection and mycobacterium ulcerans disease in Benin. AIDS 22(7):901–903. https://doi.org/10.1097/QAD.0b013e3282f7690a

Kanga JM, Kacou ED (2001) Epidemiological aspects of Buruli ulcer in Côte d'Ivoire: results of a national survey. Bull Soc Pathol Exot 94:46–51. French

Kibadi K, Panda M, Tamfum JJ, Fraga AG, Longatto Filho A, Anyo G, Pedrosa J, Nakazawa Y, Suykerbuyk P, Meyers WM, Portaels F (2008) New foci of Buruli ulcer, Angola and Democratic Republic of Congo. Emerg Infect Dis 14:1790–1792

Kondo M, Kurokawa I, Ito Y, Yamanaka K, Yamazaki T, Mizutani H (2009) Leg ulcer caused by *mycobacterium ulcerans* ssp. shinshuense infection. Int J Dermatol 48:1330–1333
Lavender CJ, Senanayake SN, Fyfe JA, Buntine JA, Globan M, Stinear TP, Hayman JA, Johnson PD (2007) First case of *Mycobacterium ulcerans* disease (Bairnsdale or Buruli ulcer) acquired in New South Wales. Med J Aust 186:62–63
Leigheb G, Cammarota T, Zavattaro E, Sarno A, Carriero A, Gambaro AC, Dossou A, Poggio F, Clemente C, Johnson RC, Sopoh G, Leigheb F (2008) Ultrasonography for the monitoring of subcutaneous damage in *mycobacterium ulcerans* infection (Buruli ulcer). Ultrasound Med Biol 34:1554–1563
Lunn HF, Connor DH, Wilks NE, Barnley GR, Kamunvi FA, Clancey JK, Bee JD (1965) Buruli (mycobacterial) ulceration in Uganda (a new focus of Buruli ulcer in Madi district, Uganda): report of a field study. East Afr Med J 42:275–288
MacCallum P, Tolhurst JC, Buckle G, Sissons H (1948) A new mycobacterial infection in man. J Pathol Bacteriol 60:93–122
Marsollier L, Robert R, Aubry J, Saint André JP, Kouakou H, Legras P, Manceau AL, Mahaza C, Carbonnelle B (2002) Aquatic insects as a vector for *mycobacterium ulcerans*. Appl Environ Microbiol 68(9):4623–4628
Marsollier L, Sévérin T, Aubry J, Merritt RW, Saint André JP, Legras P, Manceau AL, Chauty A, Carbonnelle B, Cole ST (2004) Aquatic snails, passive hosts of *Mycobacterium ulcerans*. Appl Environ Microbiol 70:6296–6298
Marsollier L, Aubry J, Milon G, Brodin P (2007) Aquatic insects and transmission of *Mycobacterium ulcerans*. Med Sci (Paris) 23:572–575
Mensah-Quainoo E, Yeboah-Manu D, Asebi C, Patafuor F, Ofori-Adjei D, Junghanss T, Pluschke G (2008) Diagnosis of *Mycobacterium ulcerans* infection (Buruli ulcer) at a treatment Centre in Ghana: a retrospective analysis of laboratory results of clinically diagnosed cases. Trop Med Int Health 13:191–198
Merritt RW, Benbow ME, Small PLC (2005) Unraveling an emerging disease associated with disturbed aquatic environments: the case of Buruli ulcer. Front Ecol Environ 3:323–331
Meyers WM, Shelly WM, Connor DH, Meyers EK (1974) Human *Mycobacterium ulcerans* infections developing at sites of trauma to skin. Am J Trop Med Hyg 23:919–923
Meyers WM, Tignokpa N, Priuli GB, Portaels F (1996) *Mycobacterium ulcerans* infection (Buruli ulcer): first reported patients in Togo. Br J Dermatol 134:1116–1121
Minime-Lingoupou F, Beyam N, Zandanga G, Manirakiza A, N'Domackrah A, Njuimo S, Eyangoh S, Cottin J, Marsollier L, Marion E, Portaels F, Le Faou A, Bercion R (2010) Buruli ulcer, Central African Republic. Emerg Infect Dis 16:746–748
N'Zi KP, N'Dri K, Aka BR, Diabate AS, Ouattara DN, Djedje AT (1998) Radiographic aspects of osteoarticular complications in Buruli ulcers. Bull Soc Pathol Exot 91:229–231. French
Nackers F, Johnson RC, Glynn JR, Zinsou C, Tonglet R, Portaels F (2007) Environmental and health-related risk factors for *mycobacterium ulcerans* disease (Buruli ulcer) in Benin. Am J Trop Med Hyg 77:834–836
Oliveira MS, Fraga AG, Torrado E, Castro AG, Pereira JP, Filho AL, Milanezi F, Schmitt FC, Meyers WM, Portaels F, Silva MT, Pedrosa J (2005) Infection with *Mycobacterium ulcerans* induces persistent infl ammatory responses in mice. Infect Immun 73:6299–6310
Oluwasanmi JO, Solanke TF, Olurin EO, Itayemi SO, Alabi GO, Lucas AO (1976) *Mycobacterium ulcerans* (Buruli) skin ulceration in Nigeria. Am J Trop Med Hyg 25:122–128
Ouoba K, Sano D, Traor A, Oudraogo R, Sakand B, Sanou A (1998) Buruli ulcers in Burkina Faso: apropos of 6 cases. Tunis Med 76:46–50
Peduzzi E, Groeper C, Schütte D, Zajac P, Rondini S, Mensah-Quainoo E, Spagnoli GC, Pluschke G, Daubenberger CA (2007) Local activation of the innate immune system in Buruli ulcer lesions. J Invest Dermatol 127:638–645
Phillips R, Horsfield C, Kuijper S, Lartey A, Tetteh I, Etuaful S, Nyamekye B, Awuah P, Nyarko KM, Osei-Sarpong F, Lucas S, Kolk AH, Wansbrough-Jones M (2005) Sensitivity of PCR targeting the IS2404 insertion sequence of *Mycobacterium ulcerans* in an assay using punch biopsy specimens for diagnosis of Buruli ulcer. J Clin Microbiol 43:3650–3656

Phillips RO, Sarfo FS, Osei-Sarpong F, Boateng A, Tetteh I, Lartey A, Adentwe E, Opare W, Asiedu KB, Wansbrough-Jones M (2009) Sensitivity of PCR targeting *Mycobacterium ulcerans* by use of fine-needle aspirates for diagnosis of Buruli ulcer. J Clin Microbiol 47:924–926

Portaels F (1989) Epidemiology of ulcers due to *Mycobacterium ulcerans*. Ann Soc Belg Med Trop 69:91–103. Review. French

Portaels F (2004) Mycobacteria and tropical regions: from the environment to man. Med Trop 64:219–220. French

Portaels F, Aguiar J, Fissette K, Fonteyne PA, De Beenhouwer H, de Rijk P, Gudnon A, Lemans R, Steunou C, Zinsou C, Dumonceau JM, Meyers WM (1997) Direct detection and identifi ca-tion of *Mycobacterium ulcerans* in clinical specimens by PCR and oligonucleotide-specific capture plate hybridization. J Clin Microbiol 35:1097–1100

Portaels F, Elsen P, Guimaraes-Peres A, Fonteyne PA, Meyers WM (1999) Insects in the transmission of *Mycobacterium ulcerans* infection. Lancet 353:986

Portaels F, Chemlal K, Elsen P, Johnson PD, Hayman JA, Hibble J, Kirkwood R, Meyers WM (2001) *Mycobacterium ulcerans* in wild animals. Rev Sci Tech 20:252–264. Review

Portaels F, Johnson C, Aguiar J, Meyers WM, Debacker M (2008a) Etude de 106 cas d'ulcères de Buruli avec atteintes osseuses trait&s à Zagnanado, B&nin. Bull de l'ALLF 23:48–50

Portaels F, Meyers WM, Ablordey A, Castro AG, Chemlal K, de Rijk P, Elsen P, Fissette K, Fraga AG, Lee R, Mahrous E, Small PL, Stragier P, Torrado E, Van Aerde A, Silva MT, Pedrosa J (2008b) First cultivation and characterization *of mycobacterium ulcerans* from the environment. PLoS Negl Trop Dis 2:e178

Pouillot R, Matias G, Mbondji Wondje C, Portaels F, Valin N, Ngoss F, Njikap A, Marsollier L, Fontanet A, Eyangoh S (2007) Risk factors for Buruli ulcer disease in Cameroon. PLoS Negl Trop Dis 1:e101

Ravisse P (1977) Skin ulcer caused by *mycobacterium ulcerans* in Cameroon: clinical, epidemiological and histological study. Bull Soc Pathol Exot Fil 70:109–124. French

Rondini S, Mensah-Quainoo E, Troll H, Bodmer T, Pluschke G (2003) Development and application of real-time PCR assay for quantification of *Mycobacterium ulcerans* DNA. J Clin Microbiol 41:4231–4237

Ross BC, Marino L, Oppedisano F, Edwards R, Robins-Browne RM, Johnson PD (1997) Development of a PCR assay for rapid diagnosis of *Mycobacterium ulcerans* infection. J Clin Microbiol 35:1696–1700

Ruf MT, Chauty A, Adeye A, Ardant MF, Koussemou H, Johnson RC, Pluschke G (2011) Secondary Buruli ulcer skin lesions emerging several months after completion of chemotherapy: paradoxical reaction or evidence for immune protection? PLoS Negl Trop Dis 5(8):e1252. https://doi.org/10.1371/journal.pntd.0001252

Silva MT, Portaels F, Pedrosa J (2007) Aquatic insects and *Mycobacterium ulcerans*: an association relevant to Buruli ulcer control? PLoS Med 4:e63

Silva MT, Portaels F, Pedrosa J (2009) Pathogenetic mechanisms of the intracellular parasite *Mycobacterium ulcerans* leading to Buruli ulcer. Lancet Infect Dis 9:699–710

Simmonds RE, Lali FV, Smallie T, Small PL, Foxwell BM (2009) Mycolactone inhibits monocyte cytokine production by a posttranscriptional mechanism. J Immunol 182:2194–2202

Sizaire V, Nackers F, Comte E, Portaels F (2006) *Mycobacterium ulcerans* infection: control, diagnosis, and treatment. Lancet Infect Dis 6:288–296

Smith JH (1970) Epidemiologic observations on cases of Buruli ulcer seen in a hospital in the lower Congo. Am J Trop Med Hyg 19:657–663

Stienstra Y, van der Werf TS, Guarner J, Raghunathan PL, Spotts Whitney EA, van der Graaf WT, Asamoa K, Tappero JW, Ashford DA, King CH (2003) Analysis of an IS2404-based nested PCR for diagnosis of Buruli ulcer disease in regions of Ghana where the disease is endemic. J Clin Microbiol 41:794–797

Stinear T, Ross BC, Davies JK, Marino L, Robins-Browne RM, Oppedisano F, Sievers A, Johnson PD (1999) Identification and characterization of IS2404 and IS2606: two distinct

repeated sequences for detection of *Mycobacterium ulcerans* by PCR. J Clin Microbiol 37:1018–1023
Toll A, Gallardo F, Ferran M, Gilaberte M, Iglesias M, Gimeno JL, Rondini S, Pujol RM (2005) Aggressive multifocal Buruli ulcer with associated osteomyelitis in an HIV-positive patient. Clin Exp Dermatol 30:649–651
Torrado E, Adusumilli S, Fraga AG, Small PL, Castro AG, Pedrosa J (2007a) Mycolactone-mediated inhibition of tumor necrosis factor production by macrophages infected with *Mycobacterium ulcerans* has implications for the control of infection. Infect Immun 75:3979–3988
Torrado E, Fraga AG, Castro AG, Stragier P, Meyers WM, Portaels F, Silva MT, Pedrosa J (2007b) Evidence for an intramacrophage growth phase of *Mycobacterium ulcerans*. Infect Immun 75:977–987
Torrado E, Fraga AG, Logarinho E, Martins TG, Carmona JA, Gama JB, Carvalho MA, Proença F, Castro AG, Pedrosa J (2010) IFN-gamma-dependent activation of macrophages during experimental infections by Mycobacterium ulcerans is impaired by the toxin mycolactone. J Immunol 184(2):947–955. https://doi.org/10.4049/jimmunol.0902717
van der Werf TS, Stinear T, Stienstra Y, van der Graaf WT, Small PL (2003) Mycolactones and *Mycobacterium ulcerans* disease. Lancet 362:1062–1064
van Zyl A, Daniel J, Wayne J, McCowan C, Malik R, Jelfs P, Lavender CJ, Fyfe JA (2010) *Mycobacterium ulcerans* infections in two horses in South-Eastern Australia. Aust Vet J 88:101–106. Review
Wallace JR, Gordon MC, Hartsell L, Mosi L, Benbow ME, Merritt RW, Small PL (2010) Interaction of *Mycobacterium ulcerans* with mosquito species: implications for transmission and trophic relationships. Appl Environ Microbiol 76:6215–6222
Walsh DS, Portaels F, Meyers WM (2008) Buruli ulcer (*Mycobacterium ulcerans* infection). Trans R Soc Trop Med Hyg 102:969–978
Walsh DS, Eyase F, Onyango D, Odindo A, Otieno W, Waitumbi JN, Bulimo WD, Schnabel DC, Meyers WM, Portaels F (2009) Short report: clinical and molecular evidence for a case of Buruli ulcer (Mycobacterium ulcerans infection) in Kenya. Am J Trop Med Hyg 81:1110–1113
Walsh DS, Portaels F, Meyers WM (2010) Recent advances in leprosy and Buruli ulcer (Mycobacterium *ulcerans* infection). Curr Opin Infect Dis 23:445–455
Walsh DS, Portaels F, Meyers WM (2011) Buruli ulcer: advances in understanding *Mycobacterium ulcerans* infection. Dermatol Clin 29:1–8
Watanabe T, Ohkusu K, Nakanaga K, Ishii N, Nakashima K, Shindo M, Yoshida Y, Yamamoto O (2010) Buruli ulcer caused by "*Mycobacterium ulcerans* subsp. shinshuense". Eur J Dermatol 20:809–810
WHO (2014) In: Portaels F (ed) Laboratory diagnosis of Buruli ulcer: a manual for health care providers. World Health Organization, Geneva, WHO/HTM/NTD/IDM/2014.1
World Health Organisation (2004) Provisional guidance on the role of specific antibiotics in the management of *Mycobacterium ulcerans* disease (Buruli ulcer). WHO/CDS/CPE/GBUI.10, Geneva, http://whqlibdoc.who.int/hq/2004/WHO_CDS_CPE_GBUI_2004.10.pdf: 42p
World Health Organization (1997) WHO joins battle against a new emerging disease, Buruli ulcer. (Press release WHO/88), Geneva
World Health Organization (2000) Buruli ulcer: *Mycobacterium ulcerans* infection. In: Asiedu K, Scherpbier RW, Raviglione M (eds) WHO/CDS/GBUI/2000.1, Geneva. 160p
World Health Organization (2001a) Diagnosis of *Mycobacterium ulcerans* disease (Buruli ulcer). WHO/CDS/CPE/GBUI/2001.4, Geneva, http://whqlibdoc.who.int/hq/2001/WHO_CDS_CPE_GBUI_2001.4.pdf: 100p
World Health Organization (2001b) Management of *Mycobacterium ulcerans* disease (Buruli ulcer). WHO/CDS/CPE/GBUI/2001.3, Geneva, http://whqlibdoc.who.int/hq/2001/WHOCDS_CPE_GBUI_2001.3.pdf: 80p
World Health Organization (2003) Buruli ulcer disease. *Mycobacterium ulcerans* infection. Wkly Epidemiol Rec 78:163–168
World Health Organization (2006) Buruli ulcer: prevention of disability (POD). WHO/CDS/NTD/GBUI/2006, Geneva, http://whqlibdoc.who.int/publications/2006/9241546816_eng.pdf: 115p

World Health Organization (2008a) Buruli ulcer: progress report, 2004–2008. Wkly Epidemiol Rec 83:145–156

World Health Organization (2008b) Prevention of disability in Buruli ulcer: basic rehabilitation-practical guide. WHO/HTM/NTD/IDM/GBUI/2008.1, Geneva, http://whqlibdoc.who.int/hq/2008/WHO_HTM_NTD_IDM_GBUI_2008.1_eng.pdf :104p

World Health organization (2009) Cotonou declaration on Buruli ulcer. WHO, Geneva. http://www.who.int/neglected_diseases/Benin_declaration_2009_eng_ok.pdf. Accessed the 18 Nov 2010

World Health Organization (2010a) Guidance on sampling techniques for laboratory-confirmation of *Mycobacterium ulcerans* infection (Buruli ulcer disease). http://www.who.int/buruli/Guidance_sampling_techniques_MU_infection.pdf. 5p

World Health Organization (2010b) Working to overcome the global impact of neglected tropical diseases: fi rst WHO report on neglected tropical diseases. WHO/HTM/NTD/2010.1, Geneva, http://whqlibdoc.who.int/publications/2010/9789241564090_eng.pdf. 184p

World Health Organization (2012) Treatment of Mycobacterium ulcerans disease (Buruli ulcer): guidance for health workers. WHO/HTM/NTD/IDM/2012.1, Geneva, http://apps.who.int/iris/bitstream/10665/77771/1/9789241503402_eng.pdf. Accessed the 6 Oct 13: 73p

World Health Organization (2015) Management of Buruli ulcer-HIV coinfection: Technical update. WHO/HTM/NTD/IDM/2015.01.Geneva. http://apps.who.int/iris/bitstream/10665/154241/1/WHO_HTM_NTD_IDM_2015.01_eng.pdf:21p

World Health Organization (2020) Ending the neglect to attain the Sustainable Development Goals: a road map for neglected tropical diseases 2021–2030

Dracunculiasis (Guinea Worm Disease)

Franklin Asiedu-Bekoe, Gautam Biswas, and John O. Gyapong

Contents

F. Asiedu-Bekoe (✉)
Public Health Division, Ghana Health Service, Accra, Ghana
e-mail: franklin.asiedubekoe@ghs.gov.gh

G. Biswas
Formerly of Department of Control of Neglected Tropical Diseases, World Health Organization, Geneva, Switzerland

J. O. Gyapong
Centre for Neglected Tropical Diseases, Institute of Health Research, University of Health and Allied Sciences, Ho, Ghana

J. O. Gyapong, B. A. Boatin (eds.), *Neglected Tropical Diseases - Sub-Saharan Africa*, Neglected Tropical Diseases, https://doi.org/10.1007/978-3-031-53901-5_3

Abstract

Dracunculiasis, also known as guinea worm disease, is caused by the nematode *Dracunculus medinensis*, which infects humans typically through drinking unsafe water. The disease occurs in the most remote and poor areas of the world. There is neither a vaccine to prevent nor medication to treat the disease. Nevertheless, the disease has been set for eradication since the 1980s and has been eliminated from many of the endemic countries through preventive measures that include behavioral change in patients and communities, self-reporting of suspected cases and animals to health workers or volunteers, filtering drinking water, drinking water from improved sources, and preventing infected individuals and animals from entering or swimming in drinking water sources. These are complemented by vector control and the provision of improved water sources. The disease has reached its lowest levels ever recorded in the drive toward eradication; however, there are current infections being recorded in animals. This chapter reviews, among other things, the epidemiology of the disease and the progress made in eradicating dracunculiasis since the eradication campaign began, the biology, the eradication strategies, current challenges to the eradication, further research for control, and the outlook for the next 10 years.

Introduction

Dracunculiasis, a parasitic disease caused by the nematode *Dracunculus medinensis*, is also known as guinea worm disease. *D. medinensis* is in the Spirurida order, an order of parasites that includes the filariae *Wuchereria bancrofti, Brugia malayi*, and *Loa loa. Dracunculus medinensis* is a 60–120-cm-long nematode worm, known to humankind since antiquity. It is on the verge of being eradicated from the world, with human cases limited to a few countries, including Central African Republic, Ethiopia, South Sudan, Mali, and Chad (Fig. 1 and Table 1). It must however be

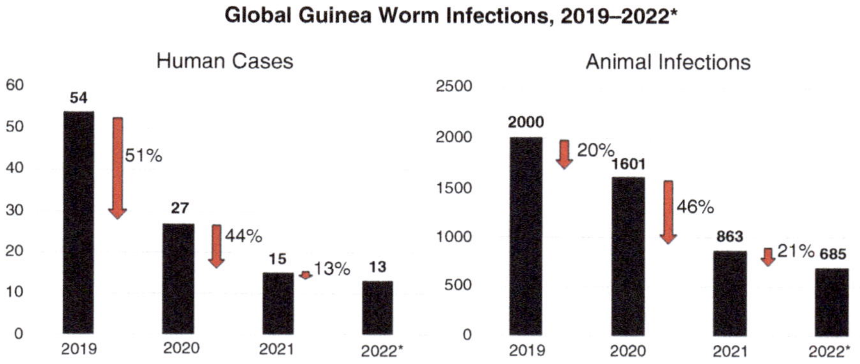

Fig. 1 Reported human and animal guinea worm infections

Table 1 Countries reporting cases of guinea worm

Guinea worm infections 2022 (provisional)		
Country	Humans	Animals
Chad	6	606
Mali	0	41
Cameroon	0	27[a]
Angola	0	7
South Sudan	5	1
Ethiopia	1	3
Central African Republic	1[a]	0
Total	13	685

[a]Apparently imported from Chad

noted that though significant cases in animals are being recorded, there is a decline in their numbers.

The first known mention of the disease was in the Turin Papyrus in the fifteenth century B.C. by the Egyptians; the disease has since been described by ancient Greek, Roman, Arab, Persian, and Indian physicians (Groove 1900). Guinea worm disease derived its common name from its prevalence in the Gulf of Guinea (Palmer and Reeder 2001). The disease has rarely been fatal, and fatality was not typically caused by the primary infection but usually in cases of secondary infection of the exit site of the parasite that leads to sepsis or tetanus (Greenaway 2004). However, morbidity is a major concern. The disease is known to cause temporary incapacitation and sometimes is associated with permanent disability (Hopkins et al. 2000; WHO 1988), resulting in loss of income and reduced school attendance among the already deprived communities and households associated with the disease. The socioeconomic consequences could be devastating. The Dogon people of Mali refer to dracunculiasis as "the disease of the empty granary" (Yoro, the empty granary; film (35 mm) and videotape 1998).

Biology and Life Cycle

Dracunculiasis is transmitted to humans typically by drinking water from stagnant ponds and wells contaminated with infected freshwater small crustaceans, copepods (*Cyclops*) (Fig. 2), the intermediate host (WHO 2012), which contains the larval stage of the parasite (*D. medinensis*). Man is the principal definitive host, but natural infections have been reported in a variety of animals, including monkeys, herbivores, domestic and wild Canidae, and Felidae (Macpherson et al. 2012).

Following ingestion of water containing infected *Cyclops*, by humans, the cyclops are killed in the stomach, upon which the infective larvae are liberated. The mature larvae penetrate the intestinal wall and migrate to the connective tissue of the host. The larvae develop inside the body. The male worm dies soon after impregnating the female, which becomes fully mature after 9–14 months, at which time it can measure over a meter long. The adult fertilized female worm (Figs. 3 and 4),

Fig. 2 Cyclops, the intermediate host for guinea worm

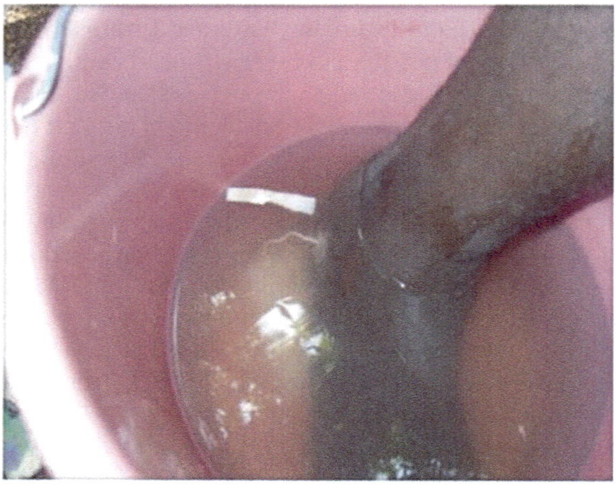

Fig. 3 A foot with an emerging worm immersed in a bucket of water (Controlled immersion) (Source: WHO)

which harbors the larvae responsible for the infection, is attracted to cool surfaces of the skin, emerging from the skin about a year after infection. The larvae are released from the uterus of the adult worm on contact with stagnant water. The first stage motile larvae are ingested by the cyclops in which they undergo development to the third larval infective stage in some 2 weeks, thus perpetuating the transmission cycle, as shown in Fig. 5 (Muller 1985).

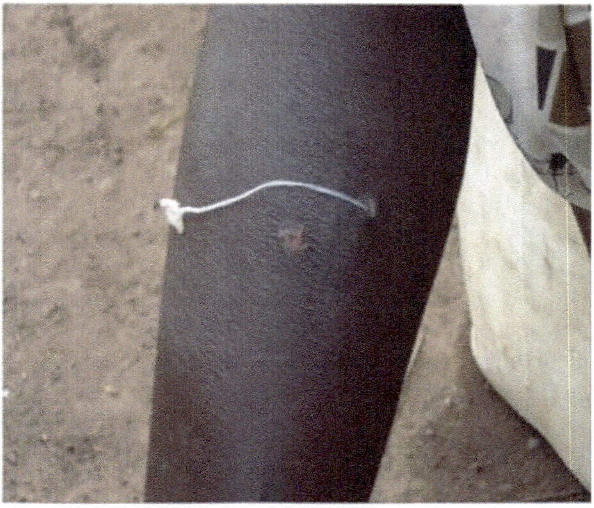

Fig. 4 An emerging worm wrapped around a stick (Source: WHO)

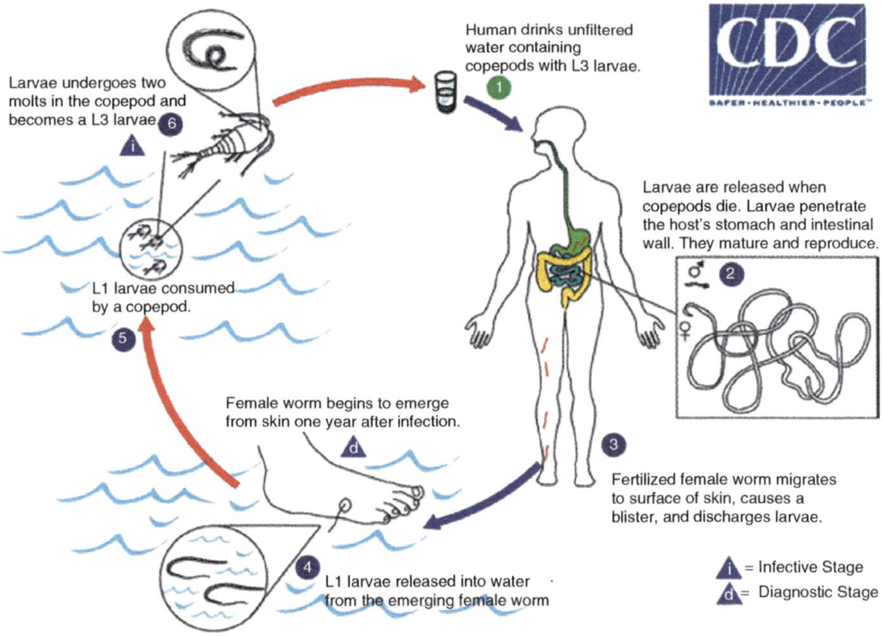

Fig. 5 Life cycle of guinea worm

Clinical Presentation

Guinea worm presents as redness and tenderness of the skin from where the adult worm would emerge, usually in the lower part of the leg. However, the disease can present with worms emerging from any part of the body, including the breast, chest, back, abdomen, testes, periorbital tissue, lungs, pancreas, or spinal cord. There could also be general allergic symptoms that subside with the discharge of the larvae (Figs. 6 and 7). The emergence of the worm, which could be more than one sometimes, is accompanied by painful edema, intense generalized pruritus, blistering, and ulceration of the area from which the worm emerges. This painful process may incapacitate the affected person for a period lasting about 8 weeks or more. If the worm is broken before it is completely extruded from the skin, this could cause an intense inflammatory reaction with pain, swelling, severe cellulitis, and other

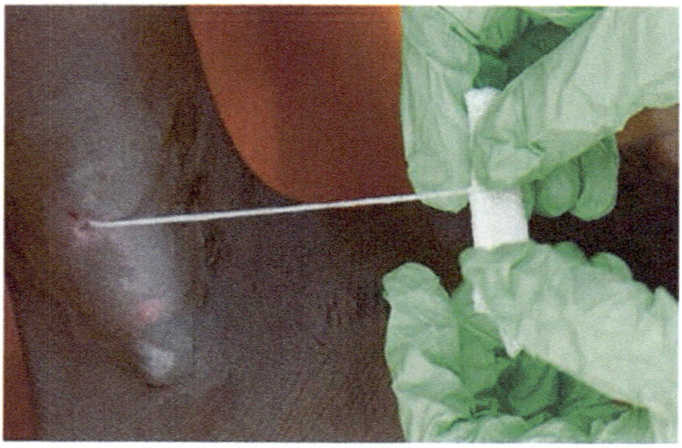

Fig. 6 Adult guinea worm from the leg *(Source: Ghana GWEP)*

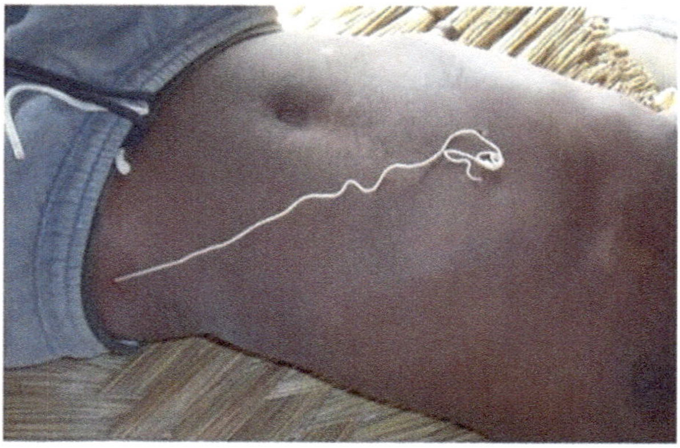

Fig. 7 Adult guinea worm from the body *(Source: Ghana GWEP)*

pyogenic infections. Chronic complications such as encapsulation of the adult worm may occur when calcified remains of the worm persist in the extremity of the patient. This can result in recurrent pain and intermittent swelling of the extremity.

In some cases, there is permanent scarring or deformity of the lower extremity, even after the worm has been extracted. Chronic pain may also persist for about 2 years or more. Infected individuals may have multiple worms exit at the same time. The success of the global campaign to eradicate the disease not only reduced the number of cases recorded by year but also reduced the load of infection everyone may bear.

In 2014, an average of 1.3 worms per case were recorded (maximum of 4 worms in one case) as compared with an average of 1.4 worms per case (maximum of 9 worms in one case) in 2013 and 1.7 worms per case (maximum of 21 worms in one case) in 2012 (WHO 2015a). There is no medicine to treat guinea worm, and neither is there a vaccine to prevent the disease (Cairncross et al. 2002; Muller 1985; Issaka-Tinogah et al. 1994; Greenaway 2004). On the other hand, the transmission cycle can be interrupted (Belcher 1985) by preventing the contamination of sources of drinking water, filtering unsafe water with fine mesh strainers before consuming or drinking water from improved sources, and by undertaking appropriate vector control interventions (Biswas et al. 2013).

The Disease Burden and Epidemiology

The distribution of the disease is determined largely by the availability of open stagnant water sources such as ponds and shallow or step wells and patients or animals infected with dracunculiasis. Artificial ponds often constitute a main source of transmission. Transmission of the disease is seasonal and depends on climatic factors. In endemic areas of Africa, there are two broad patterns: in the Sahelian zone, transmission generally occurs during the rainy season (May to September); in the humid savannah and forest zones, transmission usually peaks during the dry season (September to January). However, there are local variations to these patterns (WHO 2015b).

Guinea worm affects rural populations whose livelihood depends on subsistence farming. Since the disease causes temporary disability, lasting from a few weeks to a few months, which may prevent patients from leaving their beds while the worm emerges, the cost of lost school attendance and revenue from decreased agricultural productivity for individuals and the community can be very high (Hopkins et al. 2000; Fig. 8).

The economic loss of dracunculiasis in India alone was estimated to be 11.7 million person-days annually among 4% of the 12.2 million people living in endemic villages (Rao 1985). In 1985, researchers found that dracunculiasis had an impact on school attendance in Benin and Nigeria and on agriculture in Benin and Burkina Faso (WHO 1986, 1987). In 1987, a study in Nigeria estimated an annual loss of US$ 20 million per single crop (rice, yam, and cassava) from farmers incapacitated by the disease for an average duration of 5 weeks (WHO 1988, World Health Organization 1990a, b, 1998; Hopkins et al. 2000). The disease was therefore an

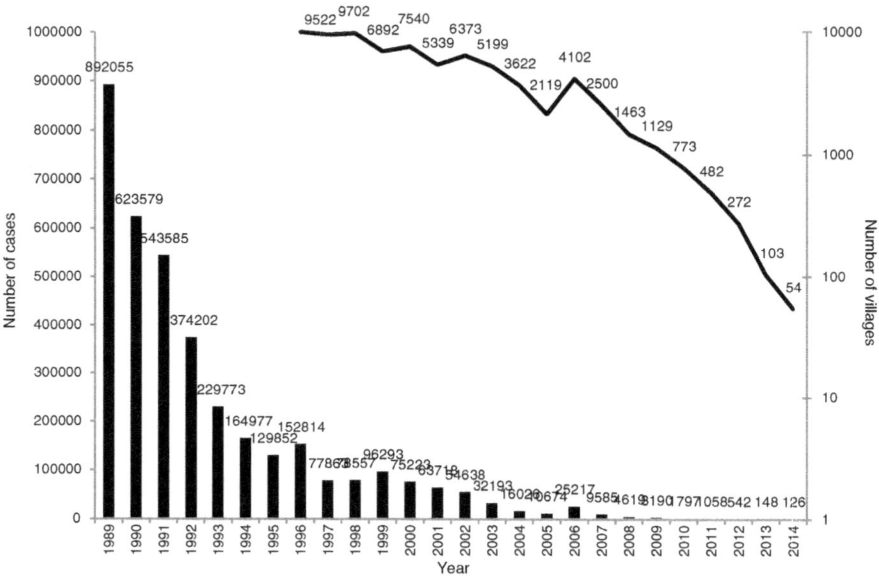

Fig. 8 Annual number of reported dracunculiasis cases and localities of case detection, worldwide, 1989–2014

important cause of socioeconomic toll on the affected communities and countries. The cost–benefit outcome of public health interventions was one of the factors that influenced the decision for eradication (Hopkins et al. 2000; Biswas et al. 2013).

In the late 1940s, more than 48 million people were estimated to be affected by the disease in Africa, India, and the Middle East (Stoll 1947). By 1976, the WHO estimated that ten million dracunculiasis cases were reported globally, with less than 0.003% of cases being reported worldwide (WHO 1982). By the 1980s, the disease was known to be endemic in 20 countries, including 17 countries in Africa, two countries in Asia, and one country in the Middle East. During the mid-1980s, an estimated 3.5 million cases were reported yearly. As a result of efforts by national Guinea Worm Eradication Programmes since the 1980s, several countries have eliminated the disease, and transmission now remains endemic only in Angola Chad, Ethiopia, Mali, and South Sudan. Since its inception, the Global Eradication Campaign has continued to progress steadily toward wiping dracunculiasis off the planet. In addition to the reduction of transmission zones to limited geographical foci, significant decreases in the incidence of dracunculiasis were noted; significantly very low cases have been recorded for close to a decade.

On the recommendation of the International Commission for the Certification of Dracunculiasis Eradication (ICCDE), WHO has now certified 200 countries, areas, and territories as Guinea worm-free. Only five endemic countries (Angola, Chad, Ethiopia, Mali, South Sudan) and one formerly endemic country (Sudan) remain to be certified (Fig. 9).

In 2022, a total of 13 human cases were identified. As of January 2023, dracunculiasis remained endemic in five countries (Angola, Chad, Ethiopia, Mali, and

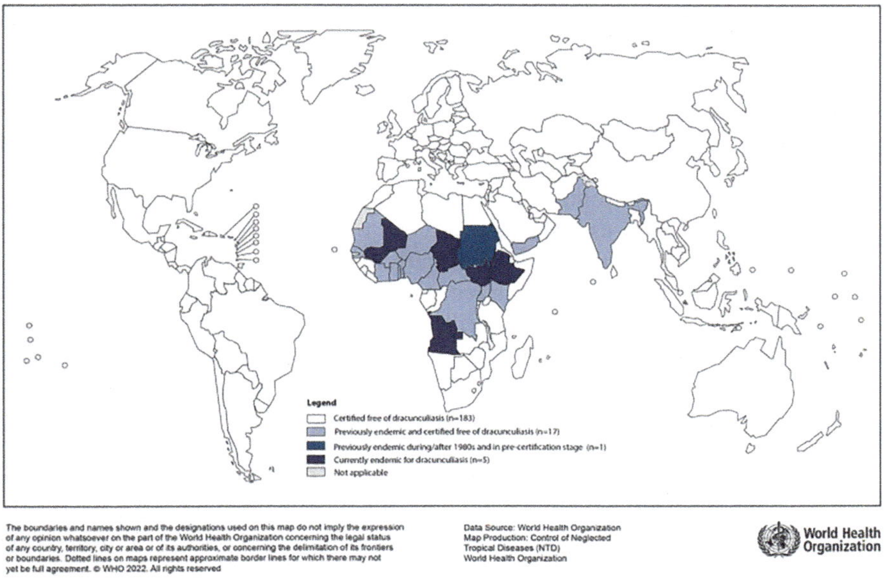

Fig. 9 Map of status of certification

South Sudan). Eradication efforts in these countries are challenged by infection in animals, the COVID-19 pandemic, civil unrest, and insecurity.

Animal infections, mostly in domestic dogs, some domestic cats, and in Ethiopia, a few baboons, have now surpassed human cases, with 863 reported animal infections in 2021 and 296 during January–June 2022. With only 15 human cases identified in 2021 and three during January–June 2022, program efforts appear to be closer to reaching the goal of eradication. However, dog infections and impeded access because of civil unrest and insecurity in Mali and South Sudan continue to be the greatest challenges for the program.

Control Tools and Eradication Strategies

The development of the dracunculiasis eradication strategies was based on evidence gathered from various studies on the disease, the effective interventions, and lessons learned from the effective smallpox eradication program as well as from the non-conclusive malaria eradication program (United States Centres for Disease Control and Prevention 1999; Grand Rounds: The Opportunity for and Challenges to Malaria Eradication 2011; Cairncross et al. 2002; Richards and Ruiz-Tiben 2011). The strategies adopted for eradication have largely withstood the scrutiny based on the evidence that 17 countries that were previously endemic in the 1980s interrupted the transmission of the disease and 15 of these countries are now certified as free of transmission by WHO (WHO 2015a; Biswas et al. 2013). The cost-effectiveness of the eradication strategies has been documented in various studies (Hopkins et al. 2000; Rao 1985; WHO 1986, 1987, 1988, 1989, 1990a, b).

The strategies address the following areas:

(a) Surveillance for prompt detection of cases and reporting.

Surveillance, which now includes animals as well, involves the community itself (supported by a supervisory structure) in detecting, containing, and reporting dracunculiasis cases (Muller 2005). The objective of surveillance and reporting is to promptly detect all cases until the absence of transmission is confirmed worldwide. As the eradication process advances, a country endemic to the disease will go through the following stages: (1) endemicity, (2) interruption of transmission, and (3) precertification. The precertification stage lasts at least 3 years, during which a country must sustain nationwide surveillance to demonstrate evidence of continued absence of transmission to be qualified for certification as dracunculiasis-free. (4) Once WHO certifies a country free of dracunculiasis, it enters post-certification surveillance, which is continued until eradication of dracunculiasis is declared globally. To increase the sensitivity of the surveillance, the ICCDE recommends that national dracunculiasis eradication programs announce a suitable cash reward for voluntary reporting of cases (WHO 1996). The current amount of the reward in endemic countries and those in the pre- and post-certification stages ranges from US$ 40 to US$1160 (WHO 2015a). A global reward scheme like the one adopted by the Smallpox Eradication Campaign will be announced by WHO when the transmission of the disease is interrupted.

(b) Case management and containment by preventing infected persons and restricting animals from wading into drinking water sources. A case of guinea worm disease is defined as "a person exhibiting a skin lesion with emergence of a guinea worm, ideally with laboratory confirmation." (WHO Collaborating Centre for Research, Training and Eradication of Dracunculiasis, CDC 2014). A case is counted only once during the calendar year, that is, when the first worm emerges from that person. Other human lesions may mimic guinea worm disease (Eberhard et al. 2010, 2015), hence making laboratory confirmation necessary during the latter stage of the eradication effort. However, any suspect guinea worm disease case should be managed as a true case until proven otherwise through laboratory diagnosis or the worms are fully expelled.

Every case detected should be treated and its further transmission prevented. The tried and tested treatment is as follows: when the worm begins slowly to extrude from the skin, the affected part is dipped in clean water or water is poured on the part to facilitate the emergence of the worm. The extruding worm is then gradually rolled out of the body using a gauze or a clean stick (matchstick) (Cairncross et al. 2002). The wound is properly bandaged to prevent secondary infection (Fig. 10).

The global eradication effort describes the management processes to be undertaken to prevent transmission of guinea worm disease from one patient to the other, and before a case of guinea worm disease is considered contained, the following must be met

- The case is detected before or within 24 h of the worm's emergence.

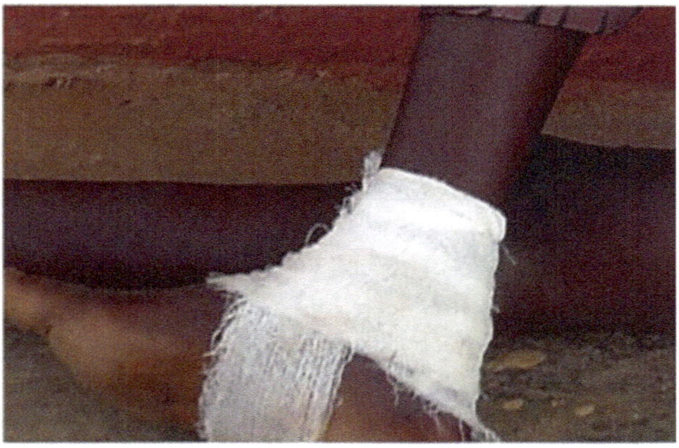

Fig. 10 Fully bandaged guinea worm disease lesion of the foot (Source: Ghana GWEP)

- The patient:
  - Did not contaminate any water source since the worm emerged.
  - Received proper care by cleaning and bandaging the wound until all the worms are fully expelled.
  - Received health education on not entering any water source
- A supervisor verified the case as dracunculiasis within 7 days (WHO 2003).
- Temephos is used if there is any uncertainty about contamination of sources of drinking water or if a source of drinking water is known to have been contaminated.

Within the context of the eradication effort, the management of a case of guinea worm disease extends beyond the individual suffering from the disease. It involves the commitment, collaboration, and collective effort of the community volunteer, the patient's family, and the community in which he lives to ensure that all the criteria mentioned above are met. This is because some criteria, such as early case detection, case confinement, effective education, and abate application, will, to a large extent, also depend on factors within the community (mostly socioeconomic and cultural).

To further ensure that the case is appropriately contained, patients are being hospitalized and treated in existing health facilities or temporary camps to prevent patient contact with unsafe drinking water sources. Health education is provided until all worms are expelled.

Painkillers, such as aspirin or ibuprofen, can help reduce pain and swelling. Antibiotic ointment can help prevent secondary bacterial infections, or antibiotic treatment may become necessary to treat secondary infection or cellulitis.

Optimal daily case management involves the following steps:

1. First, each day, the affected body part is immersed in a container of water to encourage more of the worm to come out.
2. Next, the wound is cleaned.
3. Then, gentle traction is applied to the worm to slowly pull it out. Pulling stops when resistance is met to avoid breaking the worm. Because the worm can be as long as 1 m in length, full extraction can take several days to weeks.

4. The worm is then wrapped around a rolled piece of gauze or a stick to maintain some tension on the worm and encourage more of the worm to emerge. This also prevents the worm from slipping back inside. Afterward, topical antibiotics are applied to the wound to prevent secondary bacterial infections.
5. The affected body part is then bandaged with fresh gauze to protect the site. Medicines, such as aspirin or ibuprofen, are given to help ease the pain of this process and reduce inflammation.
6. These steps are repeated every day until the whole worm is successfully pulled out.

To prevent contamination, the infected person is not allowed to enter drinking water sources throughout the treatment. This can be achieved by properly educating the patient and at the same time sensitizing his/her household to support him/her to comply.

(c) Access to improved drinking water supply.

With a strong and culturally sensitive community health education, geo-graphical access to improved drinking water sources can make a difference in guinea worm disease-affected communities. Ways and means to provide affected communities with improved drinking water supplies include the following: (1) constructing barriers to prevent humans from entering surface drinking water sources, (2) protecting hand-dug wells with walls and sinking deep bore wells, and (3) filtering surface drinking water through filters (Brieger et al. 1997; WHO 1989). Access to improved drinking water sources in communities endemic to the disease provides a sustainable solution for dracunculiasis eradication, as experienced in the Islamic Republic of Iran, Saudi Arabia, and Uzbekistan (Biswas et al. 2013). Efforts to provide safe drinking water are ongoing in the remaining endemic areas. Communities affected by the disease are prioritized for providing access to improved water sources.

(d) Vector control by treating the contaminated stagnant drinking water sources with a larvicide, temephos.

Temephos application to unsafe stagnant drinking water sources is effective in killing the cyclops, the intermediate host. Temephos application must be initiated within 14 days of case detection and then be carried out regularly at monthly intervals until no case is reported and/or at least 1 month after the risk of transmission in the locality ends. It is labor-intensive (Cairncross et al. 2002) and not always easy to carry out in some difficult-to-reach places. However, when implemented systematically—covering all eligible unsafe water sources in each endemic area—temephos application is effective in the reduction of guinea worm disease cases (Chippaux et al. 1991; Lyons 1973).

(e) Community-based health education and mobilization for increased community adherence to control interventions including filtering all unsafe drinking water through cloth filters to prevent infection. The overarching goal of health education and social mobilization is to encourage affected communities to adopt healthy behavior to prevent and ultimately interrupt disease transmission. Community volunteers and their supervisors (health staff) deliver simple and easy-to-understand messages (Cairncross et al. 2002) to community members with the aim that communities adopt the above-mentioned strategies and protect themselves from the disease and to avoid its propagation.

Modified Intervention Indices to Reflect Variable Modes of Transmission

With *D. medinensis* infections occurring in animals in three of the last four endemic countries (South Sudan is the exception), the evidence mounting to suggest that the infection is being transmitted to humans and animals not just by drinking water, but likely also by people and animals eating raw or undercooked transport hosts such as small fish (up to 2–3 inches/5–7.5 cm long) and/or raw fish guts, as well as through eating undercooked aquatic paratenic hosts such as frogs and larger fish. Guinea Worm Eradication Programs have adopted new interventions to counter the new challenges. Given this new situation, national GWEPs are encouraged to monitor a modified set of operational indicators. Among the former indicators, trained village volunteers, regular health education, and reporting by villages under active surveillance, including endemic villages, can be assumed to be at or near 100%. Coverage with cloth filters protects against contaminated drinking water, such as in Ethiopia in 2017, but not against eating an infected transport or paratenic host, which are the most common mode of infection for humans and animals in Chad and Mali (Table 2).

Table 2 Currently suggested operational indicators for GWEP

Modality	Indicator	Outcome
Reward awareness *Combined results for Village under Active Surveillance (VAS) levels I & II (endemic and high-risk villages) for reporting human and dog infections*	% of persons surveyed who are aware	Detect infections quickly
Containment of infected humans and animals	% of infected humans and animals contained or tethered	Prevent contamination
Abate coverage *Water bodies may be ineligible for Abate treatment from time to time when they become too large (>1000mx3) or dry up*	% of cumulative villages where Abate was applied in the current year and villages with infections in the current or previous year	Prevent infection and contamination
Bury fish guts	% of people surveyed in VAS level I villages with demonstrated fish gut burial practice	Prevent infection
Safe water source	%·of VAS·level I villages with at least one functioning source of safe drinking water	Prevent large point source outbreaks
Accessibility	% of VAS level I villages (endemic villages+) that are safely accessible by the program	Prevent infection

Challenges Toward Eradication of Guinea Worm Disease

The current challenges to eradication of dracunculiasis are to interrupt transmission in the few last foci of remaining countries and ensure surveillance in all other countries and areas at risk of disease reintroduction. As the disease's incidence in humans dwindles, there is a surge of the worm in animals.

The eradication campaign is paying greater attention to the following challenges to program implementation, which can be grouped into three tiers:

- Insecurity that prevents full program rollout in some parts of South Sudan, northern regions of Mali, and recently in Chad with ensuing movements of the population within and outside of national borders
- Need for continued nationwide surveillance and awareness of the reward scheme to report and investigate every suspected case
- An unusual epidemiology where many infected animals are observed, with cases in humans appearing to be sporadic

Country Specific Reports

Chad

Chad has reported six confirmed human cases of Guinea worm disease (33% contained) in January–October 2022, compared to eight human cases in the same period of 2021, which is a 25% reduction in human cases. It reported 456 infected dogs (66% contained) and 75 cats in January–October 2022, compared to 753 dogs and 65 cats in the same period of 2021 (Fig. 11). Measures to reduce

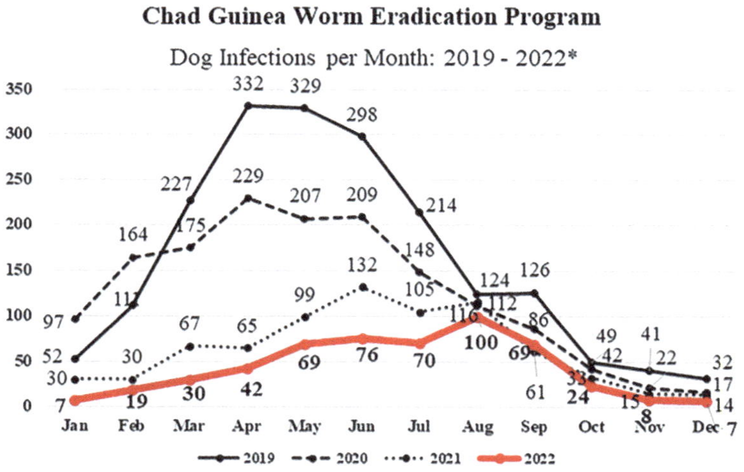

Fig. 11 Dogs infected per month (2019–2022) in Chad

transmission of guinea worm infections to and from dogs and humans are being implemented. However, this outbreak of dog infections in Chad and the need to understand this rare and unique epidemiology are the current challenges facing the dracunculiasis eradication efforts. There is a need to proactively tether all eligible dogs and cats in 2023, with special attention to the Bongor district, which borders the Guere district of Cameroon and to districts of the regions of Salamat and Moyen-Chari that border the Central African Republic. It is imperative that no dogs in Bongor and Guere districts cross the border in either direction during the transmission season in 2023.

Mali

The country continues to report indigenous transmission of dracunculiasis. Security concern in the north of the country is an impediment to program implementation. Mali has reported provisional totals of confirmed Guinea worm infections in 39 dogs and 2 cats for 2022, 63% (26/41) of which were contained. This 141% increase over the 17 animal infections reported in Mali in 2021 follows the Mali Guinea Worm Eradication Program's (MGWEP) proactive tethering of dogs and cats, which improved surveillance for Guinea worm infections starting in late 2021. Mali's limited remaining endemic area shares a similar riverine ecology and hypothesized Guinea worm transmission by eating raw or poorly cooked fish as in Chad, but with many fewer dog and human infections.

In Mali, however, some dogs are fed fish and fish guts to fatten them for human consumption, while some stray dogs that live along the river scavenge discarded small fish and fish guts. Both are at risk of Guinea worm infection and of being swept up, transported, and sold in a robust internal trade and market, which contributes to Guinea worm transmission. Despite ongoing dog infections, Mali reported no Guinea worm cases in humans in 2022 and for four consecutive years in 2016–2019. An outbreak of 29 cases at Tanzikratene in the Gao Region in 2014 was Mali's most recent common-source water-borne outbreak in humans.

South Sudan

Six confirmed Guinea worm infections (four contained) were reported in one dog and five humans, with a total of 11 worms in 2022. Four of the human cases with worm emergences between August 31 and October 2 were linked to Jarweng village in the Awerial district of Lakes State, including three persons in one family (a mother and two young sons), who were likely exposed to a common source of infection. The salient question is whether the sparse Guinea worm infections in South Sudan in recent years are being sustained by missed cases in humans, by undiscovered Guinea worm infections in animals, or both. Based on epidemiologic investigations, the program appears to have prevented forward transmission from all known Guinea worm cases for at least the last five years. Genetic analysis of emerging worms may help clarify this in 2023. Also unclear is the relative role of potential modes of transmission: via drinking water or via eating undercooked aquatic animals. The occurrence of most cases in 1–14-year-old children and 25–29-year-old women in 2007–2022 suggests Guinea worm may be being transmitted near homes.

Angola

Angola detected three human Guinea worm cases and eight infected dogs in 2018–2022, all in Cunene Province, including seven uncontained dogs reported in 2022. With WHO support, the Angola Guinea Worm Eradication Program is carrying out active surveillance in 61 villages: 39 in Namacunde, 11 in Cuanahama, and 11 in Cuvelai Municipality. Angola's January–May transmission season for Guinea worm infections coincides with the rainy season, which impedes surveillance and interventions. So far, flooding has prevented collection of a specimen from a suspected Guinea worm infection in a dog in February 2023. From January to mid-March 2023, a total of 27 rumors (three in humans and 24 in animals/dogs) have been notified by the communities. Completed investigations discovered 16 provisional animal infections. Samples from these animals will be tested at the CDC Atlanta for confirmation. Angola received a shipment of Abate in November 2022, has trained health staff to use it, and plans to start using Abate as soon as the flooding recedes, likely in April–May 2023. The program began tying up some dogs with ropes in 2023 and is exploring how best to implement proactive tethering.

Cameroon

Since 2019, Guinea worm cases and animal infections in Cameroon have occurred in a cluster of villages where extended families live on both sides of the border with Chad where indigenous transmission is believed to occur. Cameroon resumed proactive tethering of dogs in late November/early December 2022 in preparation for the season of peak Guinea worm occurrence in Extreme North Region's Guere health district, which reported 26 Guinea worm-infected dogs, all of which were *reportedly contained*, in 2022. The proactive tethering will continue until about June 2023. The peak season for Guinea worm occurrence in this area is January–April. Cameroon began proactive tethering of area dogs in December 2021; it expanded active surveillance in January 2022, including all nine villages with infected dogs, and applied Abate in villages with infected dogs. The dogs from Cameroon are believed to have been infected on the other side of the Logone River in Chad's endemic Bongor district because the families in the area live on both sides of the river. Cameroon has reported no human cases of Guinea worm disease in January–October 2022.

Ethiopia

Ethiopia reported four confirmed Guinea worm infections (two contained) with a total of 11 worms reported in 2022: two baboons, one dog, and one human. The Ethiopia Dracunculiasis Eradication Programme (EDEP) and Ethiopian wildlife authorities are discussing a protocol that will govern how they manage live baboons that are discovered to have Guinea worm infection when they are trapped for study. Since 2018, the EDEP has expanded Abate treatments and proactive tethering of dogs and cats aggressively in the limited remaining area of transmission in the Gog district of the Gambella Region.

Central African Republic

The Central African Republic has reported one confirmed case of Guinea worm disease. In 2021, a cattle herder from Chad was seen at Gordil health facility with an alleged emerged guinea worm; however, he did not agree to be hospitalized nor did he return for follow-up. The area is visited regularly by transhumance cattle herders from Chad and Sudan. No further Guinea worm infection of humans or animals was found in the locality and surrounding areas. Central African Republic has been engulfed in a multi-factorial civil war for the past decade, including the conflict in the northeast part of the country, where this case occurred, making in-country and external communications very difficult.

Need for Further Research for Additional Control Measures

Guinea worm disease is on the verge of being eradicated, with a few small pockets yet to be cleared and the infections in animals that now far outnumber human cases. The needed crucial research is under way to investigate this transmission dynamics to identify appropriate ways to accelerate the interruption of transmission in humans and animals. The following areas of research require attention to:

- Identify the implications of guinea worm infection in dogs on the transmission in humans.
- Help understand the dynamics of guinea worm infection in dogs.
- Unravel the reasons for the continued low levels of transmission in Ethiopia and appropriate ways to overcome it.
- Address the operational challenge of vector control in large water bodies.

Conclusion and Outlook

A painful and debilitating disease, dracunculiasis, is on the verge of eradication despite issues of security and current epidemiological challenges, as a result of a highly focused international public health initiative that uses inexpensive, practical interventions in target areas (Hopkins et al. 2000). The global eradication effort is led by national governments and communities, with the support of a coalition of partners, including the Carter Centre, WHO, WHO Collaborating Centre for Research, Training, and Eradication of Dracunculiasis at the US-CDC, UNICEF, and many donors.

In 2022, the global Guinea Worm Eradication Program reduced human cases by 13% to their lowest level ever, i.e., 13 cases, and reduced animal infections by 21% compared to 2021, and the World Health Organization certified the Democratic Republic of the Congo as Guinea worm-free in 2023.

The last stage of the program is the most difficult; it requires more concentrated, concerted, and focused efforts on surveillance, not only in endemic areas but also in areas free of transmission but at risk of importation from endemic areas (WHO 2015c).

Continued and increased national political support for eradication is needed to strengthen the intensification of interventions including facilitating access by health workers to areas with security concerns and epidemiological challenges. During this decade, this ancient scourge will likely be defeated. However, finishing the last mile of dracunculiasis eradication will require the full resolve of all the affected countries, supporting partners, donors, and endemic communities to make an all-out effort with zero tolerance for any slippage to achieve this target. The main challenges concern security, especially in certain areas of Mali and South Sudan, where it has implications within and beyond their borders due to population movements and the unusually high number of dogs with confirmed guinea worm infections in Chad, which poses epidemiological and biologic questions. However, these challenges could be overcome by making the best use of all opportunities while maintaining quality surveillance in guinea worm-free areas to prevent any introduction or re-establishment of new foci of transmission resulting from imported cases due to population movements (WHO 2015a).

Once eradicated, dracunculiasis will be the first parasitic human disease to be extinct. It will be the first disease eradication campaign to be carried out and successfully concluded without vaccine or medicine but solely by using public–private sector partnerships to fund and support community-level interventions and innovative incentives to empower exceptional community involvement (Barry 2006; Molyneux 2004). The program will leave a legacy of public health goods in terms of access to improved drinking water, improved surveillance, and no school absenteeism or lack of income due to incapacitation from the disease (Biswas et al. 2013).

References

Barry M (2006) Slaying the little dragons: lessons from the dracunculiasis eradication program. Am J Trop Med Hyg 75:1–2

Belcher D (1985) Opportunities for control of dracunculiasis: transmission of epidemiology. In: Ruskin FR (ed) Workshop on opportunities for control of dracunculiasis. National Academy Press, Washington, pp 1–7

Biswas G, Sankara D, Agua-Agum J, Maiga A (2013) Dracunculiasis (guinea worm disease): eradication without a drug or a vaccine. Philos Trans R Soc Lond Ser B Biol Sci 368:20120146
Brieger W, Otusanya S, Adeniyi D, Tijani J, Banjoko M (1997) Eradicating guinea worm without wells: unrealized hopes of the water decade. Health Policy Plan 12(4):354–362
Cairncross S, Muller R, Zagaria N (2002) Dracunculiasis (guinea worm disease) and the eradication initiative. Clin Microbiol Rev 15(2):223–246
Centers for Disease Control and Prevention (1999) Global disease elimination and eradication as public health strategies. Morb Mortal Wkly Rep (MMWR Suppl) 48(16–22):46–48
Chippaux J, Laniyan I, Akogbeto M (1991) Evaluation de l'efficacité du Témephos dans la lutte contre la dracunculose. Ann Soc Belg Med Trop 71:279–285
Eberhard M, Ruiz-Tiben E, Korkor A, Roy S, Downs P (2010) Emergence of onchocerca volvulus from skin mimicking dracunculiasis medinensis. Am J Trop Med 83(6):1348–1351
Eberhard M, Thiele E, Yembo G, Yibi M, Cama V, Ruiz-Tiben E (2015) Thirty-seven human cases of sparganosis from ethiopia and South Sudan caused by Spirometra Spp. Am Soc Trop Med Hyg 15(0236):350–355
Greenaway C (2004) Dracunculiasis (guinea worm disease). Can Med Assoc 170(4):495–500
Groove D (1900) A history of human helminthology. CAB International, Wallingford
Hopkins DR, Ruiz-Tiben E, Ruebush TK, Diallo N, Agle A, Withers CJ (2000) Dracunculiasis eradication: delayed, not denied. Am J Trop Med Hyg 62(2):163–168
Issaka-Tinogah A, Magnussen P, Bloch P, Yakubu A (1994) Lack of effect of ivermectin on prepa-tent guinea-worm: a single-blind, placebo-controlled trial. Trans R Soc Trop Med 88(3):346–348
Lyons G (1973) The control of guinea-worm with abate: a trial in a village of North-West Ghana. Bull WHO 49:215–216
Macpherson CN, Meslin FX, Wandeler A (2012) Dogs, zoonoses and public health. CAB International, London, pp 237–239
Grand rounds: the opportunity for and challenges to malaria eradication (2011) MMWR Morb Mortal Wkly Rep 60(15):476–480
Molyneux DH (2004) "Neglected" diseases but unrecognised successes-challenges and opportunities for infectious disease control. Lancet 364(9431):380–383
Muller R (1985) Life cycle of dracunculus medinensis. In: Workshop on opportunities for control of dracunculiasis: contributed papers. National Academy Press, Washington, pp 13–17
Muller R (2005) Guinea worm disease – the final chapter? Trends Parasitol 21(11):521
Palmer P, Reeder M (2001) Guinea worm infection (dracunculiasis). Retrieved 27 Sept 2012, from the imaging of tropical diseases. http://tmcr.usuhs.mil/tmcr/chapter27/intro.htm
Rao C (1985) Epidemiology of dracunculiasis in India. In: Workshop on opportunities for control of dracunculiasis, contributed papers, Delhi. National Academy Press, Washington, pp 25–31
Richards FO, Ruiz-Tiben E (2011) Dracunculiasis eradication and the legacy of the smallpox campaign: what is new and innovative? What is old and principled? Vaccine 29:D86–D90
Stoll N (1947) This wormy world. J Parasitol 33(1):1–18
WHO Collaborating Centre for Research, Training and Eradication of Dracunculiasis, CDC (2014) The carter centre. Retrieved 21 July 2015, from http://www.cartercenter.org/resources/pdfs/ news/health publications/Guinea worm/wrap-up/228.pdf
World Health Organization (1982) Dracunculiasis surveillance. Wkly Epidemiol Rec 57(9):65–72
World Health Organization (1986) Global surveillance summary – 1985. Wkly Epidemiol Rec 61(5):29
World Health Organization (1987) Dracunculiasis, global surveillance summary –1986. Wkly Epidemiol Rec 62(45):337
World Health Organization (1988) Dracunculiasis: global surveillance summary-1987. Wkly Epidemiol Rec 49(2):375–378
World Health Organization (1989) Dracunculiasis: global surveillance summary- 1988. Wkly Epidemiol Rec 64(39):297–302
World Health Organization (1990a) Dracunculiasis: global surveillance summary, 1989. Wkly Epidemiol Rec 65(30):229–233

World Health Organization (1990b) Dracunculiasis, global surveillance summary, 1989. Wkly Epidemiol Rec 65(30):229–236
World Health Organization (1996) Criteria for the certification of dracunculiasis eradication. Retrieved 20 July 2015, from http://whqlibdoc.who.int/hq/1996/WHO_FIL_96.188.pdf?ua=1
World Health Organization (1998) Dracunculiasis, global surveillance summary –1997. Wkly Epidemiol Rec 73(18):129–135
World Health Organization (2003) Dracunculiasis eradication: case definition, surveillance and performance indicators. Wkly Epidemiol Rec 78(37):323–328
World Health Organization (2012) Dracunculiasis (guinea-worm disease). In: Working to overcome. World Health Organization, Geneva, pp 103–106
World Health Organization (2015a) Dracunculiasis eradication: global surveillance summary, 2014. Wkly Epidemiol Rec 90(19):201–215
World Health Organization (2015b) Dracunculiasis. Retrieved 20 July 2015, from http://www.who.int/dracunculiasis/disease/disease_more/en/
World Health Organization (2015c) Dracunculiasis. Retrieved 20 July 2015, from http://www.who.int/dracunculiasis/en/
Yoro (1998) The empty granary. Film (35 mm) and videotape. Motion picture

Human African Trypanosomiasis (HAT)

Pascal Lutumba, Enock Matovu, and Boakye A. Boatin

Contents

P. Lutumba
Institut National de Recherche Biomédicale (INRB),
Kinshasa, Democratic Republic of Congo

Kinshasa University (UNIKIN), Faculty of Medicine, Tropical Medicine Department,
Kinshasa, Democratic Republic of Congo

E. Matovu
College of Veterinary Medicine, Animal Resources and Biosecurity, Makerere University,
Kampala, Uganda

B. A. Boatin (✉)
LFSCA, Accra, Ghana

J. O. Gyapong, B. A. Boatin (eds.), *Neglected Tropical Diseases - Sub-Saharan Africa*, Neglected Tropical Diseases, https://doi.org/10.1007/978-3-031-53901-5_4

Abstract

Human African Trypanosomiasis (HAT) found only in sub-Saharan Africa is caused by the parasite *Trypanosoma brucei*, which is transmitted by tsetse flies. Only two sub-species of *T. brucei* are pathogenic for humans: *T. b. gambiense* and *T. b. rhodesiense*. HAT is endemic in 36 sub-Saharan countries, and 98% of all reported HAT cases are due to *T. b. gambiense*. Fifty-five million persons in Africa are at risk of HAT. The number of HAT cases reported globally decreased fivefold in the last decade, which encouraged WHO to set a target to eliminate HAT as a public health problem by the year 2020, aiming for zero transmission by the year 2030. Tsetse flies do not lay eggs, but the female fly deposits a single mature larva in humid soil. The larva pupates and emerges as an adult fly 20 to 80 days later. A female fly produces only 3–5 larvae during her lifetime that typically lasts for 3 months making the intrinsic growth rate of tsetse populations rather low. An infected tsetse fly injects the infective form of the parasites into the mammalian host when it feeds. These parasites undergo, and can switch, their antigenic variation of their variant surface glycoprotein (VSG) coat, allowing them to escape the host immune response. This phenomenon of antigenic variation makes the development of an effective vaccine unlikely. The disease affects mainly the lymphoid system, heart, lungs, and brain, manifesting as intermittent fever, general malaise, severe headache, joint pains and muscle aches, pruritus, urticaria, or facial oedema. Lymphadenopathy is common with the classical Winterbottom's sign. Patients with the meningo-encephalitic stage suffer continuous headaches with poor response to painkillers and show more specific neurological signs of the rather typical sleep disturbances. Diagnosis of HAT is a three-step procedure: (1) screening test to identify HAT suspects, (2) confirmatory parasitological tests, and (3) staging. Treatment is based on the stage of illness, with current options being pentamidine, suramin, melarsoprol, eflornithine, fexinidazole, and the nifurtimox-eflornithine combination therapy (NECT). Two strategies are used for the reduction or interruption of HAT transmission: elimination of the parasite reservoir and vector control. This last decade has seen several breakthroughs in clinical R&D for HAT, bringing new diagnostics and drugs to patient care, but the effective and efficient implementation of these new tools in HAT control and proper treatment of patients will require further research.

Epidemiology of the Disease

Human African Trypanosomiasis (HAT) or sleeping sickness is a disease caused by the parasite *Trypanosoma brucei* and transmitted by tsetse flies belonging to the genus *Glossina*. As this vector is restricted to sub-Saharan Africa, so is the disease. Only two sub-species of *T. brucei* are pathogenic for human beings: *T. b. gambiense* and *T. b. rhodesiense*. *T. b. gambiense* causes a slowly progressive form of the disease in West and Central Africa, while *T. b. rhodesiense* is responsible for a more acute disease pattern in East and Southern Africa (WHO 2013). HAT is endemic in 36 sub-Saharan countries, of which 24 are affected by *T. b. gambiense*—representing 98% of all reported HAT cases—while 13 countries are affected by *T. b.*

rhodesiense (Franco et al. 2014; WHO 2013). WHO (2022) estimates that the population at risk is 55 million people for the period 2016–2020, with only three million people at moderate or higher risk. Three countries have more than 70% of their population living in endemic areas: the Democratic Republic of Congo (DRC) (36.2 million), Uganda (ten million), and Angola (4.7 million) (Fig. 1). Figures 2 and 3 show the geographic distribution of cases over the decade 2000–2009 and for the period 2017–2018, respectively. The Democratic Republic of Congo continued in the last 5 years to report the highest (70% of the reported cases) number of cases with an average of less than 1000 cases declared annually, followed by Angola, Central African Republic, Chad, Congo, Gabon, Guinea, Malawi, and South Sudan declaring between 10 and 100 new cases in 2019. Cameroon, Côte d'Ivoire,

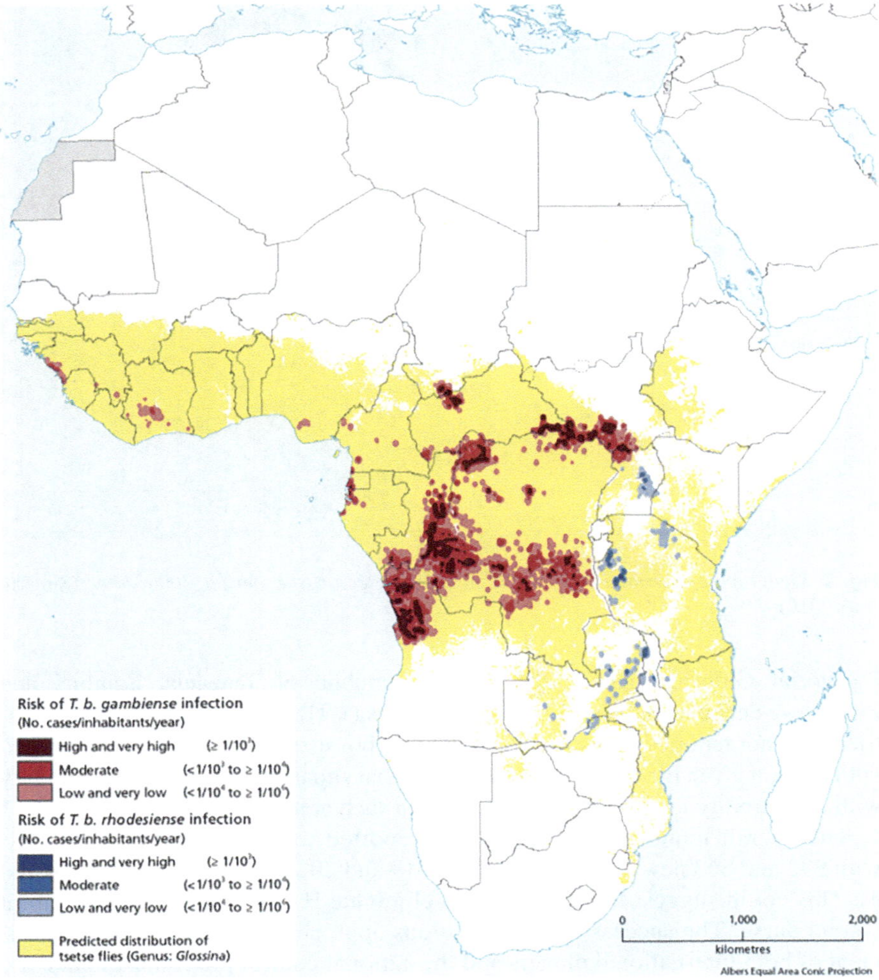

Fig. 1 *Population at risk for HAT* (Simarro et al. 2012a)

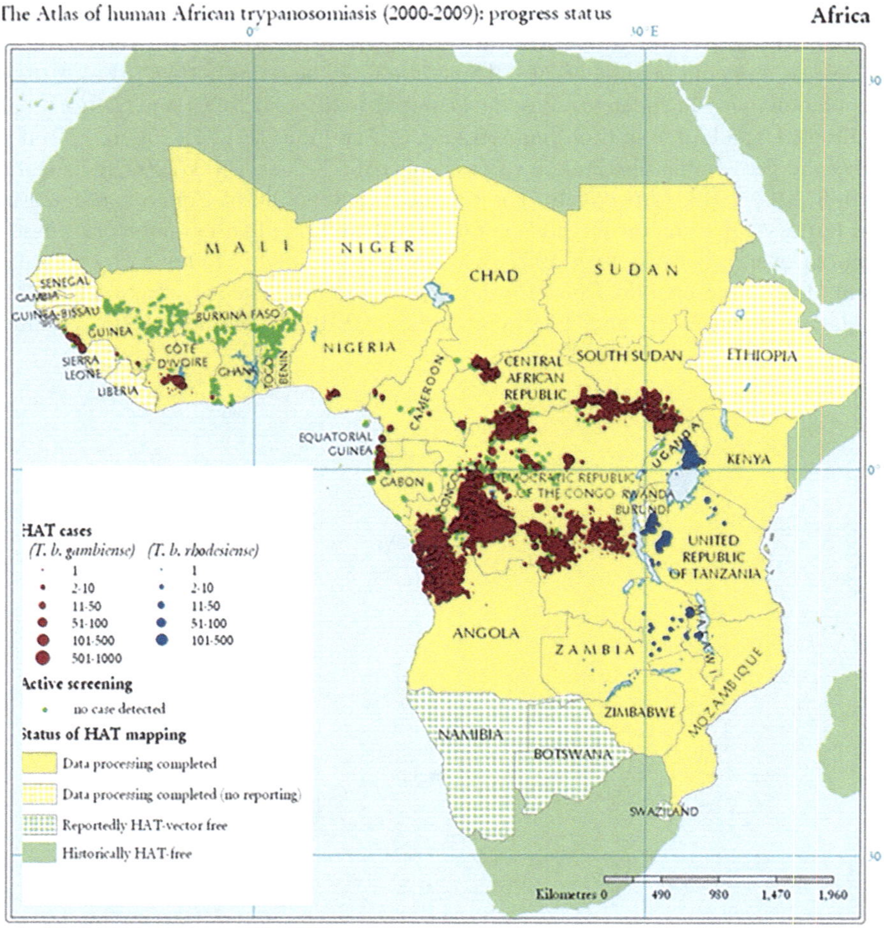

Fig. 2 *Geographic distribution of HAT cases reported in the period 2000–2009* (Simarro et al. 2010)

Equatorial Guinea, Uganda, the United Republic of Tanzania, Zambia, and Zimbabwe declared between 1 and 10 new cases (WHO 2022). Sixteen other countries have not reported cases for over a decade; however, such data must be viewed with caution given the challenges of fragile social situations as well as the difficulty with accessibility for surveillance activities in such areas.

The overall number of HAT cases is reported to have fallen considerably, with 992 and 663 new cases reported in 2019 and 2020, respectively (Figs. 4 and 5). This could suggest that the target to eliminate HAT by the year 2030 might be on course. The success of this ambitious approach will rely on the commitment of both international players and the national control programs of endemic countries.

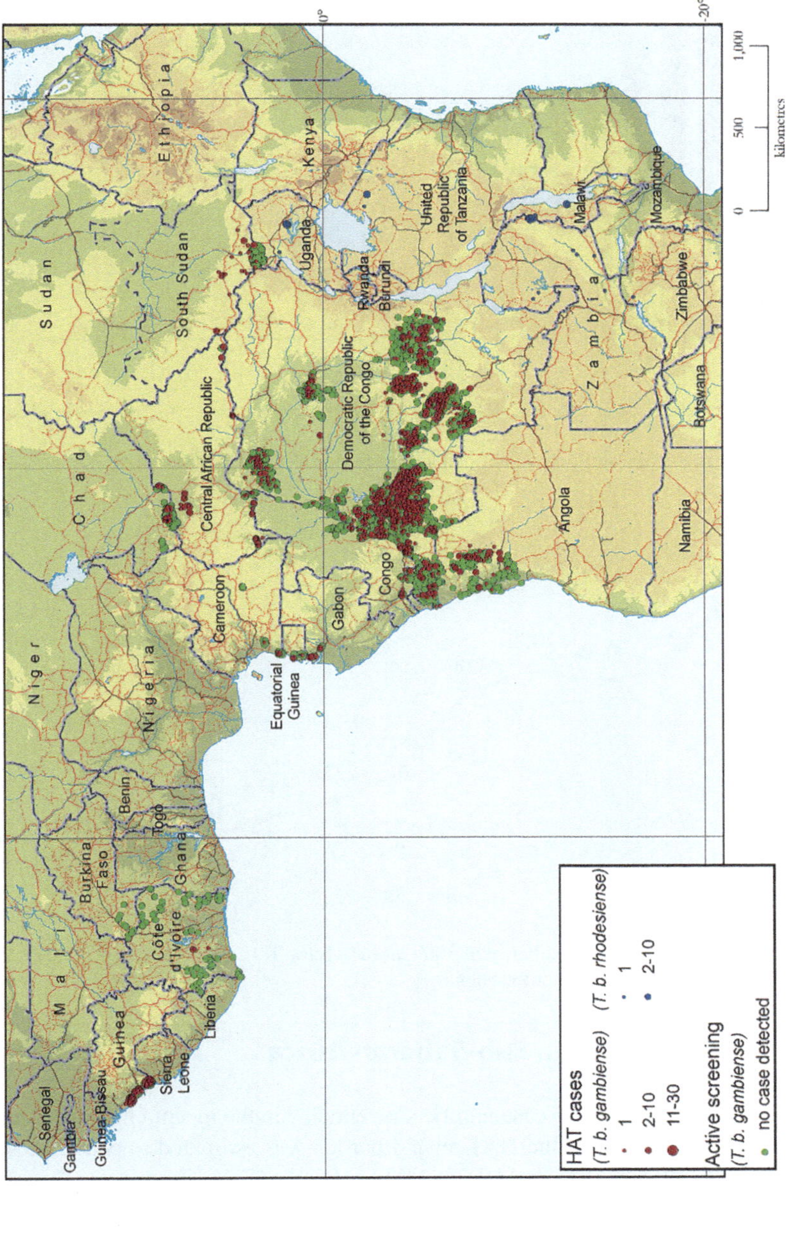

Fig. 3 *Geographic distribution of human African trypanosomiasis in the period 2017–2018.* https://doi.org/10.1371/journal.pntd.0008261.g002

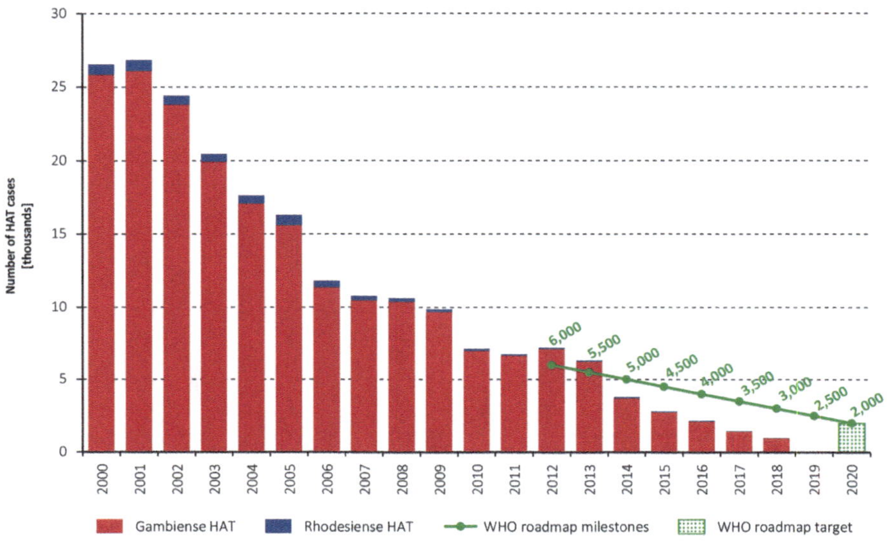

Fig. 4 Trends in the cases of human African trypanosomiasis from 2000 to 2018

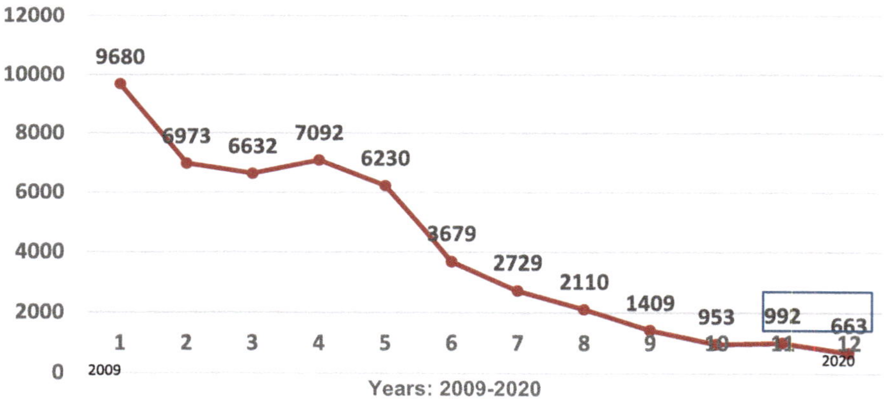

Fig. 5 HAT: New cases reported between 2009 and 2018 for T. b. gambiense and in 2019–2020 for both T. b. gambiense and T. b. rhodesiense

Public Health Impact in Sub-Saharan Africa

The decreasing trend in HAT case numbers is reflected in the recent Global Burden of Disease estimates. In 2019, the HAT-related burden was estimated to cause a loss of 82,600 (95% Confidence Interval UI 376,000 to 156,000) DALYs per year. This was a 98% reduction from the 1990 DALY estimate, when HAT was thought to cause the loss of two million DALYs per year (www.healthdata.org/results/gbd_summaries/2019/african-trypanosomiasisAfrican trypanosomiasis|Institute for Health Metrics and Evaluation(healthdata.org) 2019). Although the global HAT prevalence has decreased, sleeping sickness remains a major public health problem in communities at risk, which are mainly the rural poor. In the Democratic Republic of Congo (DRC), the most affected country, the number of years of life lost (YLL) per death caused by HAT was estimated at 27 years (Lutumba et al. 2007). Politi et al. (1995). estimated that in Uganda the number of DALYs incurred per premature death of *T. b. rhodesiense* HAT was 25. These high numbers are because HAT mainly affects adolescents and young economically active adults, and secondly, it is inevitably fatal in untreated individuals. The use of DALYs for expressing the HAT burden has been criticized since the global burden of disease rankings at the country or region level does not reflect the devastating socio-economic impact of the disease on the affected communities. HAT is a highly clustered disease, and country or regional averages may therefore be misleading. Within the HAT foci, the prevalence can be high, often around 1% and higher in epidemic contexts. In the absence of control, the prevalence can rise relatively rapidly, sometimes affecting over half of the population of certain villages. Thus, the impact of this disease can be very heavy in specific foci (Fevre et al. 2008).

Sleeping sickness patients require a great deal of supportive care from their relatives. Seeking a diagnosis and obtaining treatment is often costly and time-consuming. There have been a few attempts to quantify the full costs borne by households with a HAT patient, which includes care during hospitalization and at home, seeking a diagnosis, lost income, medical fees, transport, etc. Lutumba et al. (2007) investigated the burden of HAT in a rural community of the Democratic Republic of Congo in a retrospective household survey. The burden of HAT on households and livelihoods was high, between 1.5 and 10 months of income, even though diagnostics and HAT drugs were provided free of charge by the national control program (Lutumba et al. 2007). In the Republic of Congo-Brazzaville, the cost to households with HAT patients who were correctly diagnosed and treated came to an amount equivalent to 2.6–5 months of household income generated from agriculture (Gouteux et al. 1987). The cost-of-illness due to HAT borne by households is thus considerable and can compromise the timely uptake of HAT treatment. Households need to take their time to prepare themselves and mobilize resources, relying on the solidarity of the extended family before presenting themselves for treatment. This high cost for the household may partly explain the low participation rates at active population screening sessions organized by the national HAT control programs because people fear being identified as a HAT case and having to bear the high costs (Robays et al. 2007). Moreover, diagnosis and treatment require often long and invasive medical procedures that also act as a deterrent for persons to freely present themselves for treatment or the required post-treatment follow-up to confirm cure.

Biology

Vector and Parasite Life Cycle

Thirty-one different species of tsetse fly (genus *Glossina*; Fig. 6) can transmit HAT; these belong to one of three subgroups: fusca, palpalis, or morsitans. In general terms, species of the morsitans group are involved in the transmission of *T. b.*

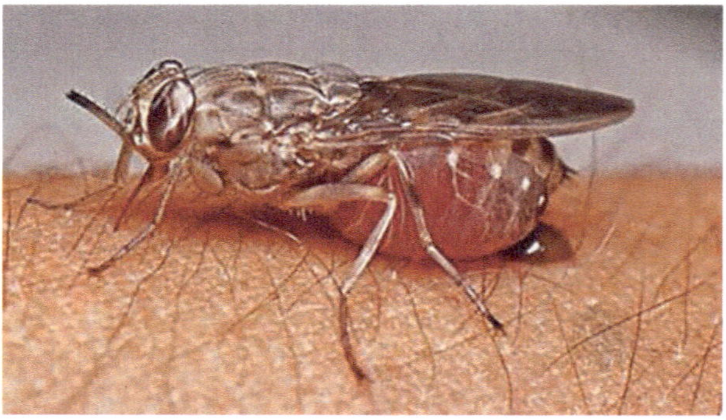

Fig. 6 Adult tsetse fly

rhodesiense, while species of the fuscipes and the palpalis group are mainly involved in the transmission of *T. b. gambiense*. The most important species involved are *G. palpalis palpalis*, *G. palpalis gambiensis*, *G. tachinoïdes*, *G. fuscipes quanzensis*, *G. fuscipes martini*, and *G. fuscipes fuscipes*. The latter is the major vector of both forms of HAT in Uganda. Tsetse flies have a peculiar reproductive cycle, and they do not lay eggs, but the female fly deposits a single mature larva in humid soil. The larva burrows into the soil, pupates, and emerges as an adult fly 20 to 80 days later (Fig. 7). A female fly will only produce 3–5 such larvae during her lifetime, which typically lasts for 3 months (2 months for males). As such, the intrinsic growth rate of tsetse populations is rather low.

When a male or female tsetse fly feeds on a *Trypanosoma*-infected mammalian host, the parasites enter the digestive tract of the fly. During the following 3–5 weeks, they undergo several differentiation steps, from dividing midgut forms to the migrating epimastigote forms, which develop in the salivary glands of the fly into the infective metacyclic forms. The latter will be injected during the next blood meal into the mammalian host (Figs. 7 and 8).

Injected into human skin, the trypanosome parasites will first proliferate at the site of the tsetse fly bite. The local inflammation reaction can lead to a trypanosomal chancre (also called "trypanome") and some local lymphadenopathy. Waves of trypanosomes then invade the bloodstream (Brun et al. 2010) and differentiate as dividing slender forms, intermediate forms, and stumpy forms. The latter do not replicate but are adapted to differentiate again into procyclic forms if taken up by a tsetse fly.

The parasites have a unique system of antigenic variation of their variant surface glycoprotein (VSG) coat, which allows them to escape the host immune response (Sternberg 2004). This coat of about ten million copies of a single VSG is strongly immunogenic, and the infected person produces a high amount of specific anti-VSG IgM antibodies that can destroy the majority of the circulating parasite population. However, the parasite can switch to the expression of different variants of VSG, and those that have switched their VSG in the meantime will survive and replicate. The variable antigens released repeatedly in the blood are responsible for a profound dysregulation of the immune response and cytokine production, resulting in

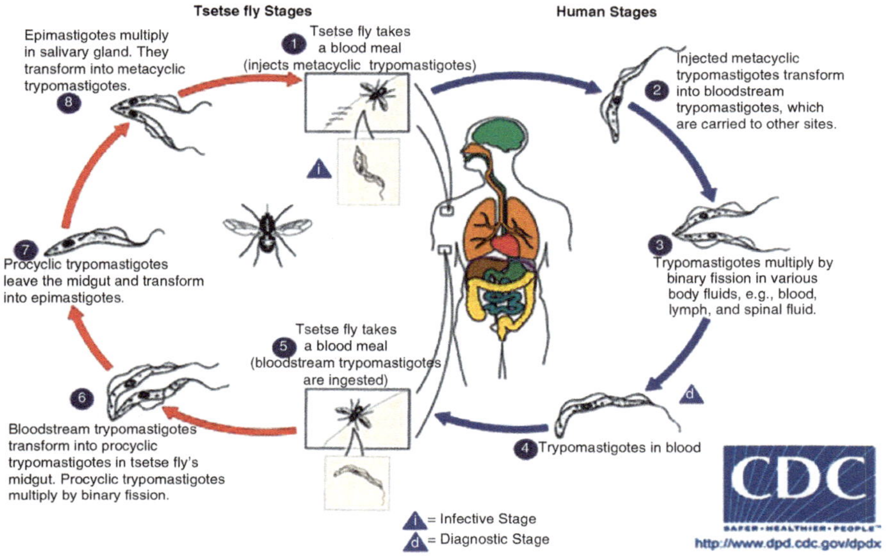

Fig. 7 Life cycle of Trypanosoma spp. (CDC 2015)

Fig. 8 Trypanosomes

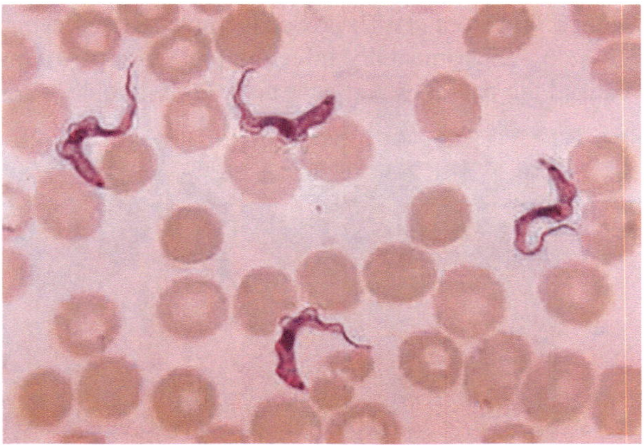

immuno-suppression. This phenomenon of antigenic variation makes the development of an effective vaccine unlikely (Magez et al. 2010).

Mode of Transmission

T. b. rhodesiense HAT is a zoonosis. Its transmission is predominantly maintained in an animal reservoir of cattle and game. In countries where the reservoir is mainly wild game, sporadic transmission to humans, including tourists visiting national parks, is occasionally reported. Countries such as Tanzania, Zambia, and Malawi have conservation areas within HAT-endemic regions. Such a complex transmission cycle involving protected animal species makes it impossible to envisage the elimination of *T. b. rhodesiense* HAT. However, in Uganda, all recognized HAT endemic regions are outside conservation areas, which makes effective control more feasible. Due to population growth, most wild animals living outside conservation areas have been decimated by hunting and other activities that alter their natural habitats. Thus, domestic animals are currently the most important reservoirs of the disease in the country (Hide et al. 1996; Hide 1999). A previous study in Busoga demonstrated that *T. b. rhodesiense* HAT is up to 5 times more likely to be transmitted through a cattle–fly–human cycle than by a human–fly–human cycle (Hide et al. 1996; Fig. 9).

Gambiense HAT, on the contrary, is maintained by the human–tsetse fly–human cycle. Very rarely, the parasite is transmitted by blood transfusion, or across the placenta, although mother-to-child transmission does occur. Other forms of possible transmission are mechanical transmission, through other blood-sucking insects (its epidemiological impact is not clear) (World Health Organisation 2018), and accidental infections in laboratories due to pricks with contaminated needles. Transmission of the parasite through sexual contact has been reported (Rocha et al. 2004).

The role of an animal reservoir in *T. b. gambiense* HAT is debated. The first indication that some *T. brucei* isolates from animals are infective to man came from Denice (1941), who infected himself with a parasite strain from a dog in Equatorial Guinea. Van Hoof (1947) proved the potential reservoir role of the pig for *T. b. gambiense* by experimentally infecting pigs with parasite strains of human origin and realized ten passages and later up to 20, without losing its infectivity for man. The

Fig. 9 Animals chronically infected with T. brucei exhibit symptoms similar to those of late stage HAT. Cattle from a *T. b. rhodesiense* endemic district of eastern Uganda was found to be co-infected with T. congolense (the cause of African animal trypanosomiasis) and *T. b. rhodesiense* and act as animal reservoirs in transmission of the acute form of HAT

same researchers were also able to show that a *T. b. gambiense* strain isolated from a human could infect goats, sheep, and dogs (Van Hoof 1947). More recent investigations using modern molecular tools have confirmed the presence of *T. b. gambiense* in pigs in Cameroon and Ivory Coast (Jamonneau et al. 2004; Nkinin et al. 2002). Infection was also detected in several primates, lizards, and rodents (Molyneux 1973) using the same technology. Eight out of 24 different wild animal species were infected with *T. b. gambiense* in a HAT focus in Cameroon, such as monkeys, ungulates, carnivores, and rodents such as *Cricetomys gambianus*, the Gambian pouched rat (Herder et al. 2002). Although it is commonly accepted that *T. b. gambiense* is maintained by the human reservoir, this does not exclude that in some foci of Gambian sleeping sickness both an animal and human reservoir may co-exist (WHO 2013). Mathematical modeling pointed repeatedly to the role of animal reservoirs in maintaining transmission in inter-epidemic periods.

Pathogenesis

The disease affects mainly the lymphoid system, heart, lungs, and brain. Gambiense HAT evolves in two stages. The first stage is haemato-lymphatic, when proliferating trypanosomes spread from the inoculation site. Parasitemia is typically low (less than 100 parasites per ml of blood) and fluctuating. In the second or late stage (also called meningo-encephalic stage), parasites will cross the blood–brain barrier (Kennedy 2013; Lundkvist et al. 2004). The parasites invade perivascular areas with subsequent infiltration in the white matter first, then later in the grey matter of the brain. The clinical syndrome is explained by the invasion of the choroid plexus, thalamus, the area postrema, and median eminence. In this second stage, signs of myocarditis (Blum et al. 2009), cerebral and meningeal edema, and punctuate hemorrhages are typically observed. Mott cells, plasma cells containing immunoglobins, are characteristic of the infection. The severity of pathological lesions correlates only partially with parasitemia (WHO 2013).

Cases of asymptomatic chronic carriers of the infection have been described (Ilboudo et al. 2014; Jamonneau et al. 2012). Some asymptomatic individuals who refused treatment and were followed up for long periods (5–15 years) were shown to eliminate the parasites from circulation coupled with a decline in the serological response. However, it is unclear whether these were cured spontaneously; these cases of asymptomatic infection are often too rare (Jamonneau et al. 2012) to enable conclusive investigation of this phenomenon.

Clinical Presentation

Signs and symptoms of HAT differ according to the stage of the disease. Symptoms in the early stage are not specific and include intermittent fever, general malaise, severe headache, joint pains, and muscle aches (Kennedy 2013). Sometimes patients have pruritus, urticaria, or facial edema. The trypanoma or chancre is the typical primary lesion in HAT caused by *T. b. gambiense,* but it is rarely observed in practice, as opposed to *T. b. rhodesiense* where the chancre is occasionally observed. Lymphadenopathy is more common: axillary and inguinal in East African HAT, while cervical lymphadenopathy is more often present in the West African type. The classical Winterbottom's sign is an enlarged gland of the posterior cervical triangle that is painless and feels firm, rubbery, and mobile.

The duration of the first stage in gambiense HAT varies from a few months to several years, while for *T. b. rhodesiense* it is a matter of weeks to a few months.

In the meningo-encephalitic stage, the patient suffers continuous headaches with poor response to painkillers and shows more specific neurological signs. These are the rather typical sleep disturbances, with daytime somnolence and nocturnal insomnia (Fig. 10) leading to a disappearance of the circadian rhythm of sleep and wakefulness, mental confusion, and a wide range of psychiatric disorders. These disorders include personality disorders, behavioral changes, and mood alterations. Other neurological symptoms include abnormal reflexes such as Babinski's sign, exaggerated osteotendinous reflexes and clonus, tone disorders (either hypertonia or hypotonia), abnormal movements, sensory disorders (Kerandel's sign), paraesthesia, and Hofman's sign. There is also a loss of sense of position, coordination disorders including ataxia and abnormal gait, convulsions, incontinence, and archaic reflexes (Kennedy 2008, 2013). Other features include amenorrhea, infertility, abortion, and wasting (Kennedy 2013).

The very diverse clinical picture mimics a range of other diseases, such as AIDS, schizophrenia, and TB, and clinicians may easily miss the diagnosis if they are not familiar with the disease. This is particularly a problem for clinicians dealing with imported cases in non-endemic areas, e.g., in urban areas. Moreover, control strategies against HAT in such urban areas are not well defined (Ebeja et al. 2003).

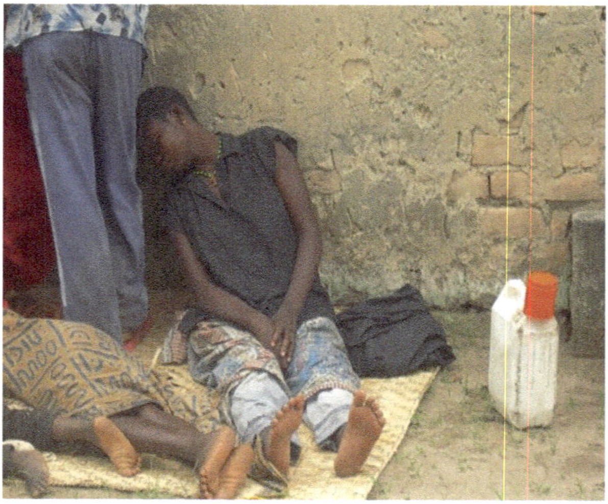

Fig. 10 Patient with HAT in the meningo-encephalitic phase

Diagnosis

HAT diagnosis is based on laboratory tests due to the absence of specific clinical signs. Because of the toxicity of the drugs in current use and the different stages of the disease, diagnosis of HAT occurs in three steps: (i) screening test to identify HAT suspects, (ii) confirmatory parasitological tests, and (iii) staging (Mitashi et al. 2012). The latter is required for clinical management as drugs that do not pass the brain–blood barrier can only be prescribed to patients in the early stage of the disease.

Screening. The most widely used screening test for Gambiense HAT is the Card Agglutination Test for Trypanosomiasis (CATT) based on the antigen Litat 1.3 (Magnus et al. 1978). Its sensitivity and specificity reportedly vary between 68.8–100% and 83.5–99.3% (Mitashi et al. 2012). Although its cost is affordable, limiting factors are the need for cold chain storage, the 50-dose format leading to loss of non-used doses, and the need for a 12 Volt power source. Since 1997, CATT has been used on a large scale in active population screening, with up to three million tests per year used by mobile teams. During the last 5 years, rapid diagnostic tests (RDTs) were developed based on the antigens Litat 1.3 and 1.5 (Buscher et al. 2013, 2014; Jamonneau et al. 2015; Sternberg et al. 2014). The first generation of these RDTs is based on native antigens and is available as individual cassette or strip tests with results available after 15–20 min. Its sensitivity and specificity vary between 89.0–99.6% and 95.0–99.0% (Buscher et al. 2013, 2014; Jamonneau et al. 2015; Sternberg et al. 2014). Second-generation RDTs based on recombinant antigens are now in clinical development (Sternberg et al. 2014) and also include antigens other than the LiTats traditionally used in the CATT and the first-generation RDTs in order to improve specificity. New, more specific tests—using protein, nucleic acid, or metabolite biomarkers (Buscher and Deborggraeve 2015)—will be needed as one gets to the "end game" and the positive predictive value of the current CATT test declines with the decrease of the burden of the disease (Barrett 2018).

Confirmation. All current HAT confirmatory tests are based on microscopic examination for the visualization of parasites, resulting in a specificity approaching 100% if carried out by trained staff. However, there is no single confirmatory test that has a satisfactory sensitivity, and in practice combinations of several are used in sequence. These include lymph node aspirates in suspects with cervical adenopathy, blood or cerebrospinal fluid for direct examination of Giemsa-stained smear, capillary tube centrifugation (CTC), quantitative buffy coat (QBC), or the mini anion-exchange centrifugation technique (mAECT). The most sensitive tests are the mAECT and CTC, but they require expertise and equipment such as a centrifuge and light source, which may not always be available in rural, HAT-endemic areas. The sensitivity of CTC varies between 44.3 and 93.0%, and its cost is estimated to be 0.76 €. The sensitivity of mAECT varies between 75.3 and 90.9%, and its cost is estimated to be 3–5 €.

Staging. In confirmed HAT patients, a lumbar puncture (Fig. 11) is done for staging of the disease. When parasites are visualized in the cerebrospinal fluid and/or the number of white blood cells is higher than $5/mm^3$ (Malvy and Chappuis 2011),

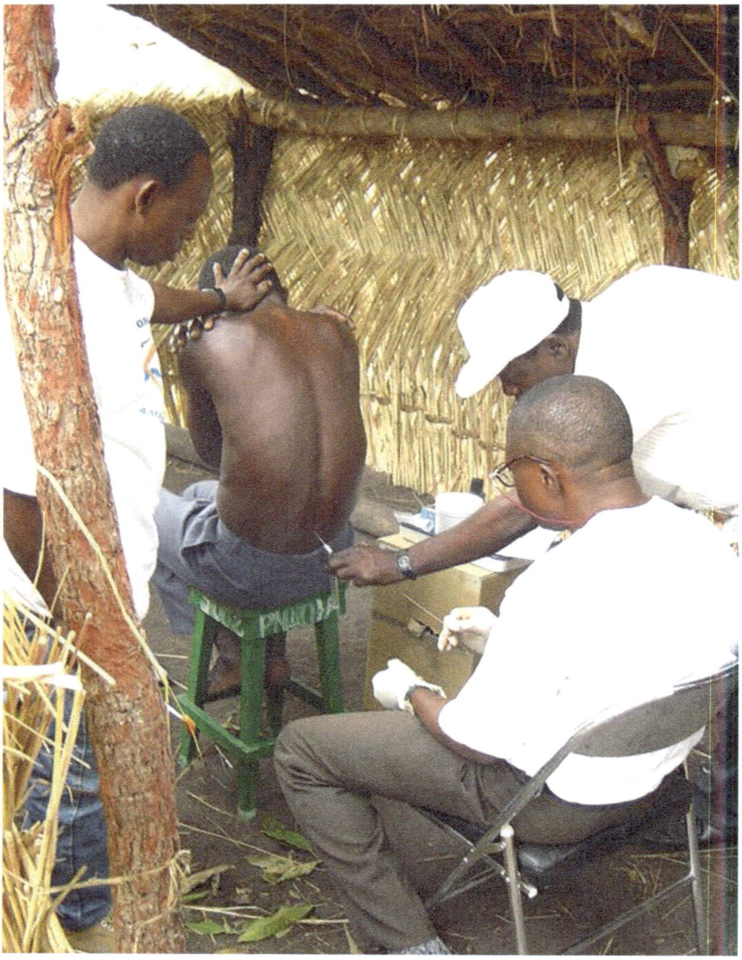

Fig. 11 Lumbar puncture for the staging of HAT during an active screening session in DRC

the patient is considered to be in the late stage of the disease. If CSF is sampled in a collection tube with a tapered tip followed by a light centrifugation and examination of the base (modified single centrifugation), this can increase the sensitivity of detection of parasitized CSF.

*Other tests***.** Molecular diagnostic tests have been developed mainly for species identification and have not been used so far in clinical diagnostic algorithms, methods recently reviewed by Buscher and Deborggraeve (2015). Differentiation between *T. b. rhodesiense* and *T. b. gambiense* by PCR is based on two specific genes: *T. b. gambiense* specific glycoprotein (TgsGP) and serum resistance-associated gene (SRA) (Simarro et al. 2012b; Urech et al. 2011). However, this specific PCR-targeted TgsGP and SRA gene is not very sensitive, as the two specific genes are present in a single copy in the trypanosome genome (Mitashi et al. 2013; Mugasa et al. 2014). PCR is limited by lack of standardization and the use of sophisticated equipment, which confine its use to reference and research laboratories. Some recent advances have tried to circumvent this. The loop-isothermal amplification of DNA (LAMP) was developed as a promising tool for field diagnosis of HAT. Unlike PCR, LAMP does not require the thermocycler for DNA amplification. The sensitivity of LAMP varies in laboratory conditions between 76.9 and 93.0% and specificity ranges from 92.8 to 100% (Mitashi et al. 2013; Mugasa et al. 2014). The cost of a LAMP reaction tube is around 5. Other simplified molecular methods include nucleic acid sequence-based amplification (NASBA) and FISH (fluorescent in situ hybridization) based on the use of fluorescent oligonucleotide probes to detect and localize specific target nucleic acid sequences. These methods require more evaluation before they can be recommended for field use. Several countries are presently evaluating LAMP alongside conventional screening and parasitological techniques in intensified efforts toward HAT elimination.

Treatment

Current treatment regimens for HAT depend on the stage of the disease but also on the infecting parasite sub-species. Currently, four drugs and one drug combination are available: pentamidine, suramin, melarsoprol, eflornithine, and the nifurtimox-eflornithine combination therapy (NECT). Pentamidine is the first-line drug for *gambiense* HAT at the first stage. It is administered at the dosage of 4 mg/kg/day for 7 days by intramuscular injection (WHO 2013). Other treatment regimens are being evaluated, such as 3 days instead of 7, but the results are not available yet. Pentamidine is a drug known since 1940 and is indicated for Gambiense HAT; however, it has also shown its effectiveness in the treatment of *rhodesiense* HAT in travelers (Simarro et al. 2012b; Urech et al. 2011). The therapeutic efficacy of pentamidine is 93–98%, and this cure rate has not changed for decades (WHO 1998). The major side effect is the immediate hypotension observed in 10% of the cases.

The first line treatment for stage I *rhodesiense* HAT is suramin, in use since 1920, and is usually administered intravenously at a dosage of 20 mg/kg (maximum 1 g). Patients receive a dose once a week for 5 weeks, which is preceded by a testing dose of 4–5 mg/kg (WHO 2013). Side effects are rare but can be severe due to early or late hypersensitivity, urticaria, or exfoliative dermatitis. The associated renal impairment is reversible once treatment ceases.

For the second-stage gambiense disease, NECT is the first-line treatment. Eflornithine is administered at 400 mg/kg/day intravenously in two intravenous perfusions (diluted in 250 physiological fluid) for 7 days in addition to oral nifurtimox at 15 mg/kg/day for 10 days. NECT effectiveness is reportedly 97.7% (Priotto et al. 2009). Major side effects have been observed in a small proportion, including neurological and psychiatric problems such as convulsions and agitations. Case fatality rates observed under NECT treatment vary from 0.5 to 1.6% (Alirol et al. 2013; Schmid et al. 2012; WHO 2013).

Second-stage *rhodesiense* HAT is still treated with melarsoprol at 22 mg/kg/day intravenously for 10 days (WHO 2013). Its effectiveness is 99%. However, its major side effect is reactive encephalopathy in 5–18% of the cases, with a lethal outcome in 10–70% of the reacting cases (Pepin and Milord 1994; WHO 1998). In Uganda and Tanzania, death rates of up to 8.4% and 9.4% of treated patients were respectively reported (WHO 2013).

In the case of treatment failure or contra-indications for certain drugs, the first-line treatment needs to be replaced by one of the above-mentioned drugs as monotherapy (WHO 2013). The basis for the therapeutic choice of second-line drugs is quite empirical since the mechanisms of treatment failure are not well understood, and the choice is left to the clinician. This is a gap that needs more particular attention.

Most drugs used for the treatment of sleeping sickness are more than 70 years old, and the choice is limited. In the 1990s, eflornithine was developed and the NECT combination regimen was adopted in 2010 after a multicenter phase III study had shown its efficacy. Besides these drugs, there are multiple products being currently evaluated at different clinical phases. Fexinidazole and benzoxaboroles (SCYX-7158), oral tablets that could be active on both stages of the disease, are the most promising and could revolutionize control and treatment strategies (Jacobs et al. 2011; Tarral et al. 2014; Torreele et al. 2010). Fexinidazole as an oral treatment for *gambiense* human African trypanosomiasis was included in 2019 in the WHO Essential Medicines List and WHO Human African Trypanosomiasis Treatment Guidelines. This molecule is indicated as the first line for the first stage and the non-severe second stage. It should be administered for 10 days within 30 min after a solid meal and under the supervision of trained medical staff. Currently, a clinical trial for its use in *rhodesiense* HAT is ongoing (WHO 2022).

In addition to providing support and technical assistance to national control programs, WHO provides the anti-trypanosome medicines free of charge to endemic countries. The drugs are made through public–private partnerships with Sanofi (pentamidine, melarsoprol, eflornithine, and fexinidazole) and with Bayer (suramin and nifurtimox; Bradley et al. 2021).

Control Tools and Strategies

There are two main strategies for the reduction or interruption of HAT transmission: (1) elimination of the parasite reservoir and (2) vector control. For the elimination of the parasite reservoir, two interventions are generally used by national control programs: active and passive screening. For gambiense HAT, where the main reservoir is human, active screening is the most effective (but the most costly) strategy because it can identify patients in an early stage while passive screening mostly identifies patients in the second stage (Hasker et al. 2011; WHO 2013).

In active screening, control programs usually deploy mobile teams generally composed of 7–8 persons. These mobile teams travel from village to village and examine the entire population using CATT as screening and a battery of confirmation tests

(Fig. 12). When HAT is confirmed, a lumbar puncture is required to determine the stage of the disease in the patient. All confirmed cases are referred to dedicated treatment centers. Early-stage patients receive treatment as outpatients, while late-stage patients require hospitalization, although in some countries such as Uganda both stages are handled as inpatients. The impact of a population screening (or active case finding) strategy on *T. b. gambiense* transmission is not well documented, and the optimal frequency of screening rounds is the subject of debate (Abel et al. 2004; Bruneel et al. 1994; Simarro et al. 1991; Van Nieuwenhove et al. 2001). WHO (1998) recommends one screening round per year in a typical endemic situation, although some national control programs perform two rounds a year. Successful active case finding must be frequently undertaken, cover a high proportion of the population at risk, use very sensitive tests, and provide sufficient capacity to treat the cases (WHO 1998). Under such circumstances, a decrease in the annual HAT detection rate to levels below 0.1% would be expected (Van Nieuwenhove et al. 2001). The reservoir of HAT cases could then be reduced to interrupt transmission for some years. For *rhodesiense* HAT, active screening is not relevant due to the acute and epidemic aspect of the disease as well as the domestic and wild animal reservoirs keeping transmission still active. Its screening activities are therefore mostly performed during epidemics to identify as many cases as possible for adequate treatment (Franco et al. 2014; WHO 2013). The "one health" approach of identifying and treating humans and domestic animals could also reduce *T. b. rhodesiense* (WHO 2013).

A consensus for passive screening strategies to control the disease has not been reached yet; however, a well-thought-out strategy could play an important role in

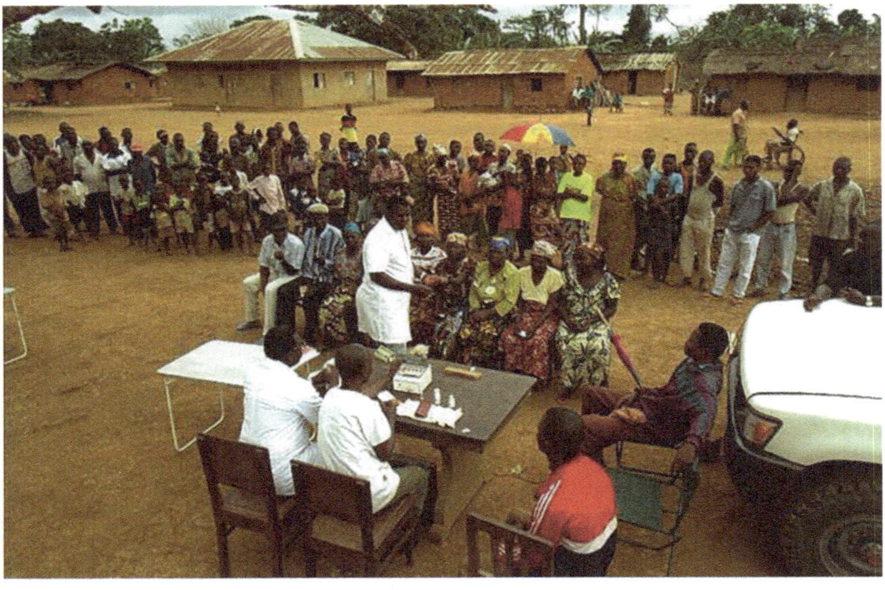

Fig. 12 *Mobile team for HAT screening at work in a village in DRC*

epidemiological surveillance and re-orientate active screening. Effective passive screening requires HAT diagnostic expertise at numerous health facilities, which is often absent. It was reported that patients present themselves more than seven times to health centers before a proper diagnosis is established. This is usually when the patient is already in the second stage of the disease (Brun et al. 2010), which is even more difficult to treat due to the associated drug toxicity complications.

Vector control against tsetse flies has long been considered an important strategy to reduce the impact of HAT on people and livestock. Early attempts to control tsetse flies were based on clearing land to remove their habitat and killing wildlife to remove tsetse hosts (Lancien 1991). Vector control interferes with the transmission of human and animal trypanosomiasis by reducing its reservoir in both humans and animals. Various strategies are used, such as traps, screens, and targeted insecticides (Abd-Alla et al. 2013; Bouyer et al. 2007; Bouyer 2008; Kgori et al. 2006; Lindh et al. 2012; Vreysen 2001); the sterile insect technique is not yet widely applicable. Currently, the most frequently used methods are traps and small insecticide-treated targets (tiny targets) (Lindh et al. 2012). However, it is noteworthy that in many countries vector control does not receive the attention it deserves to effectively suppress populations, thereby significantly hindering disease transmission.

All methods have the potential to diminish the density of tsetse flies considerably; however, their cost and the time that the area remains tsetse free varies. Despite the remarkable reduction of tsetse fly density, the efficacy of vector control in reducing transmission is not well established. A number of authors claim an effect of traps on the tsetse fly population, but most were small-scale projects that took place in low prevalence foci and these studies primarily assessed the costs and the feasibility, while the effect on transmission remained unclear (Gouteux et al. 1982, 1986; Gouteux and Lancien 1986; Gouteux and Sinda 1990; Lancien 1991; Laveissiere et al. 1994). However, mathematical modeling suggests that vector control may have a significant impact on the incidence of HAT (Artzrouni and Gouteux 1996, 2001). A study conducted by Simarro (Simarro et al. 1991) in Equatorial Guinea compared the addition of tsetse trapping to population screening in one HAT focus to exclusive population screening in two other foci. This trial showed a more rapid reduction of HAT prevalence in the focus with vector control. However, HAT was also eliminated after a single year in the two control foci with population screening only, and the paper concludes that vector control, in addition to population screening, was not cost-effective. The recently developed tiny targets were deployed at a cost of USD 85.4 per km^2 (Shaw et al. 2015). This tool is currently in evaluation for its effectiveness in reducing tsetse population and in the reduction or even elimination of HAT.

Until now, most HAT control programs in the *T. b. gambiense* foci are nearly entirely based on population screening with a very limited role for vector control. This is in sharp contrast with the *T. b. rhodesiense* area in East Africa (Welburn et al. 2001). Many argue that vector control is often not a priority in *T. b. gambiense* control programs because treatment of infected people is imperative for ethical reasons and funding for it is more easily found than for vector control. The limited budgets and the current instability in countries like South Sudan and others in central Africa make it difficult for HAT control programs to divert the scarce resources from case finding. Given these historical factors and the continued uncertainty on the effectiveness and cost-effectiveness of vector control, HAT control programs simply do not invest in vector control in the *T. b. gambiense* areas. Besides, the notion that humans are the major reservoir of gambiense HAT warrants that surveillance and treatment of identified cases take center stage. Historically, a number of foci have been brought under control solely by active case-finding programs with a very limited contribution of vector control (Abel et al. 2004; Bruneel et al. 1994; Frezil and Coulm 1977; Paquet et al. 1995; Simarro et al. 2001). However, annual active case-finding activities for more than 15 years did not bring the HAT epidemic under control in two provinces of the Democratic Republic of Congo, Bandundu and Kasai, which account now for more than half of the annually reported worldwide cases (Hasker et al. 2012). Difficulties in securing sufficient participation of the population in screening sessions are definitely one of the reasons (Robays et al. 2004). Mpanya et al. (2012) have demonstrated that sociocultural factors constitute an important barrier to participation in active screening. However, another factor might be that disease transmission may be more intense in these provinces than elsewhere, or else, the presence of an animal reservoir that is to date not fully understood.

Challenges of Program Implementation

With the intensified control efforts of recent years, the number of HAT cases has decreased considerably. However, with these decreasing numbers, the marginal cost per case detected increases, and the population loses its motivation to participate in the campaigns. This loss of efficiency of current control measures is of great concern. Moreover, since HAT control depends on external aid for over 90% (Lutumba et al. 2005), the commitment of funding agencies for HAT control is crucial and donor fatigue because of dwindling case numbers is an even bigger threat. It is well known that abandoning HAT control activities usually results in recrudescence of the disease (Hasker et al. 2012; Moore and Richer 2001). In addition to this, the populations living in an unstable socio-political context or war zones (North-East Congo, South Sudan, and Central-African Republic) or those living in inaccessible zones for various reasons are in need of alternative control strategies. In the short term, alternative, innovative, and cost-effective strategies should be established to facilitate further decrease of the HAT prevalence even in remote areas. Several alternatives have been proposed and are in the process of validation. This includes the possibility of mini-teams to perform rapid tests or CATT in a house-to-house visit mode with or without the possibility of confirmation testing. These mini-teams could travel by bike or motorcycle and be integrated into the local healthcare services. Several studies to measure the efficiency of such measures, their impact on the transmission of the disease, and further epidemiological surveillance are currently ongoing. Indeed, HAT control needs more than ever an adequate, cost-effective, and feasible system to detect as soon as possible recrudescence of the disease to undertake the required adequate actions. Other research projects investigate the process of integrating these HAT control activities within the local health centers to include screening, confirmation, treatment, and vector control.

Further Research for Policy and Control

In the recent past, there have been several breakthroughs in clinical R&D for HAT, bringing new diagnostics and drugs to patient care. The oral drug fexinidazole, for the treatment of gambiense human African trypanosomiasis, is now on the WHO Essential Medicines List. The possible use of benzoxaboroles could allow for a drastic simplification of the diagnostic algorithm to possibly eliminate the parasitological confirmation and/or staging steps. An RDT that is sufficiently sensitive and specific would allow treatment with the new safe molecules in a single step if such were the case.

Future implementation research will have to focus on the effective and efficient implementation of these new tools in HAT control and proper treatment of patients. This operational research should aim at a better integration of control measures in the health system in general and a rational and effective surveillance of HAT. The research will have to be accompanied by qualitative components that will allow the documentation on the perception and acceptability of new tools and strategies from the community and healthcare workers. The evidence generated by this research will therefore allow the political decision-makers to revise strategic documents in order to integrate the new tools and make HAT control more efficient (Franco et al. 2014).

Outlook for the Next Decade

While the number of HAT cases reported globally has been decreasing over the past decade, it is too early to consider that the problem is globally brought under control. Endemic countries as well as the donor community must be warned against the dangers of laxity in control efforts that would accompany such declines in incidence, for such would be an opportune moment for the HAT resurgence.

West African HAT affects communities in a prolonged way, as the disease tends to flare up again in the so-called historic foci several years after it was brought under control (WHO 2013). While there is consensus that HAT control should be a sustained and uninterrupted effort, it is unclear today what type of surveillance should be exerted once the incidence rates in a region have been brought to very low levels and the large-scale population screening campaigns become hugely inefficient. To eliminate HAT as a public health problem, it is thus imperative to apply enormous efforts and secure continuous support from funding agencies and the affected countries themselves. Moreover, the surveillance needs to be maintained for many years after HAT case numbers have de facto decreased to virtual zero to effectively interrupt the transmission cycle.

It would be worthwhile to undertake modeling to suggest at what level of prevalence post-control that recrudescence of the infection is unlikely or likely to occur.

The inefficiency of active case finding by the current mobile teams could be addressed by using alternative and innovative screening methods and by integrating these effective methods into the local healthcare system. However, this healthcare system will have to be sufficiently functional and capable of organizing such control activities to detect recrudescence of the disease timely, and this remains a huge challenge in countries such as DRC that are home to 80% of the current HAT burden (Mitashi et al. 2015).

References

Abd-Alla AM, Bergoin M, Parker AG, Maniania NK, Vlak JM, Bourtzis K, Boucias DG, Aksoy S (2013) Improving sterile insect technique (SIT) for tsetse flies through research on their symbionts and pathogens. J Invertebr Pathol 112(Suppl):S2–S10

Abel PM, Kiala G, Loa V, Behrend M, Musolf J, Fleischmann H, Theophile J, Krishna S, Stich A (2004) Retaking sleeping sickness control in Angola. Trop Med Int Health 9(1):141–148

Alirol E, Schrumpf D, Amici HJ, Riedel A, de Patoul C, Quere M, Chappuis F (2013) Nifurtimox-eflornithine combination therapy for second-stage gambiense human African trypanosomiasis: Medecins sans Frontieres experience in The Democratic Republic of the Congo. Clin Infect Dis 56(2):195–203

Artzrouni M, Gouteux JP (1996) Control strategies for sleeping sickness in Central Africa: a model-based approach. Trop Med Int Health 1(6):753–764

Artzrouni M, Gouteux JP (2001) A model of Gambian sleeping sickness with open vector populations. IMA J Math Appl Med Biol 18(2):99–117

Barrett MP (2018) The elimination of human African trypanosomiasis is in sight: report from the third WHO stakeholders meeting on elimination of gambiense human African trypanosomiasis. PLoS Negl Trop Dis 12(12):e0006925. https://doi.org/10.1371/journal.pntd.0006925

Blum JA, Schmid C, Burri C, Hatz C, Olson C, Fungula B, Kazumba L, Mangoni P, Mbo F, Deo K, Mpanya A, Dala A, Franco JR, Pohlig G, Zellweger MJ (2009) Cardiac alterations in human African trypanosomiasis (T. b. Gambiense) with respect to the disease stage and antiparasitic treatment. PLoS Negl Trop Dis 3(2):e383

Bouyer J (2008) Does isometamidium chloride treatment protect tsetse flies from trypanosome infections during SIT campaigns? Med Vet Entomol 22(2):140–143

Bouyer J, Stachurski F, Kabore I, Bauer B, Lancelot R (2007) Tsetse control in cattle from pyrethroid footbaths. Prev Vet Med 78(3–4):223–238. Available from: PM:17126431

Bradley M, Taylor R, Jacobson J et al (2021) Medicine donation programmes supporting the global drive to end the burden of neglected tropical diseases. Trans R Soc Trop Med Hyg 115(2):136–144. https://doi.org/10.1093/trstmh/traa167

Brun R, Blum J, Chappuis F, Burri C (2010) Human African trypanosomiasis. Lancet 375(9709):148–159

Bruneel H, van den Eeckhout A, Molisho D, Burke J, Degroof D, Pepin J (1994) Control of Trypanosoma gambiense trypanosomiasis. Evaluation of a strategy based on the treatment of serologically suspected cases with a single dose of diminazene. Ann Soc Belg Med Trop 74(3):203–215

Buscher P, Deborggraeve S (2015) How can molecular diagnostics contribute to the elimination of human African trypanosomiasis? Expert Rev Mol Diagn 15(5):607–615

Buscher P, Gilleman Q, Lejon V (2013) Rapid diagnostic test for sleeping sickness. N Engl J Med 368(11):1069–1070

Buscher P, Mertens P, Leclipteux T, Gilleman Q, Jacquet D, Mumba-Ngoyi D, Pyana PP, Boelaert M, Lejon V (2014) Sensitivity and specificity of HAT Sero-K-SeT, a rapid diagnostic test for serodiagnosis of sleeping sickness caused by Trypanosoma brucei gambiense: a case-control study. Lancet Glob Health 2(6):e359–e363

CDC (2015) Life cycle of Trypanosoma spp. http://www.cdc.gov/dpdx/trypanosomiasisAfrican/index.html.. Accessed 28 July 2015

Ebeja AK, Lutumba P, Molisho D, Kegels G, Miaka Mia BC, Boelaert M (2003) Sleeping sickness in the region of the town of Kinshasa: a retrospective analysis during the surveillance period 1996-2000. Trop Med Int Health 8(10):949–955

Fevre EM, Wissmann BV, Welburn SC, Lutumba P (2008) The burden of human African trypanosomiasis. PLoS Negl Trop Dis 2(12):e333

Franco JR, Simarro PP, Diarra A, Ruiz-Postigo JA, Jannin JG (2014) The journey towards elimination of gambiense human African trypanosomiasis: not far, nor easy. Parasitology 141(6):748–760

Frezil JL, Coulm J (1977) Present concept of strategy against sleeping sickness in the Democratic Republic of Congo. Ann Soc Belg Med Trop 57(4–5):315–322

Gouteux JP, Lancien J (1986) The pyramidal trap for collecting and controlling tsetse flies (Diptera: Glossinidae). Comparative trials and description of new collecting technics. Trop Med Parasitol 37(1):61–66

Gouteux JP, Sinda D (1990) Community participation in the control of tsetse flies. Large scale trials using the pyramid trap in The Congo. Trop Med Parasitol 41(1):49–55

Gouteux JP, Challier A, Laveissiere C, Couret D (1982) The use of screens in the anti-tsetse campaign in the forest zone. Tropenmed Parasitol 33(3):163–168

Gouteux JP, Lancien J, Noireau F, Sinda D (1986) Control of vectors by trapping and its impact on sleeping sickness in a zone of high density of glossina fuscipes quanzensis (Lefini River, People's Republic of the Congo). Trop Med Parasitol 37(2):101–104

Gouteux JP, Bansimba P, Noireau F, Frezil JL (1987) Cost of the individual treatment of Trypanosoma brucei gambiense trypanosomiasis in a focus of infection in Niari (Congo). Med Trop (Mars.) 47(1):61–63

Hasker E, Lumbala C, Mbo F, Mpanya A, Kande V, Lutumba P, Boelaert M (2011) Health care-seeking behaviour and diagnostic delays for human African trypanosomiasis in The Democratic Republic of the Congo. Trop Med Int Health 16(7):869–874

Hasker E, Lutumba P, Chappuis F, Kande V, Potet J, De Weggheleire A, Kambo C, Depoortere E, Pecoul B, Boelaert M (2012) Human african trypanosomiasis in The Democratic Republic of the Congo: a looming emergency? *PLoS.Negl.* Trop Dis 6(12):e1950

Herder S, Simo G, Nkinin S, Njiokou F (2002) Identification of trypanosomes in wild animals from southern Cameroon using the polymerase chain reaction (PCR). Parasite 9(4):345–349

Hide G (1999) History of sleeping sickness in East Africa. Clin Microbiol Rev 12(1):112–125

Hide G, Tait A, Maudlin I, Welburn SC (1996) The origins, dynamics and generation of Trypanosoma brucei rhodesiense epidemics in East Africa. Parasitol Today 12(2):50–55

Ilboudo H, Bras-Goncalves R, Camara M, Flori L, Camara O, Sakande H, Leno M, Petitdidier E, Jamonneau V, Bucheton B (2014) Unravelling human trypanotolerance: IL8 is associated with infection control whereas IL10 and TNFalpha are associated with subsequent disease development. PLoS Pathog 10(11):e1004469

Jacobs RT, Nare B, Wring SA, Orr MD, Chen D, Sligar JM, Jenks MX, Noe RA, Bowling TS, Mercer LT, Rewerts C, Gaukel E, Owens J, Parham R, Randolph R, Beaudet B, Bacchi CJ, Yarlett N, Plattner JJ, Freund Y, Ding C, Akama T, Zhang YK, Brun R, Kaiser M, Scandale I, Don R (2011) SCYX-7158, an orally-active benzoxaborole for the treatment of stage 2 human African trypanosomiasis. PLoS Negl Trop Dis 5(6):e1151

Jamonneau V, Ravel S, Koffi M, Kaba D, Zeze DG, Ndri L, Sane B, Coulibaly B, Cuny G, Solano P (2004) Mixed infections of trypanosomes in tsetse and pigs and their epidemiological significance in a sleeping sickness focus of Cote d'Ivoire. Parasitology 129(Pt 6):693–702

Jamonneau V, Ilboudo H, Kabore J, Kaba D, Koffi M, Solano P, Garcia A, Courtin D, Laveissiere, Lingue K, Buscher P, Bucheton B (2012) Untreated human infections by trypanosoma brucei gambiense are not 100% fatal. PLoS Negl Trop Dis 6(6):e1691
Jamonneau V, Camara O, Ilboudo H, Peylhard M, Koffi M, Sakande H, N'Dri L, Sanou D, Dama E, Camara M, Lejon V (2015) Accuracy of individual rapid tests for serodiagnosis of gambiense sleeping sickness in West Africa. PLoS Negl Trop Dis 9(2):e0003480
Kennedy PG (2008) Diagnosing central nervous system trypanosomiasis: two stage or not to stage? Trans R Soc Trop Med Hyg 102(4):306–307
Kennedy PG (2013) Clinical features, diagnosis, and treatment of human African trypanosomiasis (sleeping sickness). Lancet Neurol 12(2):186–194
Kgori PM, Modo S, Torr SJ (2006) The use of aerial spraying to eliminate tsetse from the Okavango Delta of Botswana. Acta Trop 99(2–3):184–199
Lancien J (1991) Campaign against sleeping sickness in south-West Uganda by trapping tsetse flies. Ann Soc Belg Med Trop 71(Suppl 1):35–47
Laveissiere C, Sane B, Meda HA (1994) Measurement of risk in endemic areas of human African trypanosomiasis in Cote d'Ivoire. Trans R Soc Trop Med Hyg 88(6):645–648
Lindh JM, Goswami P, Blackburn RS, Arnold SE, Vale GA, Lehane MJ, Torr SJ (2012) Optimizing the colour and fabric of targets for the control of the tsetse fly glossina fuscipes fuscipes. PLoS. Negl.Trop.Dis. 6, (5):e1661
Lundkvist GB, Kristensson K, Bentivoglio M (2004) Why trypanosomes cause sleeping sickness. Physiology(Bethesda) 19:198–206
Lutumba P, Robays J, Miaka Mia BC, Mesu VK, Molisho D, Declercq J, Van der Veken W, Meheus F, Jannin J, Boelaert M (2005) Trypanosomiasis control, Democratic Republic of Congo, 1993-2003. Emerg Infect Dis 11(9):1382–1388
Lutumba P, Makieya E, Shaw A, Meheus F, Boelaert M (2007) Human African trypanosomiasis in a rural community, Democratic Republic of Congo. Emerg Infect Dis 13(2):248–254
Magez S, Caljon G, Tran T, Stijlemans B, Radwanska M (2010) Current status of vaccination against African trypanosomiasis. Parasitology 137(14):2017–2027
Magnus E, Vervoort T, Van Meirvenne N (1978) A card-agglutination test with stained trypanosomes (C.A.T.T.) for the serological diagnosis of T. B. gambiense trypanosomiasis. Ann Soc Belg Med Trop 58(3):169–176
Malvy D, Chappuis F (2011) Sleeping sickness. Clin Microbiol Infect 17(7):986–999
Mitashi P, Hasker E, Lejon V, Kande V, Muyembe JJ, Lutumba P, Boelaert M (2012) Human african trypanosomiasis diagnosis in first-line health services of endemic countries, a systematic review. PLoS Negl Trop Dis 6(11):e1919
Mitashi P, Hasker E, Ngoyi DM, Pyana PP, Lejon V, Van der Veken W, Lutumba P, Buscher P, Boelaert M, Deborggraeve S (2013) Diagnostic accuracy of loopamp Trypanosoma brucei detection kit for diagnosis of human African trypanosomiasis in clinical samples. PLoS Negl Trop Dis 7(10):e2504
Mitashi P, Hasker E, Mbo F, Van Geertruyden JP, Kaswa M, Lumbala C, Boelaert M, Lutumba P (2015) Integration of diagnosis and treatment of sleeping sickness in primary healthcare facilities in The Democratic Republic of the Congo. Trop Med Int Health 20(1):98–105
Molyneux DH (1973) Isolation of Trypanosoma (Trypanozoon) brucei gambiense in rabbits by the intratesticular inoculation technique. Ann Trop Med Parasitol 67(4):391–397
Moore A, Richer M (2001) Re-emergence of epidemic sleeping sickness in southern Sudan. Trop Med Int Health 6(5):342–347
Mpanya A, Hendrickx D, Vuna M, Kanyinda A, Lumbala C, Tshilombo V, Mitashi P, Luboya O, Kande V, Boelaert M, Lefevre P, Lutumba P (2012) Should I get screened for sleeping sickness? A qualitative study in Kasai province, Democratic Republic of Congo. PLoS Negl Trop Dis 6(1):e1467
Mugasa CM, Katiti D, Boobo A, Lubega GW, Schallig HD, Matovu E (2014) Comparison of nucleic acid sequence-based amplification and loop-mediated isothermal amplification for diagnosis of human African trypanosomiasis. Diagn Microbiol Infect Dis 78(2):144–148

Nkinin SW, Njiokou F, Penchenier L, Grebaut P, Simo G, Herder S (2002) Characterization of Trypanosoma brucei s.l. subspecies by isoenzymes in domestic pigs from the Fontem sleeping sickness focus of Cameroon. Acta Trop 81(3):225–232

Paquet C, Castilla J, Mbulamberi D, Beaulieu MF, Gastellu Etchegorry MG, Moren A (1995) Trypanosomiasis from Trypanosoma brucei gambiense in the center of north-West Uganda. Evaluation of 5 years of control (1987-1991). Bull Soc Pathol Exot 88(1):38–41

Pepin J, Milord F (1994) The treatment of human African trypanosomiasis. Adv Parasitol 33:1–47

Politi C, Carrin G, Evans D, Kuzoe FA, Cattand PD (1995) Cost-effectiveness analysis of alternative treatments of African gambiense trypanosomiasis in Uganda. Health Econ 4(4):273–287

Priotto G, Kasparian S, Mutombo W, Ngouama D, Ghorashian S, Arnold U, Ghabri S, Baudin E, Buard V, Kazadi-Kyanza S, Ilunga M, Mutangala W, Pohlig G, Schmid C, Karunakara U, Torreele E, Kande V (2009) Nifurtimox-eflornithine combination therapy for second-stage African Trypanosoma brucei gambiense trypanosomiasis: a multicentre, randomised, phase III, non-inferiority trial. Lancet 374(9683):56–64

Robays J, Bilengue MM, Van der Stuyft P, Boelaert M (2004) The effectiveness of active population screening and treatment for sleeping sickness control in the Democratic Republic of Congo. Trop Med Int Health 9(5):542–550

Robays J, Lefevre P, Lutumba P, Lubanza S, Kande Betu KMV, Van der Stuyft P, Boelaert M (2007) Drug toxicity and cost as barriers to community participation in HAT control in the Democratic Republic of Congo. Trop Med Int Health 12(2):290–298

Rocha G, Martins A, Gama G, Brandão F, Atouguia J (2004) Possible cases of sexual and congenital transmission of sleeping sickness. Lancet 363(9404):247

Schmid C, Kuemmerle A, Blum J, Ghabri S, Kande V, Mutombo W, Ilunga M, Lumpungu I, Mutanda S, Nganzobo P, Tete D, Mubwa N, Kisala M, Blesson S, Mordt OV (2012) In-hospital safety in field conditions of nifurtimox eflornithine combination therapy (NECT) for T. b. Gambiense sleeping sickness. PLoS Negl Trop Dis 6(11):e1920

Shaw AP, Tirados I, Mangwiro CT, Esterhuizen J, Lehane MJ, Torr SJ, Kovacic V (2015) Costs of using "tiny targets" to control glossina fuscipes fuscipes, a vector of gambiense sleeping sickness in Arua District of Uganda. PLoS Negl Trop Dis 9(3):e0003624

Simarro PP, Sima FO, Mir M, Mateo MJ, Roche J (1991) Control of human African trypanosomiasis in Luba in Equatorial Guinea: evaluation of three methods. Bull World Health Organ 69(4):451–457

Simarro PP, Franco JR, Asumu PN (2001) Has the focus of human African trypanosomiasis in Luba, Equatorial Guinea been eradicated? Med Trop Mars 61(4–5):441–444

Simarro PP, Cecchi G, Paone M, Franco JR, Diarra A, Ruiz JA, Fevre EM, Courtin F, Mattioli RC, Jannin JG (2010) The atlas of human African trypanosomiasis: a contribution to global mapping of neglected tropical diseases. Int J Health Geogr 9:57

Simarro PP, Cecchi G, Franco JR, Paone M, Diarra A, Ruiz-Postigo JA, Fevre EM, Mattioli RC, Jannin JG (2012a) Estimating and mapping the population at risk of sleeping sickness. PLoS Negl Trop Dis 6(10):e1859

Simarro PP, Franco JR, Cecchi G, Paone M, Diarra A, Ruiz Postigo JA, Jannin JG (2012b) Human African trypanosomiasis in non-endemic countries (2000-2010). J Travel Med 19(1):44–53

Sternberg JM (2004) Human African trypanosomiasis: clinical presentation and immune response. Parasite Immunol 26(11–12):469–476

Sternberg JM, Gierlinski M, Bieler S, Ferguson MA, Ndung'u JM (2014) Evaluation of the diagnostic accuracy of prototype rapid tests for human African trypanosomiasis. PLoS Negl Trop Dis 8(12):e3373

Tarral A, Blesson S, Mordt OV, Torreele E, Sassella D, Bray MA, Hovsepian L, Evene E, Gualano V, Felices M, Strub-Wourgaft N (2014) Determination of an optimal dosing regimen for fexinidazole, a novel oral drug for the treatment of human African trypanosomiasis: first-in-human studies. Clin Pharmacokinet 53(6):565–580

Torreele E, Bourdin TB, Tweats D, Kaiser M, Brun R, Mazue G, Bray MA, Pecoul B (2010) Fexinidazole--a new oral nitroimidazole drug candidate entering clinical development for the treatment of sleeping sickness. PLoS Negl Trop Dis 4(12):e923. Available from: PM:21200426

Urech K, Neumayr A, Blum J (2011) Sleeping sickness in travelers - do they really sleep? PLoS Negl Trop Dis 5(11):e1358

Van Hoof LM (1947) Observations on trypanosomiasis in the Belgian Congo. Trans R Soc Trop Med Hyg 40(6):728–761

Van Nieuwenhove S, Betu-Ku-Mesu VK, Diabakana PM, Declercq J, Bilenge CM (2001) Sleeping sickness resurgence in the DRC: the past decade. Trop Med Int Health 6(5):335–341

Vreysen MJ (2001) Principles of area-wide integrated tsetse fly control using the sterile insect technique. Med Trop (Mars) 61(4–5):397–411

Welburn SC, Fevre EM, Coleman PG, Odiit M, Maudlin I (2001) Sleeping sickness: a tale of two diseases. Trends Parasitol 17(1):19–24

WHO (1998) Control and surveillance of African trypanosomiasis. Report of a WHO expert committee. World Health Organ Tech Rep Ser 881:I-114

WHO (2013) Control and surveillance of human African trypanosomiasis. World Health Organ Tech Rep Ser 984:1–237. www.healthdata.org/results/gbd_summaries/2019/african-trypanosomiasisAfricantrypanosomiasis|Institute for HealthMetrics and Evaluation(healthdata.org)

WHO (2022) www.who.int/news-room/fact-sheets/detail/trypanosomiasis-human-african-(sleeping-sickness

World Health Organisation (2018) Human African trypanosomiasis: The transmissioncycle. http://www.who.int/trypanosomiasis_african/disease/transmissioncycle/en/. Accessed on 20 September 2018

Leishmaniasis in Sub-Saharan Africa

Asrat Hailu, Mounir Lado, Abdelgadir Ali Bashir Ali, Ahmed Mudawi Musa, and Daniel Mwiti

Contents

A. Hailu (✉)
Department of Microbiology, Immunology & Parasitology, School of Medicine, College of Health Sciences, Addis Ababa University, Addis Ababa, Ethiopia
e-mail: asrat.hailu@aau.edu.et

M. Lado
Ministry of Health, Juba, South Sudan
e-mail: mlado@imaworldhealth.org

A. A. B. Ali
Faculty of Medicine, University of Khartoum, Khartoum, Sudan

A. M. Musa
Institute of Endemic Diseases, Faculty of Medicine, University of Khartoum, Khartoum, Sudan
e-mail: amusa@iend.org

D. Mwiti
Leishmaniasis Focal Point Person, Ministry of Health, Nairobi, Kenya

J. O. Gyapong, B. A. Boatin (eds.), *Neglected Tropical Diseases - Sub-Saharan Africa*, Neglected Tropical Diseases, https://doi.org/10.1007/978-3-031-53901-5_5

Abstract

Visceral leishmaniasis is widespread in East Africa, with several foci in Eritrea, Ethiopia, Kenya, Somalia, South Sudan, Sudan, and Uganda. Outside this subregion, cases of visceral leishmaniasis have been reported from Burkina Faso, Cameroon, Chad, Ivory Coast, Mauritania, Niger, Nigeria, and the Gambia. It causes frequent outbreaks in the savanna and forest areas where sand flies live around and in termite mounds and soil cracks. Cutaneous leishmaniasis occurs throughout western and eastern Africa. A belt runs from Mauritania, Senegal, Burkina Faso, and Mali in the West, through Niger, Nigeria, and Cameroon, to Chad, Sudan, Ethiopia, and Kenya in the East. Though their recognition as a public health problem is growing, leishmaniases remain one of the most neglected tropical diseases. Their control is still hampered by a lack of safe and efficacious drugs, simple diagnostics, effective vaccines, and vector control tools.

Introduction

Leishmaniases are a diverse group of neglected tropical diseases affecting humans in 98 countries and caused by at least 20 species of the protozoan parasite of the genus *Leishmania.* The disease manifestation ranges from self-healing cutaneous leishmaniasis (CL) to the mutilating mucocutaneous leishmaniasis or diffuse cutaneous leishmaniasis and the life-threatening visceral leishmaniasis (VL). The World Health Organization (WHO) estimates that there are approximately 50–90,000 new VL cases and 0.6–1.0 million new CL cases per year (World Health Organization 2022). The disease does not rank high either on the political or on the research agenda of many countries as it typically affects marginalized communities and the poorest of the poor (Boelaert et al. 2009).

Leishmania causing diseases in man are transmitted through the bite of a bloodsucking female sand fly. Over 90 species of sand flies are proven or suspected to transmit the diseases in various parts of the world (World Health Organization 2010). The parasites are morphologically dimorphic in their life cycle, transforming from the extracellular promastigote stage in the sand fly gut to the intracellular amastigote stage in the mammalian host. Several leishmania species are transmitted in a zoonotic cycle, with dogs or rodents as animal reservoir, but others are strictly anthroponotic.

Visceral leishmaniasis (VL) is endemic in at least 79 countries with an estimated annual incidence of 50,000–90,000 new cases per year (World Health Organization 2022), over 90% being reported from 10 countries of the Indian subcontinent, East Africa, and Brazil. Two distinct leishmania species cause VL: *Leishmania donovani* and *L. infantum/L. chagasi*, which are transmitted by sand flies of the genera *Phlebotomus* and *Lutzomyia. L. infantum/chagasi* is found mainly in the Mediterranean littoral (North Africa, South Europe, and western Middle East) and South America, whereas *L. donovani* is restricted to the Indian subcontinent and sub-Saharan Africa and occurs sporadically in isolated foci of the Arabian subcontinent. Previously, *L. archibaldi* and *L. infantum* were also considered to cause VL in Sub-Saharan Africa. However, recent molecular and evolutionary genetic analyses have annulled the existence of *L. archibaldi* as a separate species, and a revision of the taxonomy has led to a consensus that only *L. donovani* is causing VL in Sub-Saharan Africa (Jamjoom et al. 2004).

With the advent of the HIV/AIDS epidemic, coinfection of HIV and visceral leishmaniasis has become an emerging public health problem and a challenge to the clinical management of patients (Alvar et al. 2008). At least 45 countries in the world have reported cases of *HIV-leishmania* coinfections by 2021 (World Health Organization 2022).

Cutaneous leishmaniasis is caused by several species (or species complexes), mainly *L. braziliensis*, *L. guyanensis*, *L. panamensis*, *L. peruviana*, *L. mexicana*, *L. amazonensis*, *L. venezuelensis*, *L. major*, *L. tropica*, and *L. aethiopica*. Each year approximately 0.6–1.0 million cases of CL occur in at least 89 countries (World Health Organization 2022; WER 2021), the highest burden being in just 10 countries—Afghanistan, Algeria, Colombia, Brazil, Iran, Syria, Ethiopia, North Sudan, Costa Rica, and Peru (Alvar et al. 2012; World Health Organization 2022). In the Old World, major foci of anthroponotic *L. tropica* transmission stretch from India through central and western Asia into northern Africa; zoonotic transmission of *L. major* occurs from central Asia through western Asia into northern Africa, and *L. aethiopica* is confined to East Africa. However, the spectrum of skin pathology, the varied responses to treatments, the multitude of sand fly species and animal reservoirs involved in the life cycle, and the epidemiological complexities make these diseases intricately complex.

We will discuss below in more detail the epidemiological and clinical features of visceral leishmaniasis and cutaneous leishmaniasis in sub-Saharan Africa.

Part 1: Visceral Leishmaniasis in Sub-Saharan Africa

Geographical Distribution

The global distribution and endemicity of visceral leishmaniasis in tropical and subtropical areas are shown in Figure 1. Underreporting of VL is primarily attributed to the lack of diagnostic tools and surveillance mechanisms. Consequently, the actual

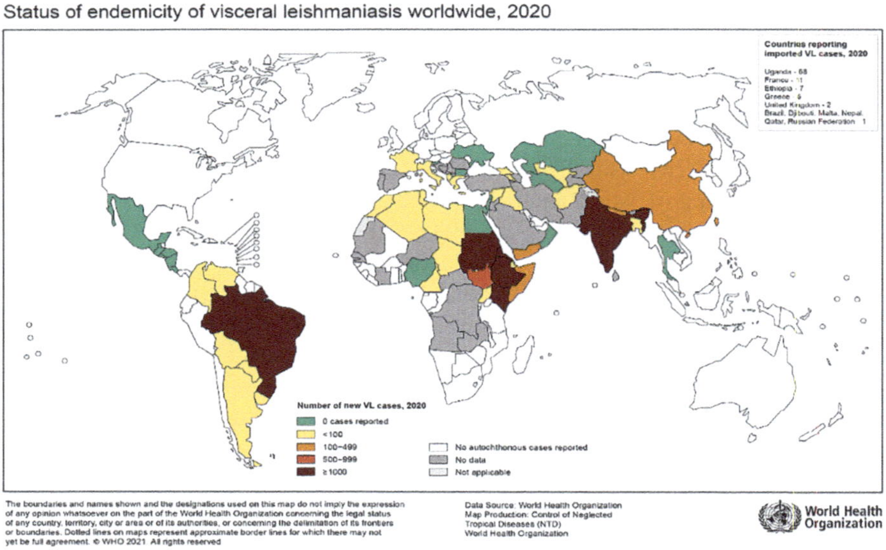

Fig. 1 Worldwide distribution and endemicity of visceral leishmaniasis in 2020

burden of VL remains unknown. Six countries of the eastern Africa region [EAR] (i.e., Eritrea, Ethiopia, Kenya, Somalia, South Sudan, and Sudan) are found in the epicenter of the VL epidemic, although the disease is also reported from Uganda, Djibouti, and western and central regions of Sub-Saharan Africa. In the west, the majority of visceral leishmaniasis cases are reported from Ivory Coast, Niger, Mauritania, Senegal, Nigeria, and Gambia (Kone et al. 2019). Visceral leishmaniasis has occurred sporadically in the N'Djamena and Lake Chad areas, with foci extending eastward throughout southern Chad (Sirol et al. 1976). The seven endemic countries of the Eastern Africa Region (Eritrea, Ethiopia, Kenya, Somalia, South Sudan, Sudan, and Uganda) as well as Cameroon, Chad, and Djibouti are now reporting cases to WHO (Global Health Observatory 2022, updated Aug. 2021).

Figure 1 shows the global distribution of visceral leishmaniasis, depicting endemicity in tropical and sub-tropical areas. Endemic countries are reporting cases of visceral leishmaniasis to WHO, but the reporting is incomplete.

Figure 2a, b show the number of visceral leishmaniasis cases reported to WHO by sub-region (Fig. 2a) and country-wise in Africa (Fig. 2b) for the period 2005–2019. Figure 2a reveals the Eastern Africa Region is presently shouldering a proportionally highest number of cases. By 2020, the Eastern Africa Region comprising Eritrea, Ethiopia, Kenya, Somalia, South Sudan, Sudan, and Uganda reported 57% of the global VL cases (WER 2021).

Visceral leishmaniasis is widespread in the Eastern Africa Region, with several foci identified in Eritrea, Ethiopia, Kenya, Somalia, South Sudan, Sudan, and Uganda. The epidemiology varies hugely with geographical locations and ecological settings and typically exhibits micro-focal variation in its clinical and epidemiological

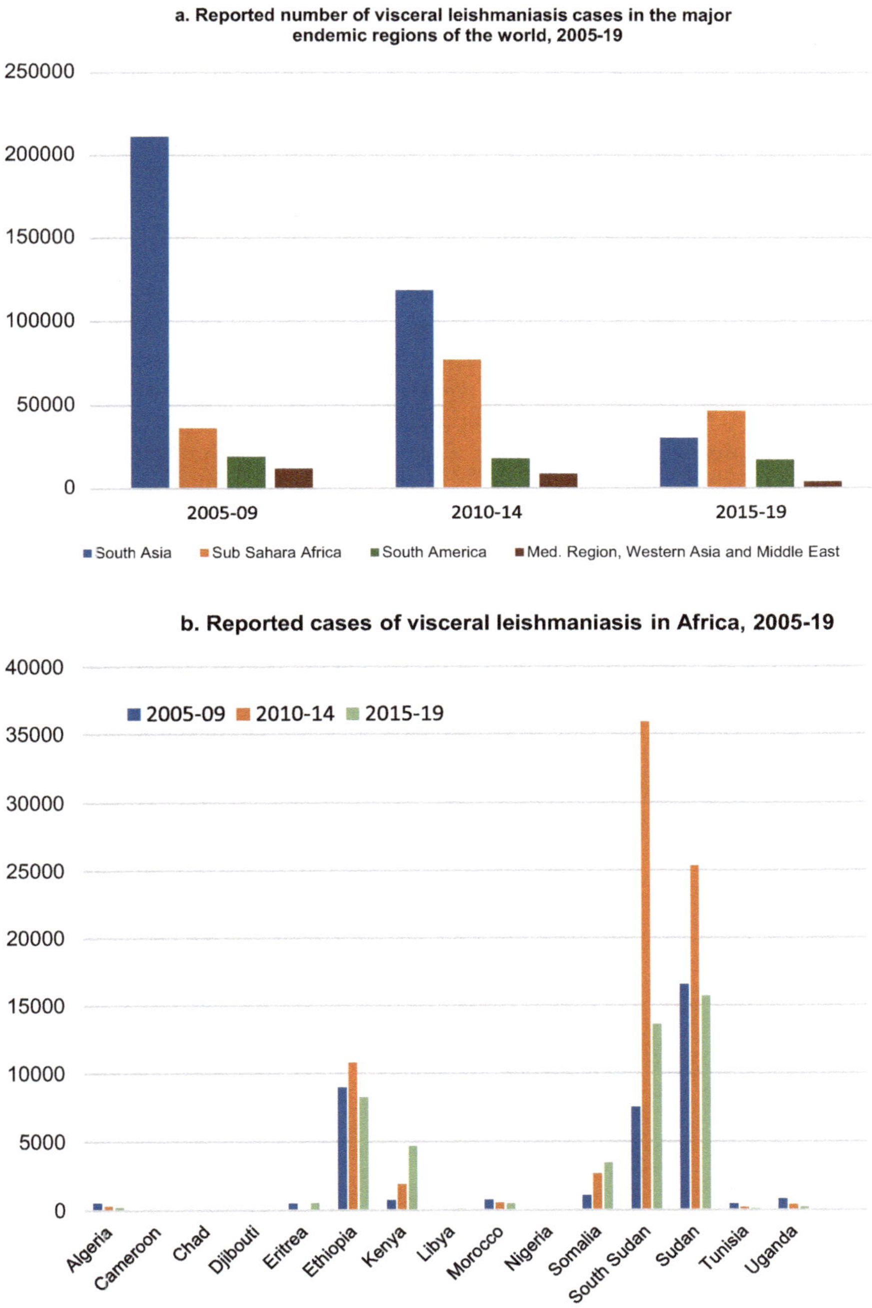

Fig. 2 (**a**). Reported number of visceral leishmaniasis cases in the major endemic regions of the world, 2005-19. (**b**). Reported cases of visceral leishmaniasis in Africa, 2005-19

characteristics. The important visceral leishmaniasis foci in the Eastern Africa Region are described below and shown in Fig. 3.

Eritrea: The eastern lowlands that abound the Red Sea littoral, which includes localities around Algena and Nakfa (Northern Red Sea Region), localities in the Southern Red Sea Region the western lowlands of Anseba region (lowlands around Keren) bordering Kassala region in eastern Sudan, and the Gash-Barka region in south west Eritrea bordering northwest Ethiopia and Kassala in eastern Sudan are localities with potential VL transmission (Ayele and Ali 1984; Hailu unpublished; Malaria Consortium 2010).

Ethiopia: The disease is widespread throughout the lowlands, being reported from over 40 isolated localities. Metema-Humera lowlands, which account for 60% of VL cases in Ethiopia, and the lowlands in South and southwest Ethiopia, e.g., Segen, Woitu, and Omo river plains, are the major endemic regions (Hailu et al. 2006). Other foci are in Borena and Guji zones (Moyale, Yabelo, Genale, Dawa) and in the Burji district—South Ethiopia. Emerging VL foci have been identified in southeast Ethiopia, e.g., in the Liben district (Guji zone of Oromia), and further east in the Somali region of Ethiopia (Afder, Gode; Wabe Shebelle area).

The 2005 outbreak in Libo Kemkem and Fogera districts (South Gondar zone) in Lake Tana Basin was curtailed after the lives of hundreds were claimed (Alvar et al. 2007). This outbreak has been attributed to the large-scale seasonal migration of highlanders to the lowlands of northwest Ethiopia—a hot spot of visceral leishmaniasis transmission. An increasing number of cases are also being reported from Tahtay Adiyabo, Badme, and Welkait districts in north Ethiopia (Hail unpublished). There also appear to be some isolated foci of visceral leishmaniasis in northeast Ethiopia (Timuga and Waja zone, middle and lower Awash valleys) and in the dry river valleys and gorges of central Ethiopia (Hailu unpublished).

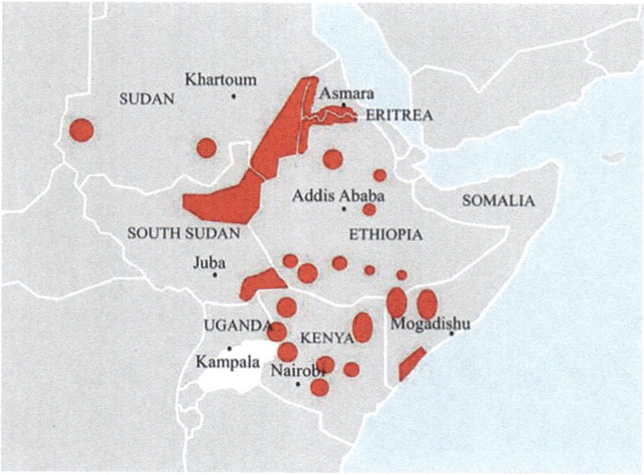

Fig. 3 Important visceral leishmaniasis foci in the Eastern Africa Region

Kenya: The main foci of visceral leishmaniasis are found in the Rift Valley counties (West Pokot, Baringo, Samburu), some localities of the northern county (i.e., Marsabit), eastern counties (Machakos, Kitui, Mwingi, Makueni), and northeastern counties (Mandera, Wajir, and Garissa) (Schaefer et al. 1994; Marlet et al. 2003; Tonui 2006). In the northeast, refugee camps around Dadaab and Hagadera have been affected. Other foci are found in Turkana and the coastal counties in the East (Tonui 2006). In Kenya, VL appears to have a wide geographic distribution, and new foci are continually being discovered.

The Republic of Djibouti: Isolated cases of visceral leishmaniasis have been reported (Pratlong et al. 2005), but the exact locations of the transmission areas are unknown.

Somalia: The geographical distribution of visceral leishmaniasis in Somalia is not very well known. An endemic locality was first described by Baruffa (1965), and cases were reported from hospitals in Mogadishu and Kismayo. Presently, it is known to be endemic in the regions of Bay, Bakool, Gedo, and western Hiiraan (Marlet et al. 2003; Raguenaud et al. 2006). Based on historically documented cases, the Middle Shabelle and Lower Juba regions are considered endemic.

South Sudan: The catchment area of the White Nile in South Sudan is the main visceral leishmaniasis transmission area and includes the northern foci (Unity, Upper Nile, and Jonglei states) and the southern foci (eastern Equatoria State). In these four states, 2.7 million people living in 28 counties are considered to be at risk (Abubakar et al. 2014). Highly affected areas are Bentiu in Unity State, Lankien in Jonglei State; and Paloich, Malakal, Ulang, Kiech Kuon, and Nasir in Upper Nile State. The Kapoeta focus is situated in eastern Equatoria (Abubakar et al. 2014). In South Sudan, epidemic hotspots are known to shift from location to location.

Sudan: Gedaref State in eastern Sudan is presently the main focus of visceral leishmaniasis in Sudan. The endemic localities in Gedaref State are located in the large expanse of forested landmass, including localities around Dinder National Park that are found north and east of the Rahad River and southwest of Atbara River. This region borders the lowlands of northwestern Ethiopia. In the 1980s, major visceral leishmaniasis outbreaks occurred in Blue Nile State (reviewed in: Osman et al. 2000) just north of Upper Nile State (South Sudan) and among displaced people in Khartoum. Other foci are found in Sennar, Darfur, and Kordofan states (Malaria Consortium 2010).

Uganda: Visceral leishmaniasis has a limited distribution in the northeastern region of the country, mainly in the Pokot region, which is contiguous with the Kenyan focus in the Pokot region (Kolaczinski et al. 2008). The majority of the patients come from Nakapiripirit District, while other adjacent districts are also reporting sporadic cases.

Transmission Pattern, Incidence, and Burden of Visceral Leishmaniasis in Sub-Saharan Africa

The transmission patterns of visceral leishmaniasis in Sub-Saharan Africa exhibit significant variation between the different ecological foci. Several field studies have witnessed varied degrees of transmission and incidence rates in different endemic areas, e.g., 2.0 cases per 1000 person-years in Kenya (Schaefer et al. 1995) and 6.9 cases per 1000 person-years in Ethiopia (Ali and Ashford 1994). Zijlstra et al. (1994) reported 38.5 cases per 1000 person-years in a community in eastern Sudan. These incidence rates reflect the varied degrees of endemicity on the one hand and the high threshold for epidemic outbreaks on the other.

In East Africa, visceral leishmaniasis is characterized by two distinct disease ecotypes, and each exhibits unique eco-epidemiology, varied disease transmission patterns, and clinical presentation. Visceral leishmaniasis in Sudan is characterized by a high frequency of lymphadenopathy and high rates of post-kala-azar dermal leishmaniasis (which sharply contrasts with the lower rates in neighboring countries). The response to treatment with various antileishmanial drugs, particularly paromomycin (Hailu et al. 2010) and liposomal amphotericin B (Khalil et al. 2014), is variable. Entomological and parasitological data support the notion that two distinct ecotypes of visceral leishmaniasis are found in the Eastern African Region.

Based on 2004–2009 reported data, the annual incidence of visceral leishmaniasis in the Eastern Africa Region was estimated between 29,400 and 56,700 (Alvar et al. 2012), with Sudan and South Sudan accounting for nearly 80% of the burden. The estimated annual incidence rates based on the 2004–2009 data are as follows: Sudan (15,700–30,300), South Sudan (7400–14,200), Ethiopia (3700–7200), Somalia (1400–2700), Kenya (610–1200), Uganda (350–520), and Eritrea (200–400).

In Sudan, since Neave reported the first case of visceral leishmaniasis in 1904 (reviewed in: Abubakar et al. 2014), the disease had seized an endemic appearance until the 1940s. The first outbreak that reached epidemic proportions was reported near the town of Melut in Upper Nile province (Stephenson 1940). The second major outbreak was reported in 1956 with reportedly "thousands of cases" in southern Fung. Subsequently, another outbreak of visceral leishmaniasis was identified in a small community of 5000 among the Jum Jum tribe and assailed the affected community with an attack rate of 10% and a case fatality ratio (CFR) of 41% (Sati 1958). Since then, VL has continued unabated in Sudan in its endemic or epidemic proportions, often exacerbated by the cumulative effects of famine and civil war (de Beer et al. 1991; Seaman et al. 1996). Seaman et al. (1996) described a devastating epidemic in South Sudan whereby during the period 1984–1994, an estimated 100,000 VL deaths occurred in a group of 280,000 people living in western Upper Nile province. Forced migration due to the civil war was a major factor in transmission, and the poor nutritional status of the population contributed to the high mortality rates. This region of South Sudan has paid a huge death toll due to visceral leishmaniasis, depicting the significance of the disease to humankind.

A recent outbreak has similarly affected the highlands of northern Ethiopia, even though the death toll was much lower than in South Sudan (Alvar et al. 2007). There is consensus among experts that over many decades, the incidence of VL in Ethiopia has increased drastically. A case in point is that of northwestern Ethiopia, where the incidence has increased tenfold in two decades from just a few hundred in the mid-1990s (numbers were 56 in 1995 and 165 in 1996). This is largely attributed to migration and new settlements driven by war or dire economic conditions in the region. The large-scale agricultural farms that attract several thousands of manual laborers, often recruited from poor farming communities of the highland fringes, are among the important factors that contribute to a high burden of the disease. Population movements or invasion of previously unoccupied localities by way of new settlement schemes or deforestation by large-scale farms have probably led to a closer association between humans and the animal reservoirs, creating a conducive environment for increased transmission. This phenomenon appears to be a consistent pattern throughout the Eastern Africa Region. Food insecurity forcing people to migrate or leading to widespread malnutrition and unusual rainfall patterns were considered to have played an important role in the emergence of visceral leishmaniasis as an important health problem in the Bakool region of Somalia (Marlet et al. 2003).

HIV-VL coinfection can also be one of the drivers of transmission, as coinfected cases are very infectious to sand flies. In Ethiopia, coinfection affects specific risk groups, e.g., military personnel and migrant workers. Presently, significant percentages (10–40%) of VL patients in northern Ethiopia suffer from coinfection.

Clinical Manifestations, Diagnosis, and Treatment of Visceral Leishmaniasis in Sub-Saharan Africa

Clinical Features, Complications, and Comorbidities

Visceral leishmaniasis patients present with prolonged but irregular fever, weight loss, hepatosplenomegaly, lymphadenopathy, pancytopenia, and hypergammaglobulinemia. Thrombocytopenia may cause uncontrollable epistaxis or bleeding. The course of the disease is insidious yet carries a high risk of mortality in the advanced stages (Fig. 4), which is characterized by emaciation, severe anemia, bleeding, jaundice, diarrhea, and severe neutropenia. Comorbid conditions—mainly superinfections (e.g., bacterial pneumonia)—and late presentations are risk factors for adverse prognosis and death. Differential diagnosis for visceral leishmaniasis must consider malaria, relapsing fever, typhoid fever, schistosomiasis, tuberculosis, brucellosis, histoplasmosis, EBV, AIDS, brucellosis, chronic hepatitis, cirrhosis, myeloproliferative diseases, lymphomas, and leukemia (Piccioni et al., 2020).

In South American visceral leishmaniasis, HIV coinfection, dyspnea, bacterial infections, and age <5 or >40–45 were identified as significant predictors of bad prognosis (Belo et al. 2014). Similarly, among northern Ethiopian patients, age above 45, HIV coinfection, edema, severe malnutrition, pneumonia, tuberculosis, and vomiting were associated with high mortality (Herrero et al. 2009). Bacterial sepsis was a common finding among northern Ethiopian visceral leishmaniasis patients, with a prevalence of 17.1% (12 out of 70) in HIV-negative and 30.8% (3 out of 13) in HIV-positive patients (Endris et al. 2014). Malnutrition and intestinal parasitosis were highly prevalent among visceral leishmaniasis patients in northern Ethiopia (Mengesha et al. 2014; Endris et al. 2014). Malaria coinfection exacerbates signs and symptoms of visceral leishmaniasis and complicates the presentation at onset. In a retrospective case-control study among

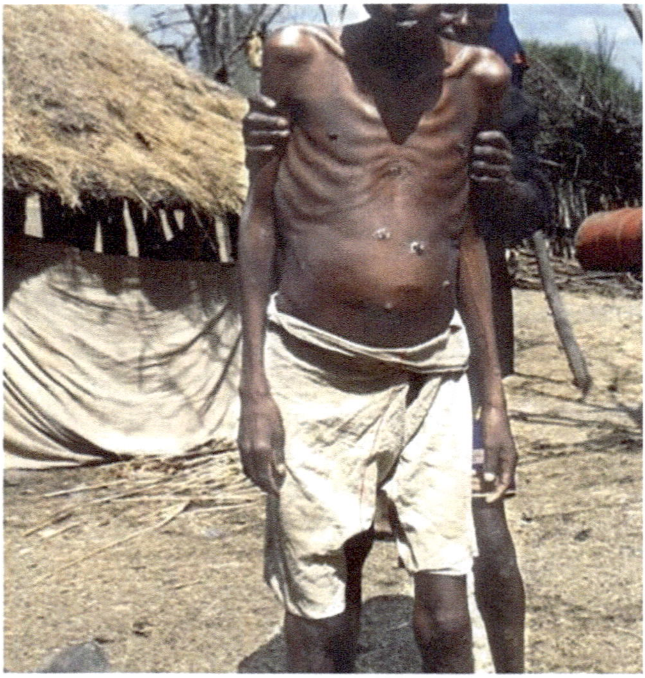

Fig. 4 Patient with advanced visceral leishmaniasis

Sudanese patients, emaciation, anemia, jaundice, and mortality were found to be positively associated with malaria coinfection (Van den Bogaart et al. 2013). Pulmonary tuberculosis, pneumonia, and diarrhea are common comorbid conditions that could lead to death. Hyperglobulinemia is a typical feature of visceral leishmaniasis that accompanies profound cellular immune suppression. The suppression of cell-mediated immunity predisposes patients to intercurrent infections with pulmonary tuberculosis and pneumonia. In Mediterranean visceral leishmaniasis, 50% of patients with HIV coinfection met the AIDS-defining criteria during the first episode of primary visceral leishmaniasis (Russo et al. 2003). There is now consensus that visceral leishmaniasis should be considered an AIDS-defining illness (stage IV diagnosis) irrespective of the CD4 count. Furthermore, HIV coinfection in visceral leishmaniasis is associated with failure of treatment and frequent relapses. Coinfected patients tend to enter a chronic carrier state despite a high parasitemic index in multiple organs.

Post-kala-azar dermal leishmaniasis (PKDL) is a dermal complication of visceral leishmaniasis, mostly occurring in treated patients. It is characterized by macular, maculopapular, and nodular rashes in the skin, especially in the exposed body parts (mainly in the face, neck, trunk, and limbs). Mucosal involvement is a common complication. Ocular involvement can manifest as blepharoconjunctivitis or pan-uveitis (El Hassan et al. 1998), with a risk of causing blindness (Khalil et al. 2011). Most PKDL lesions in sub-Saharan

Africa are self-resolving; however, severe and chronic lesions tend to be persistent, requiring treatment. The incidence of PKDL in treated visceral leishmaniasis patients can be higher than 50% in Sudanese patients, whereas it is around 5–10% in the Indian subcontinent (Ganguly et al. 2010). Interestingly, the interval between treatment for visceral leishmaniasis and the emergence of PKDL also varies between the two continents, being within 6 months in Sudan and 2–3 years in India (Zijlstra et al. 2003). PKDL is associated with *L. donovani* and hence found in Sub-Saharan Africa (mainly in Sudan and in northern Ethiopia to a lesser extent) and Asia. Khalil et al. (2013) discussed PKDL as a paradoxical IRIS in HIV-negative VL patients. PKDL-like lesions are also reported in HIV coinfected patients and are often regarded as either an atypical presentation of HIV/AIDS (Boumis et al. 2006) or as a feature of IRIS in patients on HAART (Antinori et al. 2007; Gelanew et al. 2010; Celesia et al. 2014). PKDL is not reported in HIV-negative visceral leishmaniasis patients infected with *L. infantum*.

PKDL is considered to play a role in the transmission of visceral leishmaniasis, especially in the inter-epidemic period (Zijlstra et al. 2003). Several theories and hypotheses (parasite factors, immune pathogenesis, environmental factors, etc.) have been suggested to explain the pathogenesis of PKDL.

Diagnosis of Visceral Leishmaniasis

In hospital settings, visceral leishmaniasis is diagnosed by parasitological examination of stained tissue aspirates obtained either from the spleen, bone marrow, or lymph nodes. The choice of organ for tissue aspirates varies from facility to facility, country to country, and level of expertise and experience. In Sudan, most aspirates are from lymph nodes (Babiker et al. 2007), partly due to the high frequency of lymphadenopathy. In Ethiopia, Kenya, and Uganda, tissue aspirates are often obtained either from the spleen or bone marrow. Tissue aspiration procedures are generally considered invasive and thus not used at lower health facility settings; hence, diagnosis relies on clinical diagnosis and/or serological tests, including rapid immunochromatographic tests. Nonetheless, splenic aspirates are the most sensitive to detect amastigote stages of the parasite in stained tissue. Technically, splenic aspiration requires specific skill and experience and could carry a risk of internal bleeding. Bone marrow punctures are painful to the patients and also less sensitive than splenic aspirates. Lymph node aspiration is simple and safe, but the least sensitive, aside from being painful.

In visceral leishmaniasis control programs, the diagnostic approach is well articulated in the national diagnosis and treatment guidelines and is guided by an algorithm that provides sign- and symptom-based case definition, often requiring the conduct of a serological test, usually a rapid test. When these are well organized, the approach helps reduce morbidity and mortality from visceral leishmaniasis. However, the clinical symptoms and signs of visceral leishmaniasis can still be confused with other co-endemic diseases and require reasonable clinical skills. Malaria is often co-endemic, as are schistosomiasis and brucellosis. An operational case definition, therefore, requires the exclusion of malaria.

To date, the majority of serological tests target the detection of antibodies rather than antigens in blood samples. Direct agglutination test (DAT) is a frequently used antibody test owing to its simplicity in sample collection and test performance and the visual detection of end results. DAT can be performed on blood samples collected on Whatman filter papers. However, the test still requires basic laboratory facilities and a trained technician to perform titrations of the patient's serum. In addition, test results are not immediately available, requiring at least 8 h of incubation.

Field-adapted rapid diagnostic tests have been developed and made available commercially as point-of-care tests. The development of rapid diagnostic tests has revolutionized the diagnosis of visceral leishmaniasis in control programs. Among the few rapid diagnostic tests that have evolved, the rk39 immunochromatographic (ICT) test in a dipstick or cassette format is best known and has been most extensively validated so far (Boelaert et al. 2014). Its sensitivity was found to be slightly lower in East Africa compared to the Indian subcontinent, and a negative test does not completely rule out the probability of VL. In such cases, a more sensitive test, such as the DAT, is recommended. A new immunochromatographic test, the rk28 test that combines rk39 with other recombinant antigens (rk9, rk16, and rk26), has been under clinical evaluation, with preliminary results showing problems of false sensitivity and cross-reactions with malaria.

A urine agglutination test, KATEX, has also been developed for antigen detection. Initial field trials have necessitated further improvements; a Cochrane analysis showed an overall sensitivity of only 63.6 % (Boelaert et al. 2014). Loop-mediated isothermal amplification (LAMP) is a simple molecular test that has undergone a clinical evaluation. To date, tests that aim to detect leishmanial antigens in blood samples of patients are not available even though such a test would seem to be the ideal test. Availability and access to RDTs for visceral leishmaniasis remain a challenge in resource-constrained settings.

Treatment of Visceral Leishmaniasis

Until recent years, the standard treatment regimen for primary visceral leishmaniasis in Sub-Saharan Africa was based upon a 30-day course of pentavalent antimonial compounds: sodium stibogluconate (SSG) or meglumine antimoniate (MA). Second-line treatments depended on the availability of other medications (e.g., amphotericin B deoxycholate, usually in its lipid formulation). In the past, access to and options for treatments were limited. Presently, the main drugs that are in use for treatment include one or a combination of the following: proprietary sodium stibogluconate (Pentostam®, GSK), generic sodium stibogluconate (SSG) produced by Albert David (Delhi, India), Glucantime® (meglumine antimoniate) produced by Aventis Pharma, miltefosine (Impavido®), amphotericin B deoxycholate (Fungizone, Squibb), AmBisome® produced by Gilead Scientific Inc., and paromomycin (aminosidine sulfate) produced by Gland Pharma in India. Among these, miltefosine is the only oral treatment. Not all these drugs are included in the national essential drug list of the affected countries in the region, and supplies are procured mostly by international donors.

The WHO has recommended a 17-day course of combined SSG (20 mg/kg) and paromomycin (15 mg/kg) as a first-line treatment for Eastern African visceral leishmaniasis (World Health Organization 2010). However, the 30-day course of pentavalent antimonials remains effective in Sub-Saharan Africa, and it is still used when paromomycin is not available. Paromomycin and miltefosine, when used as monotherapy, are less effective in Sub-Saharan Africa compared to India, and their use is likely to be confined to combined treatments. In Sub-Saharan Africa, AmBisome should be given in higher doses when used in monotherapy, for a total dose of 30 mg/kg in HIV-negative and at least 40 mg/kg in HIV-positive patients.

In Eastern Africa, PKDL heals spontaneously within a year in 85% of the cases (Musa et al. 2013). Treatment is indicated in patients with persistent lesions or in those where PKDL lesions are severe and disfiguring (grade II and III, WHO classifications), including those patients with ocular or oral mucosal involvement. The WHO recommendations for PKDL treatment in the region include the following options: (1) a 30–60-day course of SSG, (2) a combination of paromomycin (17 days, 11 mg/kg base per day) and SSG (17–60 days, 20 mg/kg per day), and (3) liposomal amphotericin B (20 days, 2.5 mg/kg per day). Anecdotal data show that miltefosine could be an option to treat PKDL in HIV-coinfected patients (Belay et al. 2006) and may also be used in extended doses (28–60 days) in HIV-negative PKDL patients (unpublished observations). These regimens are protracted and not ideal, and thus, new short-course treatment regimens are urgently needed. Patients with concomitant anterior uveitis and blepharoconjunctivitis need special precautions due to the high risk of inflammation that could lead to blindness during treatment (Khalil et al. 2011). A coadministration of steroid and atropine eye drops is recommended (El Hassan et al. 1998).

The treatment of HIV-positive visceral leishmaniasis cases is undoubtedly the most challenging as current treatments cannot clear parasites, and the immune system is compromised. Recurrences/relapses in these patients are the most difficult to treat, being vulnerable to drug toxicity and having low CD4 T-cell counts. HIV coinfected patients would require prolonged courses of anti-leishmanials, presupposing the availability of safe treatments. In HIV coinfected visceral leishmaniasis patients of northern Ethiopia, initial cure rates are generally high and probably as good as in HIV-negative patients. However, as definitive cure is rarely achieved, recurrence/relapse is a major challenge. A retrospective cohort study of HIV coinfected patients in northern Ethiopia found an initial cure rate of 83.8% (Abongomera et al. 2018). A randomized controlled trial in northern Ethiopia (Diro et al. 2019a) used a sequential adaptive trial design to evaluate the effectiveness of two treatment regimens: AmBisome monotherapy (40 mg/kg) and AmBisome (30 mg/kg) combined with miltefosine (100 mg/day for 28 days). At day 58, the adjusted efficacy was 55% (32–78%, 95% CI) in the monotherapy arm and 88% (79–98%, 95% CI). A prospective follow-up of 44 patients enrolled in this study revealed one-year relapse-free survival rates of 50% (35–63%, 95%CI) in 22 patients who did not receive secondary prophylaxis and 46% (30–71%, 95%CI) in 29 patients who received secondary prophylaxis (Diro et al. 2019b). This study showed that more than 50% of the patients had experienced a relapse of visceral leishmaniasis within one year, whether or not they received a secondary prophylactic treatment. It seems

HIV coinfected visceral leishmaniasis patients would require a treatment that confers a sterile cure, a subject that continues to be debated upon by the scientific community.

HIV–Visceral Leishmania Coinfection in Sub-Saharan Africa

In the northwestern lowlands of Ethiopia, high HIV infection among visceral leishmaniasis patients (up to 40%) has been reported (Alvar et al. 2008; Diro et al. 2014). Previously, Hailu and Berhe (2002) reported an HIV coinfection rate higher than 50% among patients admitted to Addis Ababa hospitals. These patients were among those diagnosed in the late 1990s and early 2000s, and 98% were soldiers or policemen that have been exposed to leishmania parasites while on a military mission in rural areas. Unpublished data show that coinfections rates in Kenya, South Sudan, and Sudan are not high, less than 3% up until 2013 (Diro et al. 2014).

The high rate of coinfection in northern Ethiopia is unique and is attributed to the seasonal migration of teenagers and young adults to the lowlands of northern Ethiopia seeking employment in commercial-scale farms, where most men are exposed to sand fly bites, while at the same time being predisposed to encounters of commercial sex. The increasing agricultural activity and the influx of huge populations to the lowlands have clearly led to an increasing burden of visceral leishmaniasis and HIV coinfection.

HIV coinfection of individuals exposed to leishmania has several untoward clinical and epidemiological consequences in both the HIV and leishmania infections:

1) HIV increases the risk of visceral leishmaniasis development in endemic areas due to the high risk of reactivation in immune-compromised individuals (Saporito et al. 2013; Monge-Maillo et al. 2014).
2) HIV coinfection may obscure the diagnoses of visceral leishmaniasis due to atypical clinical presentations. Splenomegaly is often less frequent among HIV-coinfected patients (Cota et al. 2014); dermal manifestations are common (Ritmeijer et al. 2001; Puig and Pradinaud 2003; Gelanew et al. 2011), and invasion of mucosal membranes takes place (Puig and Pradinaud 2003).
3) Visceral leishmaniasis patients with HIV coinfection are difficult to treat and at high risk of eventually relapsing (Alvar et al. 2008; Diro et al. 2014). These patients are likely to suffer from intercurrent infections/illnesses requiring multiple treatments alongside the treatments for HIV and leishmania, and the majority of the drugs have well-documented toxicity that needs to be monitored.
4) Immune reconstitution associated inflammatory syndrome could ensue during antiretroviral therapy and make the management ever challenging.
5) Early stages of HIV and leishmania infections may progress rapidly to AIDS and overt visceral leishmaniasis, respectively (Alvar et al. 2008).

6) Coinfected individuals, especially once visceral leishmaniasis is manifested, are highly infective to sand flies as they harbor high numbers of amastigotes in the skin and peripheral blood (Molina et al. 2003).

Sand Flies Transmitting Visceral Leishmaniasis in Sub-Saharan Africa

Sand fly (Fig. 5) species transmitting visceral leishmaniasis in eastern Africa include *P. orientalis* (Gebre-Michael et al. 2010; Elnaiem 2011) and *P. martini* (Perkins et al. 1988; Gebre-Michael and Lane 1996). These species are restricted to specific habitats and determine the distribution of the disease. The latter species is uniquely associated with eroded termite hills in the arid and semiarid lowlands of southern Ethiopia (Gebre-Michael and Lane 1996), the Kapoeta focus in South Sudan (Elnaiem 2011), many foci in Kenya (Mutinga 1991), Somalia (Elnaiem 2011), and Uganda (Kolaczinski et al. 2008; Elnaiem 2011).

The main visceral leishmaniasis endemic areas in Sudan, South Sudan, and north Ethiopia are found in the Acacia-Balanites forests, where *P. orientalis* is abundantly found during the drier seasons, usually between January and May. This species has been found naturally infected with *L. donovani* in Sudan (Schorscher and Goris 1992; Elnaiem et al. 1998; Hassan et al. 2008) and southwest Ethiopia (Hailu et al. 1995a). Another species involved in transmission is *P. celiae* (Gebre-Michael and

Fig. 5 Blood-fed female sand fly

Lane 1996). Suspected but unproven sand fly vector species are *P. vansomerenae* (Perkins et al. 1988; Gebre-Michael et al. 2013; Marlet et al. 2003), *P. alexandri* (Balkew et al. 1999), and *P. rodhaini* (Elnaiem et al. 2011). The latter was implied to transmit *L. donovani* between wild animal reservoir hosts in eastern Sudan.

Animal Reservoirs of Visceral Leishmaniasis in Sub-Saharan Africa

Efforts to incriminate the potential animal reservoir(s) of visceral leishmaniasis in Sub-Saharan Africa started in the 1950s and 1960s, especially in Sudan and Kenya. The work of Hoogstraal et al. (1963), and Hoogstraal and Dietlein (1964) in South Sudan demonstrated a few natural infections of rodents and small carnivores. Among rodents, natural infections were found in *Rattus rattus*, the spiny mouse (*Acomys* sp.), and *Arvicanthis niloticus* (Hoogstraal et al. 1963). Other species found with natural infections of leishmania were *Felis serval* and *Genetta* sp. (Hoogstraal and Dietlein 1964), jackals (Sixl et al. 1987), and the Egyptian mongoose (Elnaiem et al. 2001). In Kenya, among 1128 wild rodents examined, none were found infected with *L. donovani* (Githure et al. 1996). However, Kassahun et al. (2015a) found some species of rodents (*Mastomys* sp., *Arvicanthis* sp., *Gerbilliscus* sp.) that were PCR positive for L. donovani in specimens of spleen.

A high rate of natural infection was found among 69 dogs examined in Barbar El Fugara, Gedaref State in Sudan (Dereure et al. 2003). However, in 2002, low rates of serological and PCR positivity were found among 87 dogs examined near Dinder National Park in Eastern Sudan (Hassan et al. 2009). In an outbreak investigation of visceral leishmaniasis in Libo Kemkem (north Ethiopia), a few PCR-confirmed infections of domestic dogs were also found (Bashaye et al. 2009). Contrastingly, Kalayou et al. (2011) reported a high seroprevalence rate among dogs in surveys carried out in north Ethiopia. In Kenya, there was also a controversial report of natural infection of dogs (Mutinga et al. 1980). Regarding domestic animals other than dogs, serological surveys have indicated a high rate of seropositivity in donkeys, cows, sheep, and goats (Mukhtar et al. 2000).

Overall, the animal reservoir studies carried out so far remain inconclusive. Based on the field studies to date, no animal species could be singled out as a culprit. As a consequence, most authors and experts in the field of leishmaniasis prefer to consider visceral leishmaniasis in Sub-Saharan Africa as predominantly anthroponotic. Without precluding the existence of a natural source of infection in wild animals, many experts would also agree on the fact that the major epidemics of the disease in Sub-Saharan Africa are outcomes of man-to-man transmission, as has occurred in the war-ravaged South Sudan and in the intensified agricultural activities of north Ethiopia. However, the stable visceral leishmaniasis endemic regions of south Ethiopia, Kenya, and Uganda do implicate the existence of a zoonotic cycle.

Future Perspectives of Visceral Leishmaniasis Control and Research in Sub-Saharan Africa

Leishmaniasis in the Sub-Saharan Region remains a largely neglected and poorly described disease. The control of leishmaniases in Sub-Saharan Africa requires an in-depth understanding of the epidemiology to unravel factors that drive and sustain transmission. The basic biology, pathogenesis, and ecology of the diseases have many unknowns that need to be addressed. Knowledge and tools that could be used to predict outbreaks and to map the distribution of the diseases are still scanty.

Tremendous efforts have been made to eliminate visceral leishmaniasis from the Indian subcontinent, and remarkable results have been recorded. The lessons learned from this region will be vital in planning a similar elimination program in Sub-Saharan Africa. However, it is clear that the epidemiology of visceral leishmaniasis in Sub-Saharan Africa is much more complex, requiring in-depth understanding of transmission, sand fly biology, and ecology. The role of so-called asymptomatic human infection in transmission is obscure. Xenodiagnostic studies have not revealed any evidence of human reservoirs. Man-to-man transmission of visceral leishmaniasis in Sub-Saharan Africa is not an exclusive phenomenon, especially in stable endemic areas. If humans transmit visceral leishmaniasis, it is likely that superspreaders are important, and there is a need to identify them by sensitive diagnostic tool. Availability of such a tool will be vital for planning future interventions and putting in place feasible control/elimination strategies. The need for new surveillance tools transcends the agenda of diagnostic tests and will include tools to map distribution, predict outbreaks, assess levels of herd immunity, and to identify superspreaders.

Efforts to simplify diagnostic and treatment approaches should be intensified, as these are pillars of the control programs. The development of an rK39-based antibody detection test has been a major breakthrough for visceral leishmaniasis control programs, but it is certainly not a panacea. The use of these tests is restricted to patients who have a clear clinical syndrome of febrile splenomegaly and who had no visceral leishmaniasis ever before due to the well-known limitations of antibody detection. Many people who live in endemic villages develop a sub-clinical infection and clear it without ever needing treatment, but they will show a positive rK39-RDT. After treatment, visceral leishmaniasis patients will keep antibodies for a median of 9 months. Therefore, a positive test in a patient relapsing with fever is by no means diagnostic. The current antibody detection technology used in the RDTs neither allows the monitoring of treatment outcomes nor confirmation of cure or diagnosis of relapse. Moreover, the rK39-based antibody detection tests are not optimally sensitive in East Africa, and a negative test does not rule out disease. Therefore, improved rapid tests are needed, preferably based on antigen detection systems.

Availability of a diagnostic test is a crucial factor for improving access to treatments and to effectively carry out surveillance programs. Early diagnosis has a positive impact on treatment outcomes and needs to be promoted. Individuals not diagnosed at an early stage of the disease will be in the pool of superspreaders for a considerable number of days/weeks and increases costs of healthcare due to the fact that patients with advanced visceral leishmaniasis will require admission for many days or weeks.

Investment in drug development is crucially needed to fill the pipeline with novel compounds, as current treatments are toxic, difficult to administer, require many days of administration (currently 10–17 days in first-line treatment regimens), and are not readily available. The development of safe oral treatments will be game changing.

The management of HIV/leishmanial coinfection remains a challenge, the main problem being recurrences, which become refractory to treatment. This seems to be associated with the failure to reconstitute CD4 T-cell counts even when patients have achieved good viral clearance. The increasing incidence of relapses could be associated with the emergence of drug resistance, which calls for simple tools to monitor emerging drug resistance.

An effective vaccine would significantly improve visceral leishmaniasis control, as would new methods of vector control to prevent human infection. Insecticide-treated nets (ITNs) have been evaluated for the control of visceral and cutaneous leishmaniases; however, their effectiveness was shown to be variable depending on the context. There are currently no arguments for the wide-scale use of ITNs for interrupting the transmission of endemic visceral leishmaniasis in sub-Saharan Africa where transmission is largely in extradomestic environments. However, targeted deployment of ITNs in specific populations, e.g., non-immune migrants, can prevent exposure to sand fly bites. Households located near termite hills are at high risk of exposure, and the residents will benefit from sleeping under ITNs. Persons who are affected by visceral leishmaniasis but are not receiving treatment (and those at risk of relapse) should be advised to sleep under ITNs.

Part 2: Cutaneous Leishmaniasis in Sub-Saharan Africa

Species and Geographical Distribution

The paucity of epidemiological data about the exact burden and geographic distribution of the disease in most of the Sub-Saharan Africa countries (Alvar et al. 2012) is an obstacle in the global initiative to control the disease. The main challenge in reporting is the lack of diagnostic capability in the health systems of African countries as the routine diagnostic procedures for cutaneous leishmaniasis are invasive, complicated, and not sensitive enough. Notwithstanding, a recent systematic review has shown cutaneous leishmaniasis is prevalent in Niger, Nigeria, Senegal, Cameroon, Mali, Mauritania, Burkina Faso, Guinea, Sudan, Ethiopia, and Kenya (Sunyoto et al. 2018). Noting that reporting is incomplete (less than 10% in some countries), the true burden of cutaneous leishmaniasis remains unknown. Figure 6 shows the global distribution of cutaneous leishmaniasis, while Fig. 7 summarizes the 2005–2019 Global Health Observatory (GHO) data (accessed 29 May 2022) and highlights the eight high burden African countries, including Burkina Faso, Ethiopia, and Sudan among the Sub-Saharan countries.

Cutaneous leishmaniasis in Sub-Saharan Africa is caused mainly by *L. major*, *L. aethiopica*, and *L. tropica*. However, cases of cutaneous leishmaniasis associated with *L. infantum* and *L. donovani* are also reported especially from the western sub-Region of Sub-Saharan Africa. In Sub-Saharan Africa, cases of mucocutaneous and mucosal leishmaniasis have been reported from Chad, Senegal, and Sudan, while in Ethiopia numbers are considerably high.

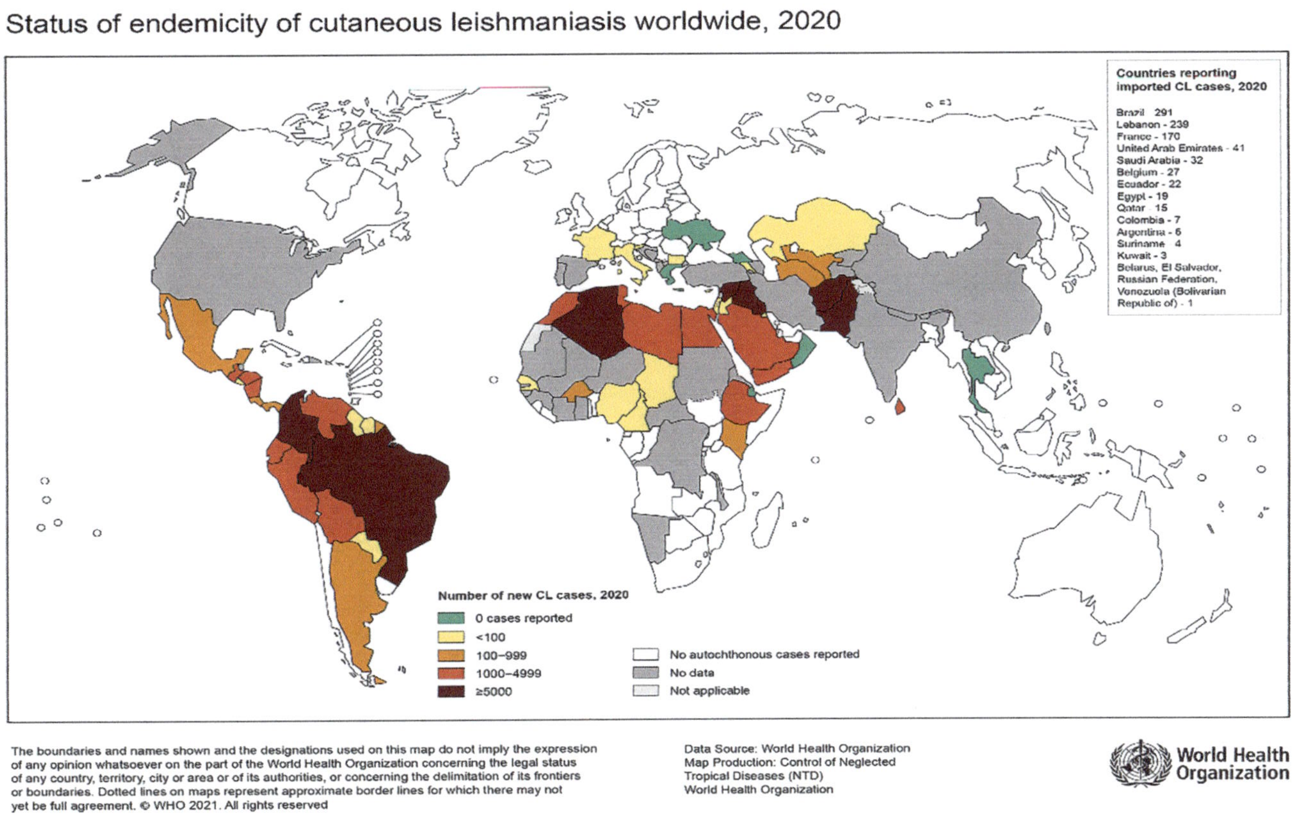

Fig. 6 Worldwide distribution and endemicity of cutaneous leishmaniasis in 2020

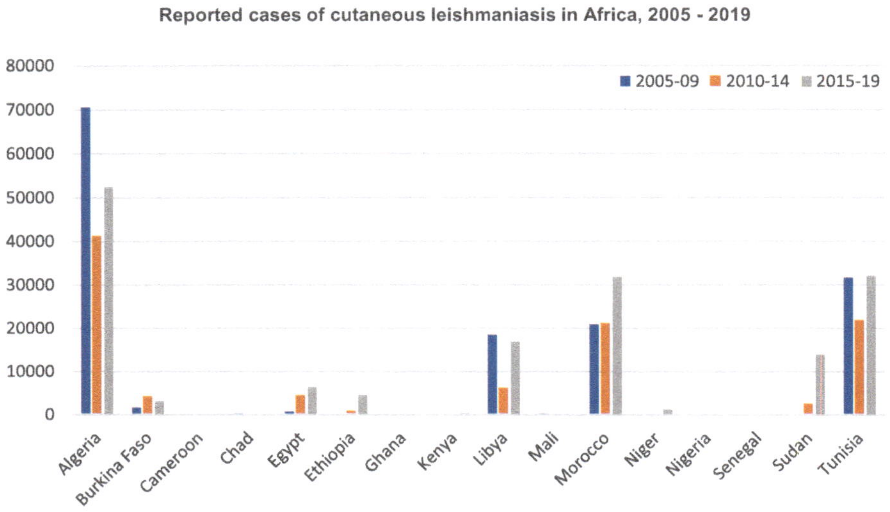

Fig. 7 High-burden countries in Africa

Cutaneous leishmaniasis caused by *L. major* (also known as zoonotic or rural cutaneous leishmaniasis) is the most widely distributed in Sub-Saharan Africa. It has been reported from Burkina Faso, Cameroon, Central African Republic, Chad, Ethiopia, Ghana, Guinea, Guinea-Bissau, Kenya, Mali, Mauritania, Niger, Nigeria, Senegal, and Sudan.

Cutaneous leishmaniasis caused by *L. aethiopica* is prevalent in Ethiopia and has also been reported in Kenya (Ngere et al. 2020). *L. aethiopica* infections are acquired from zoonotic cycles involving rock hyraxes. *L. aethiopica* causes a variety of lesion types, including localized cutaneous leishmaniasis (LCL), muco-cutaneous (MCL), and diffuse cutaneous leishmaniasis (DCL), which are prevalent in the Ethiopian highlands. Data on MCL/DCL from Kenya are limited.

Cutaneous leishmaniasis caused by *L. tropica* occurs in rural areas of Sub-Saharan Africa and is transmitted in zoonotic cycles. Urban anthroponotic cutaneous leishmaniasis, which is characteristic of *L. tropica* infections in the Middle East/Central Asia, is non-existent. Sporadic cases caused by *L. tropica* have been reported from Ethiopia and Kenya.

Transmission Pattern, Incidence, and Burden of Cutaneous Leishmaniasis in Sub-Saharan Africa

In Sub-Saharan Africa, *L. major*, *L. tropica*, and *L. aethiopica* cause zoonotic cutaneous leishmaniasis. The risk of leishmaniasis in humans is increased with the expansion of agricultural projects and irrigation systems. These development activities and ecological changes are accompanied by the intrusion of large numbers of nonimmune immigrants into an existing sylvatic cycle of leishmaniasis. Transmission to humans is favored by the practice of sleeping outdoors in areas of transmission. The risk for infection is also increased by such activities like tourism and pilgrimages. In foci of cutaneous leishmaniasis caused by *L. aethiopica* in the highlands of Ethiopia and Kenya, increased human–fly contact occurs in villages built on or near rock hills or riverbanks, which are the natural habitat of hyraxes (reservoir hosts) and the sand fly vectors (*P. longipes* and *P. pedifer)*. Spiritual establishments and monasteries are common places where rock hyraxes prevail and transmission of *L. aethiopica* takes place.

Outbreaks of cutaneous leishmaniasis due to *L. aethiopica* have been described in the northwest, central, and southern parts of Ethiopia (Seid et al. 2014). A recent outbreak has also been described in Gilgil—a southeastern region of the Kenya rift valley—where transmission was taking place in a peri-urban setting (Ngere et al. 2020). Even though the leishmania species associated with this outbreak has not been described, based on epidemiological and environmental data, *L. aethiopica* was implicated.

In Sudan, outbreaks of cutaneous leishmaniasis due to *L. major* were observed in 1976 around Shendi-Atbara area (Abdalla and Sherif 1978) and later on in Khartoum in 1986 (El-Safi and Peters 1991). In Sudan, at least five states (Northern Darfur, Southern Darfur, Northern Kordofan, El Gezira and Khartoum) are endemic to cutaneous leishmaniasis caused by *L. major* (Collis et al. 2019).

Recently, increasing numbers of cases of cutaneous leishmaniasis have been reported from Ouagadougou in Burkina Faso, where a major outbreak occurred in the period 1999–2001 that continued with high incidence until 2005 (Bamba et al. 2013). In the Volta region of Ghana, in the period 2002–2003, around 8876 cases of possible cutaneous leishmaniasis have been reported (Kweku et al. 2011).

Clinical Features, Diagnosis, and Treatment of Cutaneous Leishmaniasis in Sub-Saharan Africa

Localized Cutaneous Leishmaniasis

The clinical manifestations of cutaneous leishmaniasis vary according to species. Usually, the lesion starts as a papule or nodule (Fig. 8) at the site of inoculation and grows slowly, taking days or weeks before it reaches its final size. A crust develops centrally (Fig. 9), which may fall away, exposing an ulcer up to 5 cm in diameter with a raised edge and variable surrounding induration, which heals gradually over months or years. This often leaves a depressed scar with altered pigmentation. Even though the majority of localized lesions self-heal over a variable period of time, some localized lesions tend to be persistent. A variety of anomalous skin lesions (Figs. 10, 11 and 12), mostly non-self-healing, are occasionally found, e.g., the sporotrichoid type involving the skin lymphatics (Fig. 13). Other localized lesion types include the recidiva type, commonly associated with *L. tropica* infections, but also found in *L. aethiopica* (Fig. 14).

Lesions from *L. major* infections are painless unless complications occur and are usually severely inflamed and ulcerated, healing only after 2–8 months. In

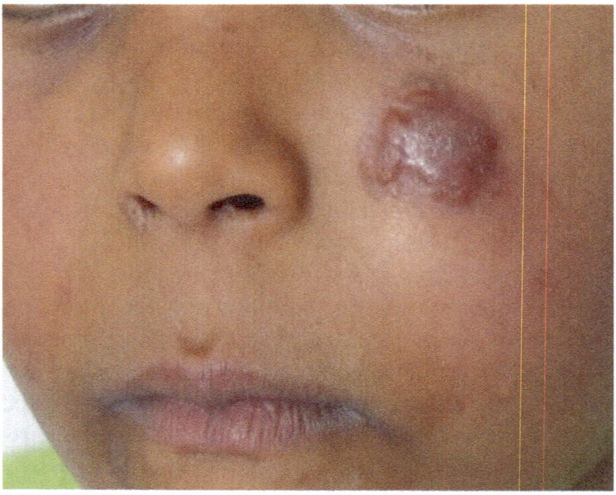

Fig. 8 Nodular lesion on the cheek

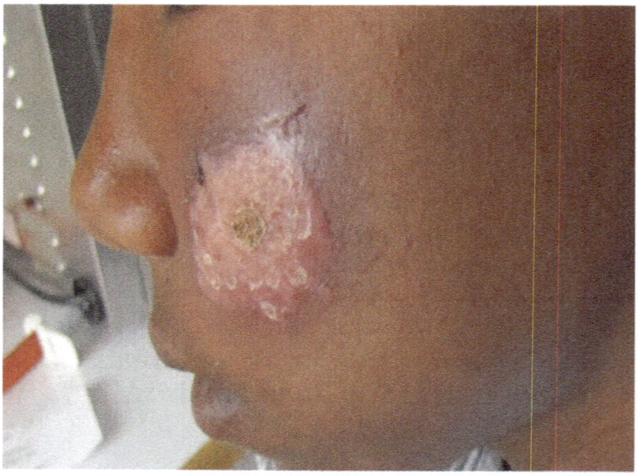

Fig. 9 Crust formation in an active LCL lesion accompanying ulceration

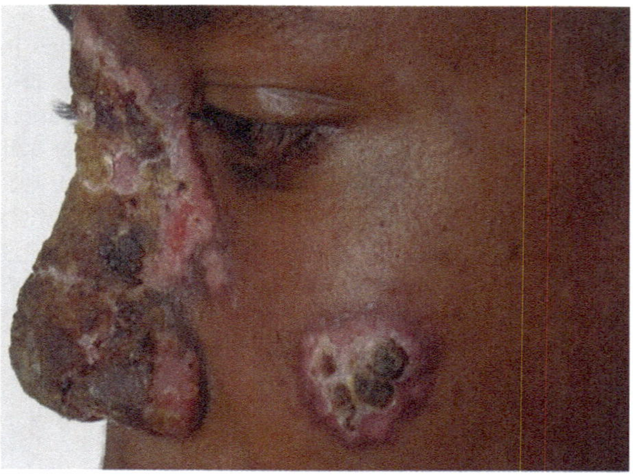

Fig. 10 Persistent and erosive lesion in the nose bridge, and an isolated ulcerated lesion with central crusts

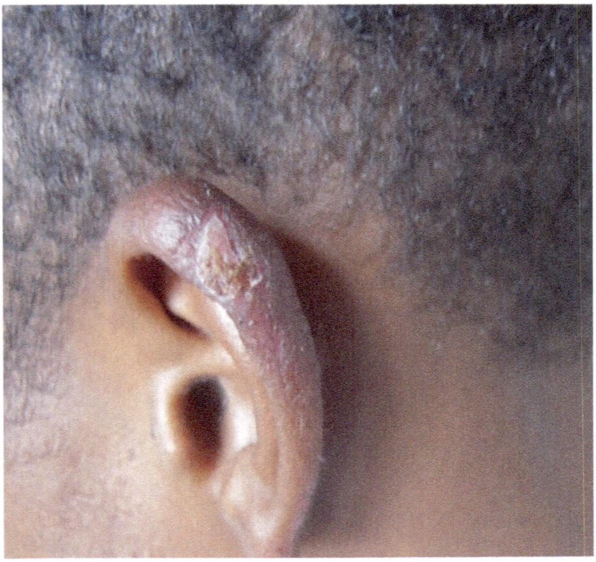

Fig. 11 Persistent erosive lesion on the ear lobe

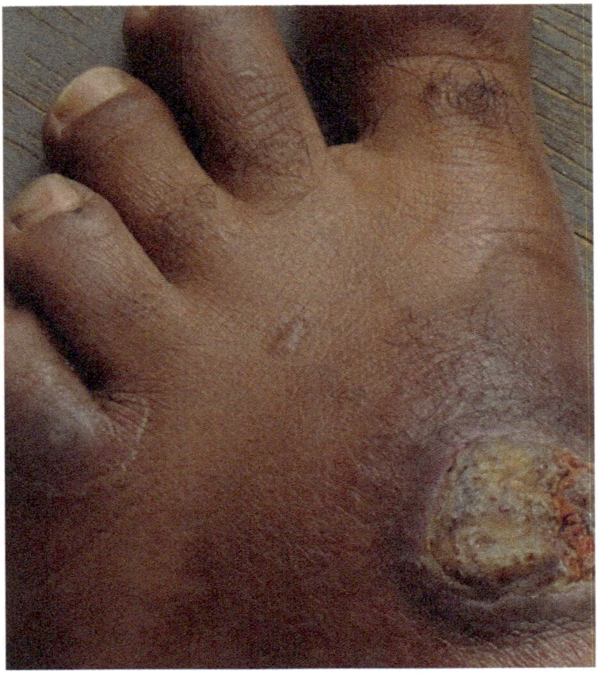

Fig. 12 Ulcerated lesion on the foot with deep wounds and crusting

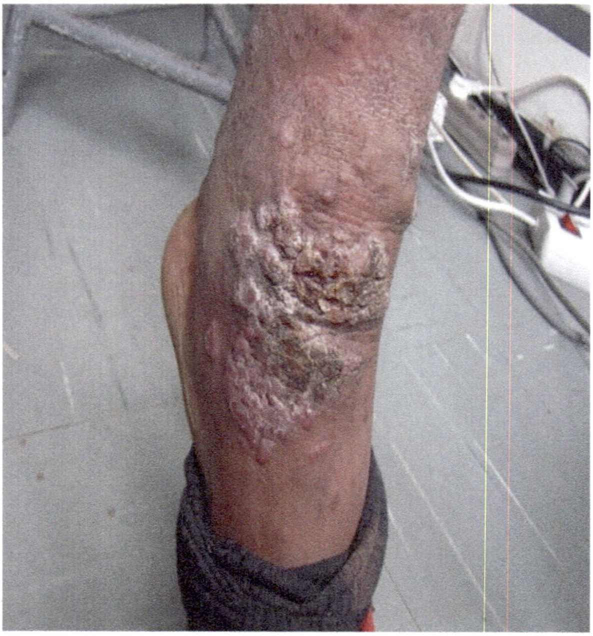

Fig. 13 Sporotrichoid type lesion

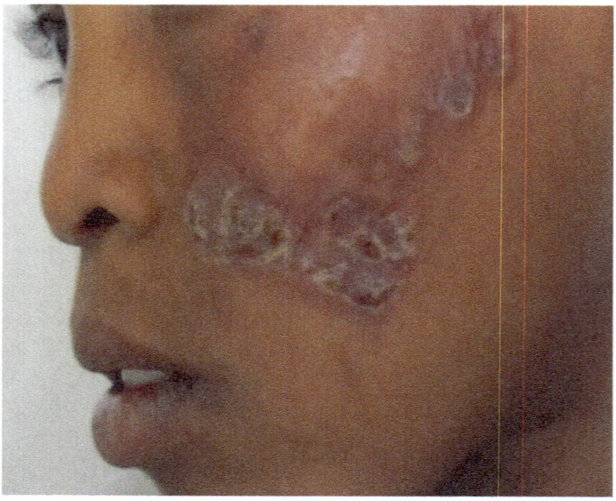

Fig. 14 Recidiva-type lesions on the cheek

non-immune individuals, multiple confluent lesions are observed. Such lesions are prone to secondary infection and are often slow to heal, leaving large, disfiguring, or disabling scars.

L. tropica causes painless, dry ulcers of the skin, which usually heal spontaneously within about 1 year, or sometimes longer, often leading to disfiguring scars. Recidiva lesions are common findings.

L. aethiopica infections tend to be infiltrative at the site of lesions, especially when lesions occur in the lip, nose, and ear lobes and produce starkly edematous

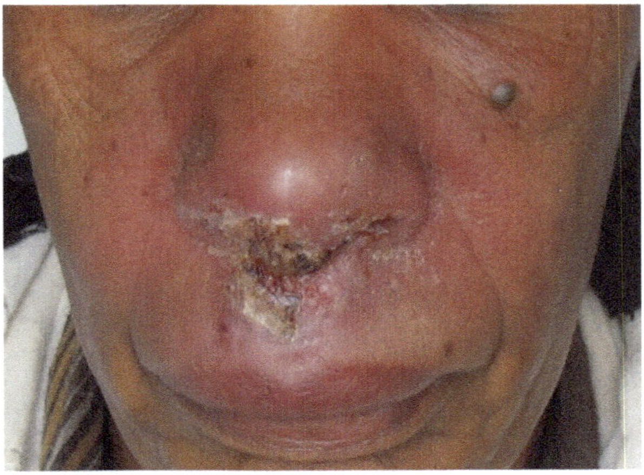

Fig. 15 Infiltrative lesions on the nose and lips

chronic inflammation of the face, often involving the lips and nose (Fig. 15). Lesions are usually localized occurring in various body sites commonly affecting the face and limbs. Ulceration is late or sometimes absent. Spontaneous healing typically takes place within 2–5 years. Some lesions are lupoid, resembling the recidiva type.

Mucocutaneous and Mucosal Leishmaniasis

Some leishmanial lesions evolve slowly and spread locally, especially when the nostrils, ear lobes, and lips are the sites of infection, ultimately causing muco-cutaneous and mucosal (oro-nasal) leishmaniasis. Mucosal infections are often accompanied by persistent inflammation of mucosal membranes and surrounding tissues.

Mucocutaneous and mucosal leishmaniasis (Fig. 16) are typical features of *L. aethiopica* infections and can be found in up to 20% of patients seeking healthcare in Ethiopia. Mucosal lesions of the buccal mucosa or larynx caused by *L. infantum*, *L. major*, and *L. tropica* may present in elderly people or people with minor forms of immunosuppression.

Diffuse Cutaneous Leishmaniasis

Diffuse cutaneous leishmaniasis in Sub-Saharan Africa is caused by *L. aethiopica* and is characterized by widely disseminated cutaneous macules, papules, nodules, or plaques, and by diffuse infiltration of the skin, especially on extensor surfaces of the limbs and on the face, where thickening of the eyebrows and ear lobes may resemble lepromatous leprosy (Fig. 17). Lesions are typically non-ulcerative and do not heal spontaneously. The lesions are refractory to treatment, requiring repeated courses of anti-leishmanials. Definitive cure is not achieved by any of the currently available treatments; thus, relapse is inevitable.

Diagnosis of Cutaneous Leishmaniasis

The differential diagnosis of cutaneous leishmaniasis is very broad as cutaneous leishmaniasis lesions resemble a number of skin conditions, including staphylococcal or streptococcal infection, mycobacterial ulcer, leprosy, fungal infection, cancer,

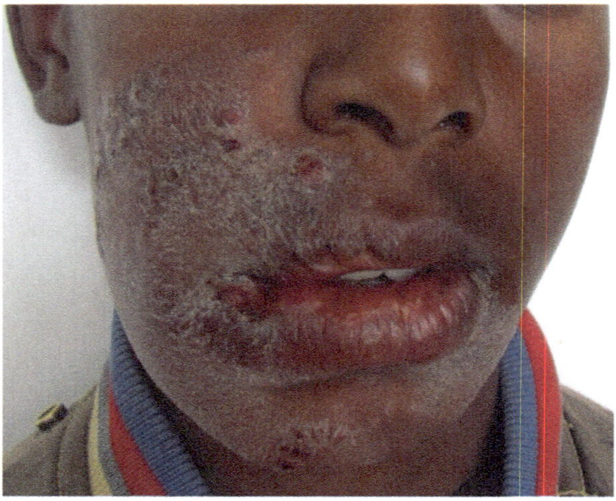

Fig. 16 Muco-cutaneous leishmaniasis in an Ethiopian patient

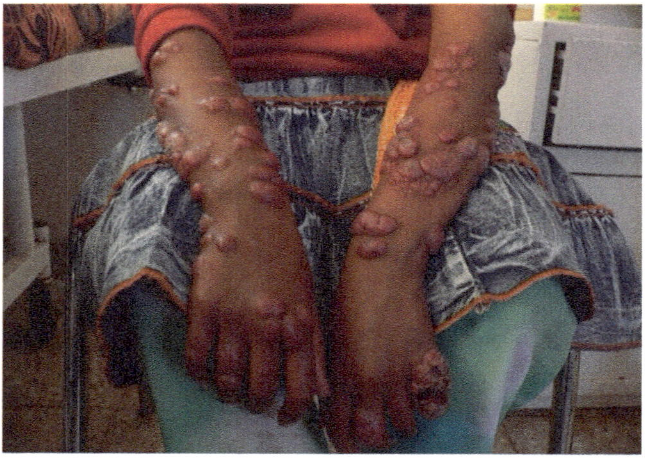

Fig. 17 Diffuse cutaneous leishmaniasis lesions spread in the hands of a young girl

sarcoidosis, and tropical ulcer. As the clinical presentation of cutaneous leishmaniasis lacks specificity, and treatment is costly and toxic, confirmatory diagnosis is necessary.

Parasitological diagnosis remains the reference standard in the diagnosis of cutaneous leishmaniasis because of its high specificity. The sensitivity varies over a wide range depending on the geographical location, species involved, and stage and type of the lesions. Multiple parasitological diagnostic tests should therefore be performed on each patient. Parasitological diagnosis can be performed by skin slit scraping of ulcerative and nodular/macular lesions, fine-needle aspiration of nodular lesions, or biopsy in all lesion types. The material obtained by any of these methods can be used for microscopic examination, culture, and molecular diagnostic techniques. Smears from MCL lesions are the least sensitive, and most sensitive in DCL lesions.

Microscopic examination of Giemsa-stained material is often the only available method at all levels of healthcare in endemic areas. Culture of the parasite in blood agar medium, e.g., NNN medium (Novy–MacNeal–Nicolle medium), gives extra benefits of diagnosis. Detection of parasite nucleic acids by molecular techniques, essentially by PCR-based methods, improves the diagnostic sensitivity and allows identification and characterization of the leishmania species. Both culture and molecular-based diagnosis require substantial laboratory infrastructure and technical expertise, limiting their use to reference laboratories.

Treatment of Cutaneous Leishmaniasis

Many different therapeutic interventions, including topical, systemic, and non-pharmacologic treatments, have been proposed for cutaneous leishmaniasis. The disease is not a life-threatening condition, and severe complications are infrequent. Cutaneous leishmaniasis due to *L. major* is usually associated with a self-cure rate of >50% at 6 months. Normally, the recommended drug or treatment approach should not induce life-threatening complications; however, in severe cases, the risk–benefit ratio is different. The treatment decision is based firstly on the risk–benefit ratio of the intervention for each patient. The use of a treatment option with a risk for severe adverse events is acceptable if the patient suffers from numerous (typically, more than four) disfiguring and complicated lesions, the size or location of which makes local therapy impossible, or if local therapy has been tried and failed. Appropriate attention to contraindications and follow-up is important. In patients with mild disease or with comorbid conditions, a safer treatment should be preferred, even if the level of evidence for efficacy is weak. It is important to clean lesions during treatment with specific anti-leishmanials and even with topical treatments as superficial secondary infections may complicate ulcerated lesions. According to the recommendations of the WHO Technical Expert Group (World Health Organization 2010), local wound care with careful follow-up is indicated for patients fulfilling the following criteria:

- Confirmed or strongly suspected infection with *L. major*
- Fewer than four lesions requiring immediate treatment
- Lesions <5 cm in diameter
- No potentially disfiguring or disabling lesion (face, joints, toes, fingers)
- No immunosuppression
- Possibility for follow-up

The patient's agreement to these options must be obtained after a full explanation of the clinical risks and the inconveniences of each of the options. If at least one criterion is absent, local therapy should be proposed. Local therapy is an attractive option, but intra-lesional treatment and, to a much lesser degree, thermotherapy cause significant discomfort.

The options of local therapy for L. major are:

- Fifteen percent paromomycin/12% methylbenzethonium chloride ointment twice daily for 20 days

- Intralesional antimonials, 1–5 ml per session plus cryotherapy (liquid nitrogen: -196 °C), both every 3–7 days (one to five sessions)
- Thermotherapy: one to two sessions with localized heat (50 °C for 30 s)
- Intralesional antimonials or cryotherapy independently, as above

*Local therapy for L. tropica, L. aethiopica, * and L. infantum**

- Fifteen percent paromomycin/12% methylbenzethonium chloride ointment, twice daily for 20 days
- Intralesional antimonials plus cryotherapy, as above
- Thermotherapy, one to two sessions with localized heat (50 °C for 30 s)
- Intralesional antimonials, alone, as above
- Cryotherapy alone, as described above

Systemic therapy in L. major

- Fluconazole, 200 mg oral daily for 6 weeks
- Pentavalent antimonials, 20 mg Sb5/kg per day intramuscularly or intravenously for 10–20 days
- Pentavalent antimonials, 20 mg Sb5/kg per day intramuscularly or intravenously plus pentoxifylline, 400 mg three times a day for 10–20 days

*Systemic therapy in L. tropica and L. infantum**

- Pentavalent antimonials, 20 mg Sb5/kg per day intramuscularly or intravenously for 10–20 days
- Pentavalent antimonials, 15–20 mg Sb5/kg per day intramuscularly or intravenously for 15 days plus oral allopurinol 20 mg/kg for 30 days (to treat leishmaniasis recidivans caused by *L. tropica*)

Systemic therapy in L. aethiopica

- Pentavalent antimonials 20 mg Sb5/kg per day intramuscularly or intravenously plus paromomycin 15 mg (11 mg base)/kg per day intramuscularly for 60 days or longer to treat mucosal and diffuse cutaneous leishmaniasis.
- Anecdotal observations indicate miltefosine could be a treatment option for complicated *L. aethiopica* infections, especially for muco-cutaneous and diffuse cutaneous leishmaniasis. This needs to be verified by well-documented studies. An in vitro study using the amastigote–macrophage model has also confirmed the high susceptibility of *L. aethiopica* to miltefosine (Utaile et al. 2013).

HIV–Cutaneous Leishmania Coinfection in Sub-Saharan Africa

Cutaneous leishmaniasis and HIV coinfection have been reported in Ethiopia (Berhe et al. 1995; Padovese et al. 2009), Burkina Faso (Guiguemdé et al. 2003), Ghana

(Lartey et al. 2006), Cameroon (Tangie et al. 2017), and Sudan (Mukhtar et al. 2010). The coinfection nearly always translates into atypical presentations, including extensive ulceration or dissemination, increased mucosal involvement, erysipeloid, sporotrichoid forms, and multiple locations. Lymphoedema of the limbs has been observed in a few patients infected with *L. aethiopica*.

Sand Flies Transmitting Cutaneous Leishmaniasis in Sub-Saharan Africa

In most of the 26 endemic countries in Sub-Saharan Africa, the vectors are unknown. However, the following sand fly species were identified as proven and suspected vectors so far: *P. duboscqi*, *P. bergeroti*, *P. longipes*, *P. pedifer*, *P. sergenti*, *P. saevus*, *P. papatasi*, *P. guggisbergi*, and *P. rossi* (Gebre-Michael et al. 2004; Hailu et al. 2006; Lawyer et al. 1991; Maroli et al. 2013). Species proven/implicated in *L. aethiopica* transmission are *P. longipes* (in Ethiopia), *P. pedifer* (in Ethiopia and Kenya), *P. saevus*, and *P. sergenti* (both species suspected to play a secondary role in isolated foci in Ethiopia). Of these, *P. sergenti* and *P. saevus* are also implicated in the transmission of *L. tropica* in Ethiopia. *P. guggisbergi* and *P. saevus* are the species associated with *L. tropica* transmission in Kenya (Lawyer et al. 1991; Owino et al. 2019). *L. major* is transmitted by *P. duboscqi* (in Kenya and southern Ethiopia) or by *P. papatasi* (in Sudan). *P. bergeroti* and *P. duboscqi* are found in sympatric association with *P. papatasi* in Sudan, and their role in *L. major* transmission is still unknown.

Animal Reservoirs of Cutaneous Leishmaniasis in Sub-Saharan Africa

Hyraxes (*Procavia capensis*, *Procavia johnstoni*, *Heterohyrax brucei*, and *Dendrohyrax arboreus*) and several rodent species (e.g., *Arvicanthis* spp., *Tatera* spp., *Xerus* spp., etc.) are known or suspected animal reservoir hosts of cutaneous leishmaniasis in Sub-Saharan African countries (Mutinga 1975; Gramiccia and Gradoni 2005; Hailu et al. 2006). In Ethiopia, some species of rodents (*Acomys* sp., *Arvicanthis* sp., and *Gerbillus* sp.) and bats (*Cardioderma cor* and *Nycteris hispida*) were found PCR-positive for *L. tropica or L. major* (Kassahun et al. 2015a, b). The African green monkey is susceptible to experimental infections with *L. major* (Lawyer et al. 1990) and *L. aethiopica* (Hailu et al. 1995b), but its role in natural transmission is not known.

Future Perspectives and Priorities of Cutaneous Leishmaniasis Control and Research in Sub-Saharan Africa

The incidence of cutaneous leishmaniasis is increasing unabatedly as no interventions are in place to prevent transmission. To date, no simplified diagnostic tests have replaced conventional procedures; and treatments are still complicated, toxic, and ineffective. With no proven preventions or control tools in place, the incidence of cutaneous leishmaniasis will keep increasing and further be aggravated by the increasing interaction of man with the environment and the zoonotic cycles of transmission.

A diagnostic test development strategy targeting antibody or antigen detection in blood specimens is not promising. Ultrasensitive test platforms to detect leishmanial antigens or DNA in skin specimens (e.g., in skin slit smears or swabs) are urgently called for.

On the treatment side, modalities that aim to deliver anti-leishmanials directly onto lesions will be novel and game changing.

L. major and *L. aethiopica* outbreaks can affect many thousands of people. Even though cutaneous leishmaniasis in Sub-Saharan Africa is transmitted in zoonotic cycles, outbreaks are often associated with man-to-man transmission in domestic settings. In high transmission settings, people can be bitten repeatedly and get re-infected. In such settings, the number of patients with multiple lesions is high and they are sources of parasites for biting sand flies. Much is desired to control cutaneous leishmaniasis during outbreaks. Surveillance and reporting systems should be in place in endemic settings to facilitate early detection of outbreaks and to investigate and untangle the factors that have triggered them. Sand fly control tools will be important to interrupt transmission, particularly during outbreaks. The sand fly control tools should aim to prevent sand flies from biting humans and from resting and feeding directly on lesions.

The transmission dynamics of cutaneous leishmaniasis is the least studied. Field data on infection intensity and incidence of the disease are not readily available. Data on sand fly biting behavior and infection rates are also lacking. Such data are crucial for modeling transmission and should be sought through research.

Conclusion

Even though leishmaniases are generally the most neglected among NTDs, cutaneous leishmaniasis is catastrophically neglected partly due to the fact that the disease is not life-threatening. The persistent localized, the mutilating mucosal/mucocutaneous, and the incurable diffuse cutaneous lesions predispose affected individuals to high risk of mental health, psychosocial morbidity, and reduced quality of life. These ill health effects are aggravated by the lack of hope to be treated.

Researchers are confronted by the many species of leishmania that cause cutaneous disease, the diverse zoonotic and anthroponotic cycles, and the many sand fly species involved in transmission. Research funds are scarce, while the needs are immense, frustrating control efforts.

The treatment of visceral leishmaniasis and PKDL cases not only reduces morbidity and mortality but also curtails transmission. Given the absence of a vaccine and effective vector control tools, it is likely that early diagnosis by passive and active surveillance using a combination of screening and confirmatory tests coupled with prompt treatment will be the cornerstone of visceral leishmaniasis control strategy in Sub-Saharan Africa for the years to come. While control efforts need to be enhanced, research needs to be carried out in deciphering transmission dynamics. It is important to underscore that the stepping point for an envisaged elimination of visceral leishmaniasis in Sub-Saharan Africa will depend on capacity building and the level of investment in research. The ministries of health/disease control programs and other stakeholders, together with academic institutions, should draw a common agenda to promote research directed toward the control of leishmaniasis. Access to tested and proven control tools, be it novel diagnostics and treatments or others, should be guaranteed by way of registration and quality-assured procurement and supply.

Diagnostics, treatments, and vaccines are undoubtedly the cornerstones of the future elimination program of leishmaniasis in Sub-Saharan Africa. It also seems that the ultimate control/elimination of leishmaniasis in the region will likely depend on the availability of vaccines and new diagnostic tools. Tools to detect superspreaders in anthroponotic cycles of visceral leishmaniasis will boost the control and elimination efforts. However, there is a lack of investments in R&D and translational research, particularly concerning vector control tools, diagnostics, and vaccines. Efforts to discover new treatments are underway, but progress is slow.

References

Abdalla RE, Sherif H (1978) Epidemic of cutaneous leishmaniasis in Northern Sudan. Ann Trop Med Parasitol 72(4):349–352. https://doi.org/10.1080/00034983.1978.11719327

Abongomera C, Diro E, Pereira AL, Buyze J, Stille K, Ahmed F, Griensven J, Ritmeijer K (2018) The initial effectiveness of liposomal amphotericin B (AmBisome) and miltefosine combination for treatment of visceral leishmaniasis in HIV co-infected patients in Ethiopia: a retrospective cohort study. PLoS Negl Trop Dis 12(5):e0006527.26. https://doi.org/10.1371/journal.pntd.0006527

Abubakar A, Ruiz-Postigo JA, Pita J, Lado M, Ben-Ismael R et al (2014) Visceral leishmaniasis outbreak in south Sudan 2009–2012: epidemiological assessment and impact of a multisectoral response. PLoS Negl Trop Dis 8(3):e2720. https://doi.org/10.1371/journal.ontd.0002720

Ali A, Ashford RW (1994) Visceral leishmaniasis in Ethiopia. IV. Prevalence, incidence and relation of infection to disease in an endemic area. Ann Trop Med Parasitol *88*(3):289–293

Alvar J, Bashaye S, Argaw D, Cruz I, Aparicio P, Kassa A et al (2007) Kala-azar outbreak in Libo Kemkem, Ethiopia: epidemiologic and parasitologic assessment. Am J Trop Med Hyg 77(2):275–282

Alvar J, Aparicio P, Aseffa A, Den Boer M, Canavate C, Dedet JP et al (2008) The relationship between leishmaniasis and AIDS: the second 10 years. Clin Microbiol Rev 21(2):334–359

Alvar J, Velez ID, Bern C, Herrero M, Desjeux P, Cano J, Jannin J et al (2012) Leishmaniasis worldwide and global estimates of its incidence. PLoS One 7(5):e35671. https://doi.org/10.1371/journal.pone.0035671

Antinori S, Longhi E, Bestetti G, Piolini R, Acquaviva V, Foschi A et al (2007) Post-kala-azar dermal leishmaniasis as an immune reconstitution inflammatory syndrome in a patient with acquired immune deficiency syndrome. Br J Dermatol 157(5):1032–1036

Ayele T, Ali A (1984) The distribution of visceral leishmaniasis in Ethiopia. Am J Trop Med Hyg 33(4):548–552

Babiker ZO, Davidson R, Mazinda C, Kipngetich S, Ritmeijer K (2007) Utility of lymph node aspiration in the diagnosis of visceral leishmaniasis in Sudan. Am J Trop Med Hyg 76(4):689–693

Balkew M, Hailu A, Berhe N, Gemetchu T (1999) Leishmaniasis in the lower Omo plains, south-west Ethiopia: the sand fly fauna. Ethiop Med J 37:31–40

Bamba S, Gouba A, Drabo MK, Nezien D, Bougoum M, Guiguemdé TR (2013) Epidemiological profile of cutaneous leishmaniasis: retrospective analysis of 7444 cases reported from 1999 to 2005 at Ouagadougou, Burkina Faso. Pan Afr Med J 14:108. https://doi.org/10.11604/pamj.2013.14.108.1140

Baruffa G (1965) Kala-azar among the nomads of the middle Webi shebeli region; first report of a confirmed endemic source of kala-azar in Somalia. Trans R Soc Trop Med Hyg 59(6):705–708

Bashaye S, Nombela N, Argaw D, Mulugeta A, Herrero M, Nieto J et al (2009) Risk factors for visceral leishmaniasis in a new epidemic site in Amhara Region, Ethiopia. Am J Trop Med Hyg 81(1):34–39

Belay AD, Asafa Y, Mesure J, Davidson RN (2006) Successful miltefosine treatment of post-kala-azar dermal leishmaniasis occurring during antiretroviral therapy. Ann Trop Med Parasitol 100(3):223–227

Belo VS, Struchiner CJ, Barbosa DS, Nascimento BW, Horta MA, da Silva ES, Werneck GL (2014) Risk factors for adverse prognosis and death in American visceral leishmaniasis: a meta-analysis. PLoS Negl Trop Dis 8(7):e2982. https://doi.org/10.1371/journal.pntd.0002982

Berhe N, Hailu A, Gemetchu T (1995) Human immunodeficiency virus and recurrence of cutaneous leishmaniasis long after healed localized cutaneous leishmaniasis due to *Leishmania aethiopica*. Trans R Soc Trop Med Hyg 89(4):400–401. https://doi.org/10.1016/0035-9203(95)90027-6

Boelaert M, Meheus F, Sanchez A, Singh SP, Vanlerberghe V, Picado A, Meessen B, Sundar S (2009) The poorest of the poor: a poverty appraisal of households affected by visceral leishmaniasis in Bihar, India. Trop Med Int Health 14(6):639–644

Boelaert M, Verdonck K, Menten J, Sunyoto T, van Griensven J, Chappuis F, Rijal S (2014) Rapid tests for the diagnosis of visceral leishmaniasis in patients with suspected disease. Cochrane Database Syst Rev 6:CD009135. https://doi.org/10.1002/14651858.CD009135.pub2

Boumis E, Chinello P, Della Rocca C, Paglia MG, Proietti MF, Petrosillo N (2006) Atypical disseminated leishmaniasis resembling post-kala-azar dermal leishmaniasis in an HIV-infected patient. Int J STD AIDS 17(5):351–353

Celesia BM, Cacopardo B, Massimino D, Gussio M, Tosto S, Nunnari G, Pinzone MR (2014) Atypical presentation of PKDL due to Leishmania infantum in an HIV-infected patient with relapsing visceral leishmaniasis. Case Rep Infect Dis 2014:370286. https://doi.org/10.1155/2014/370286

Collis S, El-Safi S, Atia AA, Bhattacharyya T, Hammad A, Den Boer M, Le H, Whitworth JA, Miles MA (2019) Epidemiological and molecular investigation of resurgent cutaneous leishmaniasis in Sudan. Int J Infec Dis 88:14–20. https://doi.org/10.1016/j.ijid.2019.08.018

Cota GF, de Sousa MR, de Mendoca AL, Patrocinio A, Assuncao LS, de Faria SR, Rabello A (2014) Leishmania-HIV co-infection: clinical presentation and outcomes in an urban area in Brazil. PloS Negl Trop Dis 8(4):e2816. https://doi.org/10.1371/journal.pntd.0002816

de Beer P, el Harith A, Deng LL, Semiao-Santos SJ, Chantel B, van Grootheest M (1991) A killing disease epidemic among displaced Sudanese population identifi ed as visceral leish-maniasis. Am J Trop Med Hyg 44(3):283–289

Dereure J, El-Safi SH, Bucheton B, Boni M, Kheir MM, Davoust B et al (2003) Visceral leishmaniasis in eastern Sudan: parasite identifi cation in humans and dogs; host-parasite relationships. Microbes Infect 5(12):1103–1108

Diro E, Lynen L, Ritmeijer K, Boelaert M, Hailu A (2014) Visceral leishmaniasis and HIV coinfection in east Africa. PLoS Negl Trop Dis 8(6):e2869. https://doi.org/10.1371/journal.pntd.0002869

Diro E, Blesson S, Edwards T, Ritmeijer K, Fikre H, Admassu H, Kibret A, Ellis SJ, Bardonneau C, Zijlstra EE, Soipei P, Mutinda B, Omollo R, Kimutai R, Omwalo G, Wasunna M, Tadesse F, Alves F, Strub-Wourgaft N, Hailu A, Alexander N, Alvar J (2019a) A randomized trial of AmBisome monotherapy and AmBisome and miltefosine combination to treat visceral leishmaniasis in HIV co-infected patients in Ethiopia. PLoS Negl Trop Dis 13(1):e0006988. https://doi.org/10.1371/journal.pntd.0006988

Diro E, Edwards T, Ritmeijer K, Fikre H, Abongomera C, Kibret A, Bardonneau C, Soipei P, Mutinda B, Omollo R, Griensven J, Zijlstra EE, Wasunna M, Alves F, Alvar J, Hailu A, Alexander N, Blesson S (2019b) Long term outcomes and prognostics of visceral leishmaniasis in HIV infected patients with use of pentamidine as secondary prophylaxis based on CD4 level: a prospective cohort study in Ethiopia. PLoS Negl Trop Dis 13(2):e0007132. https://doi.org/10.1371/journal.pntd.0007132

El Hassan AM, Khalil EA, el Sheikh EA, Zijlstra EE, Osman A, Ibrahim ME (1998) Post kala-azar ocular leishmaniasis. Trans R Soc Trop Med Hyg 92(2):177–179

Elnaiem DE (2011) Ecology and control of the sand fly vectors of Leishmania donovani in east Africa, with special emphasis of Phlebotomus orientalis. J Vector Ecol 36(Suppl 1):S23–S31. https://doi.org/10.1111/j.1948-7134.2011.00109.x

Elnaiem DA, Ward RD, Hassan HK, Miles MA, Frame IA (1998) Infection rates of Leishmania donovani in Phlebotomus orientalis from a focus of visceral leishmaniasis in eastern Sudan. Ann Trop Med Parasitol 92(2):229–232

Elnaiem DA, Hassan MM, Maingon R, Nureldin GH, Mekawi AM, Miles M, Ward RD (2001) The Egyptian mongoose, Herpestes ichneumon, is a possible reservoir host of visceral leishmaniasis in eastern Sudan. Parasitology 122(Pt 5):531–536

Elnaiem DE, Hassan HK, Osman OF, Maingon RD, Killick-Kendrick R, Ward RD (2011) A possible role for Phlebotomus (Anaphlebotomus) rodhaini (Parrot, 1930) in transmission of *Leishmania donovani*. Parasite Vectors 4:238. https://doi.org/10.1186/1756-3305-4-238

El-Safi SH, Peters W (1991) Studies on the leishmaniases in the Sudan. 1. Epidemic of cutaneous leishmaniasis in Khartoum. Trans R Soc Trop Med Hyg 85(1):44–47. https://doi.org/10.1016/0035-9203(91)90151-n

Endris M, Takele Y, Woldeyohannes D, Tiruneh M, Mohammed R, Moges F et al (2014) Bacterial sepsis in patients with visceral leishmaniasis in Northwest Ethiopia. Biomed Res Int 2014:361058. https://doi.org/10.1155/2014/361058

Ganguly S, Das NK, Barbhuiya JN, Chatterjee M (2010) Post-kala-azar dermal leishmaniasis – an overview. Int J Dermatol 49(8):921–931

Gebre-Michael T, Lane RP (1996) The roles of Phlebotomus martini and P. celiae (Diptera: Phlebotominae) as vectors of visceral leishmaniasis in the Aba Roba focus, southern Ethiopia. Med Vet Entomol 10(1):53–62

Gebre-Michael T, Balkew M, Ali A, Ludovisi A, Gramiccia M (2004) The isolation of *Leishmania tropica* and *L. aethiopica* from Phlebotomus (Paraphlebotomus) species (Diptera: Psychodidae) in the Awash Valley, northeastern Ethiopia. Trans R Soc Trop Med Hyg 98(1):64–70

Gebre-Michael T, Balkew M, Berhe N, Hailu A, Mekonnen Y (2010) Further studies on the phlebotomus sandflies of the kala-azar endemic lowlands of Humera-Metema (north-west Ethiopia) with observations on their natural blood meal sources. Parasite Vectors 3(1):6. https://doi.org/10.1186/1756-3305-3-6

Gebre-Michael T, Balkew M, Yimer M, Hailu A (2013) First record of Phlebotomus (Synphlebotomus) vansomerenae (Diptera: Psychodidae) in Ethiopia. J Med Entomol 50(3):659–663

Gelanew T, Amogne W, Abebe T, Kuhls K, Hailu A, Schonian G (2010) A clinical isolate of *Leishmania donovani* with ITS1 sequence polymorphism as a cause of para-kala-azar dermal leishmaniasis in an Ethiopian human immunodeficiency virus-positive patient on highly active antiretroviral therapy. Br J Dermatol 163(4):870–874

Gelanew T, Hurissa Z, Diro E, Kassahun A, Kuhls K et al (2011) Disseminated cutaneous leishmaniasis resembling post-kala-azar dermal leishmaniasis caused by Leishmania donovani in three patients co-infected with visceral leishmaniasis and human immuno-deficiency virus/acquired immune-deficiency syndrome in Ethiopia. Am J Trop Med Hyg 84:906–912

Githure JI, Ngumbi PM, Anjili CO, Luqalia R, Mwanyumba PM, Kinoti GK, Koech DK (1996) Animal reservoirs of leishmaniasis in Marigat, Baringo District, Kenya. East Afr Med J 73(1):44–47

Global Health Observatory (2022) Leishmaniasis. WHO, Geneva. https://www.who.int/data/gho/data/themes/topics/topic-details/GHO/leishmaniasis, updated on 17 August 2021, and accessed 29 May 2022

Gramiccia M, Gradoni L (2005) The current status of zoonotic leishmaniases and approaches to disease control. Int J Parasitol 35(11–12):1169–1180

Guiguemdé RT, Sawadogo OS, Bories C, Traore KL, Nezien D, Nikiema L, Pratlong F, Marty P, Houin R, Deniau M (2003) Leishmania major and HIV co-infection in Burkina Faso. Trans R Soc Trop Med Hyg 97(2):168–169

Hailu A (Unpublished) Visceral leishmaniasis in northern Ethiopia and Eritrea

Hailu A, Berhe N (2002) The performance of direct agglutination tests (DAT) in the diagnosis of visceral leishmaniasis among Ethiopian patients with HIV co-infection. Ann Trop Med Parasitol 96(1):25–30

Hailu A, Balkew M, Berhe N, Meredith SE, Gemetchu T (1995a) Is Phlebotomus (Larroussius) orientalis a vector of visceral leishmaniasis in south-west Ethiopia? Acta Trop 60(1):15–20

Hailu A, Negesse Y, Abraham I (1995b) *Leishmania aethiopica*: experimental infections in non-human primates. Acta Tropica 59(3):243–250. https://doi.org/10.1016/0001-706x(95)00085-s

Hailu A, Gebre-Michael T, Berhe N, Balkew M (2006) Leishmaniasis in Ethiopia. In: Kloos H, Berhane Y, Hailemariam D (eds) The ecology and epidemiology of health and disease in Ethiopia, 1st edn. Routledge, New York, pp 615–634

Hailu A, Musa A, Wasunna M, Balasegaram M, Yifru S, Mengistu G, Hurissa Z et al (2010) Geographical variation in the response of visceral leishmaniasis to paromomycin in East Africa: a multicentre, open-label, randomized trial. PLoS Negl Trop Dis 4(10):e709. https://doi.org/10.1371/journal.pntd.0000709

Hassan MM, Elamin EM, Mukhtar MM (2008) Isolation and identification of Leishmania donovani from Phlebotomus orientalis, in an area of eastern Sudan with endemic visceral leishmaniasis. Ann Trop Med Parasitol 102(6):553–555

Hassan MM, Osman OF, El-Raba'a FM, Schallig HD, Elnaiem DE (2009) Role of domestic dog as reservoir host of Leishmania donovani in eastern Sudan. Parasite Vectors 2(1):26. https://doi.org/10.1186/1756-3305-2-26

Herrero M, Orfanos G, Argaw D, Mulugeta A, Aparicio P, Parreno F et al (2009) Natural history of visceral leishmaniasis outbreak in highland Ethiopia. Am J Trop Med Hyg 81(3):373–377

Hoogstraal H, Dietlein DR (1964) Leishmaniasis in the Sudan Republic: recent results. Bull World Health Organ 31(1):137–143

Hoogstraal H, Heyneman D, Dietlein DR, Browne HG, Reid TP, van Peenen PFD, Saber AH, Rohrs LC (1963) Leishmaniasis in the Sudan Republic: epidemiological findings. Bull World Health Organ 28(2):263–265

Jamjoom MB, Ashford RW, Bates PA, Chance ML, Kemp SJ, Watts PC, Noyes HA (2004) Leishmania donovani is the only cause of visceral leishmaniasis in East Africa; previous descriptions of L. infantum and "L. archibaldi" from this region are a consequence of convergent evolution in the isoenzyme data. Parasitology 129(Pt4):399–409

Kalayou S, Tadelle H, Bsrat A, Abebe N, Haileselassie M, Schallig HD (2011) Serological evidence of Leishmania donovani infection in apparently healthy dogs using direct agglutination test (DAT) and rk39 dipstick tests in Kafta Humera, north-west Ethiopia. Transbound Emerg Dis 58(3):255–262

Kassahun A, Sadlova J, Dvorak V, Kostalova T, Frynta D, Aghova T, Yasur-Landau D, Lemma W, Hailu A, Baneth G, Warburg A, Volf P, Votypka J (2015a) Detection of *Leishmania donovani* and *L. tropica* in Ethiopian wild rodents. Acta Tropica 145:39–44

Kassahun A, Sadlova J, Benda P, Kostalova T, Warburg A, Hailu A, Baneth G, Volf P, Votypka J (2015b) Natural infection of bats with *Leishmania* in Ethiopia. Acta Tropica 150:166–170

Khalil EA, Musa AM, Younis BM, Elfaki ME, Zijlstra EE, Elhassan AM (2011) Blindness following visceral leishmaniasis: a neglected post-kala-azar complication. Trop Doct 41(3):139–140

Khalil EA, Khidir SA, Musa AM, Musa BY, Elfaki ME, Elkadaru AM et al (2013) Post-kala-azar dermal leishmaniasis: a paradigm of paradoxical immune reconstitution syndrome in non-HIV/AIDS patients. J Trop Med 2013:275253. https://doi.org/10.1155/2013/275253

Khalil EA, Weldegebreal T, Younis BM, Omollo R, Musa AM, Hailu W et al (2014) Safety and efficacy of single dose versus multiple doses of AmBisome for the treatment of visceral leish-maniasis in eastern Africa: a randomized trial. PLoS Negl Trop Dis 8(1). https://doi.org/10.1371/journal.pntd.0002613

Kolaczinski JH, Reithinger R, Worku DT, Ocheng A, Kasimiro J, Kabaterine N, Brooker S (2008) Risk factors of visceral leishmaniasis in east Africa: a case-control study in Pokot territory of Kenya and Uganda. Int J Epidemiol 37(2):344–352

Kone AK, Niare DS, Piarroux M, Izri A, Marty P, Laurens MB, Piarroux R, Thera MA, Doumbo OK (2019) Visceral leishmaniasis in west Africa: clinical characteristics, vectors and reservoirs. J Parasitol Res 9282690. https://doi.org/10.1155/2019/9282690

Kweku MA, Odoom S, Puplampu N, Desewu K, Nuako GK, Gyan B, Raczniak G, Kronmann KC, Koram K, Botero S, Boakye D, Akuffo H (2011) An outbreak of suspected cutaneous leishmaniasis in Ghana: lessons learnt and preparation for future outbreaks. Glob Health Action 4. https://doi.org/10.3402/gha.v4i0.5527

Lartey M, Adusei L, Hanson-Nortey L, Addy JH (2006) Coinfection of cutaneous leishmaniasis and HIV infection. Ghana Med J 40(3):110–112. https://doi.org/10.4314/gmj.v40i3.55264

Lawyer PG, Githure JI, Anjili CO, Olobo JO, Koech DK, Reid GDF (1990) Experimental transmission of *Leishmania major* to vervet monkeys (*Cercopithecus aethiops*) by bites of *Phlebotomus duboscqi* (Diptera: Psychodidae). Trans R Soc Trop Med Hyg 84(2):229–232

Lawyer PG, Mebrahtu YB, Ngumbi PM, Mwanyumba P, Mbugua J, Kiilu G, Kipkoech D, Nzovu J, Anjili CO (1991) *Phlebotomus guggisbergi* (Diptera: Psychodidae), a vector of *Leishmania tropica* in Kenya. Am J Trop Med Hyg 44(3):290–298. https://doi.org/10.4269/ajtmh.1991.44.290

Malaria Consortium (2010) Leishmaniasis control in eastern Africa: past and present efforts and future needs. Situation and gap analysis. Available: http://www.malariaconsortium.org/user-files/file/NTD%20Resources/VL%20EA%20Situation%20Analysis%20Fina_Janl.pdf

Marlet MV, Wuillaume F, Jacquet D, Quispe KW, Dujardin JC, Boelaert M (2003) A neglected disease of humans: a new focus of visceral leishmaniasis in Bakool, Somalia. Trans R Soc Trop Med Hyg 97(6):667–671

Maroli M, Feliciangeli MD, Bichaud L, Charrel RN, Gradoni L (2013) Phlebotomine sandflies and the spreading of leishmaniases and other diseases of public health concern. Med Vet Entomol 27(2):123–147

Mengesha B, Endris M, Takele Y, Mekonnen K, Tadesse T, Feleke A, Diro E (2014) Prevalence of malnutrition and associated risk factors among adult visceral leishmaniasis patients in northwest Ethiopia: a cross sectional study. BMC Res Notes 7:75. https://doi.org/10.1186/1756-0500-7-75

Molina R, Gradoni L, Alvar J (2003) HIV and the transmission of Leishmania. Ann Trop Med Parasitol 97(Suppl 1):29–45

Monge-Maillo B, Norman FF, Cruz I, Alvar J, Lopez-Velez R (2014) Visceral leishmaniasis and HIV coinfection in the Mediterranean Region. PLoS Negl Trop Dis 8(8):e3021. https://doi.org/10.1371/journal.pntd.0003021

Mukhtar MM, Sharief AH, el Saffi SH, Harith AE, Hiqazzi TB, Adam AM, Abdalla HS (2000) Detection of antibodies to Leishmania donovani in animals in a kala-azar endemic region in eastern Sudan: a preliminary report. Trans R Soc Trop Med Hyg 94(1):33–36

Mukhtar MM, Elamin EM, Bakhiet SM, Kheir MM, Ali AB (2010) First report on Leishmania major/HIV coinfection in a Sudanese patient. East Mediterr Health J 16(6):702–703

Musa AM, Khalil EAG, Younis BM, Elfaki MEE, Elamin MY, Adam AOA, Mohamed HAA, Dafalla MMM, Abuzaid AA, El-Hassan AM (2013) Treatment-based strategy for the management of Post-Kala-Azar Dermal Leishmaniasis patients in the Sudan. J Trop Med 2013:708391. https://doi.org/10.1155/2013/708391

Mutinga MJ (1975) The animal reservoir of cutaneous leishmaniasis on Mount Elgon, Kenya. East Afr Med J 52(3):142–151

Mutinga MJ (1991) A review of investigations on leishmaniasis vectors in Kenya. Parassitologia 33(Suppl):453–461

Mutinga MJ, Ngoka JM, Schnur LF, Chance ML (1980) The isolation and identifi cation of leishmanial parasites from domestic dogs in Machakos District of Kenya, and the possible role of dogs as reservoir hosts of kala-azar in east Africa. Ann Trop Med Parasitol 74(2):139–144

Ngere I, Boru WG, Isack A, Muiruri J, Obonyo M, Matendechero S, Gura Z (2020) Burden and risk factors of cutaneous leishmaniasis in a peri-urban settlement in Kenya, 2016. PLoS ONE 15(1):e0227697. https://doi.org/10.1371/journal.pone.0227697

Osman OF, Kager PA, Oskam L (2000) Leishmaniasis in the Sudan: a literature review with emphasis on clinical aspects. Trop Med Int Health 5(8):553–562

Owino BO, Matoke-Muhia D, Alraey Y, Mwangi JM, Ingonga JM, Ngumbi PM, Casas-Sanchez A, Acosta-Serrano A, Masiga DK (2019) Association of *Phlebotomus guggisbergi* with *Leishmania major* and *Leishmania tropica* in a complex transmission setting for cutaneous leishmaniasis in Gilgil, Nakuru county, Kenya. PLoS Negl Trop Dis 13(10):e0007712. https://doi.org/10.1371/journal.pntd.0007712

Padovese Z, Terranova M, Toma L, Barnabas G, Morrone A (2009) Cutaneous and mucocutaneous leishmaniasis in Tigray, northern Ethiopia: clinical aspects and therapeutic concerns. Trans R Soc Trop Med Hyg 103(7):707–711. https://doi.org/10.1016/j.trstmh.2009.02.023

Perkins PV, Githure JI, Mebrahtu Y, Kiilu G, Anjili C, Ngumbi PS et al (1988) Isolation of Leishmania donovani from Phlebotomus martini in Baringo district, Kenya. Trans R Soc Trop Med Hyg 82(5):695–700

Pratlong F, Debord T, Garnotel E, Garrabe E, Marty P, Raphenon G, Dedet JP (2005) First identification of the causative agent of visceral leishmaniasis in Djibouti: Leishmania donovani. Ann Trop Med Parasitol 99(1):21–25

Puig L, Pradinaud R (2003) Leishmania and HIV co-infection: dermatological manifestations. Ann Trop Med Parasitol 97(Suppl 1):S107–S114

Raguenaud ME, Jansson A, Vanlerberghe A, Deborggraeve S, Dujardin JC, Orfanos G et al (2006) Epidemiology and clinical features of patients with visceral leishmaniasis treated by an MSF clinic in Bakool region, Somalia, 2004–2006. PLoS Negl Trop Dis 1(1):e85

Ritmeijer K, Veeken H, Melaku Y, Leal G, Amsalu R et al (2001) Ethiopian visceral leishmaniasis: generic and proprietary sodium stibogluconate are equivalent; HIV co-infected patients have a poor outcome. Trans R Soc Trop Med Hyg 95:668–672

Russo R, Laguna F, Lopez-Velez R, Medrano FJ, Rosenthal E, Cacopardo B, Nigro L (2003) Visceral leishmaniasis in those infected with HIV: clinical aspects and other opportunistic infections. Ann Trop Med Parasitol 97(Suppl 1):99–105

Saporito L, Giammanco GM, De Grazia S, Colomba C (2013) Visceral leishmaniasis: host- parasite interactions and clinical presentation in the immunocompetent and in the immunocompromised host. Int J Infect Dis 17(8):e572–e576. https://doi.org/10.1016/j.ijid.2012.12.024

Sati MH (1958) Early phases of an outbreak of kala-azar in the Southern Fung. Sudan Med J 1:98–111

Schaefer KU, Kurtzhals JA, Sherwood JA, Githure JI, Kager PA, Muller AS (1994) Epidemiology and clinical manifestations of visceral and cutaneous leishmaniasis in Baringo District, Rift Valley, Kenya. A literature review. Trop Geogr Med 46(3):129–133

Schaefer KU, Kurtzhals JA, Gachihi GS, Muller AS, Kager PA (1995) A prospective seroepidemiological study of visceral leishmaniasis in Baringo District, Rift Valley Province, Kenya. Trans R Soc Trop Med Hyg 89(5):471–475

Schorscher JA, Goris M (1992) Incrimination of Phlebotomus (Larroussius) orientalis as a vector of visceral leishmaniasis in western upper Nile Province, southern Sudan. Trans R Soc Trop Med Hyg 86(6):622–623

Seaman J, Mercer AJ, Sondorp HE, Herwaldt BL (1996) The epidemic of visceral leishmaniasis in western upper Nile, southern Sudan: course and impact from 1984 to 1994. Int J Epidemiol 25(4):862–871

Seid A, Gadisa E, Tsegaw T, Abera A, Teshome A, Mulugeta A, Herrero M, Argaw D, Jorge A, Kebede A, Aseffa A (2014) Risk map for cutaneous leishmaniasis in Ethiopia based on environmental factors as revealed by geographical information systems and statistics. Geospat Health 8(2):377–387

Sirol J, Vedy J, Barabe P, Cesari C, Berger P (1976) Kala-Azar in the Republic of Chad. 6-year survey at the Central Hospital of N'Djamena (Fort-Lamy). Bull Soc Pathol Exot Fil 69(3):232–237

Sixl W, Sebek Z, Reinthaler F, Mascher F (1987) Investigations of wild animals as leishmania reservoir in south Sudan. J Hyg Epidemiol Microbiol Immunol 31(4 Suppl):483–485

Stephenson R (1940) An epidemic of kala-azar in the Upper Nile province of the Anglo-Egyptian Sudan. Ann Trop Med Parasitol 34:175–179

Sunyoto T, Verdonck K, El Safi S, Potet J, Picado A, Boelaert M (2018) Uncharted territory of the epidemiological burden of cutaneous leishmaniasis in sub-Saharan Africa – a systematic review. PLoS Negl Trop Dis 12(10):e0006914. https://doi.org/10.1371/journal.pntd.0006914

Tangie LN, Desmond A, Aminde LN, Ako AM, Halle PM (2017) Cutaneous leishmaniasis in a severely immunocompromised HIV patient in Kumbo, Northwest region of Cameroon: case report. BMC Research Notes 10(425). https://doi.org/10.1186/s13104-017-2751-1

Tonui WK (2006) Situational analysis of leishmaniases research in Kenya. Afr J Health Sci 13(1–2):7–21

Utaile M, Kassahun A, Abebe T, Hailu A (2013) Susceptibility of clinical isolates of *Leishmania aethiopica* to miltefosine, paromomycin, amphotericin B and sodium stibogluconate using amastigote-macrophage in vitro model. Exp Parasitol 134(1):68–75. https://doi.org/10.1016/l.exppara.2013.01.022

Van den Bogaart E, Berkhout MM, Nour AB, Mens PF, Talha AB, Adams ER et al (2013) Concomitant malaria among visceral leishmaniasis in-patients from Gedarif and Sennar States, Sudan: a retrospective case-control study. BMC Public Health 13:332. https://doi.org/10.1186/1471-2458-13-332

WER (2021) Global leishmaniasis surveillance: 2019–2020, a baseline for the 2030 roadmap. Weekly Epidemiological Record (WER), 3 September 2021, Vol. 96, No. 35 (pp. 401–420)

World Health Organization (2010) Control of the Leishmaniases. In: WHO Technical Report Series, 949: report of a meeting of the WHO Expert Committee on the Control of Leishmaniases, Geneva, 22–26 March 2010. World Health Organization, Geneva. http://whqlibdoc.who.int/trs/WHO_TRS_949_eng.pdf

World Health Organization (2022) Leishmaniasis Fact sheet updated 08 January 2022. Available at: https://www.who.int/news-room/fact-sheets/detail/leishmaniasis, Accessed 29 May 2022.

Zijlstra EE, el-Hassan AM, Ismael A, Ghalib HW (1994) Endemic kala-azar in eastern Sudan: a longitudinal study on the incidence of clinical and subclinical infection and post-kala-azar dermal leishmaniasis. Am J Trop Med Hyg 51(6):826–836

Zijlstra EE, Musa AM, Khalil EA, el-Hassan IM, el-Hassan AM (2003) Post-kala-azar dermal leishmaniasis. Lancet Infect Dis 3(2):87–98

Leprosy

Benedict Okoe Quao, Herman Joseph Ssekamatte Kawuma, Shimelis N. Doni, and Ekow Amankrah-Otabir

Contents

B. O. Quao (✉)
Ankaful Leprosy & General Hospital, Cape Coast, Ghana

National Leprosy Control Programme, Disease Control & Preventive Department, Ghana Health Service Public Health Division, Accra, Ghana

H. J. S. Kawuma
German Leprosy and TB Relief, Association Uganda Office, Kampala, Uganda

S. N. Doni
ALERT hospital, Addis Ababa, Ethiopia

Ethiopian Dermatology and Venereology Society, Addis Ababa, Ethiopia

E. Amankrah-Otabir
University of Cape Coast School of Medical Sciences, Cape Coast, Ghana

J. O. Gyapong, B. A. Boatin (eds.), *Neglected Tropical Diseases - Sub-Saharan Africa*, Neglected Tropical Diseases, https://doi.org/10.1007/978-3-031-53901-5_6

Abstract

Leprosy has afflicted humans since ancient times with molecular evidence showing Eastern Africa probably played a more significant role in its global spread. The tide turned against the disease in the 1980s when effective multidrug therapy (MDT) was introduced worldwide. By 2005, all countries in sub-Saharan Africa had attained the WHO elimination target of prevalence less than 1 per 10,000 population. In 2021, Africa accounted for 15% of global leprosy new case detection, with a worrying number of new cases having significant disability at diagnosis. Public stigma has also contributed to the persistence of segregated settlements for leprosy sufferers long after cure. Leprosy spread is believed to be by droplet and skin transmission, with growing evidence for zoonotic spread from infected armadillos, a species not found outside the Americas. The presentation of leprosy in sub-Saharan Africa is as elsewhere, varying between two polar states of tuberculoid and lepromatous leprosy and modified by complications, including acute reactions. No case in sub-Saharan Africa has been linked to the recently discovered *Mycobacterium lepromatosis*.

Clinical diagnosis and classification of leprosy at present are most amenable for treatment purposes, with the continuous supply of MDT being the backbone of leprosy control activities. A fixed-duration course of rifampicin, dapsone, and clofazimine remains the mainstay of treatment. There is now a greater push toward interrupting leprosy transmission with contact tracing and leprosy prevention with SDR being key strategies. Also vitally important is the needed strengthening of weak African healthcare systems within which integrated leprosy services are provided as well as increased involvement of sufferers and communities.

Leprosy control activities are now suffering from reduced attention on both global and subregional scales, and the knowledge of frontline healthcare workers regarding leprosy is dwindling. With leprosy likely to remain with us for the foreseeable future, more research, political commitment, and collaborative action are required to sustain the gains and move Africa and the rest of the world toward zero leprosy—i.e., zero disease, zero disability, and zero discrimination and stigma.

Introduction

Leprosy is a chronic mildly infectious disease that has inflicted man for several centuries and principally affects the skin and peripheral nerves and, in some cases, mucous membranes. It has historically been known for its complications of clawing and resorption of digits, which contributes to the public stigma of the disease and negative treatment meted out to sufferers in our societies. Although exact figures over time are unavailable, it is estimated that about three to four million people could be currently living with some impairment or disability arising from leprosy (World Health Organization 2021a).

The disease is also named Hansen's disease after the Norwegian physician Gerhard Armauer Hansen, who first demonstrated *Mycobacterium leprae* in infected tissue samples under the microscope in 1873 (Jay 2000). This was also the first time a bacterium was linked to a disease in man and thus an important milestone in the battle against infectious diseases.

Leprosy has existed for millennia, long before this feat, and descriptions fitting leprosy had been recorded in ancient Indian manuscripts dating as far back as 600 BC and in early Chinese medical publications around 400 BC. An alternate theory to the long-held belief of leprosy originating in India and spreading westward by returning soldiers of Alexander the Great was suggested in 2005 after molecular studies by Monot et al. (2005) pointed to a potential East African origin, more specifically from Ethiopia. With an Eastern African origin, it would also mean that colonialism and the slave trade were probably the major drivers of the worldwide spread of the disease (Monot et al. 2005).

Since ancient times, many different methods, including the use of natural oils and mercury, have been used to try to cure the disease with no notable success. The disease remained largely incurable, causing great disfigurement and disability in sufferers, and was greatly feared throughout the world. As it was believed to be highly contagious, sufferers were quarantined in leprosy colonies or "leprosaria" where they remained until their death. It was not until the 1940s, with the successful use of promin and its derivative dapsone subsequently, that an effective treatment was finally found.

From the 1940s, dapsone monotherapy became the mainstay of leprosy treatment, although this was fraught with many challenges. Therapy was long-term and associated with clinical overload and noncompliance, and soon, widespread resistance to dapsone monotherapy began to emerge. Multidrug therapy (MDT) for leprosy was introduced in the 1980s and allowed the treatment of individuals within their communities for shorter periods (World Health Organization 1982). Since 1995, the World Health Organization has supplied free effective MDT drugs worldwide, which, together with early case detection and a variety of control efforts, have contributed to a significant decline in leprosy prevalence over the past three decades.

Epidemiology of the Disease

Distribution and Burden

Leprosy is mostly found in but not limited to the tropics. It is well known that leprosy was in times past quite common in Europe far from the tropics, with sufferers saddled with clanging bells about their waists to announce their presence. Many, however, attribute the decline of the prevalence of leprosy in Europe and the rest of the advanced world more to the advent of industrialization and consequent improvements in living conditions rather than some climatic feature or drug treatment alone.

WHO statistics on leprosy, despite the challenges with the timeliness of country reporting and variations in data submitted year after year, give the best picture of the current distribution of this ancient scourge. As observed historically, the most recently released WHO figures from 2021, despite the effects of the COVID-19 pandemic, show that Southeast Asia alone continues to contribute about two-thirds of the worldwide leprosy new case detection, with numbers from Europe still virtually insignificant (Barua 2013; World Health Organization 2022). Africa and the Americas (mainly South America) follow in that order, with the WHO African region accounting for 15.1% of global new case detection in 2021 (World Health Organization 2022).

Despite the regional distribution as outlined above, the disease appears to be very focal in its distribution across countries. From historically collected data, there are some 23 global priority countries that report more than 1000 new leprosy cases yearly or have very high annual new case detection rates. Together, they account for about 95% of the new leprosy case detection worldwide annually, with only 3 of these 23 countries—India, Brazil, and Indonesia—also accounting for nearly 80%, based on pre-pandemic data (World Health Organization 2019, 2020a). Of these 23 global priority countries, 13 are African countries—Angola, Comoros, Cote D'Ivoire, Democratic Republic of Congo, Egypt, Ethiopia, Madagascar, Mozambique, Nigeria, Somalia, South Sudan, Sudan, and Tanzania (World Health Organization 2022).

In 1991, the World Health Assembly set the goal of elimination of leprosy as a public health problem by the year 2000. This goal was successfully reached through concerted worldwide efforts and the widespread use of multidrug therapy (MDT) in the year 2000, when the prevalence of leprosy fell below the set threshold of one case per 10,000 population (World Health Organization 2001a). Countries, nonetheless, continued in efforts to achieve the target also at national level, with all countries within the WHO African region achieving the target by the end of 2005 (World Health Organization, Regional Office for Africa 2012) and almost all countries globally by 2015 (Fig. 1). Translation of this apparent progress to the subnational level has, however, remained challenging with several countries still having endemic pockets such as urban slums, remote mountainous areas, and regions ravaged by conflict (World Health Organization, Regional Office for Africa 2012).

Despite the widespread reduction in the number of leprosy cases registered for treatment (or leprosy prevalence), the decline in new leprosy case detection has not been as dramatic, with only an average of 1% decline annually in the number of new leprosy cases globally observed between 2010 and the onset of the COVID-19 pandemic (Malecela and Ducker 2021). The detection of cases in under 15 year-olds is an indicator of recent transmission, so although there was a reduction in the new case detection rate among children from 10.1 in 2014 to 7.9 per million children in 2019 (World Health Organization 2021a), this still suggests that "true elimination" of leprosy had not been attained yet, and the world can expect to continue detecting new cases of leprosy for many years to come. In fact, many countries have continued to report large numbers of new leprosy cases annually (Fig. 2) and for well over a decade, there were about 200,000 leprosy cases reported annually until 2020 when the disruption of leprosy services due to the COVID-19 pandemic resulted in nearly a 40% reduction of globally reported new cases (World Health Organization 2021b).

The number of new patients with visible or Grade 2 disability, which is responsible for most of the negative social consequences such as stigma and

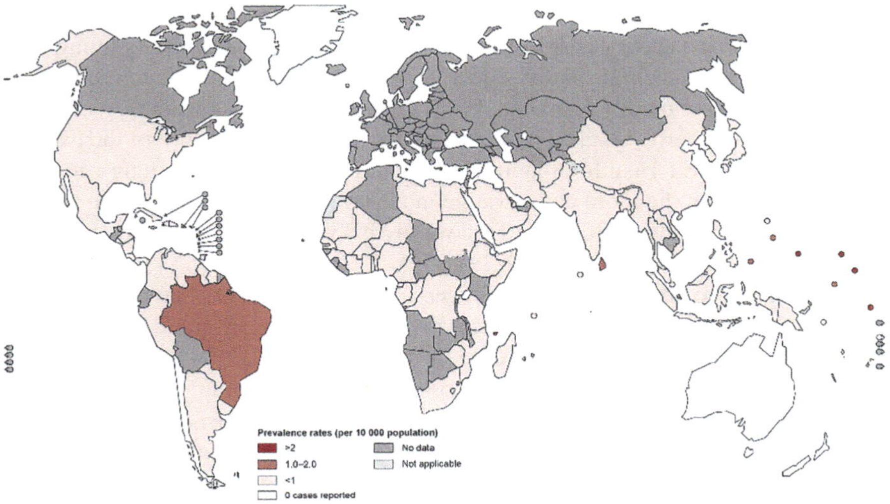

Fig. 1 Registered cases of leprosy per 10,000 population (prevalence rates), 2015 (Map produced by WHO Department of Control of Neglected Tropical Diseases and published in Global Leprosy Update 2015)

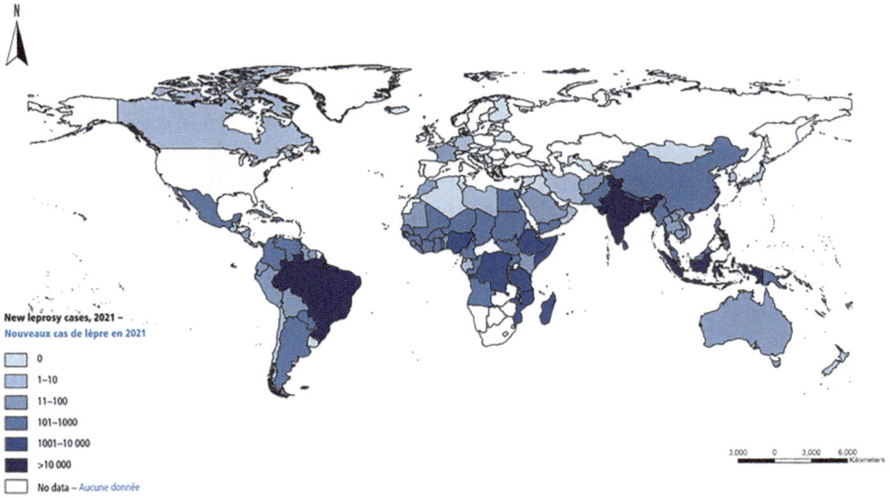

Fig. 2 Geographic distribution of newly detected leprosy cases, 2021 (Map produced by WHO Global Leprosy Programme and published in Global Leprosy Update 2021)

impoverishment, has also declined over the years. Yet in 2019 before the COVID-19 pandemic broke out, the Grade 2 Disability (G2D) rate in newly detected cases was 1.4 per million population, still above the 2020 target of <1.0 per million population

(World Health Organization 2021a). Beyond the annually reported figures on new cases with G2D, an accurate estimate of the number of persons currently living with some form of permanent disability due to leprosy who will require continuing care for disability is not readily available. From the estimated two million persons who were living with leprosy-related permanent disability around the turn of the twenty-first century (World Health Organization 1998a), the current true figure may lie anywhere between the more conservative modeled projection of about one million (Richardus and Habbema 2007) and the widely held estimate of three to –four million (World Health Organization 2021a).

In 2011, there were 25,231 new cases of leprosy detected within the WHO African region alone, with as many as 10% of them having grade 2 disabilities (World Health Organization, Regional Office for Africa 2013) while in 2021, a decade on, 3246 out of 21,201 new cases, representing about 15%, were diagnosed with G2D (World Health Organization 2022). Aside from such individuals with G2D in need of physical treatment and socioeconomic rehabilitation, the current Global Leprosy Strategy 2021–2030 with a vision of Zero Leprosy, which consists of three elements—zero disease, zero disability, and zero discrimination and stigma—has redefined the burden of leprosy to include all other persons suffering from the mental health impact of leprosy (World Health Organization 2021a).

Leprosy disease burden is thus multifaceted. It "can be measured in terms of new cases reported, the number of cases registered for treatment, the number of cases with impairments and disabilities, the number of children affected, and the number of affected people requiring rehabilitation due to consequences of the disease" (World Health Organization, Regional Office for South-East Asia 2012a). The concept of leprosy burden can be stretched to include the proportion of cases in remote and non-accessible areas, the resources available to support the leprosy program, and the impact of the stigma associated with the disease (World Health Organization, Regional Office for Africa 2012). In the Africa region, there are nine indicators that have historically been used as the "Leprosy Burden Score (LBS)" to determine the leprosy burden of each country (Tabah et al. 2016; World Health Organization, Regional Office for Africa 2012). Table 1 shows the indicators used and the weighted score given to each, while Fig. 3 shows the most recently published country distribution of leprosy burden in the WHO-AFRO from 2012 based on the LBS and comprising most of the countries within sub-Saharan Africa (Onyeze 2013). As has been done in Cameroon subsequent to this publication, countries can use the LBS to assess changing burden of leprosy at sub-national levels using available routinely collected leprosy epidemiological data (Tabah et al. 2016).

Over the last couple of years, however, there has been a shift to focusing on stopping transmission and new cases, with the WHO Global Leprosy Programme coordinating ongoing work to define new public health targets for leprosy and associated criteria and indicators to measure progress toward them. This has resulted in consensus being reached on three different phases of leprosy elimination and a new standardized approach to assessing the status of countries and sub-national geographic areas (World Health Organization, Regional Office for South-East Asia

Table 1 Score scale of leprosy burden indicators within the Africa region

Indicator	Burden score according to indicator level High (score = 2)	Medium (score = 1)	Low (score = 0)
Detection number	>1000 new cases	500–1000 new cases	<500 new cases
Prevalence rate	>2/10,000	1–2/10,000	<1/10,000
Detection rate	>20/100,000	10–20/100,000	<10/100,000
Proportion of multibacillary cases[a]	<50%	50–75%	>75%
Proportion of children[a]	>20%	10–20%	<10%
Proportion of grade 2 disability[a]	>20%	10–20%	<10%
Proportion of females[a]	<40%	>60%	30–60%
Prevalence/detection ratio	>2	1–2	<1
Grade 2 disability per 100,000 population	>1	0.5–1	<0.5
Total	≥5	3–4	0–2

WHO-AFRO (2012, 2013)
[a] Among new cases detected within the year

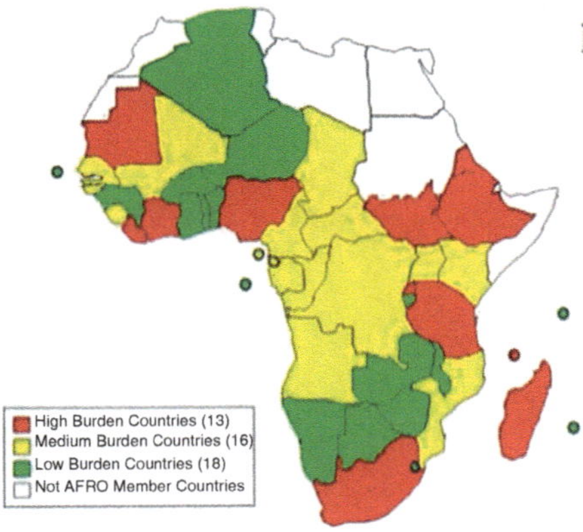

High Burden	Medium Burden	Low Burden
Comoros	Angola	Algeria
Cote d'Ivoire	Cameroon	Benin
Eritrea	Central African Republic	Botswana
Ethiopia	Chad	Burkina Faso
Lesotho	Congo	Burundi
Liberia	Democratic Republic of Congo	Cape Verde
Madagascar	Equatorial Guinea	Ghana
Mauritania	Gabon	Guinea
Nigeria	Gambia	Mauritius
Rwanda	Guinea Bissau	Malawi
South Africa	Kenya	Namibia
South Sudan	Mali	Niger
Tanzania	Mozambique	Sao Tome & Principe
	Senegal	Seychelles
	Sierra Leone	Swaziland
	Uganda	Togo
		Zambia
		Zimbabwe

Fig. 3 Leprosy burden within the WHO Africa region (Onyeze 2013)

2021). With this new paradigm targeting the achievement of zero new cases of leprosy, the focus of burden classification rests mainly on the absolute number of new leprosy cases being detected, and Fig. 4 shows the current status of countries regarding leprosy elimination in the WHO African Region based on 2021 figures.

Public Health Impact

A worrying observation about leprosy, as with most of the other neglected tropical diseases, is the predilection for the more underprivileged segments of societies, whose

Fig. 4 Leprosy status in the WHO/AFRO region (Source: WHO-AFRO)

poor living conditions and weaker immune systems make them more vulnerable to the disease. If left untreated, the disease runs a protracted course, causing damage mainly to peripheral nerves and resulting in deformities of the hands and feet as well as problems with vision. These deformities of the hands, feet, and eyes make manual tasks difficult and thus render many of those affected more impoverished since they are frequently unable to continue in their previous economic activity.

In sub-Saharan Africa, despite the progress made over the last two decades, about 35% of the population were still below the poverty line in 2019 (Beegle and Christiaensen 2019; World Bank 2022). Many of these African countries also lack well-established social welfare systems, so the effect of leprosy "maiming" a family breadwinner is profound, potentially depriving successive generations of any chance of breaking the cycle of poverty. This has been made even more grave, with the world experiencing its first increase in extreme poverty in over two decades following the broad economic shocks brought on by the COVID-19 pandemic (World Bank 2022).

Stigma surrounding leprosy also remains a significant problem for sufferers of the disease in this region, where the disease process is poorly understood by much of the population. Many continue to be shunned by other members of their community in the belief that they are cursed with a highly contagious and untreatable disease. Permanent disability such as digit clawing provides a perpetual reminder long after cure, labeling and exposing sufferers to public discrimination and isolation.

The highly stigmatized environment in sub-Saharan Africa has contributed to the persistence of segregated settlements for leprosy sufferers despite the worldwide abolishment of the concept of isolating leprosy patients. Many cured persons with residual deformities, in search of acceptance, escape from the harsh stigmatized and discriminatory setting within their former communities to set up dwellings near the former leprosaria from where they obtained treatment. In Ghana, for example, there are well-known settlements at Ho, Weija, Ankaful-fie, and Nkankyina, among many others. Although the living conditions within these settlements may not be optimal, the "privileged ones" within them are more likely to benefit from hand-outs and support from aid agencies and philanthropists than those within the general community.

At the height of the COVID-19 pandemic, there were reported interruptions of distribution of leprosy drugs and an increased vulnerability of persons affected by leprosy to experiencing mental health issues (Cavaliero and Quao 2020). Apart from the significant impact of COVID-19 on the general health systems worldwide, there have been new challenges for persons affected by leprosy. It has also been noted that pre-existing inequalities like stigma and discrimination and access to clean water experienced by marginalized populations have been made worse (Tucker et al. 2020).

Basic Biology

Infective Agents and Life Cycle

Mycobacterium leprae was historically the only known causative agent of leprosy until 2008 when a second agent *Mycobacterium lepromatosis* was linked to diffuse lepromatous leprosy (DLL), a unique form of leprosy endemic in Mexico and the Caribbean (Han et al. 2008). Through work subsequently carried out on archived and autopsied specimens and several case reports published from different locations in South East Asia and the Americas, *M. lepromatosis* has been shown to be associated with other forms of leprosy beyond DLL, including lepromatous leprosy (LL), borderline leprosy (BL), and reaction states such as erythema nodosum leprosum (ENL) and Lucio phenomenon (LP), either singly or as a co-infection with *M. leprae* (Deps and Collin 2021; Han et al. 2012).

Since ancient times, leprosy was regarded as a disease of man. Studies later confirmed that *M. leprae* occurs naturally in nine-banded armadillos native to the Americas, with infectivity rates approaching 30% in some armadillo populations, thereby establishing the armadillo species as an important natural reservoir of *M. leprae* (Walsh et al. 1981). More recently, red squirrels in the British Isles have also been shown to be a natural reservoir of both *M. leprae* and the newly discovered *M. lepromatosis* (Avanzi et al. 2016; Deps and Collin 2021). Armadillos with their much cooler body temperatures, which *M. leprae* species prefer, are believed to have acquired the agent from original European human settlers to the Americas within the last half-millennium (Truman et al. 2011). A similar anthroponotic introduction of *M. leprae* in the British Isles has also been suggested by the observed close relation at the molecular level of the British Isles red squirrel strain to the North American Armadillo strain (Avanzi et al. 2016).

Leprosy bacilli are obligate intracellular bacteria, which multiply mainly in macrophages and nerve Schwann cells and, unlike other *Mycobacterium* species, fail to grow on artificial media outside the human body (Han and Silva 2014). They are rod-shaped gram-positive organisms that are acid-fast when stained by the Ziehl Nielson (ZN) or Fite methods. Although neither agent has been cultivated in any laboratory as yet, the successful inoculation of *M. leprae* in athymic (naked) mice and armadillos has permitted some limited research into the disease. Leprosy bacilli have a life span of about 6 months with the longest doubling time of any known bacterium. In the mouse footpad, *M. leprae* has a generation time of about 12–13 days during the exponential phase of growth (Yawalkar 2009).

Disease Transmission

Even though *Mycobacterium leprae* was the first microorganism to be linked with disease causation in man, the exact mode of transmission is still not yet known and presents a major challenge to control efforts. The long incubation period of leprosy, which ranges from 2 to 10 years but can be as long as 20 years, complicates recalling of the initial exposure event, and many patients at the time of diagnosis usually have no history of apparent exposure to any known cases of leprosy. Nonetheless, the observation over the years of increased occurrence of leprosy among household contacts has always suggested a direct human-to-human spread of the disease, driving the public fear that led to the ostracization of leprosy patients in former leper colonies and leprosaria.

Despite a deficiency of direct causal evidence, the skin and nasal routes are regarded as the most probable exit and entry routes of leprosy bacilli. Untreated multibacillary leprosy patients are deemed most infectious and are known to shed millions of leprosy bacilli in their nasal secretions daily. Skin transmission was originally thought to require broken skin owing to the confinement of leprosy granulomas to deeper layers of the dermis (Yawalkar 2009), but later demonstration of leprosy bacilli on the surface of the skin of infected leprosy patients and their household contacts suggests intact skin may also play a role in transmission (Job et al. 2008). Naked mice, when exposed to leprosy bacilli by different routes, developed generalized disease only with topical nasal mucosa exposure and subcutaneous injection of bacilli (Chehl et al. 1985), giving further credence to the two routes being the likely portals of entry.

The exact mode of transfer of leprosy bacilli in man has also remained unclear, although droplet transmission is generally regarded as the most plausible. Experimental evidence with naked mice suggests that infection occurs more readily with direct instillation of leprosy bacilli on traumatized nasal mucosa or tongue compared with other modes of transfer such as inhalation or feeding (McDermott-Lancaster and McDougall 1990). Direct skin-to-skin transmission, although theoretically possible, has also not been demonstrated as yet. Indirect transmission from environmental and other non-human sources such as arthropods and water has not been proven as yet, although there is some evidence from studies with nude mice that suggests that penetrating injuries to skin contaminated with *M. Leprae* from the environment may also play a role (Job et al. 1994).

Leprosy bacilli shed into the environment are able to remain viable for periods of up to 3–5 months, depending on the prevailing physical conditions (Desikan and Sreevatsa 1995; Truman and Fine 2010), and this may be a significant factor in sustaining community transmission of leprosy. Nasal carriage of leprosy bacilli appears to play a significant role in community transmission. Indeed, nasal carriage of leprosy without any associated clinical signs has been demonstrated using PCR among contacts of persons diagnosed with leprosy as well as other non-contacts living in endemic communities, indicating exposure (Hatta et al. 1995). Nasal carriage, nonetheless, tends to be transient with clustering observed, and community carriage levels appear to be time-dependent (Hatta et al. 1995; Lavania et al. 2013). This apparent seasonal variation of PCR nasal carriage is, however, frequently absent in highly endemic pockets, where transmission is likely to be continuous (Lavania et al. 2013).

Recent evidence (Truman et al. 2011) points to the possibility of zoonotic spread of leprosy from infected nine-banded armadillos to humans, which may explain the occurrence of leprosy in individuals in the United States with no history of travel to leprosy-endemic areas. The molecular study comparing *M. leprae* bacilli isolated from an armadillo with extractions from three cases in the United States demonstrated a similar genetic strain of *M. leprae* occurring in all cases, dissimilar to any of the strains reported from elsewhere in the world (Truman et al. 2011).

Disease Presentation

The development and expression of clinical leprosy in exposed individuals are highly modulated by immunological mechanisms. In the majority of individuals exposed to leprosy bacilli, subclinical infection does not progress to clinical disease, with only about 5% of humans having genetic susceptibility to leprosy (Britton and Lockwood 2004; de Messias-Reason et al. 2009). Furthermore, in those in whom the infection becomes established, the form of leprosy they develop is dependent on their level of cell-mediated immunity, ranging between two polar states, tuberculoid leprosy (TT) and lepromatous leprosy (LL) (Ridley and Jopling 1966).

Tuberculoid leprosy is characterized by a strong cell-mediated immunity and low bacillary load with lesions limited to a few skin sites and/or nerves, while lepromatous leprosy is characterized by an absence of specific cell-mediated immunity and unchecked proliferation of leprosy bacilli resulting in diffuse lesions. The forms of leprosy in between these two extreme states are regarded as very unstable, tending to drift toward either pole depending on the individual's level of cell-mediated immunity. Both the internationally adopted 1953 Madrid Congress classification, which was purely clinical, and the later Ridley–Jopling classification recognize these intermediate forms, with the latter proposing three borderline states, namely, borderline tuberculoid (BT), mid borderline (BB), and borderline lepromatous leprosy (BL) to make up a five-group spectrum (Ridley and Jopling 1966).

There is almost always skin involvement in leprosy, accompanied most of the time by nerve enlargement. Skin lesions may be hypopigmented or erythematous and range from macules and papules to nodules, plaques, and infiltrations. Infrequently, a purely neural form of leprosy can also occur. The disease is thought to begin as indeterminate leprosy (IL), characterized by one or a few ill-defined skin lesions, usually without loss of sensation and often missed, which progresses to one of the other forms of leprosy or resolves without treatment. The skin lesions of TT tend to be well-defined macules, asymmetrically or unilaterally distributed, sometimes having slightly raised margins and usually with a dry surface and early loss of sensation over lesion from the involvement of free nerve endings in skin. The appearance and distribution of skin lesions of BT are similar to those of TT but are more numerous with the presence of satellites (Ridley and Jopling 1966; Fig. 5).

Skin lesions of LL are multiple and bilaterally and symmetrically distributed with initial erythematous macules and papules becoming plaques and nodules as the disease progresses (Ridley and Jopling 1966). LL is commonly associated with edema of feet and lower legs, and in later stages, patients may develop diffuse thickening and ridging of facial skin with thickening of nose and ears ("leonine facies") and loss of eyelashes (madarosis), keratitis, nasal ulceration, and saddle-nose deformities as well as bone and teeth involvement (Ridley and Jopling 1966). BL closely resembles LL, but its lesions differ in that they are not as shiny and are more likely to have a loss of sensation over them, and their distribution is not as symmetrical,

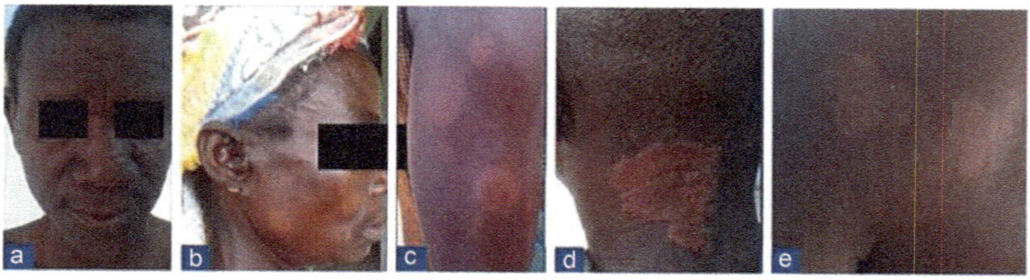

Fig. 5 (**a**) Facial papules and nodules in LL. (**b**) Hypopigmented macules and patches of BT with satellites. (**c**) BT lesions with satellites on arm. (**d**) Fairly well-defined early BT erythematous annular plaque with satellites. (**e**) Well-defined unilateral TT lesions (Pictures Source: Benedict Okoe Quao)

with sufferers sometimes presenting also with dimpled nodules and punched-out saucer-shaped plaques (Ridley and Jopling 1966). BB, as the name suggests, has balanced features of both TT and LL, tending toward symmetrical distribution with the presence of few satellites as in BT (Figs. 5 and 6).

Nerve involvement in leprosy beyond the loss of sensation over skin lesions is crucial in the development of deformities and disability, which remain long after the cure of leprosy sufferers. Neuropathy of a mixed type occurs when leprosy bacilli invade and cause inflammation in peripheral nerve trunks, composed of motor, sensory, and sympathetic nerve fibers (Lastória and de Abreu 2014). This process is often insidious in nature and involves a variety of mechanisms, including early axonal atrophy and eventual segmental demyelination (Scollard et al. 2015). It usually is silent and does not become clinically apparent until 25–30% of the nerve fiber is damaged (Malaviya 2003), unless there is painful exacerbation as occurs during a reaction, when nerve fiber damage is also accelerated. Nerve involvement may manifest as neural thickening or functional abnormalities such as loss of sensation, pain, increased sensation, muscle weakness, or loss of autonomic function such as sweating (Fig. 7). Temperature discrimination is usually the first sensory modality to be lost, while joint position sense is rarely lost because the afferents are carried via tendons rather than within the nerve trunk (Malaviya 2003).

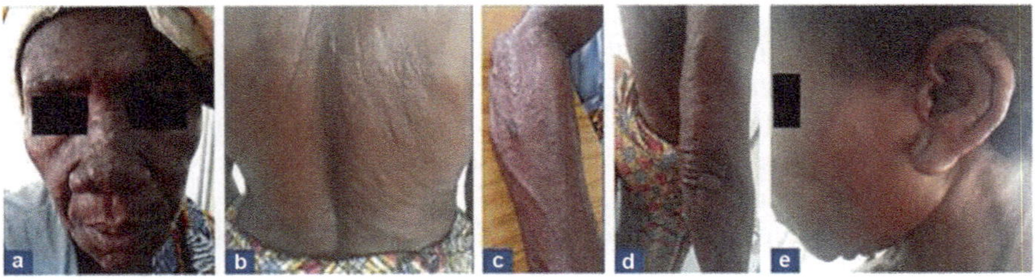

Fig. 6 Lepromatous leprosy. (**a**) "Leonine" facies with diffuse infiltration of facial skin. (**b**) Diffuse symmetrical skin nodules. (**c**) Dry skin. (**d**) Nodules at elbow. (**e**) Firm thickening of ear lobes *(Pictures Source: Benedict Okoe Quao)*

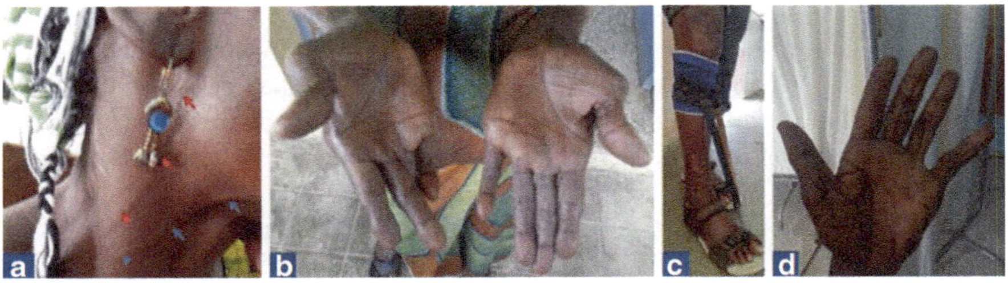

Fig. 7 Nerve involvement. (**a**) Enlarged great auricular (red arrows) and transverse cervical (blue arrows) nerves. (**b**) Ulnar nerve involvement with wasting and clawing of hands. (**c**) Correction of foot drop resulting from common peroneal nerve involvement. (**d**) Dry innervated hand *(Pictures Source: Benedict Okoe Quao)*

In TT leprosy, nerve involvement occurs early and is limited to gross and irregular enlargement of usually one peripheral nerve near the site of the skin lesion, with nerve function impairment invariably present (Ridley and Jopling 1966). The borderline forms of leprosy tend to have multiple peripheral nerves enlarged but not as irregularly and grossly thickened as occurs in TT. Nerve thickening does not occur early in pure LL, with peripheral nerves diffusely undergoing hyaline degeneration or fibrosis only in the late stages of the disease and resulting in symmetric anesthesia of hands and feet, so-called glove-and-stocking anesthesia (Ridley and Jopling 1966; Fig. 7).

About a third of leprosy sufferers may experience a reactional state at some point in the course of the disease, either as the presenting feature, during treatment or within the few couple of years after completion of drug treatment (Saunderson and ILEP Medico Social Commission 2002; World Health Organization 2020b). Triggers of the reactional state include vaccination, physical or emotional stress, pregnancy, and surgery. Irrespective of the point during the disease course when they occur, reactions always require early recognition and prompt treatment since they are periods of accelerated nerve damage. When leprosy reactions occur after cure, they may be confused with a relapse of clinical leprosy, which requires a repeat of a full course of treatment, and so there is always the need to cautiously differentiate post-treatment reactional states from relapses.

Two main types of reactions are recognized: lepra type 1 (or reversal) reaction and lepra type 2 reaction (or erythema nodusum leprosum, ENL). Isolated inflammation of nerves (or neuritis) may also occur in the absence of features of either type 1 or type 2 reactions. The so-called Lucio's phenomenon, endemic in Mexico and characterized by necrotizing ulcerating lesions, has so far not been described in African leprosy patients.

Type 1 reaction commonly occurs in BT patients but may also occur in any of the other borderline states and represents an upgrading of specific immunity against leprosy bacilli toward the TT pole. It is a type IV delayed-type hypersensitivity reaction. It may be the first presentation symptom, although some patients may only experience a type 1 reaction after initiation of treatment, usually within the first 6 months, and rarely, up to 5 years after treatment when it is usually mistaken for a relapse (Saunderson and ILEP Medico Social Commission 2002). In type 1 reaction, already existing lesions, which may have been unknown to the patient, become swollen, reddened, and warm, and there may be associated facial and acral edema. Key features of type 1 reaction are the involvement of nerves with tenderness and/or loss of function and the absence of significant systemic symptoms (Fig. 8).

Type 2 reaction, on the other hand, occurs mainly in LL patients but may occasionally occur in BL patients. The body's humoral immunity reacts to antigenic proteins from killed and decomposing leprosy bacilli, leading to the deposition of immune complexes in tissues. Most type 2 reactions will therefore occur within the first 3 years after initiation of treatment but in some cases may continue to recur even much longer after cessation of treatment (Saunderson and ILEP Medico Social

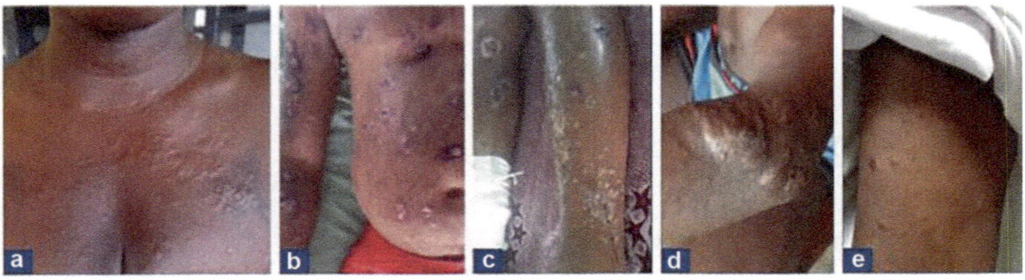

Fig. 8 Reactions in leprosy. (**a**) Tumid erythematous lesions in type 1 reaction (**b**) and (**c**) pustulating and ulcerating erythema nodusum leprosum (ENL) lesions. (**d**) Type 1 lesions complicated by bacterial skin infection with crusting. (**e**) Typical ENL lesions (Pictures Source: Benedict Okoe Quao)

Commission 2002). In ENL, patients develop painful and reddened subcutaneous nodules mainly on the limbs unrelated to pre-existing leprosy skin lesions (Fig. 8). Unlike type 1, type 2 reactions are characterized by significant systemic symptoms such as fever, general malaise, muscle and joint aches, and lymphadenopathy. Type 2 reactions may also involve other organs, such as the eyes (iritis), testes (orchitis), liver, and kidneys (deranged liver and kidney function tests), leading to permanent damage (Saunderson and ILEP Medico Social Commission 2002). The 10-item ENLIST ENL severity scale, which during its development included African patients, allows clinicians to grade the severity of the complications of ENL for treatment and monitoring purposes (Walker et al. 2017).

Long-term sequelae of the disease, which occur as permanent impairments/deformities or secondary damage to tissues resulting from these impairments/deformities, tend to persist way beyond the completion of leprosy treatment and may be the reason for seeking medical care. These problems include recurrent plantar ulcers, chronic osteomyelitis, and recurrent burn injuries of extremities, as well as poor vision resulting from any or a combination of ocular complications of the disease (Fig. 9).

Assessment of the eyes, hands, and feet, in accordance with the WHO disability grading system, allows some standardized categorization of the severity of such impairment/deformities (Brandsma and Brakel 2003; World Health Organization 1988) (Fig. 10), and this can be carried out reliably by both experienced and inexperienced testers, provided adequate training is given (Nienhuis et al. 2004). Based on the WHO International Classification of Functioning, Disability and Health (ICF), "disability" is an umbrella term for impairments, activity limitations, and participation restrictions arising from a disease or condition (World Health Organization 2001b). It has thus been suggested that, in the strict sense, based on the ICF, the WHO disability grading system assesses impairments rather than disability and should probably be renamed the "WHO Impairments Grading System" (Nienhuis et al. 2004).

In practice, the maximum impairment grade when the WHO disability grading system is applied—referring to the highest score for all six sites—is used as an indicator of the severity of impairment for the patient. It is a proxy measure for early case detection/reporting at diagnosis, with the presence of grade 2 disability suggesting late diagnosis.

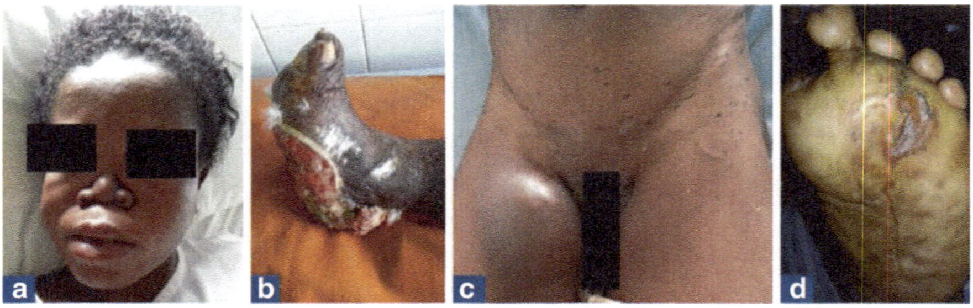

Fig. 9 (**a**) Saddle-nose deformity. (**b**) Squamous cell carcinoma complicating chronic foot ulcer. (**c**) Same patients with SCC in B with inguinal lymph node involvement. (**d**) Plantar ulcer with absent big toe *(Pictures Source: Benedict Okoe Quao)*

HANDS AND FEET	
Grade 0	No anaesthesia, no visible deformity or damage
Grade 1	Anaesthesia present, but no visible deformity or damage
Grade 2	Visible deformity or damage present
EYES	
Grade 0	No eye problem due to leprosy; no evidence of visual loss
Grade 1	Eye problems* due to leprosy present, but vision not severely affected as a result (vision: 6/60 or better; can count fingers at 6 metres)
Grade 2	Severe visual impairment (vision worse than 6/60; inability to count fingers at 6 metres); also includes lagophthalmos, iridocyclitis and corneal opacities

Fig. 10 WHO Disability Grading System for grading impairments/deformities in leprosy (Brandsma and Brakel 2003)

The individual impairment scores can also be summed up to compute an impairment sum-score known as the "EHF" score, which is used as a change indicator of impairments for the patient on treatment or being followed up after treatment.

The "Screening Activity Limitation & Safety Awareness" (SALSA) Scale, consisting of 20 items of daily activities related to three domains—mobility, self-care, and work—is also another assessment scale that can be used to assess leprosy-associated impairments, but measures activity limitations, not impairments or participation restrictions (SALSA Collaborative Study Group et al. 2007).

Diagnosis

The diagnosis of leprosy requires not only detecting the presence of the disease but also being able to accurately classify leprosy for treatment since different treatment regimens exist depending on the form of leprosy present. Improper classification may lead to overtreatment, more damage to control efforts, or undertreatment of the infection. Leprosy infection can be demonstrated and classified using different modalities, namely, clinical, bacteriological, immunological, and histopathological methods. Diagnosis for treatment and control activities is nonetheless mainly clinical. In low-resource settings, as is often the case in much of sub-Saharan Africa, this approach has proven to be highly beneficial, as it enables low-level frontline healthcare staff to receive training in the clinical diagnosis of leprosy.

The presence of at least one of three cardinal signs of leprosy is enough to make a diagnosis of leprosy (World Health Organization 2018, 2012; Fig. 11). Clearly, diagnosing leprosy in the late stages when the disease is easily recognizable is much easier than in the early stages, when the lesions are now forming. The importance of diagnosing leprosy early, however, cannot be overemphasized; late diagnosis has negative implications with regard to the continued infectivity of untreated patients within the community and the development of advanced disease with permanent deformities that will persist beyond cure by the time of initiation of treatment.

Regarding the individual criteria for diagnosing leprosy, the finding of anesthetic skin patches appears to be the most sensitive, being present in about 70% of all leprosy cases (Levis 2002). Although they generally appear much later in the course of the disease than do anesthetic skin lesions, the finding of enlarged nerves is also very useful but requires the assessor to have knowledge of the proper examination technique. In low-resource settings, teaching frontline healthcare personnel to be able to check for just ulnar and peroneal nerve enlargement will suffice since the vast majority of cases of leprosy-associated nerve enlargement will involve either or both of these nerves (Groenen et al. 2002). The demonstration of acid-fast bacilli in a slit-skin smear, despite the challenges with its sensitivity, will appear to be the most specific, with the presence of AFB almost invariably confirming leprosy (Levis 2002). Taken together, all three cardinal signs have a sensitivity of 97% with a positive predictive value of 98% (Levis 2002).

Considering the complexity of the historically available classification systems for leprosy, there was a need to have a simpler one, which could be utilized by frontline healthcare staff for treatment purposes (World Health Organization 1982). In this regard, a number of "operational classifications" of leprosy have been

Fig. 11 Cardinal signs of leprosy (WHO 2012)

Cardinal Signs of Leprosy

- Hypopigmented or reddish skin patch with definite loss of sensation
- Thickened/enlarged peripheral nerve with loss of sensation +/- muscle weakness *(Nerve thickening without any associated signs and symptoms should not be diagnosed as leprosy)*
- The presence of acid-fast bacilli (AFB) in slit skin smear

postulated over the years, with the current consensus allowing the use of clinical criteria alone, in the absence of reliable facilities for bacteriological examination of slit-skin smears (WHO Study Group on Chemotherapy of Leprosy 1994). Under the current operational classification system (World Health Organization 2012), a patient with more than five skin lesions is deemed to have multibacillary (MB) leprosy, while a patient having one to five skin lesions will be classified as having paucibacillary (PB) leprosy. When bacteriological examination is possible, all smear-positive cases are still to be classified as MB leprosy, irrespective of the number of skin lesions (World Health Organization 1988).

The current WHO classification of leprosy, using only clinical criteria, has a reasonable balance of sensitivity and specificity, approaching 90%, respectively, making it the best available option (Croft et al. 1998; Norman et al. 2004). This makes it a very convenient classification system for the treatment of leprosy in low-resource settings, with the obvious advantage of its simplicity of use. With respect to leprosy control challenges, using the number of skin lesions alone may, however, lead to a few MB patients being wrongly classified as PB leading to under treatment (Croft et al. 1998). This therefore calls for more research into newer diagnostic methods and the strengthening of health system resources, particularly in sub-Saharan Africa, in order to overcome this challenge.

The gold standard for diagnosis remains the demonstration of granulomas and acid-fast bacilli (AFB) in skin and nerve biopsies using the bacterial index in granuloma (BIG), but this is highly invasive and requires a high level of resources not readily available in many care settings (Bhushan et al. 2008). Demonstration of acid-fast bacilli in slit-skin smears (SSS) stained with the Ziehl–Neelsen method is a good proxy to biopsy, although not as sensitive since skin smears do not reach the deep layers of the dermis where acid-fast bacilli in leprosy granulomas are confined, separated from the epidermis by a subepidermal clear zone (Bhushan et al. 2008; Ridley and Jopling 1966). Furthermore, the sensitivity of bacteriological examination of slit-skin smears for acid-fast bacilli, found to range from as low as 10–50% (Levis 2002), is greatly limited by the skill level of the laboratory personnel and is another major reason why slit-skin smears are no longer routinely required to diagnose and treat or declare cure of leprosy (Fig. 12).

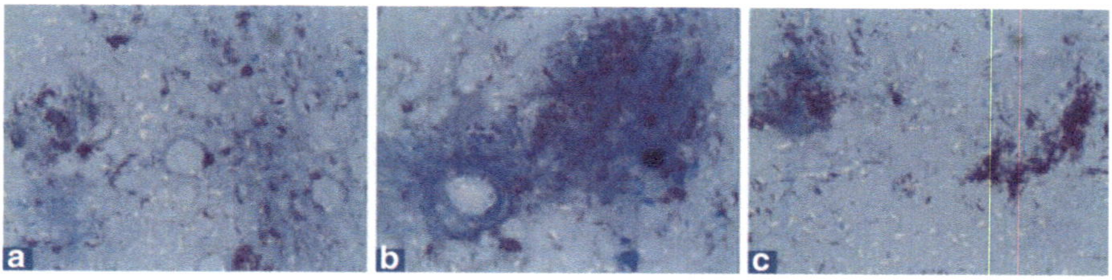

Fig. 12 Light micrographs of slit-skin smear from a patient with high bacillary load showing solid-staining rods; (**a**) and (**c**) AFB occur singly or in clumps. (**b**) AFB could be clumped together in large globi containing >100 bacilli (Credit: Smear by Mr. Arhin, Ankaful Leprosy General Hospital, Cape-Coast; Light micrograph by Dr. K. Akapko, Pathology Department, University of Cape-Coast School of Medical Sciences)

The lepromin test is an immunologically based test modality that entails the intradermal injection of inactivated leprosy bacilli and the examination of the injection site on day 3 and/or day 28 after injection. Although not routinely used in the diagnosis of leprosy, the test results do offer a clue as to the type of leprosy the individual has. The lepromin test positivity is dependent on the strength of the cell-mediated immunity to leprosy and varies from strongly positive with TT leprosy, across the whole clinical spectrum, to virtually absent with LL leprosy. There are two reactions that indicate test positivity: the so-called early reaction occurring within 2 days of intradermal injection, marked by erythema with edema and thickening of the skin, and the classic "Mitsuda" reaction occurring 3–4 weeks after and characterized by the formation of a nodule, sometimes with ulceration (Dharmendra and Lowe 2012).

The publication of the *M. leprae* Genome in 2001 (Cole et al. 2001) and the subsequent discovery of *M. lepromatosis* as a distinct species in 2008 (Han et al. 2012) are two notable achievements that have paved the way for more advanced molecular research to understand, describe, and develop diagnostics and treatments targeted at the two causative organisms. Consequently, tests to detect *M. leprae* and *M. lepromatosis* nucleic acids based on the polymerase chain reaction (PCR) have emerged as a promising diagnostic modality. Despite the high level of technology required, two recent successes in the development of new molecular diagnostic assays—first, the development of a qPCR assay based on the highly specific *M. leprae* repetitive element RLEP (Beissner et al. 2019; Braet et al. 2018) and second, the development of a molecular diagnostic assay for *M. lepromatosis* following the identification and validation of its unique repetitive element RLPM (Sharma et al. 2020)—means it may now be easier to detect and differentiate both causative organisms in clinical and surveillance settings, beyond the current application mainly in research.

Since the 1980s, when Brennan and Barrow described leprosy-specific phenolic glycolipid I (PGL-I) anti-bodies (Brennan and Barrow 1980), there has been a lot of research and development in this area to produce a rapid serological test that will be able to diagnose early/subclinical leprosy infection. A rapid diagnostic test for leprosy, suitable for use in impoverished and remote areas, as found in many parts of sub-Saharan Africa, would offer clear advantages. It could facilitate early intervention, breaking the transmission chain, and preventing deformities associated with late diagnosis of leprosy. This has led to the development of promising serological tests and lateral flow assays to detect biomarkers that could be either antibodies such as PGL-I and NDOLID (Leturiondo et al. 2019) or chemokines and chemokines like interleukin 10 (IL-10) and interferon (IFN)-γ induced protein 10 (IP-10) (Bobosha et al. 2014).

Toward the LL end of the disease spectrum, there is a good humeral response to leprosy, with multibacillary leprosy patients tending to have higher anti-PGL-I antibody titers, which can be monitored and have been shown to fall progressively after initiation of treatment (Bach et al. 1986; Miller et al. 1987). There is also evidence for the occurrence of higher anti-PGL-I antibody titers in household contacts of leprosy patients compared with apparently unexposed individuals and a higher risk of future disease conferred on the individual by the presence of anti-PGL-I antibodies (Ulrich et al. 1991). Serological tests based on PGL-I are however not sensitive, particularly for paucibacillary cases with low or absent antibodies, and call for the inclusion of multiple recombinant antigens in the development of a reliable future serological screening test (Spencer et al. 2005; Ulrich et al. 1991). Anti-PGL-I antibodies may however still be useful for monitoring epidemiological changes within communities or among contacts, although not specific enough to predict whom among exposed contacts will actually develop clinical disease (Ulrich et al. 1991).

In addition to the described diagnostic modalities, ongoing technological advancements have also opened up new possibilities, such as the employment of artificial intelligence in screening skin lesions likely to be leprosy (De Souza et al. 2021), employment of mobile phone applications by peripheral health workers in low-resource settings as pertains in many parts of sub-Saharan Africa to diagnose and treat persons affected by various skin-related neglected tropical diseases including leprosy (Mieras et al. 2018), and the use of high-resolution ultrasound to detect nerve thickening at a higher sensitivity than clinical palpation (Sreejith et al. 2021).

Despite the numerous advances made in the area of leprosy diagnostics, the recommendation by the Guidance Development Group (GDG) in WHO's first-ever evidence-based guidelines for leprosy published in 2018 is still to base the diagnosis of leprosy on clinical examination with or without slit-skin smears or pathological examination of biopsies (World Health Organization 2018). Additionally, the GDG could not also recommend any test to diagnose latent leprosy among asymptomatic contacts based on available evidence (World Health Organization 2018). Nonetheless, the Global Partnership for Zero Leprosy (GPZL) has successfully coordinated the co-creation of two target product profiles (TPP) for the desired leprosy diagnostic, which, when accepted by the WHO, will pave the way for the development of the ideal diagnostic tests to confirm the diagnosis of leprosy in individuals with clinical signs and symptoms, as well as detect *Mycobacterium leprae* infection among asymptomatic household or familial contacts of leprosy patients for treatment with appropriate prophylactic interventions (Global Partnership for Zero Leprosy 2021).

Control Tools and Strategies

Since leprosy is principally a disease of man in many parts of the world, early diagnosis and treatment of infected individuals is important to limit the spread and burden of the disease. While the provision of effective multidrug therapy (MDT) to treat identified leprosy remains a critical component of leprosy control, it is now clear more than two decades after the achievement of the global target for "elimination of leprosy as a public health problem" that passive case detection and treatment with MDT alone will be insufficient to interrupt leprosy transmission (World Health Organization 2021a). The new WHO "Global Leprosy Strategy 2021–2030" thus focuses on interrupting leprosy transmission to achieve zero autochthonous cases in contrast to previous strategies, which focused on eliminating leprosy as a public health problem (World Health Organization 2021a, 2016; World Health Organization, Regional Office for South-East Asia 2009).

The "Global leprosy strategy (GLS) 2021–2030" has been developed as one of the disease-specific strategies underpinning the WHO Roadmap for Neglected Tropical Diseases 2021–2030 and is aligned with the Sustainable Development Goals (SDGs), especially SDG 3, SDG 10, and SDG 17, which respectively deal with healthy lives and well-being, including universal health coverage (UHC), reduced inequalities, and partnerships (World Health Organization 2021a, 2020c). It is also broader in scope than previous strategies with its vision of moving toward zero leprosy, not limited to achieving and verifying elimination of transmission post-elimination surveillance, but also fully incorporating strategies to achieve zero disability and zero discrimination and stigma (World Health Organization 2021a).

Ensuring the sustainability of leprosy services necessitates their integration into the general healthcare systems of endemic countries. This integration is crucial not only for accessibility but also to eliminate stigma and ensure that individuals affected by leprosy can readily access the necessary services. Much effort is also needed to improve relatively weaker healthcare systems within the sub-Saharan African region if leprosy services are to be effectively delivered through UHC, with strengthened linkages with other public sectors such as "water, sanitation and hygiene" (WASH) and social protection. In line with this, GLS 2021–2030 represents a shift toward multi-disease integration, digitalization, and accountability while addressing some of the key issues such as human resource capacity, surveillance, and antimicrobial resistance (World Health Organization 2021a).

Compared with the previous global leprosy strategy, GLS 2021–2030 is hinged on four instead of three pillars, with a totally new pillar added seeking to scale up leprosy prevention alongside integrated active case detection (World Health Organization 2021a, 2016) (Fig. 13). While each of the four pillars is composed of different sub-pillars with several indicators, the GLS 2021–2030 headline targets for 2030 in line with the new shift of focus are as follows: 120 countries achieving

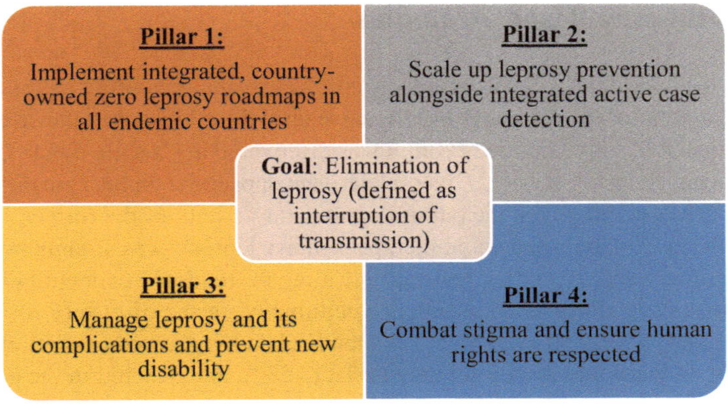

Fig. 13 Pillars of GLS 2021–2030 (Produced by Benedict Quao—Source GLS 2021–2030)

zero new autochthonous cases; 70% reduction in annual number of new cases detected with respect to 2019 figures; 90% reduction in rate per million population of new cases with G2D; and 90% reduction in rate per million children of new child cases with leprosy (World Health Organization 2021a).

While case detection has become largely passive over the years following the achievement of the global target for leprosy elimination as a public health problem, screening of household contacts remains a viable active screening method since this subset of the population has a proven higher risk of developing leprosy (Fine et al. 1997; Jesudasan et al. 1984). With established evidence now pointing to nearly a 60% reduction in future leprosy risk following prophylactic treatment with single-dose rifampicin (SDR), the newly recommended strategy to give SDR to household, neighbor, and social contacts will help improve contact tracing and screening (World Health Organization 2018; World Health Organization, Regional Office for South-East Asia 2020). The new strategy also involves undertaking integrated active case finding in targeted populations, such as hyperendemic hotspots and hard-to-reach areas, such as those coming out of conflicts, to actively identify and treat leprosy cases as well as other skin NTDs.

Regarding zero disability, in addition to encouraging early case detection before the development of deformities and disability, the current strategy continues to stress on community-based rehabilitation services alongside the intensification of self-care programs and strengthening of referral systems within an integrated health system to manage acute complications such as reactions. The incorporation of the United Nations *Principles and Guidelines for the Elimination of Discrimination against Persons Affected by Leprosy and their Family Members* (Human Rights Council 2010; UN General Assembly 2010) into national policy frameworks is a major component of the strategies directed at combating stigma and ensuring human rights are respected (World Health Organization 2021a). This is supplemented by continuing to push for greater inclusion of persons affected by leprosy in helping to frame policy and to actively participate in efforts to reduce stigma and improve the living standards of sufferers.

Chemotherapy-Based Strategies

Despite the historical use of a variety of remedies to attempt curing leprosy, a promising treatment for the disease proved evasive until the 1940s when the effectiveness of the injectable sulphonamide Promin was discovered at Carville by Dr. Guy Faget and his team (Faget et al. 1942). Its oral derivate dapsone soon became the mainstay of leprosy treatment, given the painful nature and toxicity of the former, but by the 1970s, the use of dapsone monotherapy against leprosy was fraught with widespread primary and secondary resistance leading to frequent treatment failures. The search for better alternatives eventually culminated in the discovery of multidrug therapy (MDT) for leprosy, which was then backed by a recommendation from the WHO as standard therapy for leprosy (WHO 1982), and ushered in the era of control programs for leprosy. The marked reduction in case numbers observed during the initial experiences with MDT largely contributed to the belief that leprosy could now finally be eliminated, contributing to the setting of the target for eliminating leprosy as a public health problem in 1991.

Since then, chemotherapy for leprosy has primarily comprised MDT as well as medications used for the treatment of leprosy reactions, which may occur anytime during treatment or even after. MDT utilizes three effective first-line drugs, all possessing some bactericidal properties with different mechanisms of action, given over a fixed duration with the primary goal of achieving a cure with little or no risk of relapse. This same concept underlines the use of combination therapy in the treatment of other conditions, such as tuberculosis and human immunodeficiency virus (HIV).

The three standard first-line drugs are rifampicin, clofazimine, and dapsone, with rifampicin being the pillar of the various MDT regimens owing to its far greater bactericidal activity against leprosy bacilli. Six other drugs, namely, ofloxacin, clarithromycin, minocycline, moxifloxacin, rifapentine, and bedaquiline (formerly TMC 207), have been shown to have significant bactericidal effects against leprosy bacilli and are at different stages of clinical development. Some are already being employed in second-line combinations when the use of one of the standard drugs is not possible (WHO 2012, 2018; WHO-SEARO 2012b). Though MDT regimens depending on whether multibacillary (MB) or paucibacillary (PB) leprosy is being treated have historically differed in the number of constituent drugs and duration of treatment, recent WHO guidance recommends using the same set of three drugs over the same treatment durations for both types of leprosy as before— i.e., 6 months for PB and 12 months for MB (WHO 2018) (Table 2).

Though 6 months of treatment with rifampicin and dapsone has been shown to be effective for paucibacillary leprosy with no need for extension when appropriately given (WHO Study Group on Chemotherapy of Leprosy 1994), current evidence suggests a three-drug regimen over 6 months may be superior (WHO 2018). For individuals with single-lesion PB leprosy, it is feasible to treat them with a single-dose regimen consisting of rifampicin, ofloxacin, and minocycline (WHO 1998b). Therapy for MB leprosy involves the addition of clofazimine to rifampicin and dapsone, with duration of treatment being much longer. WHO currently considers 12 months of treatment for MB leprosy as sufficient (WHO 2018, 1998a), although some countries still treat MB leprosy for up to 24 months as per the original guidelines when MDT was first introduced.

Once started on MDT, a leprosy patient is generally regarded as non-infectious since a single dose of rifampicin drastically reduces the patient's viable bacilli load to such a level as to render him/her so (World Health Organization 2009). Achieving cure, however, requires completion of the treatment course within the stipulated time. While leprosy patients may have obvious economic and physical challenges commuting to health facilities at a distance for their monthly supervised doses, this enables care providers to trace patients who start defaulting early to ensure compliance with treatment. Generally, underdeveloped sub-Saharan communities with a lack of proper address systems and inadequate transport facilities present a major challenge in this regard. Nonetheless, patients' compliance can be maintained by adequately informing them about the course of the disease and the implication of

Table 2 Standard MDT for leprosy

Weight/Age	Drug	Monthly supervised dose	Daily unsupervised dose
Paucibacillary leprosy[a]			
Duration: 6 months (completed in a maximum of 9 months)			
Adult (50–70 kg)	Rifampicin	600 mg	–
	Dapsone	100 mg	100 mg
Child (10–14 years)	Rifampicin	450	–
	Dapsone	50 mg	50 mg
Child <10 years	Rifampicin	10 mg/kg body weight	–
	Dapsone	2 mg/kg body weight	2 mg/kg body weight
Multibacillary leprosy			
Duration: 12 months (completed in a maximum of 18 months)			
Adult (50–70 kg)	Rifampicin	600 mg	–
	Clofazimine	300 mg	50 mg
	Dapsone	100 mg	100 mg
Child (10–14 years)	Rifampicin	450	–
	Clofazimine	150 mg	50 mg every other day
	Dapsone	50 mg	50 mg
Child <10 years	Rifampicin	10 mg/kg body weight	–
	Clofazimine	1 mg/kg alt days, depending on dosage	1 mg/kg alt days, depending on dosage
	Dapsone	2 mg/kg body weight	2 mg/kg body weight
Single lesion paucibacillary leprosy			
Duration: single dose			
Weight/age	Treatment		
Adult (50–70 kg)	Rifampicin (600 mg) + ofloxacin (400 mg) + minocycline (100 mg)		
Child (5–14 years)	Rifampicin (300 mg) + ofloxacin (200 mg) + minocycline (50 mg)		

Not recommended for pregnant women and children < 5 years
WHO Expert Committee on Leprosy (2012), World Health Organization (1998a, b)
[a]Current WHO recommendations prefer a three-drug regimen over a two-drug regimen for treating PB leprosy (WHO 2018)

incomplete treatment and ensuring that they can always receive MDT at health facilities within their localities.

Ensuring compliance with MDT is a crucial factor in the quality of case management since completing the whole fixed-course duration is required to cure the disease and hence reduce the potentially infective caseload in the community. It has been difficult to track the quality of case management within the subregion over the years since the MDT completion or cure rate is not regularly reported by individual countries, resulting in the use of the prevalence/detection ratio (P/D ratio) as a proxy (World Health Organization, Regional Office for Africa 2013, 2012). This is not ideal because prevalence and detection rates are influenced by other factors. Beyond strengthening the collection of data on MDT completion rates, the number of relapses and retreatments other than relapse will need to be monitored under the current strategy to maintain a high quality of leprosy control services with regard to treatment (World Health Organization 2021a).

The WHO continues to play a leading role when it comes to managing the currently available drugs for treating leprosy. Since 1995, the WHO has supplied MDT to endemic countries for free thanks to the benevolence of institutions like the Nippon Foundation and Novartis, who have committed to continue supplying these drugs for the foreseeable future. With rifampicin critical to the current MDT regimens, it is necessary to continue surveillance for rifampicin resistance even though current levels are insignificant. In this regard, the WHO surveillance network actively collects slit-skin samples from MB patients from all over the world to test for drug-resistance mutations using the PCR technique (World Health Organization, Regional Office for South-East Asia 2012b).

The mainstay of treatment for reactions and isolated neuritis is steroids (prednisolone), although clofazimine may also be used in cases of severe erythema nodusum leprosum (ENL). Thalidomide is also useful in severe or recurrent ENL but may not be readily available owing to restrictions arising from its immense teratogenic potential. Other useful adjuncts in the management of reactions include rest and analgesics such as acetylsalicylate and paracetamol, which may all but suffice in very mild cases of lepra reactions. It is important during reactions not to discontinue MDT if the patient is still undergoing treatment for leprosy.

Among the NTDs, leprosy has traditionally been grouped with the so-called case-management NTDs, requiring innovative and intensified disease management (IDDM). Since the discovery of effective treatment for leprosy, several attempts have also been made to discover an effective preventive therapy that will further accelerate efforts at eliminating the disease, using various components of MDT—i.e., dapsone (and another derivative acedapsone), single dose rifampicin, and the combination of rifampicin, ofloxacin, and minocycline (ROM) (World Health Organization, Regional Office for South-East Asia 2020). The COLEP trial, a double-blind, randomized control trial (RCT) undertaken in Bangladesh from 2002 to 2007 with single-dose rifampicin, was, however, the first study to show a significant protective effect of any preventive therapy (Moet et al. 2004; Richardus and Oskam 2015). In this COLEP trial, single-dose rifampicin given to contacts of leprosy patients resulted in a 57% reduction of future leprosy risk at 2 years compared to placebo, with the protective effect enhanced to 80% when there was a history of prior Bacille-Calmette-Guérin (BCG) vaccination for tuberculosis (Moet et al. 2008).

The first evidence-based WHO guidelines for leprosy, released in 2018, recommended that countries extend screening and treatment beyond the traditional approach of screening only household contacts. The guidelines advise screening and administering SDR to neighbors and social contacts as well, aiming to maximize the protective effect. Subsequent feasibility studies carried out within the context of the multi-country "Leprosy Post-Exposure Programme" (LPEP) study have also demonstrated that giving SDR within routine program conditions to contacts is both feasible and acceptable and results in a sustained reduction in leprosy incidence after an initial increase in case detection observed in the first year of implementing such a program (Blok et al. 2021; Richardus et al. 2021). Furthermore, the availability of such a prophylactic intervention appears to re-invigorate local leprosy control services in general, beyond the positive impact of potentially reducing community leprosy transmission (Richardus et al. 2021).

Challenges in Program Implementation

The remarkable success achieved in attaining the target for leprosy elimination as a public health problem has come at a price. In the myriad of worldwide health problems, leprosy control activities now attract less financial support and attention coupled with relative complacency on the part of national health bodies. It is hoped that with the current move toward incorporating the organization of leprosy control with other neglected tropical diseases, the much-needed visibility for leprosy control activities will be attained.

Although the push for integration enables the sharing of certain resources with leprosy control services, it also results in a decline in attention and resources allocated to dedicated leprosy services. In alignment with Pillar 1 of the GLS 2021–2030, crucial components have been identified. These include establishing political commitment with sufficient resources for leprosy within an integrated framework. Additionally, the strategy emphasizes the formation of national partnerships for zero leprosy and the development of leprosy roadmaps that actively engage all relevant stakeholders (World Health Organization 2021a). It also calls for strengthening of the primary healthcare systems, particularly within leprosy-endemic countries, so that leprosy sufferers all over sub-Saharan Africa can enjoy equitable, high-quality care wherever they may be.

It is also ironic that as we push toward the integration of leprosy services within the general healthcare systems, the knowledge of frontline healthcare workers regarding leprosy appears to be on the decline since fewer cases of leprosy are now being seen. This has not been helped by the loss of expertise and knowledge to retirement, death, and transfer of highly skilled leprosy health workers to other areas of need in the health sector. With staff capacity identified as the severest obstacle to achieving zero leprosy, it is crucial that countries build and sustain the capacity of integrated public health and clinical services staff in all aspects of leprosy prevention and care, as well as program management and laboratory capacity for microbiological analysis including slit-skin smears (World Health Organization 2021a).

The leprosy stigma also continues to be a barrier, even among frontline healthcare workers who rather prefer to engage in more lucrative activities of other health programs. In Ghana, for instance, the use of innovative "information, education and communication" (IEC) techniques are being utilized to help overcome some of these challenges that will otherwise lead to late leprosy diagnosis and consequently increase the negative public health impact of the disease. Illustrative posters highlighting the early stages of leprosy have been designed for use by frontline healthcare workers, while public education is carried out with the aid of picture banners at marketplaces and other public gatherings. In Ethiopia, as an example of what may be required to overcome such stigma, researchers went a step further to develop positive images of persons affected by leprosy, choosing to move away from showing the extremely negative aspects of the disease to depicting the lives of patients beyond the disease (Kumar et al. 2019).

Further Research for Policy and Control

Much is still not known about the exact mode of transmission of leprosy, and there remains a general paucity of knowledge regarding the recently discovered second causative agent *Mycobacterium lepromatosis*, despite recent advances made with molecular work on the new species. There is a need for more molecular studies to characterize both agents further and to investigate possible non-human sources of leprosy infection that can negatively impact control efforts. To curtail transmission and development of advanced disease, and with work advancing on finalizing TPPs for the ideal leprosy diagnostic, more research on improved diagnostic modalities toward product development is required. Finding the ideal diagnostics will help peripheral healthcare workers diagnose leprosy earlier within integrated health systems and assist in better targeting leprosy preventive measures for those with subclinical infection and more likely to develop future leprosy.

With the relatively recent breakthrough in leprosy prevention therapy using SDR through the COLEP trial, followed by the LPEP study establishing its feasibility, the translation of this recommendation now into widespread programmatic application should incorporate much research and operational learning to optimize the intervention for maximal impact on curbing leprosy transmission. More research will be required to determine other modalities that can be added to the existing SDR intervention to further increase the reduction in risk of future leprosy among contacts when it is applied. Beyond the use of prophylactic drugs as leprosy prevention, Pillar 2 of GLS 2021–2030 also calls for action on the development of existing and potential new vaccines, with the current leading candidate having been shown to be capable of dually protecting against the development of leprosy and mitigating neurological complications of leprosy (Duthie et al. 2011; Global Biodefense Staff 2018).

The clinical management of leprosy has also not changed much over the years. Drug trials involving some of the newer drugs and the formulation of other short-course combination therapies will help improve treatment quality and outcomes. More trials are also needed in the area of prevention and novel treatment of reactions to help drastically reduce the disability burden attributable to leprosy. More research is also required to improve the current operational and epidemiological tools utilized in leprosy control to help make further progress in reducing leprosy transmission as the world moves toward zero leprosy.

Conclusion and Outlook for the Next Decade

Although incidence rates of leprosy cases have continued to slowly decrease over the years, it has become clear that our efforts need to double up to make real progress toward Zero Leprosy in all its forms, i.e., zero disease (or transmission), zero disability, and zero discrimination and stigma. There is a great and urgent need for more research work to fill persisting knowledge gaps, to refine existing tools and develop new tools and interventions, and to support and guide operational actions. The momentum garnered as we approached the end of the second decade following the attainment of the global target for the elimination of leprosy as a public health problem bodes well for collective action going forward into the next decade, despite the apparent interruption due to the COVID-19 pandemic.

The world does not yet possess the magic bullet against this ancient scourge, but the new tools and advancements in recent years appear to be paving the way for bigger discoveries that will potentially drive down case numbers faster and improve the quality of care and overall patient experience in the future. The progressive use of new-age digital technology to improve surveillance for leprosy will be a key game-changer for case-detection and curbing transmission, and it is hoped that in the next decade, the current leading vaccine candidate with both therapeutic and disease prevention potential will make it to production for clinical use.

Looking ahead to the next decade, in addition to anticipated new discoveries, two noteworthy recent developments include the establishment of the Global Partnership for Zero Leprosy (GPZL), which unites stakeholders within the leprosy community for more effective collective action, and the commitments made by countries and partners through the Kigali Declaration—as the successor to the London Declaration on NTDs—which hold the promise of unlocking global potential to eradicate leprosy and other NTDs. These initiatives are crucial in sustaining country ownership, promoting integration, and fostering cross-sectoral collaboration as vital components of leprosy control in the coming years.

References

Avanzi C, del-Pozo J, Benjak A, Stevenson K, Simpson VR, Busso P, McLuckie J, Loiseau C, Lawton C, Schoening J, Shaw DJ, Piton J, Vera-Cabrera L, Velarde-Felix JS, McDermott F, Gordon SV, Cole ST, Meredith AL (2016) Red squirrels in the British Isles are infected with leprosy bacilli. Science 354:744–747. https://doi.org/10.1126/science.aah3783

Bach MA, Wallach D, Flageul B, Hoffenbach A, Cottenot F (1986) Antibodies to phenolic glycolipid-1 and to whole mycobacterium leprae in leprosy patients: evolution during therapy. Int J Lepr Mycobact Dis Off Organ Int Lepr Assoc 54:256–267

Barua S (2013) Global leprosy situation and new challenges and priorities with reference to Bangkok Summit

Beegle K, Christiaensen L (2019) Accelerating poverty reduction in Africa. World Bank, Washington, DC. https://doi.org/10.1596/978-1-4648-1232-3

Beissner M, Woestemeier A, Saar M, Badziklou K, Maman I, Amedifou C, Wagner M, Wiedemann FX, Amekuse K, Kobara B, Herbinger K-H, Kere AB, Löscher T, Bretzel G (2019) Development of a combined RLEP/16S rRNA (RT) qPCR assay for the detection of viable M. Leprae from nasal swab samples. BMC Infect Dis 19:753. https://doi.org/10.1186/s12879-019-4349-9

Bhushan P, Sardana K, Koranne RV, Choudhary M, Manjul P (2008) Diagnosing multibacillary leprosy: a comparative evaluation of diagnostic accuracy of slit-skin smear, bacterial index of granuloma and WHO operational classification. Indian J Dermatol Venereol Leprol 74:322–326. https://doi.org/10.4103/0378-6323.42892

Blok DJ, Steinmann P, Tiwari A, Barth-Jaeggi T, Arif MA, Banstola NL, Baskota R, Blaney D, Bonenberger M, Budiawan T, Cavaliero A, Gani Z, Greter H, Ignotti E, Kamara DV, Kasang C, Manglani PR, Mieras L, Njako BF, Pakasi T, Saha UR, Saunderson P, Smith WCS, Stäheli R, Suriyarachchi ND, Tin Maung A, Shwe T, van Berkel J, van Brakel WH, Vander Plaetse B, Virmond M, Wijesinghe MSD, Aerts A, Richardus JH (2021) The long-term impact of the leprosy Post-exposure prophylaxis (LPEP) program on leprosy incidence: a modelling study. PLoS Negl Trop Dis 15:e0009279. https://doi.org/10.1371/journal.pntd.0009279

Bobosha K, Tjon Kon Fat EM, van den Eeden SJF, Bekele Y, van der Ploeg-van Schip JJ, de Dood CJ, Dijkman K, Franken KLMC, Wilson L, Aseffa A, Spencer JS, Ottenhoff THM, Corstjens PLAM, Geluk A (2014) Field-evaluation of a new lateral flow assay for detection of cellular and humoral immunity against mycobacterium leprae. PLoS Negl Trop Dis 8:e2845. https://doi.org/10.1371/journal.pntd.0002845

Braet S, Vandelannoote K, Meehan CJ, Brum Fontes AN, Hasker E, Rosa PS, Lucena-Silva N, Rigouts L, Suffys PN, De Jong BC (2018) The repetitive element RLEP is a highly specific target for detection of mycobacterium leprae. J Clin Microbiol 56:e01924–e01917. https://doi.org/10.1128/JCM.01924-17

Brandsma JW, Brakel WHV (2003) WHO disability grading: operational definitions. Lepr Rev:366–373

Brennan PJ, Barrow WW (1980) Evidence for species-specific lipid antigens in mycobacterium leprae. Int J Lepr Mycobact Dis Off Organ Int Lepr Assoc 48:382–387

Britton WJ, Lockwood DN (2004) Leprosy. Lancet 363:1209–1219. https://doi.org/10.1016/S0140-6736(04)15952-7

Cavaliero A, Quao B (2020) GPZL working group 1 assesses challenges and a path forward for leprosy during COVID-19. Lepr Rev 91:421–424. https://doi.org/10.47276/lr.91.4.421

Chehl S, Job CK, Hastings RC (1985) Transmission of leprosy in nude mice. Am J Trop Med Hyg 34:1161–1166. https://doi.org/10.4269/ajtmh.1985.34.1161

Cole ST, Eiglmeier K, Parkhill J, James KD, Thomson NR, Wheeler PR, Honoré N, Garnier T, Churcher C, Harris D, Mungall K, Basham D, Brown D, Chillingworth T, Connor R, Davies RM, Devlin K, Duthoy S, Feltwell T, Fraser A, Hamlin N, Holroyd S, Hornsby T, Jagels K, Lacroix C, Maclean J, Moule S, Murphy L, Oliver K, Quail MA, Rajandream MA, Rutherford KM, Rutter S, Seeger K, Simon S, Simmonds M, Skelton J, Squares R, Squares S, Stevens K, Taylor K, Whitehead S, Woodward JR, Barrell BG (2001) Massive gene decay in the leprosy bacillus. Nature 409:1007–1011. https://doi.org/10.1038/35059006

Croft RP, Smith WCS, Nicholls P, Richardus JH (1998) Sensitivity and specificity of methods of classification of leprosy without use of skin-smear examination. Int J Lepr Other Mycobact Dis 66:445–450

de Messias-Reason I, Kremsner PG, Kun JFJ (2009) Functional haplotypes that produce Normal Ficolin-2 levels protect against clinical leprosy. J Infect Dis 199:801–804. https://doi.org/10.1086/597070

De Souza MLM, Lopes GA, Branco AC, Fairley JK, Fraga LADO (2021) Leprosy screening based on artificial intelligence: development of a cross-platform app. JMIR Mhealth Uhealth 9:e23718. https://doi.org/10.2196/23718

Deps P, Collin SM (2021) Mycobacterium lepromatosis as a second agent of Hansen's disease. Front Microbiol:12

Desikan KV, Sreevatsa N (1995) Extended studies on the viability of mycobacterium leprae outside the human body. Lepr Rev 66:287–295

Dharmendra, Lowe J (2012) The immunological skin tests in leprosy (part-II): the isolated protein antigen in relation to the classical mitsuda reaction and the early reaction to lepromin. Indian J Med Res 136:502

Duthie MS, Gillis TP, Reed SG (2011) Advances and hurdles on the way toward a leprosy vaccine. Hum Vaccin 7:13

Faget GH, Johansen FA, Ross H (1942) Sulfanilamide in the treatment of leprosy. Public Health Rep 57:1892. https://doi.org/10.2307/4584304

Fine PE, Sterne JA, Pönnighaus JM, Bliss L, Saui J, Chihana A, Munthali M, Warndorff DK (1997) Household and dwelling contact as risk factors for leprosy in northern Malawi. Am J Epidemiol 146:91–102. https://doi.org/10.1093/oxfordjournals.aje.a009195

Global Biodefense Staff (2018) Promising leprosy vaccine candidate. Glob. Biodefense. https://globalbiodefense.com/2018/06/01/promising-leprosy-vaccine-candidate/. Accessed 12 Dec 22

Global Partnership for Zero Leprosy (2021) Leprosy Diagnostic Working Group develops 2 target product profiles - Global Partnership for Zero Leprosy [WWW Document]. https://zeroleprosy.org/leprosy-diagnostic-working-group-develops-2-target-product-profiles/. Accessed 12 Mar 22

Groenen G, Saunderson PR, ILEP Medico Social Commission, International Federation of Anti-Leprosy Associations (2002) ILEP learning guide one: how to diagnose and treat leprosy, reprint ed. ILEP Learning Guide. International Federation of Anti-Leprosy Associations, London

Han XY, Silva FJ (2014) On the age of leprosy. PLoS Negl Trop Dis 8:e2544. https://doi.org/10.1371/journal.pntd.0002544

Han XY, Seo Y-H, Sizer KC, Schoberle T, May GS, Spencer JS, Li W, Nair RG (2008) A new mycobacterium species causing diffuse lepromatous leprosy. Am J Clin Pathol 130:856–864. https://doi.org/10.1309/AJCPP72FJZZRRVMM

Han XY, Sizer KC, Velarde-Félix JS, Frias-Castro LO, Vargas-Ocampo F (2012) The leprosy agents mycobacterium lepromatosis and mycobacterium leprae in Mexico. Int J Dermatol 51:952–959. https://doi.org/10.1111/j.1365-4632.2011.05414.x

Hatta M, van Beers SM, Madjid B, Djumadi A, de Wit MY, Klatser PR (1995) Distribution and persistence of mycobacterium leprae nasal carriage among a population in which leprosy is endemic in Indonesia. Trans R Soc Trop Med Hyg 89:381–385. https://doi.org/10.1016/0035-9203(95)90018-7

Human Rights Council (2010) Draft set of principles and guidelines for the elimination of discrimination against persons affected by leprosy and their family members* (No. A/HRC/15/30). United Nations General Assembly

Jay V (2000) The legacy of Armauer Hansen. Arch Pathol Lab Med 124:496–497. https://doi.org/10.5858/2000-124-0496-TLOAH

Jesudasan K, Bradley D, Smith PG, Christian M (1984) Incidence rates of leprosy among household contacts of "primary cases.". Indian J Lepr 56:600–614

Job CK, Chehl SK, Hastings RC (1994) Transmission of leprosy in nude mice through thorn pricks. Int J Lepr Mycobact Dis Off Organ Int Lepr Assoc 62:395–398

Job CK, Jayakumar J, Kearney M, Gillis TP (2008) Transmission of leprosy: a study of skin and nasal secretions of household contacts of leprosy patients using PCR. Am J Trop Med Hyg 78:518–521

Kumar A, Lambert S, Lockwood DNJ (2019) Picturing health: a new face for leprosy. Lancet 393:629–638. https://doi.org/10.1016/S0140-6736(19)30158-8

Lastória JC, de Abreu MAMM (2014) Leprosy: review of the epidemiological, clinical, and etiopathogenic aspects - part 1. An Bras Dermatol 89:205–218. https://doi.org/10.1590/abd1806-4841.20142450

Lavania M, Turankar RP, Karri S, Chaitanya VS, Sengupta U, Jadhav RS (2013) Cohort study of the seasonal effect on nasal carriage and the presence of mycobacterium leprae in an endemic area in the general population. Clin Microbiol Infect Off Publ Eur Soc Clin Microbiol Infect Dis 19:970–974. https://doi.org/10.1111/1469-0691.12087

Leturiondo AL, Noronha AB, do Nascimento MOO, de Ferreira CO, da Rodrigues FC, Moraes MO, Talhari C (2019) Performance of serological tests PGL1 and NDO-LID in the diagnosis of leprosy in a reference Center in Brazil. BMC Infect Dis 19:22. https://doi.org/10.1186/s12879-018-3653-0

Levis L (2002) The diagnosis and classification of leprosy. Int J Lepr Other Mycobact, Dis, p 70

Malaviya GN (2003) Sensory perception in leprosy-neurophysiological correlates. Int J Lepr Mycobact Dis Off Organ Int Lepr Assoc 71(2):119–124

Malecela MN, Ducker C (2021) A road map for neglected tropical diseases 2021–2030. Trans R Soc Trop Med Hyg 115:121–123. https://doi.org/10.1093/trstmh/trab002

McDermott-Lancaster RD, McDougall AC (1990) Mode of transmission and histology of M. Leprae infection in nude mice. Int J Exp Pathol 71:689–700

Mieras LF, Taal AT, Post EB, Ndeve AGZ, van Hees CLM (2018) The development of a Mobile application to support peripheral health workers to diagnose and treat people with skin diseases in resource-poor settings. Trop Med Infect Dis 3:102. https://doi.org/10.3390/tropicalmed3030102

Miller RA, Gorder DV, Harnisch JP (1987) Antibodies to phenolic glycolipid-I during long-term therapy: serial measurements in individual patients. Int J Lepr Mycobact Dis Off Organ Int Lepr Assoc 55:633–636

Moet FJ, Oskam L, Faber R, Pahan D, Richardus JH (2004) A study on transmission and a trial of chemoprophylaxis in contacts of leprosy patients: design, methodology and recruitment findings of COLEP. Lepr Rev 75:376–388

Moet FJ, Pahan D, Oskam L, Richardus JH (2008) Effectiveness of single dose rifampicin in preventing leprosy in close contacts of patients with newly diagnosed leprosy: cluster randomised controlled trial. BMJ 336:761–764. https://doi.org/10.1136/bmj.39500.885752.BE

Monot M, Honoré N, Garnier T, Araoz R, Coppée J-Y, Lacroix C, Sow S, Spencer JS, Truman RW, Williams DL, Gelber R, Virmond M, Flageul B, Cho S-N, Ji B, Paniz-Mondolfi A, Convit J, Young S, Fine PE, Rasolofo V, Brennan PJ, Cole ST (2005) On the origin of leprosy. Science 308:1040–1042. https://doi.org/10.1126/science/1109759

Nienhuis WA, van Brakel WH, Butlin CR, van der Werf TS (2004) Measuring impairment caused by leprosy: inter-tester reliability of the WHO disability grading system. Lepr Rev 75:221–232

Norman G, Joseph G, Richard J (2004) Validity of the WHO operational classification and value of other clinical signs in the classification of leprosy. Int J Lepr Mycobact Dis Off Organ Int Lepr Assoc 72:278–283. https://doi.org/10.1489/0020-7349(2004)72<278:VOTWOC>2.0.CO;2

Onyeze A (2013) Elimination of neglected tropical diseases in the African Region—an overview

Richardus JH, Habbema JDF (2007) The impact of leprosy control on the transmission of *M. leprae*: is elimination being attained? Lepr Rev 78:330–337

Richardus JH, Oskam L (2015) Protecting people against leprosy: chemoprophylaxis and immunoprophylaxis. Clin Dermatol 33:19–25. https://doi.org/10.1016/j.clindermatol.2014.07.009

Richardus JH, Tiwari A, Barth-Jaeggi T, Arif MA, Banstola NL, Baskota R, Blaney D, Blok DJ, Bonenberger M, Budiawan T, Cavaliero A, Gani Z, Greter H, Ignotti E, Kamara DV, Kasang C, Manglani PR, Mieras L, Njako BF, Pakasi T, Pandey BD, Saunderson P, Singh R, Smith WCS, Stäheli R, Suriyarachchi ND, Tin Maung A, Shwe T, van Berkel J, van Brakel WH, Vander Plaetse B, Virmond M, Wijesinghe MSD, Aerts A, Steinmann P (2021) Leprosy post-exposure prophylaxis with single-dose rifampicin (LPEP): an international feasibility programme. Lancet Glob Health 9:e81–e90. https://doi.org/10.1016/S2214-109X(20)30396-X

Ridley DS, Jopling WH (1966) Classification of leprosy according to immunity. A five-group system. Int J Lepr Mycobact Dis Off Organ Int Lepr Assoc 34:255–273

SALSA Collaborative Study Group, Ebenso J, Fuzikawa P, Melchior H, Wexler R, Piefer A, Min CS, Rajkumar P, Anderson A, Benbow C, Lehman L, Nicholls P, Saunderson P, Velema JP (2007) The development of a short questionnaire for screening of activity limitation and safety awareness (SALSA) in clients affected by leprosy or diabetes. Disabil Rehabil 29:689–700. https://doi.org/10.1080/09638280600926587

Saunderson PR, ILEP Medico Social Commission (2002) ILEP learning guide two: how to recognise and manage leprosy reactions., ILEP learning guide. International Federation of Anti-Leprosy Associations, London

Scollard DM, Truman RW, Ebenezer GJ (2015) Mechanisms of nerve injury in leprosy. Clin Dermatol Leprosy I(33):46–54. https://doi.org/10.1016/j.clindermatol.2014.07.008

Sharma R, Singh P, McCoy RC, Lenz SM, Donovan K, Ochoa MT, Estrada-Garcia I, Silva-Miranda M, Jurado-Santa Cruz F, Balagon MF, Stryjewska B, Scollard DM, Pena MT, Lahiri R, Williams DL, Truman RW, Adams LB (2020) Isolation of mycobacterium lepromatosis and

development of molecular diagnostic assays to distinguish mycobacterium leprae and *M. lepromatosis*. Clin Infect Dis 71:e262–e269. https://doi.org/10.1093/cid/ciz1121
Spencer JS, Dockrell HM, Kim HJ, Marques MAM, Williams DL, Martins MVSB, Martins MLF, Lima MCBS, Sarno EN, Pereira GMB, Matos H, Fonseca LS, Sampaio EP, Ottenhoff THM, Geluk A, Cho SN, Stoker NG, Cole ST, Brennan PJ, Pessolani MCV (2005) Identification of specific proteins and peptides in mycobacterium leprae suitable for the selective diagnosis of leprosy. J Immunol 175:7930–7938. https://doi.org/10.4049/jimmunol.175.12.7930
Sreejith K, Sasidharanpillai S, Ajithkumar K, Mani RM, Chathoth AT, Menon PS, George B, Manakkad SP, Neerackal RJ, Menon D, Mann P (2021) High-resolution ultrasound in the assessment of peripheral nerves in leprosy: a comparative cross-sectional study. Indian J Dermatol Venereol Leprol 87:199–206. https://doi.org/10.25259/IJDVL_106_20
Tabah EN, Nsagha DS, Bissek A-CZ-K, Bratschi MW, Njamnshi TN, Plushke G, Njamnshi AK (2016) The burden of leprosy in Cameroon: fifteen years into the Post-elimination era. PLoS Negl Trop Dis 10:e0005012. https://doi.org/10.1371/journal.pntd.0005012
Truman R, Fine PEM (2010) "Environmental" sources of mycobacterium leprae: issues and evidence. Lepr Rev 81:89–95
Truman RW, Busso P, Rougemont J, Kapopoulou A, Scollard DM, Cole ST (2011) Probable zoonotic leprosy in the southern United States. N Engl J Med:8
Tucker A, Cruz A, Duck M, Timalsina A (2020) Persons affected by leprosy and the COVID-19 global health crisis: a consultative calls report from GPZL's emergency response working group 2. Lepr Rev 91:425–430. https://doi.org/10.47276/lr.91.4.425
Ulrich M, Smith PG, Sampson C, Zuniga M, Centeno M, Garcia V, Manrique X, Salgado A, Convit J (1991) IgM antibodies to native phenolic glycolipid-I in contacts of leprosy patients in Venezuela: epidemiological observations and a prospective study of the risk of leprosy. Int J Lepr Other Mycobact Dis 59:405–415
UN General Assembly (2010) Promotion and protection of human rights: human rights questions, including alternative approaches for improving the effective enjoyment of human rights and fundamental freedoms (No. A/65/456/ADD.2(PARTII)). United Nations
Walker SL, Sales AM, Butlin CR, Shah M, Maghanoy A, Lambert SM, Darlong J, Rozario BJ, Pai VV, Balagon M, Doni SN, Hagge DA, Nery JAC, Neupane KD, Baral S, Sangma BA, Alembo DT, Yetaye AM, Hassan BA, Shelemo MB, Nicholls PG, Lockwood DNJ, Group, on behalf of the E.N.L.I.St (2017) A leprosy clinical severity scale for erythema nodosum leprosum: an international, multicentre validation study of the ENLIST ENL severity scale. PLoS Negl Trop Dis 11:e0005716. https://doi.org/10.1371/journal.pntd.0005716
Walsh GP, Meyers WM, Binford CH, Gerone PJ, Wolf RH, Leininger JR (1981) Leprosy - a zoonosis. Lepr Rev 52. https://doi.org/10.5935/0305-7518.19810060
WHO Study Group on Chemotherapy of Leprosy (1994) Chemotherapy of leprosy: report of a WHO study group. World Health Organization, Geneva
World Bank (2022) Poverty and shared prosperity 2022: correcting course. World Bank, Washington D.C.
World Health Organization (1982) Chemotherapy of leprosy for control programmes: report of a WHO study group. World Health Organization, Geneva
World Health Organization (1988) WHO expert committee on leprosy: sixth report. World Health Organization, Geneva
World Health Organization (1998a) WHO expert committee on leprosy: seventh report. World Health Organization, Geneva
World Health Organization (1998b) WHO model prescribing information: drugs used in leprosy (no. WHO/DMP/DSI/98.1). World Health Organization, Geneva
World Health Organization (2001a) Leprosy: global target attained [WWW Document]. https://www.who.int/news/item/16-05-2001-leprosy-global-target-attained. Accessed 11.26.22
World Health Organization (ed) (2001b) International classification of functioning, disability and health: ICF. World Health Organization, Geneva
World Health Organization (2009) Report of the Tenth Meeting of the WHO Technical Advisory Group on Leprosy Control. New Delhi, India

World Health Organization (ed) (2012) Eighth report/WHO expert committee on leprosy: [Geneva, 12–19 October 2010], WHO technical report series. World Health Organization, Geneva
World Health Organization (2016) Global Leprosy Strategy 2016–2020: Accelerating towards a Leprosy-free World
World Health Organization (2018) Guidelines for the Diagnosis, Treatment and Prevention of Leprosy
World Health Organization (2019) Global leprosy update, 2018: moving towards a leprosy-free world. Wkly Epidemiol Rec 35(36):389–411
World Health Organization (2020a) Global leprosy (Hansen disease) update, 2019: time to step-up prevention initiatives. Wkly Epidemiol Rec 36:417–440
World Health Organization (2020b) Leprosy Hansen disease: management of reactions and prevention of disabilities: technical guidance. World Health Organization, Geneva
World Health Organization (2020c) Ending the neglect to attain the sustainable development goals: a road map for neglected tropical diseases 2021–2030. World Health Organization, Geneva
World Health Organization (2021a) Towards zero leprosy. Global leprosy (Hansen's disease) strategy 2021–2030. World Health Organization, Geneva
World Health Organization (2021b) Global leprosy (Hansen disease) update, 2020: impact of COVID-19 on global leprosy control. Wkly Epidemiol Rec 36:24
World Health Organization (2022) Global leprosy (Hansen disease) update, 2021: moving towards interruption of transmission. Wkly. Epidemiol. Rec. 36
World Health Organization, Regional Office for Africa (2012) Progress Towards the Reduction of the Burden of Leprosy. Disease Prevention & Control Clusters (Neglected Tropical Disease Programme)
World Health Organization, Regional Office for Africa (2013) Leprosy: Facing the Remaining Challenges– The WHO African Region Strategic Plan for Further Reducing the Burden of Leprosy 2013–2017
World Health Organization, Regional Office for South-East Asia (2009) Enhanced global strategy for further reducing the disease burden due to leprosy (2011–2015): plan period: 2011–2015). WHO Regional Office for South-East Asia
World Health Organization, Regional Office for South-East Asia (2012a) Enhanced Global Strategy for further Reducing the Disease Burden due to Leprosy: Questions and Answers. World Health Organization, Regional Office for South-East Asia, New Delhi, India
World Health Organization, Regional Office for South-East Asia (2012b) Report of the eleventh meeting of the WHO technical advisory group on leprosy control. WHO Regional Office for South-East Asia
World Health Organization, Regional Office for South-East Asia (2020) Leprosy/Hansen disease: Contact tracing and post-exposure prophylaxis. World Health Organization. Regional Office for South-East Asia
World Health Organization, Regional Office for South-East Asia (2021) Task force on definitions, criteria and indicators for interruption of transmission and elimination of leprosy: report of the final meeting (publications no. SEA-GLP-6). World Health Organization, New Delhi
Yawalkar SJ (2009) Leprosy for medical practitioners and paramedical workers, 8th edn. Novartis Foundation for Sustainable Development, Basle

Loiasis

Joseph Kamgno, Hugues C. Nana-Djeunga,
and Marc Kouam-Kenmogne

Contents

J. Kamgno (✉)
Higher Institute for Scientific and Medical Research (ISM), Yaoundé, Cameroon

Faculty of Medicine and Biomedical Sciences, University of Yaoundé 1, Yaoundé, Cameroon
e-mail: kamgno@crfilmt.org

H. C. Nana-Djeunga
Higher Institute for Scientific and Medical Research (ISM), Yaoundé, Cameroon

M. Kouam-Kenmogne
Higher Institute for Scientific and Medical Research (ISM), Yaoundé, Cameroon

Department of Animal Production, University of Dschang, Dschang, Cameroon

J. O. Gyapong, B. A. Boatin (eds.), *Neglected Tropical Diseases - Sub-Saharan Africa*, Neglected Tropical Diseases, https://doi.org/10.1007/978-3-031-53901-5_7

Abstract

Loiasis, also known as African eye worm, is a vector-borne parasitic disease caused by the filarial nematode *Loa loa* and transmitted by the bite of tabanid vectors (*Chrysops dimidiata*) from the genus *Chrysops*. Loiasis exists exclusively in Africa, especially in forested areas of Central and West Africa where prevalence rarely exceeds 30% in endemic communities. The real burden of the disease is yet to be established, and *L. loa* remains wrongly classified as a benign disease; however, the disease has become important because of the serious adverse events which occur in some heavily infected patients after treatment with Mectizan for onchocerciasis or lymphatic filariasis. The life cycle of the parasite takes about 10 days in the vector (intermediate host) and about 3 months in humans (definite host). The passage of the adult worm in the bulbar conjunctiva is the classic manifestation of loiasis. Other manifestations are common migratory transient oedema (Calabar swellings). Both are used in the clinical diagnosis, complemented by microscopy for microfilaria. Rare but serious manifestations of *Loa loa* such as encephalitis, glomerular damage, retinal damage, endomyocardial fibrosis, albuminuria and hydrocele may also occur. Both surgical removal of adult worms and chemotherapy are used for managing *L. loa*. Currently Diethylcarbamazine (DEC), Mectizan®, mebendazole (MBZ) and albendazole (ALB) are effective in varying degrees against mf and/or adult *L. loa*, while further research is needed to improve the efficacy of ALB in the management of *L. loa*.

Introduction

Loiasis is a parasitic disease due to the filarial nematode *Loa loa* (*L. loa*) and transmitted by tabanid vectors from the genus *Chrysops*. *L. loa* is also known as the African eye worm because the most specific clinical manifestation of the disease is the migration of the adult worm under the conjunctiva of the eye of infected individuals. Although loiasis was reported as one of the most common reasons for medical consultation (Pinder 1988), its real burden has never been evaluated; it remains wrongly qualified as a benign disease. It is a neglected condition among neglected tropical diseases (NTDs), which increasingly becomes important thanks to the serious adverse events arising in some patients heavily infected with *L. loa* and treated against onchocerciasis and/or lymphatic filariasis (LF), another NTDs with dedicated control programmes.

Epidemiology of the Disease

Distribution and Burden

Loiasis was described as far back as the end of the sixteenth century in Congolese slaves taken to America. The disease disappeared from America at the end of the slavery and presently exists exclusively in Africa where it is mainly focalized in the big forest areas of Central Africa (Veletzky et al. 2020; Mogoung-Wafo et al. 2019; Boussinesq and Gardon 1997; Hawking 1977; Rodhain and Rodhain-Rebourg 1973). Some restricted foci were described in Benin, Nigeria, Northern Cameroon, Chad and in South Sudan which constitutes the west and the northern boundary of the disease. Some cases of loiasis have also been described outside these areas, but they are probably imported cases (Rodriguez et al. 2021; El Haouri et al. 2001).

Based on the recent vast surveys carried out by the African Program for Onchocerciasis Control (APOC) using the rapid mapping of *L. loa* (RAPLOA) strategy, the disease is found in two large foci with high prevalence: the west focus covering the south east of Nigeria, the south of Cameroon, Equatorial Guinea, Gabon, west of Congo, the Coastal plains of Angola, the Bas Congo in Democratic republic of Congo (DRC), the west of Central African Republic (CAR) and the south of Chad. The East focus covers the east of CAR, the south of Southern Sudan and the northeast of DRC (Zoure et al. 2011). A small focus is found at the boundary between Kenya and South Sudan.

The areas with low prevalence stretch from southwest Benin to the West of Ethiopia and from the north of Angola to the central region of Chad (Fig. 1).

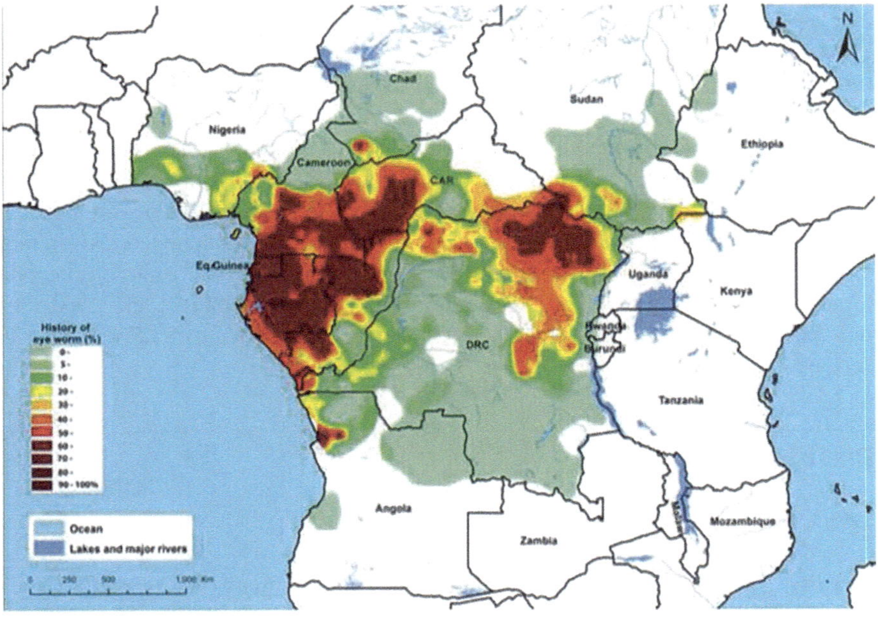

Fig. 1 Geographic distribution of *Loa loa* in Africa (© Zoure et al. 2011)

These data are in accordance with the description made previously with the confirmation that *Chrysops dimidiata* has never been found in the west of Benin. Some *Chrysops silacea* specimen were reported in Ghana, but their vectorial capacity has never been studied (Boussinesq 2006). In the endemic areas, large foci are spread in the forest areas where there are appropriate biotopes for the development of the *Chrysops* vectors. Nevertheless, there are some areas in the savannah zones with "forest galleries" where the prevalence of loiasis is very high (above 50%) (Kamgno and Boussinesq 2001). In these areas (forest galleries), the prevalence of the disease is higher than in the other forest areas probably because the installation of the infection is recent and that people are not immunologically protected from the infection as in the forest areas. Another explanation could be the concentration of vectors in "forest galleries", thus reducing the dilution phenomenon.

The burden of loiasis is understudied. However, the overall burden of disease caused by loiasis in an endemic region of Gabon, using disability-adjusted life years (DALYs) has been recently investigated, and showed to be 412.9 (95% CI: 273.9–567.7) morbidity-based DALYs per 100,000 people in rural Gabon (Veletzky et al. 2020).

Public Health Impact in Sub-Saharan Africa

Loiasis is a complex infection since the prevalence of the disease rarely exceeds 30% in endemic communities, even in areas where the vast majority of the populations presents with the infection. In Gabon, it was shown that from as early as 2 years of age, all the children have specific IgE antibodies to *L. loa* (Goussard et al. 1984). The impact of this disease is up to now not well studied. Loiasis was nonetheless described, in some endemic areas, as one of the main causes of medical consultation (Pinder 1988). Newly arrived people in endemic areas seem to develop, after infection, more clinical signs than the natives some of whom however, have very high microfilaria (mf) loads (Churchill et al. 1996; Klion et al. 1991). The interest in loiasis was renewed when serious adverse events (SAEs) were reported after administration of ivermectin (IVM) for the treatment of onchocerciasis in individuals harbouring high *L. loa* micro-filaraemia. These SAEs consist of encephalopathy usually accompanied by retinal haemorrhages (Fobi et al. 2000); the outcome can be fatal, and individuals who survive an episode of SAE present with serious sequelae (Boussinesq 2006).

Basic Biology

Life Cycle

The life cycle of *L. loa* takes place in two hosts, in humans (definitive hosts) and tabanids of the genus *Chrysops* (intermediate hosts or vectors) (Fig. 2).

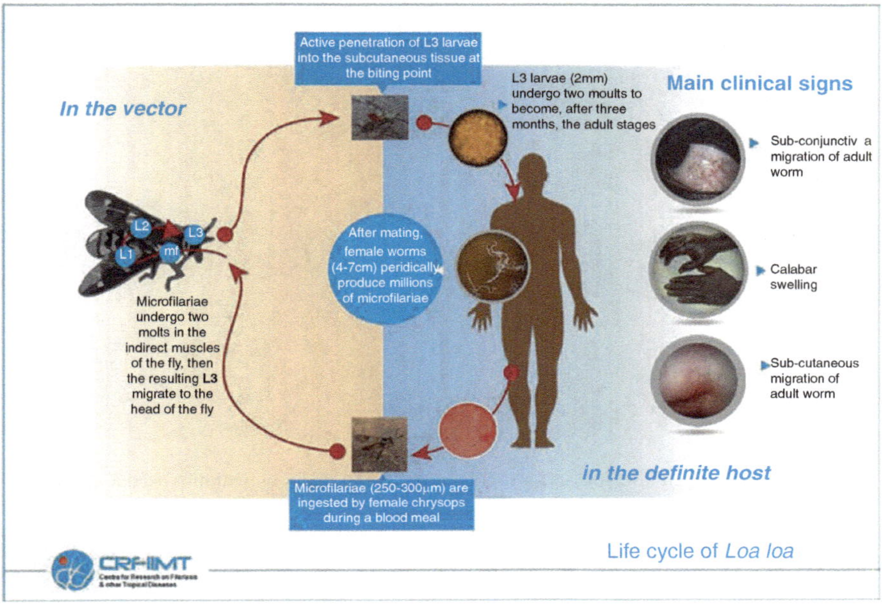

Fig. 2 Life cycle of the filarial nematode *Loa loa* (©ISM)

In the Vector

L. loa mf (Fig. 3) are ingested by female Chrysops during a blood meal taken from an infected human host. These mf cross the peritrophic membrane and the wall of the digestive tract to reach the haemocoel of the fly, particularly in the adipose tissue of the abdomen. A phagocytosis of mf by fat cells then takes place and leads to the formation of a "filarial syncytium" within which metamorphosis takes place and leads to the infective larval stage (Lebied 1957). After 10 days, the infective larvae leave the fat cells and migrate to the chest and then to the head of the fly. During a new blood meal, the *Chrysops* release mf at the biting point.

In the Definitive Host (Human)

During the blood meal, the *Chrysops* make a laceration in the epidermis and dermis, which allows the active passage of larvae into the subcutaneous tissue. These larvae undergo two moults to become, after 3 months, the adult stages. The latter move and pass through the skin into the deeper connective tissue or under the conjunctiva of the eye. Mated females (Fig. 4) produce, periodically, mf which actively accumulate in the pulmonary capillaries during the night and are distributed passively in the peripheral blood during the day (Fain and Maertens 1973; Hawking 1955).

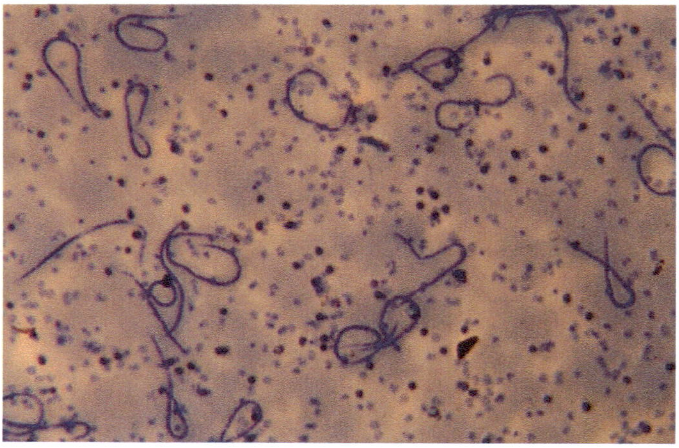

Fig. 3 Giemsa-stained thick blood film showing *L. loa* mf under optical microscope (©ISM)

Fig. 4 Adult female *L. loa* (©ISM)

Mode of Transmission

Loiasis is a vector-borne parasitic disease. *Chrysops dimidiata*, commonly called the mango fly or the mangrove fly, and *C. silacea* are the most common vectors of *L. loa* in Central Africa (Fig. 5; Kouam et al. 2013; Padgett and Jacobsen 2008).

These blood-sucking vectors have a day-biting activity which facilitates the trans-mission of the parasites from hosts to hosts due to the diurnal periodicity of their mf. Even though secondary, other *Chrysops* species are also able to transmit *L. loa* to humans (Duke 1955). *Chrysops langui* and *Chrysops centurionis* are the main vectors of simian strains of *Loa* which have a crepuscular activity.

Disease Presentation

The most classic manifestation of loiasis is the spectacular passage of the adult worm in the bulbar conjunctiva. This passage, which lasts from a few hours to over 24 h, can cause significant discomfort. It is often accompanied by severe swelling of the eyelids and the eyeball (Figs. 6 and 7). The subcutaneous migration of an adult worm within the eyelid can also occur (Fig. 8) leading to a painful "movement sensation" of the eyelid (Rodriguez et al. 2021).

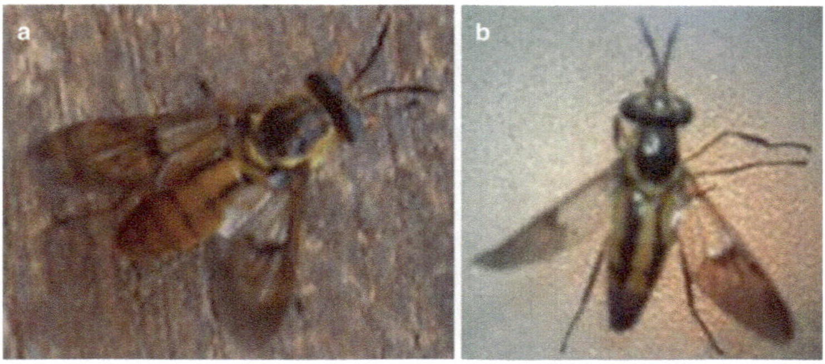

Fig. 5 *Chrysops silacea* (**a**) and *dimidiata* (**b**) (©ISM)

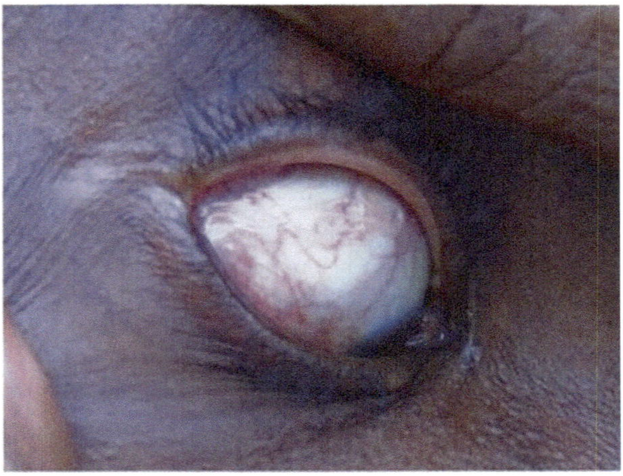

Fig. 6 Subconjunctival passage of the adult *Loa loa* in a patient suffering from loiasis (©ISM)

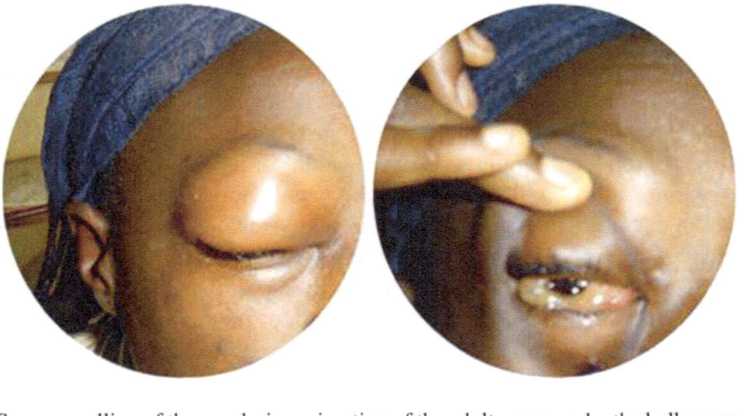

Fig. 7 Severe swelling of the eye during migration of the adult worm under the bulbar conjunctiva

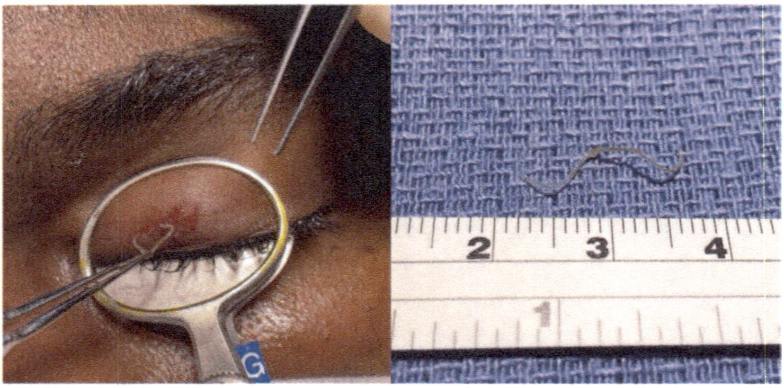

Fig. 8 Surgical extraction of male adult Loa loa worm from the left upper eyelid and gross specimen at right which measured 3.2 cm from head to tail (from Rodriguez et al. 2021)

Other manifestations of loiasis are the quite common migratory transient oedema, commonly called Calabar swellings Fig. 7 (Gruntzig and Gewalt 1979). Their occurrence is caused by episodic microfilarial laying by female adult worms.

Lysis of mf leads to an allergic reaction responsible for oedema (Fain 1978). The latter may also be linked to the release of antigenic products by the adult worm (Negesse et al. 1985). This hypothesis is supported by observations made during the migration of adult worms in the bulbar conjunctiva. Indeed, this passage is often accompanied by severe swelling of the conjunctiva and periorbital tissues (Fig. 7). Oedema due to loiasis is often associated with pruritus, arthralgia and headache (Carme et al. 1989). In an endemic area of Congo, 51% of the population had had a history of Calabar swellings, and 69% experienced the sub-conjunctiva migration of the adult worm in the last 12 months.

Apart from therapeutic accidents after treatment with DEC or IVM of patients harbouring high microfilarial loads, it seems that *L. loa* can cause spontaneously rare but serious manifestations such as encephalitis (Lukiana et al. 2006; Taiwo and Tamiowo 2007; Arrey-Agbor et al. 2018), glomerular damage, retinal damage, endomyocardial fibrosis, albuminuria and hydrocele (Languillat et al. 1977; Noireau et al. 1990). A case of pulmonary effusion due to loiasis has also been reported (Klion et al. 1992). Cases of intraocular migration of *Loa* adults or mf, as well as hemorrhagic and exudative retinopathy and filarial uveitis have been reported in Gabon. Retinal haemorrhages can cause large and sudden decrease in visual acuity and are in most cases irreversible (Védy et al. 1975). A recent study demonstrated that individuals aged 25 years'old with 30,000 mf/ml died significantly earlier than those who were amicrofilaremic, and the population-attributable fraction of mortality associated with presence of L loa microfilaraemia was 14.5% (Chesnais et al. 2017).

Despite these clinical manifestations, and epidemiological consequences, loiasis is still considered as a benign disease probably because most of the lesions described in loiasis patients are hospital-based cases. A study performed in a loiasis hyperendemic area in Cameroon aiming at describing the clinical impact of loiasis, at the

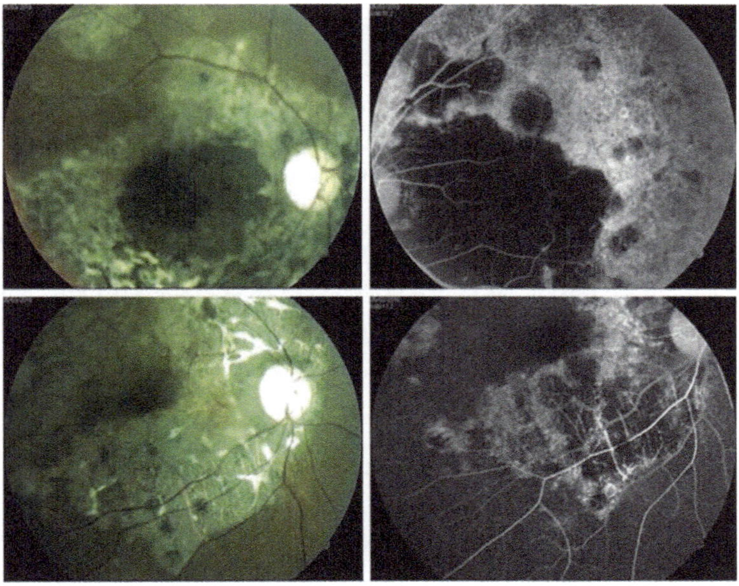

Fig. 9 Retinal lesions in individuals infected with *Loa loa* (©ISM)

community level, showed that some cardiac and ocular lesions (Fig. 9) may be attributed to loiasis (Kamgno, personal communication).

Diagnosis

The diagnosis of *L. loa* infections can be challenging, mostly because some parasite carriers are amicrofilaraemic, asymptomatic or present clinically several months or years after infection. Different diagnostic tools are available, and the choice of a particular diagnostic method depends on the resource available, the laboratory facilities and the objectives of the test (routine diagnosis, case report, epidemiological studies, etc.).

Clinical

These include Calabar swelling, visualization of adults worms migration across the eye into the subconjunctival tissues and surgical removal and identification of an adult worm under the skin. Eye worm passage under the bulbar conjunctiva, though limited to a very short timeframe (few hours to less than a week), proved to be a very useful tool in the mapping of loiasis using a rapid assessment procedure for loiasis (RAPLOA), which is based on a simple questionnaire on the history of eye worm (Zoure et al. 2011). Surgical removal and identification of adult worm for diagnosis purposes is not common but is often carried out in atypical cases of loiasis in patients living in non-endemic areas but with a history of travel in endemic areas (Richardson et al. 2012).

Microscopical/Parasitological

So far microscopy is universally used for both epidemiological data collection and medical diagnosis of human loiasis. This is based on the demonstration of mf in Giemsa-stained thin or thick blood smears (Fig. 3). Due to the diurnal periodicity of *Loa* mf, peripheral blood collection for parasitological examination should be performed between 10 AM and 4 PM. In case of low number of circulating mf, concentration techniques such as nucleospore membrane filtration, Knott's concentration, leucoconcentration and saponin lysis may increase the sensitivity of the parasitological test. A calibrated thick blood smear to quantify the number of mf per μL or ml of blood may be performed for treatment purposes or epidemiological studies.

Serological

These are of limited number, including: immunoblotting assay (Egwang et al. 1989), enzyme-linked immunosorbent assays (ELISAs) (Akue et al. 1997, 1998; Toure et al. 1999b; Gobbi et al. 2020), *L. loa* SXP-1 (LlSXP-1) ELISA (Klion et al. 2003), luciferase immunoprecipitation systems (LIPS) and rapid LIPS format (QLIPS) assays (Burbelo et al. 2008). While immunoblotting and ELISA assays show poor specificity due to cross-reactivity with other filarial infections, the serologic assays based on the *L. loa* SXP-1 antigen and others reliably distinguish loiasis from other filarial and helminthic infections.

Molecular

There are various polymerase chain reaction (PCR)-based assays available which show higher sensitivity and specificity than the previously mentioned detection tests. Conventional PCR assays (Toure et al. 1997) as well as nested PCR assays targeting the repeat three region of the gene coding for *L. loa* 15 kDa polyprotein have been developed (Toure et al. 1998, 1999a). The nested PCR was shown to have 95% sensitivity and 100% specificity (Toure et al. 1998). Three other specific nested PCRs targeting the internal transcribed spacer (ITS1) region of the DNA were developed for the differential detection of *L. loa*, *Mansonella perstans* and *Wuchereria bancrofti*, the three blood circulating mf (Jimenez et al. 2011). This same target (ITS1 region) was used by the same authors to develop species-specific polymerase chain reaction-restriction fragment length polymorphism (PCR-RFLP) protocol for the differentiation of *L. loa* and *M. perstans* from *W. bancrofti*. Real-time PCR assays were also developed to demonstrate their utility for the diagnosis of filarial infections in mobile populations (Fink et al. 2011a) and to estimate the mf burden of blood samples in comparison with calibrated microscopy (Fink et al. 2011b).

Adequacy of the Diagnostic Methods

Microscopy examination is the tool mostly used in clinic and research settings, but some infected people are amicrofilaraemic (occult loiasis) (Dupont et al. 1988). Thus for cases of occult loiasis, microscopy is not adequate. Moreover, microscopy becomes a drudgery when large amounts of samples are to be processed for a screening study.

Serological tools may be used to handle very large amount of samples more conveniently than other available tools. However, they have some major shortcomings: they cannot distinguish between active and past infections, quantify microfilaraemia or be used in treatment trials to test for the efficacy of a drug.

Molecular tools are highly sensitive and specific, readily differentiating *L. loa* from other blood circulating mf, and most importantly they successfully quantify the amount of mf in blood samples from both microfilaraemic and amicrofilaraemic individuals (Fink et al. 2011b). The ability to quantify the number of mf even in blood samples from subjects with occult loiasis is a very important asset of the molecular tools, given the relevance of *Loa* microfi larial load in area co-endemic with *Onchocerca volvulus* and *W. bancrofti* where IVM is distributed. However, steps to make these tools practical in field settings are needed.

Clinical, parasitological, serological and molecular tools have shown sufficient sensitivity and specificity. Clinically, *Loa loa* is the only worm with subconjunctival migration of the adult. Concerning the parasitological examination, *Loa loa* microfilariae (Fig. 3) have day periodicity, and in endemic area *W. bancrofti* that has the same shape has night periodicity. Serological and molecular tool development and adoption usually require two major steps. The first step focuses on the development of the tool in laboratory conditions using laboratory strains of the target organism or the target organism maintained in laboratory conditions. The following step is the applicability of the assay using field samples for which the tool is being developed. This latter step is necessary to support the ability of the tool to discriminate the target organism. It also provides evidence of the reliability of the tool. Most of the *L. Loa* diagnostic tools such as calibrated microscopy and real-time PCR (Fink et al. 2011b).

Cost-Effectiveness

One of the main advantages of microscopy is probably its cost-effectiveness in comparison with other tools. Microscopy for *L. loa* detection is well suited for resource-poor laboratories, the only expensive appliance being the microscope. Consumables (slides, capillary tubes, staining jars, etc.) and reagents (Giemsa stain, distilled water and absolute methanol) are affordable.

Serological and molecular tools require expensive equipment, consumables and reagents for their application. Even when the equipment is available, processing of large amount of samples is possible only for resource-rich laboratories due to the high cost of reagents and consumables. Nevertheless, thanks to substantial efforts made during the last decade, there are now rapid serological tools for the diagnosis of loiasis that are currently commercially available and can be used for rapid mapping of loiasis (Pedram et al. 2017; Ella et al. 2021).

The Ideal Diagnostic Method

Tools are available for the diagnosis of loiasis as a parasitic disease. Nowadays the importance of loiasis is not related to its debilitating effect as a parasitic disease but to its potential to provoke severe adverse reactions in people with high *Loa* microfilarial load (>30,000 mf/ml of blood) who take IVM for the treatment of onchocerciasis or lymphatic filariasis. Loiasis therefore stands as a hindrance to the successful implementation of onchocerciasis and lymphatic filariasis control or elimination programmes. Since mass distribution of ivermectin is community based and those in charge of administering the drug are simple lay men and women, there is a need when distributing the drug, to distinguish between those eligible for the treatment from those that are not, by using a simple, rapid, reliable and user-friendly tool. In the last years, this ideal diagnostic method - able to diagnose high *Loa* microfilarial loads within a few minutes, user-friendly and field applicable—was developed for the rapid quantitative diagnosis (D'Ambrosio et al. 2015). The tool has allowed the development of a new strategy for the treatment of onchocerciasis in loiasis co-endemic areas, the Test and not Treat strategy (Kamgno et al. 2017, 2018).

Control Tools and Strategies

Key Tools and Strategies

Two main tools are currently used for the control of loiasis: adult worm removal and chemotherapy.

The removal of adult worms usually takes place during the subconjunctival migration of the latter. This action has a very limited impact on the person being dewormed since these worms represent only a small proportion of the total worms harboured by the host.

Four chemotherapeutic agents, diethylcarbamazine (DEC), IVM or Mectizan®, mebendazole (MBZ) and albendazole (ALB)) are effective against mf and/or adult *L. loa* (see details below). Only DEC is adulticidal and can be considered as a definitive therapy for loiasis. Despite the effectiveness of these drugs on *L. loa*, no large-scale control efforts have ever been implemented in affected areas for two main reasons. As with another filarial infection (mansonellosis), it has been difficult to associate loiasis with a clear and specific clinical picture, especially at the community level. This filariasis is then still considered as a benign or less pathogenic disease which might not need as much attention as the known debilitating diseases, namely, LF, which is responsible for the enlargement of lymphatic vessels, lymphoedema, and elephantiasis and onchocerciasis for which the striking symptoms are blindness and severe dermatitis (Wanji et al. 2003). Also, it was demonstrated that people harbouring high (>8000 mf/ml) or very high (>30,000 mf/ml) microfilarial loads can experience SAEs, with sometimes neurological consequences, when treated with DEC or IVM (Gardon et al. 1997a). Therefore, individual case management has been suggested, especially for people living in endemic areas.

Pre-treatments of infected individuals aiming at lowering the *Loa* microfilarial load before usage of a more powerful or definitive therapy have been recommended. Apheresis has been suggested, but the cost and the difficulty to implement this technique seriously limit its application. Unlike IVM and DEC, ALB does not lead to a marked decrease in *Loa* microfilarial load and can therefore be used as alternative to cytapheresis pretreatment to lower microfilarial loads before usage of DEC or IVM (Klion and Nutman 2010). Unfortunately, different regimens (Klion et al. 1993; Kamgno et al. 2016 used, up to date, are not yet completely satisfactory. It is therefore important to prevent the infection or limit reinfestation.

To date, no vaccine is available to prevent becoming infected with *L. loa*, but weekly chemoprophylaxis with 300 mg DEC was found efficient for the prevention of loiasis. Vector control can also be useful in preventing the disease acquisition. Personal protection measures including insect repellents, treated or protective clothing or avoiding *Chrysops* breeding sites may prevent or reduce the risk of infection.

Effectiveness of the Tools and Strategies

There is a contrast between the efficacy and the safety of tools available for the treatment of loiasis, which renders the management of this infection very complex. In fact, the most effective drugs or tools useful in the control of *L. loa* infection are not really safe, especially when individuals are heavily infected, and the safer ones are not completely efficient. DEC and IVM are the most efficient chemotherapeutic agents against *L. loa*, but these two drugs also lead to SAEs with sometimes fatal outcome in patients harbouring high or very high microfilarial loads (Kamgno et al. 2008). In contrast, some benzimidazole (especially ALB) are safer for the treatment of *L. loa*-infected patients (adverse effects have, up to now, never been observed even in people harbouring more than 50,000 mf/ml), but their effect on microfilarial loads is either low (around 20% reduction of the initial microfilarial density) or often transitory and varies widely between patients (Boussinesq 2006; Kamgno et al. 2016).

Evidence for Choice of Tools and Strategies

Numerous literatures are available for the evaluation of the efficacy of chemotherapeutic agents on *L. loa* infection (Carme et al. 1991b; Chippaux et al. 1992; Duong et al. 1997; Gardon et al. 1997b; Garin and Garin 1951; Klion et al. 1999; Shookhoff and Dwork 1949; Tsague-Dongmo et al. 2002; Van Hoegaerden and Flocard 1985). For microfilaricidal assessment, that is, the evaluation of the efficacy of drugs on mf, the follow-up of post-treatment dynamics of the parasite density has been used (Chippaux et al. 1992; Duong et al. 1997; Gardon et al. 1997b; Kamgno et al. 2007). For macrofilaricidal effect—effect of drugs on adult worms—evidence of the efficacy of DEC was evaluated by observing degenerating adult worms under the skin (Bonnin and Moretti 1950). Direct approach to evaluate drug efficacy is difficult since sourcing and studying adult worms (including the developmental stages they are harbouring) are very tedious and appear only possible after animals' sacrifice. Then, there is a need to develop robust and easier model systems – current models are time consuming and require large quantities of drugs – as well as diagnostic test to detect the presence of viable adult worms (Grand Challenge Exploration website).

Although performed on a limited number of subjects, the efficacy of apheresis in lowering *Loa* microfilarial loads was clearly demonstrated (Abel et al. 1986; Chandenier et al. 1987; Muylle et al. 1983). A progressive decrease in microfilarial densities was observed after a three-course session of *Loa* infected subject blood purification.

Apart from the DEC-based chemoprophylaxis of *L. loa* infection which has been clearly demonstrated, the other prevention methods including drugs (IVM), personal protection (*Chrysops* repellents, treated or protective clothing, etc.), vector control or vaccines were not really evaluated.

Sustainability and Cost-Effectiveness of the Tools/Strategies

The usage of tools to control loiasis is only applicable at individual level because the evaluation of the microfilarial load of the patient is a prerequisite for the treatment of loiasis. Individuals harbouring low microfilarial loads (>8000 mf/ml) can be cured by a 21 day regimen of DEC at a dose of 8–10 mg/kg/day (Klion and Nutman 2010). This treatment, although definitive, seems appropriate only for long-term visitors living in endemic areas, and would not be sustainable in autochthones living in the endemic area because of reinfestations. This is particularly true for those subjects with high microfilarial levels for whom it is recommended to first lower their microfilarial densities by ALB whose efficacy is still questionable or by three apheresis sessions which are very expensive and difficult to implement.

Chemotherapy-Based Strategies

Evidence for Choice of Strategies

The optimal treatment strategy for loiasis relies on what is known about the efficacy and safety of the three drugs- DEC, ALB and IVM- currently used to treat it. Regarding efficacy, the only one of the three drugs for which a macrofilaricidal effect has been demonstrated is DEC, as evidenced by the finding of dead and necrotic worms in biopsy specimens from post-treatment swellings (Bonnin and Moretti 1950) and an autopsy in nonhuman primates after treatment with DEC (Duke 1990). In 10–25% of the cases, more than one course of DEC has to be given to achieve a complete cure; patients who are refractory to more than four courses are very rare (Gentilini and Carme 1981; Klion et al. 1999). DEC has an activity against the mf of *L. loa*, which are cleared from the peripheral blood within hours of DEC administration. Short courses of ALB have little effect on *L. loa* (Tabi et al. 2004; Tsague-Dongmo et al. 2002), but when given at a dose of 200 mg twice a day for 21 days, the drug has probably an embryotoxic effect (i.e. it interrupts embryogenesis in the uteri of the adult female worms) and perhaps also a macrofilaricidal effect (Klion et al. 1993). After a single dose of IVM (150 μg/kg), *Loa* microfilaraemia decreases by 70–80% within the first 3 days (Kamgno et al. 2000; Paris et al. 1991; Richard-Lenoble et al. 1988). The densities then plateau or decrease more slowly and remain at very low values up to 1 year after treatment (Gardon et al. 1997b). Whether this is due to a partial macrofilaricidal or to an embryostatic effect (preventing the release of developed mf from the uteri of the adult female worms) is not known. Monthly treatment with IVM has a cumulative effect, leading after six doses to extremely low microfilarial densities, which remain so for at least several months (Kombila et al. 1998). Besides its effects on the parasite, IVM also has a beneficial effect on the clinical manifestations of loiasis and seems to prevent the reappearance of Calabar swellings for several months (Hovette et al. 1994). Lastly, as *L. loa* does not harbour *Wolbachia* endosymbionts (McGarry et al. 2003), antibiotics (doxycycline) are useless in the treatment of loiasis. Other agents have been used in the treatment of loiasis, including suramin (Klion et al. 1994) and the benzimidazole MBZ (Van Hoegaerden et al. 1987). Suramin and the benzimidazoles have activity against the adult worms, but the widespread use of Suramin is limited by its side effects which include pancytopenia and nephrotoxicity (Klion et al. 1999), while MBZ, which has a poor and variable absorption following oral administration, has been used with only limited success for the treatment of loiasis (Van Hoegaerden et al. 1987).

Now that the reasons for using the drugs mentioned above as first-line drugs are provided, it is necessary to point out that the treatment strategy depends firstly on the risk of adverse events, which is related to the patient's *Loa* microfilarial load. The latter must mandatorily be quantified before any therapy decision, by examining a calibrated Giemsa-stained thick blood smear (50 μL) prepared between 10:00 am and 2:00 pm, i.e. when *Loa* microfilaraemia in the peripheral blood is the highest. In countries located outside the loiasis distribution area, this assessment and the resulting treatment should be conducted in specialized units or by specialized physicians. DEC and IVM can induce potentially fatal encephalopathies in persons harbouring >30,000–50,000 mf/ml of blood (Fain 1978; Gardon et al. 1997a). Functional impairment without alteration of consciousness but requiring assistance for several days can occur after DEC administration in individuals with >2000 mf/ ml (Carme et al. 1991a) and after IVM in patients with densities exceeding 8000 mf/ ml (Gardon et al. 1997a). The use of ALB in loiasis patients is usually very safe (Klion et al. 1999). Given the risk of serious adverse events after DEC or IVM treatment, the following strategy has been proposed (Boussinesq 2012):

- If the patient's microfilarial density is below 2000 mf/ml, DEC can be administered straightaway. The first course should last 3–4 weeks and start with low doses (3 or 6 mg/day) if mf are present in the blood, or 50 mg/day if the patient is amicrofilaraemic) divided into two or three doses. The dose is doubled every day until 400 mg/day (or 8–10 mg/kg/day) still divided in two to three doses. Treatment should be started in hospital, and oral antihistamines or corticosteroids may be useful in the first days to reduce the severity of side effects (pruritus, angiooedema, arthralgia, headache, fever, etc.) which occur in 50% of the cases. As stated above, several courses of DEC may be needed. If the patient is refractory to DEC, a course of ALB (200 mg twice a day for 21 days) can be undertaken.
- In subjects with *Loa* microfilarial densities between 2000 and 8000 mf/ml, it is advisable to begin treatment by using IVM (a single dose of 150 μg/kg). Mild side effects similar to those mentioned above after DEC can occur. Treatment with IVM can be repeated at intervals of 1–3 months to reduce the microfilarial loads further. When the microfilaraemia is less than 2000 mf/ml, a course of DEC can be started, using the protocol proposed above.
- When the microfilarial loads are between 8000 and 30,000 mf/ml, IVM can also be given, but a close surveillance is needed, with hospitalization during the first 3–4 days. An alternative is to start with a course of ALB (200 mg twice daily for 21 days) which reduces the *Loa* microfilaraemia by 60% at 6 months and to continue with IVM treatment with or without hospitalization depending on the microfilarial loads at that time.
- If the microfilaraemia exceeds 30,000 mf/ml, ALB is probably the best option, even if the effect of long courses of ALB has never been evaluated in patients with such high loads. Apheresis, which allows to rapidly reduce the loads by 75% after three sessions, has also been proposed. However, this technique is expensive and probably indicated only exceptionally, given the usually mild manifestations of loiasis.

Gaps in Knowledge

Usually the age of subjects enrolled in therapeutic trials for loiasis is at least 5 years old probably either because the drug under trial is contraindicated for children less than 5 years, or, since mf have a cumulative effect on their host and microfilaraemia is very low in children, subject less than 5 years old do not develop symptoms. As seen above, treatment may be given according to the situation, either as a single dose (irrespective of the body weight) or per kg of body weight, but those receiving this therapy are always adults, not children. Consequently, the effect of treatment in children below 5 years old is unknown. Because young children are not treated, they may constitute a parasite reservoir likely to prolong or hamper any control program. A contemporaneous example is the African Program for Onchocerciasis Control (APOC) for which IVM is given only to people of at least 5 years old.

Another observation is the immunologically mediated difference to infection between natives from endemic areas who are frequently microfilaraemic, but often asymptomatic, and travelers or expatriates who visit endemic areas, who are in the majority amicrofilaraemic but suffer more from annoying manifestations, such as angiooedemas (Calabar swellings) (Klion et al. 1991). The mechanisms underlying this difference in immunological response are not well understood, and there is no indication that expatriates and natives from endemic areas with similar microfilaraemias respond differently to treatment in terms of efficacy (Churchill et al. 1996).

There is no report of a variation in efficacy of any currently used filaricid related to age or sex (Gardon et al. 1997a, b; Klion et al. 1999). However, some filaricidal agents such as IVM are contraindicated for children below 5 years old. As previously mentioned, these groups of untreated individuals are excluded from treatment during control programme and might thus contribute to the maintenance of infection.

DEC is known as the treatment of choice for amicrofilaraemic patients, but some studies have demonstrated a small number of people who did not respond to DEC treatment despite multiple courses of therapy (Klion et al. 1994, 1999). This finding suggests a development of drug resistance in the parasite or may be an insufficient host immune response since DEC killing does not occur in vitro. A combination of DEC and ALB has already been successfully used after failure of DEC alone (Klion et al. 1999). Various combinations are possible, depending on the initial mf load of the infected individual. There is a common agreement on the potential of combination therapy to slow down drug resistance because of the difference in the mechanism of action among the available drugs. Some impediments to the effective implementation of drug combination are the unavailability of DEC in some *Loa-endemic* countries and the ban of IVM in the USA where patients from *Loa*-endemic areas are frequently reported.

High *Loa* microfilarial load is known to be the main cause of post-IVM SAEs, but there is some controversy about the microfilarial density required to provoke SAEs. Indeed, the word "high" has been variously described as greater than 10,000 mf per millilitre of peripheral blood (mf/ml) before treatment at a consultation organized by the Mectizan® Donation Program (MDP) in Paris in October 1995 or greater than 30,000 mf/ml in a hospital study in Southern Cameroon (Ducorps et al. 1995) or greater than 50,000 mf/ml in a community trial in Southern Cameroon (Gardon et al. 1997a). Most of the data on SAEs used for various policy recommendations is derived from SAE reports and medical records received by the MDP from onchocerciasis mass treatment programmes through a passive surveillance system. Also, it is unknown what the true numbers of SAE cases are, given the variability of reporting from the 34 onchocerciasis endemic countries that have had, or have, IVM mass treatment programmes.

Challenges of Programme Implementation

Presently, there is not a specific programme dedicated to *L. loa* infection. However, as loiasis is now a hindrance to the implementation of onchocerciasis control programme in areas of loiasis-onchocerciasis co-endemicity, it is expected that a particular attention will be paid to this infection. An important challenge of programme implementation is the unavailability of a safe drug which is practical for mass dis-tribution. ALB is the only drug which does not induce SAEs in people with high *Loa* microfilarial loads (Klion et al. 1999), but a complete course of ALB treatment lasts 21 days. The long treatment course makes the use of ALB unrealistic in loiasis control programme. The remaining two drugs, DEC and IVM, both provoke SAEs, but unlike DEC which is not suitable for mass distribution in a loiasis control programme—treatment requires several doses for several consecutive days—IVM can be useful because it has been shown to markedly decrease *L. loa* microfilaraemia 6 and 12 months after a single dose (Gardon et al. 1997b). Nevertheless, care should be taken for the proper management of SAEs that are likely to occur before using IVM as a tool for loiasis control. In this respect, *The LoaScope,* which allows for a quantitative diagnosis of loasis and to discriminate those who are at risk of serious adverse events, makes it possible today to safely treat onchocerciasis in Loa loa co-endemic areas, and to consider mass treatment of loasis. In L. loa endemic areas, the few ineligible persons that represent less than 2% could then become eligible after an ALB cure to lower the level of microfilarial density (Boussinesq 2012).

Vector control is an alternative to mass human chemotherapy. The findings that the population density of adults *Chrysops* is fairly low (~1000/km^2) and their flight range usually not great (theoretical range: <6000 m and maximum distance: 4500 m) in secondary forest were very promising for use to limit loiasis in an effective vector control programme (Chippaux et al. 2000). Some promising methods have been attempted including the use of insecticides (Dieldrin, DDT) (Williams and Crewe 1967), creation of anthropic savannah hostile for *Chrysops* development around habitations (Gordon et al. 1950) and trapping (Fain 1978).

Further Research for Policy and Control

Very promising results have been obtained with ALB, but the efficacy of this drug is still limited, for an up to now unknown reason, to a small fraction of the population (Klion et al. 1993; Kamgno et al. 2016). Research is therefore needed to identify a drug which can efficiently lower *Loa* loads before the usage of IVM or DEC for a definitive therapy. Moxidectin that has been shown to be effective in onchocerciasis needs to be tested also for its effect on loiasis. Given the observed effect of moxidectin on onchocerciasis, it would be prudent to start the first trials on Loa with very low levels of a microfilarial load to avoid SAEs.

Excess mortalities due to high microfilarial load as well as numerous cases of spontaneous encephalopathies in heavily infected individuals have been observed in endemic areas (Chesnais et al. 2017). A recent community-based trial with the aim to assess the epidemiological impact of loiasis has shown that this disease should not be a concern only because of its implication in IVM or DEC-provoking SAEs (Kamgno, personal communication). It is now clear that loiasis should not remain the so-called benign disease which poses problems only when onchocerciasis or lymphatic filariasis is targeted. Therefore, there is a need to find a way to treat loiasis not as a prerequisite in the treatment of onchocerciasis or lymphatic filariasis but because of its medical effects. So far, three drugs have been found to be efficient on mf and/or adult worms, but only DEC is adulticidal and can constitute a definitive treatment for loiasis. More research will be needed to improve on the number of medicines that can be used in loa treatment.

Outlook for the Next Decade

Inspired by the World Health Organization's Roadmap on NTDs, the 2011 London Declaration perspective was to concentrate efforts on the elimination of neglected tropical diseases (NTDs) or reductions in their impact to levels at which they are no longer considered as public-health problems by 2020. Up to now, loiasis is still considered as a benign disease and was not taken into account by the World Health Assembly (WHA) when selecting neglected tropical diseases targeted for elimination and/or eradication by 2020. It appears clear that more epidemiological impact of *L. loa* infection needs to be clarified in order to attract interest in this filariasis and to encourage the development of new effective, safe, inexpensive and tolerable medicines to treat populations at risk, especially those co-infected by lymphatic filariasis and/or onchocerciasis. Therefore, the proper understanding of the physiopathology of SAEs is urgently needed to safely treat loiasis and to achieve the goal of elimination for those diseases co-endemic with loiasis (lymphatic filariasis and onchocerciasis). Elimination of onchocerciasis and/or lymphatic filariasis is now possible in *L. loa* endemic areas thanks to the *Test and not Treat strategy* using the LoaScope that allows the exclusion from treatment of individuals who are at risk of developing SAEs (individuals harbouring high levels of *L. loa*). The TaNT strategy allows for the selection of individuals at risk of SAEs who can be treated by doxycycline against onchocerciasis and lymphatic filariasis, an antibiotic targeting endosymbiotic bacteria that is absent in *L. loa* (Holmes 2013). Research should however focus on a doxycycline regimen applicable at the community level, since the 4–6-week course regimen currently being used is very difficult to implement as community intervention (Wanji et al. 2009). Nowadays, the integrated control of NTDs is very much encouraged since drugs currently donated by pharmaceutical partners are effective against several of these infections and in addition people can be infected with more than one neglected pathogen simultaneously (Grand Challenge Exploration website). Therefore, strategies which aim at treating multiple diseases at the same time are absolutely needed, and the integrated evaluation of these diseases is awaited.

References

Abel L, Ioly V, Jeni P, Carbon C, Bussel A (1986) Apheresis in the management of loiasis with high microfilariaemia and renal disease. Br Med J 292:24

Akue JP, Hommel M, Devaney E (1997) High levels of parasite-specifi c IgG1 correlate with the amicrofilaremic state in Loa Loa infection. J Infect Dis 175:158–163

Akue JP, Hommel M, Devaney E (1998) IgG subclass recognition of Loa Loa antigens and their correlation with clinical status in individuals from Gabon. Parasite Immunol 20:387–393

Arrey-Agbor DB, Nana-Djeunga HC, Mogoung-Wafo AE, Mafo M, Danwe C, Kamgno J (2018) Case report: probable case of spontaneous encephalopathy due to Loiasis and dramatic reduction of Loa Loa Microfilariaemia with prolonged repeated courses of albendazole. Am J Trop Med Hyg 99(1):112–115. https://doi.org/10.4269/ajtmh.17-0664. Epub 2018 May 3

Bonnin H, Moretti G (1950) Preuves cliniques et biopsiques de l'action léthale d'un dérivé de la pipérazine sur la filaire Loa loa adulte. Bull Soc Pathol Exot 33:4

Boussinesq M (2006) Loiasis. Ann Trop Med Parasitol 100:715–731

Boussinesq M (2012) Loiasis: new epidemiologic insights and proposed treatment strategy. J Travel Med 19:140–143

Boussinesq M, Gardon J (1997) Prevalences of Loa Loa microfi laraemia throughout the area endemic for the infection. Ann Trop Med Parasitol 91:573–589

Burbelo PD, Ramanathan R, Klion AD, Iadarola MJ, Nutman TB (2008) Rapid, novel, specific, high-throughput assay for diagnosis of Loa Loa infection. J Clin Microbiol 46:2298–2304

Carme B, Mamboueni JP, Copin N, Noireau F (1989) Clinical and biological study of Loa Loa filariasis in Congolese. Am J Trop Med Hyg 41:331–337

Carme B, Boulesteix J, Boutes H, Puruehnce MF (1991a) Five cases of encephalitis during treatment of loiasis with diethylcarbamazine. AmJTrop Med Hyg 44:684–690

Carme B, Ebikili B, Mbitsi A, Copin N (1991b) Therapeutic trial with ivermectin in loiasis with medium and high microfi laremia. Ann Soc Belg Med Trop 71:47–50

Chandenier J, Pillier-Loriette C, Datry A, Rosenheim M, Danis M, Felix H, Nozais JP, Gentilini M (1987) Value of cytapheresis in the treatment of loaiasis with high blood microfilaria levels. Results in 7 cases. Bull Soc Pathol Exot 80:624–633

Chesnais CB, Takougang I, Paguélé M, Pion SD, Boussinesq M (2017) Excess mortality associated with loiasis: a retrospective population-based cohort study. Lancet Infect Dis 17(1):108–116. https://doi.org/10.1016/S1473-3099(16)30405-4. Epub 2016 Oct 21

Chippaux JP, Ernould JC, Gardon J, Gardon-Wendel N, Chandre F, Barberi N (1992) Ivermectin treatment of loiasis. Trans R Soc Trop Med Hyg 86:289

Chippaux JP, Bouchite B, Demanou M, Morlais I, Le Goff G (2000) Density and dispersal of the loaiasis vector Chrysops dimidiata in southern Cameroon. Med Vet Entomol 14:339–344

Churchill DR, Morris C, Fakoya A, Wright SG, Davidson RN (1996) Clinical and laboratory features of patients with loiasis (Loa Loa filariasis) in the U.K. J Infect 33:103–109

D'Ambrosio MV, Bakalar M, Bennuru S, Reber C, Skandarajah A, Nilsson L, Switz N, Kamgno J, Pion S, Boussinesq M, Nutman TB, Fletcher DA (2015) Point-of-care quantification of blood-borne filarial parasites with a mobile phone microscope. Sci Transl Med 7(286):286re4. https://doi.org/10.1126/scitranslmed.aaa3480

Ducorps M, Gardon-Wendel N, Ranque S, Ndong W, Boussinesq M, Gardon J, Schneider D, Chippaux JP (1995) Secondary effects of the treatment of hypermicrofi laremic loiasis using ivermectin. Bull Soc Pathol Exot 88:105–112

Duke BO (1955) The development of Loa in flies of the genus chrysops and the probable significance of the different species in the transmission of loiasis. Trans R Soc Trop Med Hyg 49:115–121

Duke BO (1990) Effect of diethylcarbamazine on adult Loa in monkeys. Ann Trop Med Parasitol 84:387–392

Duong TH, Kombila M, Ferrer A, Bureau P, Gaxotte P, Richard-Lenoble D (1997) Reduced Loa Loa microfi laria count ten to twelve months after a single dose of ivermectin. Trans R Soc Trop Med Hyg 91:592–593

Dupont A, Zue-N'dong J, Pinder M (1988) Common occurrence of amicrofilaraemic Loa Loa filariasis within the endemic region. Trans R Soc Trop Med Hyg 82:730

Egwang TG, Dupont A, Leclerc A, Akue JP, Pinder M (1989) Differential recognition of Loa Loa antigens by sera of human subjects from a loiasis endemic zone. Am J Trop Med Hyg 41:664–673

El Haouri M, Erragragui Y, Sbai M, Alioua Z, Louzi EMW, Sedrati O (2001) Cutaneous filariasis Loa Loa: 26 moroccan cases of importation. Ann Dermatol Venereol 128:899–902

Ella SN, Ogoussan K, Gass K, Hundley L, Diggle PJ, Johnson O, Biamonte M, Atsame J (2021) An Integrated District mapping strategy for Loiasis to enable safe mass treatment for onchocerciasis in Gabon. Am J Trop Med Hyg 106(2):732–739
Fain A (1978) Current problems of loaiasis. Bull World Health Organ 56:155–167
Fain A, Maertens K (1973) Notes sur la ponte des microfi laires chez loa loa et sur le degré de maturité des vers en migration. Bull Soc Pathol Exot 66:6
Fink DL, Fahle GA, Fischer S, Fedorko DF, Nutman TB (2011a) Toward molecular parasitologic diagnosis: enhanced diagnostic sensitivity for filarial infections in mobile populations. J Clin Microbiol 49:42–47
Fink DL, Kamgno J, Nutman TB (2011b) Rapid molecular assays for specific detection and quantitation of Loa Loa microfi laremia. PLoS Negl Trop Dis 5:e1299
Fobi G, Gardon J, Santiago M, Ngangue D, Gardon-Wendel N, Boussinesq M (2000) Ocular find-ings after ivermectin treatment of patients with high Loa Loa microfilaremia. Ophthalmic Epidemiol 7:27–39
Gardon J, GardonWendel N, DemangaNgangue D, Kamgno J, Chippaux J, Boussinesq M (1997a) Serious reactions after mass treatment of onchocerciasis with ivermectin in an area endemic for Loa Loa infection. Lancet 350:18–22
Gardon J, Kamgno J, Folefack G, Gardon-Wendel N, Bouchite B, Boussinesq M (1997b) Marked decrease in Loa Loa microfilaraemia six and twelve months after a single dose of ivermectin. Trans R Soc Trop Med Hyg 91:593–594
Garin C, Garin JP (1951) Treatment of filariasis from filaria Loa by notezine. J Med Bordeaux Sud-Ouest 128:250–252
Gentilini M, Carme B (1981) Treatment of fi lariases in a hospital setting. Complications – results. Ann Soc Belg Med Trop 61:319–326
Gobbi F, Buonfrate D, Boussinesq M, Chesnais CB, Pion SD, Silva R, Moro L, Rodari P, Tamarozzi F, Biamonte M, Bisoffi Z (2020) Performance of two serodiagnostic tests for loiasis in a non-endemic area. PLoS Negl Trop Dis 14(5):e0008187
Gordon RM, Kershaw WE, Crewe W, Oldroyd H (1950) The problem of loiasis in West Africa with special reference to recent investigations at Kumba in the British Cameroons and at Sapele in southern Nigeria. Trans R Soc Trop Med Hyg 44:11–47
Goussard B, Ivanoff B, Beville G, Garin Y (1984) Current aspects of the ELISA technic applied to the serological diagnosis of helminthiasis. Pathol Biol 32:1043–1049
Gruntzig J, Gewalt W (1979) Cameroon swelling; infection by the filaria Loa Loa. Fortschr Med 97:1932–1937
Hawking F (1955) Periodicity of microfi lariae of Loa Loa. Trans R Soc Trop Med Hyg 49:132–142
Hawking F (1977) The distribution of human filariasis throughout the world, part III. Africa Trop Dis Bull 74:8
Holmes D (2013) Loa Loa: neglected neurology and nematodes. Lancet Neurol 12:631–632
Hovette P, Debonne JM, Touze JE, Gaxotte P, Imbert P, Fourcade L, Laroche R (1994) Efficacy of ivermectin treatment of Loa Loa fi lariasis patients without microfi laraemias. Ann Trop Med Parasitol 88:93–94
Jimenez M, Gonzalez LM, Carranza C, Bailo B, Perez-Ayala A, Muro A, Perez-Arellano JL, Garate T (2011) Detection and discrimination of Loa Loa, Mansonella perstans and Wuchereria bancrofti by PCR-RFLP and nested-PCR of ribosomal DNA ITS1 region. Exp Parasitol 127:282–286
Kamgno J, Boussinesq M (2001) Hyperendemic loaiasis in the Tikar plain, shrub savanna region of Cameroon. Bull Soc Pathol Exot 94:342–346
Kamgno J, Gardon J, Boussinesq M (2000) Analysis of the prevention of post-ivermectin Loa Loa encephalopathy by administration of initial low dose. Med Trop 60:275–277
Kamgno J, Pion SD, Tejiokem MC, Twum-Danso NA, Thylefors B, Boussinesq M (2007) Randomized, controlled, double-blind trial with ivermectin on Loa Loa microfilaraemia: efficacy of a low dose (approximately 25 microg/kg) versus current standard dose (150 microg/kg). Trans R Soc Trop Med Hyg 101(8):777–785. https://doi.org/10.1016/j.trstmh.2007.03.018. Epub 2007 May 24

Kamgno J, Boussinesq M, Labrousse F, Nkegoum B, Thylefors BI, Mackenzie CD (2008) Encephalopathy after ivermectin treatment in a patient infected with Loa Loa and plasmodium spp. Am J Trop Med Hyg 78(4):546–551

Kamgno J, Nguipdop-Djomo P, Gounoue R, Téjiokem M, Kuesel AC (2016) Effect of two or six doses 800 mg of albendazole every two months on Loa Loa Microfilaraemia: a double blind, randomized, placebo-controlled trial. PLoS Negl Trop Dis 10(3):e0004492. https://doi.org/10.1371/journal.pntd.0004492. eCollection 2016

Kamgno J, Pion SD, Chesnais CB, Bakalar MH, D'Ambrosio MV, Mackenzie CD, Nana-Djeunga HC, Gounoue-Kamkumo R, Njitchouang GR, Nwane P, Tchatchueng-Mbouga JB, Wanji S, Stolk WA, Fletcher DA, Klion AD, Nutman TB, Boussinesq M (2017) A testand-not-treat strategy for onchocerciasis in Loa loa-endemic areas. N Engl J Med. 377(21):2044–2052. https://doi.org/10.1056/NEJMoa1705026

Kamgno J, Nana-Djeunga HC, Pion SD, Chesnais CB, Klion AD, Mackenzie CD, Nutman TB, Boussinesq M (2018) Operationalization of the test and not treat strategy to accelerate the elimination of onchocerciasis and lymphatic filariasis in Central Africa. Int Health 10(suppl_1):i49–i53. https://doi.org/10.1093/inthealth/ihx051

Klion AD, Nutman TB (2010) In: Guerrant RL, Walker DH, Weller PF (eds) Loiasis and Mansonella infections, in tropical infectious diseases: principles, pathogens and practice, 3rd edn. Churchill Livingstone, New York

Klion A, Massougbodji A, Sadeler B, Ottesen E, Nutman T (1991) Loiasis in endemic and nonendemic populations: immunologically mediated differences in clinical presentation. J Infect Dis 163:1318–1325

Klion AD, Eisenstein EM, Smirniotopoulos TT, Neumann MP, Nutman TB (1992) Pulmonary involvement in loiasis. Am Rev Respir Dis 145:961–963

Klion AD, Massougbodji A, Horton J, Ekoue S, Lanmasso T, Ahouissou NL, Nutman TB (1993) Albendazole in human loiasis: results of a double-blind, placebo-controlled trial. J Infect Dis 168:202–206

Klion AD, Ottesen EA, Nutman TB (1994) Effectiveness of diethylcarbamazine in treating loiasis acquired by expatriate visitors to endemic regions: long-term follow-up. J Infect Dis 169:604–610

Klion AD, Horton J, Nutman TB (1999) Albendazole therapy for loiasis refractory to diethylcarbamazine treatment. Clin Infect Dis Off Publ Infect Dis Soc Am 29:680–682

Klion AD, Vijaykumar A, Oei T, Martin B, Nutman TB (2003) Serum immunoglobulin G4 antibodies to the recombinant antigen, Ll-SXP-1, are highly specifi c for Loa Loa infection. J Infect Dis 187:128–133

Kombila M, Duong TH, Ferrer A, Perret JL, Marion MC, Nguiri C, Gaxotte P, Manfoumbi M, Richard-Lenoble D (1998) Short- and long-term action of multiple doses of ivermectin on loiasis microfilaremia. Am J Trop Med Hyg 58:458–460

Kouam KM, Tchatchueng-Mbougua JB, Demanou M, Boussinesq M, Pion SDS, Kamgno J (2013) Impact of repeated ivermectin treatments against onchocerciasis on the transmission of loiasis: an entomologic evaluation in Central Cameroon. Parasite Vector 6:283

Languillat G, Garin Y, Tursz A, Beauvais B, Lariviere M (1977) Etiological study of low fertility in eastern Gaboon. III. Filarial endemicity (Loa Loa, D. Perstans). Prevalence of microfi lariae in hydroceles (author's transl). Rev Epidemiol Sante Publique 26:273–282

Lebied B (1957) Iconography of the intrasynctial evolution of the Loa Loa in chrysops; preliminary note on the internal and external factors determining the evolutive cycle of Filariata (Scrjabin) in their intermediary hosts. Ann Soc Belg Med Trop 37:641–645

Lukiana T, Mandina M, Situakibanza NH, Mbula MM, Lepira BF, Odio WT, Kamgno J, Boussinesq M (2006) A possible case of spontaneous Loa Loa encephalopathy associated with a glomerulopathy. Filaria J 10(5):6. https://doi.org/10.1186/1475-2883-5-6

McGarry HF, Pfarr K, Egerton G, Hoerauf A, Akue JP, Enyong P, Wanji S, Klager SL, Bianco AE, Beeching NJ, Taylor MJ (2003) Evidence against Wolbachia symbiosis in Loa loa. Filaria J 2:9

Mogoung-Wafo AE, Nana-Djeunga HC, André Domche A, Fossuo-Thotchum F, Bopda J, Mbickmen-Tchana S, Djomo-Kamga H, Joseph Kamgno J (2019) Prevalence and intensity of Loa Loa infection over twenty-three years in three communities of the Mbalmayo health district (Central Cameroon). BMC Infect Dis 19:146. https://doi.org/10.1186/s12879-019-3776-y

Muylle L, Taelman H, Moldenhauer R, Van Brabant R, Peetermans ME (1983) Usefulness of apheresis to extract microfilarias in management of loiasis. Br Med J 287:519–520

Negesse Y, Lanoie LO, Neafie RC, Connor DH (1985) Loiasis: "Calabar" swellings and involvement of deep organs. Am J Trop Med Hyg 34:537–546

Noireau F, Apembet JD, Nzoulani A, Carme B (1990) Clinical manifestations of loiasis in an endemic area in The Congo. Trop Med Parasitol 41:37–39

Padgett JJ, Jacobsen KH (2008) Loiasis: African eye worm. Trans R Soc Trop Med Hyg 102:983–989

Paris L, Datry A, Durepaire R, Felix H, Gaxotte P, Danis M, Gentilini M (1991) Usefulness of ivermectin as initial treatment of loiasis. Presse Med 20:1393–1393

Pedram B, Pasquetto V, Drame PM, Ji Y, Gonzalez-Moa MJ, Baldwin RK, Nutman TB, Biamonte MA (2017) A novel rapid test for detecting antibody responses to *Loa Loa* infections. PLoS Negl Trop Dis 11(7):e0005741

Pinder M (1988) Loa Loa – a neglected filaria. Parasitol Today 4:279–284

Richard-Lenoble D, Kombila M, Rupp EA, Pappayliou ES, Gaxotte P, Nguiri C, Aziz MA (1988) Ivermectin in loiasis and concomitant O. Volvulus and M. Perstans infections. Am J Trop Med Hyg 39:480–483

Richardson ET, Luo R, Fink DL, Nutman TB, Geisse JK, Barry M (2012) Transient facial swellings in a patient with a remote African travel history. J Travel Med 19:183–185

Rodhain F, Rodhain-Rebourg F (1973) A propos de la distribution géographique de la loase. Med Mal Infect 11:8

Rodriguez L, White JM, Richards NQ, You AX, Couser NL (2021) An adult Loa Loa worm in the upper eyelid: an atypical presentation of loiasis in the United States. Case Rep Ophthalmol Med. Article ID 6630875. https://doi.org/10.1155/2021/6630875

Shookhoff HB, Dwork KG (1949) Treatment of Loa Loa infections with hetrazan. Am J Trop Med Hyg 29:589–593

Tabi TE, Befidi-Mengue R, Nutman TB, Horton J, Folefack A, Pensia E, Fualem R, Fogako J, Gwanmesia P, Quakyi I, Leke R (2004) Human loiasis in a Cameroonian village: a double-blind, placebo-controlled, crossover clinical trial of a three-day albendazole regimen. Am J Trop Med Hyg 71:211–215

Taiwo SS, Tamiowo MO (2007) Loa loa meningoencephalitis in southwestern Nigeria west. Afr J Med 26(2):156–159

Toure FS, Egwang TG, Wahl G, Millet P, Bain O, Georges AJ (1997) Species-specifi c sequence in the repeat 3 region of the gene encoding a putative Loa Loa allergen: a diagnostic tool for occult loiasis. Am J Trop Med Hyg 56:57–60

Toure FS, Kassambara L, Williams T, Millet P, Bain O, Georges AJ, Egwang TG (1998) Human occult loiasis: improvement in diagnostic sensitivity by the use of a nested polymerase chain reaction. Am J Trop Med Hyg 59:144–149

Toure FS, Leroy EM, Mavoungou E, Egwang TG (1999a) Sequence conservation of repeat 3 region of the gene coding for the 15 kDa polyprotein within human and simian Loa Loa. J Med Primatol 28:57–61

Toure FS, Mavoungou E, Deloron P, Egwang TG (1999b) Comparative analysis of two diagnostic methods of human loiasis: IgG4 serology and nested PCR. Bull Soc Pathol Exot 92:167–170

Tsague-Dongmo L, Kamgno J, Pion SD, Moyou-Somo R, Boussinesq M (2002) Effects of a 3-day regimen of albendazole (800 mg daily) on Loa Loa microfilaraemia. Ann Trop Med Parasitol 96:707–715

Van Hoegaerden M, Flocard F (1985) Mebendazole treatment of loaiasis. Lancet 1:1278

Van Hoegaerden M, Ivanoff B, Flocard F, Salle A, Chabaud B (1987) The use of mebendazole in the treatment of filariases due to Loa Loa and Mansonella perstans. Ann Trop Med Parasitol 81:275–282

Védy J, Cahuzac G, Labegorre J (1975) Manifestations oculaires atypiques des filarioses à Loa loa. Méd Armées 3:8
Veletzky L, Hergeth J, Stelzl D, Mischlinger J, Zoleko Manego R, Mombo-Ngoma G (2020) Burden of disease in Gabon caused by loiasis: a cross-sectional survey. Lancet Infect Dis 20:1339–1346
Wanji S, Tendongfor N, Esum M, Ndindeng S, Enyong P (2003) Epidemiology of concomitant infections due to Loa Loa, mansonella perstans and onchocerca volvulus in rain forest villages of Cameroon. Med Microbiol Immunol 192:15–21
Wanji S, Tendongfor N, Nji T, Esum M, Che JN, Nkwescheu A, Alassa F, Kamnang G, Enyong PA, Taylor MJ, Hoerauf A, Taylor DW (2009) Community-directed delivery of doxycycline for the treatment of onchocerciasis in areas of co-endemicity with loiasis in Cameroon. Parasite Vector 2:39
Williams P, Crewe W (1967) Studies on the control of the vectors of loiasis in West Africa. 8. An attempt to control loiasis at Kumba, West Cameroon. Ann Trop Med Parasitol 61:148–158
Zoure HG, Wanji S, Noma M, Amazigo UV, Diggle PJ, Tekle AH, Remme JH (2011) The geographic distribution of Loa Loa in Africa: results of large-scale implementation of the rapid assessment procedure for Loiasis (RAPLOA). PLoS Negl Trop Dis 5:e1210

Lymphatic Filariasis (Elephantiasis)

Yao K. Sodahlon, Monique A. Dorkenoo, and John O. Gyapong

Contents

Y. K. Sodahlon (✉)
Mectizan Donation Program, Decatur, GA, USA
e-mail: ysodahlon@taskforce.org

M. A. Dorkenoo

Université de Lomé, Faculté des Sciences de la Santé, Lomé, Togo
e-mail: monicadork@yahoo.fr

J. O. Gyapong

University of Health and Allied Sciences, PMB 31 Ho, Volta Region, Ghana
e-mail: jgyapong@uhas.edu.gh

J. O. Gyapong, B. A. Boatin (eds.), *Neglected Tropical Diseases - Sub-Saharan Africa*, Neglected Tropical Diseases, https://doi.org/10.1007/978-3-031-53901-5_8

Abstract

Sub-Saharan Africa (SSA) carries nearly one-third of the global burden of lymphatic filariasis (LF), a vector-borne and debilitating disease that afflicts the poorest population. *Wuchereria bancrofti* is the only species responsible for the disease and is mainly transmitted to humans by mosquito species belonging to *Anopheles* and *Culex*. The disease is estimated to be transmitted in 35 out of the 48 South Saharan countries with an estimated at-risk population of 432 million people. Recognizing the burden of the disease and the availability of new tools, the World Health Assembly took in 1997 the resolution WHA 50.29 calling for the elimination of LF as a public health problem. The subsequent Global Programme to Eliminate Lymphatic Filariasis was set up with the aim of eliminating the disease by 2020. Although after two decades of program implementation in SSA the initial goal has not been attained, great progress has been made. Indeed two countries (Malawi and Togo) are validated by the world Health Organization (WHO) to have achieved the elimination of LF as a public health problem; two countries (Mali and Uganda) are under post-treatment and the treatment is scaled-up in all implementation units in 23 countries. In 2019, nearly 30% of the 1736 implementation units covering 155 million people completed transmission assessment surveys (TAS) and have stopped mass treatment. The morbidity alleviation, now fully part of the validation process, is getting more attention from the country programmes and their partners.

The goal set for LF in the new WHO 2021–2030 NTD road map is that by 2030, all countries would be either validated or entirely under post treatment surveillance. If commitment is maintained at all levels, this goal is achievable as opportunities and alternative treatments including albendazole alone in *Loa loa* endemic countries and triple therapy in countries where there is no onchocerciasis, are available to accelerate LF elimination in endemic countries.

Introduction

Lymphatic filariasis (LF) is a debilitating mosquito-transmitted disease caused by the helminthes *Wuchereria bancrofti*, *Brugia malayi*, and *Brugia timori* (WHO 1992). According to recent reports, the disease is endemic in 72 countries with an estimated at-risk population of 1.39 billion people. Among the estimated 120 million people infected, 40 million suffer from the stigmatizing and disabling clinical manifestations of the disease, including 15 million who have lymphedema (elephantiasis), and 25 million men who have urogenital swelling, principally scrotal hydrocele (WHO 2010; Zeldenryk et al. 2011).

In recognition of the global burden of LF, the World Health Assembly passed the resolution WHA 50.29 in 1997, calling for collaborative efforts by member states to eliminate the disease as a public health problem (WHA 1997). In 2000, the Global Program to Eliminate Lymphatic Filariasis (GPELF) was formed in response to the goal to eliminate the disease as a public health problem by 2020. The program adopted a two-pronged strategy: (1) to interrupt transmission of the causal parasite and (2) to alleviate morbidities associated with the disease (Ottesen 2000).

In sub-Saharan Africa (SSA), the focus of this chapter, *W. bancrofti* is the only species prevalent (WHO 1992) and is mainly transmitted to human by mosquito species belonging to *Anopheles* and *Culex* (Brengues 1975; Simonsen et al. 2008). The disease is endemic in 35 out of the 48 Sub-Saharan countries with a total at-risk population estimated at 432 million people accounting for nearly one-third of the global burden (Addiss and Global Alliance to Eliminate Lymphatic 2010; WHO 2012a). LF is co-endemic with other neglected tropical diseases (NTDs) in settings of extreme poverty, making efforts to control/eliminate a key strategic component for achieving Africa's Sustainable Development Goals (SDGs) (Malecela 2019).

By 2020, following 20 years of LF elimination efforts in SSA countries, tremendous achievements had been made, however only two countries, Togo and Malawi met the 2020 elimination goals, interventions have not started in two countries, and six countries are yet to extend mass drug administration (MDA) to all endemic districts. Overall, two-thirds of the population at-risk in SSA still require mass treatment and morbidity control continues to lag behind.

In the WHO NTD 2021–2030 road map (WHO 2020a) which was endorsed by the member states during the 73rd World Health Assembly in November 2020, the new goal set for LF is to accelerate interventions to reach by 2030 the validation of LF in 81% endemic countries and 100% endemic implementations units to be under post-treatment surveillance. The achievement of these goals will require more commitment from the governments to increase investment in their national programs.

Biology, Life Cycle, Disease Presentation

Parasite

In SSA, *Wuchereria bancrofti* is the species that causes lymphatic filariasis in humans (WHO 1992). The adult worms with a length of around 4–8 cm live in the lymphatic system, especially the lymphatic vessels and the nodes. The worm lives in mutual symbiosis with *Wolbachia endobacteria* which is essential for larval development and sex determination and responsible for a major part of the inflammation leading to symptoms (Arumugam et al. 2008). The adult worm can survive 5–8 years. During this period, it releases into the lymphatic system and the blood millions of 100–150 μm long-sheathed microfilariae (L1 stage), at approximately 10,000 daily. The microfilariae (mf) are found in the peripheral blood all the time with a peak density between 9 pm and 4 am, concurrent with the period of most active feeding by the local vectors (Brengues 1975; WHO 1992). The nocturnal periodicity and the morphology of the mf are used to differentiate *W. bancrofti* infection from other species such as such us *L. loa* and some *Mansonella* species also endemic in SSA (Paily et al. 2009).

Vectors

Despite the absence of specific entomological data from many countries, the available bibliography suggests that *Anopheles gambiae s.1.*, and *An. funestus* are the major vectors of *W. bancrofti* throughout SSA and the Indian Ocean Islands (Brengues 1975; Boakye et al. 2004; Simonsen et al. 2008). The same vectors are involved in the transmission of malaria which offers a good opportunity for integrated vector control (Bockarie et al. 2009; Ashton et al. 2011). In Eastern Africa, especially in urban and semi-urban areas of the coastal areas, *Culex quinquefasciatus* is involved in the transmission of the disease (Hawking 1957; Simonsen et al. 2008). A study conducted in Ghana by Ughasi et al. (2012) reported that *Mansonia africana* and *Mansonia uniformis* are associated with the transmission of *W. bancrofti* in Ghana and possibly in other parts of West Africa.

Transmission and Life Cycle

The extrinsic life cycle of the parasite is initiated when the mf are ingested by a mosquito vector during feeding on the host blood (Fig. 1). On maturity, most of the infective L3-stage larvae migrate to the head and proboscis of the mosquito, ready to be transmitted to the host during subsequent feeding. They develop to the adult L5 stage and the period of development and the longevity of the parasites vary according to the species of the nematode and the mammalian host (Paily et al. 2009). It was demonstrated in West Africa that in the absence of reinfestation, the

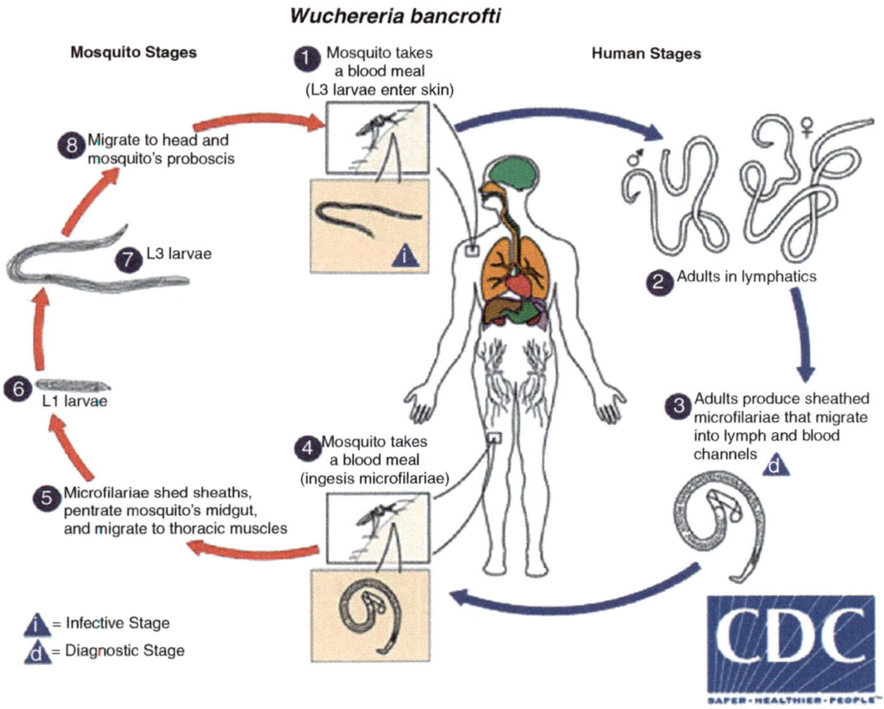

Fig. 1 Life cycle of *Wuchereria bancrofti*

microfilaremia can spontaneously disappear in about 5 years (Brengues 1975) corresponding to the life span of the adult worm.

Clinical Manifestations

With the discovery of more sensitive diagnostic tools, it is now well documented that the infection is acquired in childhood, a period during which damage to the lymphatic system takes place (Witt and Ottesen 2001; Shenoy and Bockarie 2011). Bancroftian filariasis is characterized by a wide range of clinical manifestations. The common clinical manifestations are acute adenolymphangitis (ADLA) (Figs. 2 and 3), lymphedema (Figs. 4 and 5), elephantiasis (Fig. 6), and hydrocele (Figs. 7 and 8) (WHO 1992).

These symptoms and signs often cause considerable physical, psychological, and social disabilities to the patients (Krishna Kumari et al. 2005; Zeldenryk et al. 2011) and source of income loss (Babu et al. 2002). Other reported manifestations with no public health importance are chyluria, lymphocele, scrotal lymphedema, tropical pulmonary eosinophilia, adenopathy, and hematuria (WHO 1992).

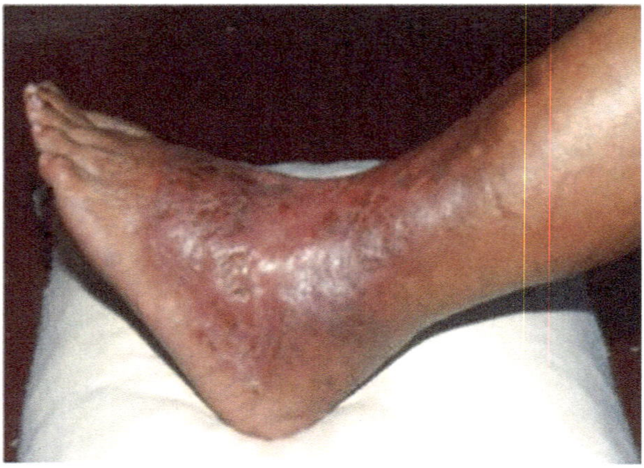

Fig. 2 Acute adenolymphangitis (ADL)

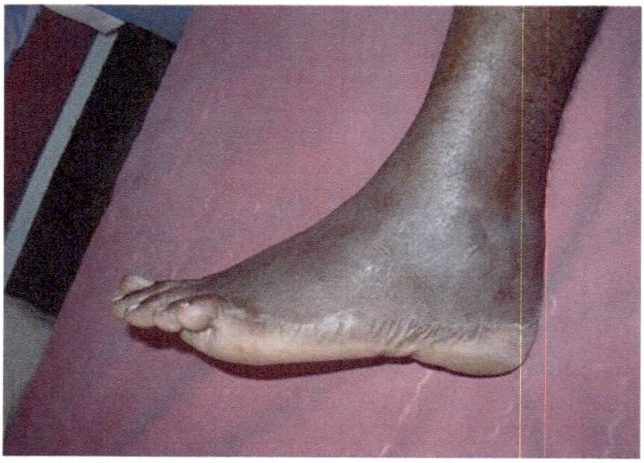

Fig. 3 Same patient recovering from ADL with peeling of skin and some residual swelling

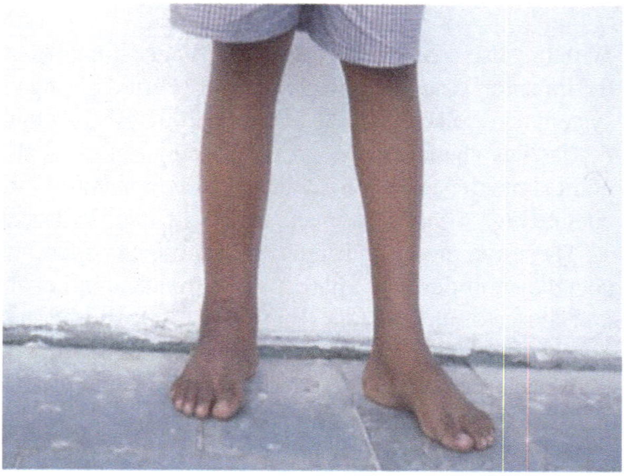

Fig. 4 Early lymphedema of the right foot

Fig. 5 Advanced lymphedema

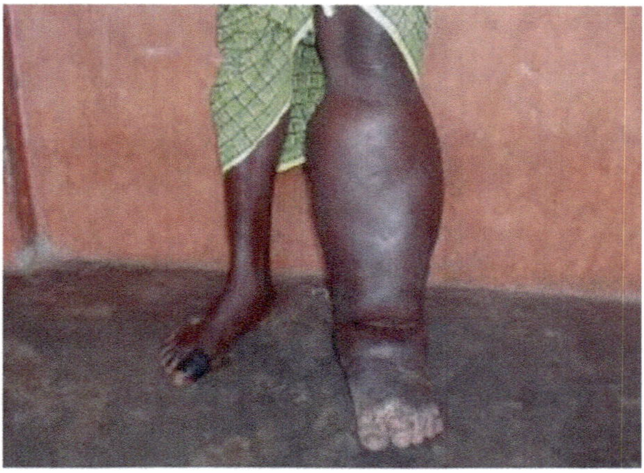

Fig. 6 Elephantiasis

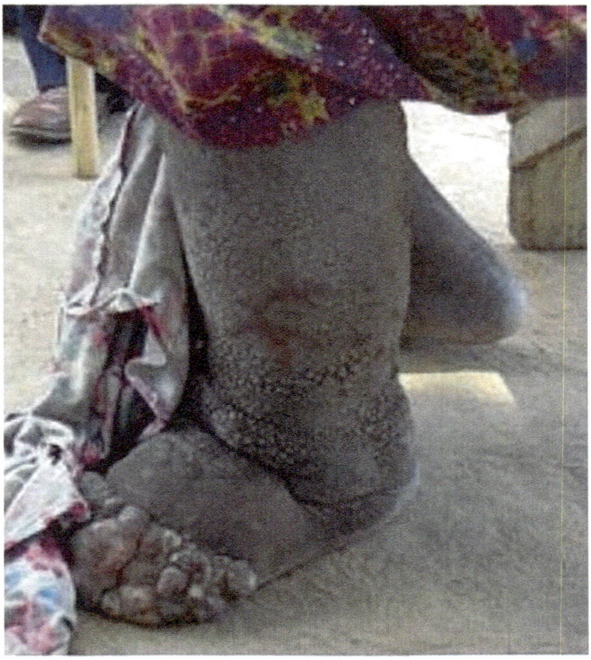

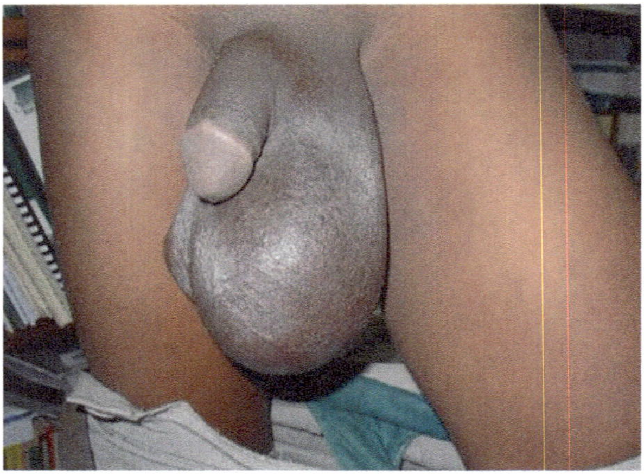

Fig. 7 Hydrocele

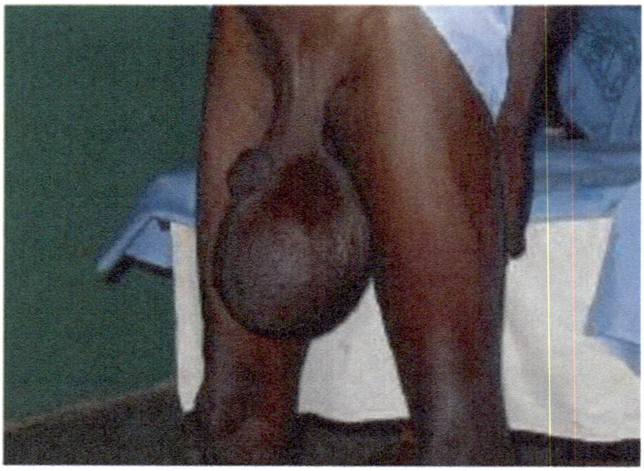

Fig. 8 Advanced hydrocele

Burden of Disease and Distribution in SSA

Distribution of the Infection

The discovery of the filarial antigen-based diagnostic test that can be used any time of the day (Weil et al. 1997) and the subsequent development of the Rapid Assessment of Geographical Distribution of Filariasis (RAGFIL) methodology by the Special Programme for Research and Training in Tropical Diseases (TDR) (Gyapong and Remme 2001) have accelerated the mapping of LF in SSA. The mapping data (Fig. 9) suggested that the disease has focal distribution, and the intensity of the infection varies by region. In the East-Southern African block, the endemicity levels are highest along the coastal areas and decline inwardly (Njenga et al. 2008; Simonsen et al. 2010). In the West-African block of countries, the pattern is variable in relation to coastal endemicity levels along the whole coast, but generally high prevalence rates are in inland countries (Gyapong et al. 1996a, b, 2002).

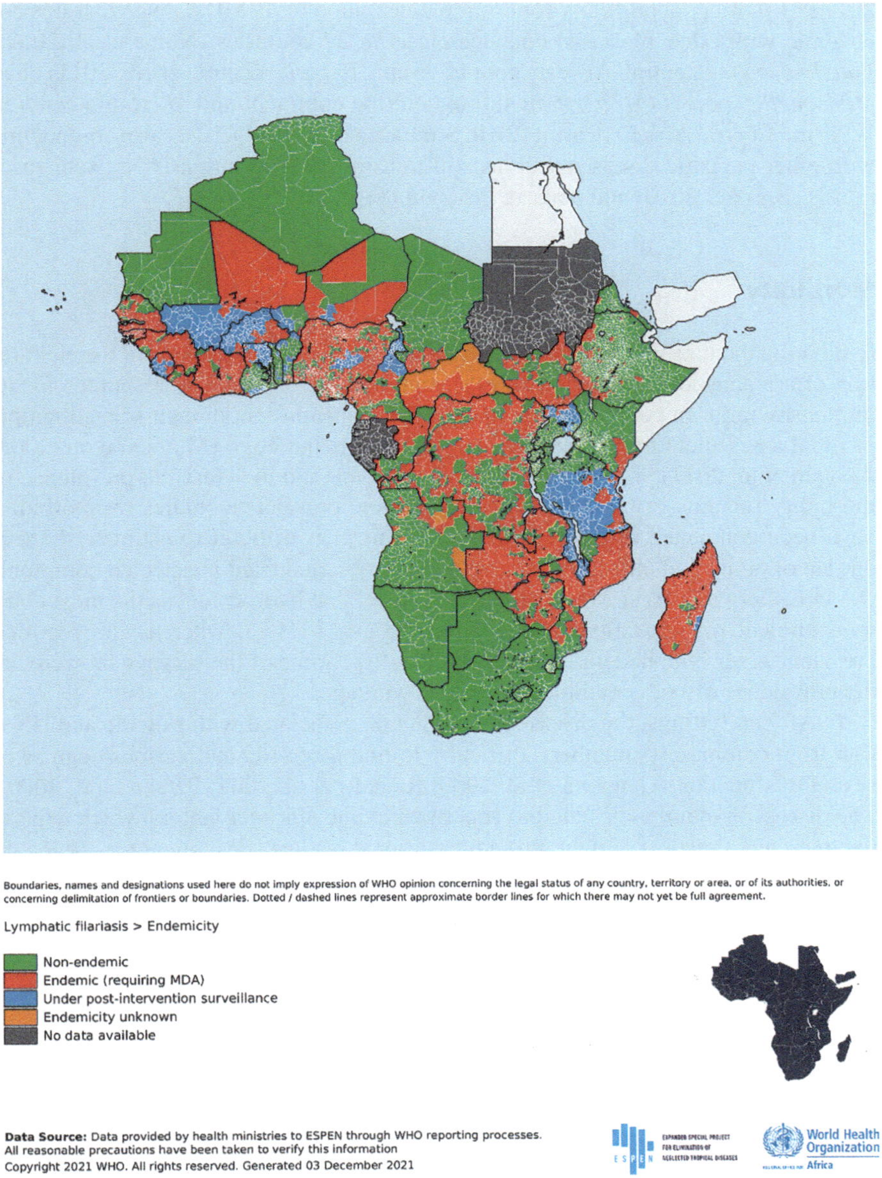

Fig. 9 Endemicity map in SSA as of Dec. 2019 (Courtesy ESPEN-Afro)

In Central Africa, the endemicity is relatively low (Richard-Lenoble et al. 1980; Kelly-Hope et al. 2018; Wanji et al. 2019).

Four countries (Burundi, Cape Verde Island, Mauritius, Rwanda, and Seychelles) have been removed from the list of endemic countries requiring intervention for elimination following the review of the endemicity data by WHO in 2011 (WHO

2011a). LF offers opportunity for co-implementation of its MDA given that it is co-endemic with other filariasis: onchocerciasis in 29 countries (Noma et al. 2002), *Loa loa* in 11 Central African countries and Nigeria (Zoure et al. 2011), and *Mansonella* species (*M. perstans* throughout the continent and *M. streptocerca* in West and Central Africa) (Carme 2010; Simonsen et al. 2011). LF is also co-endemic with other parasitic diseases such as soil-transmitted helminthiasis and schistosomiasis (Brooker 2010) and malaria (Manguin et al. 2010).

Morbidity

Estimating the morbidity prevalence in a country is the first step toward the elaboration of a management program and the mobilization of adequate resources for its implementation. In contrast to the mapping of transmission, though some attempts of rapid assessment have been tried out in Nigeria and Togo (Mathieu et al. 2008; Akogun et al. 2011), a validated standard rapid method to assess the prevalence of morbidity remains to be developed. Only a few prevalence studies on morbidity have been conducted in the region. As a result, it is difficult to estimate the real burden of clinical manifestations in SSA. Surveys conducted in selected communities in Eastern and West Africa (Table 1) showed that hydrocele was the most common clinical manifestation among adult males (7–34%), whereas the reported prevalences for lymphedema were lower at 0–9%. Some of the surveys showed that the prevalence of both conditions increases with age.

In African settings, the disease is found to be associated with ridicule and isolation from community members, difficulty to find a spouse, and various degrees of sexual dysfunction (Gyapong et al. 2000; Richard et al. 2007; Person et al. 2009). The disease is mainly attributed to supernatural and spiritual factors which explain the treatment practices sought through traditional healers (Gyapong et al. 1996a, b; Richard et al. 2007). Acute adenolymphangitis (ADL) is the major acute complication of both conditions. The incidence is estimated up to 95.9 per 1000 per annum among adults more than 10 years of age (Gyapong et al. 1996a, 2000). Individuals with lymphedema experienced more frequent acute episodes compared to those

Table 1 Prevalence of hydrocele and lymphedema in select SSA communities

Country	Hydrocele (in men) %	Lymphedema %	Authors
Ethiopia	20.3	0	Jemaneh and Kebede (1995)
Kenya	10–34	2.4–8.5	Estambale et al. (1994), Mukoko et al. (2004) and Njenga et al. (2007)
Malawi	13–17.9	1.3–3.7	Nielsen et al. (2002)
Tanzania	14.5–21.3	0.6–3.3	Simonsen et al. (1995)
Uganda	7–28	4–9	Onapa et al. (2001)
Ghana	32	3.6	Gyapong et al. (1994)
Nigeria	7–9.7	2.3–6	Okon et al. (2010) and Uttah (2011)

with hydrocele. ADL episodes ranged from one to five per annum. Each ADL episode can last on average 8.6 days and in 72.5% of the episodes, the affected individuals were incapacitated and unable to do their normal activities for an average duration of 3.7 days (Gasarasi et al. 2000).

History of LF Control in SSA

Lymphatic filariasis control in Africa began in the 1940s with two main strategies: the use of diethylcarbamazine (DEC)—commonly called Banocide or Hetrazan or notezine- and mosquito control (which included source reduction). The strategies, however, were limited to small areas and pilot projects to determine whether the strategies would work. The chemotherapeutic approach was either treatment with DEC tablets or the use of DEC fortified salt. In East Africa attempts to control LF were carried out in the Ukara Islands of Lake Victoria in Tanzania mainland (then Tanganyika) and in the Pate Islands of Kenya (Hawking 1962). In his paper "the eradication of filariasis on Ukara Island", Jordan P. explains the approach of treating all microfilaraemics to reduce transmission and eventually eliminate the disease (Jordan 1959). Similar work was also carried out in prisoners (Davis and Bailey 1969). In West Africa, similar work was done in Kerebe in the Gambia whereby DEC at doses of 5 and 2.5 mg/ kg, respectively, were given to microfilaraemia-positive persons residing in the area (McGregor et al. 1952; McGregor and Gilles 1956). Other work in West Africa includes work done in Liberia (Chlebowsky and Zielke 1980; Zielke and Chlebowsky 1980) and in Senegal (Diallo et al. 1983). In the 1970s the importance of vector control as a strategy to control LF became more evident (White 1971), and hence there was a focus on spraying source reduction (Bushrod 1979), and in some later studies even building design that would reduce man-vector contact (Kolstrup et al. 1981). The efforts were carried out at a research level and were not coordinated.

Makunduchi in Zanzibar (Maxwell et al. 1990) carried out a more coordinated and comprehensive approach, which involved killing larvae breeding in pit latrines using polystyrene beads together with the distribution of DEC. The intervention resulted in a major reduction in mosquito infectivity and a reduction of the infection in humans in a specific part of Unguja Island. Larviciding with *Bacillus sphaericus* was also carried out in open drains in the Zanzibar town (Maxwell et al. 1990). All in all, most efforts in Africa focused on small areas with an aim to gain information on what strategies would be feasible for large-scale control. It was clear, however, that although some tools were available they could not be used widely. For example, DEC, which was very effective in some areas, could not be used in onchocerciasis endemic areas for fear of ocular toxicity and the Mazzotti reaction. The Mazzotti reaction which may include anaphylactic shock is a result of the death of the microfilaria in the skin.

With more research came new tools; this included information that combinations of ivermectin and albendazole were effective in killing microfilariae (Dunyo et al. 2000; Horton et al. 2000). Then there was the development of rapid immunochromatographic tests (Weil et al. 1997), which would allow testing to be done during the day. This was a major revolution over the night blood testing which was inconvenient to the community. It was a major effort to get people to come and get tested between 9 pm and 4 am. The new test made the possibility of testing during the day a reality. The major turning point for the LF program came with the adoption in 1997 of the World Health Assembly Resolution WHA 50.29. This was followed by donation of two drugs, albendazole and ivermectin, which were to be used for the interruption of transmission of LF in areas in Africa where LF and onchocerciasis were co-endemic. The resolution urged the member states among other things: "to strengthen local programmes and their integration with the control of other diseases, particularly at community level, in order to implement simple, affordable acceptable and sustainable activities, based on community wide treatment strategies, but supplemented where feasible by vector control and improved sanitation." The resolution urged the Director General among other things "to mobilize support for global and national elimination" (WHA 1997).

This resolution was the basis for defining a Global Strategy for Control of Lymphatic Filariasis. The strategy focused on mass chemotherapy using ivermectin and albendazole in countries in Africa where LF and onchocerciasis were co-endemic and DEC and albendazole in countries non-endemic for onchocerciasis. The other part of the strategy involved morbidity management for people who already had the debilitating manifestations of the disease. The focus was on alleviating symptoms of acute filarial attack and preventing bacterial superinfection in people with lymphedema and promoting surgery for those with hydroceles (Ottesen 2000). It is against this backdrop that National Lymphatic Filariasis Elimination Programs (NLFEP) began in Africa under the auspices of the WHO Global Programme to eliminate LF.

Current Control Tools and Strategies

The control of lymphatic filariasis in SSA is currently based on the strategies adopted globally aiming to eliminate the disease as a public health problem. The two adopted strategies are (1) to interrupt transmission of the causal parasite and (2) to alleviate morbidities associated with the disease (Ottesen 2000). Tools are developed and resources made available to facilitate the implementation of these strategies in the endemic countries.

Interrupt the Transmission of the Causal Agent

The major recommended tool to interrupt transmission is once yearly treatment with a single dose of two medicines administered together to the entire eligible at-risk population: albendazole (400 mg) plus either ivermectin (150–200 mcg/kg) or DEC (6 mg/kg) for 4–6 years.

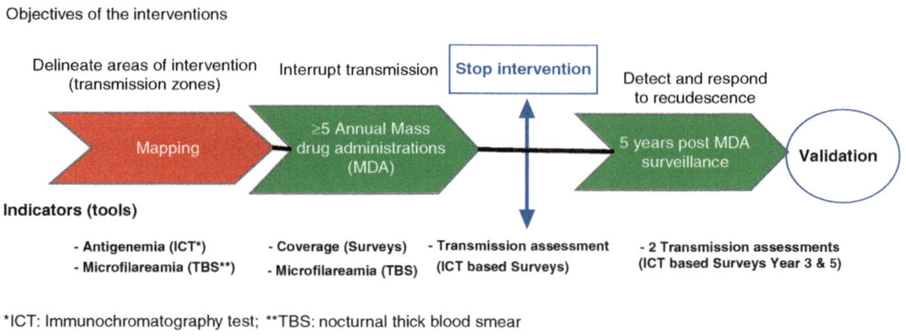

Fig. 10 Key programmatic steps toward the elimination of LF transmission

In 2017, WHO recommended triple administration of ivermectin+DEC + albendazole (IDA) once yearly to accelerate the elimination of transmission in settings where onchocerciasis is not endemic (WHO 2017a). Clinical trials demonstrated good safety and higher efficacy of the IDA compared to the association DEC+ albendazole and will likely require fewer rounds to interrupt transmission (King et al. 2018; Weil et al. 2019).

Three key steps to be followed by the national program for LF elimination are summarized in the Fig. 10.

Mapping

To identify the implementation units (IUs) that are eligible for (MDA), the mapping is conducted using rapid immunochromatographic card tests (ICT)– now replaced by filariasis test strip (FTS)- based on WHO recommended rapid method (RAGFIL) (Gyapong and Remme 2001). An IU, generally a district, is eligible for MDA if the circulating filariasis antigen (CFA) ratio is $\geq 1\%$.

The strategy served the global LF programme well as a low-cost strategy for scaling up MDA, particularly in settings where the index of suspicion that LF was present was high. Unfortunately, the strategy was less useful where programme managers lacked LF-specific information, leaving a number of countries with districts whose mapping results were not sufficiently convincing to make decisions about whether or not to begin MDA. A protocol, adapted from the transmission assessment survey (TAS), was developed based on testing of school-age children (9–14 years of age) for filarial antigen or antifilarial antibody. This confirmatory mapping protocol (also referred to as the mini-TAS) has translated into significant resource savings where the need for MDA for LF was averted based on the findings with this new protocol (Gass et al. 2017).

Cross-Reactivity of Filariais ICT Cards is reported in areas endemic for *Loa loa* in Central Africa (Bakajika et al. 2014; Wanji et al. 2015) therefore, positive cases

Table 2 Progress of LF elimination programmes in the SSA (as of Dec 2020); (WHO 2020b)

Status		Countries
Mapping	Not completed (n = 0)	None
Mass drug administration	Not required (*n* = 6)	Burundi, Cape Verde, Gambia, Mauritius, Rwanda, Seychelles,
	Not started (*n* = 2)	Gabon, Equatorial Guinea
	Scaling up (*n* = 6)	Angola, Central African Republic, Nigeria, Madagascar[a], South Sudan, Sudan,
	Full scale/ Downscaling (n = 23)	Benin, Burkina Faso, Cameroon, Chad, Comoros[a], Congo, Cote d'Ivoire, Democratic Republic Congo, Eritrea[a], Ethiopia, Ghana, Guinea, Guinea-Bissau, Kenya[a], Liberia, Mozambique, Niger, Senegal, Sierra Leone, Sao Tome & Principe[a], United Republic of Tanzania, Zambia[a], Zimbabwe[a]
Post-MDA surveillance	Countrywide (n = 2)	Mali, Uganda
Validated as a public health problem	n = 2	Malawi, Togo

[a] Countries eligible to use DEC + Albendazole or ivermectin+DEC + albendazole (IDA) for mass drug administration

must undergo a confirmatory test, generally night blood smear (Pion et al. 2016; Wanji et al. 2019).

Mass Drug Administration

The following 28 South Saharan African countries where onchocerciasis is co-endemic are eligible for the use of combination of ivermectin (donated by Merck & Co, Inc., White House Station, NJ, USA[1]) and albendazole (donated by GSK, Brentfort, UK) for LF elimination: Angola, Benin, Burkina Faso, Cameroon, Central African Republic, Chad, Congo-Brazzaville, Côte d'Ivoire, Democratic Republic of Congo, Equatorial Guinea, Ethiopia, Gabon, Ghana, Guinea, Guinea-Bissau, Malawi, Mali, Mozambique, Niger, Nigeria, Senegal, Liberia, Sierra Leone, Sudan, South Sudan, United Republic of Tanzania, Togo, and Uganda. The remaining nine countries (marked [a] in Table 2) will be using DEC-based therapy (supplied for free by Eisai Co., Ltd., Tokyo, Japan). The nine countries are also eligible to use triple therapy (ivermectin+DEC + albendazole) to accelerate the elimination of LF.

[1] Known as MSD outside USA and Canada

The Distribution Strategies

In the endemic areas, the treatment is provided door to door to individuals of 5 years old and above using a network of community volunteers. The target is to reach at least 65% of the total population. The under-5-year-old children, pregnant women, and heavily sick individuals are excluded from the treatment. In settings where LF is co-endemic with onchocerciasis, the MDAs are integrated targeting both infections. The integrated distribution strategies targeting other tropical diseases are more and more promoted in the continent (Kabatereine et al. 2010). The implementation of the recommended bi-therapy is contraindicated in areas where LF is co-endemic with *Loa loa* as it can be associated with the occurrence of severe adverse events (Twum-Danso 2003). Recent clinical trials demonstrated that albendazole in monotherapy twice per year is also effective in reducing the LF microfilaremia (Pion et al. 2015; Pion et al. 2020; Ouattara et al. 2021) therefore, WHO recommended biannual albendazole mass treatment along with vector control for malaria as a safe alternative strategy in communities where *Loa loa* is co-endemic (WHO 2017a).

Monitoring and Evaluation

The key indicators to monitor during the MDAs are the reported and surveyed geographic and therapeutic coverages and the impact of the intervention on nocturnal microfilaremia in sentinel and spot-check sites. A manual has been developed and regularly updated by WHO to guide the program managers (WHO 2011b).

Stopping MDA

A transmission assessment survey (TAS) protocol has been developed by WHO to determine when to stop MDA in a given IU (WHO 2011b, 2012b). It is estimated that following at least five rounds of MDA with effective coverage (≥65% of the total population) and a microfilaremia ≤1% in sentinel and spot-check sites, an implementation unit can be eligible to undergo the TAS.

Post-MDA Surveillance

The protocol currently recommended by WHO is to repeat the TAS 3 and 5 years following the end of MDAs. At least two additional rounds of MDAs must be implemented in an evaluation unit that does not satisfy the criteria set for a successful post MDA TAS. A country will be ready to submit a dossier to WHO for validation if all the endemic IU successfully passed the 4–5 years post-MDA surveillance and if the recommended package of care is provided in all areas in which there are patients with lymphoedema or hydrocele (WHO 2017b, 2019).

Supplemental lab-based post-treatment ongoing surveillance targeting both endemic and non-endemic districts was developed and implemented in Togo until the country was validated in 2017 (Mathieu et al. 2011; Sodahlon et al. 2013; Dorkenoo et al. 2018).

Other Interventions

Integrated Vector Management

Because the parasite of LF is transmitted mainly in SSA by *Anopheles* mosquitoes, there is a growing advocacy for the use of integrated vector management to accelerate and sustain the elimination (Manga 2002; Bockarie et al. 2009; WHO 2011c).

The recent effort to reach universal coverage with insecticide-treated long-lasting nets (ITNs/LLINs) and the implementation of indoor residual spraying (IRS) for malaria control in Africa may significantly impact LF transmission in the region (Ashton et al. 2011; Kelly-Hope et al. 2013).

Drug Development

Anti-*Wolbachia* Therapy

The anti-*Wolbachia* therapy is currently under development based on the use of antibiotics to target the *Wolbachia* endosymbionts in *W. bancrofti* (Slatko et al. 2010). However, at present the recommended antibiotic doxycycline seems unlikely to be suitable for large-scale mass chemotherapy due to the long course (6 weeks) of daily treatment required to obtain desirable effects and the fact that this drug is contraindicated in young children and pregnant or potentially preg-nant women. Research is underway to identify alternative antibiotics that exert an anti- Wolbachia effect with a shorter course of treatment and good safety (Taylor et al. 2019).

Alleviate Morbidities Associated with the Disease

Despite the availability of effective measures to treat and prevent clinical manifestations associated with lymphatic filariasis, the implementation of this second pillar is still lagging. Only a few countries are implementing lymphedema management at scale. NGDOs such as Handicap International, IMA World Health HDI, HKI etc. supported some projects in Benin, Burkina Faso, Cameroon, Ethiopia, Ghana, Madagascar, and Togo. An example of an innovative fund from Tanzania is the President Kikwete LF Fund, which was established in April 2008 to support patients suffering from the debilitating effects of lymphatic filariasis (Malecela et al. 2009). One of the first objectives of the project was to address the backlog of men needing hydrocele surgery. The total backlog at the time was 25,000 men.

The inclusion in 2017 of the morbidity management and disability prevention (MMDP) indicators in the validation dossier (WHO 2017b) and the publication by WHO of a toolkit to facilitate implementation resulted in growing interest from countries and funding partners to scale up interventions at country level. The dossier will require that national LF elimination programmes provide (1) data on the number of patients with lymphoedema (or elephantiasis) and hydrocele, (2) the number of health facilities designated to provide care, and (3) the readiness and quality of the care provided. Mobile technology using Short Messaging Service (SMS) i.e. text messages were used in Ethiopia, Ghana, and Tanzania for case finding (Mwingira et al. 2017; Martindale et al. 2018), first step toward setting up an MMDP project.

Lymphedema Management

The minimum treatment package recommended by WHO includes (1) hygiene, (2) treatment and prevention of skin lesions trough, (3) exercises, (4) elevation of the afflicted limb, and (5) the wearing of comfortable shoes. The patients are trained to implement each component of the package daily. Training materials have been developed by WHO to promote this home-based self-care in the endemic countries. The evaluation in African settings showed that the home-based management of lymphedema is well accepted by the patients and has resulted in a drastic reduction of acute attacks and improvement of the quality of life of the patients enrolled in the program (Jullien et al. 2011; Mathieu et al. 2013).

Hydrocele Management

Three techniques for the management of hydrocele are available: (1) aspiration and sclerotherapy (Khaniya et al. 2009) (2) eversion of the tunica vaginalis (Thomas et al. 2009), and (3) excision of the tunica vaginalis (Noroes and Dreyer 2010). None of them is found to be the gold standard technique or recommended by WHO. Nevertheless, the procedure with total tunical resection and closure of the wound without drainage is the one being currently promoted in SSA especially in West Africa through a project supported by a Norwegian NGDO, Health and Development International (Mante and Gueye 2011). In 6 years, the initiative enabled training of 214 surgeons in ten countries and the operation of 3000 cases. Projects of this kind will help clear the backlog of hydrocele surgery.

Validation of the Elimination of Lymphatic Filariasis As Public Health Problem

Countries seeking official acknowledgement from WHO as having met the criteria for elimination of LF as a public health problem should submit a dossier to the World Health Organization documenting the measures taken and the evidence supporting the claim. Validation criteria and process are detailed and a template developed in a manual published by WHO (2017b). Following the successful implementation of the two pillars of the elimination programme, a dossier is submitted to WHO for review by an ad hoc Regional Reviewing Authority. If the review concludes that the country meets the validation criteria, the WHO Director-General will send a letter of official acknowledgment to the country through the WHO Regional Office.

Post Validation Activities

Countries should continue to conduct post-validation surveillance for early detection and response to potential recrudescence. MMDP activities must continue and be fully integrated into the health services. Affordable and sustainable post-validation guides are yet to be developed by WHO (2017a, b).

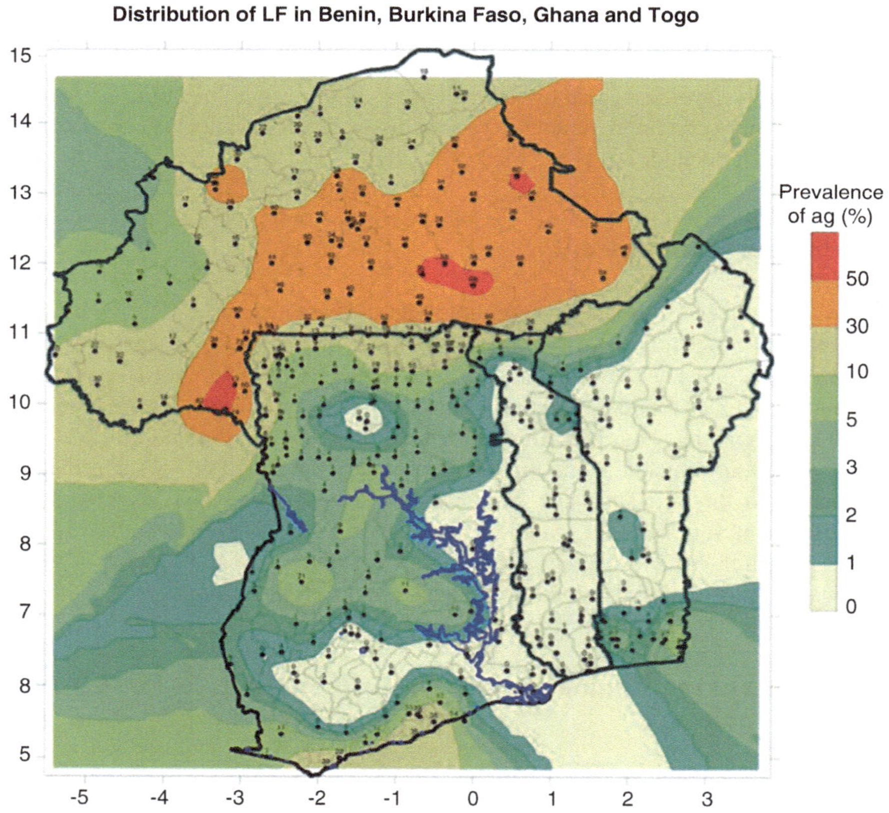

Fig. 11 LF endemicity map for West Africa (Gyapong et al. 2002)

Status of Implementation of LF Elimination Programmes in SSA

The programs always started off with mapping to find out the real extent of the infection and were based on measuring circulating filarial antigens in 100 people in an implementation unit. The methodology was based on studies on Rapid Geographical Assessment of Lymphatic Filariasis (RAGFIL) (Gyapong and Remme 2001; Gyapong et al. 2002; Fig. 11) which compared the use of the ICT test, BinaxNOW (now replaced by Filariasis Test Strip (FTS, Abbott) (Chesnais et al. 2017)) as a rapid method of determining the prevalence of LF in a particular area.

Following the development of initial country plan, NLFEP programs were launched in Africa in 2000 with three countries starting off namely, Ghana, Tanzania, and Togo. The programs initially focused on MDA scaling up implementation units year after year. The programs then added a morbidity component at different stages of the program implementation. Out of the 35 countries in South Saharan Africa where MDA may be required for LFE with at-risk population estimated at 432 million, 33 currently have functional programs including Malawi and Togo under post validation surveillance (see Table 2).

In 2019, nearly 340 million people required treatment in 32 countries. Nearly 30% of the 1736 implementation units covering 155 million people have completed transmission assessment surveys (TAS) and have stopped MDA. IDA is being implemented in three countries (Kenya, Madagascar and Sao Tome & Principe) to accelerate the elimination of LF (WHO 2020b).

The program has been implemented at the village level with strong community support. Different approaches have been tried and tested in different places, but the emphasis has been to focus on what works in the various communities. In some countries it has been the use of community drug distributors (CDDs) that were and are being used by the Onchocerciasis Programme in the community-directed treatment approach (Gyapong et al. 2001; Homeida et al. 2002). In Zanzibar, an approach of interpersonal communication COMBI (WHO 2002) was what was used where the distributors also gave intense health education on their visits. All in all, there was no one size fits all and countries had to adapt to the delivery methods that allowed them maximum efficiency with limited disruption to health system activities.

All countries had specific training programs where the distributors were trained either directly by staff at the national level or via a cascade method where the national level trained the trainers of trainers who then went on to train others. In Tanzania, for example, the national trainers trained the regional trainers. The regional trainers then trained the district leaders who then went out and trained the village health workers (Malecela et al. 2009). Messages were kept very simple throughout to ensure that key information was imparted to the distributors. These messages included what to do before the program began, including the importance of census for getting accurate numbers, what to do during MDA, and finally what to do after the MDA including ensuring that data summaries were sent to the districts. Another important component of the program is "advocacy and social mobilization," which also differs from country to country. This is the key to the success of the programs and needed to be carried out at all levels. In some cases, even within countries, social mobilization needed to be tailor-made to fit particular communities, e.g., migrants, fishermen, etc.

Challenges of Program Implementation

Like all public health programs, the LF program has many challenges as set out below.

Coverage

Coverage (geographical and treatment) is a major problem, and this is for several reasons. One of the main reasons was the lack of clear understanding as to why one should take drugs when one was not sick or showing any symptoms of the disease. There were also myths and rumors about the drugs in that the drug was being used for birth control and that it could cause erectile dysfunction. In some areas it was because the social mobilization had not been inclusive. For example, engaging religious leaders and traditional healers was helpful in some areas of Tanzania. The issue of coverage becomes an even greater concern in areas of low endemicity where the symptoms of the disease, i.e., hydrocele and elephantiasis, are less obvious, hence making it difficult to convince people of the fact that they could be carrying the disease.

Funding

One of the other key problems of getting programs to expand has been funding. Initially partners supported programs to develop a proof of concept. They then insisted that countries take on the expansion which has been very slow in some of the countries. Districts were asked to budget for LF in a number of countries where a decentralized health system existed, but this was almost impossible given the competing priorities at district level. A clear approach to funding had not been established, and while the medicines are made available and there was a clear expansion plan for the elimination, there was no corresponding domestic financial plan to go with it. The advent of the integrated NTD program has allowed new funding opportunities for preventive chemotherapy programs to be used to scale up interventions for all seven PCT diseases (Kabatereine et al. 2010; Hanson et al. 2012; Hooper et al. 2013; Wainwright et al. 2020). This in some way has helped mobilize the core funding needed to support delivery while at the same time expanding the programs at an unprecedented pace in supported countries. The challenge now is that although the different disease programs are integrated for better functioning, they need to be integrated into the health system. Emerging diseases with outbreak potential such Ebola virus disease in West Africa (Thomas et al. 2017) and currently the Covid-19 pandemic (Molyneux et al. 2020) have high potential to divert scarce resources and break the elimination dynamic. Empowering country level actors to maintain pressure on their Governments to increase domestic funding for NTDs must remain a priority (Sodahlon et al. 2020).

Capacity

Lack of capacity to implement the programs is a challenge. This has been addressed in the earlier years by training through support from the Bill Melinda Gates Grant to the Global Alliance to Eliminate Lymphatic Filariasis, which focused predominantly on training researchers at Masters and PhD level. The program was designed such that the students would work within their country programs on an issue relevant to the program research needs. This method has developed a cohort of leaders who are now leading NTD control programs in several countries in Africa. Capacity strengthening, however, includes it at the national level right down to the district level and finally to the village level where all the major work is done. Countries have adapted several capacity building models, and different programs have focused on different key areas: financial management, supply chain management, and interpersonal marketing skills. There is still a gap in the capacity needed to run these programs. It is envisaged that integrated NTD programs may be the answer to this crisis as all the single disease focal persons are now working together in one integrated program. Skills like entomology and pathology, which are still essential for LF, however, are yet to be addressed.

The Lagging Behind Countries

Another major challenge is that while great progress is being made in Africa, as shown in Table 2, there are still countries where interventions have not yet started ($n = 2$) and countries where intervention are not fully scaled up to all endemic districts ($n = 6$) (WHO 2020b). The common reasons for slow program expansion are lack of political commitment, conflict and post conflict situation. Expanding the programs to reaching full geographic scale in these countries is a major undertaking but it is also what will increase the pace of elimination on the continent.

Hotspots

There is a need to understand "hotspots," that is, areas like in Burkina Faso (Kima et al. 2019) where MDA has gone on for several years, but the reduction in transmission has been slow. While the immediate response in such circumstances is the issue of low coverage, there might be other epidemiological, climatic, vectorial factors affecting transmission. These issues have not been adequately studied. Models, however, have suggested that in areas of higher endemicity, it could be more difficult to reach the target of elimination as even within those areas, there are focal points with continued transmission despite good coverage. This opens the door for further research in this field. There might need to be a different approach to address the issues in areas of higher endemicity.

Loa Loa

Co-endemicity of LF and loiasis. An important challenge in addressing LF in Africa is the presence and level of endemicity of *Loa loa* in the Central part of Africa. Challenges are two folds (1) cross-reactivity with the antigen test making the diagnostic less reliable for LF mapping in these countries (Bakajika et al. 2014; Pion et al. 2016) (2) the administration of ivermectin to individuals with *Loa loa* may result in *Loa loa* encephalopathy, which can eventually result in death (Twum-Danso 2003). Algorithms were developed to confirm ITC-based LF mapping results in Cameroon (Beng et al. 2020). WHO has also recently recommended the use of albendazole only twice yearly in association with vector control for malaria in communities where LF is co-endemic with *Loa loa* as an alternative strategy to eliminate LF (WHO 2017a).

Indirect Impact

The Lymphatic Filariasis Programme in SSA has had far-reaching impact than just the health benefits to the victims of the infection. One important outcome from the LF programs is the ability to get commodities through it to areas "beyond the end of the road' (Haddad et al. 2008). It has increased the health systems outreach through a network, which goes even beyond the level of the village by reaching very remote settlements. The network can be used to distribute other commodities. It has created a good platform for other interventions that need to be delivered at the community level and improved intersectoral collaboration (Anonymous 2010). With the current push on the use of community health workers (there are several millions already in action), the LF program is a good example of what works and that of best practices. The approaches however will be country specific, but it is clear that community health workers can be supported to carry out several interventions. In Ethiopia, they have made great progress in maternal and child health (Medhanyie et al. 2012). The LF program has been a platform for health promotion activities and continues to be a wider platform as part of an integrated program. The program supports a sub-district health infrastructure, which is very much in line with primary health care and universal health coverage. If well utilized, it can improve access to health prevention tools, resulting in gains in health promotion and behavioral change. Other health-related impact includes reduction in scabies, reduced anemia attributable to hookworm infection, and increase in school attendance (De Rochars et al. 2004; Mohammed et al. 2012).

Further Research for Control and Elimination

The areas for research are very closely linked to the challenges faced by the program and are set out as follows:

(i) Which mechanism to apply to increase domestic funding and country ownership?

(ii) There is a need for operational research that deals with the expansion of MDA in post-conflict areas. These are also areas where sporadic MDA programs or nonexistent ones can result in untreated areas, which then adversely affects the whole concept of interrupting transmission.

(iii) The ecological determinants of transmission in "hotspots" where even after several rounds of MDA and despite good coverage there is still transmission need to be studied. A better understanding of these hotspots including issues around climate and microecology of the area are critical in understanding how to design programs to address this phenomenon.

(iv) Strategies for urban areas also need research especially since, as programs expand, large urban areas which in some cases have unplanned settlements with intense migration will need to be treated. The fundamental questions will be what approaches serve best in these areas. Research on how to attain good converge in these areas and sustain the program for the number of years it takes to interrupt transmission will be needed.

(v) Research and development for new drugs is another important area for the program in Africa. In this respect the ongoing research on the possible use of antibiotics as macrofilaricidal drugs is a priority.

(vi) As LFE moves toward the end game, xenomonitoring will become key and more and more pertinent as infections in humans decrease to levels below 1%. Xenomonitoring is regarded by many as a possible surveillance tool. Operations research to make this tool user-friendly by countries who mount surveillance will be required.

(vii) Development of sustainable and user-friendly post validation surveillance methods to be implemented through the health system.

Outlook for the Next Decade

The next decade symbolizes the countdown to the targets set forth in the new 2021–2030 NTD road map for the elimination of LF and reminds us all of the daunting task ahead of us. The focus needs to be on increasing geographical coverage especially of big countries and making domestic funds available to ensure that drug delivery is optimally carried out. This is an issue both for the endemic countries as well as their partners. As countries come close toward elimination focus on transmission assessment surveys needs to be enhanced. The integrated NTD program has built on some of the successes of the LF program in a number of countries; this should continue as it maximizes use of minimum resources. Focus on technical expertise for LF needs to remain a priority especially toward the end game. As Africa primes itself for the elimination of LF, the issue of surveillance needs, especially in post validation phase, to be taken into consideration and linked to health system surveillance activity.

There is currently a lot of enthusiasm and good will by all in the NTD fraternity, in the march toward the elimination of LF. It is therefore envisaged that coverage, domestic funding, improved treatment regimes, and social determinants will receive the needed attention. The next decade should therefore see the acceleration of the interventions leading to a rapid decline in the prevalence of and ultimately the elimination of LF from all of the endemic countries in SSA. Research to unravel the best way to manage urban filariasis will be intensified to help mop up infection in urban areas that have so far lagged behind in drug delivery.

During the next 10 years, many of the previously LF-endemic countries will be in the surveillance phase; the questions on whether or not recrudescence will actually occur and, when and if it occurs, how to manage it may be aided by the information obtained from continued modeling which can be tested with real field-generated data during the decade.

Conclusion

Lymphatic filariasis (LF) is a debilitating disease. Since WHO passed resolution WHA 50.29 in 1997, calling for collaborative efforts by member states to eliminate the disease as a public health problem, and the creation of the Global Program to Eliminate Lymphatic Filariasis (GPELF), considerable progress has been made toward elimination in several countries. The disease, however, continues to be a public health problem in many countries, especially in the very large countries, in certain urban areas and in countries with low political commitment. The donation of ivermectin, albendazole and DEC free of charge for treatment in SSA was a magnanimous gesture from the drug companies that has reduced the burden of managing this disease. Some of the good lessons learnt from the management of LF with the involvement of communities, communities at the helm of affairs in drug treatment and morbidity management, and the good practices put into play will be of practical use for other public health interventions. There is great optimism that, with the renewed commitment of the countries for the implementation of the new 2021–2030 NTD road map, that countries in SSA will be either validated or in post treatment surveillance by 2030.

References

Addiss, D. and F. Global Alliance to Eliminate Lymphatic (2010) The 6th Meeting of the Global Alliance to Eliminate Lymphatic Filariasis: A half-time review of lymphatic filariasis elimination and its integration with the control of other neglected tropical diseases. Parasit Vectors 3:100

Akogun OB, Akogun MK, Apake E, Kale OO (2011) Rapid community identification, pain and distress associated with lymphoedema and adenolymphangitis due to lymphatic filariasis in resource-limited communities of North-eastern Nigeria. Acta Trop 120(Suppl 1):S62–S68

Anonymous (2010) Community-directed interventions for priority health problems in Africa: results of a multicountry study. Bull World Health Organ 88(7):509–518

Arumugam S, Pfarr KM, Hoerauf A (2008) Infection of the intermediate mite host with Wolbachia-depleted Litomosoides sigmodontis microfilariae: impaired L1 to L3 development and subsequent sex-ratio distortion in adult worms. Int J Parasitol 38(8–9):981–987

Ashton RA, Kyabayinze DJ, Opio T, Auma A, Edwards T, Matwale G, Onapa A, Brooker S, Kolaczinski JH (2011) The impact of mass drug administration and long-lasting insecticidal net distribution on Wuchereria bancrofti infection in humans and mosquitoes: an observational study in northern Uganda. Parasit Vectors 4:134

Babu BV, Nayak AN, Dhal K, Acharya AS, Jangid PK, Mallick G (2002) The economic loss due to treatment costs and work loss to individuals with chronic lymphatic filariasis in rural communities of Orissa, India. Acta Trop 82(1):31–38

Bakajika DK, Nigo MM, Lotsima JP, Masikini GA, Fischer K, Lloyd MM, Weil GJ, Fischer PU (2014) Filarial antigenemia and Loa loa night blood microfilaremia in an area without bancroftian filariasis in the Democratic Republic of Congo. Am J Trop Med Hyg 91(6):1142–1148

Beng AA, Esum ME, Deribe K, Njouendou AJ, Ndongmo PWC, Abong RA, Fru J, Fombad FF, Nchanji GT, Amambo G, Gandjui NTV, Biholong B, Nko'Ayissi G, Mbia P, Akame J, Enyong PI, Reid SD, Tougoue JJ, Zhang Y, Wanji S (2020) Mapping lymphatic filariasis in Loa loa endemic health districts naïve for ivermectin mass administration and situated in the forested zone of Cameroon. BMC Infect Dis 20(1):284

Boakye DA, Wilson MD, Appawu MA, Gyapong J (2004) Vector competence, for Wuchereria bancrofti, of the Anopheles populations in the Bongo district of Ghana. Ann Trop Med Parasitol 98(5):501–508

Bockarie MJ, Pedersen EM, White GB, Michael E (2009) Role of vector control in the global program to eliminate lymphatic filariasis. Annu Rev Entomol 54:469–487

Brengues J (1975) La Filariose de Bancroft en Afrique de L'ouest. ORSTOM, Paris

Brooker S (2010) Estimating the global distribution and disease burden of intestinal nematode infections: adding up the numbers--a review. Int J Parasitol 40(10):1137–1144

Bushrod FM (1979) Studies on filariasis transmission in Kwale, a Tanzanian coastal village, and the results of mosquito control measures. Ann Trop Med Parasitol 73(3):277–285

Carme B (2010) Mapping of lymphatic filariasis: "to be (in English) or not to be". Med Trop (Mars) 70(5–6):425–427

Chesnais CB, Awaca-Uvon NP, Bolay FK, Boussinesq M, Fischer PU, Gankpala L, Meite A, Missamou F, Pion SD, Weil GJ (2017) A multi-center field study of two point-of-care tests for circulating Wuchereria bancrofti antigenemia in Africa. PLoS Negl Trop Dis 11(9):e0005703

Chlebowsky HO, Zielke E (1980) Studies on Bancroftian filariasis in Liberia, West Africa. III. Efficacy of repeated treatment with diethylcarbamazine and vector control on the microfilarial reservoir in a rural population. Tropenmed Parasitol 31(2):181–193

Davis A, Bailey DR (1969) The effect of salt medicated with diethylcarbamazine in bancroftian filariasis. Bull World Health Organ 41(2):195–208

De Rochars MB, Direny AN, Roberts JM, Addiss DG, Radday J, Beach MJ, Streit TG, Dardith D, Lafontant JG, Lammie PJ (2004) Community-wide reduction in prevalence and intensity of intestinal helminths as a collateral benefit of lymphatic filariasis elimination programs. Am J Trop Med Hyg 71(4):466–470

Diallo S, Bah IB, Victorius A, N'Dir O, Diouf F (1983) Effects of diethylcarbamazine on the transmission of Wucheria bancrofti in a Senegalese reservoir of lymphatic filariasis. Dakar Med 28(1):179–186

Dorkenoo MA, Bronzan R, Yehadji D, Tchalim M, Yakpa K, Etassoli S, Adjeloh P, Maman I, Sodahlon Y (2018) Surveillance for lymphatic filariasis after stopping mass drug administration in endemic districts of Togo, 2010–2015. Parasit Vectors 11(1):244

Dunyo SK, Nkrumah FK, Simonsen PE (2000) Single-dose treatment of Wuchereria bancrofti infections with ivermectin and albendazole alone or in combination: evaluation of the potential for control at 12 months after treatment. Trans R Soc Trop Med Hyg 94(4):437–443

Estambale BB, Simonsen PE, Knight R, Bwayo JJ (1994) Bancroftian filariasis in Kwale District of Kenya. I. Clinical and parasitological survey in an endemic community. Ann Trop Med Parasitol 88(2):145–151

Gasarasi DB, Premji ZG, Mujinja PG, Mpembeni R (2000) Acute adenolymphangitis due to bancroftian filariasis in Rufiji district, south east Tanzania. Acta Trop 75(1):19–28

Gass KM, Sime H, Mwingira UJ, Nshala A, Chikawe M, Pelletreau S, Barbre KA, Deming MS, Rebollo MP (2017) The rationale and cost-effectiveness of a confirmatory mapping tool for lymphatic filariasis: Examples from Ethiopia and Tanzania. PLoS Negl Trop Dis 11(10):e0005944

Gyapong JO, Remme JH (2001) The use of grid sampling methodology for rapid assessment of the distribution of bancroftian filariasis. Trans R Soc Trop Med Hyg 95(6):681–686

Gyapong JO, Magnussen P, Binka FN (1994) Parasitological and clinical aspects of bancroftian filariasis in Kassena-Nankana District, upper east region, Ghana. Trans R Soc Trop Med Hyg 88(5):555–557

Gyapong JO, Gyapong M, Adjei S (1996a) The epidemiology of acute adenolymphangitis due to lymphatic filariasis in northern Ghana. Am J Trop Med Hyg 54(6):591–595

Gyapong M, Gyapong JO, Adjei S, Vlassoff C, Weiss M (1996b) Filariasis in northern Ghana: some cultural beliefs and practices and their implications for disease control. Soc Sci Med 43(2):235–242

Gyapong M, Gyapong J, Weiss M, Tanner M (2000) The burden of hydrocele on men in Northern Ghana. Acta Trop 77(3):287–294

Gyapong M, Gyapong JO, Owusu-Banahene G (2001) Community-directed treatment: the way forward to eliminating lymphatic filariasis as a public-health problem in Ghana. Ann Trop Med Parasitol 95(1):77–86

Gyapong JO, Kyelem D, Kleinschmidt I, Agbo K, Ahouandogbo F, Gaba J, Owusu-Banahene G, Sanou S, Sodahlon YK, Biswas G, Kale OO, Molyneux DH, Roungou JB, Thomson MC, Remme J (2002) The use of spatial analysis in mapping the distribution of bancroftian filariasis in four West African countries. Ann Trop Med Parasitol 96(7):695–705

Haddad D, Cross C, Thylefors B, Richards FO Jr, Bush S, Hopkins AD, Baker SK (2008) Health care at the end of the road: opportunities from 20 years of partnership in onchocerciasis control. Glob Public Health 3(2):187–196

Hanson C, Weaver A, Zoerhoff KL, Kabore A, Linehan M, Doherty A, Engels D, Savioli L, Ottesen EA (2012) Integrated implementation of programs targeting neglected tropical diseases through preventive chemotherapy: identifying best practices to roll out programs at national scale. Am J Trop Med Hyg 86(3):508–513

Hawking F (1957) The distribution of Bancroftian filariasis in Africa. Bull World Health Organ 16(3):581–592

Hawking F (1962) A review of progress in the chemotherapy and control of filariasis since 1955. Bull World Health Organ 27:555–568

Homeida M, Braide E, Elhassan E, Amazigo UV, Liese B, Benton B, Noma M, Etya'alé D, Dadzie KY, Kale OO, Sékétéli A (2002) APOC's strategy of community-directed treatment with ivermectin (CDTI) and its potential for providing additional health services to the poorest populations. African Programme for Onchocerciasis Control. Ann Trop Med Parasitol 96(Suppl 1):S93–S104

Hooper PJ, Zoerhoff KL, Kyelem D, Chu B, Mann Flueckiger R, Bamani S, Bougma WR, Fleming F, Onapa A, Pare AB, Torres S, Traore MO, Tuinsma M, Linehan M, Baker M (2013) The effects of integration on financing and coverage of neglected tropical disease programs. Am J Trop Med Hyg 89(3):407–410

Horton J, Witt C, Ottesen EA, Lazdins JK, Addiss DG, Awadzi K, Beach MJ, Belizario VY, Dunyo SK, Espinel M, Gyapong JO, Hossain M, Ismail MM, Jayakody RL, Lammie PJ, Makunde W, Richard-Lenoble D, Selve B, Shenoy RK, Simonsen PE, Wamae CN, Weerasooriya MV (2000) An analysis of the safety of the single dose, two drug regimens used in programmes to eliminate lymphatic filariasis. Parasitology 121(Suppl):S147–S160

Jemaneh L, Kebede D (1995) Clinico-epidemiological study of lymphatic filariasis southwestern Ethiopia. Ethiop Med J 33(3):143–153

Jordan P (1959) A pilot scheme to eradicate bancroftian filariasis with diethylcarbamazine; results of the first year's treatment. Trans R Soc Trop Med Hyg 53(1):54–60

Jullien P, Some J, Brantus P, Bougma RW, Bamba I, Kyelem D (2011) Efficacy of home-based lymphoedema management in reducing acute attacks in subjects with lymphatic filariasis in Burkina Faso. Acta Trop 120(Suppl 1):S55–S61

Kabatereine NB, Malecela M, Lado M, Zaramba S, Amiel O, Kolaczinski JH (2010) How to (or not to) integrate vertical programmes for the control of major neglected tropical diseases in sub-Saharan Africa. PLoS Negl Trop Dis 4(6):e755

Kelly-Hope LA, Molyneux DH, Bockarie MJ (2013) Can malaria vector control accelerate the interruption of lymphatic filariasis transmission in Africa; capturing a window of opportunity? Parasit Vectors 6:39

Kelly-Hope LA, Hemingway J, Taylor MJ, Molyneux DH (2018) Increasing evidence of low lymphatic filariasis prevalence in high risk Loa loa areas in Central and West Africa: a literature review. Parasit Vectors 11(1):349

Khaniya S, Agrawal CS, Koirala R, Regmi R, Adhikary S (2009) Comparison of aspiration-sclerotherapy with hydrocelectomy in the management of hydrocele: a prospective randomized study. Int J Surg 7(4):392–395

Kima A, Guiguemde KT, Meda ZC, Bougma R, Serme M, Bougouma C, Drabo F (2019) Evaluation of the effect of mass drug administration against lymphatic filariasis in three health districts and public health implications: study of 12 epidemiological surveillance sites in Burkina Faso. Med Sante Trop 29(1):55–60

King CL, Suamani J, Sanuku N, Cheng YC, Satofan S, Mancuso B, Goss CW, Robinson LJ, Siba PM, Weil GJ, Kazura JW (2018) A trial of a triple-drug treatment for lymphatic filariasis. N Engl J Med 379(19):1801–1810

Kolstrup N, McMahon JE, Magayuka SA, Mosha FW, Bushrod FM, Bryan JH (1981) Control measures against Bancroftian filariasis in coastal villages in Tanzania. Ann Trop Med Parasitol 75(4):433–451

Krishna Kumari A, Harichandrakumar KT, Das LK, Krishnamoorthy K (2005) Physical and psychosocial burden due to lymphatic filariasis as perceived by patients and medical experts. Trop Med Int Health 10(6):567–573

Malecela MN (2019) Reflections on the decade of the neglected tropical diseases. Int Health 11(5):338–340

Malecela MN, Mwingira U, Mwakitalu ME, Kabali C, Michael E, Mackenzie CD (2009) The sharp end - experiences from the Tanzanian programme for the elimination of lymphatic filariasis: notes from the end of the road. Ann Trop Med Parasitol 103(Suppl 1):S53–S57

Manga L (2002) Vector-control synergies, between 'roll back malaria' and the Global Programme to Eliminate Lymphatic Filariasis, in the African region. Ann Trop Med Parasitol 96(Suppl 2):S129–S132

Manguin S, Bangs MJ, Pothikasikorn J, Chareonviriyaphap T (2010) Review on global co-transmission of human Plasmodium species and Wuchereria bancrofti by Anopheles mosquitoes. Infect Genet Evol 10(2):159–177

Mante SD, Gueye SM (2011) Capacity building for the modified filarial hydrocelectomy technique in West Africa. Acta Trop 120(Suppl 1):S76–S80

Martindale S, Mableson HE, Kebede B, Kiros FH, Tamiru A, Mengistu B, Krueger A, Mackenzie CD, Kelly-Hope LA (2018) A comparison between paper-based and m-Health tools for collating and reporting clinical cases of lymphatic filariasis and podoconiosis in Ethiopia. Mhealth 4:49

Mathieu E, Amann J, Eigege A, Richards F, Sodahlon Y (2008) Collecting baseline information for national morbidity alleviation programs: different methods to estimate lymphatic filariasis morbidity prevalence. Am J Trop Med Hyg 78(1):153–158

Mathieu E, Dorkenoo A, Otogbe FK, Budge PJ, Sodahlon YK (2011) A laboratory-based surveillance system for Wuchereria bancrofti in Togo: a practical model for resource-poor settings. Am J Trop Med Hyg 84(6):988–993

Mathieu E, Dorkenoo AM, Morgah K, Datagni M, Cantey PT, Morgah K, Harvey K, Ziperstein J, Drexler N, Chapleau G, Sodahlon Y (2013) It is Possible: Availability of Lymphedema Case Management in each Health Facility in Togo: Program Description, Evaluation, and Lessons Learned. Am J Trop Med Hyg 89(1):16–22

Maxwell CA, Curtis CF, Haji H, Kisumku S, Thalib AI, Yahya SA (1990) Control of Bancroftian filariasis by integrating therapy with vector control using polystyrene beads in wet pit latrines. Trans R Soc Trop Med Hyg 84(5):709–714

McGregor IA, Gilles HM (1956) Diethylcarbamazine control of bancroftian filariasis; follow-up of a field trial in West Africa. Br Med J 1(4962):331–332

McGregor IA, Hawking F, Smith DA (1952) The control of filariasis with hetrazan; a field trial in a rural village (Keneba) in The Gambia. Br Med J 2(4790):908–911

Medhanyie A, Spigt M, Kifle Y, Schaay N, Sanders D, Blanco R, GeertJan D, Berhane Y (2012) The role of health extension workers in improving utilization of maternal health services in rural areas in Ethiopia: a cross sectional study. BMC Health Serv Res 12:352

Mohammed KA, Deb RM, Stanton MC, Molyneux DH (2012) Soil transmitted helminths and scabies in Zanzibar, Tanzania following mass drug administration for lymphatic filariasis--a rapid assessment methodology to assess impact. Parasit Vectors 5:299

Molyneux DH, Aboe A, Isiyaku S, Bush S (2020) COVID-19 and neglected tropical diseases in Africa: impacts, interactions, consequences. Int Health 12(5):367–372

Mukoko DA, Pedersen EM, Masese NN, Estambale BB, Ouma JH (2004) Bancroftian filariasis in 12 villages in Kwale district, Coast province, Kenya - variation in clinical and parasitological patterns. Ann Trop Med Parasitol 98(8):801–815

Mwingira U, Chikawe M, Mandara WL, Mableson HE, Uisso C, Mremi I, Malishee A, Malecela M, Mackenzie CD, Kelly-Hope LA, Stanton MC (2017) Lymphatic filariasis patient identification in a large urban area of Tanzania: An application of a community-led mHealth system. PLoS Negl Trop Dis 11(7):e0005748

Nielsen NO, Makaula P, Nyakuipa D, Bloch P, Nyasulu Y, Simonsen PE (2002) Lymphatic filariasis in Lower Shire, southern Malawi. Trans R Soc Trop Med Hyg 96(2):133–138

Njenga SM, Wamae CN, Njomo DW, Mwandawiro CS, Molyneux DH (2007) Chronic clinical manifestations related to Wuchereria bancrofti infection in a highly endemic area in Kenya. Trans R Soc Trop Med Hyg 101(5):439–444

Njenga SM, Wamae CN, Njomo DW, Mwandawiro CS, Molyneux DH (2008) Impact of two rounds of mass treatment with diethylcarbamazine plus albendazole on Wuchereria bancrofti infection and the sensitivity of immunochromatographic test in Malindi, Kenya. Trans R Soc Trop Med Hyg 102(10):1017–1024

Noma M, Nwoke BE, Nutall I, Tambala PA, Enyong P, Namsenmo A, Remme J, Amazigo UV, Kale OO, Seketeli A (2002) Rapid epidemiological mapping of onchocerciasis (REMO): its application by the African Programme for Onchocerciasis Control (APOC). Ann Trop Med Parasitol 96(Suppl 1):S29–S39

Noroes J, Dreyer G (2010) A mechanism for chronic filarial hydrocele with implications for its surgical repair. PLoS Negl Trop Dis 4(6):e695

Okon OE, Iboh CI, Opara KN (2010) Bancroftian filariasis among the Mbembe people of Cross River state, Nigeria. J Vector Borne Dis 47(2):91–96

Onapa AW, Simonsen PE, Pedersen EM, Okello DO (2001) Lymphatic filariasis in Uganda: baseline investigations in Lira, Soroti and Katakwi districts. Trans R Soc Trop Med Hyg 95(2):161–167

Ottesen EA (2000) The global programme to eliminate lymphatic filariasis. Trop Med Int Health 5(9):591–594

Ouattara AF, Bjerum CM, Aboulaye M, Kouadio O, Marius VK, Andersen B, Lew D, Goss CW, Weil GJ, Koudou BG, King CL (2021) Semiannual Treatment of Albendazole Alone is Efficacious for Treatment of Lymphatic Filariasis: A Randomized Open-label Trial in Cote d'Ivoire. Clin Infect Dis 74(12):2200–2208

Paily KP, Hoti SL, Das PK (2009) A review of the complexity of biology of lymphatic filarial parasites. J Parasit Dis 33(1–2):3–12

Person B, Bartholomew LK, Gyapong M, Addiss DG, van den Borne B (2009) Health-related stigma among women with lymphatic filariasis from The Dominican Republic and Ghana. Soc Sci Med 68(1):30–38

Pion SD, Chesnais CB, Bopda J, Louya F, Fischer PU, Majewski AC, Weil GJ, Boussinesq M, Missamou F (2015) The impact of two semiannual treatments with albendazole alone on lymphatic filariasis and soil-transmitted helminth infections: a community-based study in the Republic of Congo. Am J Trop Med Hyg 92(5):959–966

Pion SD, Montavon C, Chesnais CB, Kamgno J, Wanji S, Klion AD, Nutman TB, Boussinesq M (2016) Positivity of Antigen Tests Used for Diagnosis of Lymphatic Filariasis in Individuals Without Wuchereria bancrofti Infection But with High Loa loa Microfilaremia. Am J Trop Med Hyg 95(6):1417–1423

Pion SDS, Chesnais CB, Awaca-Uvon NP, Vlaminck J, Abdou A, Kunyu-Shako B, Kuyangisa Simuna G, Tambwe JP, Weil GJ, Boussinesq M (2020) The impact of four years of semiannual treatments with albendazole alone on lymphatic filariasis and soil-transmitted helminth infections: A community-based study in The Democratic Republic of the Congo. PLoS Negl Trop Dis 14(6):e0008322

Richard SA, Mathieu E, Addiss DG, Sodahlon YK (2007) A survey of treatment practices and burden of lymphoedema in Togo. Trans R Soc Trop Med Hyg 101(4):391–397

Richard-Lenoble D, Kombila M, Carme B, Gilles JC, Delattre PY (1980) Prevalence of human filariasis with microfilaremia in Gabon. Bull Soc Pathol Exot Filiales 73(2):192–199

Shenoy RK, Bockarie MJ (2011) Lymphatic filariasis in children: clinical features, infection burdens and future prospects for elimination. Parasitology 138(12):1559–1568

Simonsen PE, Meyrowitsch DW, Makunde WH, Magnussen P (1995) Bancroftian filariasis: the pattern of microfilaraemia and clinical manifestations in three endemic communities of Northeastern Tanzania. Acta Trop 60(3):179–187

Simonsen PE, Malecela NM, Mackenzie CD (2008) Lymphatic Filariasis. Research and Control in Eastern and Southern Africa. DBL-Centre for Health Research and Development, Denmark

Simonsen PE, Pedersen EM, Rwegoshora RT, Malecela MN, Derua YA, Magesa SM (2010) Lymphatic filariasis control in Tanzania: effect of repeated mass drug administration with ivermectin and albendazole on infection and transmission. PLoS Negl Trop Dis 4(6):e696

Simonsen PE, Onapa AW, Asio SM (2011) Mansonella perstans filariasis in Africa. Acta Trop 120(Suppl 1):S109–S120

Slatko BE, Taylor MJ, Foster JM (2010) The Wolbachia endosymbiont as an anti-filarial nematode target. Symbiosis 51(1):55–65

Sodahlon YK, Dorkenoo AM, Morgah K, Nabiliou K, Agbo K, Miller R, Datagni M, Seim A, Mathieu E (2013) A success story: Togo is moving toward becoming the first sub-Saharan African Nation to eliminate lymphatic filariasis through mass drug administration and countrywide morbidity alleviation. PLoS Negl Trop Dis 7(4):e2080

Sodahlon Y, Ross DA, McPhillips-Tangum C, Lawrence J, Taylor R, McFarland DA, Krentel A, Anyaikea C, Batcho WE, Bougouma C, Dantas APC, Dhingra N, Guambe MM, Ibrahim K, Kargbo-Labour IJ, Matwale GK, Momprevil FN, Monteiro MAA, Nko'Ayissi GB, Omondi WP, Randrianirina LH, Salissou BA, Schuler HR, do Rosário ASS, Sitima LD, Togbey KM, Victor B, Malecela MN (2020) Building country capacity to sustain NTD programs and progress: A call to action. PLoS Negl Trop Dis 14(10):e0008565

Taylor MJ, von Geldern TW, Ford L, Hübner MP, Marsh K, Johnston KL, Sjoberg HT, Specht S, Pionnier N, Tyrer HE, Clare RH, Cook DAN, Murphy E, Steven A, Archer J, Bloemker D, Lenz F, Koschel M, Ehrens A, Metuge HM, Chunda VC, Ndongmo Chounna PW, Njouendou AJ, Fombad FF, Carr R, Morton HE, Aljayyoussi G, Hoerauf A, Wanji S, Kempf DJ, Turner JD, Ward SA (2019) Preclinical development of an oral anti-Wolbachia macrolide drug for the treatment of lymphatic filariasis and onchocerciasis. Sci Transl Med 11(483)

Thomas G, Richards FO Jr, Eigege A, Dakum NK, Azzuwut MP, Sarki J, Gontor I, Abimiku J, Ogah G, Jindau MY, Jiya JY, Miri ES (2009) A pilot program of mass surgery weeks for treatment of hydrocele due to lymphatic filariasis in central Nigeria. Am J Trop Med Hyg 80(3):447–451

Thomas BC, Kollie K, Koudou B, Mackenzie C (2017) Commentary: restarting NTD programme activities after the Ebola outbreak in Liberia. Infect Dis Poverty 6(1):52

Twum-Danso NA (2003) Loa loa encephalopathy temporally related to ivermectin administration reported from onchocerciasis mass treatment programs from 1989 to 2001: implications for the future. Filaria J 2(Suppl 1):S7

Ughasi J, Bekard HE, Coulibaly M, Adabie-Gomez D, Gyapong J, Appawu M, Wilson MD, Boakye DA (2012) Mansonia africana and Mansonia uniformis are vectors in the transmission of Wuchereria bancrofti lymphatic filariasis in Ghana. Parasit Vectors 5:89

Uttah EC (2011) Prevalence of endemic Bancroftian filariasis in the high altitude region of south-eastern Nigeria. J Vector Borne Dis 48(2):78–84

Wainwright E, Evans D, Rotondo L, Pou B, Yevstigneyeva V, Zoerhoff KL, Ottesen EA, Reithinger R (2020) The elimination of neglected tropical diseases: a case study exemplifying how foreign assistance funding can be catalytic in reducing the burden of major global health conditions. Clin Infect Dis 70(5):958–964

Wanji S, Amvongo-Adjia N, Koudou B, Njouendou AJ, Chounna Ndongmo PW, Kengne-Ouafo JA, Datchoua-Poutcheu FR, Fovennso BA, Tayong DB, Fombad FF, Fischer PU, Enyong PI, Bockarie M (2015) Cross-reactivity of filariais ICT cards in areas of contrasting endemicity of loa loa and mansonella perstans in Cameroon: implications for shrinking of the lymphatic filariasis map in the Central African Region. PLoS Negl Trop Dis 9(11):e0004184

Wanji S, Esum ME, Njouendou AJ, Mbeng AA, Chounna Ndongmo PW, Abong RA, Fru J, Fombad FF, Nchanji GT, Ngongeh G, Ngandjui NV, Enyong PI, Storey H, Curtis KC, Fischer K, Fauver JR, Lew D, Goss CW, Fischer PU (2019) Mapping of lymphatic filariasis in loiasis areas: A new strategy shows no evidence for Wuchereria bancrofti endemicity in Cameroon. PLoS Negl Trop Dis 13(3):e0007192

Weil GJ, Lammie PJ, Weiss N (1997) The ICT Filariasis Test: A rapid-format antigen test for diagnosis of bancroftian filariasis. Parasitol Today 13(10):401–404

Weil GJ, Bogus J, Christian M, Dubray C, Djuardi Y, Fischer PU, Goss CW, Hardy M, Jambulingam P, King CL, Kuttiat VS, Krishnamoorthy K, Laman M, Lemoine JF, O'Brian KK, Robinson LJ, Samuela J, Schechtman KB, Sircar A, Srividya A, Steer AC, Supali T, Subramanian S (2019) The safety of double- and triple-drug community mass drug administration for lymphatic filariasis: a multicenter, open-label, cluster-randomized study. PLoS Med 16(6):e1002839

WHA (1997) Resolution WHA50.29: Elimination of lymphatic filariasis as a public health problem. Fiftieth World Health Assembly, Geneva

White GB (1971) The present importance of domestic mosquitoes (Culex pipiens fatigans Wiedemann) in East Africa and recent steps towards their control. East Afr Med J 48(6):266–274

WHO (1992) Lymphatic filariasis: The disease and its control. Fifth report of the WHO Expert Committee on Filariasis. World Health Organization, Geneva

WHO (2002) The global elimination of lymphatic filariasis: the story of Zanzibar. WHO, Geneva. WHO/CDS/CPS/SMT/2002.15

WHO (2010) Progress report 2000–2009 and strategic plan 2010–2020 of the global programme to eliminate lymphatic filariasis: halfway towards eliminating lymphatic fi lariasis. WHO/HTM/NTD/PCT/2010.6

WHO (2011a) Global Programme to eliminate lymphatic filariasis: progress report on mass drug administration, 2010. Wkly Epidemiol Rec 86(35):377–388

WHO (2011b) Monitoring and epidemiological assessment of mass drug administration in the global programme to eliminate lymphatic fi lariasis: a manual for national elimination programmes. WHO/HTM/NTD/PCT/2011.4. WHO, Geneva

WHO (2011c) WHO position statement on integrated vector management to control malaria and lymphatic filariasis. Wkly Epidemiol Rec 86(13):121–127

WHO (2012a) Global programme to eliminate lymphatic filariasis: progress report, 2011. Wkly Epidemiol Rec 87(37):346–356

WHO (2012b) Transmission assessment surveys in the Global Programme to Eliminate Lymphatic Filariasis: WHO position statement. Wkly Epidemiol Rec 87(48):478–482

WHO (2017a) Guideline: Alternative mass drug administration regimens to eliminate lymphatic filariasis. WHO/HTM/NTD/PCT/2017.07. WHO, Geneva

WHO (2017b) Validation of elimination of lymphatic filariasis as a public health problem. WHO/HTM/NTD/PCT/2017.01. WHO, Geveva

WHO (2019) Global programmeto eliminate lymphatic filariasis: progress report, 2018. Wkly Epidemiol Rec 94(41):457–472

WHO (2020a) Ending the neglect to attain the Sustainable Development GoalsA road map for neglected tropical diseases 2021–2030. World Health Organization, Geneva

WHO (2020b) Global programme to eliminate lymphatic filariasis: progress report, 2019. Wkly Epidemiol Rec 95(43):509–524

Witt C, Ottesen EA (2001) Lymphatic filariasis: an infection of childhood. Trop Med Int Health 6(8):582–606

Zeldenryk LM, Gray M, Speare R, Gordon S, Melrose W (2011) The emerging story of disability associated with lymphatic filariasis: a critical review. PLoS Negl Trop Dis 5(12):e1366

Zielke E, Chlebowsky HO (1980) Studies on bancroftian filariasis in Liberia, West Africa. V. The influence of treatment with diethylcarbamazine and vector control on the transmission of Wuchereria bancrofti. Tropenmed Parasitol 31(4):444–458

Zoure HG, Wanji S, Noma M, Amazigo UV, Diggle PJ, Tekle AH, Remme JH (2011) The geographic distribution of Loa loa in Africa: results of large-scale implementation of the Rapid Assessment Procedure for Loiasis (RAPLOA). PLoS Negl Trop Dis 5(6):e1210

Onchocerciasis Part I

Boakye A. Boatin, Yankum Dadzie, and Uche Amazigo

Contents

B. A. Boatin (✉)
Institute of Parasitology, McGill University and Lymphatic Filariasis Support Centre for Africa, Accra, Ghana

Y. Dadzie
Accra, Ghana

U. Amazigo
Enugu, Enugu, Nigeria

J. O. Gyapong, B. A. Boatin (eds.), *Neglected Tropical Diseases - Sub-Saharan Africa*, Neglected Tropical Diseases, https://doi.org/10.1007/978-3-031-53901-5_9

Abstract

Onchocerciasis, also known as river blindness, is a dermic filariasis caused by the nematode *Onchocerca volvulus*. The disease is found in 30 African countries between latitudes 15°N and 14°S. About 20.9 million people were estimated infected globally (2017) of which 14.6 million had skin disease while 1.15 million had vision loss. Worldwide, onchocerciasis is a very important infectious cause of blindness, being second only to trachoma. Transmitted by the vector, *Simulium damnosum*, the adult parasite lives up to 14 years in humans, producing millions of microfilariae, which are responsible for the pathology of the disease. The disease presents incessant pruritus, acute and chronic papular onchodermatitis, and eye lesions that often end in blindness. There is evidence of protective immunity at play in these lesions. Onchocerciasis is responsible for 1.23 million (95% UI 0.765–1.82) DALYs (disability-adjusted life years) annually, with troublesome itching accounting for 60% while blindness and visual impairment account for 40% of the attributable DALYs. Individual diagnosis depended on symptomatology and parasitological tests. Parasitological tests together with rapid epidemiological mapping (nodule palpation) used for community diagnosis have produced geographical distribution of onchocerciasis in Africa. Larviciding was used successfully for control in the West African Onchocerciasis Control Programme, while the African Programme for Onchocerciasis Control and Partners have driven the control and now elimination through community-directed treatment with ivermectin in the rest of the endemic African countries. Current efforts geared toward elimination of the disease, necessitate extending operations beyond the previously covered areas with the attendant managerial, financial, and logistic challenges. Some challenges of onchocerciasis control included in the following list have become more pertinent with the change to elimination: (1) the diagnostic tool that needs to be highly specific and sensitive as in evaluations in elimination, low level infection in endemic areas is encountered. This necessitates accurate diagnosis to determine when to stop mass drug administration (MDA) and to identify areas with ongoing low transmission requiring continued MDA; (2) finding a technical and cost-effective approach to eliminating onchocerciasis in co-endemic areas with *Loa loa* without the occurrence of severe adverse reaction. Other challenges to address include (3) how to minimize noncompliance with the drug treatment; (4) the best approach to obtain sustainability of the drug distribution program; (5) how to sustain community engagement/interest in MDA; (6) discovering an acceptable macrofilaricide that can either kill or sterilize the adult; (7) how to optimize treatment regime of possible antibiotic that targets *Wolbachia endobacteria* the elimination of which will incapacitate the adult worm; and (8) how to discover early, any resistance to the current drug. All these require research, which was always central to the two erstwhile control programs.

Introduction

Onchocerciasis also known as river blindness is a dermic filariasis, which has been an important cause of blindness and debilitating skin lesions for several decades in Africa, Latin America, and a small portion in the Arabian Peninsula. The disease occurs in 30 countries in Africa (Map 1, Table 1). Onchocerciasis is a neglected tropical disease (NTD), which affects the poor, causes poverty, and is an impediment to socioeconomic development. It remains a major parasitic endemic disease especially in Central and some Eastern African countries although a lot of progress in its control has been made in West Africa.

Estimates from the WHO Expert Committee on Onchocerciasis in 1995 put the number of infected persons at about 17.7 million, about 270,000 of whom were blind and another 500,000 severely visually impaired (WHO 1995a). More recently, The Global Burden of Disease Study estimated that in 2017 there were at least 20.9 million people infected worldwide ("Global Health Division of Parasitic Diseases and Malaria Report" 2019). Over 99% of the estimated infected persons live in Africa, while about 140,000 people in the Americas are infected with the worm. These estimates call for a revision, given the remarkable progress that has been made through the erstwhile Onchocerciasis Control Programme in West Africa (OCP), the African Programme for Onchocerciasis Control (APOC), and the Onchocerciasis Elimination Programme in the Americas (OEPA). In Latin America and the Caribbean, onchocerciasis was found in Mexico, Guatemala, Colombia, Ecuador, Venezuela and Brazil, with 13 focal areas and 510,000 individuals at risk

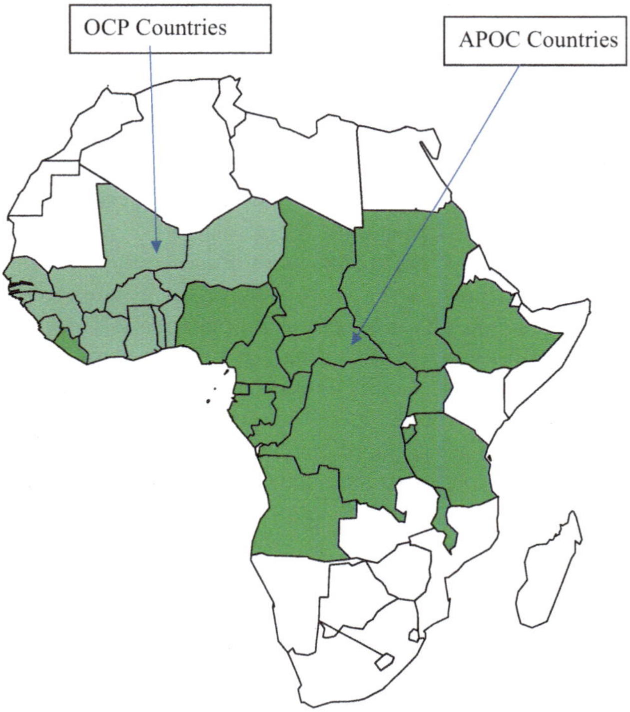

Map 1 Africa showing the countries that are endemic for onchocerciasis

Table 1 Onchocerciasis endemic countries in Africa

Angola	Ethiopia	Mozambique
Benin	Gabon	Niger
Burkina Faso	Ghana	Nigeria
Burundi	Guinea	Senega
Cameroon	Guinea Bissau	Sierra Leone
Central African Republic	Ivory Coast	South Sudan
Chad	Kenya	Sudan
Congo	Liberia	Tanzania
Democratic Rep. of Congo	Malawi	Togo
Equatorial Guinea	Mali	Uganda

of infection (Basanez et al. 2006). There were eight focal areas with evidence of interruption of transmission (Ault 2007). In 2011, Colombia became the first country in the Americas to successfully eliminate onchocerciasis (PAHO). Four countries have been verified by WHO as free of onchocerciasis after successfully implementing elimination activities for decades: Colombia (2013), Ecuador (2014), Mexico (2015), and Guatemala (2016).

Onchocerciasis is one of the most important infectious causes of blindness, being second only to trachoma in numbers of persons affected (Thylefors et al. 1995; Lewalen and Courtright 2001). Onchocercal blindness resulting from onchocercal eye disease has been esteemed to be the main public health challenge of onchocerciasis. Recently, however, severe skin disease has also been recognized as of public health significance (Hagan 1998). A global intervention effort against onchocerciasis took off nearly five decades ago in West Africa, one of the most onchocerciasis devastating areas in Africa with vector control by larviciding with its sophistication and cost as the only effective method available. The discovery of ivermectin, a microfilaricidal-cum-partial macrofilaricidal drug safe for mass treatment, which was registered in 1987 and donated by its manufacturer, Merck & Co. Inc., for as long as was required for the treatment of onchocerciasis, opened up possibilities for intervention against the disease in other areas where the disease prevailed. The concerted effort of the OCP and APOC to control the disease in the past two to three decades has produced remarkable results. Even more striking have been reports from Latin America (OEPA) of the elimination of onchocerciasis in four countries and of elimination in specific foci from some parts of Africa (Diawara et al. 2009; JAF 2011), which in Africa suggest that the disease has been or can be eliminated even with the tool in use in most areas using the right strategy and given the financial input and support. There are still challenges ahead, which are certainly not insurmountable given the progress made to date, with emphasis on research as the integral part of the interventions.

Basic Biology, Life Cycle, and Disease Presentation

Basic Biology

The Parasite

The parasite that causes human onchocerciasis is called *Onchocerca volvulus*, while that causing animal onchocerciasis is the *Onchocerca ochengi*—*O. volvulus* is a nematode with man as the only definitive known reservoir. The adult worms (Fig. 1), male (3–5 cm long) and female (30–80 cm), live in fibrous nodules found subcutaneously or deep in the connective and muscular tissues, with a tendency to overlie bony prominences. The adult worm has a reproductive life span of about 9–11 years (Plaisier et al. 1991), although it may live up to about 15 years. A female worm when fertilized produces millions of embryos called microfilariae (mfs), which themselves live for about 2 years. The mfs (up to 300 um in length) when released by the adult worm migrate from the nodules, invading the skin, eyes, and some organs but it is from the skin that the mfs are taken up when man is bitten by the vector, the blackfly. The mfs are occasionally also found in peripheral blood, urine, and sputum but are typically found in the skin and in the lymphatics of connective tissues. Different disease patterns—blinding onchocerciasis or dermal onchocerciasis—are found in different ecological and geographical areas and are associated

Fig. 1 Coiled adult onchocercal male and female worms (Source: *TDR image* library)

with different variants or strains of the parasite, each in turn associated with the different subspecies of the vector (Duke et al. 1966). *O. volvulus*, like other filarial nematodes, shares an endosymbiotic relationship with the bacterium *Wolbachia*. In the absence of *Wolbachia*, there is inhibition of worm development, embryogenesis and fertility are arrested, and worm viability diminishes (Hoerauf et al. 2008).

Vector and Transmissions

Onchocerciasis is transmitted by the female of small blackflies (Fig. 2) of the genus *Simulium*. The main vectors are the *Simulium damnosum s.l.* Of these *S. damnosum s.s.* and *sirbanum* are mostly responsible for transmission in savannah areas, while *S. yahense*, *S. squamosum*, *S. soubrense*, and *S. sanctipauli* subspecies are found more in the forest areas in Africa. Other complexes are *S. neavei* also found in Africa and *S. ochraceum*, *S. metallicum*, *S. exiguum*, and *S. oyapockense* in the Americas (Crosskey 1990; Dunbar 1968; Vajime and Dunbar 1975). More recently, *Simulium dentulosum* was found to be the main vector of human onchocerciasis within the Kakoi-Koda focus at the western slopes of the rift valley above Lake Albert, DRC. This is the first time that *S. dentulosum* has been implicated as a vector of onchocerciasis (Post et al. 2022). The larvae and pupae of *Simulium damnosum* require rapidly flowing well-oxygenated streams and rivers to develop (Blacklock 1926). Therefore, *Simulium* in general tends to stay and feed close to the breeding sites (fast-flowing river bodies at areas or points where there are rocks and/or overhanging vegetation and water vegetation) where they lay their eggs. *S. damnosum* has been found to fly long distances up to 400–500 km assisted by seasonal winds in West Africa (Garms et al. 1979; Baker et al. 1990). *Simulium damnosum s.l.* has a wide distribution in Africa and Yemen and is the most important vector of the disease. *S. neavei*, the aquatic stages of which are found on freshwater crabs, on the other hand, transmits *O. volvulus* in East Africa. The association of *S. neavei* and the freshwater crabs has been exploited successfully in the eradication of onchocerciasis transmitted by *S. neavei* from some foci in Kenya and Uganda in East Africa.

Fig. 2 Blackfly (*Simulium damnosum*) taking a blood meal (*TDR image library*)

The vectorial capacity of the blackfly has been an important factor in the propagation of transmission in different ecosystems. In this respect, blackfly vectors in the Americas are less efficient in transmitting the parasite than those in Africa. In Africa, the forest vectors are less efficient transmitters than the savannah blackflies. As a general rule, blackfly species are best adapted to transmit corresponding *O. volvulus* parasite strains from their original region (Duke et al. 1991).

Life Cycle

The life cycle of *O. volvulus* begins when an infected blackfly deposits infective *O. volvulus* larvae into the human host when it takes a blood meal. These larvae develop into "viviparous" mature adult worms in about a year, and the fertilized female worms release mfs (unsheathed) into the skin some 10–15 months after infection. The mfs then migrate throughout the body in the dermal layer, from where they can be taken up by the blood-feeding blackflies. The mfs migrate from the blackfly's midgut through the hemocoel to the thoracic muscles. There the mfs develop into first-stage larvae (L1) and subsequently into third-stage infective larvae (L3). The third-stage infective larvae migrate to the blackfly's proboscis and can infect another human when the fly takes a blood meal, thus completing the life cycle of the parasite. Development of the mfs from the L1 larvae through L3 infective larvae spans a period of 10–12 days (Fig. 3).

Disease Presentation

Clinical Presentation

The clinical manifestations of onchocerciasis can occur as early as 1–3 years after infection, when the adult worms begin producing mfs but can be variable and often occur later. Three major clinical presentations are recognized. These are predominantly dermal, lymphatic, and ocular in character and are as a result of host tissue

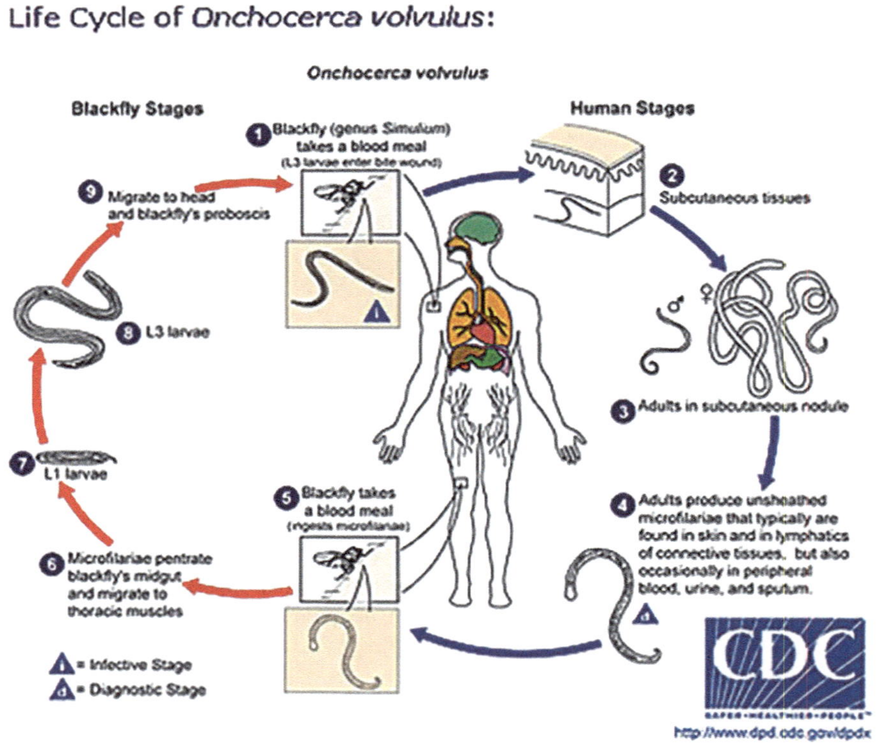

Fig. 3 Life cycle of *Onchocercal volvulus*. CDC: http.www.dcd.cdc.gov/dpdx

inflammatory reactions to the dead microfilariae. Symptoms and signs include severe itching, skin rashes and palpable nodules, and lymphatic involvement including lymphadenopathy, hanging groin, lymphedema, and eye lesions.

Dermal Lesions

The first signs that a person who has been infected with the *O. volvulus* parasite may include fever, neuralgic pain in joints, and transient rashes on the trunk (Fig. 4) and face. Some infected individuals may present with very intense itching. In others, there are disfiguring skin lesions. These result from a combination of inflammatory skin tissue reaction to dead mfs and often secondary bacterial infection of damaged skin from excessive scratching provoked by microfilaria-induced intense skin irritation. Severe "troublesome itching" is one of the most important symptoms of onchocerciasis (Brieger et al. 1998; WHO 1995b). This may affect more than 50% of the populations in hyperendemic communities. In lightly infected persons, onchocerciasis skin lesions may present as acute onchodermatitis, which is the most common symptom of the disease. Chronic onchodermatitis develops as a result of prolonged heavy microfilarial infection, when other skin manifestations—atrophy and depigmentation—do occur (Murdoch et al. 2002; Fig. 5).

The clinical classification and grading system for onchocerciasis skin disease (OSD) was not clear until recently when Murdoch et al. (1993) instituted a system that

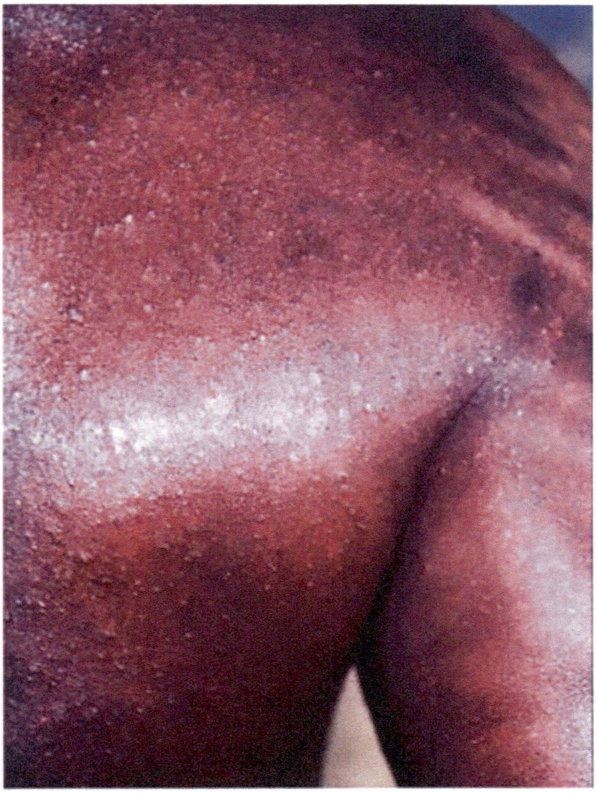

Fig. 4 Skin rash (acute papular onchodermatitis). (Source: www.stanford. edu)

defines and provides appropriate descriptions of the skin changes and lesions associated with *O. volvulus* infection: acute papular onchodermatitis (APOD), chronic papular onchodermatitis (CPOD), lichenified onchodermatitis (LOD), skin atrophy (ATR), depigmentation (DPM, also known as leopard skin), troublesome itching (pruritus), and lymphatic involvement (LYM) plus hanging groin (HG). Murdoch's scheme has provided a platform for and the basis for comparisons in prevalence surveys, psychosocial and economic studies, and selection of clinical groups of patients for detailed immunological and genetic studies (Murdoch 2010).

Ocular Lesions

Mfs migrate from the skin to enter the eyes where both dead and live mfs have been found in different eye tissues, with the slit lamp microscope mfs can be seen in the cornea, the anterior chamber, and occasionally in the vitreous. However, it is the dead mfs that cause the ocular morbidity (Hall and Pearlman 1999; Pearlman and Hall 2000) and that result in the visual impairment, including permanent blindness. Visual loss normally occurs from acute and chronic ocular disease of both the anterior segment (sclerosing keratitis and iridocyclitis) and posterior segments (optic nerve disease, optic atrophy, and choroidoretinitis). Blindness appearing in the late stage of the disease is the most serious consequence of onchocerciasis (Dadzie et al.

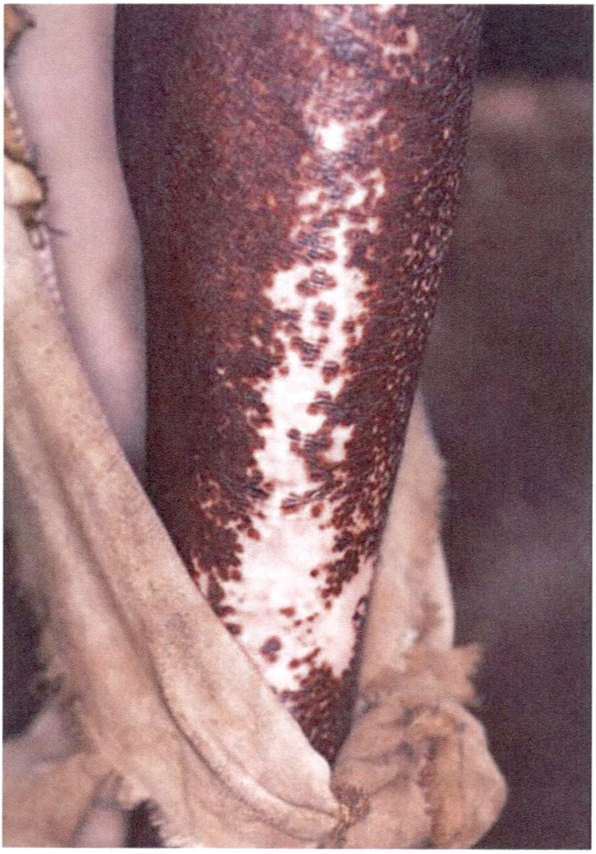

Fig. 5 Hypopigmentation (leopard skin) (TDR *image library*)

1986). This normally occurs in both eyes (Fig. 6). Blindness caused by onchocerciasis, which was associated with the "proverbial" child leading his blind parent on a long stick, as well as acute and chronic onchocercal skin disease, have become scenes of the past, as control measures with ivermectin make a significant dent in the prevalence and incidence of the disease.

The severity of onchocercal ocular disease varies considerably between geographical zones. Blindness from onchocerciasis is extensive in hyperendemic populations in the West African savannah, while virtually no or little blindness is found in forest villages with a comparable intensity of infection. On the other hand, skin manifestations tend to be the main forms of disease in the forest areas. In Yemen and Central Sudan, the major disease manifestation is Sowda, a pruritic skin condition affecting usually one limb with a swelling. These differences may be attributed to the different vector–parasite complexes with strains of *O. volvulus* that differ in their pathogenicity (Garner et al. 1973; Zimmeman et al. 1992).

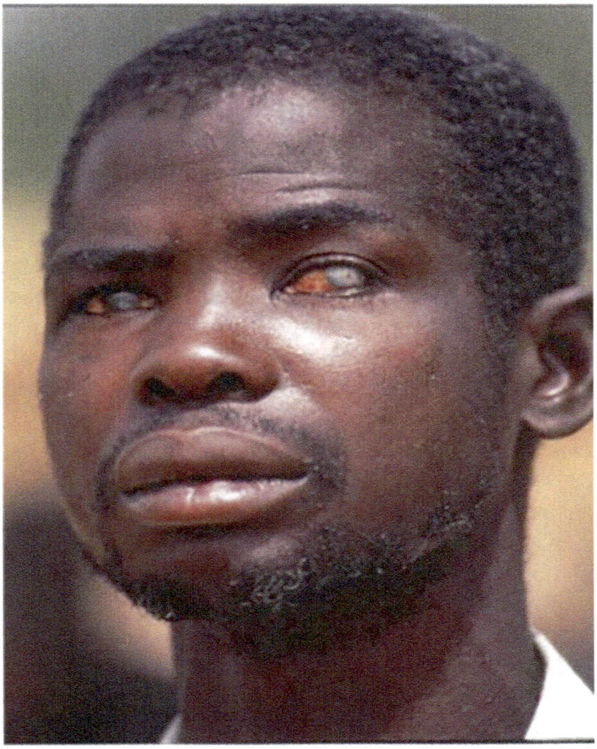

Fig. 6 Blindness from onchocerciasis (*TDR image library*)

Other Manifestations

Systemic manifestations such as low body weight and diffuse musculoskeletal pain have been described. Some evidence suggests that onchocerciasis may present with neurohormonal involvement. Evidence also suggests that onchocerciasis is a risk factor for epilepsy (Druet-Cabanac et al. 1999; Pion et al. 2009; Colebunders et al. 2018; Nelson et al. 2020) and may be responsible for a type of dwarfism (Kipp et al. 1996) known as hypo-sexual dwarfism (Nakalanga syndrome) in certain areas. A recent study investigated the association between onchocerciasis and increased risk of epilepsy and found that epilepsy prevalence increased, on average, by 0.4% for each 10% increase in onchocerciasis prevalence.

Immunity in Onchocerciasis

Onchocerciasis like other helminth infections tends to be long-lasting and shows high population prevalence. Part of the reason for this could be the fact that the onchocercal parasite has a long life span. It could also be due to the nature of the parasites in producing chronic and long-term infection through suppression of host immunity. Repeated infection in individuals living in the endemic areas results in cumulative infection, which appears to drive the eventual symptoms and clinical manifestation. Studies on immunity in onchocerciasis have spanned several decades and have looked at both pre- and posttreatment immune conditions in onchoendemic communities. There is evidence of protective immunity in onchocerciasis, an assertion that is derived principally from the existence of noninfected individuals (endemic normals who are putatively immune (PI) living in endemic or hyperendemic areas). Antibody and minimal cellular proliferative responses to parasite antigen reflect the systemic immune response of patients with onchocerciasis. Local and systemic immune mechanisms to contain inflammation (e.g., blocking antibodies, downregulating cytokines) are prominent in infected patients (Ottesen 1995). Soboslay et al. (1997) also showed that distinct states of *O. volvulus* infection correlate with a particular cellular and humoral immune response, while Elson et al. (1995) suggest that immunity to *Onchocerca volvulus* may in part be mediated by an antigen-specific Th I-type response. Additionally, concomitant protective immunity of infected humans prevents most of the newly acquired L3 infections from developing. Recent evidence suggests that *Wolbachia* endobacteria (symbionts of arthropods and filarial nematodes) contain lipopolysaccharides that are released with the death of mfs and contribute to the inflammatory pathology associated with the disease (Taylor and Hoerauf 1999; Saint Andre et al. 2002; Taylor 2003; Taylor et al. 2002, 2005). A lot of progress has been made in understanding the nature of the immunity in onchocerciasis. More clarity is still needed with the host–parasite immune relationships, i.e., the type of protective immunity against onchocerciasis, given that the existence or induction of immunity in the treated, those that have been merely exposed, or infected individuals, could have an impact on long-term (drug treatment) control measures.

Geographical Distribution and Epidemiological Patterns

Onchocerciasis in Africa is found between latitudes 15° N and 14° S. It is present in Senegal in the west and in all the countries in between up to Ethiopia in the east and from Mali in the north to Malawi in the south (Map 1).

Geographical Distribution

The geographical distribution and the disease pattern of onchocerciasis vary between different zones. The vector–parasite complexes and clinical characteristics of the disease pattern differ in a number of ways between the Latin American and the African disease pattern as well as between geoclimatic zones within Africa and the Latin American foci (Duke 1981). These differences are attributable to parasite strains and their pathogenicity and the transmitting species of *Simulium* vectors, their biting behavior, flight range, and vectorial capacities.

Epidemiological Patterns of Disease

Two broad clinical–epidemiological patterns of the disease are present in Africa: in West Africa particularly, the clinical pattern of onchocercal disease found in the savanna is blinding while that found in the forest is non-blinding. The differences between savannah and rain forest forms of onchocerciasis in West Africa have also been identified through antigenic (Bryceson et al. 1976) and other immunological techniques (Lobos and Weiss 1985; Cianchi et al. 1985).

In general, the primary manifestations of onchocerciasis in the forest belts in West Africa are severe skin disease. In West Africa, blindness rates are significantly higher in savannah communities with mf prevalence greater than 60% (hyperendemic) compared to communities with similar prevalence in the rain forest where blindness rates are low and hardly reach 2% of the total population, (Anderson et al. 1976; Dadzie et al. 1989, 1990). In the most hyperendemic villages in the savannah areas, over 10% of the population may be blind due to onchocerciasis. In Central Africa, the pattern is less clear, with blinding as well as less blinding onchocerciasis occurring in both savannah and forest belts. Onchocercal skin disease is most prevalent in East Africa, while blindness due to onchocerciasis is rare (Brieger et al. 1998; WHO 1995b). While onchocercal ocular disease and blindness are more prominent in savannah regions, onchodermatitis or onchocercal skin disease (OSD) has a higher prevalence in forest areas, possibly because of differences in *O. volvulus* strains. The importance of OSD as a contributor to the disease burden of onchocerciasis has been recognized relatively recently (Murdoch 2010) with troublesome itching, accounting for 60% of DALYs (disability-adjusted life years) attributable to onchocerciasis.

Endemicity Levels and Severity of Disease

Onchocerciasis is a disease that largely manifests itself after many years of repeated infections (cumulative infections). How severe the disease is, therefore, depends on the intensity of infection, i.e., the level of mfs in the skin and elsewhere in the body, and the duration of infection. The location of the villages in which onchocerciasis is found is also very important. The closer the location to the rivers with blackfly breeding sites, the more important the disease is, in that virtually everybody above the age of 20 years will be infected and harbor dozens of adult worms that produce microfilariae. As one moves further away from the river, where the vector density is less, the prevalence and intensity of infection decline in relation to the distance to the vector breeding sites. Although the level of intensity of infection in an individual drives the manifestation of the diseases in the person, it is the community microfilaria load (CMFL) concept of a community level of infection, or level of endemicity, which is more important in the epidemiology of onchocerciasis and which becomes a very important monitoring factor in any control measure. In a control area where the drug, ivermectin

(Mectizan®, Merck & Co.), is used, the CMFL in the community will fall faster than the prevalence of infection. There is also a linear relationship between the prevalence of onchocercal skin disease, eye disease, and blindness, as well as the level of endemicity of the community (Remme et al. 1986). In the West African savannah, onchocerciasis becomes a major public health problem in hyperendemic communities, which have a prevalence of infection greater than 60% and the community microfilarial load (CMFL) exceeding ten microfilaria per skin snip (Remme et al. 2008). In such communities, blindness may affect more than 5% of the population. Such a situation is a threat to the survival of the village itself. It has been recognized that the fear of the disease led to the depopulation of many fertile river valleys in the West African savannah. In this part of the world, onchocerciasis was therefore both an important public health problem and a major obstacle to socioeconomic development. Following many years of control effort in the West African savannah area, the high levels of blindness due to onchocerciasis are no longer encountered, due to the absence of new infection, a fact which is manifested in children that were born after the start of control operations and who have been spared of infection, the disease and its consequent blindness. Additionally, the older people who were blind from onchocerciasis more often would have died out. This state of affairs has encouraged repopulation of hitherto desolate areas.

Disease Burden

Morbidity and Mortality

Two main epidemiological methods or indices—the original parasitological method of skin snip used during the period of the OCP and recently the Rapid Epidemiological Mapping (REMO) based on nodule surveys (Ngoumou et al. 1994), both set out in more detail under the section on diagnostics—have been used to obtain the estimates of onchocerciasis infection as reported by Zouré et al. (2014) and Noma et al. (2002) in which worldwide, over 150 million people are at risk of acquiring *O. volvulus* infection the largest majority from 30 countries in tropical Africa. A more recent estimates from The Global Burden of Disease Study (Global Health, Division of Parasitic Diseases and Malaria 2019) reported 14.6 million of onchocerciasis patients had skin disease, while 1.15 million had vision loss and the DALYs is presented in Fig. 7. Over the period from 2017 till present, remarkable progress has been made, which has impacted on the prevalence and incidence of the infection, as many countries intensify their efforts toward the elimination of the disease.

Retrospective studies from the OCP suggested that the life expectancy of the blind was greatly reduced and that the mortality rate in blind adults on the average was three to four times greater than in the population who were fully sighted. More recent studies in the Cameroon also confirm that blindness gave rise to a significant increase in mortality among the blind. Detailed analysis of data from OCP showed

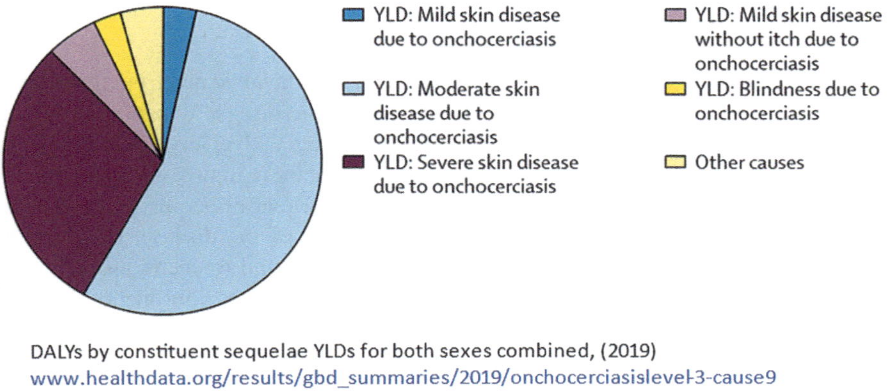

Fig. 7 The global burden of onchocerciasis in 2019 was 1.23 million (95% UI 0.765–1.82) DALYs. The constituent sequelae YLDs for both sexes combined (2019)

that the excess mortality due to onchocerciasis was 0.05 deaths per thousand per year (Little et al. 2004). Additional analysis of the relationships between microfilarial load and blindness incidence and microfilarial load and excess human mortality carried out in the same OCP period from 1975 to 2002 arrived at the conclusion that the excess relative risk of mortality is density dependent and more elevated in children.

Other Disease Burden

Patients with heavy microfilaria load may have infected lymph nodes, which may lead to mild or generalized lymphadenopathy, hanging groin, and scrotal elephantiasis with hanging groin. These could be a challenge for work in the peasants engaged almost exclusively in subsistence farming. Onchocercal skin disease has been shown to be associated with a variety of psychosocial and economic effects. The disease leads to stigmatization of affected persons and their families. Unsightly lesions from acute and chronic papular dermatitis limited the chances of adolescent girls finding marriage partners (Amazigo 1994). Other sociocultural aspects of skin disease—including people worried that the skin disease would affect their ability to interact socially, having fear of being ostracized, having a feeling of low self-esteem, and children who are likely to be distracted in school due to constant itching (Amazigo and Obikeze 1991; Amazigo 1993)—have now been recognized.

In the first decade of the OCP period, a rapid repopulation of the areas that were previously abandoned for fear of the disease took place, with the consequent impact of population pressure on the environment. Controlled resettlement with its economic dimensions at regulating the population resettlement in these oncho-free areas had to be put in place as, for example, in countries such as Burkina Faso and Ghana to ease this pressure.

Economic Impact

The constant and intense itching from onchocerciasis, the nuisance from the bites of the blackfly, and the impairment of vision due to the disease conspire to impact negatively on productivity, lead to diminished earnings, adversely affect supply of labor, and significantly reduce agricultural output. All these may result in the migration of communities away from fertile arable land for fear of acquiring the disease and its associated physical and psychosocial effects of the disease. On average, persons who suffer from onchodermatitis have been found to spend an additional US$8.10 over a 6-month period in comparison with their non-onchodermatitis counterparts from the same community and to spend an additional US$6.75 seeking health care over the same period.

Diagnosis

Until the recent past, the diagnosis of onchocerciasis was needed (1) for the diagnosis of the disease in individual patients, (2) to determine prevalence and endemicity of the disease over broad geographic areas for the planning of large-scale control programs such as in the OCP and APOC, (3) to assess impact of control measures, (4) to determine when to terminate mass drug administration (MDA), (5) for continued surveillance (Denery et al. 2010), and (6) for the purpose of validating the elimination of the disease.

Diagnosis of onchocerciasis is also required for specific epidemiological research including that for onchocercal infection in the blackfly vector and distinguishing between forest- and savannah-type infections. Parasitic diagnosis by microscopic identification following incubation of skin snips in saline can, however, be difficult especially in light infections. Improved or new diagnostic tools, which are very sensitive and highly specific, are therefore required. Box 1 shows the various diagnostic tests for onchocerciasis, many of which are not used routinely.

Box 1

- Parasitological detection of microfilariae by skin biopsy.
- Detection of microfilariae by ocular examination.
- Detection of adult worm in subcutaneous nodules.
- Skin test immediate hypersensitivity.
- Delayed hypersensitivity.
- Rapid diagnosis: nodule palpation.
- Assessment of leopard skin.
- Antibody detection complement fixation.
- Passive hemagglutination.
- Precipitation tests (gel diffusion).
- Indirect immunofluorescence.

- Radioimmunoassay (RIA).
- Radioimmunoprecipitation.
- Enzyme-linked immunosorbent assay (ELISA).
- Antigen/parasite gel diffusion product detection.
- DNA probe PCR.

Parasitological Diagnosis (Skin Snip Examination)

The demonstration of microfilariae in skin biopsies (WHO 1995a, b) is the gold standard and definitive diagnosis of onchocerciasis (Vincent et al. 2000). It is the most appropriate for measuring the intensity of infection (Kale 1978). Parasitological examination of skin snip for *O. volvulus* has, until recently, been the most widely used diagnostic method. The test is carried out by taking skin snips or biopsies from the right and left iliac crests using a sterilized corneoscleral punch. The biopsies are placed in physiological saline at 37 °C to incubate for 24 h and examined for the presence of microfilariae (mf), which are counted under a microscope. This technique has been fully described by Taylor and Green (1989). The motile mf may be counted under low-power microscopy, and expressed as the arithmetic mean mf per mg skin density (WHO 1987; Borsboom et al. 2003; Remme 2004; Ozoh et al. 2007; Mas et al. 2006; Lipner et al. 2006). The mf count is also used to determine community microfilarial load (CMFL) by control programs (Remme et al. 1986).

Skin snips should be bloodless to avoid contaminating the snips with other blood-borne microfilariae, and disposable instruments are preferred where there is a risk of transmission of blood-borne pathogens.

Microscopic demonstration of microfilariae is 100% specific for onchocerciasis, but it has the following drawbacks: (1) too low a sensitivity to be useful in areas of low transmission or for monitoring drug efficacy (Udall 2007; Ayong et al. 2005; Vincent et al. 2000), or for detection of early disease with a small microfilarial load (Enk 2006); (2) invasiveness that limits its use in epidemiological surveys, because of high nonparticipation rate—either due to pain it inflicts and the inconvenience (Boatin et al. 2002)—of people in endemic areas, which in the end could bias survey results (Tekle et al. 2012; Diawara et al. 2009), giving an underestimation of the disease risk or overestimation of the impact of a control program; (3) time-consuming in view of the nature of the disease, which is found in remote areas "at the end of the road"; and (4) does not allow for the detection of prepatent infection, an important factor in the detection of reinfection post-vector control.

Skin Tests

There are two categories of skin tests: (1) skin hypersensitivity reaction as a result of parasite response to topically applied diethylcarbamazine citrate (DEC), the so-called "Mazzotti" patch test and (2) skin hypersensitivity in response to intradermal injection of parasite extracts—both not currently used. A positive Mazzotti test (1948a) is the presence of local pruritus and a papular rash of the skin under the patch containing DEC cream. This test gave a 92% positive reaction in microfilariae-positive individuals in the Sudan (Stingl et al. 1984). The Patch test has been used in the OCP area in West Africa to determine possible renewal of transmission of *O. volvulus* in areas where the transmission had been eliminated by vector control (Toe et al. 2000; Ozoh et al. 2007). The DEC patch test is effective for detection of infection in children aged less than 15 years and can therefore be very useful for detecting recrudescence of infection in areas in which infection is believed to have been eliminated within the APOC area. It provides a noninvasive alternative to the skin snip although the latter remains the most suitable for accurate diagnosis (Boatin et al. 2002).

A new DEC patch (a transdermal application) was developed by TDR in collaboration with Lohmann Therapy System (LTS), Germany—the LTS-2—with the purpose of improving on the reliability and standardization of the DEC patch test and for ease of use (TDR 2008). The LTS-2 test has proved from field trials in Mali and Senegal in 2008 to be a simple and noninvasive surveillance tool and highly specific in the onchocerciasis foci in West Africa for early detection of recrudescence of transmission and infection in an area of low prevalence of infection. The two main operational challenges of DEC patch test are a failure of some people to return to have their patch examined post-application and some patches that had gotten partially or completely detached before they could be re-examined (Diawara et al. 2009). Other major problems associated with the test are that of nonspecificity and cross-reactions that occur in individuals with other filariases due to the extensive antigenic similarity between these species of filarial worms. Where prevalence is low, PCR and DEC patch are more sensitive than skin snip, and the DEC patch provides a good alternative to skin snip alone in low prevalence areas (Boatin et al. 2002). It may, however, be hampered by varying degrees of sensitivity in different geographic areas.

Nodule Palpation: Detection of the Adult Worm

Adult worms are normally located in nodules, which are easily palpable in infected patients. The discovery of a coil of hair-like white worms in an excised subcutaneous or intramuscular nodule in the appropriate epidemiologic setting is diagnostic of onchocerciasis on visual inspection. Most nodules have only three to five adults in them and less in hypoendemic areas. However, nodules containing as many as 50 adult worms have been found and liberated from the tissue by collagenase digestion.

The prevalence of onchocercal nodules in adult males, identified through simple nodule palpation at specified parts of the body, has also been shown to be reliable for community diagnosis of onchocerciasis (Whitword et al. 1999). Mass distribution programs need simpler and noninvasive methods to determine the population at high risk and those that require mass treatment with ivermectin (Edungbola et al. 1987). The rapid epidemiological mapping of onchocerciasis (REMO) tool developed by TDR and OCP as a simple and rapid test to determine communities at risk of onchocerciasis and requiring mass treatment (Ngoumou et al. 1994; Noma et al.

2002) has been an answer to this requirement, and it has been used extensively to map the distribution of onchocerciasis and sites for mass drug treatment in APOC countries (Noma et al. 2002; Gemade et al. 1998; Mace et al. 1997).

There are also frequently additional nodules that are located deeper in the tissue and are more difficult to detect. These conditions may lead to a substantial underestimation of nodules in women who have more subcutaneous fat and may have inhibitions about having a full examination. Nodules appear to have "preferred sites" that can vary according to geographical location.

A limitation of REMO is that (1) it uses nodule palpation, which has poor sensitivity and specificity in low prevalence communities (Table 2); (2) nodule palpation provides little information about the state of the adult worm, which is of particular interest regarding effects of drugs on the adult worm in drug trials; and (3) subcutaneous nodules may not always be palpable and other types of nodular tissue such as lipomas, lymph nodes, and cysts can frequently be mistaken to be of onchocercal nodules. Despite its limitations, REMO surveys carried out between 1997 and 2010 in 14,000 villages have produced geographical distribution of onchocerciasis in sub-Saharan Africa, which is also useful data for delineation of transmission zones.

Other Diagnostic Tests

Slit Lamp Technique

Microfilariae can be observed directly, using the slit lamp, in the cornea and the anterior chamber of the eye (Enk 2006), and the examination allows the microfilarial load in the eye to be counted. The slit lamp technique has a high specificity, but it is less commonly used for epidemiological purposes or by control programs as it requires expensive equipment and skilled personnel to perform it.

Table 2 Comparison of available diagnostic technique

Test	Specificity city	Sensitivity	*Ochengi* interferes	Throughput	Cost	Application
Skin snip	–>100%	Low	No	Low	Low	Field
Nodule palpation	Moderate	Low	No	High	Low	Field
Skin PCR	–>100%	–>100%	No	Low	High	Laboratory
Scratch PCR	–>100%	–>100%	No	Low	High	Laboratory
DEC patch	Variable	variable	No	Low	Low	Field
Ov 16 ELISA	–>100%	+60 (%)	???	High	Medium	Laboratory
Fly dissection	Low	Low	Yes	Low	Medium	Field
Pool screen PCR	–>100%	–>100%	No	High	Varies	Laboratory

Adapted from WHO/APOC meeting on feasibility of onchocerciasis elimination consultation, 2009

Ultrasound Examination

Adult worms in suspected nodules may be detected through ultrasound (Homeida et al. 1986; Poltera et al. 1987), but this has not yet been fully tested.

Current Diagnostic Tests

The current potential diagnostic tests are still in a research stage and would continue to be so for some time to come. The WHO Guidelines for Stopping Mass Drug Administration (MDA) and Verifying the Elimination of Human Onchocerciasis published in 2016 spells out the attributes and disadvantages of the current immunological tests against the historical parasitological and REMO tests together with the basis for the recommendation of the current diagnostic tests for use.

Immunological Tests

Antibody response to onchocercal infection can be detected in infected humans at 3 months and at more than 12 months before parasitologically detectable infection (Lobos et al. 1991a). This observation would make a viable immunological test, a useful tool for control and elimination programs.

Immunological diagnostic tests are based on the detection of antibodies specific to filarial helminths in infected patients or, more recently, to detection of the antigen. Some of these tests, such as complement fixation tests, have been available for many years but have now been replaced by more sensitive enzyme-linked immunosorbent assay (ELISA) tests. The most common antigen used to stimulate an immune response has been from an extract of *Dirofilaria immitis*.

The limitations of immunological tests for *O. volvulus* include cross-reactivity with other filarial parasite species and non-filarial helminth infections. While being a highly sensitive tool (Table 2), antibodies persist in persons who are no longer infected with living parasites. This limitation renders a number of immunodiagnostic tests only useful when applied on people coming into an endemic area from a non-endemic area and are therefore presumed to be seronegative. The concept would be susceptible for use as a surveillance tool on a large scale by control programs.

Detection of Parasite Antibody in Body Fluids

These tests detect antibody bound to a parasite extract and can be visualized using a secondary antibody conjugated to a radioisotope (radioimmunoassay) or an enzyme (ELISA test).

Antigen Detection Tests

A difficulty with antigen detection tests is that antigens from filarial helminths are commonly cross-reactive and therefore of low specificity (Cabrera and Parkhouse 1986) necessitating the need for extracts of the parasite that are species specific.

ELISA-based assays are more sensitive than the skin snip and are therefore useful for detecting early infections in young children born after transmission has been suppressed. The finding of early infections in such children could also serve as a warning sign for recrudescence of infection.

Monoclonal antibodies have been used to define onchocercal-specific antigens for diagnostic purposes. Promising results have been obtained but such tests are not yet available in a suitable form for field use. The use of monoclonal antibodies provides an advantage as they can be reliably produced in sufficient quantities with reproducible quality. The advent of recombinant DNA technology allows cloned antigens to be produced in sufficient quantities to overcome these problems.

Recombinant Antigens

Recombinant DNA technology may make the use of recombinant antigens more feasible for diagnosis of onchocerciasis; however, the difficulty of distinguishing between past and current infections remains a big challenge.

Detection of Parasite Antigen in Body Fluids

The detection of parasite products in the circulation or other body fluids such as urine provides conclusive evidence of an ongoing infection. The development of successful antigen detection assays for onchocerciasis has, however, proven technically difficult.

Ov 16 ELISA Test

Ov 16 ELISA serological test carried out in children aged <10 years (born after the control measures commenced and presumably have not been infected) as recommended by the WHO Guidelines, as well as currently in children aged 5–9 years as recommended by the OTS, is the serological test applied to determine whether transmission of onchocerciasis has been interrupted to enable stopping MDA. The Ov 16 ELISA assay has high specificity and throughput (Table 2). This assay uses a recombinant antigen of *O. volvulus* to measure the prevalence of immunoglobulin G4 (IgG4) antibodies (Lobos et al. 1991a, b). The test is done in the laboratory on dried blood spots (DBS) collected in the field. Two forms of ELISA serological tests are in use. One is based on alkaline phosphatase-based enzyme (used by the CDC and by OEPA) and the other horseradish-based enzyme (used in the RDT test and developed by Smith College originally). The diagnostic characteristics of the ELISA tests used by the OEPA consists of 99.9% specificity and 40% sensitivity (Richards et al. 2001). The OEPA ELISA test has been used to stop MDA in some in a number of endemic foci in Africa, viz. in Sudan (Higazi et al. 2013), Uganda, and Nigeria. The standardization of the ELISA tests are currently ongoing in three centers, viz., CDC and Kenya Medical Research Institute (KEMRI) in Kenya and Centre for Research on Filariasis and Other Tropical Diseases (CRFilMT) in Cameroon. RDT, which was designed to be carried out on blood from finger prick in the field, turned out not to be sufficiently sensitive. On the other hand, it has been found recently that the test carried out on DBS in the lab produces test results, which are comparable to ELISA tests.

Polymerase Chain Reaction (PCR) Tests

Definitive diagnosis of *O. volvulus* infection required the identification of the parasite in either the skin or palpation of subcutaneous nodules, but the gold standard parasitological method of microscopic examination of skin biopsies has the limitation of poor sensitivity (see Table 1). The polymerase chain reaction (PCR)-based assay is significantly more sensitive than skin snips and nodule palpation (Lustigman and Abraham 2009) and has the added advantage of diagnosing active disease state. The PCR technique uses species-specific probes for the "Oncho-150" repeat to diagnose *O. volvulus* infection by detecting parasite DNA in routine skin snips (Lustigman and Abraham 2009). As a highly sensitive DNA test, PCR test can detect low-level infections compared to the microscopic skin snip examination. The PCR assay is also used with pool screening to assess the frequency of *O. volvulus* infections in the pool of blackflies collected for the process of stopping MDA or carrying out surveillance (Pischke et al. 2002; Rodriguez-Perez et al. 2006). By amplifying the DNA, PCR assays can be sensitive enough to allow a single infective larva or microfilaria to be identified using DNA probes (Meredith et al. 1991). In recent times, a piloting of new blackfly qPCR protocols (that are said to provide the best combination of species-specificity, sensitivity, simplicity, and cost–effectiveness) and their accompanying quality assurance and quality control procedures has been recommended. The PCR assay is limited by its (1) lack of immediate quantification of skin microfilariae and (2) requirement for considerable laboratory infrastructure.

DNA Tests

Among the more recent developments, progress has been made toward producing probes for the detection of parasite DNA. DNA probe-based assays rely on the detection of specific sequences in the parasite's genome and not the host's response to the presence of the parasite. This means that a DNA probe-based assay cannot be used for some important applications for which an antibody-based assay is well suited, such as in detecting prepatent infections.

DNA Diagnosis of Individual Patients

Detection of parasite DNA in routine skin snips by PCR amplification of O-150, an *O. volvulus*-specific 150-bp repeated genomic DNA family, is one of the most sensitive techniques for detecting low-level infections in individual patients. A challenge with DNA probes as with other molecular techniques is to produce tests suitable for field use, as at present these tests are only suitable for research use in laboratories. There are also questions regarding the persistence and distribution of parasite DNA in the skin after ivermectin treatment has been used or after the natural death of that adult worm as this could limit the usefulness of the test.

Challenges for Appropriate Diagnostic Tools

Though onchocerciasis has a relatively well-defined distribution, there are overlapping clinical symptoms with other filarial parasites such as *Mansonella streptocerca*, *Mansonella perstans*, *M. ozzardi*, and *Loa loa* that it may be co-endemic with differentiating onchocercomas from lipomas, fibromas, lymph nodes, cysticerci, dermoid cysts, foreign body granulomas, and ganglia could also be challenging.

Onchocercal chorioretinitis needs to be differentiated from choroiditis due to toxoplasmosis, while the occasional isolated onchocercal optic atrophy needs to be differentiated from optic neuropathy, or nutritional optic atrophy. For these reasons and because of overlaps, a reliable differential diagnostic technique is required for making clinical distinctions with other conditions. While antigen tests could provide effective differential diagnosis, they are not yet sufficiently sensitive or easily used under field conditions.

Establishing new tools capable of identifying the presence of recrudescence is another challenge. The (DEC) patch skin showed promising results but although it was used to an advantage in isolated situations, its usefulness might perhaps have been overtaken by recent newer diagnostic tools. A rapid antigen detection test, similar to that used for diagnosis of lymphatic filariasis, would be preferable as it would detect active infections (Richards et al. 2001).

The challenges regarding diagnostics for onchocerciasis at present include the need for a field-accessible, highly specific, and sensitive diagnostic tool capable of detecting current infections, a diagnostic tool that can be used to assess the elimination status in transmission zones by monitoring the interruption of transmission.

Treatment

The management of onchocerciasis posed a considerable challenge in the past when the only drugs available for use were suramin and diethylcarbamazine. Suramin is a macrofilaricide (Ashburn et al. 1949), which has to be given by injection for several weeks. It is also toxic, causing damage to the kidneys. Although it has been used in limited mass treatment (Dawood 1978; Rougemont et al. 1980, 1984), there is hardly any place, presently, for its use in the treatment of onchocerciasis (Awadzi et al. 1995). The piperazine derivative diethylcarbamazine (DEC), however, has only a microfilaricidal action. This drug was used before not only for treatment but also to aid individual diagnosis—the so-called Mazzotti reaction (Mazzotti 1948b). The intense pruritus and severe reactions ("Mazzotti reaction") of fever, headache, rash, and edema and the possibility of causing "ocular inflammatory reaction," which could lead to irreversible ocular damage, also limited the use of DEC in the individual and on a large scale (Awadzi and Gilles 1980). These two drugs are therefore no longer recommended for the treatment of onchocerciasis (WHO 2001, 1995a).

Ivermectin (Mectizan®)

Ivermectin (Mectizan®, Merck & Co.) is currently the drug of choice for the treatment of onchocerciasis. It is a semisynthetic macrocyclic lactone derived from *Streptomyces avermitilis*, which was registered for the treatment of human onchocerciasis in October 1987 in France. Ivermectin is safe and effective when given orally at the standard dose of 150 mcg/kg body weight (Brown and Neu 1990). A single dose provides long-lasting suppression of microfilaridermia (Greene et al. 1985), making it suitable for community-based distribution. It is also currently the only microfilaricide that is suitable for large-scale onchocerciasis treatment (Awadzi et al. 1985; De Sole et al. 1989; Remme et al. 1989; Prod'hon et al. 1991; Whitworth et al. 1991; Collins et al. 1992). Ivermectin has mild adverse effects, which are non-ocular (Brown and Neu 1990). For practical reasons and ease of administration of ivermectin at the community level by community distributors, the tablets (3 mg strength) are given based on height as surrogate for weight as follows: 1 tablet for the 90–119 cm height group; 2 tablets for the 120–140 cm group; 3 tablets for the 141–158 cm group; and 4 tablets for the >158 cm group (Crompton 2006).

Effect of Ivermectin on the Onchocerca Parasite

As a microfilaricide, ivermectin suppresses the skin microfilariae. Ivermectin paralyzes intra-uterine microfilariae that degenerate and form a bolus blocking the anterior uterus, thus preventing insemination of oocytes that transform into degenerate ova. Repeated dosage of ivermectin also causes slow attrition of adult worms (Duke et al. 1990). Plaisier et al. (1995) suggested an irreversible decline in female adult worm microfilaria production by about 30% per treatment. Although this assertion is up for debate (Turner et al. 2013), it would suggest that if this is the case, then repeated treatment may ultimately eliminate the adult worm reproduction, provided there are no new infections. Gardon et al. (2002) report from their study that persistent three-monthly treatments with ivermectin would seem to kill more worms than annual treatments. Current findings (Diawara et al. 2009) also suggest that long-term repeat ivermectin treatment seems to have eliminated the infection from certain foci.

Other drugs that are receiving attention for the treatment of onchocerciasis are moxidectin, a macrocyclic lactone, which has completed clinical trials in adults for onchocerciasis.

Doxycycline and Other Antibiotics

Given that ivermectin is currently the only suitable drug for large-scale treatment of onchocerciasis, it is only prudent that other drugs that could replace and/or serve as adjunct to ivermectin be explored. This will be necessary in the event of the mfs developing resistance to ivermectin. In this respect, a macrofilaricide capable of eliminating the adult worm will be a useful additional product, help reduce the duration of treatment, and facilitate the prospects for the elimination of onchocerciasis.

Currently, no effective macrofilaricide, which is simple to administer as a single or double dose, is available for use against onchocerciasis (Mackenzie and Geary 2011).

Wolbachia endobacteria are symbionts of arthropods and filarial nematodes including *O. volvulus* and are essential for the fertility of *O. volvulus*. Following the earlier studies in Ghana (Hoerauf et al. 2000, 2001) that treatment of *O. volvulus*-infected individuals with doxycyline (DOX) at 100 mg daily for 6 weeks resulted in an almost complete depletion of *Wolbachia* by the 11th month, considerable interest has been shown in the potential use of DOX in the treatment of onchocerciasis, especially because of the finding that DOX inhibits embryogenesis in female *O. volvulus*, resulting in an apparent permanent elimination of skin mfs. These encouraging results have generated enthusiasm for considering *Wolbachia* endobacteria as targets for the treatment of onchocerciasis. Several regimens of DOX, (200 mg/day for 4 and 6 weeks and 100 mg/day for 5 weeks) have been tried. The current thinking following meta-analysis of the available data on trials of regimen suggests that DOX at 200 mg/day for 4 weeks can be considered as standard anti-*Wolbachia* treatment (Specht et al. 2023). The question of the feasibility of use of DOX in the field condition given the relatively long duration for treatment with the drug has been explored in one study by Wanji et al. (2009). In this study it was shown that community distributors of DOX obtained good coverage and compliance in areas endemic for both onchocerciasis and loiasis but this ought to be replicated elsewhere.

A search for additional drugs (antibiotics) that are active against *Wolbachia* without the contraindications in the use of DOX (contraindicated in children below 9 years old and pregnant women) and equally effective when given for shorter courses of treatment is a priority.

Moxidectin

Moxidectin (MOXI), a milbemycin compound, has a similar chemical structure to ivermectin. Although it has not been proven to be macrofilaricidal, studies by Trees et al. (2000) showed that it could induce sustained abrogation of embryogenesis in filarial animal models. The attraction to MOXI is that it has a longer half-life than ivermectin (Cotreau et al. 2003), a single dose provides longer duration of suppression of microfilaria than IVM (Opoku et al. 2018) with a prospect of being useful in interrupting transmission even when applied once annually. The drug has been in use in animal health for quite a while and has received FDA approval for treatment in humans at a dose of 8 mg. It is currently undergoing clinical trials for pediatric use. Although MOXI has been used successfully in animal health when there has been ivermectin resistance, it is unclear if MOXI will be effective against onchocerciasis in the event of ivermectin resistance in humans.

Vaccines

An effective vaccine in the arsenal against onchocerciasis would be an important addition given that effective vaccines are still the most economical and efficient tools to control/eliminate infectious diseases. Development of an effective vaccine against onchocerciasis has been the focus of a research program that was supported by the Edna McConnell Clark Foundation from 1985 to 1999 (Lustigman et al. 2002), and three recombinant antigens capable of inducing protective immunity to *O. volvulus* have been identified (Abraham et al. 2001). At the moment, if a vaccine that is safe and effective with 50% prevention could be developed, it could be used in combination with any of an approved chemotherapeutic agent. A lot of emphasis has been put on the effort to find a macrofilaricide against onchocerciasis; however, the interest in vaccine development against onchocerciasis has been rekindled lately, through the research work of The Onchocerciasis Vaccine for Africa (TOVA) and a further support of these activities will be in order (Lustigman and Abraham 2009).

Onchocerciasis Interventions

Vector control (1974–2002) and later drug treatment were used in the effort to control onchocerciasis and eliminate it as a disease of public health importance. There has now been a complete shift to elimination of the transmission of the disease through MDA. This confidence in the current strategy was encouraged by the successes in elimination in the Colombia focus and elsewhere in Latin America, certain foci in Uganda, and similar findings in West Africa (Diawara et al. 2009).

Vector Control (Historical Control Efforts)

Many efforts have been used in the past both in Africa and Latin America to control onchocerciasis through an attack on the vector. A catalogue of such efforts is as follows:

- In the Chiapas focus in Mexico, this involved clearing vegetation and applying plant extracts, Paris green, and creosote to the vector breeding sites. This attempt was unsuccessful.
- In Africa, *S. neavei* and onchocerciasis were successfully eradicated from a very small focus, Riana focus, in Kenya by simply clearing riverine forest.
- Wide availability of DDT ushered in new campaigns against blackflies that succeeded in eradicating *S. neavei* and hence onchocerciasis from the Kodera Valley in Kenya.
- Ghana started vector control on a small scale beginning in the Upper East Region of the country in the Black Volta valley and later in the Lower Volta valley in the south, the latter to protect the workers at the Akosombo hydroelectric dam construction site.
- During the 1960s, the Institut français de Recherche scientifique pour le Développement en Coopération (ORSTOM), previously Office de Recherche scientifique et technique Outre-Mer, and the Organisation de Coordination et de

Coopération pour la Lutte contre les Grandes Endémies (OCCGE) carried out a series of vector control studies in Ivory Coast, Burkina Faso, and Mali.
- Other early control efforts took place on the Comoé River in Burkina Faso and the Farako area of southeast Mali.

Ground Larviciding

Ground larviciding is effective in small, easy-to-reach river basins and was used to an advantage in Uganda. This was tried with DDT in a limited area (200 km^2) in Guatemala, but although the larvae of the fly were eliminated in the treated streams, the impact on the fly population was disappointing. Similarly, in Mexico, ground larviciding was practiced for about 6 years and then abandoned because it was deemed rather labor intensive. In Africa, *S. neavei* was successfully eradicated from the Kodera valley in Kenya through wide ground application of DDT. Ground larviciding has also been used successfully in small river basins for nuisance control and for the protection for personnel in the agricultural industry. More recently, ground larviciding has been used successfully in Uganda through the auspices of APOC and the Ugandan government. In very small foci and where the vector has been confirmed to be *S. neavei*, it would be profitable to undertake ground larviciding. This approach will eliminate the vector in such a focus within a short time (2–3 years) and the disease altogether in a time frame, shorter than drug treatment will. Elimination of the vector in Bioko Island was successfully (Traoré et al. 2009; Meyer et al. 2009) carried out by APOC and partners.

Aerial Larviciding (Large-Scale Vector Control)

The most comprehensive and extensive approach to the control of onchocerciasis and its elimination as a disease of public health importance in West Africa was by the Onchocerciasis Control Programme of West Africa (OCP), which was launched in 1974. This initially covered seven countries, Benin, Ghana, Côte d'Ivoire, Mali, Niger, Togo, and Burkina Faso, which later included the so-called extension areas of Guinea-Conakry, Guinea-Bissau, Senegal, and Sierra Leone. The program ran from 1974 to 2002 using aerial larviciding to control the vector. The objective of vector control was to apply larvicides to kill the vector at the laval stage at their breeding sites in water for a long enough period, being 14 years, for the parasite, *Onchocerca volvulus*, to die out of the human population or reservoir. Aerial larvicides were sprayed weekly into the mapped-out breeding sites in the river courses. Larvicides that were used included temephos (AgrEvo, France), a cheap and efficient organophosphorous insecticide with insignificant impact on nontarget fauna, pyrethroids, and biological *Bacillus thuringiensis* serotype H14–*B.t. H14* and other organophosphorous insecticides. These insecticides (larvicides) were used in rotation, taking the river flow into account (Guillet et al. 1995) to restrict emergence of insecticide resistance in vectors, to minimize adverse impact on nontarget organisms, and to have a cost-effective approach to the activity. The duration of larviciding was, at the start of the program, planned to last for a period of 20 years, as the life span of the adult female *O. volvulus* had been estimated to be up to 18 years.

However, this period was reduced to 14 years when studies on adult worm and modeling estimated the longevity to be on the average 10–12 years. With the strategy of larviciding combined with ivermectin treatment (1989/1990), the duration of larviciding reduced further to 12 years, a change that was informed by output from model predictions. Aerial larviciding as a mode of control, however, is relatively costly because of the heavy infrastructure, logistics, and insecticides that are required. However, an economic study on the activity estimated it to be cost-effective with an economic return of investment estimated to be 25%.

The impact of vector control in the OCP area has been well documented in several publications and reports. The salient features following vector control were (1) the rapid decline of human infection prevalence after 5–6 years into larviciding and (2) the absence of new infections in the areas concerned, indicating interruption of transmission. The effectiveness of larviciding in the OCP was assessed through periodic entomological and parasitological evaluations and the impact on morbidity (eye disease) as compared to baseline data. Vector control achieved a virtual interruption of transmission as observed in the zero annual transmission potentials (ATPs) recorded at the fly-catching points. This was observed in most parts of the original OCP area with the exception of a few isolated foci. Similarly, the CMFLs declined in a linear manner, reaching virtually zero after 10 years of control, followed by the predicted and accelerated fall in the prevalence of infection as predicted using the ONCHOSIM simulation model. The incidence of infection in children, a surrogate measure for new infections, was reduced by 100%. Ocular onchocerciasis became rare in the area. Any lesions found now are in the elderly from earlier infection, and the incidence of eye lesions due to onchocerciasis is nil in children born since the program began in 1975. Other post-control indices show that (1) more than 40 million people in the 11 OCP countries are now considered free from the risk of infection and eye lesions, (2) more than 1.5 million people are no longer infected, and (3) more than 600,000 cases of blindness have been prevented. 16 million children born since the program began are free of onchocerciasis. The socioeconomic impact has also been dramatic: 25 million hectares of fertile land in the river valleys were made available for resettlement and agriculture (Remme et al. 2008).

To maintain the achievements of the control program, the primary objective was to mount an effective epidemiological surveillance to detect early infection and control it before any disease recurrence. The OCP operated almost solely as a vertical program until the control activities were devolved to the participating countries. The countries have made efforts to integrate the epidemiological surveillance activities into other ongoing activities in the countries. This strategy of integrated surveillance activities is vital for sustainability of post-elimination surveillance of onchocerciasis.

Ivermectin Distribution Strategies

The donation of ivermectin (Mectizan®, Merck & Co), the drug of choice for the control of onchocerciasis, led to the development of innovative distribution mechanisms. Ivermectin mass chemotherapy was introduced following its discovery as an antihelminthic drug and its donation by the manufacturers, Merck

& Co, for as long as it is needed. At the onset (following its donation), ivermectin complemented vector control but gradually it became the only cost-effective tool (Crump et al. 2012; Cheke and Garms 2013) for the control of onchocerciasis in Latin America, Yemen, and sub-Saharan Africa and has proven to be an effective tool followed by the recent declaration of the elimination of onchocerciasis in Columbia (PAHO 2013). Distribution of ivermectin in most endemic areas was annual in Africa but is now given biannually in the majority of endemic countries while certain countries deliver as many as three or four times a year depending on their peculiar circumstances. The aim in all cases is to accelerate the elimination of the disease (Abiose et al. 2000; Sauerbrey 2008; WHO/APOC 2009).

Mobile Distribution

The mobile distribution method was the first strategy used (by the OCP) when ivermectin became available for mass treatment in 1988. The mobile teams were made up of trained health workers who visited communities in a military-like fashion, called on the community leader, and informed the leader about the date for the distribution of ivermectin to the eligible community members. Ivermectin distributions were carried out by the teams and they monitored for serious adverse events (Meredith et al. 2012). Mobile distribution by the OCP was very efficient and kept unrivalled good records of treatments; however, achieving high coverage was problematic since only members of the community available at the time of the team's visit benefited from the treatment, and inaccessible villages could be missed out. Mobile distribution method was unsustainable largely due to high personnel and logistic costs (Meredith et al. 2012).

Community-Based Treatment with Ivermectin (CBTi)

The community-based treatment with ivermectin (CBTi)—the Non-Governmental Development Organization's (NGDO) approach to the delivery of ivermectin—was introduced to improve access of remote populations to ivermectin and improve coverage, cost-effectiveness, and sustainability (Remme 2004). In CBTi, health workers mobilize and inform the community of the processes of drug distribution that had to be undertaken and ensure that community members undertake the required activities under their supervision (Crump et al. 2012). Community members, selected by the community leader or the health workers, distribute the drug supervised by health workers. Ivermectin, for those eligible individuals who were not available at the time of distribution, is sometimes kept with the drug distributor for a limited period and given to health workers when they return. However, the communities were not allowed to make any modifications to the processes or decisions on the activities. They simply carried out what the health workers instructed them to do.

Community-Directed Treatment with Ivermectin (CDTi)

The goal of the CDTi approach to ivermectin distribution was to fulfil the strategy of APOC, which was to set up sustainable delivery system for ivermectin with the objective to control onchocerciasis as a public health problem. To address the challenges of access, coverage, and goals of APOC, in sub-Saharan Africa, where health systems lack structures in remote rural areas, the TDR Task Force for Onchocerciasis Operational Research launched a multicountry study on community-directed treatment with ivermectin (CDTi) in 1995. The search for a more sustainable and cost-effective strategy resulted in the CDTi strategy. The results were striking and interesting—that communities can plan and implement drug distributions—achieving and sustaining high treatment coverage (WHO/TDR 2008, 2010). The CDTi was adopted as the principal strategy for the control of onchocerciasis by the Joint Action Forum, the governing body of the African Programme for Onchocerciasis Control (APOC) in 1997 (Amazigo et al. 2012).

CDTi, also called community-directed intervention (CDI) strategy, is a bottom-up initiative to public health-care delivery that empowers communities to take care of their own health. CDTi is considered a forerunner of community-directed intervention. In both strategies, affected communities are actively engaged in the design, drug supply chain, implementation, monitoring, and reporting of the delivery of health-care intervention. While the CDTi strategy was applied from 1997 for the delivery of ivermectin only, the CDI, a strategy also developed through implementation research, is applicable to multiple and integrated interventions including home-management of malaria.

Under the community-directed intervention, health workers approach a community to discuss the health and socioeconomic consequences for an individual and the community at large of a particular disease and the means of controlling it. The community is informed about the availability of donated medicines. The health workers respect the traditions of the community as well as the leadership structure and take note of important members and community groups that are likely allies in achieving community ownership of control strategies and tools.

Health workers and other partners jointly schedule visits to the communities, and trusting an old saying, "the villagers may know less, but they understand more," and providing them with access to the tools and knowledge that they lack, and they often exceed our expectations on the outcome. Hence, at scheduled meetings with community leaders and members, the health workers explain the importance of building a partnership between the community and health providers; they define the roles and responsibilities of external partners. In these meetings, community members understand they have the authority to make collective decisions—selection of drug distributors, changing and appointing new distributors and taking care of monetary and nonmonetary incentives of drug distributors. The community assumes the responsibility for the planning and management of the distribution of drugs under the guidance and supervision of health professionals. The community holds meetings and decides the period and mode of distribution (house to house or distribution from a central place) and collection of drugs from the nearest health facility.

Without external interference, community members select community-directed distributors (CDDs) from among their rank. Experiences of countries have shown that when given the authority to make such decisions, communities select peers, who they trust will deliver the drugs. The health workers and NGO personnel train the CDDs on census taking of households, treatment eligibility, exclusion criteria, and storage and safety of drugs as appropriate. The drug distributors are also trained to treat minor side effects, referrals of severe adverse effects, and record keeping (WHO 2010). In a recent publication on CDDs as "foot-soldiers" for the elimination of NTDs, detailed tasks, training of CDDs, and time CDDs put into their work are described (Amazigo et al. 2021).

During training, health workers emphasize the importance, to individual and community health, of accurate recording of the number of eligible persons treated, refusals and ineligibles, timely (Amazigo 1999) reporting back to the health staff, and the return of unused ivermectin. The community-directed distributors are retrained every year in the first 3 years, and new CDDs selected by the community are trained. This reduces the gap from attrition. The community is encouraged to take on the responsibility of reviewing the success of administration of ivermectin, motivation of distributors, and collection of drug supplies, through community self-monitoring (CSM). The CSM is carried out 1–3 months after distribution, and community report on how to improve the next treatment is provided to the health services.

As a major contribution, CDTi has built local capacity of five million (4,917,295) people as CDDs in sub-Saharan Africa by 2013 (Table 2). It has strengthened weak health systems, and has proven effective for the delivery of multiple health interventions including the distribution of insecticide-treated bed nets, home management of malaria and vitamin A supplementation, and distribution of other antifilarial drugs such as albendazole and Zithromax (Mutalemwa et al. 2009; Meredith et al. 2012). The CDTi empowers the communities to own, plan, and implement distribution programs, thus making the activity sustainable. CDTi has therefore been adopted by countries for the control of other neglected tropical diseases. The CDTi strategy has been shown to be a feasible and cost-effective approach for sustained ivermectin delivery (WHO/APOC 2010). The CDTi as other social innovations strategies by communities in rural Africa demonstrates the social and economic values of community engagement in the delivery of health-care services (Tiangco et al. 2023).

The major challenge with the CDTi strategy is that it is not a "quick fix." The implementation of an efficient CDTi requires patience, time, and resources at the onset when the community and health services have to be provided with the needed capacity to run the program. The initial periods of intensive community mobilization and capacity development costs are high, especially in conflict-torn countries (Amazigo et al. 2002). However, after an initial 2–3 years when the community health system support structures are in place, ivermectin distribution runs with minimal inputs from outside partners.

Through the CDTi approach, APOC prior to its closure in 2015 enabled the partner countries to deliver 194 million treatments from 180,000 communities on ivermectin treatment in the APOC endemic countries in 2018. Maps 2 and 3 show one example of the impact of the treatment on the prevalence of onchocerciasis and

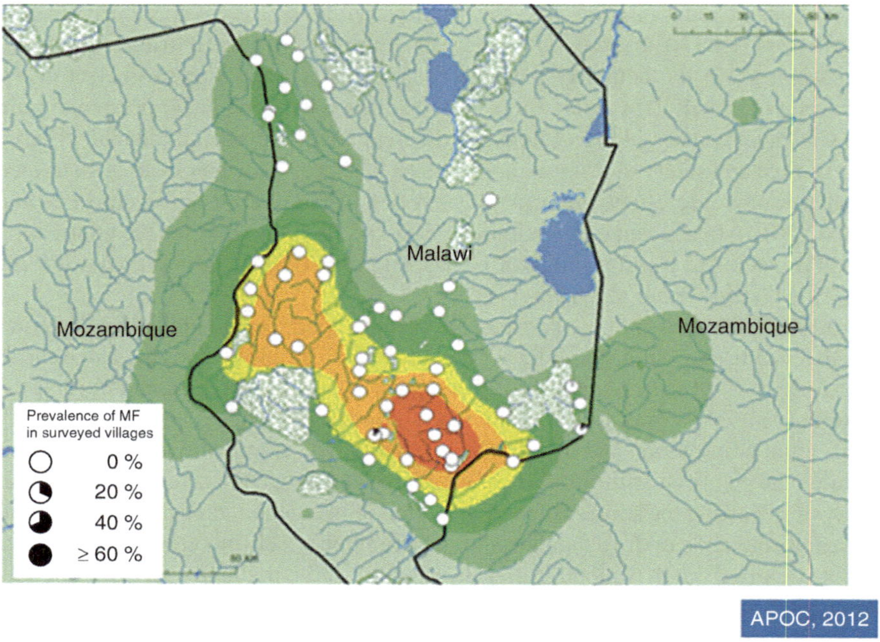

Map 2 Pre-control nodule prevalence in Malawi

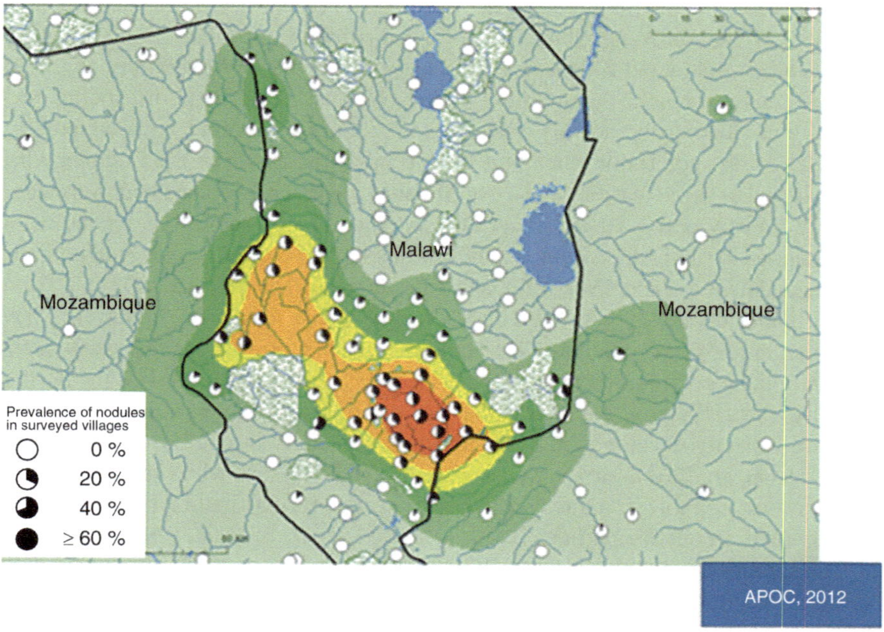

Map 3 Post-control (after 12 years) nodule prevalence in Malawi

epidemiological evaluations undertaken for the period 2008–2014 to assess progress in 12 APOC countries in CDTi areas with ≥6 years treatment showed a significant decline in onchocerciasis infection (Tekle et al. 2016).

From 1999 and 2010, CDDs had distributed almost 2 billion (1,816,138,200) ivermectin tablets in 16 countries (WHO JAF 2011). The current figures of drug donations and distribution are set out elsewhere in the book: *"The roles of medicine donation programmes in the control of and elimination of NTDs in sub-Saharan Africa."*

From Control to Elimination of Onchocerciasis in Africa

Both the OCP and APOC were conceived primarily as control programs with the principal goal to eliminate onchocerciasis as a disease of public health importance in Africa. OCP used a vector control strategy in the effort to interrupt transmission for long enough periods so that the adult worm parasite would die out from the infected population. APOC approached the control through the establishment of a sustainable drug delivery system. The OEPA has used only drug treatment strategy against onchocerciasis and reported successes with interruption of transmission in several foci in the Americas. There have also been reports of interruption of transmission in certain foci in Africa, in Senegal and Mali (Diawara et al. 2009), in Uganda (Katabarwa et al. 2012) and Sudan (Zarroug Isam et al. 2016), and several isolated foci in some of the APOC countries (Table 3). In the light of findings in

Table 3 Distribution of CDDs and HWs trained and available between 2000 and 2013

Subregion	Country	Period of APOC data submission[a]	Cumulative number of CDDs trained	Cumulative number of CDDs available[b]	Cumulative number of HWs trained
West Africa	Benin	2011–2013	29,410	29,410	837
	Burkina Faso	2011–2013	5151	5293	381
	Cote d'Ivoire	2008–2013	31,087	34,188	1949
	Ghana	2008–2013	32,880	14,443	9697
	Guinea	2011–2013	25,641	22,079	521
	Guinea Bissau	2008–2013	3219	2524	82
	Liberia	2000–2013	104,638	99,913	5601
	Mali	2011–2013	24,978	24,946	1466
	Nigeria	2000–2013	1,521,595	1,350,791	268,222
	Senegal		Not Available	Not Available	Not Available
	Sierra Leone[c]	2009–2013	93,972	80,448	4737
	Togo	2011–2013	15,773	15,733	1395

(continued)

Table 3 (continued)

Subregion	Country	Period of APOC data submission[a]	Cumulative number of CDDs trained	Cumulative number of CDDs available[b]	Cumulative number of HWs trained
Total West Africa			*1,888,344*	*1,679,808*	*294,888*
Central Africa	Angola	2005–2012	18,630	18,971	2098
	Burundi	2005–2013	67,734	68,042	1732
	Cameroon	2000–2013	338,130	359,269	31,593
	CAR	2000–2012	62,867	20,450	5221
	Chad	2000–2013	42,515	95,197	2773
	Congo	2001–2013	23,383	23,344	2271
	DRC	2001–2013	829,581	882,704	60,197
	Eq. Guinea	2000–2013	1748	1748	150
	Gabon	2002–2003	0	0	0
Total Central Africa			*1,384,588*	*1,469,725*	*106,055*
East Africa	Ethiopia	2001–2013	628,787	648,469	54,848
	Malawi	2000–2013	118,129	121,780	23,966
	South Sudan	2000–2013	78,986	79,992	5614
	Sudan	2000–2013	23,508	22,813	9847
	Tanzania	2000–2013	129,658	130,066	10,591
	Uganda	2000–2013	676,920	628,528	49,934
Total East Africa			1,655,988	1,631,648	154,800
Grand total			4,928,920	4,781,181	555,743

Source: WHO/APOC [22,23] and S1 data on the data used for this table

[a]The data collection period varies for country according to the year a country started to submit data to APOC

[b]Cumulative number of CDDs available may also include those present before the training from 2000 to 2013

[c]Sierra Leone data for CDDs were only available from 2009 to 2013. Data for HWs were from 2009 to 2013

APOC, African Programme for Onchocerciasis Control; CAR, Central African Republic; CDD, community-directed distributor; DRC, Democratic Republic of the Congo; HW, health worker; WHO, World Health Organization

Latin America (Gustavsen et al. 2011) and in Africa (Traoré et al. 2012), APOC, before it closed down, reviewed and modified its objective to include elimination of the disease. The elimination of onchocerciasis with ivermectin, which was not considered possible in Africa during the OCP and APOC eras, is now the strategy that has been adopted for Africa (Table 4).

A full and detailed text on onchocerciasis elimination is set out in the next chapter, "Onchocerciasis Part 2: Elimination."

Table 4 Results of epidemiological evaluations in 2008–2012

Conclusion of evaluation	Number of sites	Number of projects	Population (millions)
Elimination probably already achieved	20	12	17.6
Close to elimination	8	6	4.2
Ontrack, but still some years to go	12	11	17.4
Total with satisfactory progress	40	29	39.2
Unsatisfactory Progress	5	5	5.4
Toral Evaluated	45	34	44.7

Source: APOC 2012

Integration of Delivery Strategies for the Elimination of NTDs Including Onchocerciasis by 2030

Until late 1990s, the onchocerciasis and lymphatic filariasis (LF) control programs addressing individual NTDs worked as vertical programs with disease-specific funding streams. There is a consensus that integration of NTD programs improves greater efficiency, cost-effectiveness, and program coverage in co-endemic areas and in settings without the challenges of co-endemicity with *Loa loa*. Multiple advantages and key elements of successful NTD programming—careful planning, country leadership, engagement of stakeholders and disease experts, and setting clear objectives for successful integrated surveys for NTDs—have all been highlighted. In compliance with the guidelines of the WHO NTD 2030 elimination road map and considering the goal of "leaving no one behind," programs should give special attention to integration of systematic and sustained community engagement (Krentel et al. 2018) and morbidity management within wider framework of integrated health systems.

Integration of mass drug administration (MDA) for delivery of preventive chemotherapy (PCT) for the elimination of lymphatic filariasis, onchocerciasis, schistosomiasis, soil-transmitted helminthiasis, and trachoma in sub-Saharan Africa is being encouraged. The justifiable rationale is the considerable geographical overlap of these (NTDs). There is a need for each program to develop a comprehensive approach and focus on processes that will consistently improve geographical and treatment coverage while ensuring that all MDA activities are maximally integrated and affordable by the country in the absence of external donor financing.

Furthermore, understanding the costs of integrated NTD program without the contributions of communities and external partners and without compromising program quality remains a challenge now and in future in managing any recurrence of high prevalence levels. A recent study found mixed evidence regarding what aspects of the programs to integrate and how to operationalize integration and expected outcomes (Deardorff et al. 2018). For instance, multisectoral integration of Water, Sanitation and Hygiene (WASH) and deworming is difficult to measure and may result in poor and noneffective synergies.

Other fundamental challenges bedeviling integration of NTDs are differences in the definition of integration and implementation experiences of program managers even among those focusing on the five PCT alone.

Recent documents have reported differences in the definition of effectiveness among a variety of NTD stakeholder groups, including multilateral organizations, funding partners, national Ministry of Health (MOH) teams, district MOH teams, rural health workers, and community members participating in NTD programs. Also, there is insufficient information from countries' field operations regarding what aspects of the five PCT programs to integrate and how to operationalize integration for maximum benefits.

The challenges of using the CDTi strategy have been well elucidated including undermining "fragile and overstretched" health systems by continuing the endorsement and funding of vertical programs (Allen and Parker 2012). To minimize this challenge while maintaining CDD trained workforce, the partners are promoting the use of the CDTi strategy integrated in the national drug delivery system. Before the call for integration of NTD medicines delivery, 3.6 million CDDs treated 47.6 million persons cumulatively with other NTD and non-NTD interventions in nine countries (Nigeria, Uganda, Tanzania, Sierra Leone, Mali, Liberia, DRC, Cameroon, and Burundi) in 2012 (Amazigo et al. 2021). This trend can improve with investments in effective integration and documentation by program managers.

Overreliance upon volunteers (CDDs) without sufficient investment in building their capacity has resulted in low coverage. Indeed, one wonders why policy decision-makers are reluctant to listen to and engage communities and to ask what questions communities wish to be answered in order to remove barriers to field integration of NTDs as recommended by WHO (2022), a new road map of specific disease targets to control and eliminate NTDs by 2030.

The SARS-COV2 pandemic has moved community engagement to the center stage of public health and of recent considerations related to sustainability of healthcare programs. The experiences of program managers on integrated NTD control even before the SARS-COV2 pandemic is very limited (Kabatereine et al. 2010) and lack details on processes that could promote country and community ownership and sustainability for long-term benefits, as well as cost implications. There is no doubt a well implemented integration of at least the five PCT diseases can achieve elimination by 2030 as envisaged in the WHO road map.

Response of *Onchocerca volvulus* to Ivermectin

Ivermectin was registered for treatment against onchocerciasis in October 1987. Since then, millions of the tablets have been used on a large-scale treatment for onchocerciasis covering a period of over three decades. As for the case of any drug, which is used in large quantities over a long period, there has been concern of the possibility of the emergence of *resistance to ivermectin*. Taylor and Green (1989) suggest, however, that resistance is unlikely because of the ineffective transmission of resistance through a vector-dependent life cycle of *O. volvulus*. Hitherto, no drug resistance to ivermectin had been seen after years of intensive veterinary usage of the drug as prophylaxis against dog heartworm (*Dirofilaria immitis*). Some reports (Bowman 2012), however, suggest that there might be evidence of the development of resistance by *Dirofilaria immitis* to macrocyclic lactones (ML). More recent ML controlled studies showed that (loss of efficiency) strains of *D. immitis* established infections in dogs despite chemoprophylaxis with oral *ivermectin* or injectable *moxidectin* (Bourguinat et al. 2015). Additionally, reports from Ghana have shown persistent *O. volvulus* microfilaridemias in several subjects following multiple treatments with ivermectin (Awadzi et al. 2004a, b; Osei-Atweneboana et al. 2007). Remme et al. (2007), however, attributed the persistence of microfilaria in the skin in the Ghana subjects to the presence of new infections in the area concerned. Other studies (Bourguinat et al. 2007, 2008; Eng and Prichard 2005; Eng et al. 2006; Ardelli et al. 2006) have reported genetic selection by ivermectin on the genome of *O. volvulus*. In a comparative study in Cameroon, Abong et al. (2020) showed a significant reduction in the ability of IVM to clear microfilaria 30 days after treatment in communities, which had been under IVM treatment for over 20 years and those with fewer number of years IVM treatment.

The question of whether or not there is (1) a possibility of the development of ivermectin resistance, (2) an already present ivermectin resistance in humans, (3) emerging resistance, and (4) mere nonresponsiveness or atypical or suboptimal response of the adult worm to ivermectin has not yet been adequately answered. This calls for further work and vigilance given that ivermectin is at present the sole available and recommended drug for the treatment of onchocerciasis and the mainstay for the elimination of transmission on onchocerciasis.

Further Research in the Management of Onchocerciasis

The current management of onchocerciasis in sub-Saharan Africa is centered on the march to its elimination. Research in new intervention tools, effective diagnostics, revised strategies, monitoring and evaluation, and surveillance techniques and tools is required to support this objective. These, as well as other areas of research, are set out in detail in "Onchocerciasis Part 2: Elimination" in this book.

Additionally, community empowerment and ownership have been key in the CDTi strategy for onchocerciasis control and, by extension, its elimination. There is therefore the need for research on the best ways to improve on community empowerment, participation, and ownership in the programs.

The CDTi strategy has been very appropriate in getting large quantities of ivermectin to vast areas that need it. Research will be needed to find yet other ways of distribution of ivermectin and/or other new drugs in terms of regularity and frequency to respond to any future refractoriness/population fatigue to CDTi.

Challenges

Management of Ivermectin Supplies

Ivermectin is currently being given through MDA on a biannual and in some cases four times a year basis. It is also being given in all areas endemic for onchocerciasis. This calls for larger volumes and better forecasting of ivermectin supplies—forecasting optimum supplies and efficient management (storage) by the receiving countries to forestall unnecessary expiration of drugs due to inability to utilize all the supply in a timely manner.

Cross-border Issues

In a 2016 article—the authors concluded: *"Cross-border foci are common in Africa as rivers often form natural borders between countries while they also provide the breeding sites for the vector in onchocerciasis endemic areas. Simulium damnosum s.l. is a powerful fly that can travel many (and sometimes more than 20) kilometers in search of a bloodmeal without respecting national boundaries Hence cross-border coordination and synchronization of treatment and evaluation activities will be critical for onchocerciasis elimination in such areas."*

Residual Infections

The residual infection levels in different foci and especially in cross-border communities sharing common borders either internally or internationally has been an issue in onchocerciasis control and is likely to pose an important challenge for onchocerciasis elimination. The inextricable link between program operations and research should be explored to determine for each country, all foci with onchocerciasis and those with a decline in prevalence of pre-control endemicity level.

The Future

The previous two regional onchocerciasis control programs in Africa started off with the main objective to ensure that onchocerciasis is eliminated as an important public health problem throughout Africa. This was largely achieved in large tracts of the affected areas. The onchoendemic countries have now embarked on a program of total elimination of the disease from Africa. There is every optimism for the elimination of onchocerciasis altogether from SSA if given the financial and technical support. It can be envisaged that with a policy of onchocerciasis elimination from SSA, emphasis on efforts of an integrated approach for onchocerciasis and lymphatic filariasis elimination would be more cost-effective and of mutual benefit.

Conclusion

Through the erstwhile onchocerciasis control efforts, there has been an appreciable dent in the public health importance of the disease in the endemic countries. The distribution of the disease in SSA is changing as there is "shrinkage" of the onchocerciasis map as a result of the earlier vector control activities of the OCP and the impact of the very wide and large volumes of distribution of ivermectin through the auspices of APOC and its partners. Onchocerciasis, however, still remains a major parasitic disease in some of the endemic countries in Africa and therefore still requires attention. There is clear evidence to show that onchocerciasis may and can after all be eliminated altogether with the current means of treatment, an observation which has driven the shift in paradigm from control of the disease to the elimination of the transmission. Some challenges persist for the elimination programs, which need to be addressed to pave the way for rapid progress. The culture of research, particularly operational research, which played an important role in the OCP and APOC programs in the past, needs to be continued to shore up the elimination programs.

References

Abiose A et al (2000) Onchocerciasis control strategies. Lancet 356(9240):1523
Abong RA, Amambo GN, Chounna Ndongmo PW et al (2020) Differential susceptibility of onchocerca volvulus microfilaria to ivermectin in two areas of contrasting history of mass drug administration in Cameroon: relevance of microscopy and molecular techniques for the monitoring of skin microfilarial repopulation within six months of direct observed treatment. BMC Infect Dis 20:726
Abraham D, Leon O, Leon S, Lustigman S (2001) Development of a recombinant antigen vaccine against infection with the filarial worm Onchocerca volvulus. Infect Immun 69(1):262–270
Allen T, Parker M (2012) Will increased funding for neglected tropical diseases really make poverty history? Lancet 379:1097–1098
Amazigo U (1999) Community selection of ivermectin distributors. Community Eye Health 12(31):39–40
Amazigo UO (1993) Onchocerciasis and women's reproductive health: indigenous and biomedical concepts. Trop Dr 23:149–151
Amazigo UO (1994) Detrimental effects of Onchocerciasis on marriage age and breast feeding. Trop Geogr Med 46(5):322–325
Amazigo UV, Leak SGA, Zoure HGM et al (2021) Community-directed distributors—the "foot soldiers" in the fight to control and eliminate neglected tropical diseases. PLoS Negl Trop Dis https://doi.org/10.1371/journal.pntd.0009088
Amazigo UV et al (2002) The challenges of community-directed treatment with ivermectin (CDTI) within the African Programme for Onchocerciasis Control (APOC). Ann Trop Med Parasitol 96:41–58
Amazigo UV et al (2012) Community-driven interventions can revolutionise control of neglected tropical diseases. Trends Parasitol 28(6):231–238
Amazigo UO, Obikeze DS (1991) Socio-cultural factors associated with prevalence and intensity of onchocerciasis and onchodermatitis among adolescent girls in rural Nigeria. SER/TDR Project Report, World Health Organization, Geneva and challenges. Trends Parasitol 17(12):558–555
Anderson J, Fuglsang H, Marshall DC (1976) Studies on onchocerciasis in the United Cameroon Republic III. A four year follow-up of six forest and six savannah villages. The incidence of ocular lesions. Trans Trop Med Hyg 79:362–373
Ardelli BF, Guerriero SB, Prichard RK (2006) Ivermectin imposes selection pressure on P-glycoprotein from *Onchocerca volvulus*: linkage disequilibrium and genotype diversity. Parasitology 132(3):375–386
Ashburn LL, Burch TA, Brady FJ (1949) Pathologic effects of suramin, hetrazan and arsenamide on adult Onchocerca volvulus. Bol Oficina Sanit Panam 28:1107–1117

Ault SK (2007) Pan American Health Organization's Regional Strategic Framework for addressing neglected diseases in neglected populations in Latin America and the Caribbean. Mem Inst Oswaldo Cruz 102(Suppl 1):99–107

Awadzi K, Attah SK, Addy ET, Opoku NO et al (2004a) Thirty-month follow-up of sub-optimal responders to multiple treatments with ivermectin, in two onchocerciasis-endemic foci in Ghana. Ann Trop Med Parasitol 98(4):359–370

Awadzi K, Boakye DA, Edwards G, Opoku NO et al (2004b) An investigation of persistent microfilaridermias despite treatments with ivermectin, in two onchocerciasis-endemic foci in Ghana. Ann Trop Med Parasitol 98(3):231–249

Awadzi K, Dadzie KY, Schulz-Key H, Haddock DRW (1985) The chemotherapy of onchocerciasis. X. Assessment of four single dose treatment regimes of MK-933 (ivermectin) in human onchocerciasis. Ann Trop Med Parasitol 79:63–78

Awadzi K, Gilles HM (1980) The chemotherapy of onchocerciasis. III. A comparative study of DEC and metrifonate. Ann Trop Med Parasitol 79:199–210

Awadzi K, Hero M, Opoku NO, Addy ET, Büttner DW, Ginger CD (1995) The chemotherapy of onchocerciasis. XVIII. Aspects of treatment with Suramin. Trop Med Parasitol 46:19–26

Ayong LS, Tume CB, Wembe FE, Simo G, Asonganyi T, Lando G, Ngu JL (2005) Development and evaluation of an antigen detection dipstick assay for the diagnosis of human onchocerciasis. Trop Med Int Health 10(3):228–233

Baker RHA, Guillet P, Sékétéli A et al (1990) Progress in controlling the reinvasion of windborne vectors into the western area of the Onchocerciasis Control Programme in West Africa. Philos Trans R Soc Lond Ser B Biol Sci

Basanez MG, Pion SD, Churcher TS, Breitling LP, Little MP, Boussinesq M (2006) River blindness: a success story under threat? PLoS Med 3(9):e371

Blacklock BD (1926) The development of onchocerca volvulus in Simulium damnosum. Ann Trop Med Parasitol 20:1–48

Boatin BA, Toe L, Alley ES, Nagelkerke NJD, Borsboom G, Habbema JDF (2002) Detection of *Onchocerca volvulus* infection in low prevalence areas: a comparison of three diagnostic methods. Parasitology 125:545–532

Borsboom GJJM, Boatin BA, Naglekerke NJD, Agoua H, Akpoboua KLB, Soumbey EW et al (2003) Impact of ivermectin treatment on onchocerciasis transmission: Assessing the empirical evidence that repeated ivermectin mass treatments may lead to elimination/eradication in West Africa. Filaria J 2(8). https://doi.org/10.1186/1475-2883-2-8

Bourguinat C et al (2007) Genetic selection of low fertile *Onchocerca volvulus* by ivermectin treatment. PLoS Negl Trop Dis 1(1):e72

Bourguinat C et al (2008) P-glycoprotein-like protein, a possible genetic marker for ivermectin resistance selection in *Onchocerca volvulus*. Mol Biochem Parasitol 158(2):101–111

Bourguinat C, Lee ACY, Lizundia R, Blagburn BL, Liotta JL et al (2015) Macrocyclic lactone resistance in *Dirofilaria immitis*: failure of heartworm preventives and investigation of genetic markers for resistance. Vet Parasitol 210(3–4):167–178

Bowman DB (2012) Heartworm, macrocyclic lactones, and the specter of resistance to prevention in the United States. Parasit Vectors 5:138

Brieger WR, Awedoba AK, Eneanya CI, Hagan M, Ogbuagu KF, Okello DO, Ososanya OO, Ovuga EB, Noma M, Kale OO, Burnham GM, Remme JH (1998) The effects of ivermectin on onchocercal skin disease and severe itching: results of a multicentre trial. Trop Med Int Health 3:951–961

Brown KR, Neu DC (1990) Ivermectin -clinical trials and treatment schedules in onchocerciasis. Rev Acta Leidensia 59:169–175

Bryceson AD, van Veen KS, Oduloju AJ, Duke BO (1976) Antigenic diversity among *Onchocerca volvulus* in Nigeria, and immunological differences between onchocerciasis in the savanna and forest of Cameroon. Clin Exp Immunol 24(10):168–176

Cabrera Z, Parkhouse RM (1986) Identification of antigens of *Onchocerca volvulus* and *Onchocerca gibsoni* for diagnostic use. Mol Biochem Parasitol 20(3):225–231

Cheke RA, Garms R (2013) Indices of onchocerciasis transmission by different members of the Simulium damnosum complex conflict with the paradigm of forest and savanna parasite strains. Acta Trop 125:43–52

Cianchi R, Karam M, Henry MC, Villani F, Kumlien S, Bullini L (1985) Preliminary data on the genetic differentiation of *Onchocerca volvulus* in Africa (Nematoda: filaroidea). Acta Trop 42:341–351

Colebunders R, Siewe FJ, Hotterbeek A (2018) Onchocerciasis associated epilepsy, an additional reason for strengthening onchocerciasis elimination programs. Trends Parasitol 34(3):208–216

Collins RC, Gonzales-Peralta C, Castro J et al (1992) Ivermectin: reduction in prevalence and infection intensity of *Onchocerca volvulus* following bi-annual treatment in five Guatemalan communities. Am J Trop Med Hyg 47:156–169

Cotreau MM et al (2003) The antiparasitic moxidectin: safety, tolerability, and pharmacokinetics in humans. J Clin Pharmacol 43(10):1108–1115

Crompton D.W.T World Health Organization (2006) Preventive chemotherapy in human helminthiasis: a manual for health professionals and programme managers. World Health Organization, Geneva

Crosskey RW (1990) The natural history of blackflies. Wiley, Chichester

Crump A, Morel CM, Omura S (2012) The onchocerciasis chronicle: from the beginning to the end? Trends Parasitol 28:280–288. S1471-4922(12)00069-4. https://doi.org/10.1016/j.pt.2012.04.005

Dadzie KY, Remme J, Baker RHA, Rolland A, Thylefors B (1990) Ocular onchocerciasis and intensity of infection in the community. III. West African rain-forest foci of the vector Simulium sanctipauli. Trop Med Parasitol 41:376–382

Dadzie KY, Remme J, Rolland A, Thylefors B (1986) The effect of 7–8 years of vector control on the evolution of ocular onchocerciasis in West African savannah. Trop Med Parasitol 37:263–270

Dadzie KY, Remme J, Rolland A, Thylefors B (1989) Ocular onchocerciasis an intensity of infection in the community; II West African rain-forest foci of the vector Simulium yahense. Trop Med Parasitol 40:348–354

Dawood MS (1978) Field treatment of onchocerciasis in The Sudan with Suramin preparations and Banocide. World Health Organization, Alexandra

De Sole G, Remme J, Awadzi K, Accorsi S, Alley E, Ba O, Dadzie K, Giese J, Karam M, Keita F (1989) Adverse reactions after mass treatment of onchocerciasis with ivermectin. Combined results from eight community trials. Bull World Health Organ 67:707–719

Deardorff KV, Rubin Means A, Ásbjörnsdóttir KH, Walson J (2018) Strategies to improve treatment coverage in community-based public health programs: a systematic review of the literature. PLoS Negl Trop Dis 12(2):e0006211. https://doi.org/10.1371/journal.pntd.0006211

Denery JR, Nunes AA et al (2010) Neglected tropical disease research and development. The World Health Organization's 2010 report dedicated to neglected tropical diseases ... and open source collaboration methods for drugs

Diawara L, Traoré MO, Badji A, Bissan Y, Doumbia K, Goita SF, Konaté L, Mounkoro K, Sarr MD, Seck AF, Toé L, Tourée S, Remme JH (2009) Feasibility of onchocerciasis elimination with ivermectin treatment in endemic foci in Africa: first evidence from studies in Mali and Senegal. PLoS Negl Trop Dis 3:e497

Druet-Cabanac M, Preux PM, Bouteille B, Bernet-Bernady P, Hopkins AD (1999) Onchocerciasis and epilepsy: a matched case-control study in the Central African Republic. Am J Epidemiol 149(6):565–570

Duke BOL (1981) The six diseases of WHO. Onchocerciasis Br Med J (Clin Res Ed) 283(6297):961–962. https://doi.org/10.1136/bmj.283.6297.961

Duke BOL, Zea-Flores G, Castro J et al (1990) Effects of multiple doses of ivermectin on adult Onchocerca volvulus. Am J Trop Med Hyg 43(6):657–664. https://doi.org/10.4269/ajtmh.1990.43.657. PMID: 2267970

Duke BOL, Zea-Flores G, Castro J, Cupp EW, Muñoz B (1991) Comparison of the effects of a single dose and of four six-monthly doses of ivermectin on adult *Onchocerca volvulus*. Am J Trop Med Hyg 45:132–137

Duke BOL, Lewis DJ, Moore PJ (1966) Transmission of forest and Sudan-savanna strains of Onchocerca volvulus, from Cameroon, by Simulium damnosum from various West African bioclimatic zones. Ann Trop Med Parasitol 60(3)

Dunbar RW (1968) Nine cytological segregates in the Simulium damnosum complex. Bull Wld Hlth Org 40:974–979

Edungbola LD, Alabi TO, Oni GA et al (1987) 'Leopard skin' as a rapid diagnostic index for estimating the endemicity of African onchocerciasis. Int J Epidemiol 16(590):594

Elson LH, Calvopiña M, Paredes W, Araujo E, Bradley JE et al (1995) Immunity to onchocerciasis: putative immune persons produce a Th1-like response to Onchocerca volvulus. J Infect Dis 171(3):652–658

Eng JK et al (2006) Ivermectin selection on beta-tubulin: evidence in *Onchocerca volvulus* and *Haemonchus contortus*. Mol Biochem Parasitol 150(2):229–235

Eng JK, Prichard RK (2005) A comparison of genetic polymorphism in populations of *Onchocerca volvulus* from untreated- and ivermectin-treated patients. Mol Biochem Parasitol 142(2):193–202

Enk CD (2006) Onchocerciasis—river blindness. Clin Dermatol 24:176–180

Gardon J, Boussinesq M, Kamgno J et al (2002) Effects of standard and high doses of ivermectin on adult worms of *Onchocerca volvulus*: a randomized controlled trial. Lancet 360:203–210

Garms R, Walsh JF, Davies JB (1979) Studies on the reinvasion of the onchocerciasis control Programme in the Volta River basin by Simulium damnosum s.l. With emphasis on the south-western areas. Trop Med Parasitol 30:345–362

Garner A, Duke BOL, Anderson J (1973) A comparison of the lesions produced in the cornea of the rabbit eye by microfilariae of the forest and Sudan-savanna strains of onchocerca volvulus from Cameroon. Tropenmed Parasit 24:385–396

Gemade EI, Jiya JY, Nwoke BE, Ogunba EO, Edeghere H, Akoh JI, Omojola A (1998) Human onchocerciasis: current assessment of the disease burden in Nigeria by rapid epidemiological mapping. Ann Trop Med Parasitol 92(Suppl 1):79–83

Global Health Division of Parasitic Diseases and Malaria report (2019). www.healthdata.org/results/gbd_summaries/2019/onchocerciasis-level-3-cause

Greene BM, Taylor HR, Cupp EW, Murphy RP, White AT, Schulz-key H, D'Anna SA, Newland HS, Goldsmith LP, Auer C, Hanson AP, Freeman SV, Reber EW, Williams PN (1985) Comparison of ivermectin and diethylcarbamazine in the treatment of onchocerciasis. N Engl J Med 313:133–138

Guillet P, Seketeli A, Alley ES et al (1995) Impact of combined large-scale ivermectin distribution and vector control on transmission of onchocerca volvulus in The Niger basin, Guinea. Bull World Health Org 73(2):199–205

Gustavsen K et al (2011) Onchocerciasis in the Americas: from arrival to (near) elimination. Parasit Vectors 4(1):205. https://doi.org/10.1186/1756-3305-4-205

Hagan M (1998) Onchocercal dermatitis: clinical impact. Ann Trop Med Parasitol 92(Suppl 1):S85–S96

Hall LR, Pearlman E (1999) Pathogenesis of onchocercal keratitis (river blindness). Clin Microbiol Rev 12:445–453

Higazi TB, Zarroug IMA, Mohamed HA, ElMubarak WA, Deran TCM, Aziz N, Katabarwa M, Hassan HK, Unnasch TR, Mackenzie CD, Richards F, Hashim K (2013) Interruption of *Onchocerca volvulus* transmission in the Abu Hamed Focus, Sudan. Am J Trop Med Hyg Online 89:51–57

Hoerauf A, Mand S, Adjei O, Fleischer B, Büttner DW (2001) Depletion of Wolbachia endobacteria in Onchocerca volvulus by doxycycline and microfilaridermia after ivermectin treatment. Lancet 357:1415–1416

Hoerauf A, Specht S, Buttner M (2008) *Wolbachia* endobacteria depletion by doxycycline as antifilarial therapy has macrofilaricidal activity in onchocerciasis: a randomized placebo-controlled study. Med Microbiol Immun 197(3):295–311
Hoerauf A, Volkman L, Hamelmann C, Hamelmann C, Adjei O, Autenrieth IB, Fleischer B, Büttner DW (2000) Endosymbiotic bacteria in worms as targets for a novel chemotherapy in filariasis. Lancet 355:1242–1243
Homeida M, Mackenzie C, Williams H, Ghalib H (1986) The detection of onchocercal nodules by ultrasound technique. Trans R Soc Trop Med Hyg 80:570–571
Joint African Forum Ministerial meeting, Communique (2011) Kuwait City Dec 20, African Programme for Onchocerciasis Control Meeting Report 11
Kabatereine NB, Malecela M, Lado M, Zaramba S, Amiel O, Kolaczinski JH (2010) How to (or Not to) integrate vertical programmes for the control of major neglected tropical diseases in Sub-Saharan Africa. PLoS Negl Trop Dis 4(6):e755. https://doi.org/10.1371/journal.pntd.0000755
Kale O (1978) Small-scale trials of six drugs against *Onchocerca volvulus*. Tropenmed Parasitol 29(2):163–167
Katabarwa MN, Walsh F, Habomugisha P, et al (2012) Transmission of onchocerciasis in Wadelai focus of northwestern Uganda has been interrupted and the disease eliminated. J Parasitol Res 2012:748540. https://doi.org/10.1155/2012/748540
Kipp W, Burnham G, Bamuhiiga J, Leichsenring M (1996) The Nakalanga syndrome in Kabarole District, Western Uganda. Am J Trop Med Hyg 54(1):80–83
Krentel A, Gyapong M, Ogundahunsi O, Amuyunzu-Nyamongo M, McFarland DA (2018) Ensuring no one is left behind: urgent action required to address implementation challenges for NTD control and elimination. PLoS Negl Trop Dis 12(6):e0006426
Lewalen S, Courtright P (2001) Blindness in Africa: present situation and future needs. Br J Ophthalmol 85:897–903
Lipner EM, Dembele N, Souleymane S, Alley WS, Prevots DR, Toe L, Boatin B, Weil GJ, Nutman TB (2006) Field applicability of a rapid-format anti-Ov-16 antibody test for the assessment of onchocerciasis control measures in regions of endemicity. J Infect Dis 194(2):216–221
Little MP, Breitling LP, Basanez MG, Alley ES, Boatin BA (2004) Association between microfilarial load and excess mortality in onchocerciasis: an epidemiological study. Lancet 363(9420):1514–1521
Lobos E, Weiss N (1985) Immunochemical comparison between worm extracts of *Onchocerca volvulus* from savanna and rain forest. Parasite Immunol 7(3):333–347
Lobos E, Weiss N, Karam M, Taylor HR, Ottesen EA, Nutman TB (1991a) An immunogenic *Onchocerca volvulus* antigen: a specific and early marker of infection. Science 25:1603–1605
Lobos E, Weiss N, Karam M, Taylor RH, Ottesen EA (1991b) An immunogenic Onchocerca volvulus antigen: a specific and early marker of infection. Science 251(5001):1603–1605
Lustigman S, Abraham D (2009) Onchocerciasis (Chapter 67). In: Barrett AT, Stanberry LR (eds) Vaccines for biodefense and emerging and neglected diseases. Academic Press, London, pp 1379–1400. ISBN: 10: 0123694086
Lustigman S, James ER, Tawe W, Abraham D (2002) Towards a recombinant antigen vaccine against Onchocerca volvulus. Trends Parasitol 18(3):135–141
Mace J-M, Boussinesq M, Ngoumou P et al (1997) Country-wide rapid epidemiological mapping of onchocerciasis (REMO) in Cameroon. Ann Trop Med Parasitol 91(4):379–391
Mackenzie CD, Geary TG (2011) Flubendazole: a candidate macrofilaricide for lymphatic filariasis and onchocerciasis field programs. Expert Rev Anti-Infect Ther 9(5):497–501
Mas J, Ascaso C et al (2006) Reduction in the prevalence and intensity of infection in *Onchocerca volvulus* microfilariae according to ethnicity and community after 8 years of ivermectin treatment on the Island of Bioko. Equatorial Guinea Trop Med Int Health 11(7):1082–1091
Mazzotti L (1948a) The feasibility of using the allergic reactions which follow the administration of hetrazan as an auxiliary means of diagnosis of onchocerciasis. Rev Inst Salud Enferm Trop Mex 9:935

Mazzotti L (1948b) Posibilidad de utilizar como medio diagnostico auxiliar en la onchocercosis, las reacciones alérgicas consecutivas a la administración del 'Hetrazan'. Rev Inst Salubridad Enfermedadas Tropicales 38:235–237

Meredith S, Cross C, Amazigo U (2012) Empowering communities in combating river blindness and the role of NGOs: case studies from Cameroon, Mali, Nigeria, and Uganda. Health Res Policy Syst 10:16. https://doi.org/10.1186/1478-4505-10-16

Meredith SE, Lando G, Gbakima AA, Zimmerman PA, Unnasch TR (1991) *Onchocerca volvulus*: application of the polymerase chain reaction to identification and strain differentiation of the parasite. Exp Parasitol 73:335–344

Meyer RR, Mas J, McCall PJ, Post RJ, Zouré H, Noma M, Yaméogo L, Sékétéli AV, Amazigo UV (2009) The elimination of the onchocerciasis vector from the Island of Bioko as a result of larviciding by the WHO African Programme for Onchocerciasis Control. Acta Trop 111(3):211–218

Murdoch ME (2010) Onchodermatitis. Curr Opin Infect Dis 23(2):124–131

Murdoch ME, Asuzu MC, Hagan M et al (2002) Onchocerciasis: the clinical and epidemiological burden of skin disease in Africa. Ann Trop Med Parasitol 96(3):283–296

Murdoch ME, Hay RJ, Mackenzie CD, Williams JF, Ghalib HW, Cousens S et al (1993) A clinical classification and grading system of the cutaneous changes in onchocerciasis. Br J Dermatol 129(3):260–269

Mutalemwa P et al (2009) Community directed approach beyond ivermectin in Tanzania: a promising mechanism for the delivery of complex health interventions. Tanzan J Health Res 11(3):116–125

Nelson J, Fodjo S, Remme JHF, Preux P-M, Colebunders R (2020) Meta-analysis of epilepsy prevalence in West Africa and its relationship with onchocerciasis endemicity and control. Int Health 00:1–11. https://doi.org/10.1093/inthealth/ihaa012

Ngoumou P, Walsh JF, Mace JM (1994) A rapid mapping technique for the prevalence and distribution of onchocerciasis: a Cameroon case study. Ann Trop Med Parasitol 88(5):463–474

Noma M, Nwoke BEB, Nutall I, Tambala PA, Enyong P, Namsenmo A, Remme J, Amazigo UV, Kale OO, Seketeli A (2002) Rapid epidemiological monitoring of Onchocerciasis (REMO): its application by the African Programme for Onchocerciasis Control (APOC). Ann Trop Med Parasit 96(Suppl 1):S29–S40

Opoku NO, Bakajika DK, Kanza EM et al (2018) Single dose moxidectin versus ivermectin for *Onchocerca volvulus* infection in Ghana, Liberia, and the Democratic Republic of the Congo: a randomized, controlled, double-blind phase 3 trial. Lancet 392(10154):1207–1216

Osei-Atweneboana MY, Eng JK, Boakye DA, Gyapong JO, Prichard RK (2007) Prevalence and intensity of Onchocerca volvulus infection and efficacy of ivermectin in endemic communities in Ghana: a two-phase epidemiological study. Lancet 369(9578):2021–2029

Ottesen EA (1995) Immune responsiveness and the pathogenesis of human onchocerciasis. J Infect Dis 171(3):659–671

Ozoh G, Boussinesq M, Zoung-Kanyi Bissek AC et al (2007) Evaluation of the valuation of the diethylcarbamazine patch to evaluate onchocerciasis endemicity in Central Africa. Trop Med Int Health 12(1):123–129

PAHO (2013) Global Health Observatory (GHO) data. Onchocerciasis. www.who.int/entity/gho/neglected_diseases/onchocerciasis

Pearlman E, Hall LR (2000) Immune mechanisms in Onchocerca volvulus-mediated corneal disease (river Blindness). Parasite Immunol 22:625–631

Pion SDS, Kaiser C, Boutros-Toni F, Cournil A, Taylor MM et al (2009) Epilepsy in onchocerciasis endemic areas: systematic review and meta-analysis of population-based surveys. PLoS Negl Trop Dis 3(6):e461

Pischke S, Butter DW, Liebau E, Fischer P (2002) An internal control for the detection of Onchocerca volvulus DNA by PCR-ELISA and rapid detection of specific PCR products by DNA detection test strips. Trop Med Int Health 7(6):526–531

Plaisier AP, Alley ES, Boatin BA, Van Oortmarssen GJ, Remme H, De Vlas SJ, Bonneux L, Habbema JD (1995) Irreversible effects of ivermectin on adult parasites in onchocerciasis patients in the Onchocerciasis Control Programme in West Africa. J Infect Dis 172(1):204–210

Plaisier AP, Van Oortmarssen GJ, Remme J, Habbemma JD (1991) The reproductive lifespan of Onchocerca volvulus in West African savanna. Acta Trop 48:271–284

Poltera AA, Reyna O, Zea Flores G et al (1987) Detection of skin nodules in onchocerciasis by ultrasound scans. Lancet 329(8531):505

Post RJ, Laudisoit A, Mandrso M, Lakoe T et al (2022) Identification of the onchocerciasis vector in the Kakoi-Koda focus of the Democratic Republic of Congo. journals.plos.org November 4, 2022. https://doi.org/10.1371/journal.pntd.0010684

Prod'hon J, Boussinesq M, Fobi G, Prod'hum JM, Enyong P, Lafluer C, Quillévéré (1991) Lutte contre onchocercose par ivermectin; résultats d'une campagne de masse au Nord-Cameroun. Bull World Health Organ 69:443–450

Remme J, Baker RHA, De Sole G, Dadzie KY, Adams MA, Alley ES, Avissey HSK, Walsh JF (1989) A community trial of ivermectin in the onchocerciasis focus of Asubende, Ghana. I. Effect on the microfilarial reservoir and the transmission of Onchocerca volvulus. Trop Med Parasitol 40:367–374

Remme JH, Amazigo U, Engels D, Barryson A, Yameogo L (2007) Efficacy of ivermectin against Onchocerca volvulus in Ghana. Lancet 370(9593):1123–1124

Remme JHF (2004) Research for control: the onchocerciasis experience. Trop Med Int Health 9(2):243–254. https://doi.org/10.1046/j.1365-3156.2003.01192.x

Remme JHF, Boatin B, Boussinesq M (2008) Helminthic diseases: onchocerciasis and loiasis. In: International encyclopedia of public health, vol 3, 1st edn. Academic, San Diego, pp 339–351

Remme JHF, Dadzie KY, Karam M et al (1986) The effects of 11 years of simulium control in OCP on ABR's, ATP's, prevalence and intensity of *O. volvulus* infections and on the incidence and prevalence of eye. Parasitic Diseases Programme. World Health Organization, Geneva

Richards FO, Boatin B, Sauerbrey M, Sékétéli A (2001) Control of onchocerciasis today: status and challenges. Trends Parasitol 17(12):558–563

Rodriguez-Perez MA, Lugo-Rodriguez L, Lizarazo-Ortega C, Unnasch TR (2006) Entomological and serological assessment of Onchocerca volvulus transmission in the northern Chiapas focus (Mexico). Acta Entomologica Serbica 11(Suppl):51–59

Rougemont A, Hien M, Thylefors B, Prost A, Rolland A (1984) Traitement de l'onchocercose par la Suramin à faibles doses progressive dans les collectivités hyperendémiques d'Afrique occi-dentale. 2. Résultats cliniques, parasitologiques et opthalmologique en de transmission con-trôlée. Bull World Health Organ 62:261–269

Rougemont A, Thylefors B, Ducan M, Prost A, Ranque PH (1980) Traitement de l'onchocercose par la Suramin à faibles doses progressive dans les collectivités hyperendémiques d'Afrique occidentale. 1. Résultats parasitologique et surveillance opthalmologique en zone de transmission non interrompue. Bull World Health Org 58:919–122

Saint Andre A, Blackwell NM, Hall LR, Hoerauf A, Brattig NW, Volkmann L, Lustigman S, James ER, Tawe W, Abraham D (2002) Towards a recombinant antigen vaccine against *Onchocerca volvulus*. Trends Parasitol 18(3):135–141

Sauerbrey M (2008) The onchocerciasis elimination program for the Americas (OEPA). Ann Trop Med Parasitol 102(Suppl 1):25–29

Soboslay P, Geiger SM, Weiss N et al (1997) The diverse expression of immunity in humans at distinct states of Onchocerca volvulus infection. Immunology 90:592–599

Specht S., Kamgo J, Geary T (2023) Antifilarial chemotherapy: current option for humans in humans and animal filariases: landscape, challenges and control, 1st edn. Ed. By Ronald Kaminsky and Timothy Geary ©2023 WILEY-VCH GmbH. Published 2023 by WILEY-VCH GmbH

Stingl P, Ross M, Gibson DW et al (1984) A diagnostic patch test for onchocerciasis using topical diethylcarbamazine. Trans R Soc Trop Med Hyg 78:254–258

Taylor HR, Green BM (1989) The status of ivermectin in the treatment of human onchocerciasis. Am J Trop Med Hyg 41:460–466

Taylor MJ (2003) Wolbachia in the inflammatory pathogenesis of human filariasis. Ann N Y Acad Sci 990:444–449

Taylor MJ, Bandi C, Hoerauf A (2005) *Wolbachia* bacterial endosymbionts of filarial nematodes. Adv Parasitol 60:245–284

Taylor MJ, Ford L, Hise AG, Lass JH, Diaconu E, Pearlman E (2002) The role of endosymbiotic Wolbachia bacteria in the pathogenesis of river blindness. Science 295:1892–1895

Taylor MJ, Hoerauf A (1999) *Wolbachia* bacteria of filarial nematodes. Parasitol Today 15:437–442

Tekle AH, Elhassan E, Isiyaku S, Amazigo UV, Bush S, Noma M, Cousens S, Abiose A, Remme JH (2012) Impact of long-term treatment of onchocerciasis with ivermectin in Kaduna State, Nigeria: first evidence of the potential for elimination in the operational area of the African Programme for Onchocerciasis Control. Parasites Vectors 5:28

Tekle AH, Zouré HGM, Noma M et al (2016) Progress towards onchocerciasis elimination in the participating countries of the African Programme for Onchocerciasis Control: epidemiological evaluation results. Infect Dis Poverty 5:66. https://doi.org/10.1186/s40249-016-0160-7

Thylefors B, Negrel AD, Pararajasegaram R, Dadzie KY (1995) Global data on blindness. Bull World Health Organ 73:115–121

Tiangco PMP, Mier Alpano JD, Cruz JRB et al (2023) Community engagement self-monitoring (CE-SM) strategy for social innovations in health: pilot implementation in The Philippines. BMJ Innov 0:1–7. https://doi.org/10.1136/bmjinnov-2022-001049

Toe L, Adjami AG, Boatin BA et al (2000) Topical application of diethylcarbamazine to detect onchocerciasis recrudescence in west Africa. Trans R Soc Trop Med Hyg 94:519–525. https://doi.org/10.1016/S0035-9203(00)90073-7

Traoré MO, Sarr MD, Badji A, Bissan Y et al (2012) Proof-of-principle of onchocerciasis elimination with ivermectin treatment in endemic foci in Africa: final results of a study in Mali and Senegal. PLoS Negl Trop Dis 6:e1825. [PMC free article] [PubMed]

Traoré S, Wilson MD, Sima A, Barro T, Diallo A, Aké A, Coulibaly S, Cheke RA, Meyer RR, Mas J, McCall PJ, Post RJ, Zouré H, Noma M, Yaméogo L, Sékétéli AV, Amazigo UV (2009) The elimination of the onchocerciasis vector from the Island of Bioko as a result of larviciding by the WHO African Programme for Onchocerciasis Control. Acta Trop 111(3):211–218

Trees AJ, Graham SP, Renz A, Bianco AE, Tanya V (2000) *Onchocerca ochengi* infections in cattle as a model for human onchocerciasis: recent developments. Parasitology 120(Suppl):S133–S134

Turner HC, Churcher TS, Walker M, Turner HC, Churcher TS, Walker M, Osei-Atweneboana MY, Prichard RK, Basanez M'a-G (2013) Uncertainty surrounding projections of the long-term impact of ivermectin treatment on human onchocerciasis. PLoS Negl Trop 7(4):e2169

Udall DN (2007) Recent updates on onchocerciasis: diagnosis and treatment. Clin Infect Dis 44:53–60

Vajime CG, Dunbar RW (1975) Chromosomal identifications of eight species of the sub-genus *Edwardsellum damnosum* Theobald (Diptera: Simuliidae). Tropenmed Parasit 26:111–138

Vincent JA, Lustigman S, Zhang S, Weil GJ (2000) A comparison of newer tests for the diagnosis of onchocerciasis. Ann Trop Med Parasitol 94(3):253–258

Wanji S, Tendongfor N, Nji T et al (2009) Community-directed delivery of doxycycline for the treatment of onchocerciasis in areas of co-endemicity with loiasis in Cameroon. Parasit Vectors 2(1):39. https://doi.org/10.1186/1756-3305-2-39

Whitword JAG, Gemade E II (1999) Independent evaluation of onchocerciasis rapid assessment methods in Benue State, Nigeria. Trop Med Int Health 4(1):26–30

Whitworth JAG, Gilbert CE, Mabey DM, Maude GH, Morgan D, Taylor DW (1991) Effects of repeated doses of ivermectin on ocular onchocerciasis community-based trial I Sierra Leone. Lancet ii:1110–1103

WHO/APOC (2010) Bull World Health Organ 88(7):509–518. Published online 2009 Dec 1. https://doi.org/10.2471/BLT.09.069203

WHO JAF (2011) Seventeenth Session of the Joint Action Forum-JAF17: The World Health Organization Year 2011 Progress Report, 1st September 2010–31st August 2011

WHO/TDR (2008) CDI report for major problems in Africa. World Health Organization, Geneva

WHO/TDR (2010) Community-directed interventions for priority health problems in Africa: results of a multicountry study. Bull World Health Organ 88(7):509–518

World Health Organization (1987) WHO expert committee on onchocerciasis: third report. Technical series No 852. WHO, Geneva. www.who.int/entity/tdr/publications/documents/cdi_report_08.pdf

World Health Organization (1995a) Expert Committee on onchocerciasis control, Technical Report Series no. 852. WHO, Geneva

World Health Organization (1995b) The importance of onchocercal skin disease: report of the Pan-African Study Group on Onchocercal Skin Disease, Document TDR/AFR/RP/95. WHO, Geneva

World Health Organization (2001) Criteria for certification of interruption of transmission/elimination of human onchocerciasis. World Health Organization, Geneva, Document No: WHO/CDS/CEE/2001.18a

World Health Organization/African Programme for Onchocerciasis Control (2009) Informal consultation on elimination of onchocerciasis transmission with current tools in Africa—"Shrinking the Map", Ouagadougu

World Health Organization (2022) Elimination of human onchocerciasis progress report 2021: Weekly epidemiological record 18 November 2022, 97th YEAR No 46, 2022, 97, 591–598. http://www.who.int/wer

Zarroug Isam MA, Hashim K et al (2016) The first confirmed elimination of onchocerciasis focus in Africa: Abu Hamed, Sudan. Am J Trop Med Hyg 95(5):1037

Zimmeman PA, Dadzie KY, De Sole G, Remme J, Alley ES, Unnasch TR (1992) Onchocerca volvulus DNA probe classification correlates with epidemiologic patterns of blindness. J Infect Dis 165:964–968

Zouré HGM, Noma M, Tekle AH et al (2014) The Geographical distribution of onchocerciasis in the 20 participating countries of the African Programme for Onchocerciasis Control: (2) pre-control endemicity levels and estimated number infected. Parasit Vectors 7:326

Onchocerciasis Part 2: Elimination

Yankum Dadzie, Boakye A. Boatin, and Uche Amazigo

Contents

Y. Dadzie (✉)
GOEC, Neglected Tropical Disease Department , Min of Health Ghana, Accra, Ghana

B. A. Boatin
Institute of Parasitology, McGill University and Lymphatic Filariasis Support Centre for Africa, Accra, Ghana

U. Amazigo
Pan-African Community Initiative on Education and Heatlh PACIEH, Enugu, Enugu State, Nigeria

J. O. Gyapong, B. A. Boatin (eds.), *Neglected Tropical Diseases - Sub-Saharan Africa*, Neglected Tropical Diseases, https://doi.org/10.1007/978-3-031-53901-5_10

Abstract

The International Task Force for Disease Eradication (ITFDE) defines elimination as the reduction to zero of the incidence of infection caused by a disease specific pathogen in a defined geographic area, as a result of deliberate efforts, followed by a continued action to maintain the zero incidence. Disease elimination necessitates the availability of an effective diagnostic tool and an intervention tool that can reduce the incidence of the disease specific pathogen to zero. The diagnostic tool, Ov 16 ELISA assay and ivermectin, a microfilaricide cum macrofilaricide appear to have met the two conditions. Since its registration in 1987 endemic countries have used ivermectin as a control tool given its presumed solely microfilaricidal effect. The reports at the turn of the millennium of ivermectin eliminating human onchocerciasis in four countries in the Americas and in some foci in Africa engendered a change in objective from control to elimination of onchocerciasis. The strategies and activities toward elimination have included: Onchocerciasis elimination mapping; enhanced ivermectin MDA implementation; application of coverage surveys as essential M and E and emphasis on entomological evaluations. WHO Guidelines for Stopping Mass Drug Administration (MDA) and Verifying the Elimination of Human Onchocerciasis guide the elimination process with the Onchocerciasis Technical Advisory Subgroup (OTS) advising on the technical aspects. Efforts are required to address the challenges for the elimination of onchocerciasis which include, variable technical capacity at national level, especially onchocerciasis entomological skills, setting up of special laboratories for processing Ov 16 ELISA serological and O-150 PCR pool screening tests, long-term investment in community engagement, safe MDA without SAEs in oncho and loa loa co-endemic areas and safe MDA in conflict areas. Further research is required to address diagnostics, treatment, improved intervention methods, and compliance. Although progress with the elimination march is slow there is optimism that onchocerciasis elimination in the endemic countries can and will be achieved.

Introduction

The International Task Force for Disease Eradication (ITFDE) defines elimination as the reduction to zero of the incidence of infection caused by a disease specific pathogen in a defined geographic area, as a result of deliberate efforts, followed by a continued action to maintain the zero incidence. Disease elimination can be accomplished if two conditions are met, viz. (1) availability of a diagnostic tool that can detect or identify the disease in all its forms and stages and (2) availability of an intervention tool that can be used to reduce the incidence of the disease specific pathogen to zero. These two conditions appear to have been met in (1) the diagnostic tool, Ov 16 ELISA assay based on the use of the recombinant *Onchocerca volvulus* antigen to measure Immunoglobulin G4 (IgG4) antibodies, (Lobos et al. 1991a, b) and (2) ivermectin, a macrocyclic lactone which kills microfilariae of *Onchocerca volvulus*, affects the viability and reproductivity of the adult female worm, and thus is able to eliminate onchocerciasis when administered in repeated doses at regular intervals for a period of time.

Ivermectin has been used for the treatment of human onchocerciasis since its registration in 1987 with the main objective of eliminating onchocerciasis as a disease of public health importance. The elimination of onchocerciasis with ivermectin was considered unlikely when it was first reported (Aziz et al. 1982) to be capable of being used in the management of onchocerciasis, as it was simply seen as a microfilaricide, like diethylcarbamazine (Hawking 1978; Rougement 1982). A decade after its use, a meeting of experts, convened to review the question of eliminability of onchocerciasis with ivermectin, (Dadzie et al. 2003) still assessed it as only capable of eliminating onchocerciasis under limited conditions, such as that found in the Americas where the disease was mild, existed in small, isolated foci and was transmitted by vectors with low vectorial capacity. It was not until at the turn of millennium that the capacity of ivermectin to eliminate human onchocerciasis was demonstrated in several endemic areas, viz., in the Americas,- in Colombia (Nicholls et al. 2018) Ecuador (Lovato et al. 2014) Mexico (Rodríguez-Pérez et al. 2010) and Guatemala (Lindblade et al. 2007 ; Gonzalez et al. 2009) and in Africa by the demonstration of the proof of principle of elimination in Africa in foci in Mali and Senegal (Diawara et al. 2009, Traoré et al. 2012) and in Sudan (Higazi et al. 2013), Nigeria (Tekle et al. 2012) and Uganda (Katabarwa et al. 2012).

With the prospects of possible elimination of onchocerciasis with drug use, there is now a paradigm shift from just controlling the disease to eliminating the disease altogether. This is likely to be achieved in Africa provided coverage is optimum and there is further extension of the treatment to other areas that have not been included in the current treatment, e.g., some areas where there is coendemicity of onchocerciasis and *Loa loa* and onchocerciasis hypoendemic areas.

These events in South America and Africa, and the prospects of elimination of onchocerciasis led to an important revision of the view held on what can be done with ivermectin against onchocerciasis in Africa. In 2011 the Joint Action Forum of APOC passed a resolution to modify the objective of APOC to include Elimination of Onchocerciasis in all areas in Africa where it is possible. However, the formal launch of the elimination of onchocerciasis in Africa occurred with the closure of APOC in 2015, followed by the creation of the Expanded Special Programme for Elimination of Neglected Tropical Diseases (ESPEN) by WHO AFRO and the publication of WHO Guidelines for Stopping Mass Drug Administration (MDA) and Verifying the Elimination of Human Onchocerciasis in 2016. The final step taken was WHO's recommendations to all onchocerciasis endemic countries in Africa to change their goal of intervention against onchocerciasis from Onchocerciasis

Control to Elimination and to create National Onchocerciasis Elimination Expert Committees to guide them in the pursuit of the new objective. The WHO also set up the Onchocerciasis Technical Advisory Subgroup on Onchocerciasis (OTS) to advise it to steer the process of the change to Elimination of Onchocerciasis and to provide guidance in the development of technical approaches and strategies that endemic countries must adopt to ensure a smooth implementation of elimination of onchocerciasis. The WHO Guidelines recommended two indicators to apply to demonstrate that interruption of onchocerciasis has occurred before MDA could be stopped. They are:

1. to demonstrate entomologically an infectivity rate of less than one infective L3 larva in the head of one blackfly among 2000 blackflies tested by O-150 PCR pool screening at the upper level of 95% confidence interval (CI) and,
2. to demonstrate a seroprevalence of <0.1% in children aged 5–9 years at the upper bound of 95% CI.

Change to Elimination

Elimination of onchocerciasis entails a national onchocerciasis elimination programme changing its strategies of implementation to include:

1. assessment and identification of all the areas in the country where transmission occurs and subjecting all those areas to MDA. In Africa, generally, such areas will include (a) untreated areas in the country that have been classified as hypoendemic and non-endemic when the objective of the intervention was exclusively onchocerciasis control, (b) Lymphatic Filariasis (LF) and onchocerciasis co-endemic areas where LF MDA is being stopped after successful LF Transmission Assessment Survey (TAS), and (c) taking steps to develop, in some 11 countries that harbour onchocerciasis and loa loa co-endemic areas, plans to map out the endemicity of onchocerciasis and the prevalence and intensity of *loa loa* in the coendemic areas to be able to ensure safe implementation of MDA with ivermectin.
2. review and ensure that implementation of MDA in all areas achieve the maximum possible therapeutic coverage which is 100% of all who are eligible to be treated, which translates to close to 80% of the population.
3. Monitoring and evaluation (M&E) to ensure effective progress towards the endpoint when MDA can be stopped.
4. Entomological skills, which up until recently have received less attention and been relegated to the back seat, need to be brought up to the front seat to assume their role in elimination activities and in particular the preparation for and implementation of stopping MDA.
5. Countries would like to consider setting up laboratories to carry out the numerous PCR pool screening tests and the ELISA tests that are associated with various stages of the elimination programme.

Strategies for Onchocerciasis Elimination

Onchocerciasis Elimination Mapping

The first action to address, in the change to onchocerciasis elimination is to identify all the areas which need to be put on MDA. This is to ensure that no area transmitting onchocerciasis is left untreated, as such areas would or could constitute sources of new infection to the areas that have been cleared of infection. This process has been termed Onchocerciasis Elimination Mapping (OEM). It normally starts with conducting a desk top review of the country to identify and exclude all areas where onchocerciasis is unlikely to occur, a process termed onchocerciasis exclusion mapping. The currently recommended threshold at which OEM should trigger MDA is 2% seroprevalence in adults aged 20 years and above. Since currently available diagnostic does not have the ideal diagnostic performance characteristics, (Biamonte et al. 2022), a statistical threshold rule of 5% has been recommended to address the deficient performance of the current ELISA diagnostic test in use. To operationalize the use of the statistical threshold, it is recommended to use the district as the implementation unit. The objective of the exercise is to determine which of the districts among those that have not been treated to date contain communities with a seroprevalence above the statistical threshold and will therefore need to start MDA.

OEM is implemented in two stages. The first stage is to purposively identify five first line villages in the district from which 100 adults aged 20 years or more will be serologically tested per village by RDT or ELISA on DBS. The district qualifies to be put on MDA if the seroprevalence results of one or more of the villages is 5% or more. If seroprevalence results in none of the identified first line villages attains 5% the second stage mapping will be implemented. This consists of random selection of 20 villages from the district, out of which 50 adults aged 20 years and above per village will be sampled for serological tests by RDT or ELISA on DBS. The district qualifies for MDA if the seroprevalence results in two villages or more is 5% or more, or at least one village tests seroprevalence of 10% or more. It should be noted that this approach is provisional and could change with respect to the availability of the ideal diagnostic tool for use and the confirmation of the process itself which is currently under evaluation.

Areas Co-endemic with LF

In onchocerciasis and LF co-endemic areas under LF MDA, which are going to be evaluated for LF TAS, it is recommended that OEM is planned to run concurrently with the LF TAS. This is basically a setting whereby onchocerciasis treatment has occurred with the implementation of LF MDA and so the objective of the assessment is to determine whether LF MDA has been adequate to interrupt onchocerciasis transmission at the focus or there is a need to continue MDA for onchocerciasis should cessation of LF MDA be triggered by the results of LF TAS. In the event onchocerciasis MDA is not required, a stop MDA would need to be carried out.

Such an approach is in line with the requirements on National NTD programmes to show integration or coordination of NTD activities whenever an M & E is being conducted for any disease to show cost effective use of resources.

Areas Co-endemic with Loa Loa

Since severe adverse events (SAEs) with death were first reported in 1991 (Boussinesq et al. 1998) in the Cameroon and later also in the DRC following administration of ivermectin to the populations in areas co-endemic for onchocerciasis and loa loa, MDA in such co-endemic areas is only carried out under the Guidelines issued by the Mectizan Expert Committee (MEC) of Mectizan Donation Program (MDP) and the Technical Consultative Committee (TCC) of APOC (June 2004). The Guidelines which require active surveillance after MDA together with the availability of prompt hospital support for SAEs, restrict MDA implementation to only oncho and loa co-endemic areas where oncho endemicity is either hyper- or meso-endemic.

It has been imperative with the change in the objective of onchocerciasis intervention to elimination, to address the issue of being able to carry out MDA in all onchocerciasis and loa loa co-endemic areas without the occurrence of SAEs. The Test and not Treat (TnT) strategy employing a LoaScope appears to have made enormous strides towards addressing the problem (Kamgno et al. 2017). TnT requires the determination of loa loa Mf level in the blood of all individuals in a co-endemic area, with the view to identify and avoid treating individuals with very high level of infection with loa loa over a threshold, currently set at 20,000 Mf/ml with ivermectin. This requirement is assured using LoaScope, a device coupled with a smartphone which takes short videos of blood samples and provides information on the density of Mf loa in 3 min. MDA with ivermectin carried out employing TnT with LoaScope have demonstrated that the population coverage increases due to the confidence evoked in the population with the safety assured and with the occurrence of nil SAEs. TnT with LoaScope has also been used to confirm the safety of carrying out MDA without the need for TnT when the prevalence of loa infection density greater than 20,000 Mf/ml in the community is less than 1%. Mapping strategies are being developed to enable identifying onchocerciasis and loa loa co-endemic areas with high density of loa loa infection which require special care and attention during MDA intervention in order to avoid the occurrence of SAEs.

Monitoring and Evaluation Strategies

The OTS current M&E recommendation is a test to indicate that the point of stopping MDA has been reached. The process consists of selecting up to three first line villages located near productive blackfly breeding sites in the transmission area to test up to 300 children aged 5–9 years with Ov 16 ELISA. When the seroprevalence is $\leq$0.1% then the area is probably ready for conducting "stop MDA" as elaborated in the WHO 2016 Guidelines. The test can be carried out when an effective MDA has been carried out for a period of 10 years. When the result is negative the test can be repeated after 3–4 years or every 3–4 years until the stop MDA indicative results are attained to warrant undertaking "stop MDA."

Coverage Surveys

Modelling has demonstrated that the length of time required for the elimination of onchocerciasis with ivermectin MDA depends on the precontrol endemicity, the level of coverage of MDA and the frequency of MDA implementation per annum (Winnen et al. 2002). This suggests that when the precontrol endemicity is known and the coverage is optimal the required length of time for MDA and hence the time point for stop MDA can be estimated from modelling. Coverage surveys which have been developed to assess the exact coverage of MDA can be undertaken as a monitoring tool to assess the correctness and completeness of the coverage reports provided from the field. The results of the coverage surveys allow faults in the coverage report to be detected to allow corrective measures to be undertaken to ensure that MDAs are optimally implemented for interruption of transmission to occur at the expected time. As a rule, therefore, elimination programmes must make special efforts to ensure high coverage of MDA in all endemic communities in order to avoid having to prolong MDA over what is necessary to reach the endpoint. Increasing annual MDA to biannual frequency helps to accelerate the attainment of the endpoint for stop MDA but ensuring high level of coverage is the first important action to take before consideration to increasing the number of MDAs per annum.

Entomological Evaluations

Implementation of onchocerciasis elimination requires complementary entomological input whilst the process of demonstrating the elimination of onchocerciasis requires even more of entomological input. Entomological skills and know-how are necessary for identifying blackfly breeding sites, the basis for the selection of first line villages for OEM and for selecting villages for M&E. Entomological evaluation data such as blackfly biting rates, Annual Transmission Potential, identification of peak transmission season, fly species determination, fly parity assessment are important information required to guide the elimination process. Entomological assessment of the infectivity level of blackflies must be undertaken when a programme area or a transmission zone is identified for stop MDA. For MDA to be stopped in a transmission zone, the infectivity level must be demonstrated to be less than 1 L3 in the head of one blackfly among 2000 female blackflies, an indication that interruption of transmission of onchocerciasis has occurred. *This index together with a serological test of seroprevalence of ≤0.1% in children aged 5–9 years make* up the process required to be demonstrated to decide whether MDA can be stopped as required by the WHO Guidelines for Stopping Mass Drug Administration and Verifying the Elimination of Human Onchocerciasis.

Assessment of infectivity levels in blackflies requires entomologists to identify and select blackfly breeding sites and determine the appropriate blackfly catching points. The entomologists then train vector collectors on how to collect blackflies on their exposed legs in a sitting position, at specific period of the day at the specified catching points and on how to label and preserve the caught blackflies in provided reagents. The preserved blackflies are collected at intervals for storage in the lab until PCR pool screening is undertaken. It is therefore clear that endemic nations need to have entomologists and entomologist-technicians to ensure that these important entomological activities can be undertaken to demonstrate that elimination of onchocerciasis has occurred.

Demarcation of the Country into Transmission Zones

The demarcation of the country into transmission zones allows (i) the progress of the elimination process to be better assessed and (ii) to carry out the process of stop MDA in sections of the country as and when appropriate. The demarcation can follow the main river courses in the country whilst considering such factors as the ecology and vector species distribution when known. This approach is informed and justified by the natural history of onchocerciasis transmission determined by the dispersion of the biting adult female blackfly vectors of the Simuliidae family. Adult female blackfly vectors of the Simuliidae family concentrate around their breeding sites at specific points in the basins of fast flowing rivers or streams characterized by the existence of rocks or rocks interspersed with overhanging and aquatic vegetation where the adult female blackflies lay their eggs and the pre-adult stages of the blackfly develop. This contrasts with the administrative convenience of MDA implementation from the districts.

Cross Border Collaboration Between Neighbouring Countries

The era of onchocerciasis control did not require any stringent corporation by neighbouring countries as to how and when they carried out control activities at either side of their common borders. With Onchocerciasis elimination, which involves the maintenance of zero incidence of onchocercal infection after elimination of onchocerciasis in a transmission zone, neighbouring countries have to collaborate in the implementation of MDA with ivermectin at their neighbouring areas on either side of their common border. The countries must share technical information on the progress of the elimination process. This is to ensure that one border area does not get into the state of surveillance, be it post treatment or post elimination surveillance whilst the other side of the border is implementing MDA with poor coverage and thus constituting a potential source of re-infection to the side across the border where MDA has been stopped. Neighbouring countries therefore need to collaborate closely across their common border areas in the implementation of ivermectin MDA and share technical information on activities being carried out or planned to be carried out at the common border areas.

Integration and Cross-Cutting Approaches

Implementation of onchocerciasis elimination should be carried out as much as possible not as a stand-alone activity but should be planned and integrated with other NTD activities. Such areas of integration may include the Monitoring and Evaluation of, for example, onchocerciasis and lymphatic filariasis together. Planning of activities in general must consider and examine the feasibility of integrating with other NTD activity to save cost and enhance efficiency.

Stop MDA

The WHO published in 2016, Guidelines for Stopping Mass Drug Administration and Verifying the Elimination of Human Onchocerciasis, a document that sets out what is required of onchocerciasis endemic countries to demonstrate to qualify to be certified as having eliminated onchocerciasis from their countries (Fig. 1).

Onchocerciasis elimination process by MDA with ivermectin can be divided into three phases.

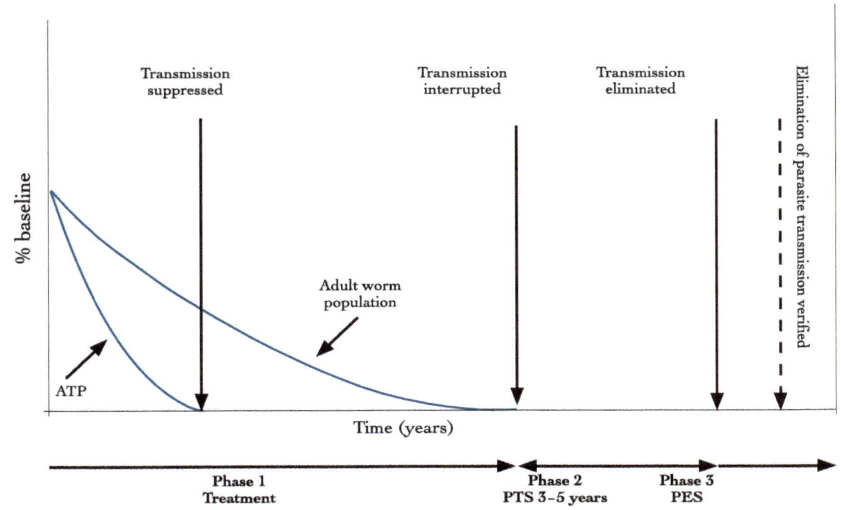

ATP, annual transmission potential; PES, post-elimination surveillance; PTS, post-treatment surveillance

Fig. 1 Phases of Onchocerciasis eliminationCourtesy: Guidelines for stopping mass drug administration and verifying elimination of human onchocerciasis: criteria and procedures. World Health Organization (2016). https://apps.who.int/iris/handle/10665/204180. Document number WHO/HTM/NTD/PCT/2016.1 I

Phase 1

The first phase is the period of intervention when ivermectin MDA is given to the eligible population residing in all the areas in the country identified to show seroprevalence of >2% in adults aged 20 years and above, for long enough periods until pre-stop MDA test results indicate that the point of stop MDA has been reached. This can encompass the entire country, in the event it is so, as in a small country or in different parts of the country, categorized into transmission zones, at different times successively.

At this point the two tests to demonstrate interruption of transmission of onchocerciasis to enable stop MDA are carried out. They consist of:

(a) demonstrating in the transmission zone concerned that Ov 16 ELISA seroprevalence in children aged 5–9 years is <0.1% at the upper bound of 95% CI. The recommended procedure consists of sampling 3000 children aged 5–9 years from the Transmission Zone. One half of the children are selected by convenience sampling from purposively selected first line villages and the other half by random sampling from randomly selected villages from non-first line villages from the Transmission Zone.
(b) demonstrating entomologically that the infectivity level in the transmission zone is less than 1 L3 in the head of one out of 2000 female blackflies at the upper bound of the 95% CI, assessed by O-150 PCR pool screening in at least 6000 blackflies collected on human baits from identified breeding sites within the Transmission Zone.

Meeting these two conditions enables the decision to stop MDA according to the 2016 WHO Guidelines, and hence to pass to phase 2 of the elimination process. In the event any one of the tests fails to meet the criteria, MDA will continue. To address the diagnostic deficiency of the ELISA serological test in use, the WHO Guidelines instructs that when less than ten children are found positive, they should be skin snipped, and the snips examined by O-150 PCR to determine whether there is an infection or not. In the event the results from the skin snip PCR are negative, the positive seroprevalence results will be interpreted to mean exposure to infection and the children considered uninfected and having passed the test.

Phase 2

The second phase starts when the decision is made to stop MDA. This triggers the post treatment surveillance (PTS) which is the activity of phase 2. PTS is assured entomologically and follows the procedure described above to demonstrate that annual infectivity levels after stopping MDA remains ≤1 L3/2000 blackflies for 3–5 years. At the end of the 3–5 years of successful PTS, i.e., after demonstration of sustained interruption of transmission, serological and entomological tests described above are repeated. When the two tests pass the set out criteria, it is a confirmation that interruption of transmission of onchocerciasis has been sustained, and it signals that onchocerciasis transmission in the Transmission Zone has been eliminated.

Phase 3

The third phase starts after it has been demonstrated at the end of phase 2 that the transmission of onchocerciasis has been eliminated, triggering the start of post elimination surveillance (PES). Phase 3 therefore will ensure PES which will be an entomological activity, at least at the moment, demonstrating at intervals that the blackfly infectivity in the Transmission Zone is below the threshold for transmission of onchocerciasis infection, viz. ≤1L3/2000 female blackflies.

Country Certification for Elimination of Onchocerciasis

When the process of demonstrating the elimination of onchocerciasis as described above has been accomplished successfully in all the Transmission Zones in the country, it means that the country has reached the stage where it can apply for certification of onchocerciasis elimination. The country is then required, with the guidance of its National Onchocerciasis Elimination Expert Committee, to put together a Dossier on the records of all the onchocerciasis elimination activities that have been carried out to reach the stage of post elimination surveillance of onchocerciasis in the entire country. The dossier is then submitted to the WHO HQ through the WHO Country Representative. The WHO in turn sets up an international verification team to verify the state of the elimination of onchocerciasis in the country and submit their findings and recommendations to the WHO which in turn issues the Certificate of Elimination of Onchocerciasis in the Country signed by the WHO Director General.

Challenges

- About 7 years have elapsed since the WHO recommended to onchocerciasis endemic countries in Africa to change their goal of intervention of onchocerciasis from control to elimination. Progress in the process of the change among endemic countries has however varied. The variations may be related to different levels of development of the health systems on which depends the basis for the rate of progress generally. This may be seen in terms of variable technical capacity generally which will account for some delays in progress, but lack of onchocerciasis entomological skills, particularly in most of the countries is a challenge which needs to be addressed if rapid progress is to be made.
- A need for review of the serological indicator for demonstrating interruption of transmission and hence STOP MDA, which is given in the WHO Guidelines as Ov 16 ELISA seroprevalence ≤0.1% in children aged less than 10 years. This

threshold has been noted to be too stringent (Gass 2018) whilst modelers have demonstrated that the threshold for stopping MDA could be seroprevalence of ≤2% in children aged 5–9 years at the upper 95% confidence interval (Coffeng et al. 2019; Hamley et al. 2020). The OTS has recommended operational research into it which are ongoing.
- Another challenge is for the countries to set up special laboratories for processing Ov 16 ELISA serological and O-150 PCR pool screening tests required to process the high volume of test materials associated with elimination processes.
- The usefulness of the current M & E test is limited only to signaling the attainment of the point at which stop MDA can be initiated. This means that impact assessment of the progress of MDA is not possible until the moment when there is a need to test to stop MDA.
- The current intervention tool appears to be adequate to enable reaching the endpoint of elimination in most parts of Africa. However, there might be some challenges in their application to bring some holoendemic areas in the Cameroon and possibly in the DRC and South Sudan to the endpoint necessitating additional intervention tools to ensure and accelerate the attainment of the endpoint. Additionally, there is the challenge of continuous proper application of the CDI strategy, in particular, community engagement (CE) for programmes to achieve the objectives.
- Undertaking MDA safely in areas where onchocerciasis is co-endemic with loa loa i.e., without the occurrence of SAE's. This has largely been addressed, but still needs to be put in operation generally in all countries concerned.
- The currently available method for post elimination surveillance is technically intensive and expensive, making its sustainability a challenge for countries as most of the funding is from external sources. There is therefore a need for a simple, inexpensive easy to operate post elimination surveillance tool.
- Political conflicts, and lack of governments' interest in NTDs continue to pose a challenge in the elimination of onchocerciasis in some countries.

Further Research

Diagnostics

Although the current IgG4 serological test has been used to demonstrate elimination of onchocerciasis in countries in the Americas and some foci in Africa, its performance characteristics are not satisfactory. There is therefore a need to improve the diagnostic performance characteristics and/or to develop a new diagnostic tool which is highly specific and very sensitive to ensure smooth implementation of the processes of elimination of onchocerciasis. The Target Product Profile (TPP) for the diagnostic tool required for carrying out oncho elimination mapping has been set at 99.8% specificity and 60% sensitivity whilst that required for use to stop MDA is set at 99.8% specificity and 89% sensitivity (Biamonte et al. 2022). Developing target product profiles of diagnostics to meet the needs of the programmes is more desirable and effective. Programmes' needs are confirmatory diagnostic(s) that can be used in low-prevalence settings to assist with mapping, MDA stopping decisions and for use in surveillance. Development of a test that could be used to exclude infection in an individual is a priority. There is also a need to prepare a diagnostic strategy for identification of communities and districts with high *loa loa* infection intensity.

Intervention

Once-to-twice-yearly ivermectin MDA is effective at breaking transmission but takes 10–15 years or more. There are also cases, e.g., holoendemic communities where biannual MDA may be insufficient to reach the endpoint of elimination. Furthermore, MDA with ivermectin cannot be safely used in settings in which loiasis and onchocerciasis are co-endemic and is even not justifiable when onchocerciasis in the co-endemic area is hypoendemic. This is an important limiting factor for the progress of onchocerciasis elimination. Even though TnT combined with Loascope application addresses largely the main obstacle to the treatment in the onchocerciasis and loa loa co-endemic areas, the cost involved is prohibitive and has slowed down its implementation. There is therefore a need for a search for a macrofilaricide which can kill or sterilize the adult worm and thus shorten the duration required for the elimination of onchocerciasis with the current tool in the difficult areas. Although it has been shown that doxycycline administered to *O. volvulus* infected individuals inhibits embryogenesis in the *O. volvulus* adult female worm by way of depletion of the symbiont endobacteria *Wolbachia* with resultant elimination of microfilariae from the skin of the infected, the mode of its application, daily dose over 6 weeks, does not appear ideal for mass administration. There is, therefore, a need for the search for an effective antibiotic which can be given in simpler doses and for shorter periods to deplete *O. volvulus* female adult worms of *Wolbachia*. An available macrofilaricide or an antibiotic effective in depleting *Wolbachia* from *O. volvulus* may also be applied effectively in areas of hypoendemic onchocerciasis co-endemic with loa loa without the occurrence of SAEs to address the problem that has prevented the use of ivermectin MDA in such areas.

Compliance

Studies are required to address compliance including systematic non-compliance to improve MDA coverage in some areas.

Ivermectin Resistance

Although ivermectin resistance does not appear to be a problem currently, there is a need to search for a tool for use to detect ivermectin resistance and prepare for the way of addressing it should ivermectin resistance occur. The evaluation of

Table 1 Top ten research priority areas for helminthiases (Top six research areas for onchocerciasis)

Area for research	Specifics of research
Diagnostic	Improve available diagnostic tests, specifically their sensitivity, specificity, and ability to measure infection intensity and detect drug resistance
	Sensitivity and specificity are mostly important to enable diagnosis of infection at low prevalence in elimination settings
Intervention tools	Optimize existing intervention tools to maximize impact
	Develop novel control tools to improve impact and sustainability
	The tools include pharmaceuticals and vaccines, the search for new or existing antibiotics to be used against *Wolbachia*, and the clinical trials and studies on moxidectin on paediatric efficacy and safety
Modelling	Develop and refine models to investigate the relationships between infection and morbidities to aid programs aiming to reduce the burden of disease (elimination of public health problem)
	Increase the use and application of epidemiological models to aid monitoring and evaluation (M&E) and surveillance, the design of cost-effective sampling protocols, and the monitoring of intervention efficacy including drug resistance
Monitoring and evaluation	Standardize and validate methodologies and cost-effective protocols for monitoring and evaluation (M&E) settings
Host–parasite interactions	Investigate modulation of host–parasite interactions at population and within-host levels, including impact on host immune response of concurrent infection, impact of parasite control interventions on such host–parasite interactions, and host's ability to seroconvert upon vaccination
Biology	Elucidate the effect of ivermectin on *O. volvulus* reproductive biology and quantify the patterns of coverage and compliance in treated communities

moxidectin use in children is therefore welcome as it will make moxidectin fully available in the event ivermectin resistance should occur. The WHO Technical Report Series 972 (2012) on Research Priorities for helminth infections aptly sets out some of the areas that require research in onchocerciasis. Table 1 is an adaptation from the Annex 5 of the research priorities listed as the "top ten research priority areas for helminthiases" recommended by the Disease Reference Group on helminths (TDR).

The Future

The start of the millennium saw the elimination of onchocerciasis in four countries of the Americas and in several foci in Africa demonstrating the feasibility of elimination of onchocerciasis both in the Americas and Africa with ivermectin MDA. This has prompted the WHO to publish "Guidelines for Stopping Mass Drug Administration and Verifying the Elimination of Human Onchocerciasis in 2016" and "Ending the neglect to obtain Sustainable Development Goals; a roadmap for neglected tropical diseases 2021 – 2030 (WHO 2020)".

Of the 37 endemic countries in the world, (6) in the Americas, (1) in Arabian Peninsula and (30) in Africa, 12 countries are foreseen to eliminate onchocerciasis by 2030. Though the process of the change to elimination of onchocerciasis has started rather slowly, with the approaches that have been put in place- including the formation of the OTS within the NTD Scientific Steering Committee by the WHO to oversee the review and development of evidence-based scientific strategies, tools to support elimination of onchocerciasis and the emphasis on operational research,- the progress should eventually improve in momentum. It is hoped that the support and unflinching interest of partners and stakeholders will continue to address and meet the financial obligations which is key to the achievement of onchocerciasis elimination. The future can be described as promising, granted that no extreme unexpected technical challenge would appear to derail the ongoing process.

A few countries have been slow in embracing and working towards the elimination of onchocerciasis. These countries have problems due to inadequate technical capacity and weak health systems, absence of political will and support to NTD elimination programmes and others may be involved in political conflicts which may obstruct progress. Support to these countries will have to be addressed to fit individual needs in order that the future timelines may be achieved.

Conclusion

The history of large-scale intervention against onchocerciasis which started in 1975 in West Africa as vector control and later extended into all endemic countries in Africa, the Americas and Yemen in the Arabian Peninsula has been as intriguing as informative. Contributing to this intriguing history was the discovery of ivermectin with its registration for use to treat human onchocerciasis in 1987, and its donation for as long as it is needed by the manufacturer, Merck & Co Inc. Building on the knowledge accumulated from small scale interventions that were carried out in the early 50s in different areas in Africa, (mainly in the West and East) and in the Americas, (particularly in Guatemala and Mexico), onchocerciasis intervention, which was foremost to control morbidity, has evolved into the elimination of transmission by MDA with ivermectin informed by successes in this direction. This has led to the decision to aim for world-wide elimination of transmission of the disease pathogen as outlined in the WHO roadmap for the elimination of NTD diseases, WHO (2020) *"Ending the neglect to obtain Sustainable Development Goals; a roadmap for neglected tropical diseases 2021 – 2030 (WHO,* 2020*)".*

There is now a wealth of scientific understanding of the disease, but there are still some uncertainties in the knowledge which need to be addressed. There is the need for the development of new and/or improvement in the diagnostic tool.

Operational research has been at the base of the progress of onchocerciasis control in the past and is equally well accepted today as a crucial component in the quest for onchocerciasis elimination and must be continued.

No end date for onchocerciasis elimination from all countries in the world has been set. This is a cautious move given the state of different levels of progress in intervention in Africa, and the presence of a few ongoing political conflicts in countries which generally impede the implementation of health interventions. However, there is no doubt that in the absence of any insurmountable technical challenges, world-wide elimination of onchocerciasis is attainable.

References

Aziz MA, Diop I, Diallo S, Lariviere M, Porta ML (1982) Efficacy and tolerance of ivermectin in human onchocerciasis. Lancet 320(8291):171–173

Biamonte MA et al (2022) Onchocerciasis: target product profiles of in vitro diagnostics to support onchocerciasis elimination mapping and mass drug administration stopping decisions. PLoS Negl Trop Dis 16(8):e0010682. https://doi.org/10.1371/journal.pntd.0010682

Bousinnesq M, Gardon JA, Gardon-Wendel N, Kamgno J, Ngoumou P, Chippaux J-P (1998) Three probable cases of Loa Loa encephalopathy following ivermectin treatment for onchocerciasis. Am J Trop Med Hyg 58(4):461–469

Coffeng LE, Stolk WA, Golden A, de Los ST, Domingo GJ, de Vlas SJ (2019) Predictive value of Ov16 antibody prevalence in different subpopulations for elimination of African onchocerciasis. Am J Epidemiol 188(9):1723–1723

Dadzie Y, Neira M, Hopkins D (2003) Final report of the conference on the eradicability of onchocerciasis. Filaria J 2(1):2

Diawara L, Traoré MO, Badji A, Bissan Y, Doumbia K, Goita SF, Konaté L, Mounkoro K, Sarr MD, Seck AF, Toé L, Tourée S, Remme JH (2009) Feasibility of onchocerciasis elimination with ivermectin treatment in endemic foci in Africa: first evidence from studies in Mali and Senegal. PLoS Negl Trop Dis 3:e497

WHO (2020) Ending the neglect to attain the sustainable development goals. A road map for neglected tropical diseases 2021–2030. Available from: https://www.who.int/publications/i/item/9789240010352

Gass KM (2018) Rethinking the serological threshold for onchocerciasis elimination. PLoS Negl Trop Dis 12:e0006249

Gonzalez RJ, Cruz-Ortiz N, Rizzo N, Richards J, Zea-Flores G, Domínguez A et al (2009) Successful interruption of transmission of *Onchocerca volvulus* in the Escuintla-Guatemala focus, *Guatemala*. PLoS Negl Trop Dis 3(3):e404. https://doi.org/10.1371/journal.pntd.0000404

Hamley JID, Walker M, Coffeng LE, Milton P, de Vlas SJ, Stolk WA et al (2020) Structural uncertainty in onchocerciasis transmission models influences the estimation of elimination thresholds and selection of age groups for Seromonitoring. J Infect Dis 221(Suppl 5):S510–S5S8

Hawking F (1978) Diethylcarbamazine. *A review of the literature with special reference to its pharmacodynamics, toxicity and use in the therapy of onchocerciasis and other filarial infections.* WHO/ONCHO/78.142

Higazi TB, Zarroug IMA, Mohamed HA, ElMubarak WA, Deran TCM, Aziz N, Katabarwa M et al (2013) Interruption of Onchocerca volvulus transmission in the Abu Hamed focus, Sudan. Am J Trop 89(1):51–57

Kamgno J, Pion SD, Chesnais CB, Bakalar MH, D'Ambrosio MV, Mackenzie CD et al (2017) A test-and-not-treat strategy for onchocerciasis in Loa loa-endemic areas. N Engl J Med 377(21):2044–2052

Katabarwa M, Walsh F, Habomugisha P, Lakwo TL, Agunyo S et al (2012) Transmission of onchocerciasis in Wadelai focus of northwestern Uganda has been interrupted and the diseases eliminated. J Parasitol Res 2012:748540. https://doi.org/10.1155/2012/748540

Lindblade KA, Arana B, Zea-Flores G et al (2007) Elimination of Onchocerca volvulus transmission in the Santa Rosa focus of Guatemala. Am J Tropical Med Hygiene 77(2):334–341. https://doi.org/10.4269/ajtmh.2007.77.334. PMID: 17690408

Lobos E, Weiss N, Karam M, Taylor HR, Ottesen EA, Nutman TB (1991a) An immunogenic Onchocerca volvulus antigen: a specific and early marker of infection. Science 25:1603–1605

Lobos E, Weiss N, Karam M, Taylor RH, Ottesen EA (1991b) An immunogenic Onchocerca volvulus antigen: a specifi c and early marker of infection. Science 251(5001):1603–1605

Lovato R, Guevara A, Guderian R, Proaño R, Unnasch T, Criollo H et al (2014) Interruption of infection transmission in the onchocerciasis focus of Ecuador leading to the cessation of ivermectin distribution. PLoS Negl Trop Dis 8(5):e2821. https://doi.org/10.1371/journal.pntd.0002821

Mectizan Expert Committee and Technical Consultative Committee (2004) Recommendations for the treatment of onchocerciasis with Mectizan in areas co-endemic for onchocerciasis and loiasis. http://www.who.int/apoc/publications/englishmectccloarecs-june04.pdf. opens in new tab

Nicholls RS, Duque S, Olaya LA et al (2018) Elimination of onchocerciasis from Colombia: first proof of concept of river blindness elimination in the world. Parasites Vectors 11:237. https://doi.org/10.1186/s13071-018-2821-9

Rodríguez-Pérez MA, Unnasch TR, Domínguez-Vázquez A, Morales-Castro AL, Peña-Flores GP, Orozco-Algarra ME, Arredondo-Jiménez JI, Richards F Jr, Vásquez-Rodríguez MA, Rendón VG (2010) Interruption of transmission of Onchocerca volvulus in the Oaxaca focus, *Mexico*. Am J Trop Med Hyg 83(1):21–27. https://doi.org/10.4269/ajtmh.2010.09-0544. PMID: 20595472; PMCID: PMC2912570

Rougement A (1982) Ivermectin for onchocerciasis. Lancet 20:1982

Tekle AH, Elhassan E, Isiyaku S, Amazigo UV, Bush S, Noma M, Cousens S, Abiose A, Remme JH (2012) Impact of long-term treatment of onchocerciasis with ivermectin in Kaduna state, Nigeria: first evidence of the potential of elimination in the operational area of the African programme for onchocerciasis control. Parasites Vectors 5:28

Traoré MO, Sarr MD, Madji A, Bissan Y et al (2012) Proof of principle of onchocerciasis elimination with ivermectin treatment in endemic foci in Africa: final results of a study in Mali and Senegal. PLoS Negl Trop Dis 6:e1825. [PMC free article] [PubMed]

Winnen M, Plaisier AP, Alley ES, Nagelkerke NJD, van Oortmarssen G, Boatin BA, Habbema JDF (2002) Can ivermectin mass treatments eliminate onchocerciasis in Africa? Bull World Health Organ 80(5)

World Health Organization (2016) Document number WHO/HTM/NTD/PCT/2016.1 I. https://apps.who.int/iris/handle/10665/204180

World Health Organization, TDR (1996) *Guidelines for analysis of REMO data using GIS*. TDR/TDF/COMDT/98.3. WHO, Geneva

Podoconiosis: Endemic Non-filarial Elephantiasis

Kebede Deribe, Wendemagegn Enbiale, Fasil Tekola-Ayele, and Gail Davey

Contents

K. Deribe (✉) · G. Davey
Centre for Global Health Research, Brighton and Sussex Medical School, University of Sussex, Brighton, UK

W. Enbiale
Bahir Dar University, Dermatovenerology Department, College of Medicine and Health Sciences, Bahir Dar, Ethiopia

Amsterdam UMC, University of Amsterdam, department of dermatology, Amsterdam Institute for Infection and Immunity (AII), Academic Medical Centre, Amsterdam, Netherlands

F. Tekola-Ayele
Epidemiology Branch, Division of Population Health Research, Division of Intramural Research, Eunice Kennedy Shriver National Institute of Child Health and Human Development, National Institutes of Health, Bethesda, MD, USA

J. O. Gyapong, B. A. Boatin (eds.), *Neglected Tropical Diseases - Sub-Saharan Africa*, Neglected Tropical Diseases, https://doi.org/10.1007/978-3-031-53901-5_11

Abstract

Podoconiosis is an endemic, non-filarial lymphoedema of the lower limb which needs to be differentiated from lymphatic filariasis (LF), lymphoedema of systemic disease and other causes of localized lymphoedema. The disease, affecting genetically susceptible individuals who go barefoot, is linked to long-term exposure to red clay soil. Interactions between genetic and environmental factors trigger an inflammatory response that leads to lymphoedema and fibrosis. Patients with podoconiosis develop signs and symptoms in the second and third decades of life with the disease being common up to the sixth decade. The disease is mostly bilateral but asymmetrical and mostly limited to below the knees. Globally, there is current evidence of podoconiosis in 17 countries across tropical Africa, central and South America and Southeast Asia. In Africa, podoconiosis has been reported in 12 countries where 114.5 million people live. Tropical African countries bear the highest disease burden with about one and a half million people living with podoconiosis in Ethiopia, a further 40,000 in Cameroon and 7000 in Rwanda. The key strategies for podoconiosis control are prevention of contact with soil (primary prevention) and lymphoedema morbidity management (secondary and tertiary prevention). The main challenge faced in podoconiosis control is lack of awareness that the condition exists and that it is different from LF and other main causes of lymphoedema in the tropics. Mapping of podoconiosis is progressing, and development of a global atlas of podoconiosis is critical. Evidence from the past 5 years suggests that podoconiosis is amenable to public health interventions such as footwear and hygiene-based morbidity management, which reduce acute clinical episodes. The development of a global strategy is important to galvanize the momentum achieved over the past 5 years. Mapping of disease must be accompanied by development of national control programmes to deliver preventive and treatment interventions. Controlling podoconiosis is achievable given the absence of a biological agent or vector also needing control, the relatively small global scale of the problem, and the safe means of podoconiosis prevention and control.

Introduction

Podoconiosis is the lesser-known of the two major forms of tropical lymphoedema. An endemic, non-filarial lymphoedema of the lower limb, podoconiosis is linked to long-term exposure to red clay soil (Davey et al. 2007a). The disease affects genetically susceptible individuals who often go barefoot (Price 1990; Tekola Ayele et al.

2012a). It causes bilateral, often below-knee lymphoedema of the lower limb. The term podoconiosis was derived from the two Greek words *podos* and *konos,* which mean foot and dust, respectively, and imply that the disease is caused by exposure of the feet to irritant clay soil (Price 1990; Davey et al. 2007b).

Epidemiology

Distribution and Burden

Using an 'evidence consensus' approach (Deribe et al. 2019a), there is evidence of podoconiosis in 17 countries (12 in Africa, three in Latin America, and two in Asia) and absolute consensus on presence in six of these, all in Africa (Fig. 1). The 12 African countries in which there is evidence of presence are: Burundi, Cameroon, Cape Verde, Ethiopia, Equatorial Guinea, Kenya, Nigeria, São Tomé and Príncipe, Sudan, Tanzania, Rwanda, and Uganda. In terms of population potentially affected, in 2020, there were 114.5 million people living in areas suitable for the occurrence of podoconiosis across Africa (Fig. 1; Deribe et al. 2020a).

So far, only four countries (Cameroon, Ethiopia, Kenya, and Rwanda) have conducted nationwide mapping of podoconiosis as part of developing the global atlas of the disease. In Ethiopia, 1.5 million people were estimated to be living with podoconiosis across the country (Deribe et al. 2017), while in Cameroon a further 40,000 and in Rwanda 7000 people with podoconiosis were estimated (Deribe et al. 2018a; 2019b). In these countries the burden of podoconiosis is lower than initial empirical estimates, although it is widespread geographically. The distribution within countries varies significantly; in Cameroon, of the 76 villages surveyed, 44

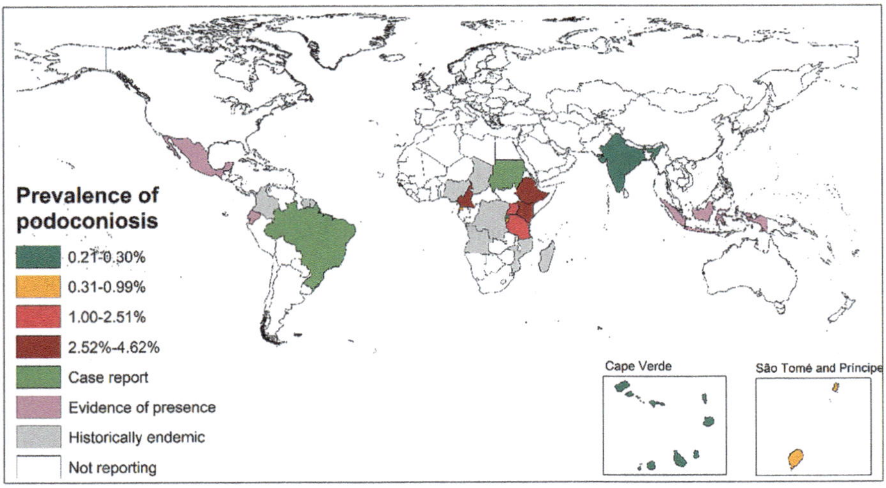

Fig. 1 Countries in which podoconiosis is endemic or has been described (Adapted from Deribe et al. 2019a)

had no cases of podoconiosis. In the remaining villages, the prevalence ranged from 0.3% to 4.9% (Deribe et al. 2018b). In Kenya, the national average prevalence of podoconiosis was 0.3%. Of 48 villages surveyed, only 10 had cases of podoconiosis with prevalence ranging from 0.32% to 3.23% (Sultani et al. 2021).

A scalable mapping approach utilizing mobile phone and geographic information system technologies has been developed and applied in these four countries. In addition to community-based surveys, the mapping approach uses remote sensing advances to produce datasets that define areas at risk of podoconiosis using a combination of climate, environmental and ecological parameters. The approach enables clear definition of the geographical distribution of the disease, and estimation of the population at risk and the number of individuals affected, which helps prioritize implementation efforts.

The mean age at first noticing the swelling is 25 years (range 5-87 years) and about 12.0% had noticed swelling when younger than 10 years of age. A nationwide mapping from Ethiopia (Deribe et al. 2015) confirmed that the disease is more common in women (male to female ratio 1 to 1.4) and this slight female predominance is also supported by a recent meta-analysis which concluded that the odds of podoconiosis risk among women was 1.15 times that among men (Berhe et al. 2021). A nationwide study in Cameroon also confirmed a similar finding (male to female ratio 1 to 1.3) (Deribe et al. 2018b).

Public Health Impact

In addition to the health impacts described below, podoconiosis has severe social and economic consequences. According to a study in southern Ethiopia in an area with 1.7 million residents, the annual economic cost of podoconiosis was more than 16 million USD per year (Tekola et al. 2006), which when extrapolated to the country as a whole suggests costs of more than 200 million USD per annum for Ethiopia. People with podoconiosis were found to be half as productive as those without podoconiosis but with the same occupations. Podoconiosis patients lose 45% of their economically productive time because of morbidity associated with the disease (Tekola et al. 2006). Total direct costs of podoconiosis amounted to the equivalent of US$143 per patient per year. In Ethiopia alone, it is estimated that podoconiosis causes 172,073 disability-adjusted life-years (DALYs) to be lost annually (Deribe et al. 2020). The majority (97.4%) of these DALYs were due to chronic lymphoedema, while 2.6% were attributable to ADL episodes.

Studies have documented that 77.4–97% of patients have experienced acute dermatolymphangioadenitis (ADLA)–a recurrent inflammatory swelling of the lymphoedematous limb at least once per year (Alemu et al. 2011; Molla et al. 2012a; Tekola Ayele et al. 2013). People with podoconiosis become bedridden during acute attacks. On average, most patients have five or more episodes of ADLA and hence lose 25 productive days per year (Molla et al. 2012a).

The social impact of podoconiosis is also significant. Qualitative studies in southern Ethiopia have shown that podoconiosis is considered the most stigmatising health problem in endemic areas (Gebrehanna 2005; Tora et al. 2011). Social stigma against people with podoconiosis is rife, patients being excluded from school; denied participation in local meetings, churches, and mosques; and excluded from marriage with unaffected individuals (Gebrehanna 2005; Yakob et al. 2008; Tekola et al. 2009; Tora et al. 2011; Molla et al. 2012b). In a study conducted in northern Ethiopia, 13% of patients mentioned that they had experienced one or more forms of social stigmatisation at school or church or in the market place (Molla et al. 2012b). In southern Ethiopia, 55.8% of community members showed stigmatising attitudes towards social interactions with podoconiosis patients (Yakob et al. 2008). The misconception that podoconiosis is caused by hereditary factors and is unrelated to environmental factors leads to the assumption that it cannot be avoided (Ayode et al. 2012). Among rural youth, accurate knowledge about gene–environment influences and rejection of infectious causes of podoconiosis were associated with less stigmatizing attitudes (Engdawork et al. 2020).

In a recent study among people with podoconiosis and healthy neighbourhood controls, the healthy controls had much higher quality of life scores than podoconiosis-affected individuals in all domains of quality of life. Overall, patients had almost seven times the risk of lower than average quality of life scores than controls. Patients' quality of life was worse than that of patients with other neglected tropical diseases (NTDs) such as soil-transmitted helminths (Mousley et al. 2013). Studies have clearly shown an association between depression and podoconiosis. A study in Ethiopia documented a prevalence of depression of 12.6% among people with podoconiosis compared to 0.7% among healthy neighbours (Bartlett et al. 2015), and another in Rwanda reported depressive symptoms in 68% of people affected by podoconiosis (Semrau et al. 2020a). The overall impact of podoconiosis on the life of an affected person has been recently documented using ethnographic film (Cremers et al. 2020).

Until recently, podoconiosis was chiefly assumed only to lead to morbidity rather than direct mortality. However, follow-up of nearly 700 podoconiosis patients in a trial for a period of 1 year enabled calculation of mortality, which was found to considerably exceed that of a comparison cohort in a similar setting in northern Ethiopia. The overall Standardised Mortality Rate for the study period was 6.0 (95% CI 3.6 to 9.4) (Masraf et al. 2020).

Basic Biology

Gene-Environment Interactions

The pathogenesis of podoconiosis is not yet clearly understood. Based on existing evidence, the most accepted hypothesis is that podoconiosis is caused by mineral particle-induced inflammation on a background of genetic susceptibility (Davey et al. 2007a). Interactions between genetic and environmental factors trigger an inflammatory response that leads to lymphoedema and fibrosis (Davey et al. 2007a). It is hypothesised that mineral particles that penetrate bare skin are engulfed by macrophages in the lower limb lymphatics and induce an inflammatory response in the lymphatic vessel lumina. This is followed by fibrosis and obstruction of the vessel lumen leading to oedema of the lower leg, which progresses to elephantiasis (Price 1972b). The stratum corneum of the lower leg and foot of patients with podoconiosis is significantly less hydrated than that of unaffected controls (Ferguson et al. 2013). Little is known of the molecular pathogenesis of podoconiosis, but early evidence suggests that transforming growth factor 3-1 (TGF 3-1) may play a role (Addisu et al. 2010). Below, we describe the pathogenesis of podoconiosis using the gene-environment interaction model.

Genetics

Not all barefoot people who live in areas of red clay soil develop podoconiosis. In the early 1970s, Price reported familial aggregation of podoconiosis in Ethiopia, Rwanda, and Burundi (Price 1976b). These observations and subsequent epidemiological studies in Ethiopia hinted at high heritability of podoconiosis. Reportedly, a third to one-half of patients have other podoconiosis-affected close relatives or family members: 48% in Illubabor resettlement schemes, western Ethiopia (Kloos et al. 1992); 38% in East and West Gojjam Zones, northern Ethiopia (Molla et al. 2012b); 34% in Bedele Zuria, western Ethiopia (Tekola Ayele et al. 2013); and 30% in Midakegn, central Ethiopia (Geshere Oli et al. 2012). These figures may be underestimated because patients may not disclose the presence of podoconiosis in blood relatives to protect their family from social stigmatisation.

Price performed segregation analysis on 80 families with more than one affected child. The study suggested the possibility of a genetic factor with an autosomal recessive mode of inheritance and an estimated risk genotype frequency of 15–40% (Price 1972a). Price also hypothesised that individual differences in the tissue handling of absorbed minerals play a role in development of full-blown podoconiosis (Price 1990).

A pedigree study conducted in 2005 in 59 multigenerational families with multiple affected members in southern Ethiopia presented evidence for a genetic basis to podoconiosis. The sibling recurrence risk ratio was 5.07 (i.e. the sibling of an affected person is at five times increased risk of developing podoconiosis when compared to a person in the general population), and the heritability of podoconiosis was estimated to be 63% (SE 0.069, $p = 1 \times 10^{-7}$) (i.e. 63% of the variance in occurrence or non-occurrence of podoconiosis is accounted for by genetics). Segregation analysis showed the most parsimonious model to be that of an autosomal codominant major gene, with age and history of use of footwear as significant environmental covariates. The study illustrated that both genetic and environmental factors contribute to the pathophysiology of podoconiosis (Davey et al. 2007a).

A genome-wide comparison of the frequency of genetic variants between podoconiosis cases and unaffected controls from southern Ethiopia revealed that genetic variants in the HLA locus (a genomic region on chromosome 6) confer susceptibility to podoconiosis. Specifically, variants in or near the class II HLA genes, namely, *HLA-DQA1*, *HLA-DRB1* and *HLA-DQB1*, were found to be associated with increased susceptibility to podoconiosis. The study findings were further corroborated with a family-based association test. Subsequent high-resolution sequence-based HLA typing showed that HLA-DRB1*0701 and HLA-DQA1*0201 are risk variants for podoconiosis. The study suggested that podoconiosis is a T-cell-mediated inflammatory condition (Tekola Ayele et al. 2012a). These findings were reproduced in a larger study involving three ethnic groups from Ethiopia (Gebresilase et al. 2020). A recent gene expression analysis of 579 immune-related genes was performed on RNA extracted from archived formalin-fixed and paraffin-embedded tissue blocks of femoral lymph node and skin samples collected in 1970 to 1976 from podoconiosis patients and unaffected controls. The study has identified 69 genes that are differentially expressed between podoconiosis patients and controls. Integrated analysis of the differentially expressed genes has revealed significant biological functions related to local immune activation (Alcantara et al. 2020).

Environment

Some of the environmental factors that facilitate formation of the red clay soils containing the putative inorganic particles that trigger the inflammatory response leading to podoconiosis have been described. Podoconiosis-endemic areas have characteristic red clay soils, elevation >1000 m above sea level and annual rainfall >1000 mm (Price 1974b).

The exact causal agent in red clay soil areas has not yet been identified. A range of studies has suggested that mineral particles present in red clay soils play a role in the pathogenesis of podoconiosis (Price 1972b, 1976a, 1990; Fyfe and Price 1985; Frommel et al. 1993; Abrahams 2002). Electron microscopy microanalysis examination of lower limb lymphatic tissues of barefoot people from Ethiopia indicated the presence of microparticles containing the elements found in clays with some difference in the amount between affected and unaffected individuals (Price and Henderson 1978). Moreover, podoconiosis-endemic areas were free of filariasis; footwear had a protective effect; birefringent silica particles were found in the lymph node macrophages; and the dermal content of silicon, aluminium, iron and trace metals was consistent with local soils, leading to the postulation that podoconiosis was a soil-induced disease (Price 1972b, 1976a, 1990). Price suggested that silicate particles cause subendothelial oedema, endolymphangitis, collagenisation and obliteration of the lymphatic lumen (Price 1976a). More recent modelling of environmental factors associated with podoconiosis suggested higher prevalence in areas with higher levels of smectite, quartz and mica in the soil (Molla et al. 2014).

Disease Presentation

The clinical picture and course of podoconiosis varies depending on the duration of disease, type of swelling, personal foot care and other unknown factors. The majority of patients develop signs and symptoms of podoconiosis in the second and third decades of life. Disease is bilateral but asymmetrical and the swelling is mostly limited to below the knees (Davey et al. 2007b; Alemu et al. 2011; Geshere Oli et al. 2012; Molla et al. 2012a; Tekola Ayele et al. 2013).

The early symptoms of podoconiosis include an on-and-off burning sensation in the foot following day-long barefoot exposure on the farm or field and itching on the dorsum of the foot around the base of the toes. The key early signs of podoconiosis are thickening of the skin over the anterior and dorsum of the foot (Fig. 2a), splaying of the forefoot (Fig. 2b), and transient plantar and lower leg pitting oedema that disappears after overnight rest (with leg elevation). The lichenification and thickening can render the skin, particularly overlying the first toe web space, stiff and unable to be pinched (positive Stemmer's sign). Warty and papillomatous growths usually involve the dorsum of foot in the anterior one third in a 'slipper' distribution accentuated over the lateral side of the foot and the heel (Figs. 2a and 3a). This hyperkeratotic and warty or velvety growth looks like 'moss' but it is rough to touch. Moss frequently resolves in the areas subjected to pressure from footwear among patients who have started to wear shoes (Price 1983, 1984, 1990; Figs. 3a, b).

Following recurrent episodes of cellulitis or ADLA, the leg diameter progressively increases and lymphoedema becomes persistent. In advanced cases, clinical presentation can vary greatly. Conventionally, three main forms of lymphoedema are distinguished in podoconiosis (Price 1990; Tekola-Ayele and Embiale 2014):

1. Water bag type—subcutaneous oedema that is soft to rubbery to touch and pits with pressure; it has little dermal fibrosis. In extreme cases, the swelling may have a flask appearance with the neck around the knee and wider base on the foot. The foot skin will have a smooth surface, with occasional lymphorrhea which attracts flies and is foul smelling. Often there is loss of normal hair. With time, the foot and lower leg become large and flabby. In about one third of these patients, multiple nodules can be seen among those who wear shoes, the nodules become smooth, and their surfaces may overlap (Fig. 4a, b). Swelling may lead to pillowy oedema presenting with redundant skin folds forming a band-like swelling arranged laterally in front of the ankle and at the dorsal junction of the

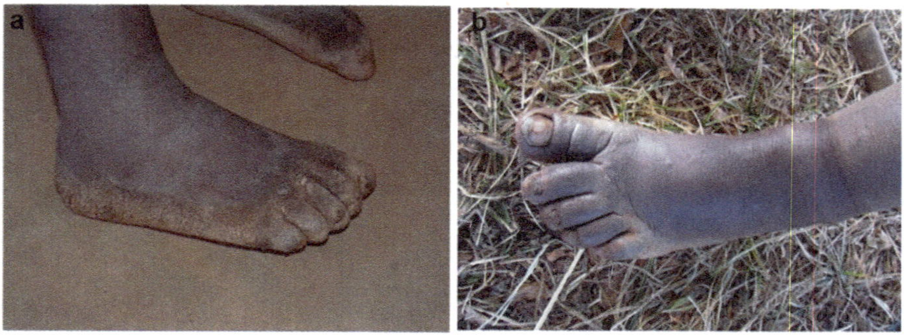

Fig. 2 (**a**, **b**) Early stage

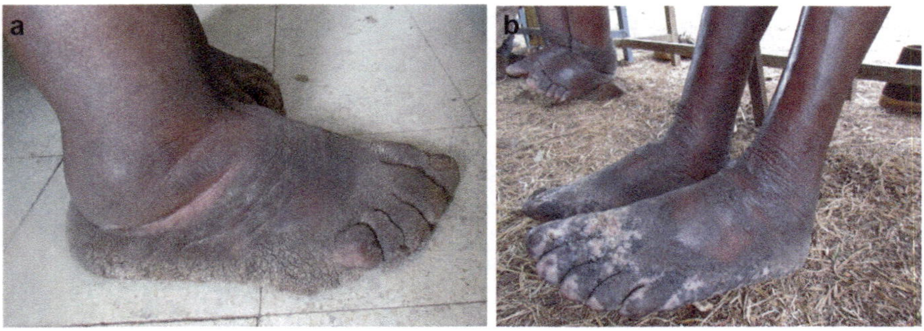

Fig. 3 (**a**, **b**) Warty hyperkeratosis or velvety papules

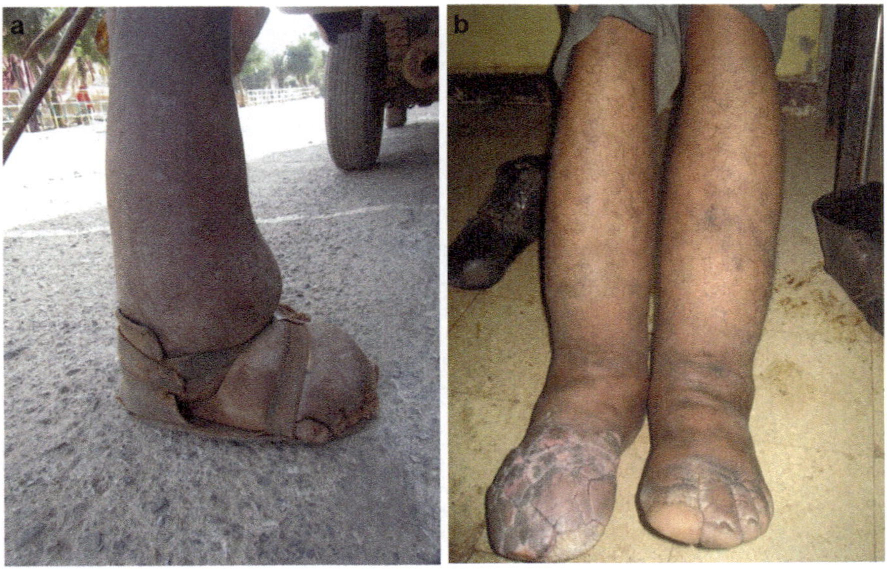

Fig. 4 (**a**, **b**) Water bag lymphoedema

toe and forefoot. Ballooning over the toes is another presentation. The disability to the patient is due to the great size of the limb, its heaviness, and the foul-smelling discharge.

2. Fibrotic/sclerotic or leathery lymphoedema—sclerosis governs the changes in the skin and sub cutis, which become woody hard, grossly thickened and does not pit. The skin may be hyperpigmented and adherent to the underlying deep tissue. The overlying epidermis on the foot takes on a sandpaper-like appearance which eventually, due to increasing hyperkeratosis, takes on the so-called 'mossy' appearance. Under areas of compression, such as a sandal strap, the skin remains smooth. There will often be multiple smooth and hyperkeratotic surface nodules on the ankles or lower legs which are rubbery in consistency. The stiff,

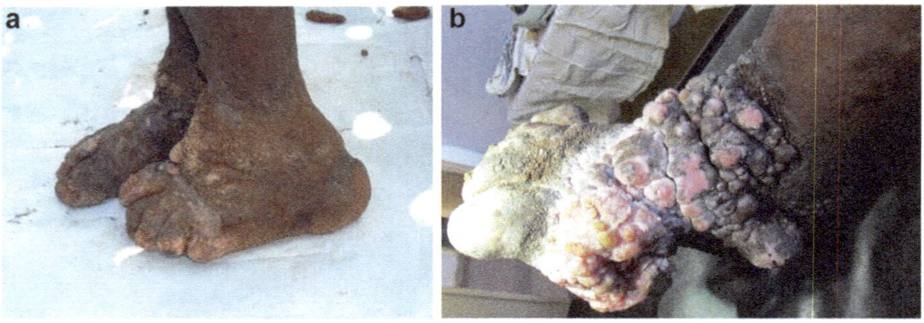

Fig. 5 (**a**, **b**) Ankylosis and maceration

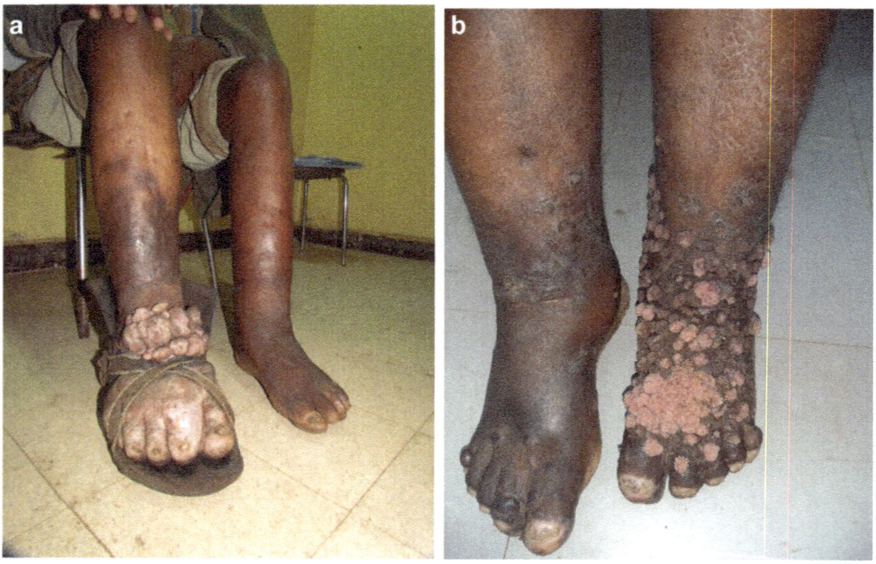

Fig. 6 (**a**, **b**) Fibrotic and Mixed lymphoedema

sclerotic nature of the altered skin especially at the ankle compromises the normal flexibility of the ankle and toe and makes it vulnerable to cracking and trauma, in addition to ankylosis of the joints (Fig. 5a).

3. Mixed lymphoedema—characterized by a limb grossly swollen below the knee with irreversible, non-pitting oedema, and areas of sclerotic change commonly on the shins. There may be variation in the compressibility of the lymphoedema below and above the ankle (Fig. 6a, b).

Late-stage disease is characterised by fusion of the interdigital spaces and ankylosis of the inter-phalangeal and ankle joints. It may also present with maceration, oozing and, rarely, with bone resorption (Figs. 5a, b) (Cohen 1960; Price 1983, 1984, 1990).

Associated Clinical Features

(1) Nodules—these occur in all types of advanced lymphoedema. Nodules can start as smooth-surfaced skin folds and progress to tumorous growth which considerably inhibits use of footwear (Enbiale et al. 2021).
(2) Inter-digital and skin fold maceration—this presents as whitish, wet patches in the interdigital spaces, with occasional fissuring. The maceration is associated with oozing (lymphorrhea) and often with bacterial or fungal infections. Lymphorrhea with microbial super-infection leads to the foul smell (Fig. 5b).
(3) Acute lymphangioadenitis or cellulitis or ADLA—is characterised by fever, chills, rigors and hot, reddened, and tender swelling of the limb which leads to loss of productivity. It represents bacterial infection in the dermis, lymphatic channel, or lymph node. It is the most disabling problem associated with podoconiosis and requires prevention strategies and emergency medical interventions. The ADLA frequency is increased in those with advanced stage of podoconiosis and those with associated interdigital fungal infections.

A system for grading the clinical stages of podoconiosis has been developed and validated. The system has five stages, with the first two stages defining disease presentation that can easily be reversed with currently implemented treatment packages (Tekola et al. 2008). Comparison of staging systems used in oncology, LF and podoconiosis suggests that a universal system might lead to better management and reporting in research (Douglass and Kelly-Hope 2019).

Diagnosis

There is currently no point-of-care diagnostic tool for podoconiosis, which is therefore a diagnosis of clinical exclusion. The commonest differential diagnoses of podoconiosis are lymphatic filariasis (LF), lymphoedema of systemic disease, recurrent cellulitis of the legs and deep fungal infection. Although there are point-of-care diagnostic tests for lymphatic filariasis infection, these are not very sensitive in established filarial infection among advanced cases.

The differentiation of podoconiosis from lymphatic filariasis uses a panel approach, including clinical history, physical examination, antigen, and antibody tests. The swelling of podoconiosis starts in the foot and progresses upwards (Price 1990), whereas the swelling in filarial elephantiasis starts elsewhere in the leg. Podoconiosis lymphoedema is bilateral, asymmetric, usually confined to below the level of the knees and unlikely to involve the groin (Davey et al. 2007b). In contrast, lymphatic filariasis is commonly unilateral and extends above the knee, usually with groin involvement. In addition to the clinical history and physical examination, an antigen-based Alere Filariasis Test Strip (FTS) can also help to further distinguish between the two lymphoedemas, although the majority of lymphatic filariasis patients are also negative for the antigen-based test. Studies suggest that antibody-based tests such as the Wb123 test may help define the filaria transmission status of a community in areas where few and possibly outlying cases of lymphoedema are found and may help to reduce misclassification of patients.

Systemic causes of lymphoedema can be ruled out by taking a proper history and physical examination of other organ systems. Some hereditary lymphoedemas can be excluded since most of the time they are unilateral, include above the knee involvement and may have other vascular anomalies associated with the lymphoedema.

Control Tools and Strategies

The key strategies for podoconiosis control are prevention of contact with irritant soil (primary prevention) and lymphoedema morbidity management (secondary and tertiary prevention).

Primary Prevention of Podoconiosis

Prevention of contact with the (still uncharacterised) mineral trigger encompasses:

- Behavioural measures (using footwear, daily foot washing)
- Environmental measures (replacing earth floors with concrete, paving roads)
- Lifestyle changes (shift from subsistence farming to alternative income-generating activities).

These measures are all predicated on the assumptions that the trigger(s) of podoconiosis are only found in certain irritant volcanic soils, that prolonged contact between bare feet and these soils is necessary for disease initiation, and that passage of the trigger through the skin of the feet can be prevented using standard footwear.

Extensive work by EW Price in the 1970s showed fine soil particles to be closely associated with disease prevalence in southern Ethiopia. The prevalence of podoconiosis among people living on 'tropical red soil' averaged 5.38%, which decreased to 2.83% at the limits of this soil and further to 0.79% among people living 25 km away from the edge of the soil (Price 1974b). Similar observations relating disease prevalence to soil type were made in Kenya, Rwanda, Burundi, and north-west Tanzania (Price 1976a).

Evidence that prolonged contact between bare feet and irritant soils comes from studies in which age of onset is recorded. In community surveys, Kloos et al. noted that most patients reported disease onset in their teens and early twenties (Kloos et al. 1992), and a national mapping study in Ethiopia concluded that about 12.0% had noticed swelling when younger than 10 years of age (Deribe et al. 2015). A study in Uganda demonstrated higher odds of podoconiosis in those who never wore shoes at work or at home (Kihembo et al. 2017).

Evidence for the effectiveness of the behavioural, environmental and lifestyle measures listed above is not yet available. The fact that interest in podoconiosis has only been rekindled in the past decade, together with the long latent period of podoconiosis has meant that long-term follow-up studies of prevention have not been launched. However, recent behavioural research has explored the relationship between beliefs about cause of disease and preventive behaviours (Ayode et al. 2012) in endemic communities in Ethiopia. This study demonstrates that people with a nuanced view of the causes of podoconiosis (who are able to consider heritability as a necessary but not sufficient cause) are more likely to perceive themselves as being able to control the disease and are also more likely to espouse positive preventive behaviours, including shoe use and foot washing. Research more closely focussed on barriers to use of shoes revealed disease-specific barriers (fit of shoe, fear of labelling) and more general, structural level barriers related to poverty (price and shoe quality) (Ayode et al. 2013). Studies have shown that youth have the potential to disseminate culturally tailored education programmes that explain the gene x environment basis of podoconiosis (Engdawork et al. 2018).

Secondary and Tertiary Prevention of Podoconiosis

Secondary and tertiary prevention of podoconiosis is currently based on lymphoedema management (foot hygiene, compression, exercises, and elevation) and use of shoes to reduce exposure to irritant soil. The foot hygiene 'package' has evolved independently in several sites in Ethiopia, and while the basic elements at each site are similar, local variations exist. The fundamental elements of this lymphoedema management 'package' are:

- Soaking the feet in water
- Washing with soap
- Drying
- Use of skin moisturizer
- Compression therapy with short stretchable bandage and leg elevation for those with water bag lymphoedema
- Exercises and elevation
- Use of socks and shoes

Antiseptic (Savlon, dilute bleach, salt, or *neem* tree [*Azadirachterindica*] extract) may be added to the soaking basin. In some sites, the soap is made with *neem* extract. Once dry, Vaseline or Whitfield's ointment, cooking oil or neem-infused oil may be used. There is as yet no evidence to support the efficacy of one alternative over another. Early stages of podoconiosis lymphoedema can be reversed and those with water bag type of lymphedema will have a control compatible to normal socio-economic position.

The GoLBeT trial provided evidence that the lymphoedema management package outlined above is effective in reducing the frequency and duration of ADLA (Negussie et al. 2018). A trial in southern Ethiopia demonstrated that 2% glycerine added to a reduced volume of washing water resulted in improved skin barrier function (Brooks et al. 2017). A cluster randomised trial in northern Ethiopia demonstrated greater reduction in frequency of acute attacks among patients following an enhanced management package, which included lymphatic stimulation (Douglass et al. 2020). The LE-Doxy trial underway among podoconiosis patients in Cameroon is testing the impact of 6 weeks' doxycycline therapy as an adjunct to the standard package above (Horton et al. 2020).

Patients with redundant skin, fibrous nodules, or pillowy oedema (Fig. 7a, b) on the foot and leg tend not to respond so well to this therapy since the nodules are too hard to be reduced by compression bandaging. Surgical removal of the nodules in these patients allows them to use custom designed shoes and to control disease progress with the lymphoedema management package described above. A recent follow-up study has demonstrated that surgical nodulectomies can be performed with satisfactory healing rates, encouraging lack of complications, and leading to significant improvement in quality of life in low resource settings (Yeshanehe et al. 2017; Enbiale et al. 2021).

There are many similarities between lymphoedema morbidity management for podoconiosis and for filarial lymphoedema, making cross-disease lymphoedema management (which does not distinguish the cause) attractive in countries such as Cameroon, Uganda, and Ethiopia in which both NTDs are found. At present, the main differences between management regimes for the two types of lymphoedema are that bandaging is recommended in podoconiosis (where it is particularly effective in soft 'water-bag'-type swelling) but is not in filarial lymphoedema, due to safety concerns. Recent implementation research has demonstrated the acceptability, feasibility, and effectiveness of integrating care across types of lymphoedema. A holistic physical and mental health care package for people with lymphoedema caused by LF, leprosy or podoconiosis was successfully integrated into routine primary health care services in northern Ethiopia, and its effectiveness and cost-effectiveness have been demonstrated (Semrau et al. 2020b; Hounsome et al. 2020).

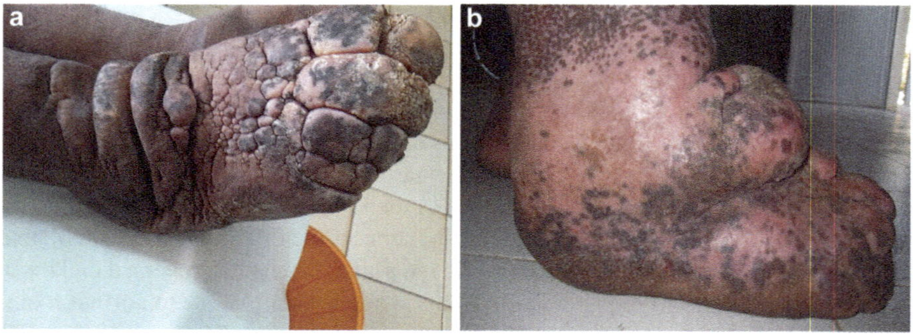

Fig. 7 (**a, b**) Pillowy oedema

Programmatic Aspects

Many features of morbidity management for podoconiosis mean that most treatment can be carried out by patients themselves, with little intervention from health-care professionals. Expert patients (patients who have been trained to successfully manage their condition and assist others in doing so) can be trained to guide treatment for uncomplicated lymphoedema (Davey and Burridge 2009). Health professional input is important in cases of diagnostic uncertainty, to manage ADLA, to evaluate for nodulectomy if prominent nodules prevent shoe wearing and to identify coexisting pathologies including jiggers, chromoblastomycosis and carcinoma.

Content analysis based on a podoconiosis programme in northern Ethiopia identified 16 key steps to establishment of a sustainable, community-led programme (Tomczyk et al. 2012). However, with the drive towards Universal Health Coverage, interest in cross-disease programmes has increased. The feasibility of burden assessment across tropical lymphoedemas (Kebede et al. 2018) and the successful development of a cross-disease package of physical, mental health and psychosocial care for people with lower-limb lymphoedema, have recently been described (Tesfaye et al. 2021).

Challenges of Programme Implementation

Intervention is still so new in podoconiosis that challenges exist chiefly in programme initiation rather than continuation. The key challenge faced is lack of awareness that the condition exists and is different from lymphatic filariasis, the other main cause of lymphoedema in the tropics. This lack of awareness is evident among health professionals (Bayisenge et al. 2020; Churko et al. 2021), academics (Fowler et al. 2020) and Ministry of Health and World Health Organization staff. There appear to be many reasons for the lack of awareness of podoconiosis, including lack of a positive diagnostic test (see Sect. "Diagnosis"), a phenotype very similar to filarial lymphoedema and lack of understanding of simple prevention and treatment strategies.

As for most neglected tropical diseases, podoconiosis primarily affects remote, rural populations. These populations are commonly underserved in terms of health facilities and personnel and have little voice at regional or national level. In addition, access to both water and shoes (necessary for disease prevention and treatment) is lower in these communities. Framing podoconiosis as an issue of human rights (Shahvisi et al. 2018) draws attention to these structural inequities and enables the duties of governments and the international community to be clearly identified.

Finally, although a global voice for podoconiosis now exists (Foot*work*, the International Podoconiosis Initiative) (Davey et al. 2012), this organisation is still young and relatively small. A strategy review in 2021 aligned Footwork's goals more clearly with Universal Health Coverage and the WHO NTD Road Map 2021-30 (WHO 2020).

Further Research for Policy and Control

Critical questions remain before a global strategy for podoconiosis elimination may be developed (Deribe et al. 2020c). First, estimating the global burden of podoconiosis is vital. Such data provide policy makers, programme planners and clinical practitioners clear burden estimates which help priority setting and decision making. Strong country-level data spur engagement of endemic countries' medical and public health communities and prompt national implementation. Second, the development of a framework and tools for surveillance, monitoring and evaluation with key programme indicators is important. Implementation research on active surveillance based on environmental, behavioural, and climatic risk factors and mandatory passive surveillance is also necessary. Third, point-of-care diagnosis for podoconiosis is important. Exploring immunological and other biological markers will be important in addressing this gap.

Finally, a costing tool for prevention and management of podoconiosis is an important area of future research. Such a tool will help programme planers to accurately estimate programme costs and facilitate proper implementation and will help advocacy and resource mobilisation.

The combination of understanding the distribution of podoconiosis, and optimising treatment and prevention activities will help to guide policy and practice in endemic countries and ultimately the elimination of podoconiosis.

Outlook for the Next Decade

The past 15 years have been marked by changes in the way podoconiosis is viewed by endemic country policy makers, global funding organisations and researchers. In February 2011, the World Health Organization included podoconiosis in its list of NTDs. Based partly on evidence of effective care integrated across lower limb NTDs (Semrau et al. 2020b), WHO now includes podoconiosis in its 'skin-centred' strategy for NTDs (Asiedu et al. 2020). Initiatives including Foot*work*, the International Podoconiosis Initiative and the Ethiopian National Podoconiosis Action Network have been formed with the shared vision of a world free of podoconiosis. The UK Wellcome Trust, the UK Medical Research Council, the UK National Institute of Health Research, and the US National Institutes of Health have funded research on genetics, geospatial, behavioural and morbidity management aspects of podoconiosis in Ethiopia, Rwanda, and Cameroon.

Controlling podoconiosis is achievable because to date, no biological agent or vector involved in podoconiosis has been identified, the global scale of the problem is relatively small, the means for podoconiosis prevention and control are safe and additional strategies for prioritisation of resources such as use of family health history (Tekola Ayele et al. 2012b) and risk maps can be implemented.

Realisation of these plans over the next decade needs a renewed model as part of the global movement to eradicate other NTDs. The currently uncoordinated efforts by small-scale programmes on podoconiosis must be integrated into a framework led by endemic country governments. This must be backed by robust public health policies, by sustainable resource allocation and by inclusion of podoconiosis in the formal education curricula of health professionals at all levels. Government support is only one step. Public-private partnerships must be strengthened and should match public resources (of government and research institutions) with private financial investments. As evidenced above, management of podoconiosis can be integrated with other forms of lymphoedema and skin problems, and services should be widely accessible to patients in primary healthcare facilities.

Basic science and applied research must continue to drive and improve upon existing prevention and treatment modalities. For example, despite the effectiveness of lymphoedema management against early clinical disease, no effective treatment beyond nodulectomy is available to reverse debilitating, advanced stage disease. Given the fact that many patients present with advanced disease, investment in cutting-edge therapeutic solutions is crucial and may be responsible in the future for considerable improvements in quality of life of patients and in the overall social capital and economy. Finally, global-level advocacy and funding will play key roles in shaping the pathway towards a world free of podoconiosis.

Conclusion

- Podoconiosis is a preventable, treatable, non-infectious form of tropical lymphoedema.
- Prevalence >1% has been documented in several countries in sub-Saharan Africa.
- The condition appears to arise from mineral particle-induced inflammation on a background of genetic susceptibility.
- No diagnostic tool for podoconiosis is available, so differentiation from the lymphoedema of lymphatic filariasis is based on clinical examination and exclusion of microfilariae.
- Prevention of disease is through protection of the feet from exposure to irritant clay soils.
- Simple foot washing, use of emollients, compression bandaging, and use of socks and shoes decreases the frequency of debilitating ADLA attacks.

Acknowledgements GD is supported by the UK National Institute of Health Research (grant 200140).

References

Abrahams PW (2002) Soils: their implications to human health. Sci Total Environ 291(1–3):1–32

Addisu S, El-Metwally TH, Davey G, Worku Y, Titheradge MA (2010) The role of transforming growth factor-beta1 and oxidative stress in podoconiosis pathogenesis. Br J Dermatol 162(5):998–1003

Alcantara DR, Jones CI, Altmann DM, Boyton RJ, Haniffa M, Newport MJ (2020) Multiplexed gene expression analysis of HLA class II-associated podoconiosis implicates chronic immune activation in its pathogenesis. Trans R Soc Trop Med Hyg 114:926–936

Alemu G, Tekola Ayele F, Daniel T, Ahrens C, Davey G (2011) Burden of podoconiosis in poor rural communities in Gulliso woreda, West Ethiopia. PLoS Negl Trop Dis 5(6):e1184

Asiedu K, Marks M, Hay RJ (2020) Podoconiosis, skin-NTDs and global health. Trans R Soc Trop Med Hyg 114:887–888
Ayode D, McBride C, de Heer H, Watanabe E, Gebreyesus T, Tadele G, Tora A, Davey G (2012) The association of beliefs about heredity with preventive and interpersonal behav-iors in communities affected by podoconiosis in rural Ethiopia. Am J Trop Med Hyg 87(4):623–630
Ayode D, McBride CM, de Heer HD, Watanabe E, Gebreyesus T, Tora A, Tadele G, Davey G (2013) A qualitative study exploring barriers related to use of footwear in rural highland Ethiopia: implications for neglected tropical disease control. PLoS Negl Trop Dis 7(4):e2199
Bartlett J, Deribe K, Tamiru A, Amberbir T, Medhin G, Malik M, Hanlon C, Davey G (2015) Depression and disability in people with podoconiosis: a comparative cross-sectional study in rural northern Ethiopia. Int Health 8:124–131
Bayisenge U, Schurer J, Wong R, Amuguni H, Davey G (2020) Podoconiosis in Rwanda: knowledge, attitudes and practices among health professionals and environmental officers. PLoS Negl Trop Dis 14(10):e0008740
Berhe B, Legese H, Mardu F, Tesfay K, Adhanom G, Kahsay T, Belay G, Negash H (2021) Epidemiology and sex differences of podoconiosis in Ethiopia: a systemic review and meta-analysis. Heliyon 7(4):e05446
Brooks J, Ersser SJ, Cowdell F, Gardiner E, Mengistu A, Matts PJ (2017) A randomized controlled trial to evaluate the effect of anew skincare regimen on skin barrier function in those withpodoconiosis in Ethiopia. Brit J Derm 177:1166
Churko C, Asfaw MA, Tunje A, Girma E, Zerdo Z (2021) Knowledge, attitude, practice, and associated factors of health professionals towards podoconiosis in Gamo zone, Ethiopia, 2019. J Foot Ankle Res 14:31
Cohen LB (1960) Idiopathic lymphoedema of Ethiopia and Kenya. East Afr Med J 37:53–74
Cremers AL, Visser BJ, Getahun Z, Borku M, Meskele E, Ahmed J, van Vugt M, Birnie E (2020) Using ethnographic film in tackling podoconiosis. Trans R Soc Trop Med Hyg 114:896–898
Davey G, Burridge E (2009) Community-based control of a neglected tropical disease: the mossy foot treatment & prevention association. PLoS Negl Trop Dis 3(5):e424
Davey G, GebreHanna E, Adeyemo A, Rotimi C, Newport M, Desta K (2007a) Podoconiosis: a tropical model for gene-environment interactions? Trans R Soc Trop Med Hyg 101(1):91–96
Davey G, Tekola F, Newport MJ (2007b) Podoconiosis: non-infectious geochemical elephantiasis. Trans R Soc Trop Med Hyg 101(12):1175–1180
Davey G, Bockarie M, Wanji S, Fuller C, Fox L, Mycoskie M, Gruin M, Tsegaye A, Ayele FT, Newport M (2012) Launch of the international podoconiosis initiative. Lancet 379(9820):1004
Deribe K, Brooker SJ, Pullan RL, Sime H, Gebretsadik A, Assefa A, Kebede A, Hailu A, Rebollo MP, Shafi O, Bockarie MJ, Aseffa A, Reithinger R, Cano J, Enquselassie F, Newport MJ, Davey G (2015) Epidemiology and individual, household and geographical risk factors of podoconiosis in Ethiopia: results from the first nationwide mapping. Am J Trop Med Hyg 92:148–158
Deribe K, Cano J, Giorgi E, Pigott DM, Golding N, Pullan RL, Noor AM, Cromwell EA, Osgood-Zimmerman A, Enquselassie F, Hailu A, Murray CJL, Newport MJ, Brooker SJ, Hay SI, Davey G (2017) Estimating the number of cases of podoconiosis in Ethiopia using geostatistical methods. Wellcome Open Res 4(2):78
Deribe K, Cano J, Njouendou AJ, Eyong ME, Beng AA, Giorgi E, Pigott DM, Pullan RL, Noor AM, Enquselassie F, Murray CJL, Hay SI, Newport MJ, Davey G, Wanji S (2018a) Predicted distribution and burden of podoconiosis in Cameroon. BMJ Glob Health 3(3):e000730
Deribe K, Beng AA, Cano J, Njouendo AJ, Fru-Cho J, Awah AR, Eyong ME, Chounna Ndongmo PW, Giorgi E, Pigott DM, Golding N, Pullan RL, Noor AM, Enquselassie F, Murray CJL, Brooker SJ, Hay SI, Enyong P, Newport MJ, Wanji S, Davey G (2018b) Mapping the geographical distribution of podoconiosis in Cameroon using parasitological, serological, and clinical evidence to exclude other causes of lymphedema. PLoS Negl Trop Dis 2(1):e0006126
Deribe K, Simpson H, Cano J, Pigott DM, Weaver ND, Cromwell EA, Brady OJ, Pullan RL, Noor AM, Argaw D, Murray CJL, Brooker SJ, Hay SI, Newport MJ, Davey G (2019a) Mapping the global distribution of podoconiosis: applying an evidence consensus approach. PLoS Negl Trop Dis 13(12):e0007925

Deribe K, Mbituyumuremyi A, Cano J, Mbonigabe JB, Giorgi E, Ruberanziza E, Bayisenge U, Leonard U, Bikorimana JP, Rucogoza A, Turate I, Rusanganwa A, Pigott DM, Pullan RL, Noor AM, Enquselassie F, Condo JU, CJL M, Brooker SJ, Hay SI, Newport MJ, Davey G (2019b) Geographical distribution and prevalence of podoconiosis in Rwanda: a cross-sectional country-wide survey. Lancet Glob Health 7(5):e671–e680

Deribe K, Negussu N, Newport MJ, Davey G, Turner HC (2020) The health and economic burden of podoconiosis in Ethiopia. Trans R Soc Trop Med Hyg 114(4):284–292

Deribe K, Simpson H, Pullan RL, Bosco MJ, Wanji S, Weaver ND, Murray CJL, Newport MJ, Hay SI, Davey G, Cano J (2020a) Predicting the environmental suitability and population at risk of podoconiosis in Africa. PLoS Negl Trop Dis 14(8):e0008616

Deribe K, Mackenzie CD, Newport MJ, Argaw D, Molyneux DH, Davey G (2020c) Podoconiosis: key priorities for research and implementation. Trans R Soc Trop Med Hyg 114:889–895

Douglass J, Kelly-Hope L (2019) Comparison of staging systems to assess lymphedema caused by cancer therapies, lymphatic filariasis, and Podoconiosis. Lymphatic Res Biol 17(5):550–556

Douglass J, Hailekiros F, Martindale S, Mableson H, Seife F, Bishaw T, Nigussie M, Meribo K, Tamiru M, Agidew G, Kim S, Betts H, Taylor M, Kelly-Hope L (2020) Addition of lymphatic stimulating self-care practices reduces acute attacks among people affected by moderate and severe lower-limb lymphedema in Ethiopia, a cluster randomized controlled trial. J Clin Med 9:4077

Enbiale W, Verdonck K, Gebeyehu M, van Griensven J, de Vries HJC (2021) Surgical debulking of podoconiosis nodules and its impact on quality of life in Ethiopia. PLoS Negl Trop Dis 15(1):e0009053

Engdawork K, McBride CM, Ayode D, Allen CG, Davey G, Tadele G (2018) Rural youths' understanding of gene x environmental contributors to heritable health conditions: the case of podoconiosis in Ethiopia. PLoS Negl Trop Dis 12(9):e0006763

Engdawork K, Davey G, Ayode D, McBride CM, Tadele G (2020) A cross-sectional survey to assess the risk factors associated with stigmatizing attitudes towards patients with podoconiosis among rural youth in Southern Ethiopia. Trans R Soc Trop Med Hyg 114:995–1002

Ferguson JS, Yeshanehe WE, Matts PJ, Davey G, Mortimer PS, Fuller LC (2013) Assessment of skin barrier function in podoconiosis: measurement of stratum corneum hydration and Transepidermal water loss. Br J Dermatol 168(3):550–5544

Fowler K, Wampande LM, Gebreselassie A, Bayisenge U, Uwase C, de Oliveira A, Schurer JM (2020) Trans R Soc Trop Med Hyg 114:899–907

Frommel D, Ayranci B, Pfeifer HR, Sanchez A, Frommel A, Mengistu G (1993) Podoconiosis in the Ethiopian rift valley. Role of beryllium and zirconium. Trop Geogr Med 45(4):165–167

Fyfe NC, Price EW (1985) The effects of silica on lymph nodes and vessels – a possible mechanism in the pathogenesis of non-fi larial endemic elephantiasis. Trans R Soc Trop Med Hyg 79(5):645–651

Gebrehanna E (2005) The social burden of podoconiosis in Wolaita zone. MPH thesis. Addis Ababa University, Addis Ababa

Gebresilase T, Finan C, Suveges D, Tessema TS, Aseffa A, Davey G, Hatzikotoulas K, Zeggini E, Newport MJ, Tekola-Ayele F (2020) Replication of HLA class II locus association with susceptibility to podoconiosis in three Ethiopian ethnic groups. Sci Rep 11:3285

Geshere Oli G, Tekola Ayele F, Petros B (2012) Parasitological, serological, and clinical evidence for high prevalence of podoconiosis (non-filarial elephantiasis) in Midakegn district, Central Ethiopia. Trop Med Int Health 17(6):722–726

Horton J, Klarmann-Schulz U, Stephens M, Budge PJ, Coulibaly Y, Debrah A, Debrah LB, Krishnasastry S, Mwingira U, Ngenya A, Wanji S, Weerasooriya M, Yahathugoda C, Kroidl I, Deathe D, Majewski A, Sullivan S, Mackenzie C, Nutman TB, Shott JP, Weil G, Ottesen E, Hoerauf A (2020) The design and development of a multicentric protocol to investigate the impact of adjunctive doxycycline on the management of peripheral lymphoedema caused by lymphatic filariasis and podoconiosis. Parasites Vect 13:155

Hounsome N, Kinfe M, Semrau M, Ali O, Tesfaye A, Mengiste A, Bremner S, Ahmed A, Fekadu A, Davey G (2020) Economic assessment of a community-based care package for peoplewith

lower limb disorder caused by lymphatic filariasis, podoconiosis and leprosy in Ethiopia. Trans R Soc Trop Med Hyg 114:1021–1034
Kebede B, Martindale S, Mengistu B, Kebede B, Mengiste A, Hailekiros F, Tamiru A, Davey G, Kelly-Hope LA, Mackenzie CD (2018) Integrated morbidity mapping of lymphaticfilariasis and podoconiosis cases in 20 co-endemic districts of Ethiopia. PLoS Negl Trop Dis 12(7)
Kihembo C, Masiira B, Lali WZ, Matwale GK, Matovu JKB, Kaharuza F, Ario AR, Nabukenya I, Makumbi I, Musenero M, Zhu BP, Nanyunja M (2017) Risk factors for Podoconiosis: Kamwenge District, Western Uganda, 2015. Am J Trop Med Hyg 96(6):1490–1496
Kloos H, Kello AB, Addus A (1992) Podoconiosis (endemic non-filarial elephantiasis) in two resettlement schemes in western Ethiopia. Trop Dr 22:109–112
Masraf H, Azemerau T, Molla M, Jones CI, Bremner S, Ngari M, Berkley JA, Kivaya E, Fegan G, Tamiru A, Kelemwork A, Lang T, Newport MJ, Davey G (2020) Trans R Soc Trop Med Hyg 114:1035–1037T
Molla YB, Tomczyk S, Amberbir T, Tamiru A, Davey G (2012a) Podoconiosis in east and west gojam zones, northern Ethiopia. PLoS Negl Trop Dis 6(7):e1744
Molla YB, Tomczyk S, Amberbir T, Tamiru A, Davey G (2012b) Patients' perceptions of podoconiosis causes, prevention and consequences in east and west Gojam, northern Ethiopia. BMC Public Health 12:828
Molla YB, Wardrop NA, Le Blond JS, Baxter P, Newport MJ, Atkinson PM, Davey G (2014) Int J Health Geog 13:24
Mousley E, Deribe K, Tamiru A, Davey G (2013) The impact of podoconiosis on quality of life in northern Ethiopia. Health Qual Life Outcomes 11:122
Negussie H, Molla M, Ngari M, Berkley JA, Kivaya E, Njuguna P, Fegan G, Tamiru A, Kelemework A, Lang T, Newport MJ, McKay A, Enquoselassie F, Davey G (2018) Lymphoedema management to prevent acute dermatolymphangioadenitis in podoconiosis in northern Ethiopia (GoLBeT): a pragmatic randomised controlled trial. Lancet Glob Health 6:e795–e803
Price EW (1972a) A possible genetic factor in non-filarial elephantiasis of the lower legs. Ethiop Med J 10(3):87–93
Price EW (1972b) The pathology of non-filarial elephantiasis of the lower legs. Trans R Soc Trop Med Hyg 66(1):150–159
Price EW (1974b) The relationship between endemic elephantiasis of the lower legs and the local soils and climate. Trop Geogr Med 26(3):225–230
Price E (1976a) The association of endemic elephantiasis of the lower legs in East Africa with soil derived from volcanic rocks. Trans R Soc Trop Med Hyg 4:288–295
Price EW (1976b) Endemic elephantiasis of the lower legs in Rwanda and Burundi. Trop Geogr Med 28:283–290
Price EW (1983) Endemic elephantiasis: early signs and symptoms, and control. Ethiop Med J 21(4):243–253
Price EW (1984) Pre-elephantiasic stage of endemic nonfilarial elephantiasis of lower legs: 'podoconiosis'. Trop Dr 14(3):115–119
Price E (1990) Podoconiosis: non-filarial Elephantiasis. Oxford Medical Publications, Oxford
Price EW, Henderson WJ (1978) The elemental content of lymphatic tissues of barefooted people in Ethiopia, with reference to endemic elephantiasis of the lower legs. Trans R Soc Trop Med Hyg 72(2):132–136
Semrau M, Davey G, Bayisenge U, Deribe K (2020a) High levels of depressive symptoms among people with lower limb lymphoedema in Rwanda: a cross-sectional study. Trans R Soc Trop Med Hyg 114:974–982
Semrau M, Ali O, Deribe K, Mengiste A, Tesfaye A, Kinfe M, Bremner SA, Hounsome N, Kelly-Hope LA, MacGregor H, Taddese HB, Banteyerga H, Haile Mariam D, Negussu N, Fekadu A, Davey G (2020b) EnDPoINT: protocol for an implementation research study to integrate a holistic package of physical health, mental health and psychosocial care for podoconiosis, lymphatic filariasis and leprosy into routine health servicein Ethiopia. BMJ Open 10:e037675
Shahvisi A, Meskele E, Davey G (2018) A human right to shoes? Establishing rights and duties in the prevention and treatment of Podoconiosis. Health Human Rights J 20(1):53–65

Sultani HM, Okoyo C, Kanyi HM, Njenga SM, Omondi WP, Ayagah I, Buliva M, Ngere I, Gachohi J, Muli J, Newport MJ, Deribe K (2021) Mapping of Podoconiosis cases and risk factors in Kenya: a Nationwide cross-sectional study. Am J Trop Med Hyg. https://doi.org/10.4269/ajtmh.21-0248

Tekola Ayele F, Adeyemo A, Finan C, Hailu E, Sinnott P, Burlinson ND, Aseffa A, Rotimi CN, Newport MJ, Davey G (2012a) HLA class II locus and susceptibility to podoconiosis. N Engl J Med 366(13):1200–1208

Tekola Ayele F, Adeyemo A, Rotimi CN (2012b) Using a "genomics tool" to develop disease prevention strategy in a low-income setting: lessons from the podoconiosis research project. J Community Genet 3(4):303–309

Tekola Ayele F, Alemu G, Davey G, Ahrens C (2013) Community-based survey of podoconiosis in BedeleZuria woreda, Southwest Ethiopia. Int Health 5(2):119–125

Tekola F, Mariam DH, Davey G (2006) Economic costs of endemic non-filarial elephantiasis in Wolaita zone, Ethiopia. Trop Med Int Health 11(7):1136–1144

Tekola F, Ayele Z, Mariam DH, Fuller C, Davey G (2008) Development and testing of a de novo clinical staging system for podoconiosis (endemic non-filarial elephantiasis). Trop Med Int Health 13(10):1277–1283

Tekola F, Bull S, Farsides B, Newport MJ, Adeyemo A, Rotimi CN, Davey G (2009) Impact of social stigma on the process of obtaining informed consent for genetic research on podoconiosis: a qualitative study. BMC Med Ethics 10(13):13

Tekola-Ayele F, Embiale WY (2014) Podoconiosis: tropical lymphedema of the lower legs. In: Dermatology and allergology—principles and practice, 1st edn. iConcept Press Ltd, Hong Kong, p 201. https://pdfs.semanticscholar.org/df19/c491c922dba13a894cc8f87db4ba-ec5df89d.pdf

Tesfaye A, Semrau M, Ali O, Kinfe M, Tamiru M, Fekadu A et al (2021) Development of an integrated, holistic care package for people with lymphoedema for use at the level of the primary health care unit in Ethiopia. PLoS Negl Trop Dis 15(4):e0009332

Tomczyk S, Tamiru A, Davey G (2012) Addressing the neglected tropical disease podoconiosis in northern Ethiopia: lessons learned from a new community podoconiosis program. PLoS Negl Trop Dis 6(3):e1560

Tora A, Davey G, Tadele G (2011) A qualitative study on stigma and coping strategies of patients with podoconiosis in Wolaita zone, Southern Ethiopia. Int Health 3(3):176–181

WHO (2020) Ending the neglect to attain the sustainable development goals: a road map for neglected tropical diseases 2021–2030. World Health Organization, Geneva

Yakob B, Deribe K, Davey G (2008) High levels of misconceptions and stigma in a community highly endemic for podoconiosis in Southern Ethiopia. Trans R Soc Trop Med Hyg 102(5):439–444

Yeshanehe WE, Tamiru A, Fuller LC (2017) Surgical nodulectomies can heal in patients with lymphoedema secondary to podoconiosis in resource-poor settings. Br J Dermatol 177(4):e128–e129

Scabies

Wendemagegn Enbiale, Ashenafi Ayalew,
and Henry J. C. de Vries

Contents

W. Enbiale (✉)
Bahir Dar University, College of Medicine and Health Sciences, Dermatovenerology Department, Bahir Dar, Ethiopia

A. Ayalew
Ethiopian Public Health Institute, Addis Ababa, Ethiopia

H. J. C. de Vries
Amsterdam UMC, University of Amsterdam, Department of Dermatology, Amsterdam, The Netherlands

Institute for Infection and Immunity (AII), location Academic Medical Centre, Amsterdam, Netherlands & Center for Sexual Health, Department of Infectious Diseases, Public Health Service Amsterdam, Amsterdam, The Netherlands

J. O. Gyapong, B. A. Boatin (eds.), *Neglected Tropical Diseases - Sub-Saharan Africa*, Neglected Tropical Diseases, https://doi.org/10.1007/978-3-031-53901-5_12

Abstract

Scabies is a Neglected Tropical Disease (NTD) and is endemic in many resource-poor settings, predominantly in tropical regions and under crowded living conditions. It involves a skin infestation of the mite *Sarcoptesscabiei*var *hominis* and transmission occurs via person-to-person contact. Scabies is one of the most common dermatological conditions with a point prevalence of 100–200 million cases globally. The disease causes considerable loss of quality of life, itching and can lead to severe secondary bacterial infections and immune-mediated disease. WHO added scabies to the list of neglected tropical diseases in 2017. Effective topical and oral treatments exist, but re-infestation is common. Recent scientific evidence suggest that scabies is amenable to population-level control, particularly through mass drug administration.

Epidemiology and Burden

Scabies is caused by infestation of the skin with a microscopic female mite (*Sarcoptesscabiei var. hominis*) (Heukelbach and Feldmeier 2006). It is a contagious disease characterized by itch and a skin eruption of variable severity. Scabies occurs in all countries, and all populations are theoretically at risk of infestation, but its distribution is not uniform. The risk is considerably higher among people living in crowded, impoverished conditions, because of greater skin-to-skin contact and lack of access to effective treatment. The burden is greater in low-income and middle-income countries where access to effective treatment is often inadequate and population crowding increases the opportunities for transmission, whereas in high-income settings most cases are sporadic, and the predominant public health issue is the management of outbreaks in institutions such as hospitals and residential care facilities for older people (Mounsey et al. 2016).

Scabies was added to the portfolio of the World Health Organization (WHO) Department of Control of Neglected Tropical Diseases (NTDs) in 2017, because of the burden of scabies and its complications, particularly in areas with limited access to healthcare and because of potential new public health control strategies (World Health Organization 2020). Globally the total number of infected people is estimated to be around 100–200 million cases of scabies at a time, with 455million incident cases annually, but further efforts are needed to assess the burden more precisely (Karimkhani et al. 2017a, b). Recent estimates of the prevalence of infection range from 02% to 71% (Romani et al. 2015). In some regions, especially sub-Saharan Africa, a general population prevalence of 33.5% has been reported, with 40% of children having associated secondary bacterial infection (impetigo) (Enbiale and Ayalew 2018). Globally, scabies is estimated to cause approximately 3.8 million DALYs, causes an age-adjusted morbidity burden like *Haemophilus influenzae*, type B meningitis, and lymphoid leukemia. This makes it one of the highest-burden NTDs (GBD 2016 DALYs and HALE Collaborators 2017). Furthermore, the estimate does not include the associated complications of infestation.

In other resource-poor settings where baseline prevalence is lower or not properly documented, drought or change in sociopolitical conditions have been associated with scabies epidemics. An example is the outbreak in Ethiopia that has been in progress since 2015 and has become widespread in most of regions of the country and more than one million people are estimated to be affected each year (Enbiale et al. 2020a). Population crowding frequently leads to high transmission, and outbreaks are common within schools, prisons, and fragile settings such as refugee camps (Chandler and Fuller 2019).

There is little data on the profile and magnitude of scabies in sub-Saharan Africa. A systematic review of scabies prevalence studies published between 1985 to 2015 included only five African countries (Egypt, Tanzania, Ethiopia, Nigeria and Cameroon with prevalence ranging 0.2–9.2%) highlighting the paucity of prevalence data in Africa (Romani et al. 2015). Since 2015 some community based prevalence data of scabies in Sub-Saharan Africa appeared; a community based screening of about 9, 057, 427 people in Ethiopia in 2018 identified 875,890 (9.7%) cases of scabies (Enbiale et al. 2020a). Other recent community-based scabies prevalence studies in Liberia, Malawi and Nigeria reported prevalences of 9.3%, 17.2% and 65% respectively (Galván-Casas et al. 2021; Collinson et al. 2020; Ugbomoiko et al. 2018). Based on the limited Sub-Saharan African data in 2015, the age-standardised scabies DALY burden per 100,000 people ranges from 40 to 60 (Karimkhani et al. 2017a, b and Galván-Casas et al. 2021).

Complications Associated with Scabies Infestation

Scabies infestation results in breaks in the skin due to scratching and therefore increases the risk of secondary bacterial skin infections. The risk is augmented by the inhibitory effects of infestation on host complement pathways, thereby promoting bacterial colonization (Swe et al. 2014). Scabies infestation is known to be a major risk factor for bacterial skin infection caused by *Streptococcus pyogenes* and *Staphylococcus aureus* and is therefore considered a risk factor for more serious infections, including cellulitis, necrotizing fasciitis, and sepsis. Several studies also showed that outbreaks of scabies are a major risk factor for acute post-streptococcal glomerulonephritis (Mason et al. 2016; Romani et al. 2015). A growing body of evidence also implicates impetigo caused by *S. pyogenes* in the pathogenesis of rheumatic fever and rheumatic heart disease (Thornley et al. 2018). In general patients in tropical areas with scabies and Staphylococcus aureus bacteraemia have a higher mortality than those without scabies (Lynar et al. 2017).

Parasite Life Cycle and Transmission

Sarcoptesscabiei var. *hominis*, the human itch mite, is in the arthropod class Arachnida, subclass Acari, family Sarcoptidae. The mites burrow into the upper layer of the skin but never below the stratum corneum. The burrows appear as tiny raised serpentine lines that are grayish or skin-colored and can be a centimeter or more in length. Other species of scabies mites (Sarcoptes scabiei var canis, Sarcoptic mange) may cause infestations in other mammals, such as domestic cats, dogs, pigs, and horses. It should be noted that the species of mites found on other animals may cause a mild infestation in humans with temporary itching due to dermatitis; however, there is no evidence they multiply on the human host (Arlian et al. 1988; Ljunggren 2005).

The different developmental stages of *S. scabiei* consist of egg, larva, protonymph, tritonymph and adult. This life-cycle is typical of that for other a stigmatid mites (Fig. 1). There is considerable difference in the reported durations of the life-cycle of *S. scabiei var. hominis*. Reported durations of the total life-cycle for scabies mites from humans are 7–21 days. In the historic published literature, it is suggested that females produce 40–50 or more eggs over a life span of 26–40 days. A systematic in vivo study of the life-cycle of *S. scabiei var. canis* using a rabbit model reported eggs incubate and hatch in 3–4 days (90% of the hatched mites die). The larvae with three pair of legs migrate to the skin surface and burrow into the intact stratum corneum to make short burrows, called molting pouches (3–4 days) and larvae molt into four pair of legs nymphs, which molt once into larger nymphs before becoming adults. Mating takes place once, and the female is fertile for the rest of her life, the male dies soon after mating. The female makes a serpentine burrow using proteolytic enzymes to dissolve the stratum corneum of the epidermis, laying eggs in the process; she continues to lengthen her burrow and lay eggs for the rest of her life, surviving 1–2 months. Transmission of impregnated females mites from person-to-person occurs through direct or indirect skin contact (Arlian et al. 1988; Arlian and Morgan 2017).

When scabies mites infest a host, they first must penetrate the stratum corneum of the epidermis of the skin. Mites on the epidermis secrete a fluid (presumably saliva) that forms a pool around their body. It appears that the stratum corneum is dissolved (lysed) and the mite sinks into a depression in the skin. As it sinks, the legs seem to move in a tortoise like fashion (digging, crawling/swimming motion) and the action of the legs propels the mite forward forming its characteristic serpentine burrow, laying eggs in the process. After the impregnated female burrows into the skin, she remains there and continues to lengthen her burrow and lay eggs for the rest of her life (1–2 months) (Arlian and Morgan 2017).

Scabies usually is spread by direct, prolonged, skin-to-skin contact with a person who has scabies. Contact generally must be prolonged; a quick handshake or hug usually will not spread scabies. Scabies is spread easily to sexual partners and household members. Early human experiment showed that fomites such as clothing and exposure to mites shed onto surfaces such as floors or bedding were unlikely to cause infection (Mellanby 1941). The exception is that patients with crusted scabies can shed numerous skin flakes containing thousands of mites. An infected person can spread scabies even if he or she has no symptoms. Humans are the source of infestation and animals do not spread human scabies. In classic scabies the average person harbors only 5–12 mites, but those with crusted scabies can harbour more than a million mites and shed thousands. On a person, scabies mites can live for as long as 1–2 months. Off a person, scabies mites usually do not survive more than 48–72 h (Mellanby 1941).

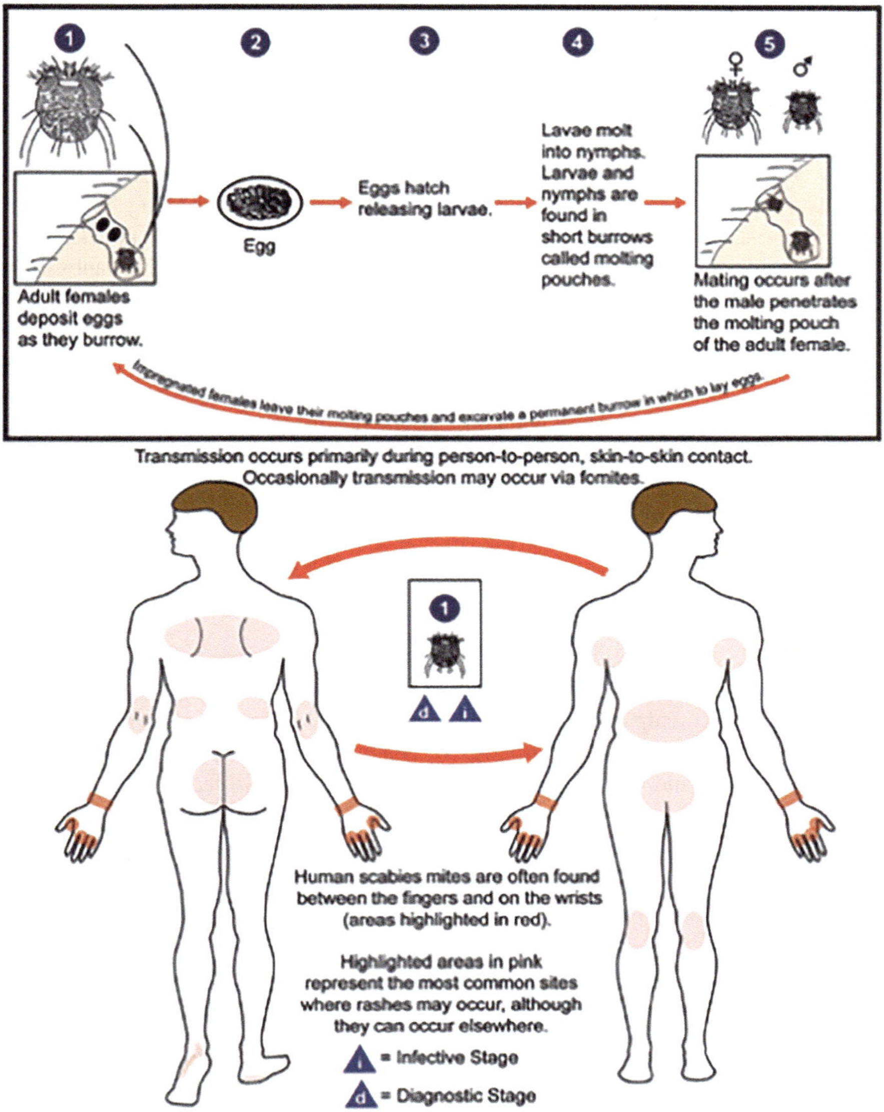

Fig. 1 ***Sarcoptesscabiei* life cycle:** Females deposit 2–3 eggs per day as they burrow under the skin. Eggs are oval and 0.10–0.15 mm in length and egg hatch and the larvae migrate to the skin surface and burrow into the intact stratum corneum to construct almost invisible, short burrows called molting pouches. After the larvae molt, the resulting nymphs. This form molts into slightly larger nymphs before molting into adults. Larvae and nymphs may often be found in molting pouches or in hair follicles and look like adults, only smaller. Adults are round, sac-like eyeless mites. Females are 0.30–0.45 mm long and 0.25–0.35 mm wide, and males are slightly more than half that size. Mating occurs after the active male penetrates the molting pouch of the adult female (CDC 2020)

Disease Presentation

If a person has never had scabies before, symptoms usually appear 2–6 weeks after initial exposure. If a person has had scabies in the past, symptoms could occur 1–4 days following mite exposure. The typical presentation of scabies is itching (pruritus), especially at night, and a rash of pimples and nodules. The rash may contain tiny blisters (vesicles) and scales. The common sites of itching are in the webbing between the fingers, wrist, elbow, armpit, shoulder blades, breasts, waist, genitals, buttocks, knees and ankles. Scabies in adults usually spares the face, palms, and soles (Mellanby 1944; Anderson and Strowd 2017). It is not known why certain parts of the body are more commonly infected than others. Burrows are sometimes seen on the skin with the help of dermatoscopy or magnifying glass (Micali et al. 2011). Skin sores and secondary bacterial skin infections can occur due to scratching of affected areas (Fig. 2). In communities affected by social and natural disasters, health professionals should maintain a high index of suspicion that undiagnosed skin rashes and conditions may be scabies. If left untreated, symptoms may continue indefinitely. After receiving appropriate treatment, symptoms (especially itching) may continue for 3–4 weeks but will gradually disappear.

Scabies in Young Children

In young children the scabies rash and itch are often generalized. Even a child's palms, soles, and scalp can be affected in contrast to the mainly intertriginous localization of lesions in older patients (Fig. 3). The most common presenting lesions are papules, nodules and vesiculopustules. Secondary bacterial infection (impetiginization) and eczematization are common (Fig. 4). Infants and children who have scabies are very irritable and often do not want to eat or sleep. The itch can keep them awake at night (Paller 1993).

Crusted Scabies

Crusted scabies is a severe form of scabies and develops because of uncontrolled proliferation of the mites in people who have a weak immune system due to a medical condition, neurologic problem, and malnutrition in the elderly. Itching which is a hallmark of classic scabies is minimal or can be absent in crusted scabies. A common sign of crusted scabies is localized or widespread horny plaques on the skin, palm, and sole and even dystrophic nails (Fig. 5). These crusts tend to be thick, crumble easily when touched, and look greyish in colour. Sometimes the crusts can be localized to one or a few areas of the body such as the scalp, back, or feet. The lesions over the extensor surface show fissuring. When the plaques are very large and thick, the whole surface may be broken up into slabs by these fissures. On removal of the crust, the undersurface is smooth, red, moist, and velvety in appearance. Patients with crusted scabies shed thousands of mites daily and they are highly infectious (Roberts et al. 2005, Karthikeyan 2009). As a result, airborne outbreaks can arise, especially in hospital settings and care institutions. These outbreaks are often extremely difficult to manage since the treatment of crusted scabies is challenging.

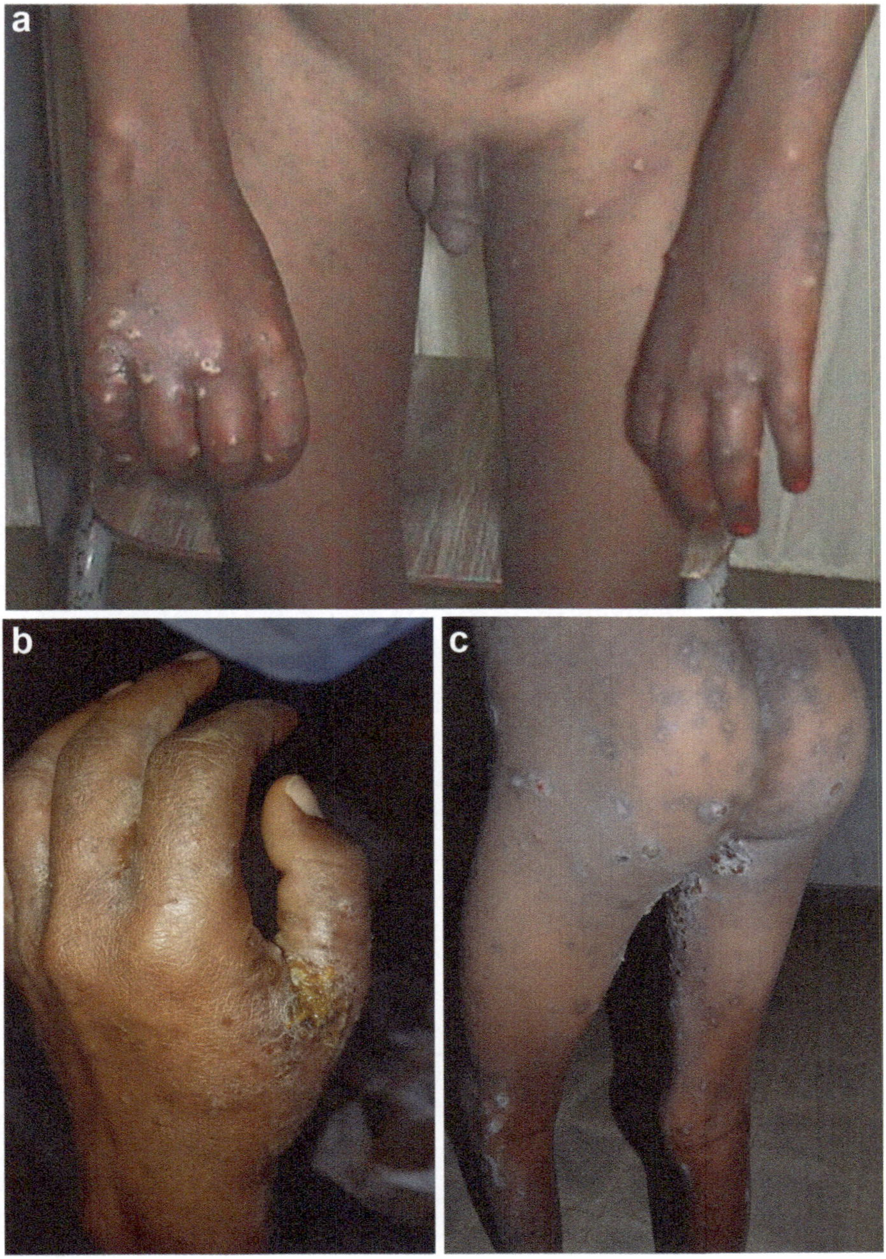

Fig. 2 (**a**) Papules, pustules and excoriations on interdigital space of right hand (**b**) Papules, pustules and yellowish crust on the dorsum of hands and fingers (**c**) Thick crust, excoraition and post inflammatory dispigmentation in the buttock and legs

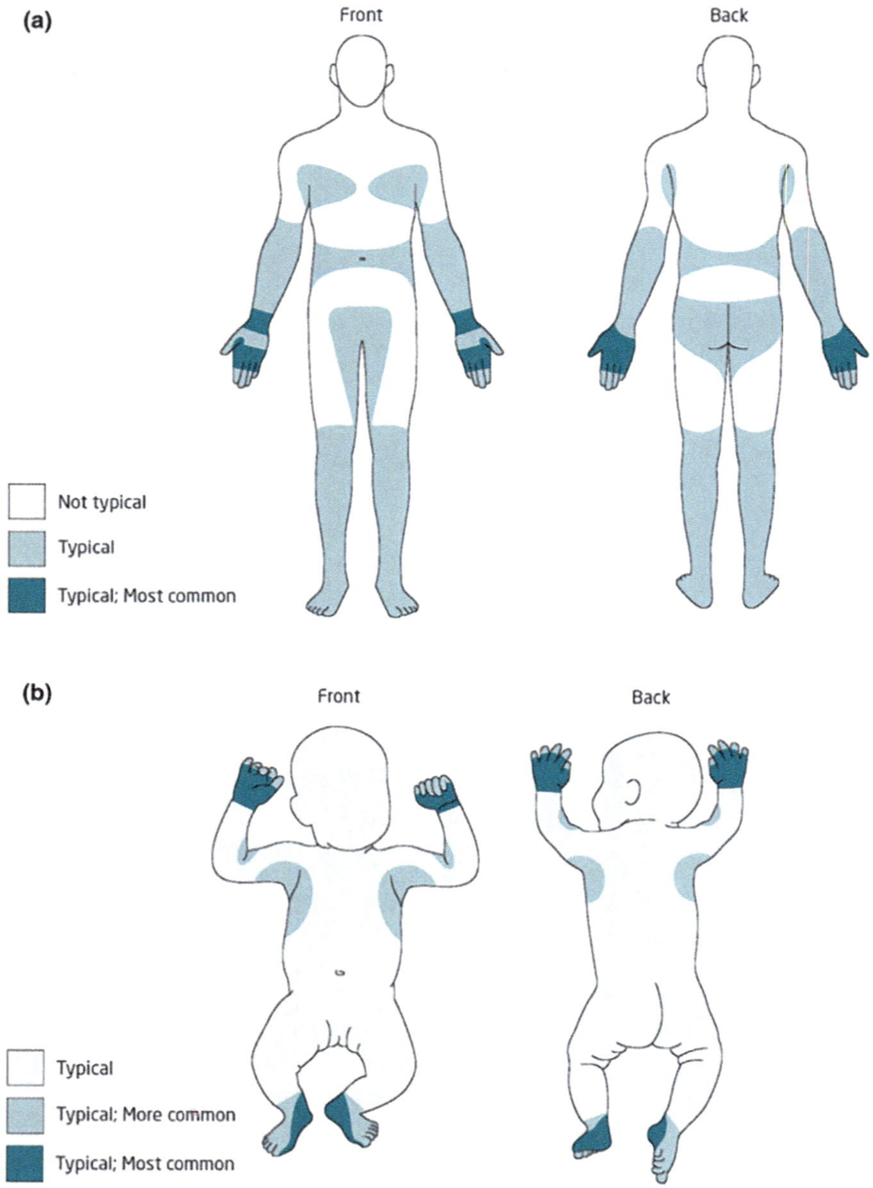

Fig. 3 Scabies typical body distribution in adult and children (Engelman et al. 2020)

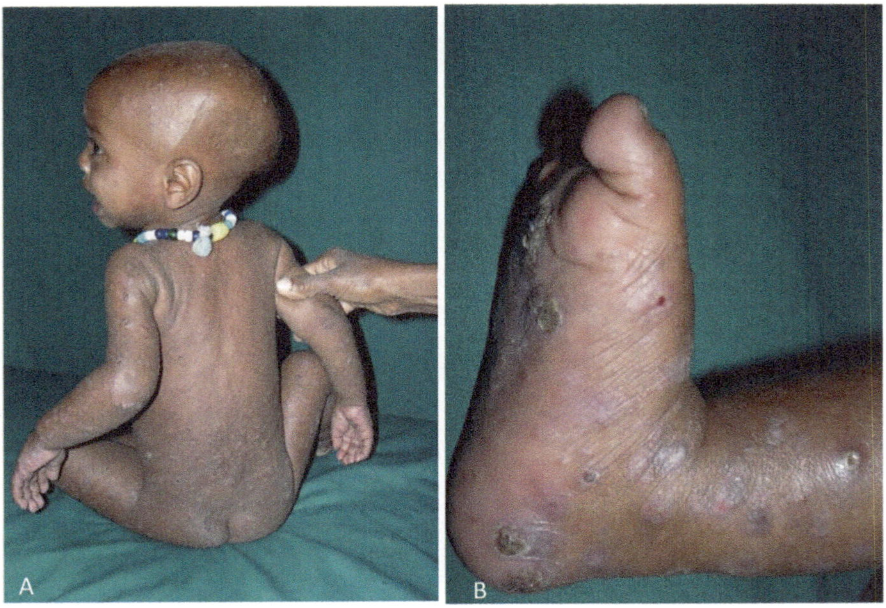

Fig. 4 (**a**) papules and excoriations on background of dry skin on scalp, trunk, and extremity. (**b**) erythematous papules, pustules on the sole of foot

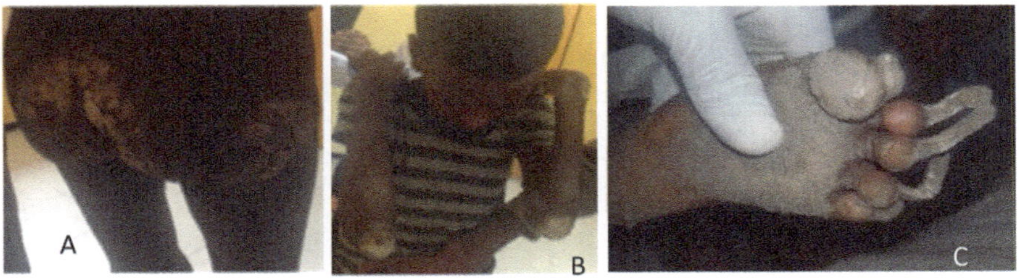

Fig. 5 Psoriasiform hyperkeratotic plaques on the buttocks in a patient with crusted scabies. (**b**) Hyperkeratotic scaly plaques with fissures on the extensor (elbow and wrist) arm. (**c**) Dirty looking hyperkeratotic crust on the sole of foot anteriorly and hyperkeratotic and elongated nails with significant deformity

Diagnosis

Clinical examination remains the mainstay of diagnosis of scabies especially for the primary health care unit and in most health care settings. The International Alliance for the Control of Scabies (IACS) has published consensus criteria for the diagnosis of scabies (Engelman et al. 2019) as follows:

IACS Consensus Criteria

A. Confirmed scabies

At least one of:

1. A1: Mites, eggs or feces on light microscopy of skin samples
2. A2: Mites, eggs or feces visualized on individual using high-powered imaging device
3. A3: Mite visualized on individual using dermoscopy

B. Clinical scabies

At least one of:

1. B1: Scabies burrows
2. B2: Typical lesions affecting male genitalia
3. B3: Typical lesions in a typical distribution and two history features

C. Suspected scabies

One of:

1. C1: Typical lesions in a typical distribution and one history feature
2. C2: Atypical lesions or atypical distribution and two history features

History features

1. H1: Itch
2. H2: Close contact with an individual who has itch or typical lesions in a typical distribution

The B1criteria of the IACS consensus criteria is problematic since burrows can be difficult to see in dark skin individuals and need specialized expertise.

To improve access for patients with scabies, widespread training of nurses or community-based health workers (CHW) is essential. A study in Ethiopia has showed that the CHWs can make reliable clinical diagnosis of scabies in 98.3% of cases (validated by a dermatologist/expert control group) (Enbiale et al. 2020a). Development of both train-the-trainer and quality assurance interventions were key factors for the success and reliability of surveys.

The standard technique for the diagnosis of scabies consists in the microscopic identification of the mite, eggs, or feces from scales obtained by skin scraping (Lacarrubba et al. 2001). Since patients are usually only infested by a few mites, repeated tests in different areas are often needed for a conclusive diagnosis and the sensitivity is low. Moreover, this technique is time-consuming and not suitable in a busy practice and not feasible for the primary health care and low resource settings (Neynaber and Wolff 2008). Other possible diagnostic procedures include the burrow ink test, the adhesive tape test, and the polymerase chain reaction (PCR)-based method for detecting *S. scabiei* DNA in skin scrapings (Fukuyama et al. 2010). The detection of *S. scabiei* antibodies (which are produced before the onset of clinical symptoms) through the enzyme-linked immunosorbent assay (ELISA) is a promising, although not a currently available diagnostic tool (Zheng et al. 2016). Non-invasive skin imaging techniques such as videodermatoscopy, dermoscopy and in vivo reflectance confocal microscopy have been found suitable for the in vivo identification of burrows and scabies mites (Lacarrubba et al. 2010, Micali et al. 2011 and Lacarrubba et al. 2015). The authors concluded that the advantages of using the specialized devices include rapid noninvasive mass-screening, post-therapeutic follow-up, no physical risk, and minimization of complications associated with a missed scabies diagnosis (Micali et al. 2016). Yet, these novel non-invasive skin imaging techniques need further field-based validation. Their economic and t logistic feasibility in the hands of mid and lower level health workers.

In summary, no "field-friendly" rapid test is currently available for diagnosing scabies. Scabies can be diagnosed definitively by microscopic examination of skin scrapings, but this is impractical in the field and insensitive. Scabies usually presents as typical skin lesions and itch, although it can take many forms. Therefore, a clinical diagnosis is the preferred method in resource limited setting (World Health Organization 2020).

Control Tools Strategies

The first and foremost essential element of any control strategy is the establishment of the burden of the disease. Subsequently, the tools for control strategies include three important parts: Surveillance, intensive case management (for low endemic settings) and community-based management (for high endemic settings).

Surveillance

Continuous community-based surveillance is essential for early detection, control, and prevention of scabies in the community. Once an outbreak is confirmed, there is a need to develop a communication strategy, multi-sectored coordination, and response to control the spread of the outbreak. The detection of cases at the health facility and at the community level is a cost effective strategy to determine the distribution of the disease and identify the high-risk areas and high-risk group. Based on these distribution data, a tailored intervention can be planned and commenced. The burden is also captured in the routine clinic reporting system of diseases from health facilities other than the surveillance system. These two (surveillance and report from health facilities) have their own pros and cons as the surveillance may suffer from incompleteness. Monitoring the weekly incidence can trigger aggressive contact tracing and initation of early response, whereas routine lower frequency disease reporting may not be helpful to detect outbreaks and trigger prompt effective responses.

Intensive Disease Management

Intensified case management, contact tracing and contact treatment are instrumental to control the disease in lower prevalence settings. To break the chain of infestations and ongoing transmission, the management of contacts is as important as the management of cases. The health care worker, who is providing care and the surveillance focal person should work closely to identify and list contacts at household, school and working areas. Contact tracing can be challenging. For instance, the identification of contacts of an index case with multiple sexual partners can be hampered by stigma and taboo. The effectiveness of the treatment is also affected by the adherence of the patient or care giver to treatment protocol/guidelines and the adherence of the case/contact to the drug (compliance).

The drug of choice for intensive disease management can be topical or systemic. The most commonly used topical drugs of choice for the individual treatment of scabies are permethrin, Benzyl benzoate, sulfur, and crotamiton, as well as the systemic option, Ivermectin.

Permethrin

Permethrin lotion/cream 5% is applied once for at least 8 h (before going to bed at night and rinsed off in the morning) and repeated after 1 week.

The complete skin surface below the chin should be treated, including the genital, intergluteal space, palm, and soles and under the nails. In infants (up to the age of 2 years) the head and scalp should also be treated. The application should be repeated after 1 week.

Dosage of Permethrin; Full tube of 30 ml for adults. Half tube for children.

Side effect of Permethrin; Allergic contact dermatitis and rarely irritation.

Benzyl Benzoate

For adult, Benzyl benzoate lotion (BBL) 25% once per day for 3 days or for children above 2 years old 12.5% BBL once per day for 3 days. BBL is contra-indicated in children below 2 years old. A systematic review on efficacy of topical anti scabies showed that BBL is less effective than Permethrin (Thadanipon et al. 2019).

All the skin below the chin should be treated, including the genital, buttocks, palm, and soles and under the nails.

Dosage: One bottle of 100 ml BBL will be enough for 3 days whole body application for adult. The lotion is left on the applied surfaces for 1 day before it is washed off.

Side effect of BBL; Irritation, Allergic contact dermatitis and rarely, Neurotoxicity

Sulfur

For children under 2 years, pregnant and breast-feeding women 5 to 10% sulfur topical application once per day for 3 days.

All the skin below the chin should be treated, including the genital, buttock, palm, and soles and under the nails. Treat the head and scalp also in infants (up to the age of 2 years).

Dosage: For children under 10 years 1 tube (50gm) of 5% sulfur is used for 3 days whole body application and left on for 1 day before washing off. For children above 10 years old and adult 2 tubes of 10% sulfur will be needed for 3 days whole body applications.

Side effect of Sulfur; messy, malodorous, irritation and allergic contact dermatitis.

Ivermectin

Ivermectin oral 200 mg/kg oral once to be repeated after 1–2 weeks.

All household contacts should be treated at the same time even if asymptomatic.

Apart from treatment, hygienic measures should be imposed. Clothing, bedding, and linens used by an index case during the 3–5 days prior to and during treatment should be washed or cleaned and dried in the sun. Another option is to place all potentially infested clothes and linens in a closed plastic bag for 4 days.

In patients with severe pruritis a sedative antihistamine can be added, and the patient must be informed that the pruritus can last for up to 4 weeks after successful treatment (post scabies itch).

Community Based Management

This is the strategy to be implemented in high prevalence settings. Mass Drug Administration (MDA) is recommended where the population-based prevalence is estimated to be $\geq$10%.

Estimates of prevalence for determining where to start MDA could be derived by rapid mapping, although a formal prevalence assessment could also be used (World Health Organization 2020).

The preferred drug for a population-based intervention or MDA is systemic ivermectin for those without any contraindications (children younger than 5 or below 15 kg, pregnant and lactating mothers) and topical 5% permethrin for children younger than 5 years old, pregnant, and lactating mothers (Enbiale et al. 2020a, b). The target threshold for stopping MDA should be pilot-tested in embedded operational research but the current expert consensus is that the MDA can be repeated annually or biannually till the population prevalence is <2% (World Health Organization 2020; Engelman et al. 2021).

As with intensive case management, contact tracing and treatment are essential also in community-based management interventions. The process demands a dedicated team coupled with proper recording and administrative tools. This is usually challenging in low- and middle-income countries.

Disease control efforts can only be effective when concerted effort is in place. Engagement of the society including the government are vital for a successful and effective outcome. To keep momentum and to ensure sustainability of the control efforts, the engagement and active participation of political leaders and non-health sectors will be essential components of the strategy.

Further Research for Policy and Control

Several thematic areas that require further research, including disease burden, treatment, and diagnostics have been identified.

- Under disease burden, research is needed to determine the correlation between "true prevalence" and "rapid mapping prevalence," as well as the consistency between prevalence in schools and in communities.
- The impact of MDA (Mass Drug Administration) where community prevalence is less than 10%, the relationship between scabies and impetigo, and severe complications in different populations and settings also need further study.
- Research is required to assess the safety and dosage of ivermectin in children weighing less than 15 kg, the efficacy of 150 μg/kg and of 200 μg/kg of ivermectin for MDA against scabies, and the effectiveness of a single dose of ivermectin and a single application of permethrin in MDA.
- The efficacy of moxidectin for the treatment and community control of scabies also needs to be evaluated.
- There is a need to validate the simplified IACS diagnostic criteria for use by front-line and mid-level health care providers in clinics and to develop diagnostic tools, such as serological testing, to standardize monitoring and evaluation.

Conclusion

Scabies is one of the NTD of the skin, caused by ectoparasites that are transmitted by skin-to-skin contact. It is extremely common, and all populations are at risk of scabies but children living in developing countries are particularly susceptible to the disease. The disease can lead to secondary bacterial infection—a major risk factor for acute glomerulonephritis and chronic renal failure. Clinical examination remains the mainstay of the diagnosis of scabies.

The essential elements of the control strategy in Scabies are surveillance, intensive case management (for low endemic settings) and community-based management (for high endemic settings).

Current evidence suggests that MDA with ivermectin or permethrin can be effective in reducing the prevalence of scabies and its associated complications, such as impetigo and post-streptococcal glomerulonephritis. The use of MDA has been shown to be more effective than treating only clinically diagnosed cases of scabies. However, there is still a need for further research to optimize MDA implementation and to determine the most effective and sustainable strategies for scabies control.

References

Anderson KL, Strowd LC (2017) Epidemiology, diagnosis, and treatment of scabies in a dermatology office. J Am Board Fam Med 30(1):78–84

Arlian LG, Morgan MS (2017) A review of Sarcoptesscabiei: past, present, and future. Parasit Vectors 10(1):297

Arlian LG, Estes SA, Vyszenski-Moher DL (1988) Prevalence of Sarcoptesscabiei in the homes and nursing homes of scabietic patients. J Am Acad Dermatol 19(5,Pt1):806–811

CDC (2020) Global Health, Division of parasitic diseases and malaria, Life cycle image and information. https://www.cdc.gov/parasites/scabies/biology.html#. Accessed 8 September 2021

Chandler DJ, Fuller LC (2019) A review of scabies: an infestation more than skin deep. Dermatology 235(2):79–90

Collinson S, Timothy J, Zayzay SK, Kollie KK, Lebas E, Candy N et al (2020) The prevalence of scabies in Monrovia, Liberia: a population-based survey. PLoS Negl Trop Dis 14(12):e0008943

Enbiale W, Ayalew A (2018) Investigation of a scabies outbreak in drought-affected areas in Ethiopia. Tropical Med Infectious Dis 3:114

Enbiale W, Baynie TB, Ayalew A, Gebrehiwot T, Getanew T, Ayal A, Ayalew M, De Vries HJ, Takarinda K, Manzi M, Zachriah R (2020a) "Stopping the itch": mass drug administration for scabies outbreak control covered for over nine million people in Ethiopia. J Infect Dev Ctries 14(6.1):28S–35S

Enbiale W, Ayalew A, Gebrehiwot T, Mulu Y, Azage M, Zachariah R et al (2020b) Does mass drug administration for community-based scabies control works? The experience in Ethiopia. J Infect Dev Ctries 14(06.1):78S–85S

Engelman D, Cantey PT, Marks M, Solomon AW, Chang AY, Chosidow O, Enbiale W, Engels D, Hay RJ, Hendrickx D, Hotez PJ, Kaldor JM, Kama M, Mackenzie CD, McCarthy JS, Martin DL, Mengistu B, Maurer T, Negussu N, Romani L, Steer AC (2019) The public health control of scabies: priorities for research and action. Lancet 394(10192):81–92

Engelman D, Yoshizumi J, Hay RJ, Osti M, Micali G, Norton S et al (2020) The 2020 international Alliance for the control of scabies consensus criteria for the diagnosis of scabies. Br J

Dermatol. https://doi.org/10.1111/bjd.18943. A, children aged > 2 years and adults; B, infants aged <2 years
Engelman D, Marks M, Steer AC, Beshah A, Biswas G, Chosidow O et al (2021) A framework for scabies control. PLoS Negl Trop Dis 15(9):e0009661
Fukuyama S, Nishimura T, Yotsumoto H, Gushi A, Tsuji M, Kanekura T, Matsuyama T (2010) Diagnostic usefulness of a nested polymerase chain reaction assay for detecting Sarcoptesscabiei DNA in skin scrapings from clinically suspected scabies. Br J Dermatol 163:892–894. pmid:20560958
Galván-Casas C, Mitjá O, Esteban S, Kafulafula J, Phiri T, Navarro-Fernández Í et al (2021) A facility and community-based assessment of scabies in rural Malawi. PLoS Negl Trop Dis 15(6):e0009386
GBD 2016 DALYs and HALE Collaborators (2017) Global, regional, and national disability-adjusted life-years (DALYs) for 333 diseases and injuries and healthy life expectancy (HALE) for 195 countries and territories, 1990–2016: a systematic analysis for the global burden of disease study 2016. Lancet 390:1260–1344
Heukelbach J, Feldmeier H (2006) Scabies. Lancet 367:1767–1774
Jouret G, Bounemeur R, Presle A, Takin R (2016) La gale hyperkératosique [Crusted scabies: a review]. Annales de dermatologie et de venereologie 143(4):251–256
Karimkhani C, Colombara DV, Drucker AM, Norton SA, Hay R, Engelman D, Steer A, Whitfeld M, Naghavi M, Dellavalle RP (2017a) The global burden of scabies: a cross-sectional analysis from the global burden of disease study 2015. Lancet Infect Dis 17(12):1247–1254
Karimkhani C, Colombara DV, Drucker AM, Norton SA, Hay R, Engelman D et al (2017b) The global burden of scabies: a cross-sectional analysis from the global burden of disease study 2015. Lancet Infect Dis 17:1247–1254
Karthikeyan K (2009) Crusted scabies. Indian J Dermatol Venereol Leprol 75(4):340–347
Lacarrubba F, Musumeci ML, Caltabiano R, Impallomeni R, West DP, Micali G (2001) High-magnification videodermatoscopy: a new non-invasive diagnostic tool for scabies in children. Ped Dermatol 18:439–441
Lacarrubba F, D'Amico V, Nasca MR, Dinotta F, Micali G (2010) Use of dermatoscopy and videodermatoscopy in therapeutic follow-up: a review. Int J Dermatol 49:866–873. pmid:21128914
Lacarrubba F, Verzì AE, Micali G (2015) Detailed analysis of in vivo reflectance confocal microscopy for Sarcoptesscabiei hominis. Am J Med Sci 350:414
Ljunggren EL (2005) Molecular analysis of Sarcoptesscabiei. Dept. of Biomedical Sciences and Veterinary Public Health, Swedish University of Agricultural Sciences, Uppsala
Lynar S, Currie BJ, Baird R (2017) Scabies and mortality. Lancet Infect Dis 17:1234
Mason DS, Marks M, Sokana O et al (2016) The prevalence of scabies and impetigo in the Solomon Islands: a population-based survey. PLoS Negl Trop Dis 10:e0004803
Mellanby K (1941) Transmission of scabies. Br Med J 2(4211):405–406. https://doi.org/10.1136/bmj.2.4211.405
Mellanby K (1944) The development of symptoms, parasitic infection, and immunity in human scabies. Parasitology 35(4):197
Micali G, Lacarrubba F, Massimino D, Schwartz RA (2011) Dermatoscopy: alternative uses in daily clinical practice. J Am Acad Dermatol 64:1135–1146. pmid:21292346
Thornley S, Marshall R, Jarrett P, Sundborn G, Reynolds E, Schofield G (2018) Scabies is strongly associated with acute rheumatic fever in a cohort study of Auckland children. J Paediatr Child Health 54:625–632
Micali G, Lacarrubba F, Verzì AE, Chosidow O, Schwartz RA (2016) Scabies: advances in non-invasive diagnosis. PLoS Negl Trop Dis 10(6):e0004691. https://doi.org/10.1371/journal.pntd.0004691
Mounsey KE, Murray HC, King M, Oprescu F (2016) Retrospective analysis of institutional scabies outbreaks from 1984 to 2013: lessons learned and moving forward. Epidemiol Infect 144(11):2462–2471
Neynaber S, Wolff H (2008) Diagnosis of scabies with dermoscopy. CMAJ 178:1540–1541. pmid:18519900

Paller AS (1993) Scabies in infants and small children. Semin Dermatol 12(1):3–8
Roberts LJ, Huffam SE, Walton SF, Currie BJ (2005) Crusted scabies: clinical and immunological findings in seventy-eight patients and a review of the literature. J Infect 50:375–381
Romani L, Steer AC, Whitfeld MJ, Kaldor JM (2015) Prevalence of scabies and impetigo worldwide: a systematic review. Lancet Infect Dis 15(8):960–967
Swe PM, Reynolds SL, Fischer K (2014) Parasitic scabies mites and associated bacteria joining forces against host complement defence. Parasite Immunol 36:585–593
Thadanipon K, Anothaisintawee T, Rattanasiri S, Thakkinstian A, Attia J (2019) Efficacy and safety of antiscabietic agents: a systematic review and network meta-analysis of randomized controlled trials. J Am Acad Dermatol 80(5):1435–1444
Ugbomoiko US, Oyedeji SA, Babamale OA, Heukelbach J (2018) Scabies in resource-poor communities in Nasarawa state, Nigeria: epidemiology, clinical features and factors associated with infestation. Trop Med Infect Dis 3(2):59
World Health Organization (2020) WHO informal consultation on a framework for scabies control: World Health Organization regional Office for the Western Pacific: Manila, Philippines, 19–21 February 2019: meeting report. World Health Organization, Geneva
Zheng Y, He R, He M, Gu X, Wang T, Lai W et al (2016) Characterization of Sarcoptesscabiei cofilin gene and assessment of recombinant cofilin protein as an antigen in indirect-ELISA for diagnosis. BMC Infect Dis 16:21. pmid:26801761

Schistosomiasis

Tony Danso-Appiah, Mustapha Immurana,
Alfred Kwesi Manyeh, William Kofi Anyan,
and John O. Gyapong

Contents

T. Danso-Appiah
Department of Epidemiology and Disease Control, School of Public Health, College of Health Sciences, University of Ghana, Accra, Ghana

M. Immurana · A. K. Manyeh · J. O. Gyapong (✉)
Institute of Health Research, University of Health and Allied Sciences, Ho, Ghana
e-mail: jgyapong@uhas.edu.gh

W. K. Anyan
Noguchi Memorial Institute for Medical Research (NMIMR), University of Ghana, Accra, Ghana

J. O. Gyapong, B. A. Boatin (eds.), *Neglected Tropical Diseases - Sub-Saharan Africa*, Neglected Tropical Diseases, https://doi.org/10.1007/978-3-031-53901-5_13

Abstract

Schistosomiasis, a neglected tropical disease, is caused by the blood fluke that resides in the blood vessels of the human hosts. It presents as an acute, but mostly chronic, illness and is commonly found in the least developed countries having water resource development projects, like dams and irrigation system, and lacking good basic facilities as well as good healthcare systems. Forty countries in sub-Saharan Africa (SSA) were endemic for schistosomiasis in 2010. Of the five known species, *Schistosoma haematobium*, *S. mansoni*, *S. japonicum*, *S. intercalatum*, and *S. mekongi*, that can infect humans, it is *S. haematobium* which causes urogenital schistosomiasis and *S. mansoni* which causes intestinal schistosomiasis that have been responsible for most of the disease and a substantial public health and economic burden in SSA. Children, women, and farmers in rural communities who depend on water contact for recreational, domestic, or occupational activities are most vulnerable to the infection. Cross-border movements from unstable and conflict zones in SSA have contributed to the spread of the disease to previously non-endemic foci, and emigration from rural areas into the cities for economic opportunities has introduced the disease into some urban areas. In 2019, schistosomiasis caused around 1.64 million and 1.23 million disability-adjusted life years (DALYs) and years lived with disability (YLDs), respectively, with at least 80% of the burden concentrated in SSA. Annual loss from disability due to schistosomiasis in Africa was estimated to be nearly half a billion US dollars, approximately 70% of the global cost. The life cycle involves intermediate planorbid freshwater snail hosts—*Bulinate* species *and Biomphalaria* species—in the transmission of the infection with sexual and asexual stages. Man is the definitive host. Pathology and clinical morbidity of schistosomiasis are caused by eggs trapped in tissues of various organs and may present as hematuria in urogenital schistosomiasis and abdominal pain, diarrhea, and bloody stool in intestinal schistosomiasis. Long-term complications from schistosomiasis include urinary tract infections, bladder calcification, hydronephrosis/hydroureter, kidney failure, lesions of the liver, portal vein, and spleen, leading to periportal fibrosis, hepatomegaly, splenomegaly, pipe-stem portal fibrosis, ascites, nodules in the vulva, and bladder cancer. Microscopic examination for parasite eggs in the urine or stool is considered definitive for the diagnosis of the infection. Control of schistosomiasis involves health education, safe water supply, mollusciding, environmental management, chemotherapy (praziquantel as drug of choice) or combination of these measures. Currently, schistosomiasis control is mainly by mass drug administration (MDA) with anthelminthics, and the ongoing debate is whether it can be controlled or eliminated using the current approach.

Epidemiology of the Disease

Geographic Distribution

Schistosomiasis is mostly tropical and one of the waterborne parasitic diseases in the world caused by the blood fluke, a group of flatworms that reside in the blood vessels in the human hosts. It presents as an acute, but mostly chronic illness. Due to its geographic and demographic distribution, the disease is listed as one of the neglected tropical diseases (NTDs). It is commonly found in countries among the least developed whose health systems face difficulties to provide basic care at the primary health level (Chitsulo et al. 2000) (Table 1). Five species, namely, *Schistosoma haematobium*, *S. mansoni*, *S. japonicum*, *S. intercalatum*, and *S. mekongi*, can infect humans, but the three with public health importance are *S. haematobium* (mostly endemic in Africa and the Middle East), *S. mansoni* (common in the tropics and subtropics), and *S. japonicum* (mainly found in the People's Republic of China and the Philippines) (WHO 2012). *S. haematobium* causes urogenital schistosomiasis, while *S. mansoni* and *S. japonicum* cause intestinal schistosomiasis.

The main schistosome species infecting animals are *S. mattheei, S. bovis*, and *S. currassoni* (Table 1). There is evidence of increasing hybridization events of schistosome species between human and animal, leading to new zoonotic infections in recent years (Aula et al. 2021).

Schistosomiasis is found commonly in low-income countries. The largest burden of the disease, i.e., 93% making about 207 million of the world's disease burden, is found in sub-Saharan Africa (SSA). Countries in SSA with the highest prevalence are Nigeria, Tanzania, Ghana, Mozambique, and the Democratic Republic of Congo, and studies have shown that these five countries contribute ~78 million cases of schistosomiasis (Onasanya et al. 2021).

Schistosomiasis commonly affects the poor rural dwellers and disadvantaged urban or peri-urban settlements with limited access to portable water, poor sanitation, and hygiene services (Hotez et al. 2009). The disease is also more

Table 1 Schistosome species and their geographical distribution

	Species	Intermediate host	Definitive hosts	Site of infection	Geographical distribution
Intestinal schistosomiasis	*S. mansoni*	*Biomphalaria* spp.	Humans, rodents	Intestinal mesenteric veins	Sub-Saharan Africa, Madagascar, the Middle East, the Caribbean, South America
Intestinal schistosomiasis	*S. intercalatum* and *S. guineensis*	*Bulinus* spp.	Humans, nonhuman primates (excluding apes)	Intestinal mesenteric veins	Central Africa, West Africa, Madagascar
Urogenital schistosomiasis	*S. mattheei*	*Bulinus* spp.	Cattle, sheep, goats	Intestinal mesenteric veins	Southeastern and Central Africa
Animal intestinal schistosomiasis	*S. curassoni*	*Bulinus* spp.	Cattle, sheep, goats	Intestinal mesenteric veins	West Africa
Animal intestinal schistosomiasis	*S. bovis*	*Bulinus* spp.	Cattle, goats, sheep, horses, camels, pigs	Intestinal mesenteric veins	North, East, West and Central Africa, the Middle East and Mediterranean (Europe) region

prevalent in fishing and agriculture dominant communities where direct interactions with fresh water increase the risk of contracting the disease. Water-related domestic activities such as washing clothes, fetching water, and performing religious activities such as water baptism and ablution as well as recreational water activities in infested water increase the risk of infection of schistosomiasis.

In SSA, the two main species infecting humans are *Schistosoma haematobium*, the cause of urogenital schistosomiasis, and *S. mansoni*, which causes intestinal schistosomiasis. Other species like *S. intercalatum* and *S. guineensis* also cause intestinal schistosomiasis but are less prevalent as shown in Table 1 (Fig. 1).

Demographic Distribution

A major feature of the infection is that it shows a characteristic aggregated distribution and overdispersion of worms. In a typical endemic setting, it has been estimated that 20% of the population usually carry approximately 80% of the total infection; the remaining 80% of the population harbor few worms that constitute only about 20% of the overall population of worm load (Anderson and May 1992; Woolhouse et al. 1991). The infection also displays age-intensity profile that is typically convex, with prevalence and intensity of infection rising in childhood,

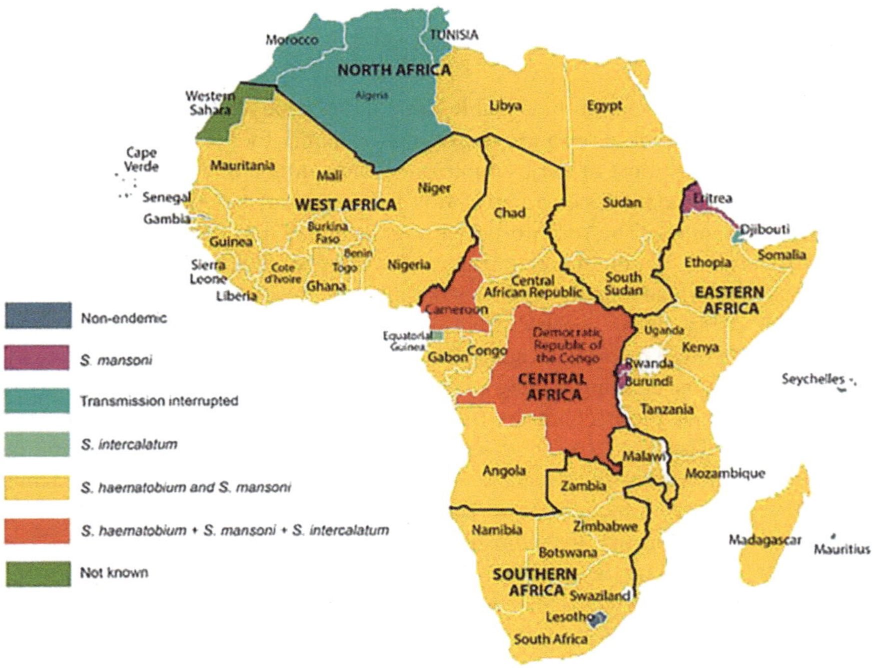

Fig. 1 Distribution of schistosome infections in Africa (Aula et al. 2021)

peaking in adolescence and decreasing in adulthood to a relatively stable phase (King et al. 1988; Mitchell et al. 2008). Children, women, and farmers in disadvantaged rural communities without access to adequate sanitation or clean water and who depend on water contact for recreational, domestic, or occupational activities are most vulnerable to the infection. Poverty is a major determinant of schistosomiasis, but attempts to improve the livelihood of poor farmers through the creation of water-related economic schemes have in turn led to the promotion of transmission of the infection (Chitsulo et al. 2000; Poda et al. 2004; Steinmann et al. 2006).

Urogenital schistosomiasis and intestinal schistosomiasis are responsible for most of the disease in SSA. In some countries, there is an overlap in distribution, resulting in mixed infection in majority of the infected persons. *S. intercalatum* which causes another form of intestinal schistosomiasis is endemic in six countries in Central and West Africa, namely, Democratic Republic of Congo, Gabon, Cameroon, Central African Republic, Chad, and Sao Tome. It has recently been reported in travelers returning from Mali, West Africa, but this is yet to be officially confirmed. This chapter focuses on the two common schistosome types in SSA, namely, *S. haematobium* and *S. mansoni*, with reference to the other species where necessary.

Impact of Conflict, Migration, and Tourism on the Spread of the Disease

Cross-border movements from unstable and conflict zones in SSA have contributed to the spread of the disease to previously non-endemic foci (Chitsulo et al. 2000). As the economy of most of these countries is concentrated in the urban centers and cities, there has been a trend of emigration from rural areas into the cities for economic opportunities. The rural-to-urban migration has introduced the disease into some urban areas. In non-endemic countries in Europe, such as Denmark, England, Germany, Netherlands, Spain, and Switzerland, the prevalence of imported schistosomiasis is rising, partly because of an increase in travelling to endemic areas and partly to emigration from these areas. The rise in the number of imported cases of schistosomiasis is creating a new diagnostic and management challenge to the health systems of these countries (Grobusch et al. 2003; Bierman et al. 2005; Helleberg and Thybo 2010). Travellers embarking on ecotourism and diving tours in endemic destinations in Africa are the most at risk (Schwartz et al. 2005; Kinkel et al. 2012). However, records show that in 2005, two thirds of all imported cases of schistosomiasis to Europe were mostly immigrants and refugees who had fled conflicts and wars in some parts of SSA (Hatz 2005).

Some studies reported schistosomiasis prevalence ranging from 1% to 76% among travelers and migrants who were infected while visiting SSA (Clerinx and Van Gompel 2011). There have been reported first cases of locally acquired urinary schistosomiasis in recent years caused by *S. haematobium* in Corsica, France. These cases were believed to be likely imported by persons who were infected in West Africa (Boissier et al. 2015, 2016).

Life Cycle and Mode of Transmission

Schistosomiasis has a complex life cycle that involves an intermediate snail host and sexual and asexual stages (Fig. 2). Man is the definitive host and some species of the planorbid freshwater snails are the intermediate hosts in the transmission of the infection: *Bulinus truncatus* for *S. haematobium* and *Biomphalaria pfeifferi* for *S. mansoni*. The asexual reproduction occurs in the snail, whereas the sexual reproduction takes place in the human host (Hoffmann and Dunne 2003). The worms do not multiply in the human host. People infected with *S. haematobium* worms excrete the parasite eggs in their urine and *S. mansoni* eggs in feces. Under optimal conditions, the eggs hatch to release miracidia (larvae) which swim to locate and penetrate a specific freshwater snail, the intermediate host. Within the snail, the miracidia undergo a two-staged developmental process to produce the infective larvae called cercariae within 4–6 weeks. Cercariae are phototropic and their release from the snail is triggered by sunlight, mostly around midday, in the tropics, which tends to coincide with the peak hours of water contact activities by children and women. It has been reported that a snail infected by one miracidium can release thousands of cercariae every day for months (Gryseels et al. 2006). Upon release into the water,

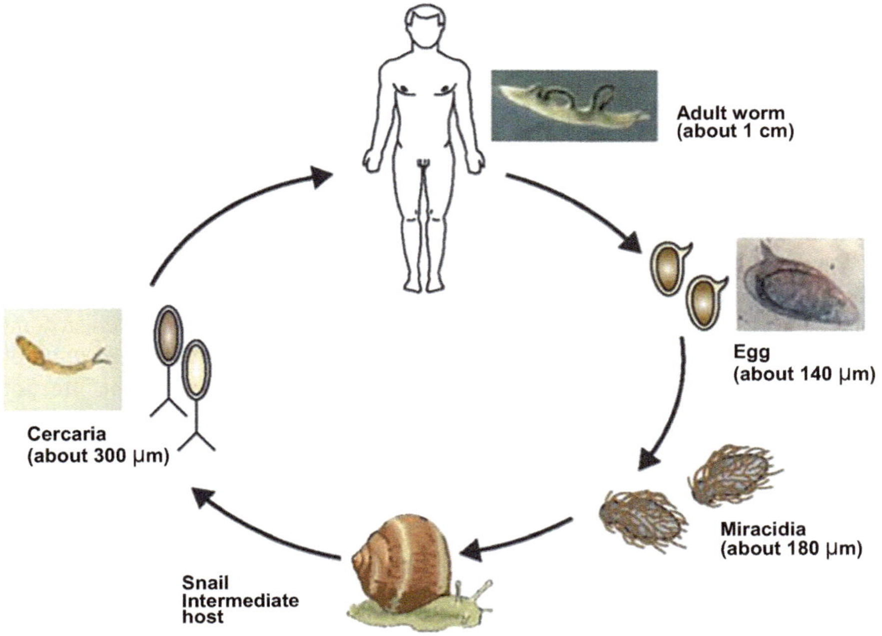

Fig. 2 Life cycle of schistosome worm and mode of transmission of schistosomiasis (Courtesy: Hoffmann and Dunne *Genome Biology* 2003)

the cercariae can swim for up to 72 h when they must locate and penetrate the skin of the human host. Their infectivity potential slows and plateaus around 9 h post-release (Whitfield et al. 2003). The infection is acquired when the infective cercariae penetrate a person's skin upon contact with contaminated water bodies. Upon entering the skin of the human host, the cercariae shed their tail and become schistosomulae which remain in the skin for up to 2 days before locating a post-capillary venule and migrate through the venous system, the right chamber of the heart, the lungs, the mesenteric arteries, and the liver via the portal vein.

Eight to ten days post-penetration of the host's skin, the worm migrates to the liver sinusoids, develops an oral sucker, and begins to feed on red blood cells. Schistosomes undergo developmental processes throughout their migratory phases until finally locating and residing in the venous plexus of the urinary bladder (*S. haematobium* worms) or superior mesenteric veins draining the large intestine (*S. mansoni* worms). A schistosome worm reaches maturity in 6–8 weeks when a male and a female worm pair up in a groove formed by the male called the gynecophoric channel. Here, the male worm holds the longer and slender female where mating takes place, and the female worms begin to produce a large number of eggs (≈300 eggs per day) (Davis 2009). An adult worm usually lives for 3–5 years, but some can live up to 30 years (Warren 1982) and produce up to 600 billion schistosomes during their lifetime (Gryseels et al. 2006). Schistosome eggs can survive for up to 7 days.

Pathology and Disease Presentation

Pathology and clinical morbidity of schistosomiasis are caused by eggs trapped in tissues but not the worms themselves, and symptoms and signs of the disease are location specific. Within the human body, eggs produced by the worms are moved progressively across the walls of blood vessels toward the lumen of the bladder and ureters for excretion with the urine (*S. haematobium*) or the intestine to be excreted with the feces (*S. mansoni*). In the process, a substantial number, up to half, of the eggs released by the worm become trapped in tissues and initiate immune-induced inflammatory reactions that lead to pathology and morbidity (Richter 2003; King and Dangerfield-Cha 2008). Although most individuals with few schistosome worms, especially adults, remain asymptomatic, about 80% of infected children present with one or more symptoms of the disease in the early stages of the infection (Mott et al. 1983; Gryseels and Polderman 1987; Olds and Dasarathy 2000). The main signs and symptoms related to schistosomiasis are summarized in Table 2.

Katayama Syndrome

The acquisition of cercariae may be followed by an initial rash or itchy skin at the site of penetration. This may be followed by an acute illness called Katayama's syndrome within 4–8 weeks following exposure to the infection that may result in a systemic hypersensitivity reaction characterized by fever, fatigue, muscle aches, abdominal pain, diarrhea, nonproductive cough, lymphadenopathy, and eosinophilia. There may also be pain in the upper right part of the abdomen just below the rib cage. Katayama syndrome is most common with *S. mansoni* and *S. japonicum* infections. It mainly occurs following primary infection in individuals with low or no protective immunity, such as tourists visiting an endemic area for the first time. Katayama syndrome is thought to be triggered by maturation and migration of schistosomula within the host (Zuidema 1981; Istre et al. 1984). The patient recovers spontaneously after 2–10 weeks, but some develop persistent and more serious disease. If symptoms do not abort, it can lead to other complications such as weight loss, dyspnea, diarrhea, diffuse abdominal pain, widespread rash, hepatosplenomegaly, and toxemia.

Early Symptoms of an Established Disease

The main early symptom of established infection with *S. haematobium* is blood in the urine (hematuria) or terminal blood after urination, appearing 10–12 weeks after the infection. This may be accompanied by painful urination (dysuria). There is evidence that parasite-induced pathological changes of the bladder and abnormal pyelon dilation of at least one of the kidneys are common in children in the early stages of the infection (Brouwer et al. 2003). This suggests that development of pathology begins far earlier in children than previously thought. For intestinal

Table 2 Schistosomiasis-related pathology and clinical presentation

Stage of symptom	*S. haematobium*	*S. mansoni*	Both (nonspecific)
Katayama syndrome[a]	Itchy skin	Itchy skin	Fever
	Nephropathy	Abdominal pain	Fatigue
		Diarrhea (with or without blood)	Muscle ache
		Nephropathy	Nonproductive cough
			Lymphadenopathy
			Eosinophilia
Early[b]	Hematuria	Diarrhea	Anemia
	Dysuria	Blood in stool/bloody diarrhea	Fatigue
		Abdominal pain	
		Gastrointestinal bleeding	
		Hepatomegaly	
		Splenomegaly	
Late stage[c]	Bladder wall thickening	Colonic polyps	Anemia
	Bladder wall calcification	Cirrhosis of the liver	Fatigue
	Bladder stones	Pipe-stem portal fibrosis	Nutritional deficiencies
	Secondary bacterial infection	Portal hypertension	Poor growth
	Higher risk for ectopic pregnancies	Varicose veins	Memory loss
	Infertility	Hematemesis	Slower reaction time
	Bladder abnormal pyelon dilation	Hemorrhage from rupturing varicose veins	Lower scores in cognitive ability
	Distal ureter hydronephrosis	Hemorrhagic shock	Low birth weight
	Pyelonephritis	Gallbladder cancer	
	Anuria	Liver cancer	
	Hydroureter		
	Kidney failure		
	Squamous cell carcinoma		

Source: Adapted from Danso-Appiah (2009)
Hematuria (blood in urine), dysuria (painful urination), hepatomegaly (enlargement of the liver), splenomegaly (enlargement of the spleen), anuria (no urine excretion)
[a] Occurs a few days or weeks after exposure to the infection
[b] May occur a few weeks or months after the infection
[c] Insidious and may take a long time depending on the intensity of infection

schistosomiasis due to *S. mansoni*, the early symptoms are nonspecific but usually include abdominal pain, diarrhea (which can be bloody), and blood in stool (Gryseels et al. 1992; Lengeler et al. 2002). These symptoms are caused by cellular granulomatous inflammatory reactions around eggs trapped in the intestinal tissues. In spite of the fact that a large number of eggs are deposited in the small intestine, most severe infection-related lesions such as colonic polyps occur in the large intestine (Cheever et al. 1978). It is commonly accepted that intensity and duration of infection determine severity of the disease, but some recent studies have challenged this assumption and suggested that the mere presence of the infection and not necessarily intensity determines morbidity (King and Dangerfield-Cha 2008), although further evidence is warranted to confirm or refute this. Inflammatory reactions in the liver lead to hepatosplenic schistosomiasis, which can manifest as hepatomegaly and splenomegaly within a couple of months from heavy infections or many years after light infections (Fig. 3). It should be noted that chronic schistosomiasis affects mainly individuals with long-standing or repeated infections.

Late-Stage Symptoms and Complications

Long-term complications caused by schistosomiasis are mostly insidious. For *S. haematobium* infection, granulomatous changes and ulcers of the bladder wall and ureter can lead to bladder obstruction, dilatation, secondary urinary tract infections, and subsequent bladder calcification. Photomicrography of the bladder in

Fig. 3 Hepatosplenomegaly in schistosomiasis patient

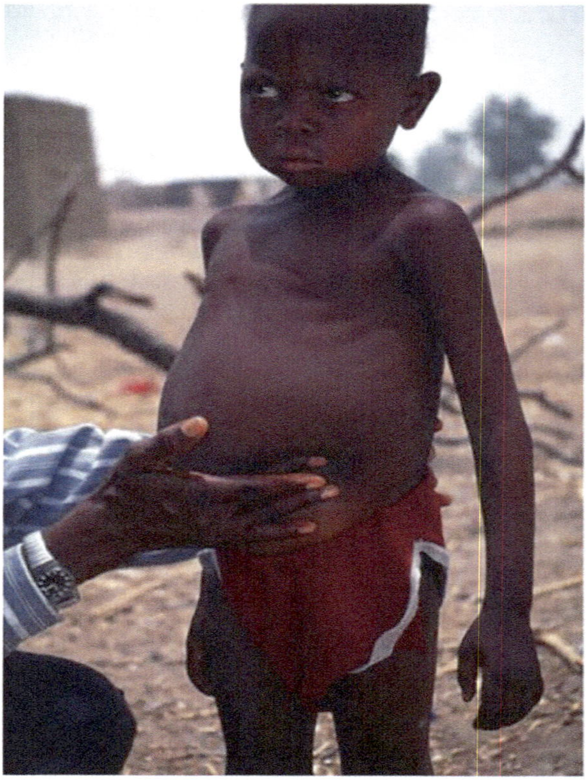

S. haematobium infection may show clusters of the parasite eggs with intense eosinophilia (Fig. 4). Diffused or localized wall thickening of the bladder and distal ureter hydronephrosis or hydroureter may eventually lead to kidney failure (Kardorff and Dohring 2001; WHO 2002a). In females, genital lesions resulting from the infection may present with vaginal bleeding, pain during sexual intercourse, and nodules in the vulva. In men, urogenital schistosomiasis can induce pathology of the seminal vesicles, prostate, and other organs that may decrease male vitality and fertility. Urogenital schistosomiasis is also associated with increased risk of bladder cancer. A review on this topic showed that bladder carcinoma is ranked seventh on the list of most common cancers worldwide in men with the highest incidence occurring among men in Egypt (Murta-Nascimento et al. 2007), which may be related to *S. haematobium* infection (Jordan 2000).

Structural changes or damage resulting from complications of the infection may allow intravascular shift of worms between locations. The superior mesenteric veins have numerous anastomoses with the veins draining the internal genital organs and the vulval area toward the inferior vena cava. The structural damage to valves allows unrestricted migration of worms to the genital organs where they deposit eggs to initiate genital organ pathology (Poggensee and Feldmeier 2001). There are also communications between the spermatic ducts with the superior and inferior mesenteric veins allowing eggs to move into the semen (Arean 1956). In females, the linkage between the ovarian and uterine plexus through anastomoses allows free movement of eggs to the cervix and the vagina thereby affecting the female reproductive organs (Arean 1956; Camara 1959). In recent studies, evidence has emerged that urogenital schistosomiasis is a potential risk factor for HIV infection, especially in women with schistosomiasis-related lesions (WHO 2010a; Mbabazi et al. 2011). There is also an increased risk of hepatocellular carcinoma and mortality from co-infection of schistosomiasis and hepatitis and HIV or malaria.

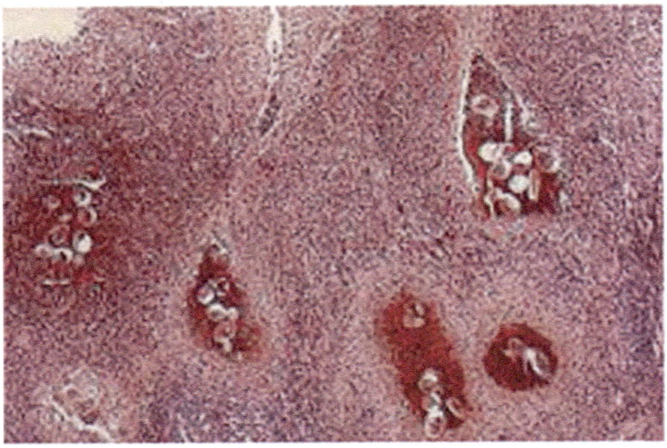

Fig. 4 Photomicrography of the bladder in *S. haematobium* infection, showing clusters of the parasite eggs with intense eosinophilia (Source: Centre for Disease Control (CDC))

The late-stage complications commonly associated with *S. mansoni* infection include lesions of the liver, portal vein, and spleen, leading to periportal fibrosis, hepatomegaly, splenomegaly, pipe-stem portal fibrosis, and ascites (Hatz 2001; Richter 2003). The fibrotic lesions in turn lead to liver cirrhosis that progressively occludes the portal system giving rise to Symmers' pipe-stem periportal fibrosis and portal hypertension. The portal hypertension eventually causes enlargement of the hepatic arteries that leads to varicose veins (esophageal varices). These may rupture and cause heavy blood loss, hemorrhagic shock and finally death. The patient may also suffer repeated episodes of variceal bleeding, the primary cause of death in hepatic schistosomiasis (Andersson and Chung 2007).

Male and Female Genital Schistosomiasis

Male Genital Schistosomiasis

Male genital schistosomiasis (MGS) was first reported in 1911 (Madden 1911) and is termed as the presence of *S. haematobium* eggs in the male genital organs and fluids. The awareness of MGS and its effects especially in schistosomiasis endemic areas is often due to misdiagnosis as STI (Kayuni et al. 2018; Leutscher et al. 2000, 2008).

Signs and symptoms of MGS include painful urination, erectile dysfunction, enlarged genital organs, painful ejaculation, irregular ejaculations, hermatospermia, prostatitis, epididymitis (which could be mistaken for tuberculosis and associated funiculitis) and infertility (Kayuni et al. 2018). Men with *S. haematobium* eggs lodged in their bladder are more likely to develop bladder cancer at a younger age than uninfected males (El-Bolkainy et al. 1981; Kitinya et al. 1986). Additionally other studies have shown that *S. haematobium* exposed cells injected into mice with no immune system caused development of tumors similar to those found in bladder cancer (Botelho et al. 2009).

Female Genital Schistosomiasis

Female genital schistosomiasis (FGS) is the presence of eggs of *S. haematobium* in the vagina and cervix. It is among the most neglected of the NTDs and affects 56 million women and girls in SSA, Middle East, and other parts of the globe (Aula et al. 2021; Christinet et al. 2016; Downs et al. 2011).

Recent research has shown that *S. haematobium* eggs penetrate the female urogenital system, resulting in uterine enlargement, menstrual disorders, cervicitis, and infertility (Nour 2010). FGS in pregnancy presents with symptoms such as anemia and may lead to low birth weight and increased infant and maternal mortality rates (Aula et al. 2021; Ben-Chetrit et al. 2015).

S. haematobium infection in women and girls causes immune response which leads to chronic inflammatory granulomatous lesions, genital epithelial bleeding, and sandy patches in the cervix and vagina. If the infection is left untreated for long, it can lead to infertility and also an easy entry point for the HIV virus (Colley et al. 2014; Downs et al. 2017; Kjetland et al. 2006, 2012). Studies have shown that while more than 70% of HIV infections worldwide occur in Africa, HIV remains a major health challenge on the continent and an important confounder for schistosome infection in SSA (Aula et al. 2021).

FGS has also been associated with ectopic pregnancies as a consequence of blockage of the fallopian tubes (Aminu et al. 2014; Okonofua et al. 1990). Women and girls suffering from FGS can recover through the administration of PZQ if the infection is treated sufficiently early (Ogunniyi et al. 1994; Okonofua et al. 1990).

FGS is mostly misdiagnosed due to different transformation stages of the *S. haematobium* parasite and immune response in the affected tissues. In asymptomatic conditions, FGS may be mistaken for sexually transmitted infections (STIs) or cervical cancer (Christinet et al. 2016; Kjetland et al. 2014). Stigma and misconception about STIs have also contributed to the misdiagnosis of FGS or reluctance of young women to present to clinicians when experiencing clinical manifestation of FGS.

The presence of an adult worm infection in the female reproductive system is usually asymptomatic. The deposition of *S. haematobium* eggs along the female urogenital tract, including the cervix and vagina, however, causes a hypersensitive immune response, resulting in scarring and fibrosis in the female genital tract, ovaries, and fallopian tubes. The eggs may look like papillary white lesion, resulting in thickening, nodular lesions, and adhesions which ultimately cause obstruction of the fallopian tubes. The consequential fibrosis and blockage causes infertility (Bentefouet et al. 2017; Ogunniyi et al. 1994).

Public Health Impact of the Disease

Risk of Infection, Morbidity, and Mortality Estimates

Schistosomiasis is mostly recurrent, and a person may be reinfected repeatedly even when regular treatment is provided, creating a long-lasting public health, social, and economic burden in endemic populations (WHO 2002b; King and Dangerfield-Cha 2008). Although most individuals with schistosomiasis show few or no early symptoms, significant morbidity can develop if untreated. Current statistics suggest that the burden due to schistosomiasis may even be comparable to that of malaria (Hotez et al. 2009). Globally, nearly 240 million people are infected with schistosomiasis, while 700 million people are at risk of acquiring the infection (WHO 2023a). Thus, using estimates by the WHO (2002a, 2023a), it can be said that currently, of those with infection, at least 139 million manifest symptoms and signs of the disease, while close to 23 million experience long-term infection-related complications. The statistics are staggering as it shows that up to 90% or more of the global total estimate of people requiring treatment for schistosomiasis live in Africa (WHO 2023b).

From mathematical modelling, it was predicted that the two schistosome worms commonly found in SSA may contribute to the deaths of more than a quarter of a million people annually from disease-specific or infection-related complications (van der Werf et al. 2003). However, given that it is uncommon for schistosomiasis to directly result in the death of the infected person, mortality estimates provide only little insight about the public health impact of the disease. Measures such as disability-adjusted life years (DALYs) and years lived with disability (YLDs) are considered to provide a better picture of the disease burden of schistosomiasis.

FGS has also been connected with increased risk of HIV infection in women and girls. This is as the result of the production of genital mucosal lesions surrounding the eggs (Downs et al. 2017; Kjetland et al. 2012). The concurrent infection of *S. haematobium* and HIV may lead to increased disease pathology. HIV infection may lead to an increased chance of contracting schistosomiasis (Aula et al. 2021). Schistosomiasis infection is suspected to cause increasing disease advancement and death in persons infected with HIV through increasing the HIV RNA load in the blood plasma (Mbabazi et al. 2011; Wall et al. 2018).

Approximately 71 million people experience hematuria, and annually, nearly 18 million infected persons suffer from urinary bladder pathology (WHO 2020). Kidney failure related deaths due to *S. haematobium* total nearly 150,000 yearly (van der Werf et al. 2003; WHO 2020). There is an estimated 54 million individuals infected with *S. mansoni* with around 4 million people experiencing diarrhea and 8.5 million hepatomegaly (WHO 2020), with deaths resulting from hematemesis estimated to be around 130,000 per annum (van der Werf et al. 2003; WHO 2020).

Bladder cancer in males is one of the leading serious complications of prolonged *S. haematobium* infection. It is estimated that 3–4 cases per 100,000 bladder cancer infections are associated with *S. haematobium* infection (Shiff et al. 2006). Also, 275,000 people globally are diagnosed with bladder cancer annually and 108,000 people die of the disease. Studies have shown that bladder cancer caused by squamous cell carcinoma which occurs in schistosomiasis-endemic African countries is as a result of untreated chronic MGS (Abol-Enein 2008; Heyns and van der Merwe 2008).

Disability-Adjusted Life Years (DALYs) and Years Lived with Disability (YLDs)

In 2019, at the global level, the DALYs and YLDs associated with schistosomiasis were estimated to be 1.64 million and 1.23 million, respectively, with at least 80% of the burden concentrated in SSA (Global Burden of Disease Collaborative Network [GBD] 2019). Nonetheless, there is the opinion that the GBD figures are underestimated, because, for instance, they do not include most acute but usual clinical outcomes of the disease (King and Galvani 2018). Additionally, underestimation could arise given the difficulty in establishing and quantifying cause-specific deaths due to schistosomiasis. For example, exposure to the infection usually takes place in childhood, but severe-chronic disease occurs later in adulthood making it difficult to establish the link between infection and long-term complications. Another reason for the underestimation of the burden of the disease is a limitation inherent in the DALYs approach itself, as the DALYs systematically undervalue chronic diseases, including schistosomiasis (King and Bertino 2008; Utzinger et al. 2009). Researchers and public health practitioners, therefore, disagree on the current morbidity and mortality assessment methods used for the estimation of burden attributable to schistosomiasis.

Subtle Morbidity and Other Schistosomiasis-Related Complications

Heavy infection in schoolchildren may be associated with short-term memory loss, slower reaction time, lower scores in some tests of cognitive ability (eye-hand coordination, language and communication, personal-social-emotional, foundation of learning, and overall general development domains) and poor growth (Kasambala et al. 2022; WHO 2002b). Urinary schistosomiasis has been found to be associated with end-stage renal failure and hypertension (Antwi et al. 2014). Iron deficiency anemia and other nutritional deficiencies have also been linked with heavy infection (Awasthi et al. 2003). The social and economic implications of schistosomiasis are thought to be even greater (Wright 1972; WHO 2012). In Burkina Faso, Brazil, Egypt, Sudan, and Tanzania, researchers found substantial reduction in labor productivity of working adults and high school absenteeism of children suffering from schistosomiasis. For example, in Burkina Faso, subsistence farmers who belonged to the poorest households (relative to those from wealthier households) have been found to be heavily affected by schistosomiasis, resulting in 32–45% average loss in yield (Rinaldo et al. 2021). A drop in labor output by as much as 35% of adults infected with schistosomiasis has also been reported in Egypt (Salehe and Hassan 2012). Similar results have been reported from studies investigating adults working in sugar plantations in Tanzania and Sudan that showed a decrease in work output by up to 15% in workers infected with schistosomiasis (Salehe and Hassan 2012). Further studies from Tanzania showed that apart from spending more time in seeking treatment and hence having fewer working hours per month, irrigation rice farmers suffering from schistosomiasis also spent much more money from their harvest on seeking health care (Salehe and Hassan 2012). A group of researchers from Nigeria found similar findings with reduced worker productivity, cash income, rates of land clearing, and farm size of those with the disease (Umeh et al. 2004). Generally, individuals suffering from schistosomiasis in rural areas show severely impaired work capacity and performance due to lethargy (WHO 2002b).

For schistosomiasis-related complications, evidence is also emerging from recent research that urogenital schistosomiasis among sexually active women is associated with a threefold increased risk of contracting HIV (Brodish and Singh 2016). Co-infection of schistosomiasis and hepatitis and HIV or malaria can increase the risk of hepatocellular carcinoma and risk of mortality. Genital lesions resulting from the infection may present with vaginal bleeding, pain during sexual intercourse, and nodules in the vulva. In men, urogenital schistosomiasis can induce pathology of the seminal vesicles, prostate and other organs that may decrease male vitality and fertility.

Economic Implications of Schistosomiasis

In the early 1970s, some staggering results from the analyses of the economic impact of schistosomiasis in terms of disability and loss of work capacity were published in the Bulletin of the WHO (Wright 1972). The main objective of the analysis was to quantify resource loss attributable to reduced productivity. The analysis estimated the total global annual loss to be greater than 600 million US dollars (US$641, 790, 130) excluding the cost of public health programs, medical care, or compensation for illness (Wright 1972). Annual loss from complete and partial disability due to schistosomiasis in Africa alone was estimated to be nearly half a billion US dollars (US$445, 866, 945), constituting approximately 70% of the global cost. This value excluded Mauritius with an estimated cost of US$755, 480. Southeast Asia accounted for US$16, 527, and 275; Southwest Asia US$118, 143, and 675; and the Americas US$60, 496, and 755. These costs estimates were based on an estimated 124, 905, and 800 prevalence of the infection in those days. The authors acknowledged that the data were based on prevalence estimates obtained from less sensitive diagnosis of a single stool (for intestinal schistosomiasis) or single urine (for urogenital schistosomiasis) examinations which could lead to serious underestimation of the actual number of cases. With current estimates of nearly 240 million people having the infection, far higher than estimates used in this calculation (around 124 million), it is expected that the cost will be far higher. Recently, in Burkina Faso, it has been found that schistosomiasis, that remains untreated, is associated with 0.8% loss in gross domestic product (GDP) (Rinaldo et al. 2021). Among Asian and African countries, the cost of interventions to control schistosomiasis has been found to range from US$0.06 to US$4.46 per person (Salari et al. 2020). The implication is that if the cost of treating schistosomiasis is to be born by individuals or their households, there is a higher likelihood of catastrophic health expenditure since most schistosomiasis-affected individuals are poor.

Diagnosis of the Infection and Disease

Parasitological Diagnosis

Microscopic examination for parasite eggs in the urine (for urogenital schistosomiasis) or stool (for intestinal schistosomiasis) is considered definitive for the diagnosis of the infection. For urogenital schistosomiasis, the diagnostic technique involves a nucleopore membrane of pore size 12 μm secured in a Swinnex chamber through which a standard 10 ml volume of urine is passed to trap the eggs if present in the urine. This method helps quantify egg count and determine the intensity of infection which is very useful in epidemiological investigations. For intestinal schistosomiasis, the Kato-Katz technique (Katz et al. 1972), which remains the gold standard, is used and requires the examination of a single or multiple stool specimens. The Kato-Katz technique is both specific and sensitive and able to quantify eggs into different intensity levels. It also has low operational cost and can be used in settings with minimal infrastructure (Rabello 1992). The sensitivity of parasitological diagnosis that requires microscopy decreases considerably when egg excretion is low, particularly in low endemicity areas or after chemotherapy.

Rectal and bladder biopsy for all schistosome species is more sensitive and occasionally done when repeated stool or urine examinations are negative for schistosome eggs (Rabello 1992). Rectal biopsies are particularly useful when conducting clinical trials of new treatments when it is necessary to observe egg morphology in tissues. The disadvantage is that biopsies are invasive and cumbersome to perform and not recommended for mass field application (Allan 2001).

Immunological Diagnosis

The underestimation of true prevalence of infection resulting from less sensitive parasitological diagnostic tools led to the development of immunodiagnostic techniques, which are more sensitive and simpler to perform. Monoclonal antibody-based technique to detect schistosome specific by-products is gaining prominence in the diagnosis of the infection (Bosompem et al. 1996, 2004a; Polman et al. 2001, 2002). Reagent strips are applied for the detection of schistosomiasis—related blood and proteins released in the urine, but this technique is unable to quantify the infection. Studies on monoclonal antibody-based dipsticks have shown promising results (Legesse and Erko 2007, 2008; Coulibaly et al. 2011). However, in immunodiagnostic techniques that involve the detection of the antibodies, there is a lack of specificity and it is impossible to accurately evaluate the real impact of chemotherapy because it is associated with high positive rates for detection of schistosomal antibodies over a long period of time (Balahbib et al. 2017). The WHO recommended the use of these methods in low prevalence areas as well as in evaluating and monitoring control program efficiency where disease elimination is the main goal (Adriko et al. 2014). More recently, the FLOTAC has evolved as a novel technique for the detection and quantification of *S. mansoni* eggs in stools (Glinz et al. 2010; Cringoli et al. 2013).

Polymerase Chain Reaction (PCR)

Polymerase chain reaction (PCR) is increasingly being used in some reference laboratories in Europe, and it is hoped that this can be assessed for evidence of test accuracy and cost-effectiveness for application in endemic settings (Sandoval et al. 2006; Cnops et al. 2012; Enk et al. 2012).

Clinical Diagnosis

Clinically, schistosomiasis is diagnosed based on blood in the urine of the patient (for *S. haematobium* infection) and blood in stool, diarrhea, and abdominal pain (for *S. mansoni* infection). Diagnosis based on presence of blood in urine is reliable in children but less so in adults where blood in the urine may be caused by other diseases than urogenital schistosomiasis, for example, sexually transmitted diseases. For intestinal schistosomiasis, diarrhea, abdominal pain, and splenohepatomegaly are nonspecific and could be due to other causes such as malaria. Also, blood in stool can be caused by dysentery but not necessarily schistosomiasis.

Mass Screening

For mass or field control, questionnaires have been suggested as operationally cost-effective for rapid screening of the infection in high-endemic communities. Questionnaire-based assessment relies on self-reporting of symptoms; it is not reliable as recall bias can be high. For urogenital schistosomiasis, blood in urine and painful urination show moderate-to-high sensitivity, depending on age and prevalence of the infection (Bogoch et al. 2012). In high-prevalence areas, sensitivity and specificity tend to be high, whereas in areas with low prevalence of the infection, sensitivity and specificity of this method are low. For intestinal schistosomiasis, the main signs and symptoms (blood in stool, diarrhea, and abdominal pain) are neither sensitive nor specific (Danso-Appiah et al. 2004, 2010), and questionnaire-based diagnosis is less encouraging (Raso et al. 2005). Another major challenge is that questionnaire application is influenced by age and gender of the infected person.

Control Strategies and Challenges of Implementation

The goal of schistosomiasis control, like any infectious disease, is to prevent new infections or transmission of the infection, usually by interrupting the parasites' life cycle. As this has been difficult to achieve, control guidelines have changed several times over the years. Historically, schistosomiasis control measures implemented across SSA have centered around two main strategies, namely, (1) transmission control and (2) morbidity control, using one or more of the following measures: health education, safe water supply, molluscicing, environmental management, chemotherapy, or combination of measures.

Transmission Control

Molluscicing

The application of molluscide formed the primary control option prior to the introduction of safe and effective antischistosomal drugs in the 1970s (WHO 1985; Sturrock 2001). The objective was to interrupt the transmission of infection by targeting the intermediate snail host. Mass application of molluscide has been conducted in various settings in SSA, particularly where outbreaks of schistosomiasis have occurred because of the creation of dams (Shiff et al. 1973; Chu 1978). In Ghana, for example, control measures adopted following the outbreaks of schistosomiasis after the construction of the Akosombo Dam focused primarily on chemical (molluscide) control targeting the intermediate host snails. Generally, large-scale molluscide control proved to be very expensive and impractical in the long term and also raised several operational and health concerns (Chu 1978; Sturrock 2001). Nevertheless, with the control programs moving from control to elimination of schistosomiasis, the WHO recommends integration of PC and molluscidng activities (WHO 2022).

Environmental Management

Environmental management attempts to reduce the snail population and thereby transmission of the infection. It has generally not been used as one of the approaches undertaken in SSA due to the high cost and paucity of information about the exact areas where this can be implemented. Positive experiences with this have been mainly documented in Morocco and Egypt, outside SSA. There is the need for the identification of areas where environmental management for snail control can be intensified in the SSA to improve effort toward control of the infection.

Health Education

Health education focuses on promoting good hygiene and sanitation, especially among school-age children and caregivers, discouraging practices such as bathing in streams and indiscriminate disposal of human waste that underpin spread of the infection. The main objective of this strategy is to decrease the number of eggs reaching and contaminating water bodies thereby reducing transmission and acquisition of the infection in endemic communities. The impact of most health education programs appears to be encouraging in the short term (Asaolu and Ofoezie 2003; Mascie-Taylor et al. 2003; Guo et al. 2005; Minamoto et al. 2012) but are unsustainable in the long run (Minamoto et al. 2012). A systematic review that pooled data to evaluate the effect of health education showed that health education helps to sustain the benefits of other control measures such as chemotherapy (Asaolu and Ofoezie 2003). The review also found that health education is effective, easy, and cheap to deliver and will remain the only tool for creating enabling environment for both chemotherapy and sanitation to succeed.

Safe Water Supply

The goal of safe water supply is to reduce the frequency of water contact by poor rural inhabitants. However, in some settings in SSA, water contact may not be avoidable as this serves as a source of occupation (fishing) and is also the only means rural inhabitants can conduct their domestic activities such as washing clothing or fetching water for drinking, as well as children swimming in streams infested with the intermediate host snails responsible for transmission of the infection. A study examining the impact of improved water supply and sanitation facilities on schistosomiasis and other NTDs by analyzing 144 independent studies and calculating disease-specific median reduction levels of the infection found marked reductions in diarrhea (26%), trachoma (27%), and ascariasis (29%) (Esrey et al. 1991). The median reduction of new infection with schistosomiasis was 77% and dracunculiasis was 78%, with overall child mortality decreasing by 55% (Esrey et al. 1991). The authors concluded that properly maintained water supply alone was highly effective for controlling NTDs. In investigating the value of individual household water supplies in curtailing the transmission of schistosomiasis in Saint Lucia, households in five villages were provided with their own water supply, as well as community laundry shower units between 1970 and 1975 (Jordan et al. 1982). Evaluation of the impact of water supply showed a marked reduction in the incidence of *S. mansoni* infections. Chemotherapy was introduced thereafter for 3 years (1975–1977) with oxamniquine to all the infected persons. The results showed a further reduction in the transmission of the infection to a very low level which remained low for over 4 years, with no sign of an increase in infection rates despite reservoirs of the infection and poor level of sanitation in the villages remaining unchanged after the treatment (Jordan et al. 1982). The authors concluded that properly maintained water supply alone appeared to be effective in keeping transmission of infection low and that effectiveness improved further when safe water supply was combined with chemotherapy.

Drug Treatment

Chemotherapy focuses on the human host and aims at killing the worms residing in the human body, thereby reducing the worm burden and infection-related pathology and morbidity. Various medicines have been tried and abandoned because of poor effect or adverse events. Egypt was the first country in Africa to use the antimonies, one of the earliest antischistosomal medicines developed after the First World War, around 1918. However, the low compliance associated with the long course of administration, pain from the injections, and the serious side effects meant that the antimonies were limited to only hospital patients (Shekhar 1991; Cioli et al. 1995; Wu and Halim 2000). Treatment-associated death rates were very high, estimated at 3 per 1000 patients treated (Jordan 2000; Sturrock 2001). Ghana also tried the antimonies in the 1960s when schistosomiasis became a big public health problem following the construction of the Akosombo Dam across the Volta River, but discontinued its use few years later because of serious adverse events (Wolfe et al. 1966).

The 1970s marked the turning point in schistosomiasis control when safe and effective medicines were introduced onto the market shifting control emphasis to chemotherapy of the human host (Cioli et al. 1995). Praziquantel (PZQ), a broad-spectrum, safe and easy-to-administer medicine immediately became the treatment of choice for schistosomiasis. The WHO endorsed selective treatment (i.e., identification of high-prevalence communities, screening the people, and treating those identified to have the infection) as a more cost-effective option (WHO 1985). PZQ was delivered through small-to-medium or large-scale selective programs across endemic countries in SSA, funded by foreign donor organizations. The objective was to reduce or interrupt transmission of the infection; morbidity was a secondary objective.

Selective treatment programs were implemented mainly as "vertical" strategies where control decisions were mainly the responsibility of the donor agencies and expatriate experts. The outcome of this approach showed short-term promise in reducing prevalence of the infection. It was expected that any initial gains would lead to a takeover of the maintenance phase by national health authorities (Bruun et al. 2008), but transmission and reinfection rates usually remained unchanged in most treated areas. Lack of sustainability after withdrawal of donor funds and the expatriates at the end of a project was a major problem (Chitsulo et al. 2000; Engels et al. 2002). The high cost of PZQ at the time before expiry of its patent made universal coverage impossible.

Morbidity Control

Realizing the minimal impact and lack of sustainability of selective treatment, the WHO issued a guideline endorsing morbidity control as a new and cost-effective approach (WHO 1985). This approach was to limit coverage to symptomatic individuals or those with heavy work load and likely to suffer complications from the infection. Morbidity control was delivered mainly through "vertical" strategies. After about 10 years of its adoption, the WHO Expert Committee reviewed the strategy in the early 1990s and agreed that morbidity control was feasible and effective (WHO 1993). However, the Expert Committee identified that a leading role of the existing health service was necessary for its success and sustainability. It was acknowledged that if the patients themselves recognize signs and symptoms of the disease and self-report to a clinic or hospital ("passive case finding") for an appropriate treatment, most severe infection-related pathology and complications may be averted. Passive case finding by local clinics looks promising because it is based on the primary healthcare concept already embraced by many endemic countries in Africa (Danso-Appiah et al. 2004, 2010).

Other potential success signs of passive case finding are that (1) control within the regular health service will strengthen the local health services including material and human resource; (2) related activities can be easily combined in an integrated concept, for example, the control of schistosomiasis can be combined with other helminthiases of public health relevance such as soil-transmitted helminths (STHs) or malaria as these diseases tend to coexist; and (3) many parts of Africa where the highest burden of the disease is concentrated did not benefit from the donor-supported vertical control programs (WHO 1998).

The morbidity control strategy was reviewed in 1998 by program managers, scientists, and public health experts from various endemic countries who reset the agenda for future control activities (WHO 1998). The Expert Committee recognized the leading role of the WHO in developing control strategies since its establishment after the Second World War but acknowledged that in SSA only a few countries had ongoing schistosomiasis control programs as at 1990 (WHO 1998). It noted that PZQ was often not available at the peripheral health facilities where it is needed most. At the same time, major challenges were identified: (1) how to make PZQ widely available to communities where the medicine is needed most at an affordable cost; (2) how to organize and finance the distribution of PZQ to those who need it in a sustainable way; and (3) how to adjust control measures to the varying distribution pattern and public health relevance of schistosomiasis in different communities.

It was also felt that drug supply from a strong centralized purchasing system with distribution through the existing health services according to local needs would facilitate decentralized planning and implementation of control strategies (WHO 1998). However, paucity of information on whether availability of PZQ alone in health facilities would influence rate of hospital uptake was a worry to policy makers. Therefore, the Expert Committee recommended the need to build local capacity and strengthen existing health systems, with main emphasis on integration of control and decentralized decision-making and service delivery. Schistosomiasis is a focal problem and often its importance is diluted when control activities are planned and managed at the national centralized level.Schistosomiasis was not acknowledged as a serious problem in preschool children until recently. Therefore, children under 5 years have been excluded from mass population treatment programs (Stothard and Gabrielli 2007; WHO 2012). A major reason for their exclusion is that there is limited information about the safety of PZQ in this age group. Also, the dose pole used for determining the dose of PZQ, recommended by the WHO and used in these control programs, only works for those over 94 cm in height. There are also no readily available syrup or pediatric formulations of PZQ.

A WHO-sponsored meeting that reviewed the results of studies on the treatment of schistosomiasis in preschool-age children in several countries in Africa (WHO 2010b) confirmed results from earlier publications (Bosompem et al. 2004b; Odogwu et al. 2006; Johansen et al. 2007; Stothard and Gabrielli 2007) that (1) schistosomiasis can represent a significant public health problem in children aged less than 5 years; (2) in preschool-age children living in areas where endemicity for schistosomiasis is high, levels of infection are comparable with those of school-age children; (3) prevalence of the infection may exceed 50%, and pathological lesions may be detectable by ultrasonography; and (4) administration of PZQ to preschool-age children is acceptable, safe, and efficacious in the context of individual case management and in group settings, such as communicated in the studies reported (WHO 2012).

Based on the above considerations, a series of recommendations were formulated. Among these, it was recommended that preschool-age children should be regarded as a high-risk group in areas endemic for schistosomiasis and that the WHO should consider formally recommending the use of PZQ in preschool-age children in areas where schistosomiasis is endemic. However, the studies reviewed in this report had some limitations in design and methodology. Recognizing the existence of knowledge, operational, regulatory, and evidence gaps in existing data, it was re-emphasized that (1) treatment of this age group should be within an appropriate healthcare setting (WHO 2012), (2) systematic review(s) involving preschool-age children, or well-designed multicountry randomized controlled trials (RCTs) using a standard protocol are warranted to contribute further evidence, and (3) a pharmacokinetic study investigating PZQ is warranted (WHO 2011). The case for health facility-based passive case reporting is strengthened given that this is the only way for preschoolers to benefit from treatment of their infection.

Chemotherapy

At present, chemotherapy is the cornerstone of schistosomiasis control and applied in endemic countries in SSA. PZQ remains the drug of choice for control and treatment programs of schistosomiasis. Pressure on this drug is growing, and it is expected that this will increase exponentially given the endorsement of the implementation of mass drug administration (MDA) in endemic countries across SSA. The risk of PZQ resistance remains a big threat and its prevention requires adequate and continued monitoring. A well-designed systematic review and meta-analysis of clinical and laboratory evidence are highly warranted to inform on the drug's performance, dynamics, and resistance.

Metrifonate

Metrifonate is an effective antischistosomal drug introduced around 1952, initially as an insecticide (Lorenz et al. 1955) before it was approved for use in humans in the 1960s (Snellen 1981). It is one of only a few treatments where clinical trials were initiated before detailed testing in animals were concluded (Aldridge and Holmstedt 1981). It has been used extensively given at a dose of 7.5–10 mg/kg three times at 14-day intervals (Feldmeier and Doehring 1987). The only limitation is that it is active against only *S. haematobium*. Adverse effects are mainly associated with cholinergic stimulation and include fatigue, muscular weakness, tremor, sweating, salivation, fainting, abdominal colic, diarrhea, nausea, vomiting, and bronchospasm. Its use was limited after the publication that it was inferior clinically, economically, and operationally to PZQ (Feldmeier and Chitsulo 1999). The chemistry, pharmacological properties, and therapeutic efficacy as well as adverse events in clinical human use have been discussed in detail elsewhere (Cioli 2000; Utzinger and Keiser 2004).

Oxamniquine

Oxamniquine belongs to the same group of drugs as hycanthone with activity against only *S. mansoni*. It was introduced onto the market for clinical use in the early 1970s and has since been used widely with good effect, particularly in Brazil where it has been a major component of the national schistosomiasis control program (NSCP). It acts by dislodging the worms from the mesenteric veins to the liver and destroying the male worms. Although oxamniquine allows the female worms to return to the mesentery, they are made sterile and are no longer able to produce eggs. In South America (mainly Brazil) and West Africa, a single lower dose of 15 mg/kg is given to adults and 20 mg/kg to children. In other countries of Africa and the Arabian Peninsula, higher doses are given, the total dose varying from 30 to 60 mg/kg (Foster 1987). Differences in the susceptibility of parasites to the drug seem to account for the variation in dosage. Adverse events are mild and transient and include dizziness with or without drowsiness, headache, and gastrointestinal effects such as nausea, vomiting, and diarrhea. Few serious events such as seizures have been reported, but all were associated with higher doses. Oxamniquine's use has declined considerably in favor of PZQ, even in Brazil, where it has long been preferred to PZQ.

Praziquantel

Praziquantel is a broad-spectrum antischistosomal drug effective against all schistosome worms. PZQ is distributed mainly as 600 mg tablet, with two or three grooves for ease of breaking. In China, it is distributed as 200 mg pills, while some manufacturers provide syrup formulation containing 600 mg/5 ml (e.g., epiquentel from EIPICO). The shelf life of PZQ is usually 4 years in temperate climates and 3 years in hot humid environments. When administered orally, the drug is rapidly and almost completely absorbed, appearing in the blood within 15 min and attaining peak concentration 1–2 h after treatment (Valencia et al. 1994). Approximately 80% of the drug is cleared within 24 h of treatment mainly through urine (Cioli et al. 1995). There are many brands and generic forms in the system including Distocide manufactured by Shin Poong in Korea, Bilharzid from Egypt, and Prazitel from Kenya (Cioli and Pica-Mattoccia 2003). There are other brands produced in the Netherlands and Malta. PZQ is administered orally at a standard single dose of 40 mg/kg body weight and well tolerated. Most common adverse events are mild and transient, mostly related to the gastrointestinal tract: abdominal pain, nausea, vomiting, anorexia, and diarrhea. Currently, it is the only drug on the WHO Model List of Essential Medicines for treating schistosomiasis. Although its effectiveness is unequivocal, PZQ is less effective against immature schistosome worms (Sabah et al. 1986). Therefore, it has been suggested that a combination therapy with drugs of unrelated mechanisms of action and targeting the different developmental stages of the schistosomes may improve therapeutic efficacy and slow the development of resistance. Drugs with potential for combination with PZQ include the artemisinin derivatives, metrifonate, and oxamniquine.

Other Medicines

Several drugs have been tried for the treatment of schistosomiasis and later abandoned because of poor effect or adverse events. These include the *antimonials*, *niridazole*, *lucanthone*, *hycanthone*, *oltipraz*, *cyclosporin A*, and *levamisole*, extensively reviewed by Cioli et al. (1995). More recently, the efficacy of *myrrh* (*Mirazid*) in the treatment of intestinal schistosomiasis was tested in Egypt at a single daily dose of 300 mg for 3 days, and virtually no activity was observed as almost all the treated participants failed treatment at 3–6 weeks (Barakat et al. 2005; Botros et al. 2005).

Status of Medicines Used for Treating Schistosomiasis

In the late 1990s, amidst the argument that metrifonate acts against only *S. haematobium* coupled with clinical, economic, and operational suitability when compared with PZQ (Feldmeier and Chitsulo 1999), this medicine was withdrawn from the WHO Model List of Essential Medicines. The application of oxamniquine (which acts against only *S. mansoni*) has declined considerably, and in areas where it was formally used as the drug of choice, it is being replaced with PZQ (Cioli 2000; Beck et al. 2001; Reich and Fenwick 2001; Utzinger and Keiser 2004; WHO 2012). Oxamniquine has subsequently been removed from the WHO Model List of Essential Medicines.

At present, there is no effective and acceptable antischistosomal vaccine (Utzinger et al. 2007; Bergquist et al. 2008), and evidence on the artemisinins are inconclusive (Utzinger et al. 2007; Utzinger et al. 2010; Liu et al. 2011; del Villar et al. 2012; Wikman-Jorgensen et al. 2012; Danso-Appiah et al. 2013). Therefore, PZQ remains the only drug available for the treatment of millions of persons suffering from schistosomiasis. At the same time, PZQ use in SSA has increased enormously following the endorsement of Resolution 54.19 at the World Health Assembly (WHA) in the early millennium. Endemic countries in SSA have been encouraged to adopt MDA as their national control strategy. PZQ distribution is expected to increase exponentially, raising fears about potential resistance emerging against this medicine. Given the aforementioned, it is critical to find and appraise available data for evidence on resistance against PZQ.

Resistance to Praziquantel

In the early 1990s, PZQ treatment in Senegal (West Africa) resulted in unacceptably low cure rates, raising concerns about possible emergence of resistance against PZQ (Gryseels et al. 1994; Stelma et al. 1995). Meanwhile, analysis of *S. mansoni* strains obtained from patients treated with PZQ but not cured in the areas where low cure rates had been registered in Senegal, as well as in Egypt, yielded isolates less susceptible to higher doses of the drug. Around the same time, laboratory experiments in mice yielded isolates of *S. mansoni* strains from Senegal that were less sensitive to PZQ (Fallon et al. 1995). The development of resistance against PZQ was cited as one of the main reasons for the observed low drug performance. Other reasons given were the very high pretreatment intensity of infection observed in Senegal, the presence of immature worms less susceptible to PZQ, and the rapid acquisition of a large number of new infections immediately following treatment. This prompted the European Commission to set up the European Commission-Concerted Action on PZQ in the late 1990s (Renganathan and Cioli 1998) to review reports of low efficacy of trials from Senegal and Egypt. However, it was difficult to distinguish between normal performance and failure of PZQ as it was not possible to analyze data quantitatively. Notwithstanding, the EC-Concerted Action on PZQ advocated continuous monitoring and evaluation of resistance to PZQ.

To ascertain the availability of evidence on clinical resistance against PZQ, various electronic databases and sources were searched for systematic reviews and meta-analysis. The search retried only one meta-analysis of observational studies that has assessed this question quantitatively (Danso-Appiah and De Vlas 2002). The meta-analysis used robust Bayesian approach of evidence synthesis, accounting for various risks of biases and confounding effects. Although no formal quality assessment of the included studies was conducted, rigorous standardization measures were used to ensure comparability across studies and that biases and errors were minimized.The results showed that the low cure rates observed in Senegal was largely the result of high pretreatment intensity of infection in persons treated there. The relatively sensitive diagnostic technique used in the studies from Senegal to some extent contributed to the observed low cure rates.

However, after correcting for pretreatment intensity and sensitivity of diagnosis, cure rates from Senegal remained somewhat lower than expected (Fig. 5). Various studies on biological and methodological aspects were reviewed to help explain the Senegal situation (Geerts and Gryseels 2001; Gryseels et al. 2001; Cioli 2000; Doenhoff 1998).

The meta-analysis concluded that although high pretreatment intensity of infection, sensitive diagnostic methods and presence of high baseline numbers of immature worms in treated patients explained most of the low performance of PZQ in Senegal, the suspicion about resistance could not be ruled out completely. This is because tools and methods for assessing resistance were not adequate or unavailable.

Combination Therapy

As far as it is known, no new medicines are being developed although some scientists suggest that the combination of PZQ and metrifonate or PZQ and oxamniquine would be potentially cost-effective treatment. The rationale is that PZQ and metrifonate or PZQ and oxamniquine are independently effective against *S. haematobium* and *S. mansoni* infections that their targets of action in the parasite are not linked. Therefore, combination may improve therapeutic efficacy and slow or prevent the development of drug resistance. However, both oxamniquine and metrifonate are active against single schistosome species, while in most endemic settings in SSA, there is an overlap between *S. haematobium* and *S. mansoni* (Anyan et al. 2019). As it is operationally less suitable within the MDA PC strategy, pharmaceutical companies may be less motivated to invest in the development of these combination therapies.

Amoscanate

Another potentially useful drug is amoscanate, a broad-spectrum anthelminthic medicine with activity against all the major schistosome species (Striebel 1976). It was discovered before the 1970s and showed effect against some systemic parasites

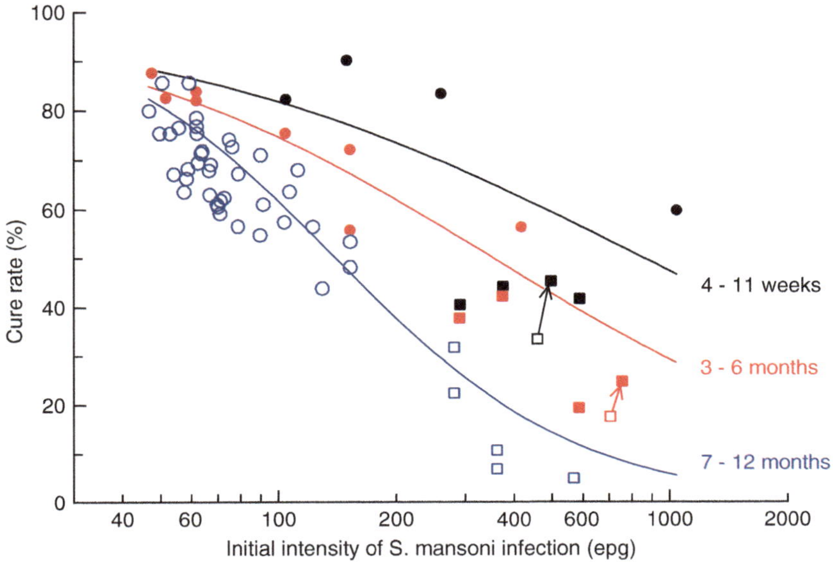

Fig. 5 The impact of pretreatment intensity of *Schistosoma mansoni* infection, diagnostic sensitivity, and follow-up time on cure rate using the standard PZQ dose of 40 mg/kg body weight in population-based, selective chemotherapy. Intensity of infection (logarithmic scale) is the geometric mean of egg count per gram of feces (epg) for individuals positive for *S. mansoni* eggs. Cure rate denotes percentage of treated subjects who appeared negative for *S. mansoni* eggs at follow-up. The dots represent different follow-up time categories: 4–11 weeks; 3–6 months; 7–12 months. Data points located on same vertical axis indicate successive follow-ups of a treatment in the same community. *Squares* represent observations from Senegal; *circles*, studies from other locations. *Arrows* illustrate shifts from published observations from Senegal based on duplicate stool specimens (*open squares*) to the corresponding cure rate and intensity from a single stool specimen. For the three follow-up time categories, curves give best fitting associations of *cure rate* and *ln epg* through cure rate = $100/(1 + e^{-Lp})$, where $Lp = \alpha + f3 \ln epg$.. α and f3 coefficient values are 4.7 and −0.70 (4–11 weeks follow-up); 5.0 and −0.85 (3–6 months); 6.9 and −1.40 (7–12 months). Coefficients α and f3 are estimated by generalized linear mixed modelling (GLMM). Out of the 13 data points from Senegal, 12 are located below the respective lines. Thus, even when initial intensity, follow-up time and sensitivity of diagnosis are accounted for, cure rates from Senegal are consistently lower than expected (Reproduced from Danso-Appiah and De Vlas (2002), *Trends in Parasitology*)

including filariae and gastrointestinal nematodes such as hookworms. It was tested extensively in China from locally produced brand called "nithiocyaminum" (Bueding et al. 1976; Striebel 1976). Toxicity was low in animal models, but mutagenicity tests in bacteria gave positive results. Also, mutagenic metabolites were constantly found in the urine of mammals administered this treatment (Batzinger and Bueding 1977). Coupled with concerns about liver toxicity, amoscanate was abandoned when PZQ became available (Cioli et al. 1995). Given its broad-spectrum schistosomicidal properties, new and well-designed research that will investigate structural modifications of the chemical structure to decrease liver toxicity is warranted (Cioli et al. 1995).

Artemisinin Derivatives

The artemisinins act against immature schistosomes (schistosomulae) that are less sensitive to PZQ (Le et al. 1982, 1983; Utzinger et al. 2003a, b, 2007); the invasive and adult worms are less susceptible to the artemisinins. Adverse effects are minor and last for less than 24 h. The combination of artemisinin with existing drugs effective against other stages of the parasite such as PZQ may improve efficacy and slow or prevent the development of resistance against PZQ. Therefore, combination therapy with the artemisinins should be pursued vigorously given that current evidence is inconclusive.

Supportive Evidence for Diagnostic Tools and Treatment

To investigate whether application of diagnostic tests and techniques in sub-Saharan Africa is supported by systematic reviews and other forms of evidence synthesis, relevant electronic databases including MEDLINE, LILACS, DARE, and the Cochrane Collaboration Databases were searched in September 2013. The reference lists of published studies were also searched, and experts in the field of schistosomiasis were contacted for additional or unpublished systematic reviews that might have been missed through the search. The search identified only one published systematic review protocol which intends to assess rapid screening and diagnostic tests for schistosomiasis in endemic areas, but data are not yet available (Ochodo et al. 2012) and six published systematic reviews of varying qualities assessed treatment options and effects (Danso-Appiah et al. 2008, 2013; Liu et al. 2011; del Villar et al. 2012; Wikman-Jorgensen et al. 2012; Stothard et al. 2013).

Given that the search did not retrieve any systematic review or carefully synthesized evidence on diagnostic accuracy, it is not possible to establish whether available evidence supports the use of current diagnostic tools and methods used in SSA. The WHO recently commissioned a systematic review and meta-analysis to assess the circulating cathodic antigen (CCA) test in the diagnosis of schistosome infections which may provide valuable information (Danso-Appiah et al. 2014, submitted). The six systematic reviews on treatment options and effects retrieved are of variable quality and thus reliability. Detailed assessment as to whether they support current treatment options and policies in SSA is beyond the remit of this study. Therefore, a new and well-designed study is warranted to examine this.

The Millennium Control Strategies

With the exceptions of Egypt and Morocco, hardly any observable progress was made in the control of schistosomiasis in the African continent up to the 1990s (Utzinger et al. 2009). The reasons for this were attributed to the absence of safe and easy-to-apply medicines prior to the 1970s and high price of available medicines after their introduction onto the market. The expiry of patents of PZQ in the early 1990s allowed the production of brands leading to a sharp fall in the price of this medicine from about US$1 at 1990 to less than UD$0.06 per tablet in the early millennium. PZQ use has since increased considerably. One other major reason for the surge in PZQ distribution may be due to evidence that dispelled fears about the likelihood of resistance against PZQ (Doenhoff 1998; Cioli 2000; Gryseels et al. 2001; Danso-Appiah and De Vlas 2002) together with increased advocacy for control of NTDs (Hotez et al. 2007; Fenwick et al. 2009; Garba et al. 2009; Utzinger et al. 2011).

The adoption of Resolution WHA54.19 in 2001 urged member states to treat at least 75% of schoolchildren infected with schistosomiasis and high-risk individuals or entire "at-risk" populations (e.g., school-age children) through mass distribution of PZQ, usually without prior diagnosis—an approach termed "preventive chemotherapy" (PC). This is the millennium control strategy endorsed by WHO and applied in many endemic countries. Usually, PZQ at a standard single oral dose of 40 mg/kg is used.

In 2002, WHO convened an Expert Committee meeting to develop operational guidelines to translate the recommendations of WHA54.19 into concrete action with emphasis on morbidity control (WHO 2002b). The strategy had a goal of control of morbidity to be achieved through large-scale distribution of PZQ to populations at risk, using defined thresholds of prevalence as criteria for selecting the appropriate interval of retreatment. Schools were identified as the most efficient delivery entry point. On the other hand, if school enrolment is low, community-based interventions are to be implemented to reach out to children not attending school. The adoption of Resolution WHA54.19 has led to thousands of school-age children in countries in SSA to receive multiple doses of PZQ initially through the Schistosomiasis Control Initiative (SCI) which started in 2003 and funded by the Bill and Melinda Gates Foundation (Fenwick et al. 2006, 2009; Linehan et al. 2011) and more recently through the integrated control approach of NTDs supported by the United States Agency for International Development (USAID), the British Department for International Development (DFID), and the Global Network for NTDs and its partners (Hotez 2009, 2011).

New evidence that emerged indicating morbidity is not necessarily determined by the intensity of infection led to a revised version of the operational details of PC in 2006 (WHO 2006a, b). The target population was expanded to include all adults in high-risk areas where prevalence of infection in school-age children is 50%, as well as special-risk groups such as people occupationally exposed to risk of infection. The targets for moderate—and low-risk areas were also revised (Table 3).

With new evidence that preschool-age children in endemic areas may be at a similar risk of infection and morbidity (WHO 2011), the recommendation is that children aged up to 5 years who are excluded from population-based mass chemotherapy should be treated in health facilities as part of national control programs. The 2006 guidelines came out with additional component about coordinated use of anthelminthic medicines to control and eliminate four main helminth infections, namely, lymphatic filariasis, onchocerciasis, schistosomiasis, and soil-transmitted helminths (STHs), which usually coexist and for which the medicines used for their treatment can safely be co-administered (Utzinger and Keiser 2004; Olsen 2007; Hotez 2009).

Table 3 Recommended treatment strategy for schistosomiasis

Category	Baseline prevalence among school-age children	Action to be taken	
High-risk community	50% by parasitological methods (both) or 30% by questionnaire for history of hematuria (for urinary schistosomiasis)	Treat all school-age children (enrolled and not enrolled) once a year	Also treat adults considered to be at risk (from special groups to entire communities living in endemic areas)
Moderate-risk community	10% but <50% by parasitological methods (both), or <30% by questionnaire for history of hematuria (for urinary schistosomiasis)	Treat all school-age children (enrolled and not enrolled) once every 2 years	Also treat adults considered to be at risk (special groups only)
Low-risk community	<10% by parasitological methods (for both urinary and intestinal schistosomiasis)	Treat all school-age children (enrolled and not enrolled) twice during their primary schooling age (e.g., once on entry and once on exit)	Praziquantel should be available in dispensaries and clinics for treatment of suspected cases

Adapted from WHO (2012)

Political Will, Functional Health Service, and Sustainable Control

Most countries that have been successful in schistosomiasis control have used multifaceted approach combining various control measures together with political will. For example, Japan, China, Iran, Laos, Mauritius, Morocco, Saudi Arabia, Tunisia, the endemic Caribbean Islands, Brazil, and Venezuela have been successful in their control effort because they made functional health services key component of the control of schistosomiasis. Also, their governments showed political will and commitment. Therefore, an adequately resourced or functional health service can play a crucial role as control is far more effective when placed in the context of a general health system (WHO 2002b). Although in terms of cost-effectiveness and sustainability the concept which underpins this strategy is sound, the process has been slow in SSA due to lack of material, human resources. and political will. The experiences of schistosomiasis control from Egypt and Uganda have been presented to highlight the importance of multifaceted approach as a way to successful and sustainable control.

Egypt is endemic for both urinary and intestinal schistosomiasis. Implementation of a National Schistosomiasis Control Program started in 1977, through the primary healthcare system. First, pilot projects were implemented using either snail control or chemotherapy with little success. From 1988, PZQ was made available free of charge for all diagnosed cases of schistosomiasis in all governmental health facilities. Then, from 1997 to 2002, mass treatment campaigns without prior diagnosis in communities with a prevalence of infection above 20% were actively embarked on. As the program progressed and the disease prevalence decreased, the threshold for mass chemotherapy was changed. As the infection levels went down, the 1997 threshold of prevalence for mass treatment was reduced from >10% in 1999, 5% in 2000, >3.5% in 2002 to >3% in 2003. Since 2003, mass chemotherapy has been applied only in hotspots (microfocal control).

Available statistics show that before the mass chemotherapy campaigns in 1996 (Fig. 6), there were 168 villages with a prevalence of >30%, 324 villages with a prevalence of 20–30%, and 654 villages with a schistosomiasis prevalence of 10–20%. However, during the program evaluation by the end of 2010, only 20 villages in the whole country had prevalence more than 3–5%, and none had a prevalence of >10%. The prevalence of both *S. mansoni* and *S. haematobium* has consistently decreased under sustained treatment effort supported by political will.

Tracing Egypt's success story, the prevalence of the two species was 38.6% and 35%, respectively, in 1983, reduced to 11.9% and 5% in 1996 and to 2.7% and 1.9% in 2002. Mass chemotherapy began to be targeted on "hotspot" areas in 2003 when the prevalence of *S. mansoni* was 2.6% and that of *S. haematobium* was 1.7%, but by 2010, these were 0.3% and 0.4%, respectively (WHO 2011).The achievements in Egypt were made possible after treating a total of 38.8 million and a further nine million people with PZQ. In parallel to the national program and functional health service as a key component, attention was also focused on snail control, universal access to potable water to almost every household, and significant increase in access to adequate sanitation.

SCI and other partners began the support to Uganda from 2003. Like Egypt, control began with a pilot phase in 2003, in one subregion in each of the 18 most

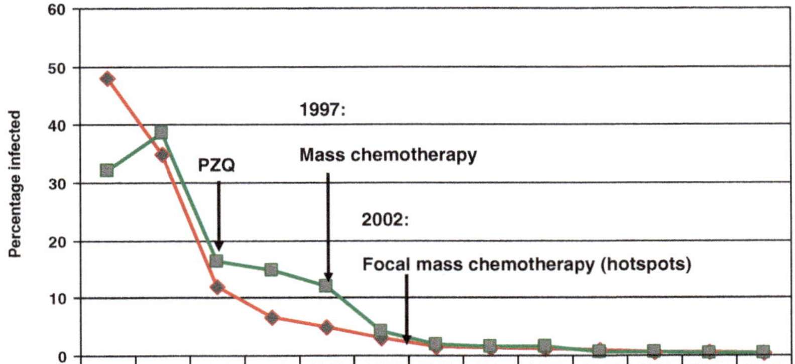

	1935	1983	1988	1993	1996	2000	2004	2005	2006	2007	2008	2009	2010
S.haematobium	48	35	11.9	6.6	5	3	1.6	1.4	1.2	0.9	0.6	0.5	0.4
S.mansoni	32	38.6	16.4	14.8	11.9	4.2	1.9	1.6	1.5	0.6	0.6	0.4	0.3

Fig. 6 The impact of mass chemotherapy with praziquantel on schistosomiasis control in Egypt (Adapted from WHO (2012). Schistosomiasis: progress report 2001–2011 and strategic plan 2012–2020)

affected districts, mainly mass treatment and health education. Activities included planning, advocacy, training of teachers and supervisors, selection and training of community drug distributors, health education, supervision, monitoring of treatment side effects, and reporting of treatment coverage and drug accountability.

During the pilot phase, 400,000 people were treated and this increased to 1.4 million in 2004, and 3 million in 2005. Annual meetings were held regularly at national level to review constraints and plan the way forward. In 2006, two million people were treated in 27 of the 38 endemic districts. In addition, health center-based treatment was conducted alongside mass chemotherapy in 11 of the districts. The main aspects monitored were program performance and the impact of treatment campaigns, severe adverse events, and drug utilization and efficacy. The impact of yearly mass treatment with PZQ on infection status and morbidity of *S. mansoni* infection in school-age children was evident in several ways. Anemia was also reduced. In adults, yearly treatment significantly reduced both the prevalence and the intensity of *S. mansoni* infection between 2003 and 2006, as well as the incidence and severity of fibrosis (WHO 2011).

Similar success stories have been reported from countries selected and supported for mass PZQ distribution as part of the SCI (2003).

Control or Elimination

From a recent survey for NTD disease-specific experts' opinions on the role of MDA for the elimination of lymphatic filariasis, onchocerciasis, schistosomiasis, soil-transmitted helminths (STHs), and trachoma, most experts in lymphatic filariasis, onchocerciasis, and trachoma thought elimination was the appropriate goal of treatment programs with MDAs. For schistosomiasis and STHs, the majority of experts thought that MDA programs were intended for control (Keenan et al. 2013) and that MDA plus other control measures would be required for their elimination.

In a debate in PLoS Medicine titled *Which new approaches to tackling NTDs show promise*? (Spiegel et al. 2010), while some experts held the view that there has been too much focus on the biomedical mechanisms and drug development for NTDs at the expense of attention to the social determinants of disease and that the weak health systems and poor socio-environmental conditions that cause and/or perpetuate NTDs have largely been ignored, others argue that the best return on investment will continue to be MDA for NTDs.

In the same paper, one school of thought suggested that the designation of a set of tropical diseases as "neglected" has indeed provoked interest in strategies for their control and research on new tools for alleviation of their burden. This initiative, however, has also exposed the over-medicalization of contemporary tropical disease control strategies. A primary example is the emphasis on drug administration alone to alleviate the burden of schistosomiasis and STHs, two groups of disease that would benefit more from a combination of MDA and other control methods. Others of contrary views advance the points that in terms of both health impact and cost-effectiveness, few other interventions can rival MDA for NTDs, particularly schistosomiasis. Furthermore, MDA is increasingly being recognized for its beneficial effects on strengthening health systems, improving economic development and achieving the Millennium Development Goals (Spiegel et al. 2010). The successful control of blindness in SSA through free community-directed treatments with ivermectin of the Onchocerciasis Control Program (OCP) and the African Programme for Onchocerciasis Control (APOC) is cited as positive examples. Furthermore, there is considerable epidemiological overlap of NTDs, and in more than 75% of countries in SSA at least six of the seven these diseases coexist in the same community and that there is ample evidence about the cost-effectiveness of simultaneous MDA to tackle these diseases.

Outlook of Control in the Next Decade

Schistosomiasis is widely recognized as a disease that is socially determined, but most programs have ignored the key issues about understanding the social and behavioral factors that underpin transmission and control of the disease. Therefore, social determinants of schistosomiasis should be considered as vital in designing strategies and policies for prevention and control (Bruun et al. 2008). Given the lack of suitable and appropriate water supplies in rural areas, the rural folk have no alternative but to be in constant contact with schistosomiasis infested water bodies for their daily use. There is, therefore, the suggestion that preventive programs could be updated and made relevant to real-life situations in remote rural areas by using insights from social scientists including medical anthropologists, health geographers, sociologists, and others whose skills enable them to explore the social context of schistosomiasis transmission and control at the micro level, in the setting in which the infection is transmitted and where efforts are made to control it (Bruun et al. 2008).

Mass distribution of PZQ to endemic countries in SSA as set out in The WHA Resolution 54.19 has been adopted by the endemic countries. Caution is, however, called for in (1) what is solely large-scale anthelminthic drug administration to the exclusion of preventive strategies for control which include access to clean water, improved sanitation, and hygiene in an integrated and sustainable disease control programs (Utzinger et al. 2003a, b) and (2) the disconnection between engineering projects and community-led initiatives in the developing world on clean water and sanitation and the vertical "drugs only" programs (Singer and Castro 2007; Utzinger et al. 2009).

Utzinger et al. (2009) predict PZQ treatment needs in SSA and estimate that approximately 128 million school-age children would need to be given PZQ. This amounts to 384 million tablets of PZQ every year for treatment in Africa alone. Although PZQ has now been made available free of charge by multilateral collaboration between donor organizations and pharmaceutical companies to poor endemic countries in SSA, issues of supplies, cost, whether free PZQ alone will be enough to achieve the objectives of schistosomiasis control and elimination, and possible drug resistance need to be kept in mind and addressed effectively.

Experiences from schistosomiasis control as, for example, in Burundi and Mali where the German Technical Cooperation (GTZ) funded large-scale vertical control programs with PZQ show that these programs were not sustainable in the long run. Usually, gains in prevalence reductions were offset by rapid reinfection after withdrawal of external funds and technical capacity at the end of the programs given practical and operational problems which resurface in the absence of adequate funding. These operational problems are not unique to Burundi and Mali. They exist in many of the endemic countries presently and will need further attention in the future. These include availability of material, human resource, operational funds and infrastructure, political will to sustain the benefit achieved with chemotherapy, availability of reliable data, and adequate technical capacity for delivery, monitoring, and evaluation of national control programs. Any one of these or a combination of some of them has led to the situation where after nearly 10 years of MDA campaign only 18 countries were yet to start MDAs as at 2010, and just over half (23) of the 42 schistosomiasis endemic countries in SSA had national control programs, mostly of limited scale. Only about five countries had reached all their at-risk populations, but prevalence rates were still 100% in some endemic settings. In some conflict and postconflict areas, e.g., the Democratic Republic of Congo, the actual schistosomiasis problem situation is still unknown. The next decade should see an amelioration of some of these challenges given the current goodwill and enthusiasm of ministries of health, partners, and donors to eliminate several of the NTDs including schistosomiasis.

The current WHO monitoring guideline which has not been specific on how parasitological and morbidity monitoring should be conducted should see updating together with the production of guidelines on how to assess drug sensitivity and timely detection of emergence of resistance.

Very few partners provide financial support for research and development; despite the devastating public health impact of NTDs and their effect on development (Hotez et al. 2009; Manderson et al. 2009; McCoy et al. 2009; Hotez and Pecoul 2010), there is, however, a cautious optimism that such a situation could change in the coming decade (Hotez and Pecoul 2010).

References

Abol-Enein H (2008) Infection: is it a cause of bladder cancer? Scand J Urol Nephrol 79–84

Adriko M, Standley CJ, Tinkitina B, Tukahebwa EM, Fenwick A, Fleming FM et al (2014) Evaluation of circulating cathodic antinging (CCA) urine-cassette assay as a survey tool for *Schistosoma mansoni* in different transmission setting within Bugiri District. Uganda Acta Trop 136:50–57

Aldridge WN, Holmstedt B (1981) Metrifonate and dichlorvos: theoretical and practical aspects History and scope of the conference. Acta Pharmacol Toxicol (Copenh) 49(Suppl 5):3–6

Allan RW (2001) The Johns Hopkins Hospital, Department of pathology. Johns Hopkins Microbiol Newsl 20:30

Aminu MB, Abdullahi K, Dattijo LM (2014) Tubal ectopic gestation associated with genital schistosomiasis: a case report. Afr J Reprod 18:144–146

Anderson RM, May RM (1992) Infectious diseases of humans: dynamics and control. Oxford University Press, Oxford

Andersson KL, Chung RT (2007) Hepatic schistosomiasis: current treatment options. Gastroenterology 10(6):504–512

Antwi S, Aboah KEK, Saprong CKG (2014) The unacknowledged impact of urinary schistosomiasis in children: 5 cases from Kumasi, Ghana. Ghana Med J 48(4):228–233. Available from https://vizhub.healthdata.org/gbd-results/

Anyan WK, Abonie SD et al (2019) Concurrent *Schistosoma mansoni* and *Schistosoma haematobium* infections in a peri-urban community along the Weija dam in Ghana: a wake up call for effective National Control Programme. https://doi.org/10.1016/j.actatropica.2019.105116

Arean VM (1956) Lesions caused by Schistosoma mansoni in the genitourinary tract of men. Am J Clin Pathol 26(9):1010–1021

Asaolu SO, Ofoezie IE (2003) The role of health education and sanitation in the control of helminth infections. Acta Trop 86(2–3):283–294

Aula OP, McManus DP, Jones MK, Gordon CA (2021) Schistosomiasis with a focus on Africa. Trop Med Infect Dis 6(3):109. https://doi.org/10.3390/tropicalmed6030109

Awasthi S, Bundy DAP, Savioli L (2003) Helminthic infections. BMJ 327:431–433

Balahbib A, Amarir F, Corstjens PL, de Dood CJ, van Dam GJ, Hajli A et al (2017) Selecting accurate post-elimination monitoring tools to prevent reemergence of urogenital schistosomiasis in Morocco: a pilot study. Infect Dis Poverty 6:75. https://doi.org/10.1186/s40249-017-0289-z

Barakat R, Elmorshedy H, Fenwick A (2005) Efficacy of myrrh in the treatment of human schistosomiasis mansoni. Am J Trop Med Hyg 73(2):365–367

Batzinger RP, Bueding E (1977) Mutagenic activities in vitro and in vivo of five antischistosomal compounds. J Pharmacol Exp Ther 200(1):1–9

Beck L, Favre TC, Pieri OS, Zani LC, Domas GG, Barbosa CS (2001) Replacing oxamniquine by praziquantel against Schistosoma mansoni infection in a rural community from the sugarcane zone of Northeast Brazil: an epidemiological follow-up. Mem Inst Oswaldo Cruz 96(Suppl):165–167

Ben-Chetrit E, Lachish T, Mørch K, Atias D, Maguire C, Schwartz E (2015) Schistosomiasis in pregnant travelers: a case series. J Travel Med 22:94–98

Bentefouet TL, Thiam M, Gaye AM, ElWardi A, Gueye L, Cisse ML (2017) Case report of tubo-ovarian Bilharziasis presented with pelvic pain and secondary infertility case rep images. Obstet Gynecol 3:19–22

Bergquist R, Utzinger J, McManus DP (2008) Trick or treat: the role of vaccines in integrated schistosomiasis control. PLoS Negl Trop Dis 2(6):e244

Bierman WF, Wetsteyn JC, van Gool T (2005) Presentation and diagnosis of imported schistosomiasis: relevance of eosinophilia, microscopy for ova, and serology. J Travel Med 12(1):9–13

Bogoch II, Andrews JR, Dadzie ERK, Utzinger J (2012) Simple questionnaire and urine reagent strips compared to microscopy for the diagnosis of *Schistosoma haematobium* in a community in northern Ghana. Trop Med Int Health 17(10):1217–1221

Boissier J, Mone H, Mitta G, Bargues MD, Molyneux D, Mas-Coma S (2015) Schistosomiasis reaches Europe. Lancet Infect Dis 15:757–758

Boissier J, Grech-Angelini S, Webster BL, Allienne JF, Huyse T, Mas-Coma S, Toulza E, Barre-Cardi H, Rollinson D, Kincaid-Smith J et al (2016) Outbreak of urogenital schistosomiasis in Corsica (France): an epidemiological case study. Lancet Infect Dis 16:971–979

Bosompem KM, Ayi I, Anyan WK, Nkrumah FK, Kojima S (1996) Limited field evaluation of a rapid monoclonal antibody-based dipstick assay for urinary schistosomiasis. Hybridoma 15:443–447

Bosompem KM, Owusu O, Okanla EO, Kojima S (2004a) Applicability of a monoclonal antibody-based dipstick in diagnosis of urinary schistosomiasis in the central region of Ghana. Trop Med Int Health 9(9):991–996

Bosompem KM, Bentum IA, Otchere J, Anyan WK, Brown CA et al (2004b) Infant schistosomiasis in Ghana: a survey in an irrigation community. Trop Med Int Health 9(8):917–922

Botelho M, Oliveira P, Gomes J, Gartner F, Lopes C, da Costa JMC, Machado JC (2009) Tumourigenic effect of Schistosoma haematobium total antigen in mammalian cells. Int J Exp Pathol 90:448–453

Botros S, Sayed H, El-Dusoki H, Sabry H, Rabie I, El-Ghannam M et al (2005) Efficacy of mirazid in comparison with praziquantel in Egyptian *Schistosoma mansoni* -infected school children and households. Am J Trop Med Hyg 72(2):119–123

Brodish PH, Singh K (2016) Association between Schistosoma haematobium exposure and human immunodeficiency virus infection among females in Mozambique. Am J Trop Med Hyg 94(5):1040

Brouwer KC, Ndhlovu PD, Wagatsuma Y, Munatsi A, Shiff CJ (2003) Epidemiological assessment of Schistosoma haematobium-induced kidney and bladder pathology in rural Zimbabwe. Acta Trop 85(3):339–347

Bruun B, Aagaard-Hansen J, Watts S (2008) The social context of schistosomiasis and its control. An introduction and annotated bibliography. Special Programme for Research & Training in Tropical Diseases (TDR) sponsored by UNICEF/UNDP/World Bank/WHO

Bueding E, Batzinger R, Petterson G (1976) Antischistosomal and some toxicological properties of a nitrodiphenylaminoisothiocyanate (C9333-Go/CGP 4540). Experientia 32(5):604–606

Camara A (1959) Female genital schistosomiasis. Rev Ginecol Obstet (Sao Paulo) 105:271–280

Cheever AW, Kamel IA, Elwi AM, Mosimann JE, Danner R, Sippel JE (1978) *Schistosoma mansoni* and *S. haematobium* infections in Egypt. III. Extrahepatic pathology. Am J Trop Med Hyg 27(1 Pt 1):55–75

Chitsulo L, Engels D, Montresor A, Savioli L (2000) The global status of schistosomiasis and its control. Acta Trop 77:41–51

Christinet V, Lazdins-Helds JK, Stothard JR, Reinhard-Rupp J (2016) Female genital schistosomiasis (FGS): from case reports to a call for concerted action against this neglected gynaecological disease. Int J Parasitol 46:395–404

Chu KY (1978) Trials of ecological and chemical measures for the control of *Schistosoma haematobium* transmission in a Volta Lake village. Bull World Health Organ 56(2):313–322

Cioli D (2000) Praziquantel: is there real resistance and are there alternatives? Curr Opin Infect Dis 13(6):659–663

Cioli D, Pica-Mattoccia L (2003) Praziquantel. Parasitol Res 90(Supp 1):S3–S9

Cioli D, Pica-Mattoccia L, Archer S (1995) Antischistosomal drugs: past, present ... and future? Pharmacology 68:35–85

Clerinx J, Van Gompel A (2011) Schistosomiasis in travellers and migrants. Travel Med Infect Dis 9(1):6–24

Cnops L, Tannich E, Polman K, Clerinx J, Van Esbroeck M (2012) Schistosoma real-time PCR as diagnostic tool for international travellers and migrants. Trop Med Int Health 17(10):1208–1216

Colley DG, Bustinduy AL, Secor WE, King CH (2014) Human schistosomiasis. Lancet 383(9936):2253–2264

Coulibaly JT, Knopp S, N'Guessan NA, Silué KD, Fürst T et al (2011) Accuracy of urine circulating cathodic antigen (CCA) test for Schistosoma mansoni diagnosis in different settings of Côte d'Ivoire. PLoS Negl Trop Dis 5(11):e1384. https://doi.org/10.1371/journal.pntd.0001384

Cringoli G, Rinaldi L, Albonico M, Bergquist R, Utzinger J (2013) Geospatial (s)tools: integration of advanced epidemiological sampling and novel diagnostics. Geospat Health 7(2):399–404

Danso-Appiah A (2009) Treatment effects and integrated morbidity control of schistosomiasis. Ipskamp Publishers, Enschede 1–202

Danso-Appiah A, De Vlas SJ (2002) Interpreting low praziquantel cure rates of Schistosoma mansoni infections in Senegal. Trends Parasitol 18(3):125–129

Danso-Appiah A, De Vlas SJ, Bosompem KM, Habbema JD (2004) Determinants of health-seeking behaviour for schistosomiasis-related symptoms in the context of integrating schistosomiasis control within the regular health services in Ghana. Trop Med Int Health 9(7):784–794

Danso-Appiah A, Utzinger J, Liu J, Olliaro P (2008) Drugs for treating urinary schistosomiasis. Cochrane Database Syst Rev 16(3):CD000053. https://doi.org/10.1002/14651858.CD000053.pub2

Danso-Appiah A, Stolk WA, Bosompem KM, Otchere J, Looman CW et al (2010) Health seeking behaviour and utilization of health facilities for schistosomiasis-related symptoms in Ghana. PLoS Negl Trop Dis 4(11):e867. https://doi.org/10.1371/journal.pntd.0000867

Danso-Appiah A, Olliaro PL, Donegan S, Sinclair D, Utzinger J (2013) Drugs for treating Schistosoma mansoni infection. Cochrane Database Syst Rev 2:CD000528. https://doi.org/10.1002/14651858.CD000528.pub2

Danso-Appiah et al (2014) Systematic review and meta-analysis of the circulating cathodic antigen (CCA) test in the diagnosis of schistosome infection (Submitted)

Davis A (2009) Schistosomiasis. In: Cook GC, Zumla AI (eds) Manson's tropical diseases, Section 11, Chapter 82, 22nd edn. Saunders Elsevier, pp 1431–1469. International edition. Printed in China

del Villar PL, Burguillo FJ, López-Abán J, Muro A (2012) Systematic review and meta-analysis of artemisinin based therapies for the treatment and prevention of schistosomiasis. PLoS One 7(9):e45867. https://doi.org/10.1371/journal.pone.0045867

Doenhoff MJ (1998) Is Schistosomicidal chemotherapy sub-curative? Implications for drug resistance. Parasitol Today 14(10):434–435

Downs JA, Mguta C, Kaatano GM, Mitchell KB, Bang H, Simplice H, Fitzgerald DW et al (2011) Urogenital schistosomiasis in women of reproductive age in Tanzania's Lake Victoria region. Am J Trop Med Hyg 84(3):364–369

Downs JA, Dupnik KM, van Dam GJ, Urassa M, Lutonja P, Kornelis D et al (2017) Effects of schistosomiasis on susceptibility to HIV-1 infection and HIV-1 viral load at HIV-1 seroconversion: a nested case-control study. PLoS Negl Trop Dis 11:e0005968

El-Bolkainy MN, Mokhtar NM, Ghoneim MA, Hussein MH (1981) The impact of schistosomiasis on the pathology of bladder carcinoma. Cance 48:2643–2648

Engels D, Chitsulo L, Montresor A, Savioli L (2002) The global epidemiological situation of schistosomiasis and new approaches to control and research. Acta Trop 82(2):139–146

Enk MJ, Silva OG, Rodrigues NB (2012) Diagnostic accuracy and applicability of a PCR system for the detection of Schistosoma mansoni DNA in human urine samples from an endemic area. PLoS One 7(6):e38947. https://doi.org/10.1371/journal.pone.0038947

Esrey SA, Potash JB, Roberts L, Shiff C (1991) Effects of improved water supply and sanitation on ascariasis, diarrhoea, dracunculiasis, hookworm infection, schistosomiasis, and trachoma. Bull World Health Organ 69(5):609–621

Fallon PG, Sturrock RF, Niang AC, Doenhoff MJ (1995) Diminished susceptibility to praziquantel in a Senegalese isolate of Schistosoma mansoni. Am J Trop Med Hyg 53:61–62

Feldmeier H, Chitsulo L (1999) Therapeutic and operational profiles of metrifonate and praziquantel in *Schistosoma haematobium* infection. Arzneimittelforschung 49(7):557–565

Feldmeier H, Doehring E (1987) Clinical experience with metrifonate. Review with emphasis on its use in endemic areas. Acta Trop 44(3):357–368

Fenwick A, Rollinson D, Southgate V (2006) Implementation of human schistosomiasis control: challenges and prospects. Adv Parasitol 61:567–622

Fenwick A, Webster JP, Bosque-Oliva E, Blair L, Fleming FM et al (2009) The schistosomiasis control initiative (SCI): rationale, development and implementation from 2002–2008. Parasitology 136(13):1719–1730

Foster R (1987) A review of clinical experience with oxamniquine. Trans R Soc Trop Med Hyg 81(1):55–59

Garba A, Touré S, Dembelé R, Boisier P, Tohon Z et al (2009) Present and future schistosomiasis control activities with support from the schistosomiasis control initiative in West Africa. Parasitology 136(13):1731–1737

Geerts S, Gryseels B (2001) Anthelmintic resistance in human helminths: a review. Trop Med Int Health 6(11):915–921

Glinz D, Silué KD, Knopp S, Lohourignon LK, Yao KP et al (2010) Comparing diagnostic accuracy of Kato-Katz, Koga agar plate, ether-concentration, and FLOTAC for Schistosoma mansoni and soil-transmitted helminths. PLoS Negl Trop Dis 4(7):e754. https://doi.org/10.1371/journal.pntd.0000754

Global Burden of Disease Study 2019 Results. Seattle, United States: Institute for Health Metrics and Evaluation (IHME). Available from https://vizhub.healthdata.org/gbd-results/

Grobusch MP, Mühlberger N, Jelinek T, Bisoffi Z, Corachán M et al (2003) Imported schistosomiasis in Europe: sentinel surveillance data from Trop Net Europe. J Travel Med 10(3):164–169

Gryseels B, Polderman AM (1987) The morbidity of schistosomiasis mansoni in Maniema (Zaire). Trans R Soc Trop Med Hyg 81(2):202–209

Gryseels B, Polderman AM, Engels D (1992) Experiences with the control of schistosomiasis mansoni in two foci in Central Africa. Mem Inst Oswaldo Cruz 87(Suppl 4):187–194

Gryseels B, Nkulikyinka L, Engels D (1994) Impact of repeated community-based selective chemotherapy on morbidity due to schistosomiasis mansoni. Am J Trop Med Hyg 51:634–641

Gryseels B, Mbaye A, De Vlas SJ, Stelma FF, Guissé F et al (2001) Are poor responses to praziquantel for the treatment of Schistosoma mansoni infections in Senegal due to resistance? An overview of the evidence. Trop Med Int Health 6(11):864–873

Gryseels B, Polman K, Clerinx J, Kestens L (2006) Human schistosomiasis. Lancet 368(9541):1106–1118

Guo JG, Cao CL, Hu GH, Lin H, Li D, Zhu R, Xu J (2005) The role of 'passive chemotherapy' plus health education for schistosomiasis control in China during maintenance and consolidation phase. Acta Trop 96(2–3):177–183

Hatz CF (2001) The use of ultrasound in schistosomiasis. Adv Parasitol 48:225–284

Hatz CF (2005) Schistosomiasis: an underestimated problem in industrialized countries? J Travel Med 12(1):1–2

Helleberg M, Thybo S (2010) High rate of failure in treatment of imported schistosomiasis. J Travel Med 17(2):94–99

Heyns CF, van der Merwe A (2008) Bladder cancer in Africa. Can J Urol 15:3899–3908

Hoffmann KF, Dunne DW (2003) Characterization of the *Schistosoma* transcriptome opens up the world of helminth genomics. Genome Biol 5:203

Hotez PJ (2009) Mass drug administration and integrated control for the world's high-prevalence neglected tropical diseases. Clin Pharmacol Ther 85:659–664

Hotez P (2011) Enlarging the "audacious goal": elimination of the world's high prevalence neglected tropical diseases. Vaccine 29(Suppl 4):D104–D110. https://doi.org/10.1016/j.vaccine.2011.06.024

Hotez PJ, Pecoul B (2010) "Manifesto" for advancing the control and elimination of neglected tropical diseases. PLoS Negl Trop Dis 4(5):e718. https://doi.org/10.1371/journal.pntd.0000718

Hotez P, Raff S, Fenwick A, Richards F Jr, Molyneux DH (2007) Recent progress in integrated neglected tropical disease control. Trends Parasitol 23(11):511–514

Hotez PJ, Fenwick A, Savioli L, Molyneux DH (2009) Rescuing the bottom billion through control of neglected tropical diseases. Lancet 373:1570–1575. https://doi.org/10.1016/S0140-6736(09)60233-6

Istre GR, Fontaine RE, Tarr J, Hopkins RS (1984) Acute schistosomiasis among Americans rafting the Omo River Ethiopia. JAMA 251:508–510

Johansen MV, Sacko M, Vennervald BJ, Kabatereine NB (2007) Leave children untreated and sustain inequity! Trends Parasitol 23(12):568–569

Jordan P (2000) From katayama to the Dakhla Oasis: the beginning of epidemiology and control of bilharzia. Acta Trop 77(1):9–40

Jordan P, Unrau GO, Bartholomew RK, Cook JA, Grist E (1982) Value of individual household water supplies in the maintenance phase of a schistosomiasis control programme in Saint Lucia, after chemotherapy. Bull World Health Organ 60(4):583–588

Kardorff R, Dohring E (2001) Ultrasound diagnosis of bilharziasis. Ultraschall Med 22(3):107–115

Kasambala M, Mduluza T, Vengesai A, Mduluza-Jokonya T, Jokonya L, Midzi H, Mukaratirwa S et al (2022) Effect of Schistosoma haematobium infection on the cognitive functions of preschool age children and benefits of treatment from an endemic area in Zimbabwe. BMC Infect Dis 22(1):1–9

Katz N, Chaves A, Pellegrino J (1972) A simple device for quantitative stool thick-smear technique in schistosomiasis mansoni. Rev Inst Med Trop Sao Paulo 14(6):397–400

Kayuni S, Lampiao F, Makaula P, Juziwelo L, Lacourse EJ, Reinhard-Rupp J, Stothard JR et al (2018) A systematic review with epidemiological update of male genital schistosomiasis (MGS): a call for integrated case management across the health system in sub-Saharan Africa. Parasite Epidemiol Control 4:e00077

Keenan JD, Hotez PJ, Amza A, Stoller NE, Gaynor BD et al (2013) Elimination and eradication of neglected tropical diseases with mass drug administrations: a survey of experts. PLoS Negl Trop Dis 7(12):e2562

King CH, Bertino A-M (2008) Asymmetries of poverty: why global burden of disease valuations underestimate the burden of neglected tropical diseases. PLoS Negl Trop Dis 2(3):1–10

King CH, Dangerfield-Cha M (2008) The unacknowledged impact of chronic schistosomiasis. Chron Illn 4(1):65–79

King CH, Galvani AP (2018) Underestimation of the global burden of schistosomiasis. Lancet 391(10118):307–330

King CH, Keating CE, Muruka JF, Ouma JH, Houser H et al (1988) Urinary tract morbidity in schistosomiasis haematobia: associations with age and intensity of infection in an endemic area of Coast Province. Kenya Am J Trop Med Hyg 39(4):361–368

Kinkel HF, Dittrich S, Bäumer B, Weitzel T (2012) Evaluation of eight serological tests for diagnosis of imported schistosomiasis. Clin Vaccine Immunol 19(6):948–953

Kitinya JN, Laurèn PA, Eshleman LJ, Paljärvi L, Tanaka K (1986) The incidence of squamous and transitional cell carcinomas of the urinary bladder in northern Tanzania in areas of high and low levels of endemic Schistosoma haematobium infection. Trans R Soc Trop Med Hyg 80:935–939

Kjetland EF, Ndhlovu PD, Gomo E, Mduluza T, Midzi N, Gwanzura L, Gundersen SG et al (2006) Association between genital schistosomiasis and HIV in rural Zimbabwean women. AIDS 20:593–600

Kjetland EF, Leutscher PDC, Ndhlovu PD (2012) A review of female genital schistosomiasis. Trends Parasitol 28:58–65

Kjetland EF, Hegertun IE, Baay MF, Onsrud M, Ndhlovu PD, Taylor M (2014) Genital schistosomiasis and its unacknowledged role on HIV transmission in the STD intervention studies. Int J STD AIDS 25:705–715

Le WJ, You JQ, Yang YQ, Mei JY, Guo HF, Yang HZ et al (1982) Studies on the efficacy of artemether in experimental schistosomiasis. YaoXue Xue Bao 17(3):187–193

Le WJ, You JQ, Mei JY (1983) Chemotherapeutic effect of artesunate in experimental schistosomiasis. YaoXue Xue Bao 18(8):619–621

Legesse M, Erko B (2007) Field-based evaluation of a reagent strip test for diagnosis of *Schistosoma mansoni* by detecting circulating cathodic antigen in urine before and after chemotherapy. Trans R Soc Trop Med Hyg 101(7):668–673

Legesse M, Erko B (2008) Field-based evaluation of a reagent strip test for diagnosis of schistosomiasis mansoni by detecting circulating cathodic antigen (CCA) in urine in low endemic area in Ethiopia. Parasite 15(2):151–155

Lengeler C, Utzinger J, Tanner M (2002) Questionnaires for rapid screening of schistosomiasis in sub-Saharan Africa. Bull World Health Organ 80(3):235–242

Leutscher P, Ramarokoto CE, Reimert C, Feldmeier H, Esterre P, Vennervald BJ (2000) Community-based study of genital schistosomiasis in men from Madagascar. Lancet 355:117–118

Leutscher PD, Ramarokoto CE, Hoffmann S, Jensen JS, Ramaniraka V, Randrianasolo B, Christensen N (2008) Coexistence of urogenital schistosomiasis and sexually transmitted infection in women and men living in an area where Schistosoma haematobium is endemic. Clin Infect Dis 47:775–782

Linehan M, Hanson C, Weaver A, Baker M, Kabore A et al (2011) Integrated implementation of programs targeting neglected tropical diseases through preventive chemotherapy: proving the feasibility at national scale. Am J Trop Med Hyg 84(1):5–14

Liu R, Dong HF, Guo Y, Zhao QP, Jiang MS (2011) Efficacy of praziquantel and artemisinin derivatives for the treatment and prevention of human schistosomiasis: a systematic review and meta-analysis. Parasitol Vectors 4:201. https://doi.org/10.1186/1756-3305-4-201

Lorenz W, Henglein A, Schrader G (1955) The new insecticide 0.0-dimethyl-2,2,2-trichloro-1-hydroxyethyl phosphate. J Am Chem Soc 77:2554

Madden F (1911) Two rare manifestations of bilharziosis. Lancet 178:754–755

Manderson L, Aagaard-Hansen J, Allotey P, Gyapong M, Sommerfeld J (2009) Social research on neglected diseases of poverty: continuing and emerging themes. PLoS Negl Trop Dis 3:e332. https://doi.org/10.1371/journal.pntd.0000332

Mascie-Taylor CGN, Karim R, Karim E, Akhtar S, Ahmed T et al (2003) The cost-effectiveness of health education in improving knowledge and awareness about intestinal parasites in rural Bangladesh. Econ Hum Biol 1(3):321–330

Mbabazi PS, Andan O, Fitzgerald DW, Chitsulo L, Engels D, Downs JA (2011) Examining the relationship between urogenital schistosomiasis and HIV infection. PLoS Negl Trop Dis 5(12):e1396. https://doi.org/10.1371/journal.pntd.0001396

McCoy D, Kembhavi G, Patel J, Luintel A (2009) The Bill & Melinda Gates Foundation's grant-making programme for global health. Lancet 373:1645–1653

Minamoto K, Mascie-Taylor CGN, Karim E, Moji K, Rahman M (2012) Short- and long-term impact of health education in improving water supply, sanitation and knowledge about intestinal helminths in rural Bangladesh. Public Health 126(5):437–440

Mitchell KM, Mutapi F, Woolhouse ME (2008) The predicted impact of immunosuppression upon population age- intensity profiles for schistosomiasis. Parasite Immunol 30(9):462–470

Mott KE, Dixon H, Osei-Tutu E, England EC (1983) Relation between intensity of *Schistosoma haematobium* infection and clinical haematuria and proteinuria. Lancet 1(8332):1005–1008

Murta-Nascimento C, Schmitz-Drager BJ, Zeegers MP, Steineck G, Kogevinas M, Real FX et al (2007) Epidemiology of urinary bladder cancer: from tumor development to patient's death. World J Urol 25(3):285–295

Nour NM (2010) Schistosomiasis: health effects on women. Rev Obstet Gynecol 3:28–32

Ochodo EA, Spek B, Reitsma JB, Van Lieshout L, Polman K et al (2012) Rapid screening and diagnostic tests for human schistosomiasis in endemic areas. Cochrane Database Syst Rev 2:CD009579. https://doi.org/10.1002/14651858.CD009579

Odogwu SE, Ramamurthy NK, Kabatereine NB, Kazibwe F, Tukahebwa E, Webster JP, Fenwick A, Stothard JR (2006) Schistosoma mansoni in infants (aged <3 years) along the Ugandan shoreline of Lake Victoria. Ann Trop Med Parasitol 100(4):315–326

Ogunniyi SO, Nganwuchu AM, Adenle MA, Dare FO (1994) Pregnancy following infertility due to pelvic schistosomiasis—a case report. West Afr J Med 13:132–133

Okonofua FE, Ojo OS, Odunsi OA, Odesanmi WO (1990) Ectopic pregnancy associated with tubal schistosomiasis in a Nigerian woman. Int J Gynaecol Obstet 32:281–284

Olds GR, Dasarathy S (2000) Schistosomiasis. Curr Treat Options Infect Dis 2:88–99

Olsen A (2007) Efficacy and safety of drug combinations in the treatment of schistosomiasis, soil-transmitted helminthiasis, lymphatic filariasis and onchocerciasis. Trans R Soc Trop Med Hyg 101(8):747–758

Onasanya A, Bengtson M, Oladepo O, Van Engelen J, Diehl JC (2021) Rethinking the top-down approach to schistosomiasis control and elimination in sub-Saharan Africa. Front Public Health 9:622809. https://doi.org/10.3389/fpubh.2021.622809

Poda JN, Wango SP, Sorgho H, Dianou D (2004) Recent evolution of schistosomiasis in the water project of Sourou in Burkina Faso. Bull Soc Pathol Exot 97(1):15–18

Poggensee G, Feldmeier H (2001) Female genital schistosomiasis: facts and hypotheses. Acta Trop 79(3):193–210

Polman K, Stelma FF, De Vlas SJ, Sow S, Fathers L et al (2001) Dynamics of egg counts and circulating antigen levels in a recent *Schistosoma mansoni* focus in northern Senegal. Trop Med Int Health 6:538–544

Polman K, Stelma FF, Le Cessie S, De Vlas SJ, Falcao-Ferreira ST et al (2002) B. Evaluation of the patterns of Schistosoma mansoni infection and re-infection in Senegal, from faecal egg counts and serum concentrations of circulating anodic antigen. Ann Trop Med Parasitol 96:679–689

Rabello AL (1992) Parasitological diagnosis of schistosomiasis mansoni: faecal examination and rectal biopsy. Mem Inst Oswaldo Cruz 87(Suppl 4):325–331

Raso G, Utzinger J, Silué KD, Ouattara M, Yapi A et al (2005) Disparities in parasitic infections, perceived ill health and access to health care among poorer and less poor schoolchildren of rural Côte d'Ivoire. Trop Med Int Health 10:42–57

Reich MR, Fenwick A (2001) Schistosoma haematobium. N Engl J Med 344(15):1170

Renganathan E, Cioli D (1998) An international initiative on praziquantel use. Parasitol Today 14(10):390–391

Richter J (2003) The impact of chemotherapy on morbidity due to schistosomiasis. Acta Trop 86(2–3):161–183

Rinaldo D, Perez-Saez J, Vounatsou P, Utzinger J, Arcand JL (2021) The economic impact of schistosomiasis. Infect Dis Poverty 10(1):134

Sabah AA, Fletcher C, Webbe G, Doenhoff MJ (1986) *Schistosoma mansoni*: chemotherapy of infections of different ages. Exp Parasitol 61:294–303

Salari P, Fürst T, Knopp S, Utzinger J, Tediosi F (2020) Cost of interventions to control schistosomiasis: a systematic review of the literature. PLoS Negl Trop Dis 14(3):e0008098

Salehe FS, Hassan SN (2012) Socio-economic effects of schistosomiasis on. Irrigation rice growers in Morogoro, Tanzania. J Exp Agric 2(3):395–406

Sandoval N, Siles-Lucas M, Pérez-Arellano JL, Carranza C, Puente S et al (2006) A new PCR-based approach for the specific amplification of DNA from different Schistosoma species applicable to human urine samples. Parasitology 133(Pt 5):581–587

Schwartz E, Kozarsky P, Wilson M, Cetron M (2005) Schistosome infection among river rafters on Omo River Ethiopia. J Travel Med 12:3–8

Shekhar KC (1991) Schistosomiasis drug therapy and treatment considerations. Drugs 42(3):379–405

Shiff CJ, de Clarke V, Evans AC, Barnish G (1973) Molluscicide for the control of schistosomiasis in irrigation schemes. A study in Southern Rhodesia. Bull World Health Organ 48:299–307

Shiff C, Veltri R, Naples J, Quartey J, Otchere J, Anyan W et al (2006) Ultrasound verification of bladder damage is associated with known biomarkers of bladder cancer in adults chronically infected with Schistosoma haematobium in Ghana. Trans R Soc Trop Med Hyg 100:847–854

Singer BH, Castro MC (2007) Bridges to sustainable tropical health. Proc Natl Acad Sci USA 104:16038–16043

Snellen WM (1981) Therapeutic properties of metrifonate. Acta Pharmacol Toxicol (Copenh) 49(Suppl V 5):114–117

Spiegel JM, Dharamsi S, Wasan KM, Yassi A, Singer B, Hotez PJ et al (2010) Which new approaches to tackling neglected tropical diseases show promise? PLoS Med 7(5):e1000255. https://doi.org/10.1371/journal.pmed.1000255

Steinmann P, Keiser J, Bos R, Tanner M, Utzinger J (2006) Schistosomiasis and water resources development: systematic review, meta-analysis, and estimates of people at risk. Lancet Infect Dis 6:411–425

Stelma FF, Talla I, Sow S, Kongs A, Niang M et al (1995) Efficacy and side effects of praziquantel in an endemic focus of Schistosoma mansoni. Am J Trop Med Hyg 53:167–170

Stothard JR, Gabrielli AF (2007) Schistosomiasis in African infants and preschool children: to treat or not to treat? Trends Parasitol 23:83–86

Stothard JR, Sousa-Figueiredo JC, Navaratnam AM (2013) Advocacy, policies and practicalities of preventive chemotherapy campaigns for African children with schistosomiasis. Expert Rev Anti-Infect Ther 11(7):733–752

Striebel HP (1976) 4-Isothiocyanato-4′-nitrodiphenylamine (C 9333-Go/CGP 4540), an anthelminthic with an unusual spectrum of activity against intestinal nematodes, filariae and schistosomes. Experientia 32(4):457–458

Sturrock RF (2001) Schistosomiasis epidemiology and control: how did we get here and where should we go? Mem Inst Oswaldo Cruz 96(Suppl):17–27

Umeh JC, Amali O, Umeh EU (2004) The socio-economic effects of tropical diseases in Nigeria. Econ Hum Biol 2:245–263

Utzinger J, Keiser J (2004) Schistosomiasis and soil-transmitted helminthiasis: common drugs for treatment and control. Expert Opin Pharmacother 5(2):263–285

Utzinger J, Keiser J, Xiao SH, Tanner M, Singer BH (2003a) Combination chemotherapy of schistosomiasis in laboratory studies and clinical trials. Antimicrob Agents Chemother 47:1487–1495
Utzinger J, Bergquist R, Xiao SH, Singer BH, Tanner M (2003b) Sustainable schistosomiasis control—the way forward. Lancet 362:1932–1934
Utzinger J, Xiao SH, Tanner M, Keiser J (2007) Artemisinins for schistosomiasis and beyond. Curr Opin Investig Drugs 8(2):105–116
Utzinger J, Raso G, Brooker S, De Savigny D, Tanner M et al (2009) Schistosomiasis and neglected tropical diseases: towards integrated and sustainable control and a word of caution. Parasitology 136(13):1859–1874
Utzinger J, N'goran EK, Caffrey CR, Keiser J (2010) From innovation to application: social-ecological context, diagnostics, drugs and integrated control of schistosomiasis. Acta Trop 120(Suppl 1):S121–S137
Utzinger J, N'goran EK, Caffrey CR, Keiser J (2011) From innovation to application: social-ecological context, diagnostics, drugs and integrated control of schistosomiasis. Acta Trop 120(Suppl 1):S121–S137
Valencia CI, Catto BA, Webster LT Jr, Barcelon E, Ofendo-Reyes R (1994) Concentration time course of praziquantel in Filipinos with mild Schistosoma japonicum infection. Southeast Asian J Trop Med Public Health 25(2):409–414
van der Werf MJ, de Vlas SJ, Brooker S, Looman CW, Nagelkerke NJ, Habbema JD et al (2003) Quantification of clinical morbidity associated with schistosome infection in sub-Saharan Africa. Acta Trop 86(2–3):125–139
Wall KM, Kilembe W, Vwalika B, Dinh C, Livingston P, Lee YM et al (2018) Schistosomiasis is associated with incident HIV transmission and death in Zambia. PLoS Negl Trop Dis 12:e0006902
Warren KS (1982) Schistosomiasis: host-pathogen biology. Infect Dis 4(4):771–775
Whitfield PJ, Bartlett A, Khammo N, Clothier RH (2003) Age-dependent survival and infectivity of Schistosoma mansoni cercariae. Parasitology 127(Pt 1):29–35
WHO (1985) The control of Schistosomiasis: report of WHO expert committee. Technical Report Series No. 728, Geneva
WHO (1998) Report of the WHO Informal Consultation on schistosomiasis control, Geneva 2–4 Dec, 1–44
WHO (2002a) Death and DALY estimates for 2004 by cause for WHO member states: persons, all ages. World Health Organization Retrieved 2009-11-12
WHO (2002b) Prevention and control of schistosomiasis and soil-transmitted helminthiasis, Geneva, World Health Organization, (WHO Technical Report Series, No. 912):1–57
WHO (2006a) Preventive chemotherapy in human helminthiasis. World Health Organization, Geneva, Also available from http://whqlibdoc.who.int/publications/2006/9241547103_eng.pdf
WHO (2006b) Preventive chemotherapy in human helminthiasis. Coordinated use of anthelminthic drugs in control interventions: a manual for health professionals and programme managers. World Health Organization, Geneva
WHO (2010a) Schistosomiasis, fact sheet no 115; February. World Health Organization Available at http://www.who.int/mediacentre/facesheets/fs115/en/index.html. Accessed 5 Oct 2010
WHO (2010b) Report of a Meeting to Review the Results of Studies on the treatment of Schistosomiasis in Preschool age Children, Geneva, 13–14 Sept. WHO/HTM/NTD/PCT/2011.7
WHO (2011) Report of an informal consultation on schistosomiasis control, Geneva, 30 March–1 April 2011
WHO (2012) Schistosomiasis: progress report 2001–2011 and strategic plan 2012–2020. World Health Organization, 1–82
WHO (2020) Current estimated total number of individuals with morbidity and mortality due to Schistosomiasis Haematobium and S. Mansoni infection in Sub-Saharan Africa. Retrieved from Pretoria, South Africa
WHO. World Health Organization (2022) Launch new guidelines for the control and elimination of human schistosomiasis

WHO. World Health Organization (1993) The control of schistosomiasis. Second report of the WHO expert committee. WHO Tech Rep Ser 830:1–86

Wikman-Jorgensen PE, Henríquez-Camacho CA, Serrano-Villar S, Pérez-Molina JA (2012) The role of artesunate for the treatment of urinary schistosomiasis in schoolchildren: a systematic review and meta-analysis. Pathog Glob Health 106(7):397–404

Wolfe MS, Quansah FO, Agbemadzo F (1966) Suppressive management of urinary schistosomiasis with antimony dimercaptosuccinate (TWSb) in Ghana. J Trop Med Hyg 69(10):217–223

Woolhouse M, Taylor P, Matanhire D, Chandiwana SK (1991) Acquired immunity and epidemiology of *Schistosoma haematobium*. Nature 351:757–759

World Health Organization (2023a) Schistosomiasis (Bilharzia): overview. Available from: https://www.who.int/health-topics/schistosomiasis#tab=tab_1

World Health Organization (2023b) Schistosomiasis: key facts. Available from: https://www.who.int/news-room/fact-sheets/detail/schistosomiasis

Wright WH (1972) A consideration of economic impact of schistosomiasis. Bull World Health Organ 47(5):559–565

Wu GY, Halim MH (2000) Schistosomiasis: progress and problems. World J Gastroenterol 6(1):12–19

Zuidema PJ (1981) The Katayama syndrome: an outbreak in Dutch tourists to the Omo National Park. Ethiopia Trop Geogr Med 33:30–35

Soil-Transmitted Helminthiasis

Michael D. Wilson, Dziedzom K. de Souza, Jewelna Akorli, and Irene Ayi

Contents

Abstract

The human burden of soil-transmitted helminthiasis is highest in sub-Saharan Africa (SSA) than in other parts of the world. The major parasites that contribute to the global burden, *Ascaris*, *Trichuris*, hookworm, schistosomes, and

M. D. Wilson (✉) · D. K. de Souza · J. Akorli · I. Ayi
Department of Parasitology, Noguchi Memorial Institute for Medical Research, University of Ghana, Legon, Ghana
e-mail: MWilson@noguchi.ug.edu.gh

J. O. Gyapong, B. A. Boatin (eds.), *Neglected Tropical Diseases - Sub-Saharan Africa*, Neglected Tropical Diseases, https://doi.org/10.1007/978-3-031-53901-5_14

Strongyloides, and even less known ones such as *Oesophagostomum bifurcum* and *Ternidens deminutus* (a.k.a. "false hookworm") all occur in SSA. In this chapter, the symptoms, and consequences of infection by these parasites including retarded growth, reduced work capacity, poor school performance, adverse cognitive development, and nutritional and reproductive health problems are described for each parasite. The most at-risk groups for the diseases they cause are also specified. Why then is SSA carrying the highest burden? The answer is the diseases' association with ignorance or lack of awareness, poor personal and environmental hygiene, poverty, impoverished health services, and economic development. These factors are common to these helminths' infections and also explain to a large degree the skewed high prevalence of the diseases in rural African communities. The parasites all share similar basics of the life cycle except for *Strongyloides stercoralis* which in addition is also capable of autoinfection (i.e., multiplication) in human and existing as free living and productive adults in the environment; attributes which make it difficult to control. Treatment and challenges with mass drug administration (MDA) with antihelminthics, the urgent need for more sensitive diagnostic tools than the existing ones for scenarios of low-intensity infections, and the likelihood of resistance developing with the MDA are highlighted. There are opportunities arising from recent technological advancements such as the availability of genome data of 22 helminth species to be exploited for novel drugs and vaccine design and of IT tools, remote sensing data, and global resource platforms for disease mapping, ecology and epidemiology studies, operational research, and implementation of control programs. What it entailed in conducting a successful and sustainable STH control in a hitherto deprived community in Ghana is presented as a case study.

Introduction

Helminthiases are generally infections by intestinal nematodes, of which the roundworm (*Ascaris lumbricoides*), the whipworm (*Trichuris trichiura*), and the hookworms (*Ancylostoma duodenale* and *Necator americanus*) are the most widespread species (Bethony et al. 2006; Brooker et al. 2006a). There are other helminth infections such as *Strongyloides stercoralis* and the even less known *Oesophagostomum bifurcum* and *Ternidens deminutus* [the "false hookworm"] (Hotez and Kamath 2009). The World Health Organization estimates that more than 1.5 billion people, or 24% of the world's population, are infected with soil-transmitted helminthiasis (STH), with the greatest numbers occurring in sub-Saharan Africa, the Americas, China, and East Asia (WHO 2020). The high prevalence of these infections is closely associated with poverty, poor environmental hygiene, and impoverished health services. The WHO, in its NTD 2021–2030 roadmap, targets the elimination of STHs as a public health problem (defined as <2% proportion of soil-transmitted helminth infections of moderate and heavy intensity due to *Ascaris lumbricoides, Trichuris trichuria*, *Necator americanus*, and *Ancylostoma duodenale*) in 96% of disease endemic countries (WHO 2020).

Children and pregnant women are particularly vulnerable to STH infection which decreases work capacity (Crompton and Stephenson 1990) and fitness, and especially in the case of children, it influences their nutritional status, causing growth retardation (Adams et al. 1994) and reduced learning ability (Nokes et al. 1992).

Helminth infections produce a wide range of symptoms including intestinal manifestations (diarrhea, abdominal pain), general malaise, and weakness that may affect working and learning capacities and impair physical growth. Hookworms cause chronic intestinal blood loss that result in iron-deficiency anemia. Some other features of helminth infections include growth stunting, rectal prolapse, and chronic dysentery. These parasitic infections can also adversely affect cognitive development in childhood (Crompton and Nesheim 2002).

Infection intensity is a key factor in understanding the morbidity of STH. Although light infections are often asymptomatic, heavy infections cause an array of morbidities, including dietary deficiencies and delayed physical and cognitive development.

Additionally, T. *trichiura* infections contribute to iron-deficiency anemia (Bethony et al. 2006; Brooker et al. 2006a; Zimmermann and Hurrell 2007).

According to the 2010 estimates, the global burden due to STH ranges between 43 and 128 disability-adjusted life-years (DALYs) per 100,000 populations (Murray et al. 2012). These diseases are predominant in rural areas and also in some poor urban areas in sub-Saharan Africa, Asia, and Latin America which are rated as low-income countries. Long-term disability and poverty can result from any of these neglected tropical diseases (Hotez et al. 2006; Molyneux et al. 2005; Lammie et al. 2006).

Globally, it is estimated that more than 1.5 billion people, or 24% of the world's population, are infected with soil-transmitted helminth infections (WHO 2020). As many as 1.2 billion individuals might be infected with *A. lumbricoides*, close to 800 million with *T. trichiura*, and more than 700 million with hookworm (Bethony et al. 2006; Lammie et al. 2006). The actual prevalence of *Strongyloides* infections is not known due to the difficulty in diagnosis, but it is estimated that 30–100 million people may be infected worldwide (Hotez and Kamath 2009; Olsen et al. 2009). It is estimated that 267 million preschool-aged children and over 568 million school-aged children reside in areas where these parasites are intensively transmitted and are in need of treatment and preventive interventions (WHO 2020).

About one-third of the world's population is infected with either one or more species of STH (Warren et al. 1993), and this is a major concern for nutritional health workers since these infections impair children's growth and development. Several studies have shown associations between helminth infection and undernutrition, iron-deficiency anemia, stunted growth, poor school attendance, and poor performance in cognition tests (Simeon et al. 1994; Savioli et al. 1992; Nokes).

About one-third of the world's population is infected with either one or more species of STH (Warren et al. 1993), and this is a major concern for nutritional health workers since these infections impair children's growth and development. Several studies have shown associations between helminth infection and undernutrition, iron-deficiency anemia, stunted growth, poor school attendance, and poor performance in cognition tests (Simeon et al. 1994; Savioli et al. 1992; Nokes and Bundy 1994; Evans and Guyatt 1995; WHO 2002). Children are more susceptible to the ill effects of parasitic attacks, at the age when their need for nutrients is high. In young children, physical and mental development may be affected by malabsorption, blood and protein loss, and diarrhea generated often by the presence of several types of worms in the gut. Many parasites interfere with the process of intestinal absorption of nutrients. They feed on the nutrients, depriving the child of its sources of nutrition. The child is thus thrown into a state of acute and chronic malnutrition.

Morbidity due to STH is relatively easy to control with simple intervention measures. Better sanitation practices reduce transmission to a minimum. However, another approach called preventive chemotherapy is to treat children or whole populations at risk on a regular basis to reduce infection rates (WHO 2020). The targeted STH diseases are ascariasis, trichuriasis, and schistosomiasis.

Helminth infections affect millions of people, and the prevalence and intensity of *T. trichiura* and *A. lumbricoides* are also greatest in school-aged children (Bundy and Cooper 1989). For these reasons and safety, coupled with the cost-effectiveness of anthelminthic delivery to children in this age group, most research and investigations target school-aged children (Anderson 1980). It has been argued that targeting delivery of antihelminthics to school-aged children by taking advantage of the existing education infrastructure and administrative system can be one of the most cost-effective approaches in minimizing the intensity of infections with both major intestinal nematodes and urinary schistosomes in many developing countries.

While school-based approach to control of STH may be cost-effective, the challenge is with the treatment of non-enrolled children. Non-enrolled children form a significant proportion of rural communities in sub-Saharan Africa, and they are potentially sources of reinfection of STH and schistosomiasis to their enrolled counterparts, who benefit from the school-based interventions. The enrolment rate in Africa has been estimated to be approximately 68% and variations between countries exist (Bundy and Guyatt 1995). Although the enrolment ratios are expected to have increased since 1995, the number of non-enrolled children still remains high. Therefore, a sizeable number of children are not covered by school-based interventions. In some communities in Ghana, for example, enrolment of children aged 6–15 years in 1999 was estimated to be 37% (Fentiman et al. 1999). This, however, may have increased substantially with the introduction of the capitation grant policy and the National School Feeding Program in 2005 (Osei-Fosu 2011). The capitation grant covers the extra cost and levies (such as examination, facilities, management, security charges, games, and sports) that parents usually pay as "school fees" in public schools. There is, however, recent evidence suggesting community-wide treatments to be more effective in reducing hookworm prevalence and intensity than school-based treatment (Pullan et al. 2019).

Basic Biology, Life Cycle, and Disease Presentation

The STHs share a basic biology and life cycle represented in the simplified diagram (Fig. 1). The adult worms in a human shed many eggs which are passed into the environment through feces. The eggs develop into infective embryo, which are ingested with contaminated food and water or ingested from contaminated hands. From the stomach, they develop into adults and lay more eggs.

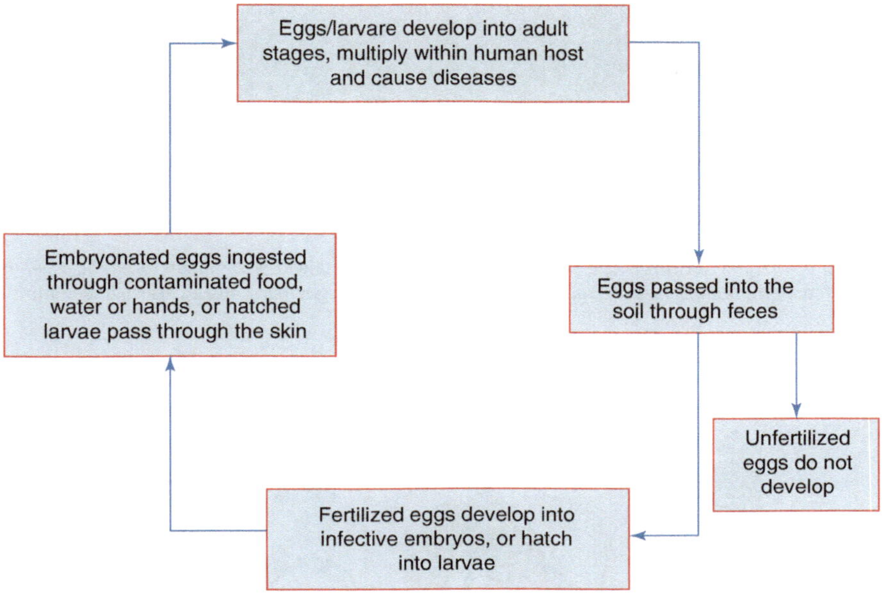

Fig. 1 A generalized transmission cycle of STHs

Roundworms

Ascaris lumbricoides is the largest nematode parasite of humans. Adult females can grow up to 20–35 cm, while adult males can measure up to 15–30 cm (Fig. 2). Adult worms live in the lumen of the small intestine. A female may produce approximately 200,000 eggs per day, which are passed with the feces. Unfertilized eggs may be ingested but are not infective. The eggs develop into embryos and become infective after 18 days to several weeks, depending on the environmental conditions (optimum: moisture, temperature, and shaded soil). The infective eggs usually enter the body through contaminated water or food or on fingers placed in the mouth after the hands have touched a contaminated object. After infective eggs are swallowed, the larvae hatch, invade the intestinal mucosa, and are carried via the portal and then systemic circulation to the lungs. In the lungs, they mature further after 10–14 days. They then penetrate the alveolar walls, ascend the bronchial tree to the throat, and are swallowed. They develop into adult worms in the small intestine. A typical life cycle (Fig. 3) takes between 2 and 3 months, from ingestion of the infective eggs to oviposition by the adult female. Adult worms can live for 1–2 years.

Ascariasis usually results in no symptoms. Symptoms, however, may present in the form of light abdominal discomfort. Heavy parasite loads can cause intestinal blockage and growth retardation in children. Other symptoms such as cough may be present and are due to migration of the worms through the body.

The life cycle of *S. stercoralis* (Fig. 4) is rather more complex than that of other nematodes, in that it has a free-living stage and a parasitic stage. In addition to this, it has the potential for autoinfection and multiplication within the host. In the free-living cycle, the rhabditiform larvae passed in the stool follow one of two cycles.

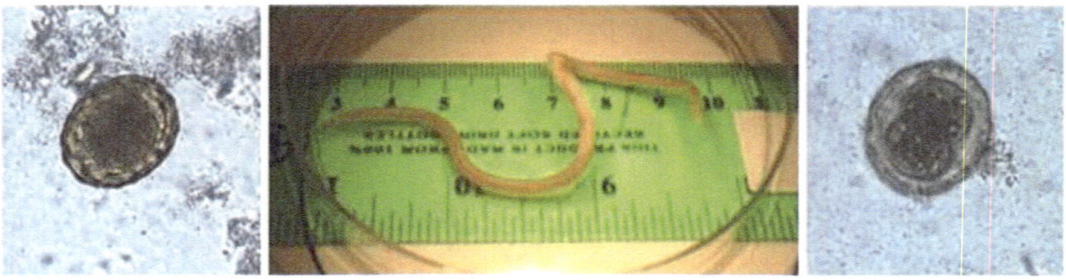

Fig. 2 *Left/right:* fertilized eggs of *A. lumbricoides* in unstained wet mounts of stool. *Center*: adult female *A. lumbricoides*. Credit: DPDx. http://www.cdc.gov/parasites/ascariasis/index.html

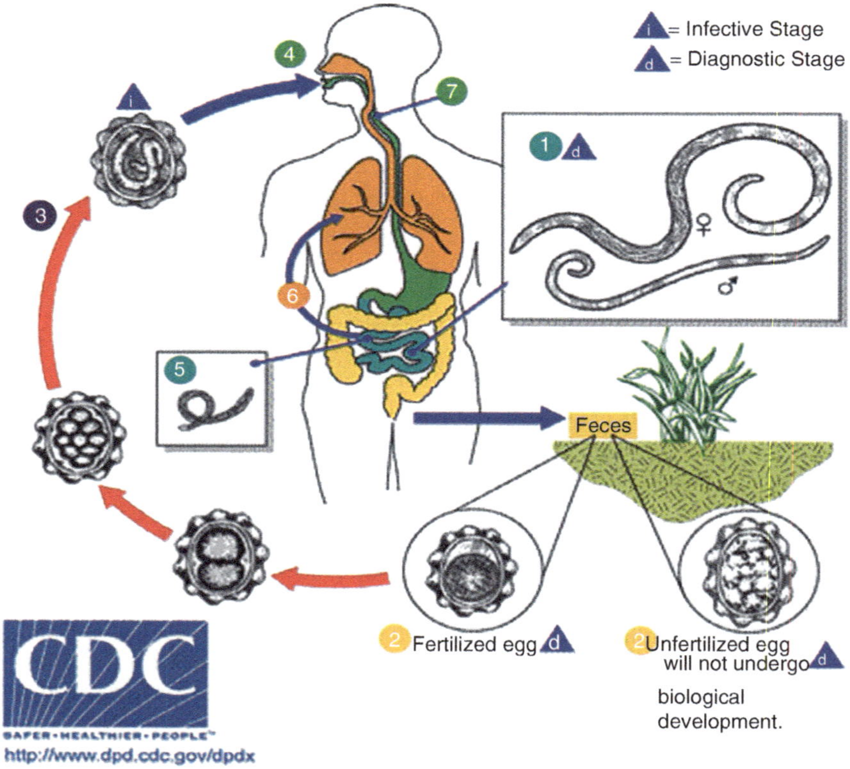

Fig. 3 Life cycle of *Ascaris lumbricoides*

In the first cycle, they molt twice and become infective filariform larvae. In the second cycle, they molt four times and become free-living adult males and females. Males grow to only about 0.9 mm (0.04 in.) in length, and females can grow from 2.0 to 2.5 mm. The males and females (Fig. 5) mate and produce eggs from which rhabditiform larvae hatch. The hatched rhabditiform larvae develop either into a new generation of free-living adults or into infective filariform larvae.

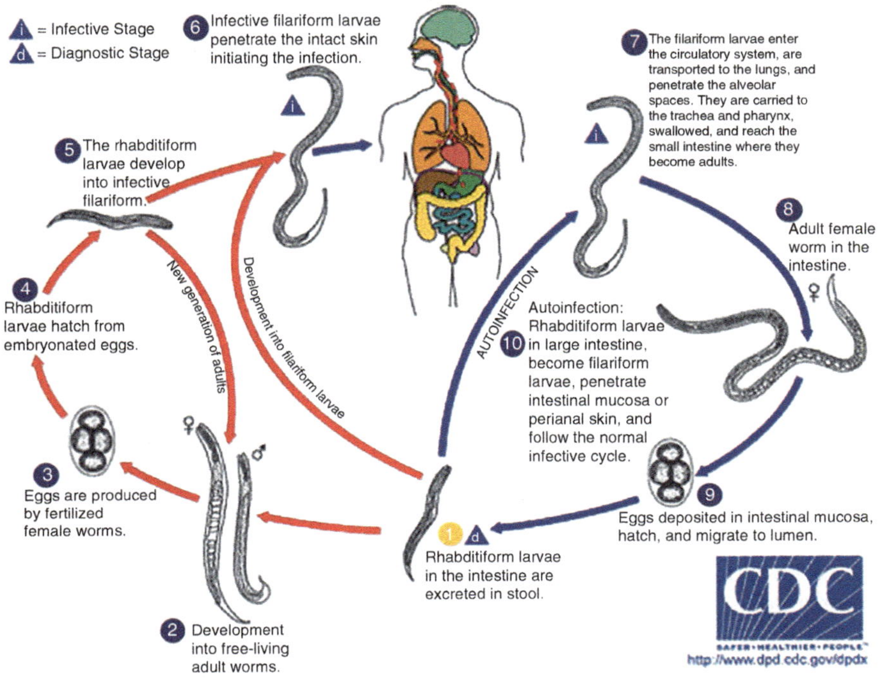

Fig. 4 Life cycle of *Strongyloides stercoralis*

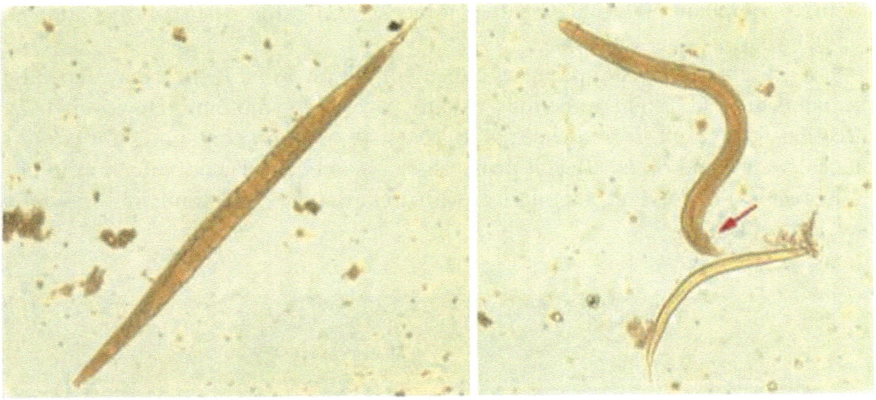

Fig. 5 *Left*: adult free-living female *S. stercoralis*. Notice the row of eggs within the female's body. *Right*: free-living adult male *S. stercoralis*, showing a spicule (*red arrow*). A smaller rhabditiform larva lies adjacent to the adult male. Credit: DPDx. http://www.dpd.cdc.gov/dpdx/HTML/ImageLibrary/Strongyloidiasis_il.htm

In the parasitic cycle, the filariform larvae in contaminated soil penetrate through the skin and are transported to the lungs. From there, they enter the alveolar spaces and are carried through the bronchial tree to the pharynx, swallowed, and then reach the small intestine. In the small intestine, they molt twice and become adult female worms. The adult females live in the epithelium of the small intestine and produce eggs by parthenogenesis. The eggs hatch into rhabditiform larvae, which either can be passed in the stool or can cause autoinfection. In autoinfection, the rhabditiform larvae become infective filariform larvae, which can penetrate either the intestinal mucosa (internal autoinfection) or the skin of the perianal area (external autoinfection); In either case, the filariform larvae may follow the previously described route, being carried successively to the lungs, the bronchial tree, the pharynx, and the small intestine where they mature into adults, or they may disseminate widely in the body.

Whipworm

Trichuris trichiura, also called the human whipworm, is another species of roundworm. The adult whipworms measure approximately 4 cm in length (Fig. 6). The whipworm life cycle is depicted in Fig. 7. They live permanently in the cecum and ascending colon. The female sheds 3000–20,000 eggs a day. These are passed with the stool, into the soil, where they embryonate and become infective in 15–30 days. When ingested through soil-contaminated hands or food, the eggs hatch in the small intestine. The larvae mature and establish themselves as adults in the colon, with the anterior portions threaded into the mucosa. The females begin to oviposit 60–70 days after infection. The life span of the adults is about 1 year.

The presentation of symptoms as a result of whipworm infection may depend on the parasite load. Light infections usually have no symptoms. However, heavy infections can result in frequent, painful passage of stool, containing a mixture of mucus, water, and blood. Rectal prolapse can also occur. Heavy infection in children results in severe anemia, growth retardation, and impaired cognitive development.

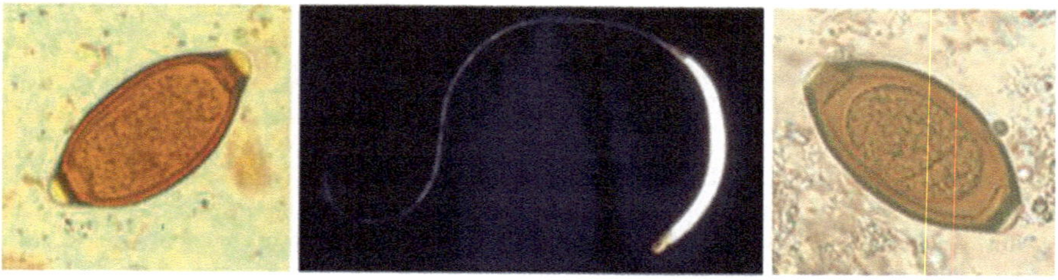

Fig. 6 Egg of *T. trichiura* in an iodine-stained wet mount. *Right*: egg of *T. trichiura* in an unstained wet mount. *Center:* micrograph of an adult female *Trichuris* human whipworm that is approximately 4 cm long. Credit: DPDx, PHIL. http://www.cdc.gov/parasites/whipworm/

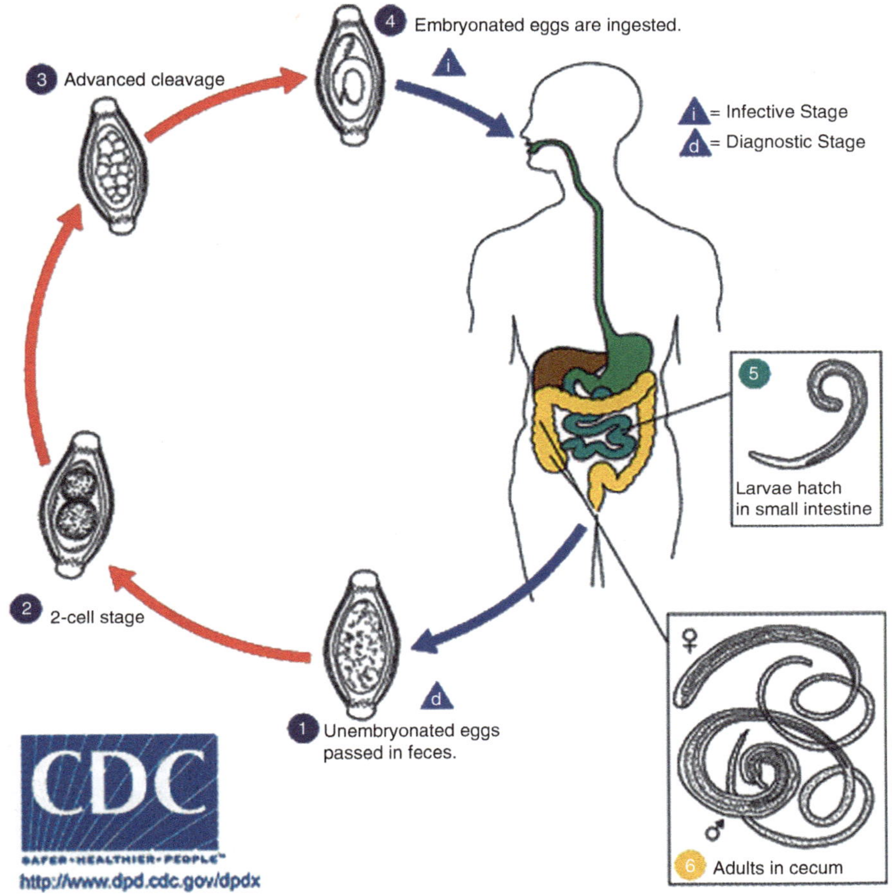

Fig. 7 Life cycle of *Trichuris trichiura*

Hookworms

Human hookworm infections are the second most common helminth infections after ascariasis. They are usually due to the nematode species, *Ancylostoma duodenale* and *Necator americanus*. The life cycle of the hookworms differs from the other intestinal helminths, in their mode of transmission (Fig. 8). Adult worms live in the lumen of the small intestine. They produce eggs that are passed in the stool. Within 1–2 days, and under favorable environmental conditions, the eggs hatch into rhabditiform (first stage) larvae (Fig. 9). The larvae grow in the feces and/or the soil and become infective filariform (third stage) larvae, after 5–10 days. These infective larvae can survive 3–4 weeks in favorable environmental conditions. On contact with the human host, the larvae break through the skin and are carried through the blood vessels to the heart and then to the lungs. They enter the pulmonary alveoli,

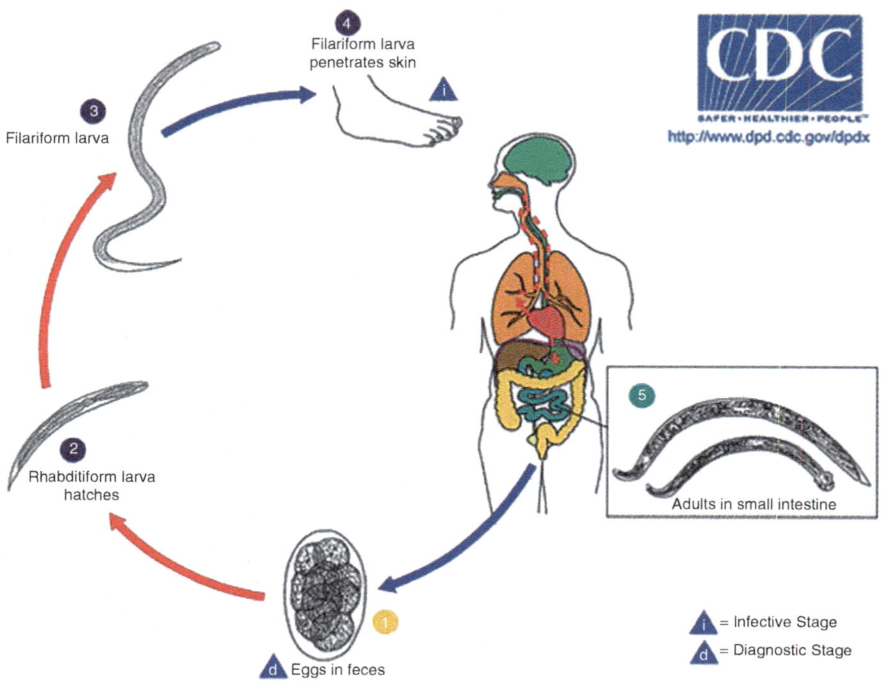

Fig. 8 Life cycle of *Ancylostoma duodenale* and *Necator americanus*

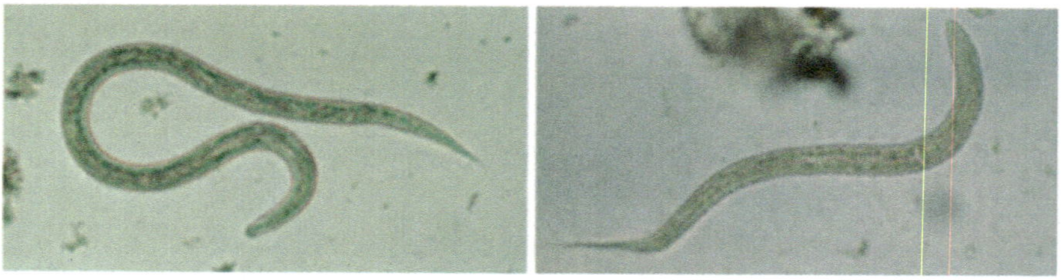

Fig. 9 *Left*: filariform (L3) hookworm larva in a wet mount. *Right*: hookworm rhabditiform larva (wet preparation). Credit: DPDx. http://www.cdc.gov/parasites/hookworm/

up the bronchial tree, and to the pharynx and are swallowed. The larvae reach the small intestine, where they reside and mature into adults. They attach to the intestinal wall with resultant blood loss by the host. The life span of most adult worms is 1–2 years but may reach several years.

Some *A. duodenal* larvae, following penetration of the host skin, can become dormant (in the intestine or muscle). In addition, infection by *A. duodenale* may probably also occur by the oral and transmammary route. *Necator americanus,* however, requires a transpulmonary migration phase.

The disease presentation is primarily iron-deficiency anemia (a result of blood loss at the site of intestinal attachment of the adult worms), in addition to gastrointestinal and nutritional/metabolic symptoms. Cardiac complications and respiratory symptoms can be observed during the cardiac and pulmonary migration of the larvae. In addition, local skin manifestations ("ground itch") can occur during penetration by the filariform (L3) larvae.

Epidemiology of Soil-Transmitted Helminths in Sub-Saharan Africa

The transmission of STHs is closely linked to poverty, poor personal hygiene and sanitation. The geographic distribution of roundworms, whipworms, and hookworms is worldwide, in areas with warm, moist climates, and is widely overlapping. Infection occurs worldwide and is most common in tropical and subtropical areas where sanitation and hygiene are poor. Both *N. americanus* and *A. duodenale* are found in Africa, Asia, and the Americas. *Necator americanus* predominates in the Americas and Australia, while only *A. duodenale* is found in the Middle East, North Africa, and southern Europe. STH infections are usually a result of ingestion of parasite eggs through contaminated food or hands or the active penetration on the skin by larvae in the soil. While STH infections occur predominantly in rural areas, with the absence of proper human waste disposal systems, the social and environmental conditions in many unplanned urban slums in developing countries provide ideal conditions for their transmission (Crompton and Savioli 1993). Helminth infections disproportionately affect marginalized, low-income, and resource-constrained regions of the world (Lustigman et al. 2012a). An equilibrium state exists between parasitic populations, host populations, and the environment. STHs are overdispersed or aggregated in human populations such that most people in a community will have either light-intensity infections, while only a small number will have heavy-intensity infections (Schad and Anderson 1985). Individuals with heavy infections suffer most of the clinical consequences of STHs and are the major source of infection for the rest of the community (WHO 2002), although their proportion in the community may be small. If environmental and/or behavioral conditions are not changed at the same time that chemotherapy program is being implemented, the prevalence and intensity of infection will tend to return to original pretreatment levels through reinfection.

The epidemiology of helminth infections is influenced by several factors, including environmental and social ecology, age, parasite population structure and composition, geographic and household clustering, multiple parasite infections, chemotherapeutic factors, and current diagnostic abilities (Hotez et al. 2008; Lustigman et al. 2012a). There is considerable geographical variation in the occurrence of infections (Brooker et al. 2009). The intensity of infection also varies with age. *A. lumbricoides* and *T. trichiura* are among the heaviest and most frequent infections in children aged 5–15 years, with a decline in intensity and frequency in adulthood (Gilles 1996). On the other hand, hookworm infections show a steady increase in intensity with age, peaking in adulthood (Bethony et al. 2002). Individuals infected with STHs have also been shown to cluster at the household levels (Forrester et al. 1988), especially with regard to familial predisposition to heavy infection with *A. lumbricoides* and *T. trichiura* (Forrester et al. 1990).

Morbidity due to STH infections is a result of worm burden (number of eggs per gram of feces). Thus, disease prevalence is commonly combined with the intensity of infection to assess the epidemiological situation for STH infections and to classify communities into transmission categories. This enables the determination of the appropriate strategies for treatment and control (WHO 2006).

In sub-Saharan Africa, recent geospatial analysis of prevalence and intensity of STHs in children between 2000 and 2018 suggested a reduction in the prevalence in children aged 5–14 years, from 44% in 2000 to 13% in 2018 (Sartorius et al. 2021). These achievements were driven by sustained delivery of preventive chemotherapy, improvement in water sanitation and hygiene (WASH), and economic development. The study also revealed that 25% implementation units still had moderate-to-heavy prevalence estimates of infection exceeding the 2% target threshold. These were largely concentrated in Nigeria, Democratic Republic of the Congo, Ethiopia, Cameroon, Angola, Mozambique, Madagascar, Equatorial Guinea, and Gabon (Sartorius et al. 2021).

Poor Sanitation, Overcrowding, and Helminth Infections

In common with many other parasitic infections, STH infections flourish in impoverished areas characterized by inadequate sanitation and overcrowding. It is commonly assumed that *A. lumbricoides* and *T. trichiura* are more prevalent in urban areas, whereas hookworm is more often found in rural areas (Crompton and Savioli 1993). However, comparable data on STH infections in urban and rural settings are remarkably few, and those that exist indicate a more complicated picture. Studies which surveyed similar age groups and socioeconomic areas indicate that the prevalence of *A. lumbricoides* and *T. trichiura* differs between urban and rural communities but in no systematic manner. By contrast, hookworm appears to be equally prevalent in both urban and rural settings. The precise reasons for the urban-rural dichotomies for *A. lumbricoides* and *T. trichiura* are as yet unclear. Differences in prevalence of *A. lumbricoides* and *T. trichiura* in urban and rural areas may reflect differences in sanitation or population density. Socioeconomic differences will also play an important role. It is clear that further work is needed to resolve these issues.

The parasites that cause helminth infections are naturally present in the environment. However, the high prevalence of intestinal parasite infections in certain parts of the world is closely correlated with poverty and poor environmental hygiene, namely, (a) the lack of safe water supply, (b) contamination of the environment by human waste, (c) lack of shoes, and (d) poor personal hygiene. Improved latrines have been shown to be associated with a reduced prevalence of STH infections (Steinbaum et al. 2019).

Unlike viruses, bacteria, fungi, and protozoa, these parasites do not multiply in the human host. Reinfection can therefore occur only as a result of new contact with the contaminated environment. The World Health Organization asserts that school-aged children are an important group in areas where STHs are prevalent, children are disproportionately infected, and the "worm load" in infected children is often greater than in adults (WHO 1997). Also, children with heavy worm infections are more likely to contaminate the environment and thereby increase the risk of infection to others.

Helminth Infections, Anemia, and Undernutrition

Iron-deficiency anemia is the most prevalent nutritional deficiency worldwide (Stoltzfus et al. 1997; Viteri 1994; Zetterstrom 2004; Van Den Broek 2003; Stephenson et al. 2000). More than 90% of affected individuals live in the developing world, where helminth infections are highly prevalent, and the parasites are endemic (Stephenson et al. 2000). Helminths are known to be significant contributors to the overall anemia burden in the developing world (Stephenson et al. 2000; Traore' et al. 1998). The health effects associated with anemia are most pronounced in children and women of reproductive age (Viteri 1994; Crompton and Whitehead 1993). In addition to the morbidity in these groups, anemia generates profound physiological and economic costs in the general population (Stephenson et al. 2000). The amount of blood loss that results from iron-deficiency anemia depends on the degree of infection (Stephenson et al. 1985; Stoltzfus et al. 1996), as well as the dietary intake of iron (Roche and Layrisse 1966; Tatala et al. 1998), and also the presence of other parasitic infections that can cause blood loss or hemolysis, such as malaria and *T. trichiura* infection (Stoltzfus et al. 1997; Robertson et al. 1992; Olsen et al. 1998). School-aged children are most vulnerable to iron-deficiency anemia caused by parasitic infection because they tend to harbor the heaviest worm loads in communities (Bundy 1988). STH infections are associated with decreased appetite and low food intake (Stephenson et al. 1993), resulting in decreased growth rate, poor fitness, decreased activity, and poor cognitive function. Micronutrient losses and nutrient malabsorption due to ascariasis and blood loss due to trichuriasis can lead to iron deficiency, iron-deficiency anemia, and poor growth rate (Stephenson et al. 1993, 2000). An analysis of the association between hookworm and anemia in school-aged children in Zanzibar showed that 25% of all anemia cases, 35% of iron-deficiency anemia cases, and 73% of severe anemia cases could be attributed to hookworm infection (Stoltzfus et al. 1997). It has been observed that hookworm contributes more than schistosomiasis or malaria to iron-deficiency anemia in school-aged children (Stephenson et al. 1985; Stephenson 1987). Other studies in sub-Saharan African countries have revealed similar trends between STH infections, anemia, and malnutrition (Osazuwa et al. 2011; Brooker et al. 1999).

Burden of Disease and Distribution in Sub-Saharan Africa

Figures 10 and 11 show the distribution and prevalence of STHs in Africa, respectively. In 2018, it was estimated that 35 million children aged 5–14 years were infected with at least one STH. It was also estimated that 2.9 million children harbor

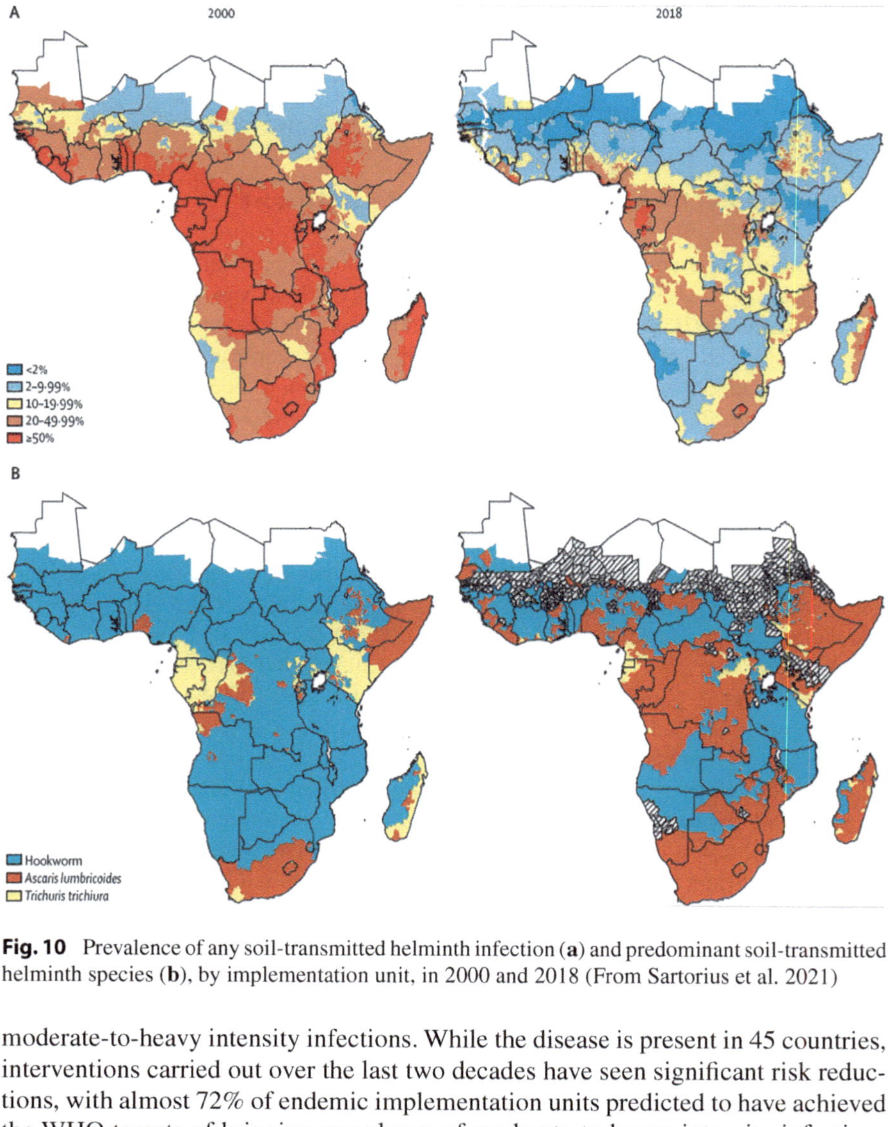

Fig. 10 Prevalence of any soil-transmitted helminth infection (**a**) and predominant soil-transmitted helminth species (**b**), by implementation unit, in 2000 and 2018 (From Sartorius et al. 2021)

moderate-to-heavy intensity infections. While the disease is present in 45 countries, interventions carried out over the last two decades have seen significant risk reductions, with almost 72% of endemic implementation units predicted to have achieved the WHO targets of bringing prevalence of moderate-to-heavy intensity infections in children to less than 2%.

It is estimated that between one-quarter and one-third of SSA's population is affected by one or more STH infections (Stephenson et al. 1990). Children, especially school-aged children, are the most infected, with half (89 million) of the estimated population of school-aged children infected with hookworm, ascariasis, trichuriasis, or some combination of these STH infections (Brooker et al. 2006a; Hotez and Kamath 2009). The high-intensity infections observed in children accounts for the physical and mental deficits, reduced school performance, and the

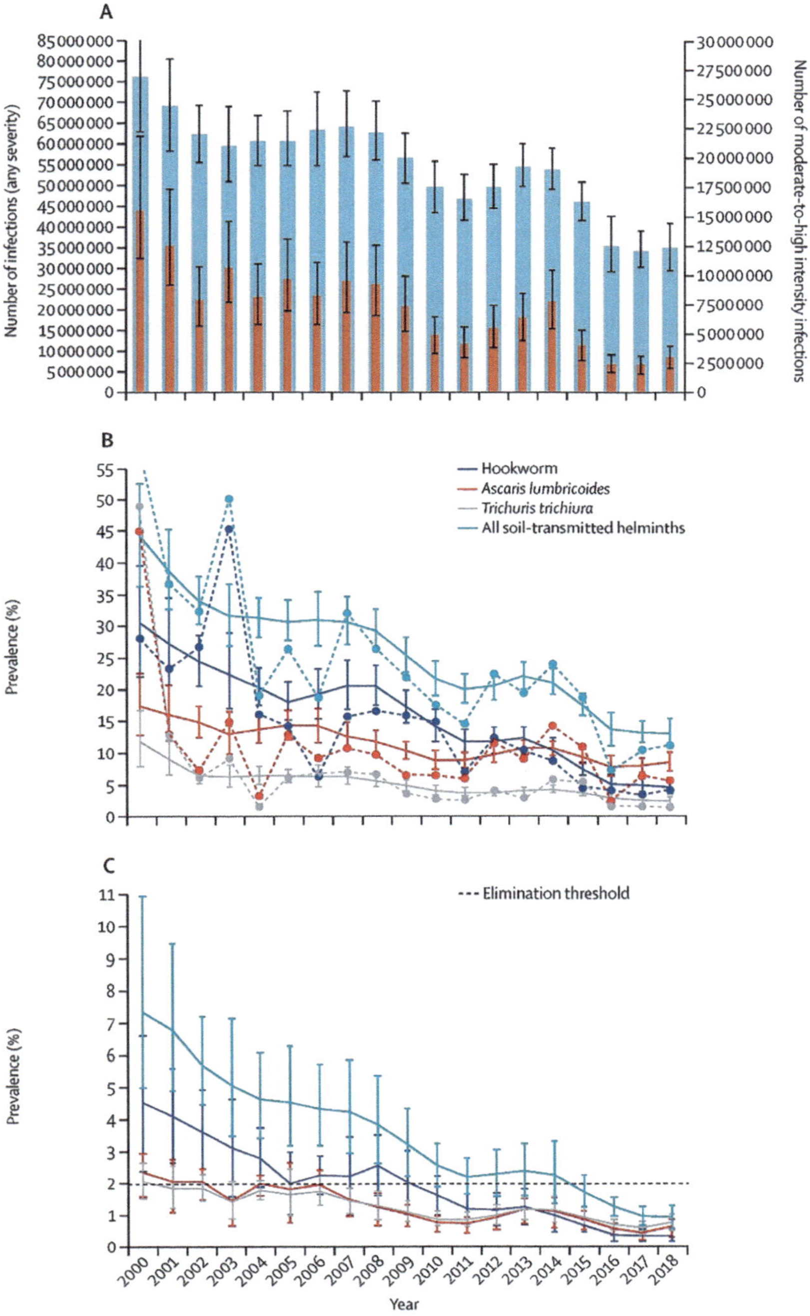

Fig. 11 Prevalence of any soil-transmitted helminth infections (**a**) and of moderate-to-high intensity infections (**b**), and the absolute number of prevalent cases (**c**), in children aged 5–14 years in sub-Saharan Africa, 2000–18 (From Sartorius et al. 2021)

high disease burden in SSA (Hotez and Kamath 2009; Stephenson et al. 1990; Miguel and Kremer 2003).

The morbidity associated with STH infections is primarily the effect on nutritional status and nutrient deficiencies in infected people (Hall et al. 2008). In children, STHs result in cognitive impairment. In severe cases, surgical interventions for intestinal and biliary obstructions are required. The specific signs associated with STH infections are intestinal bleeding and anemia (Stoltzfus et al. 1996), malabsorption of nutrients (Crompton and Nesheim 2002; Solomons 1993), competition for micronutrients (Curtale et al. 1993), impaired growth (Taren et al. 1987), loss of appetite and reduction of food intake (Stephenson et al. 1993), diarrhea (Callender et al. 1998), reduction in fluency and memory (Nokes et al. 1992; Kvalsvig et al. 1991), intestinal and biliary obstructions (De Silva et al. 1997), and rectal prolapse (Montresor et al. 2002).

Hookworms are the most common STH infections and the most common NTD in SSA, accounting for up to one-third of the total burden from NTDs in SSA (Hotez and Kamath 2009). It is estimated that 29% of SSA's population (198 million) are infected with hookworm, including 40–50 million school-aged children (Brooker et al. 2006a, b). Nearly one-third of the world's hookworm infections occur in SSA, with the greatest number of cases occurring in Nigeria (38 million) and the Democratic Republic of the Congo (31 million), followed by Angola, Ethiopia, and Côte d'Ivoire (10–11 million) (Hotez and Kamath 2009). The distribution of hookworm in SSA reveals a high prevalence and intensity, particularly in the coastal regions (Mabaso et al. 2004) and areas of extremely high temperatures (where land surface temperatures exceed 37 °C–40 °C) (Behnke et al. 2000; Brooker and Michael 2000; Brooker et al. 2000, 2002). In SSA, both *N. americanus* and *A. duodenale* are endemic, with *N. americanus* being the predominant species (Albonico et al. 2008). Iron-deficiency anemia in school-aged and preschool-aged children is a principal symptom of hookworm infection (Stoltzfus et al. 1997; Albonico et al. 1998; Brooker et al. 1999; Hotez et al. 2004). It has been estimated that 35% of iron-deficiency anemia and 73% of severe anemia in school-aged children in Zanzibar were attributable to hookworm (Stoltzfus et al. 1997). In addition to its effects on children, hookworms have also been recognized as an important cause of anemia and morbidity in women of reproductive age in SSA, especially among pregnant women (Hotez et al. 2004; Crompton 2000). At any given time, almost seven million pregnant women in SSA (up to one-third of pregnant women in the region) are infected with hookworm (Brooker et al. 2008), thus reflecting the importance of hookworm infections as a cause of maternal and child anemia (Hotez and Kamath 2009).

Like hookworm infections, the highest prevalence and intensity of *Ascaris* and *Trichuris* infections occur in school-aged children (Brooker et al. 2006a), with an estimated 173 million and 162 million people being infected in SSA with *Ascaris* and *Trichuris,* respectively. Of these, 36 million school-aged children are infected with ascariasis and 44 million with trichuriasis. For both infections, the largest number of cases occurs in Nigeria (being the most populous African country), where coinfections with hookworm are common (Dada-Adegbola et al. 2005). Tens of millions of cases are also found in Ethiopia, DRC, and South Africa (Hotez and Kamath 2009). The distribution of *Ascaris* and *Trichuris* infections in SSA is patchy, with the highest prevalence occurring in equatorial Central and West Africa, eastern Madagascar, and southeast Africa (Brooker et al. 2006a). High prevalence rates of ascariasis and trichuriasis are also found in urban areas compared with rural areas (Brooker et al. 2006a); thus, it has been suggested that increased urbanization in SSA may promote the emergence of ascariasis and trichuriasis in the future (Hotez and Kamath 2009).

Available Tools and Control Strategies

The control of STHs requires, principally, the assessment of prevalence, intensity, and distribution of infections and the use of chemotherapy to reduce and clear parasite load in infected individuals or the whole community. These approaches are based on different tools and strategies that have been tested and implemented over the years.

In assessing the prevalence and intensity of STHs, the most widely used approach is the Kato-Katz (K-K) technique (Katz et al. 1972), which has been recommended by the World Health Organization (Montresor et al. 1998). Worm burden is commonly measured by the number of eggs per gram (EPG) of feces (Montresor et al. 1998). The estimated prevalence and intensity of infections are then used to classify communities into low-risk (<20%), moderate-risk (≥20% but <50%), and high-risk (>50%) transmission categories, which enable the determination of the appropriate treatment strategy for a community (WHO 2006). The K-K method has however been shown to lack sensitivity if only a single stool sample is examined and especially in areas with high proportions of light-intensity infections (Booth et al. 2003). A small number of helminths' eggs, unequally excreted over days and patchily distributed in stool, can occasionally not be detected in the small amount of stool (i.e., 41.7 mg) examined with the K-K, hence negatively impacting on the method's sensitivity (Knopp et al. 2008). Thus, for more accurate estimations of parasite intensity (especially in areas with light-intensity infections), examination of multiple stool samples instead of a single one is advocated, in combination with other diagnostic methods (Knopp et al. 2008). These other tools for STH diagnosis include the McMaster test (Levecke et al. 2011), the Koga agar plate (KAP) method (Koga et al. 1991), FLOTAC method (Knopp et al. 2009; Cringoli 2006), FECPAK^{G2} (Ayana et al. 2018), and the molecular methods (Verweij et al. 2007). A comparison of the diagnostic performance of a single Kato-Katz (reference method) with that of other microscopy-based methods (duplicate Kato-Katz, Mini-FLOTAC, and FECPAK^{G2}) and qPCR for the detection and quantification of STH infections in three drug efficacy trials in Ethiopia, Lao PDR, and Tanzania showed that a single Kato-Katz is underestimated the intensity of infection in the community and that diagnostic specific thresholds to classify intensity of infection are warranted for Mini-FLOTAC, FECPAK^{G2}, and qPCR (Cools et al. 2019). The Kato-Katz is also the only microscopy-based method that meets the minimal TPP diagnostic criteria for application in the planning, monitoring, and evaluation phase of an STH program, while the qPCR is the only method that could be considered in the phase that aims to seek confirmation for cessation of program (Cools et al. 2019).

In addition to determining the prevalence and intensity of infections, tools such as the geographical information system (GIS) and remote sensing (RS) have been used to better understand helminth ecology and epidemiology, to develop low-cost ways to identify target populations for treatment. GIS and RS have been used to develop the Global Atlas of Helminth Infections (GAHI), and to estimate the number of infections in school-aged children in sub-Saharan Africa (Brooker et al. 2000, 2009; Brooker and Michael 2000). The GAHI provides a global information resource on the distribution of STH and schistosomiasis, and can be accessed at http://www.thiswormyworld.org/. The information in the database enables users to visualize the following: (1) maps showing the prevalence of infection based on survey data; (2) a predictive risk map showing the probability that infection prevalence warrants MDA, according to recommended WHO thresholds; and (3) a control planning map showing which districts require MDA treatment or where further surveys would be helpful in defining risk (Brooker et al. 2010). Coupled with mathematical modeling for helminth infections (Basanez et al. 2012), it is expected that data collection efforts will take advantage of the current opportunities to integrate multidimensional data, providing a more scientific input into programmatic activities and improving the decision-making in STH control.

Interventions for the control of STHs require the treatment of whole communities through mass drug administration (MDA) and/or individual chemotherapy, supplemented with improved sanitation, environmental conditions, and education (Prichard et al. 2012). MDA programs for STHs are most often targeted at school-aged children who are most at risk of acquiring heavy infection and developing associated morbidity. They are also most accessible for intervention through school-based programs.

Thus, the World Health Assembly resolution WHA54.19, passed in 2001, set the global target of treating by the year 2010 at least 75% of all school-aged children at risk of morbidity from STHs and schistosomiasis (Lustigman et al. 2012a). The current strategic plan was aimed at eliminating STHs as a public health problem in children, by the year 2020 (WHO 2012). In highly endemic areas, MDA programs may also be directed at whole communities. The focus of MDA programs for STHs is to achieve long-term reduction in infection prevalence and intensity, and consequently of the associated morbidity, rather than aiming for the elimination of infection (Prichard et al. 2012). The health outcomes aimed for in STH programs and other NTDs were closely related to those of the health-related Millennium Development Goals 1, 2, and 6 of poverty reduction, improvements in education, and combating diseases, respectively (Molyneux and Malecela 2011; Lustigman et al. 2012a). However, following the pledge to eliminate NTDs through the Sustainable Development Goals, especially Goal 3 which calls for the need to ensure health and well-being for all and ending the epidemics of AIDS, tuberculosis, malaria, and other communicable diseases by 2030, there is an urgent call for action (Smith and Taylor 2016; Addisu et al. 2019).

The current WHO recommended medicines for the treatment of STH—albendazole (400 mg) and mebendazole (500 mg)—are effective (Maipanich et al. 1997; Steinmann et al. 2011), inexpensive, and easy to administer by nonmedical personnel (e.g., teachers). They have been through extensive safety testing and have been used in millions of people with few and minor side effects. Both albendazole and mebendazole for the treatment of STHs are donated through WHO to disease control program in all endemic countries for the treatment of all school-aged children (WHO 2020) (see chapter on Medicines donation program, NTDs in SSA in this edition). Ivermectin has also been found to be effective for the control of uncomplicated *S. stercoralis* (Buonfrate et al. 2019; Hailu et al. 2020).

Challenges of Program Implementation

The STH control program is not without challenges. A challenge to program implementation is the lack of resources required to effectively carry out diagnosis, mapping, treatment, monitoring, and evaluation. While albendazole and mebendazole for STH control are donated by GlaxoSmithKline and Johnson & Johnson, respectively (Prichard et al. 2012), MDA still requires the costly prerequisite for mass screening and diagnosis. Very little funding, through the financial support from Foundations, government, and agencies, is available to support the operational costs of control programs, and these are directed at short-term objectives (Lustigman et al. 2012a). Integrating various programs therefore has the advantage of reducing the operational costs associated with MDA.

Another important challenge is in monitoring and evaluation of the control programs, especially in the light of the level of sensitivity of various diagnostic tools at low-intensity infections. There are major deficiencies in intervention and diagnostic tools that are appropriate to the changing requirements of large-scale preventative chemotherapy strategies (Boatin et al. 2012). Various diagnostic tools have their usefulness and drawbacks, and their use should be determined by the needs, requirements, and resources available. For instance, the FLOTAC method is more sensitive than the K-K particularly for hookworm eggs but requires a centrifuge and is relatively low throughput (McCarthy et al. 2012). The McMaster test on the other hand is not suitable for community studies where intestinal schistosomiasis is coendemic (McCarthy et al. 2012). Molecular methods can only determine parasite presence and genetic markers for anthelminthic resistance (Schwenkenbecher et al. 2007; Diawara et al. 2009), but they cannot determine parasite intensity. The K-K method, which has become the gold standard in STH diagnosis, lacks sensitivity if only single stool samples are examined, especially in areas with high proportions of light-intensity infections (Booth et al. 2003; Nikolay et al. 2014). A small number of helminth eggs, unequally excreted over days and patchily distributed in stool, can occasionally not be detected in the small amount of stool (i.e., 41.7 mg), examined with the K-K hence negatively impacting on the method's sensitivity (Knopp et al. 2008). Thus, for more accurate estimations of parasite intensity (especially in areas with light-intensity infections), examination of multiple stool samples instead of a single one is advocated (Knopp et al. 2008). The programmatic implications of this therefore lead to more manpower and time that must be used toward the tail end of STH control programs.

There are also limitations to MDA programs based on single-dose anthelmintic monotherapy for STHs (Keiser and Utzinger 2008). Single-dose therapy with either albendazole (ABZ) or mebendazole (MBZ) is currently highly effective against *A. lumbricoides*; this is not the case for hookworms and whipworms (Prichard et al. 2012). Research is therefore needed to optimize the dose regimes of existing drugs used for the treatment of STHs (Geary et al. 2010). Another limitation to the MDA programs is the identification of genetic polymorphisms in β-tubulin, the marker for benzimidazole resistance in livestock parasites *T. trichiura* and *N. americanus* (Prichard et al. 2012). This partly explains that the low/variable efficacy of ABZ or MBZ in *T. trichiura* and *N. americanus* may be due to benzimidazole-resistant genotypes in these parasites in some populations (Diawara et al. 2009). This development, if not properly understood, may threaten the sustainability of control efforts against STHs. Thus, routine monitoring and evaluation should be essential components for STH control programs. Unfortunately, most health-care systems do not yet prioritize surveillance of NTDs. Thus, outreach is needed to extend to and involve cooperating health services, health professionals, environmental health officers, and communities that can organize themselves for surveillance (Lustigman et al. 2012a). In addition, modeling tools may help forecast the change in prevalence of soil-transmitted helminths during control programs (Montresor et al. 2013).

Mass treatments further face the challenges of sustainability. Noncompliance and "patient fatigue" to MDA programs have been reported for lymphatic filariasis control programs (Boyd et al. 2010; El-Setouhy et al. 2007). Community involvement and ownership of control activities are also crucial for successful programs (Lustigman et al. 2012a). There may be further "donor fatigue" to sustained funding if initial successes are not maintained (Lustigman et al. 2012a). Finally, there are many social, political, and environmental factors affecting STH control (Gazzinelli et al. 2012). Control programs must be supplemented by improvements in the availability of potable water, sanitation, and hygiene (WASH); improved socioeconomic development; and environmental sustainability, through multidisciplinary studies encompassing the various stakeholders. These must principally be accompanied with behavior change if the gains made are to be sustainable.

Recent Developments in STH Research and Control

Human Gut Microbiome Research

The association of the human microbiome with several infectious and noninfectious diseases has rapidly become the focus of advanced research. The human gut microbiome is a complex, dense community of bacteria with versatility that is influenced from birth through life by age (Ursell et al. 2012), diet (Savage 1977), lifestyle (Rubel et al. 2020), and disease (Manor et al. 2020). Intestinal parasitic pathogens, such as hookworms, whipworms, and tapeworms share a common niche with gut microbes which make their interactions more closely defined. These interactions are crucial for the survival of the parasites and microbes in the human host. The long co-evolutionary persistence of these parasites with the human host has been attributed to the helminths' ability to immune-modulate the host, a function that is postulated to be the contribution of the gut microbiota (Giacomin et al. 2015).

Identification of microbial biomarkers of helminth infection: In intestinal tract disorders such as inflammatory bowel diseases (IBD), the role of the microbiome in immune-modulation and disease progression has been extensively studied (Glassner et al. 2020). Transfer of some bacterial species from IBD individuals (human or mice) to healthy mice induces inflammation and cause disease in recipients (Schaubeck et al. 2016; Britton et al. 2019; Ohkusa et al. 2003), providing a clear association of bacteria with IBD. While no similar clear-cut links between soil-transmitted helminthiasis and gut microbes have been established, several studies have investigated the dynamics of bacterial composition during helminth infection to associate signature microbes to infection (Gause and Maizels 2016; Lee et al. 2014; Gordon et al. 2020; Rubel et al. 2020). However, findings across studies are inconsistent, while some studies report no significant differences in bacteria taxa between infected and non-infected persons (Cooper et al. 2013; Cantacessi et al. 2014; Jenkins et al. 2018), others report observing differences in bacteria genera such as *Akkermansia*, *Bifidobacterium*, *Prevotella*, and *Treponema* (Lee et al. 2014; Kay et al. 2015). To address these issues and contribute to establishing definite involvement of specific candidate bacteria for further consideration in STH interventions, methods need to be standardized. This should include but is not limited to sample collection, processing, accurate metadata reporting, and collating all these information on a more accessible repository in regional and/or global catalogues. An example of such an initiative has begun with the Mosquito Microbiome Consortium (Dada et al. 2021). Putting data in one place can help appreciate what drives differences between studies and what could be common bacteria species across different helminth infections. For example, *Prevotella* is demonstrated to increase in *S. haematobium* and *Trichuris* infection in Malaysia (Ramanan et al. 2016) and Zimbabwe (Kay et al. 2015), respectively.

Microbiome association with disease pathophysiology: Among the many questions that remain around helminth infections, one is understanding disease risk factors. Intestinal parasite infections have mainly been associated with environmental (Stromberg 1997), socioeconomic (Mehraj et al. 2008; Igore et al. 2020), and hygienic conditions (Gizaw et al. 2018; Njambi et al. 2020). Although these factors significantly correlate with disease epidemiology, they do not explain pathophysiological differences in infection status and diseases outcomes observed among individuals. A recent study has shed light on the connection between microbiome and susceptibility to *S. mansoni*. Using transinfection of a mouse model with the microbial community from the human gut, it has been demonstrated that the baseline microbiome is important for determining infection load and immune regulation in response to the parasite (Cortés et al. 2020). *Ruminococcus 1*, which has mucin-degrading bacterial species, and microbial taxa such as *Butyricicoccus* (Devriese et al. 2017) and *Intestinimonas* (Bui et al. 2015) which play key roles in the production of short-chain fatty acids (SCFA) were negatively associated with infection burden. While in vivo animal studies such as the one described are available for *S. mansoni* (Bui et al. 2015; Jenkins et al. 2018) and *Trichuris* (Li et al. 2012; Wu et al. 2012; Holm et al. 2015; Broadhurst et al. 2012), the lack of an animal model for the predominant intestinal helminth, *N. americanus*, has challenged similar investigations to be done.

Immunomodulation through helminth-modified host microbiome: Helminths are known to suppress host inflammatory responses but the mechanisms by which they do this is largely undetermined (Giacomin et al. 2015). Modulation of the immune system by persisting helminth infections has been shown to influence responses of hosts to other co-infections (Degarege et al. 2012; Harris et al. 2009; Resende et al. 2007) and be associated with reduced allergies (Yazdanbakhsh et al. 2001). The interplay between helminth infections, microbiome and host immunity has been largely linked to T-helper cells (Grencis 2015; Jirků et al. 2021) but the proposed mechanisms vary (Brosschot and Reynolds 2018).

Microbiome association with treatment outcome: Albendazole and praziquantel have been accepted drugs for the treatment of hookworms and schistosome infections, respectively. Recent studies have raised concerns about reduced cure rate (Humphries et al. 2013) and possible drug resistance. Considering efforts to understand the etiology of STH infections with microbiome interactions, recent observational and interventional studies have begun to investigate microbial composition pre-and post-treatment (Martin et al. 2018; Sharpton et al. 2020). In one of these studies where a placebo was offered to infected control individuals, no differences in the microbiome at the phylum levels were observed between albendazole and placebo-treated participants pre- and post-treatment. However, in treated individuals who remained infected, significant differences were attributed to an alteration in *Bacteroidetes* with an observed increase in *Trichuris* infection (Martin et al. 2018).

Computer-Aided Anthelmintic Drug Discovery

The current drugs used for STH treatment are old, introduced for human use in the early 1980s, and they have been used in millions of patients; thus, the likelihood of a full-blown resistance developing in future exists. The benzimidazoles used for treatment and MDAs target the β-tubulin genes of the parasites, and already mutations conferring resistance have been reported widely in veterinary nematodes (Kaplan 2004) and are also present in low levels in human parasites. There are very few new drugs for human health in the pipeline, mainly because R&D costs that are associated with conventional approaches are huge and are economically disincentive to pharmaceutical companies. The conventional approaches to drug discovery are very challenging and expensive, are a time-consuming process, need several hands, and take approximately 15 years to get approval (Roses 2008). It is also characterized by a low rate of therapeutic discovery. Moreover, the number of approved drugs is steadily declining because of these challenges, with toxicity accounting for approximately one-third of withdrawals (Guengerich 2011), and the rest mostly due to early-stage terminations that are related to suboptimal efficacy and safety issues (Ferri et al. 2013). Repurposing of existing drugs for helminth treatment can overcome some of the limitations of costs. Another less expensive approach is using in silico drug design methods to screen compounds of natural products in databases against parasite targets. The publication of the genomes of major helminths, e.g., *N. Americanus* (Tang et al. 2014), *S. haematobium* (Young et al. 2012), and *S. mansoni* (Berger et al. 2021), and the diverse structural genomic data combined with high-performance biocomputing capacity are key in expediting the identification of novel anthelmintics.

Natural products have traditionally been a source of compounds for medical diagnosis and treatment (Veeresham 2012). Currently, natural products and their derivatives account for over one-third of new molecular entities approved by the US FDA (Patridge et al. 2016). Furthermore, the structural and chemical diversity of natural product-derived compounds serve as rich sources of scaffolds for the discovery of potentially novel drug (Cragg and Newman 2013; Lahlou 2013). Chemoinformatic natural product molecules in databases including the Traditional Chinese Database (Chen 2011), AfroDB the African Natural Product Database (Ntie-Kang et al. 2013), and North African Natural Product Database (Ntie-Kang et al. 2017) collectively contain over 30,000 freely accessible unique natural product compounds. Also, the development of machine learning algorithms would facilitate the rapid screening and discovery of even more novel molecules using any identified compounds as scaffolds. The approach here is targeting physiological processes that are unique to the parasite with compounds.

Computational design of novel anti-hookworm drugs circumventing the benzimidazole-associated β-tubulin mutations. The emergence of drug resistance against albendazole and mebendazole and their reduced efficacies necessitate the need for new drugs. Agyapong et al. (2021) utilized pharmacoinformatics techniques to screen a library of 3390 compounds African natural compounds against a modelled structure of *Necator americanus* β-tubulin protein through docking to binding pocket to predict ZINC compounds: ZINC95486082 and ZINC95486052 and euphohelionon as potential inhibitory molecules of the hookworm *Necator americanus* β-tubulin gene. The three compounds were predicted to form no interactions with the resistance-associated mutations Phe167, Glu198, and Phe200 and were pharmacologically profiled to be drug-like demonstrating that the compounds are potential anthelmintic drug leads. Novel critical residues comprising Leu246, Asn247, and Asn256 were also predicted for binding. Euphohelionon was then used as a template for the de novo fragment-based design of five compounds which were labelled A1, A2, A3, A4, and A5: with four of them exhibiting characteristics indicating easy synthesis. All the five new molecules docked firmly in the binding pocket of the β-tubulin with no binding interactions with the three known resistance-associated mutation residues. The usefulness of these compounds can only be determined after in vitro and in vivo assays and if they should pass the tests would serve as templates from which novel anthelmintics may be designed specifically to circumvent the resistance to current benzimidazole.

Machine learning (ML) approaches to anti-helminth compounds prediction. The leveraging of the ever-expanding datasets on open databases and existing algorithms have also been used to predict anthelmintic molecules. Agyapong et al. (2021) have recently developed a proteochemometric model (PCM) that predicts the bioactivity of small compounds against the protein receptors. The development of the model explored the activity of statin, a new antischistosomal drug (Rojo-Arreola et al. 2014) against schistosomulae. The PCM model was trained on 437 inhibitors of the β-tubulin protein from different schistosomiasis-related databases; PubChem (Kim 2016), ChEMBL (Gaulton et al. 2012), DrugBank (Wishart et al. 2018), BindingDB (Gilson et al. 2016), Protein Data Bank (PDB) (Burley et al. 2017), and SchistoDB a helminth-specific databases (Zerlotini et al. 2013), and datasets of antischistosomal molecules from high-throughput screening (HTS) experiments of the identified anti-*S. mansoni* molecules (Gaba et al. 2014). Proteochemometric modelling (PCM) Webserver for predicting the bioactivity between small compounds and tubulin receptors that uses a PCM-based Support Vector Machine (SVM) model to predict the inhibitory activity of chemical compounds against tubulin proteins is available on a server.

The ML approaches for drug discovery could speed up the discovery of novel therapeutics for neglected tropical diseases with increased collaborative efforts by the scientific community.

Establishment of African Strain of *N. Americanus*-Hamster Model

The first African strain of *N. americanus* into a laboratory rodent (hamster) has been established at Cappello's laboratory at Yale University. This involved adapting strains of the parasite from Ghana into Golden Syrian hamsters through a collaborative effort with Wilson's laboratory at NMIMR. This model in addition to the existing *Ancylostoma* model also at Yale University provides access to sustainable quantities of parasite proteins and life cycle stages (eggs, larvae, adults) and offers opportunity to study host-pathogen interactions in vivo, as well as for the development of serologic tests with adequate sensitivity/specificity for use as a monitoring tool for hookworm infections. Establishment of animal models in endemic countries laboratories will contribute to making this objective achievable early.

Further Research for Control and Elimination

The Disease Reference Group on Helminth Infections (DRG4), established in 2009 by the Special Program for Research and Training in Tropical Diseases (TDR), has outlined the research priorities and gaps associated with the control of helminth infections. Basic helminth research has been hampered by the need for operational control activities which attract most of the funding available (Lustigman et al. 2012a; Moran et al. 2009). It has been suggested that research into helminth genomics and the use of bioinformatics tools may enhance the functional analysis of genes and the proteins they encode and the identification of candidate antigens for vaccine development (Lustigman et al. 2012a).

The drugs used for helminth control are limited. In the case of STHs, there are four main antihelminthics on the World Health Organization model list of essential medicines for the treatment and control of STH, and these include albendazole, mebendazole, levamisole, and pyrantel pamoate (Utzinger and Keiser 2004; WHO 2006). However, albendazole and mebendazole are the main drugs used, because of their efficacy (Prichard et al. 2012). In Kenya, for example, it was found that the antihelminthic drug levamisole was highly effective against *A. lumbricoides* but ineffective against hookworms and *T. trichiura*; hence, levamisole was exchanged with albendazole (Magnussen et al. 1997).

The rather small number of drugs against STH infections poses concerns should there be development of anthelminthic resistance following years of treatment (Olliaro et al. 2011; Vercruysse et al. 2011), and in view of the fact that despite the benefits to treated populations, helminth infections have been resilient and persistent in their host populations (Prichard et al. 2012). Currently, single-dose treatment with albendazole and mebendazole is effective against *A. lumbricoides*, but not hookworms and whipworms (Prichard et al. 2012). In many endemic areas, STH infections are coendemic and polyparasitism is not uncommon. Thus, the use of single-dose treatments with suboptimal responses against some parasites may pose challenges to control programs. Furthermore, genetic polymorphisms in the benzimidazole drug resistance marker (β-tubulin) have been detected in *T. trichiura* and *N. americanus* (Diawara et al. 2009, 2013). Thus, one priority research need is to undertake well-designed dose-finding studies to optimize the dose regimes of existing drugs used for the treatment of STHs (Geary et al. 2010).

Another research need is to investigate the development of resistance to benzimidazole, for monitoring and evaluation purposes. Recently, molecular assays have been developed to detect putative resistance genetic changes in *A. lumbricoides*, *T. trichiura*, and hookworms (Diawara et al. 2013). However, these new assays need to be validated. A third research need is the formulation of new drug combinations, novel drugs, and vaccines. The use of new drug combinations will help increase the spectrum, effectiveness, and convenience of drug administration while reducing the development of resistance. Currently, new antihelminthics for STHs (e.g., tribendimidine, nitazoxanide, monepantel) do exist, but these lack in efficacy compared to current drugs (Prichard et al. 2012; Hu et al. 2009). It has also been suggested that new veterinary antihelminthics (e.g., emodepside, derquantel) should be evaluated for potential use against helminth parasites of humans (Prichard et al. 2012; Olliaro et al. 2011; Keiser and Utzinger 2010). Similarly, there have been some advances in vaccine development against hookworm and some other helminth infections (Prichard et al. 2012).

Being principally diseases of the poor, there are many social, cultural, political, and environmental factors that affect the success of STH control programs (Gazzinelli et al. 2012). While chemotherapy is considered the most effective control tool by far, it is recommended that a more holistic, integrated approach be used. Research is thus needed to understand the range of social, cultural, and environmental factors and conditions involved in helminth transmission, spread, persistence, and control, such as the role of poverty, environmental and climatic change, and active and broadly based community participation, intersect oral collaboration, and equitable access to health services. All of these must be considered in developing and scaling up acceptable, cost-effective, and sustainable control and elimination programs.

Finally, there is the need for establishing capacity in disease-endemic countries (DECs), in order to undertake the various research needs facing the national and global helminth control programs. The research infrastructure in DECs is at the mercy of international donor agencies, major funding bodies, and academic institutions from the developed world, who contribute to the creation of (not always equitable) North-South "partnerships" (Osei-Atweneboana et al. 2012). Thus, while most funding for control implementation programs go to DECs, most funding for research and development remains in developed countries. Furthermore, there is an outflow of trained personnel to developed countries, because of low salaries, unattractive work conditions and enabling environments, and low government support for research activities. Thus, comprehensive approaches to human resource management as well as a systemic approach to capacity building, including recognition of the importance of developing a strong research culture in DECs, are needed.

Outlook for the Next Decade

The London Declaration on Neglected Tropical Diseases (Uniting to Combat NTDs 2012) in January 2012 has led to increased commitment on the part of the international community to provide support for the global efforts to control or eliminate several NTDs by 2020. The global partners committed to:

- Sustain, expand, and extend programs that ensure the necessary supply of drugs and other interventions to help eradicate Guinea worm disease and help eliminate by 2020 lymphatic filariasis, leprosy, sleeping sickness (human African trypanosomiasis), and blinding trachoma.
- Sustain, expand and extend drug access programs to ensure the necessary supply of drugs and other interventions to help control schistosomiasis, soil-transmitted helminthes, Chagas disease, visceral leishmaniasis, and river blindness (onchocerciasis). Advance R&D through partnerships and provision of funding to find next-generation treatments and interventions for neglected diseases. Enhance collaboration and coordination on NTDs at national and international levels through public and private multilateral organizations to work more efficiently and effectively together.
- Enable adequate funding with endemic countries to implement NTD programs necessary to achieve these goals, supported by strong and committed health systems at the national level.
- Provide technical support, tools, and resources to support NTD-endemic countries to evaluate and monitor NTD programs.

These goals would determine the global agenda for the control of STH infections and other NTDs, for the next decade and beyond. The pharmaceutical companies providing the drugs used for NTDs have pledged to donate drugs for as long as they are needed. The UK's Department for International Development (DFID) announced a fivefold increase in their aid for NTDs (http://endtheneglect.org/2012/01/uk-announces-five-fold-increase-in-funding-for-ntds/). The recent funding cuts from the UK government in April 2021 (https://unitingtocombatntds.org/news/our-response-to-the-uks-cuts-to-foreign-aid/), however, means the support toward control activities will be significantly hampered unless other partners step in to fill the gaps. There has also been funding support from other sources such as the USAID (http://www.neglecteddiseases.gov/funding/index.html) and the Bill & Melinda Gates Foundation (http://www.gatesfoundation.org/Media-Center/Press-Releases/2009/01/Bill-and-Melinda-Gates-Urge-Global-Leaders-to-Maintain-Foreign-Aid), toward the 2020 goals, all in the effort to help control and greatly reduce the burden of the most prevalent neglected diseases that affect the world's poorest populations.

Despite these ambitious targets and donations, the 2020 elimination goals could not be achieved. The WHO has recently set six new targets for STHs by the year 2030 (WHO 2020). These include:

- Achieving and maintaining elimination of STH morbidity in pre-school and school age children.
- Reducing the number of tablets needed in preventive chemotherapy for STH.
- Increasing domestic financial support to preventive chemotherapy for STH.
- Establishing an efficient STH control program in adolescent, pregnant, and lactating women.
- Establishing an efficient strongyloidiasis control program in school-aged children.
- Ensuring universal access to at least basic sanitation and hygiene by 2030 in STH-endemic areas.

Conclusion

Currently, many sub-Saharan African countries have completed the STH mapping exercise. Thus, expanded programs for STH control should be in their full state within the next couple of years and especially with regard to current approaches aiming at integrating the control of co-endemic NTDs. The improvements in the cartography of helminth infections and updated estimates of the public health burden of STH (Pullan et al. 2014; Sartorius et al. 2021) coupled with the scale-up of control efforts and the free donation of drugs by pharmaceutical companies should result in significant reductions of STH infections. High levels of coverage will be required if MDA is to drive the parasite below the breakpoint under which transmission will be eliminated (Anderson et al. 2014). Further, the addition of long-term measures such as WASH interventions is believed to contribute to lowering the basic reproductive number and facilitate the ability of MDA to interrupt transmission (Anderson et al. 2014). However, further research is warranted to determine the magnitude of benefit from WASH interventions for STH control, through multisectorial, integrated intervention packages that are tailored to social-ecological contexts (Strunz et al. 2014).

Given the advanced state of helminth genomics and the mapping of the genome sequences of 22 species of helminths, including most or all of the significant STHs (Brindley et al. 2009), it is expected that the search for novel drugs and vaccines will be intensified. These advances will also lead to a better understanding of the human-parasite immune interactions. For example, higher infection burden, as a result of reinfection, has been reported in some areas where previous MDA campaigns have not resulted in transmission interruption (Narain et al. 2004). Thus, a better understanding of host-parasite immune relationships at play at the molecular level and at different life cycle stages within the host is important not only to make more precise predictions about the eventual success of the specific elimination efforts but also for monitoring and evaluation purposes, alerting MDA programs of potential challenges that might arise from altered immunity in treated communities (Lustigman et al. 2012b).

References

Adams EJ, Stephenson LS, Latham MC, Kinoti SN (1994) Physical activity and growth of Kenyan school children with hookworms, *Trichuris trichiura* and *Ascaris lumbricoides* infections are improved after treatment with albendazole. J Nutr 124:1199–1206

Addisu A, Adriaensen W, Balew A, Asfaw M, Diro E, Garba Djirmay A, Gebree D, Seid G, Begashaw H, Harries AD, Hirpa Adugna A, Ayalew Jejaw Z, Kamau EM, Kelbo T, Manzi M, Medebo Daniel D, Moloo A, Olliaro P, Owiti P, Reeder JC, Senkoro M, Takarinda K, Terry R, Timire C, Tucho S, Tweya H, Wendemagegn Y, Verdonck K, Vogt F, van Henten S, van Griensven J, Worku B, Zolfo M, Zachariah R; Ethiopia SORT IT Neglected Tropical Diseases Group (2019) Neglected tropical diseases and the sustainable development goals: an urgent call for action from the front line. BMJ Glob Health 4(1):e001334. https://doi.org/10.1136/bmjgh-2018-001334. Erratum in: BMJ Glob Health 4(2):e001334corr1

Agyapong O, Asiedu SO, Kwofie SK, Miller WA III, Parry CS, Sowah RA, Wilson MD (2021) Molecular modelling and de novo fragment-based design of potential inhibitors of beta-tubulin gene of *Necator americanus* from natural products. Informatics Med Unlocked 26:100734. https://doi.org/10.1016/j.imu.2021.100734

Albonico M, Stoltzfus RJ, Savioli L, Tielsch JM, Chwaya HM, Ercole E, Cancrini G (1998) Epidemiological evidence for a differential effect of hookworm species, Ancylostoma duodenale or Necator americanus, on iron status of children. Int J Epidemiol 27:530–537

Albonico M, Allen H, Chitsulo L, Engels D, Gabrielli AF, Savioli L (2008) Controlling soil-transmitted helminthiasis in pre-school-age children through preventive chemotherapy. PLoS Negl Trop Dis 2:e126

Anderson RM (1980) The dynamics and control of direct life cycle helminth parasites. Biomathematics 39:278–322

Anderson R, Truscott J, Hollingsworth TD (2014) The coverage and frequency of mass drug administration required to eliminate persistent transmission of soil-transmitted helminths. Philos Trans R Soc B: Biol Sci 369:20130435

Ayana M, Vlaminck J, Cools P, Ame S, Albonico M, Dana D, Keiser J, Manly H, Matoso LF, Mekonnen Z, Montresor A, Correa-Oliveira R, Rinaldi L, Sayasone S, Sowersby SJ, Tesfaye L, Vercruysse J, Mirams G, Levecke B (2018) Modification and optimization of the FECPAK^{G2} protocol for the detection and quantification of soil-transmitted helminth eggs in human stool. PLoS Negl Trop Dis 12(10):e0006655

Basanez MG, Mccarthy JS, French MD, Yang GJ, Walker M, Gambhir M, Prichard RK, Churcher TS (2012) A research agenda for helminth diseases of humans: modelling for control and elimination. PLoS Negl Trop Dis 6:e1548

Behnke JM, De Clercq D, Sacko M, Gilbert FS, Ouattara DB, Vercruysse J (2000) The epidemiology of human hookworm infections in the southern region of Mali. Trop Med Int Health 5:343–354

Berger DJ, Crellen T, Lamberton PHL, Allan F, Tracey A, Noonan JD, Kabatereine NB, Tukahebwa EM, Adriko M, Holroyd N, Webster JP, Berriman M, Cotton JA (2021) Whole-genome sequencing of Schistosoma mansoni reveals extensive diversity with limited selection despite mass drug administration. Nat Commun 12:4776

Bethony J, Chen J, Lin S, Xiao S, Zhan B, Li S, Xue H, Xing F, Humphries D, Yan W, Chen G, Foster V, Hawdon JM, Hotez PJ (2002) Emerging patterns of hookworm infection: influence of aging on the intensity of Necator infection in Hainan Province. People's Republic of China. Clin Infect Dis 35:1336–1344

Bethony J, Brooker S, Albonico M, Geiger M, Loukas A, Diemert D, Hotez PJ (2006) Soil-transmitted helminth infections: ascariasis, trichuriasis, and hookworm. Lancet 367:1521–1532

Boatin BA, Basanez MG, Prichard RK, Awadzi K, Barakat RM, García HH, Gazzinelli A, Grant WN, McCarthy JS, N'Goran EK, Osei-Atweneboana MY, Sripa B, Yang GJ, Lustigman S (2012) A research agenda for helminth diseases of humans: towards control and elimination. PLoS Negl Trop Dis 6:e1547

Booth M, Vounatsou P, N'goran EK, Tanner M, Utzinger J (2003) The influence of sampling effort and the performance of the Kato-Katz technique in diagnosing Schistosoma mansoni and hookworm co-infections in rural Côte d'Ivoire. Parasitology 127:525–531

Boyd A, Won KY, Mcclintock SK, Donovan CV, Laney SJ, Williams SA, Pilotte N, Streit TG, Beau De Rochars MVE, Lammie PJ (2010) A community-based study of factors associated with continuing transmission of lymphatic filariasis in Leogane, Haiti. PLoS Negl Trop Dis 4:e640

Brindley PJ, Mitreva M, Ghedin E, Lustigman S (2009) Helminth genomics: the implications for human health. PLoS Negl Trop Dis 3:e538

Britton GJ, Contijoch EJ, Mogno I, Vennaro OH, Llewellyn SR, Ng R, Li Z, Mortha A, Merad M, Das A, Gevers D, McGovern DPB, Singh N, Braun J, Jacobs JP, Clemente JC, Grinspan A, Sands BE, Colombel JF, Dubinsky MC, Faith JJ (2019) Microbiotas from humans with inflammatory bowel disease alter the balance of gut Th17 and RORγt+ regulatory T cells and exacerbate colitis in mice. Immunity 50(1):212–224.e4

Broadhurst MJ, Ardeshir A, Kanwar B, Mirpuri J, Gundra UM, Leung JM, Wiens KE, Vujkovic-Cvijin I, Kim CC, Yarovinsky F, Lerche NW, McCune JM, Loke P (2012) Therapeutic helminth infection of macaques with idiopathic chronic diarrhea alters the inflammatory signature and mucosal microbiota of the colon. PLoS Pathog 8(11):e1003000

Brooker S, Michael E (2000) The potential of geographical information systems and remote sensing in the epidemiology and control of human helminth infections. Adv Parasitol 47:245–288

Brooker S, Peshu N, Warn PA, Mosobo M, Guyatt HL, Marsh K, Snow RW (1999) The epidemiology of hookworm infection and its contribution to anemia among pre-school children on the Kenya coast. Trans R Soc Trop Med Hyg 93:240–246

Brooker S, Rowlands M, Haller L, Savioli L, Bundy DAP (2000) Towards an atlas of human helminth infection in sub-Saharan Africa: the use of geographical information systems (GIS). Parasitol Today 16:303–307

Brooker S, Beasley M, Ndinaromtan M, Madjiouroum EM, Baboguel M, Djenguinabe E, Hay SI, Bundy DA (2002) Use of remote sensing and a geographical information system in a national helminth control programme in Chad. Bull World Health Organ 80:783–789

Brooker S, Clements AC, Bundy DA (2006a) Global epidemiology, ecology, and control of soil-transmitted helminth infections. Adv Parasitol 62:221–261

Brooker S, Clements AC, Hotez PJ, Hay SI, Tatem AJ, Bundy DAP, Snow RW (2006b) The co-distribution of plasmodium falciparum and hookworm among African schoolchildren. Malar J 5:99

Brooker S, Hotez PJ, Bundy DA (2008) Hookworm-related anaemia among pregnant women: a systematic review. PLoS Negl Trop Dis 2:e291

Brooker S, Kabatereine NB, Smith JL, Mupfasoni D, Mwanje MT, Ndayishimiye O, Lwambo NJ, Mbotha D, Karanja P, Mwandawiro C, Muchiri E, Clements AC, Bundy DA, Snow RW (2009) An updated atlas of human helminth infections: the example of East Africa. Int J Health Geogr 8:42

Brooker S, Hotez PJ, Bundy DAP (2010) The global atlas of helminth infection: mapping the way forward in neglected tropical disease control. PLoS Negl Trop Dis 4:e779

Brosschot TP, Reynolds LA (2018) The impact of a helminth-modified microbiome on host immunity. Mucosal Immunol 11(4):1039–1046

Bui T, Ritari J, Boeren S et al (2015) Production of butyrate from lysine and the Amadori product fructoselysine by a human gut commensal. Nat Commun 6:10062

Bundy DAP (1988) Population ecology of intestinal helminth infections in human communities. Philos Trans R Soc Lond B 321:405–420

Bundy D, Cooper E (1989) Trichuris and trichuriasis in humans. Adv Parasitol 28:107–173

Bundy DAP, Guyatt HL (1995) The health of school-age children; report of a workshop. Parasitol Today 11:166–167

Buonfrate D, Salas-Coronas J, Muñoz J, Maruri BT, Rodari P, Castelli F, Zammarchi L, Bianchi L, Gobbi F, Cabezas-Fernández T, Requena-Mendez A, Godbole G, Silva R, Romero M, Chiodini PL, Bisoffi Z (2019) Multiple-dose versus single-dose ivermectin for Strongyloides stercoralis infection (Strong Treat 1 to 4): a multicentre, open-label, phase 3, randomised controlled superiority trial. Lancet Infect Dis 19(11):1181–1190

Burley SK, Berman HM, Kleywegt GJ, et al (2017) Protein Data Bank (PDB): the single global macromolecular structure archive. In: Methods in molecular biology. Humana Press, pp 627–641

Callender JE, Walker SP, Grantham-Mcgregor SM, Cooper ES (1998) Growth and development four years after treatment for the trichuris dysentery syndrome. Acta Paediatr 87:1247–1249

Cantacessi C, Giacomin P, Croese J, Zakrzewski M, Sotillo J, McCann L, Nolan MJ, Mitreva M, Krause L, Loukas A (2014) Impact of experimental hookworm infection on the human gut microbiota. J Infect Dis 210(9):1431–1434

Chen CY-C (2011) TCM database@Taiwan: the world's largest traditional Chinese medicine database for drug screening in silico. PLoS One 6(1):e15939. https://doi.org/10.1371/journal.pone.0015939

Cools P, Vlaminck J, Albonico M, Ame S, Ayana M, José Antonio BP, Cringoli G, Dana D, Keiser J, Maurelli MP, Maya C, Matoso LF, Montresor A, Mekonnen Z, Mirams G, Corrêa-Oliveira R, Pinto SA, Rinaldi L, Sayasone S, Thomas E, Verweij JJ, Vercruysse J, Levecke B (2019) Diagnostic performance of a single and duplicate Kato-Katz, Mini-FLOTAC, FECPAKG2 and qPCR for the detection and quantification of soil-transmitted helminths in three endemic countries. PLoS Negl Trop Dis 13(8):e0007446

Cooper P, Walker AW, Reyes J, Chico M, Salter SJ, Vaca M, Parkhill J (2013) Patent human infections with the whipworm, Trichuris trichiura, are not associated with alterations in the faecal microbiota. PLoS One 8(10):e76573

Cortés A, Clare S, Costain A, Almeida A, McCarthy C, Harcourt K, Brandt C, Tolley C, Rooney J, Berriman M, Lawley T, MacDonald AS, Rinaldi G, Cantacessi C (2020) Baseline gut microbiota composition is associated with *Schistosoma mansoni* infection burden in rodent models. Front Immunol 11:593838

Cragg GM, Newman DJ (2013) Natural products: a continuing source of novel drug leads. Biochim Biophys Acta - Gen Subj 1830(6):3670–3695. https://doi.org/10.1016/j.bbagen.2013.02.008

Cringoli G (2006) FLOTAC, a novel apparatus for a multivalent faecal egg count technique. Parassitologia 48:381–384

Crompton DW (2000) The public health importance of hookworm disease. Parasitology 121(Suppl):S39–S50

Crompton DW, Nesheim MC (2002) Nutritional impact of intestinal helminthiasis during the human life cycle. Annu Rev Nutr 22:35–59

Crompton DW, Savioli L (1993) Intestinal parasitic infections and urbanization. Bull World Health Organ 71:1–7

Crompton DWT, Stephenson LS (1990) Hookworm infections, nutritional status, and productivity. In: Schad GA, Warren KS (eds) Hookworm disease. Taylor and Francis Ltd., London

Crompton DW, Whitehead RR (1993) Hookworm infections and human iron metabolism. Parasitology 107(Suppl):S137–S145

Curtale F, Tilden R, Muhilal VY, Pokhrel RP, Guerra R (1993) Intestinal helminths and risk of anaemia among Nepalese children. Panminerva Med 35:159–166

Dada N, Jupatanakul N, Minard G, Short SM, Akorli J, Villegas LM (2021) Considerations for mosquito microbiome research from the mosquito microbiome consortium. Microbiome 9(1):36

Dada-Adegbola HO, Oluwatoba AO, Falade CO (2005) Prevalence of multiple intestinal helminths among children in a rural community. Afr J Med Med Sci 34:263–267

De Silva NR, Guyatt HL, Bundy DAP (1997) Worm burden in intestinal obstruction caused by Ascaris lumbricoides. Trop Med Int Health 2:189–190

Degarege A, Legesse M, Medhin G, Animut A, Erko B (2012) Malaria and related outcomes in patients with intestinal helminths: a cross-sectional study. BMC Infect Dis 12:291

Devriese S, Eeckhaut V, Geirnaert A, Van den Bossche L, Hindryckx P, Van de Wiele T, Van Immerseel F, Ducatelle R, De Vos M, Laukens D (2017) Reduced mucosa-associated Butyricicoccus activity in patients with ulcerative colitis correlates with aberrant Claudin-1 expression. J Crohns Colitis 11(2):229–236

Diawara A, Drake LJ, Suswillo RR, Kihara J, Bundy DA, Scott ME, Halpenny C, Stothard JR, Prichard RK (2009) Assays to detect beta-tubulin codon 200 polymorphism in Trichuris trichiura and Ascaris lumbricoides. PLoS Negl Trop Dis 3:e397

Diawara A, Schwenkenbecher JM, Kaplan RM, Prichard RK (2013) Molecular and biological diagnostic tests for monitoring benzimidazole resistance in human soil-transmitted helminths. Am J Trop Med Hyg 88:1052–1061

El-Setouhy M, Abd Elaziz KM, Helmy H, Farid HA, Kamal HA, Ramzy RMR, Shannon WD, Weil GJ (2007) The effect of compliance on the impact of mass drug administration for elimination of lymphatic filariasis in Egypt. Am J Trop Med Hyg 77:1069–1073

Evans DB, Guyatt HL (1995) The cost-effectiveness of mass drug therapy for intestinal helminths. PharmacoEconomics 8:14–22

Fentiman A, Hall A, Bundy D (1999) School enrolment patterns in rural Ghana: a comparative study of the impact of location, gender, age, and health on children's access to basic schooling. Comp Educ 35:331–349

Ferri N, Siegl P, Corsini A, Herrmann J, Lerman A, Benghozi R (2013) Drug attrition during preclinical and clinical development: understanding and managing drug-induced cardiotoxicity. Pharmacol Ther 138(3):470–484. https://doi.org/10.1016/j.pharmthera.2013.03.005

Forrester JE, Scott ME, Bundy DA, Golden MH (1988) Clustering of *Ascaris lumbricoides* and *Trichuris trichiura* infections within households. Trans R Soc Trop Med Hyg 82:282–288

Forrester JE, Scott ME, Bundy DAP, Golden MHN (1990) Predisposition of individuals and families in Mexico to heavy infection with *Ascaris lumbricoides* and *Trichuris trichiura*. Trans R Soc Trop Med Hyg 84:272–276

Gaba S, Jamal S, Drug Discovery Consortium OS, Scaria V (2014) Cheminformatics models for inhibitors of Schistosoma mansoni Thioredoxin glutathione reductase. Sci World J 1–9. https://doi.org/10.1155/2014/957107

Gaulton A, Bellis LJ, Bento AP et al (2012) ChEMBL: a largescale bioactivity database for drug discovery. Nucleic Acids Res 40:D1100–D1107. https://doi.org/10.1093/nar/gkr777

Gause WC, Maizels RM (2016) Macrobiota - helminths as active participants and partners of the microbiota in host intestinal homeostasis. Curr Opin Microbiol 32:14–18

Gazzinelli A, Correa-Oliveira R, Yang GJ, Boatin BA, Kloos H (2012) A research agenda for helminth diseases of humans: social ecology, environmental determinants, and health systems. PLoS Negl Trop Dis 6:e1603

Geary TG, Woo K, Mccarthy JS, Mackenzie CD, Horton J, Prichard RK, De Silva NR, Olliaro PL, Lazdins-Helds JK, Engels DA, Bundy DA (2010) Unresolved issues in anthelmintic pharmacology for helminthiases of humans. Int J Parasitol 40:1–13

Giacomin P, Croese J, Krause L, Loukas A, Cantacessi C (2015) Suppression of inflammation by helminths: a role for the gut microbiota? Philos Trans R Soc B 3702014029620140296:20140296

Gilles HM (1996) Soil-transmitted helminths (geohelminths). In: Cook GC (ed) Manson's tropical diseases. Saunders, London

Gilson MK, Liu T, Baitaluk M et al (2016) BindingDB in 2015: a public database for medicinal chemistry, computational chemistry, and systems pharmacology. Nucleic Acids Res 44:D1045–D1053. https://doi.org/10.1093/nar/gkv1072

Gizaw Z, Adane T, Azanaw J, Ayenew A, Daniel H (2018) Childhood intestinal parasitic infection and sanitation predictors in rural Dembiya, Northwest Ethiopia. Environ Health Prev Med 23:26

Glassner KL, Abraham BP, Quigley EMM (2020) The microbiome and inflammatory bowel disease. J Allergy Clin Immunol 145:16–27

Gordon CA, Krause L, McManus DP, Morrison M, Weerakoon KG, Connor MC, Olveda RM, Ross AG, Gobert GN (2020) Helminths, polyparasitism, and the gut microbiome in The Philippines. Int J Parasitol 50(3):217–225

Grencis RK (2015) Immunity to helminths: resistance, regulation, and susceptibility to gastrointestinal nematodes. Annu Rev Immuno 33:201–225

Guengerich FP (2011) Mechanisms of drug toxicity and relevance to pharmaceutical development. Drug Metab Pharmacokinet 26(1):3–14

Hailu T, Nibret E, Amor A, Munshea A, Anegagrie M (2020) Efficacy of single dose Ivermectin against Strongyloides stercoralis infection among primary school children in Amhara National Regional State. Infect Dis (Auckl) 13:1178633720932544

Hall A, Hewitt G, Tuffrey V, De Silva N (2008) A review and meta-analysis of the impact of intestinal worms on child growth and nutrition. Matern Child Nutr 4:118–236

Harris JB, Podolsky MJ, Bhuiyan TR, Chowdhury F, Khan AI, Larocque RC, Logvinenko T, Kendall J, Faruque AS, Nagler CR, Ryan ET, Qadri F, Calderwood SB (2009) Immunologic responses to Vibrio cholerae in patients co-infected with intestinal parasites in Bangladesh. PLoS Negl Trop Dis 3(3):e403

Holm JB, Sorobetea D, Kiilerich P, Ramayo-Caldas Y, Estellé J, Ma T, Madsen L, Kristiansen K, Svensson-Frej M (2015) Chronic Trichuris muris infection decreases diversity of the intestinal microbiota and concomitantly increases the abundance of Lactobacilli. PLoS One 10(5):e0125495

Hotez PJ, Kamath A (2009) Neglected tropical diseases in sub-Saharan Africa: review of their prevalence, distribution, and disease burden. PLoS Negl Trop Dis 3:e412

Hotez PJ, Brooker S, Bethony JM, Bottazzi ME, Loukas A, Xiao S (2004) Hookworm infection. N Engl J Med 351:799–807

Hotez PJ, Bundy DAP, Beegle K, Brooker S, Drake L, De Silva N, Montresor A, Engels D, Jukes M, Chitsulo L, Chow J, Laxminarayan R, Michaud C, Bethony J, Correa-Oliveira R, Shuhua X, Fenwick A, Savioli L (2006) Helminth infections: soil-transmitted helminth infections and schistosomiasis. In: Jamison DT, Breman JG, Measham AR, Alleyne G, Claeson M, Evans DB, Jha P, Mills A, Musgrove P (eds) Disease control priorities in developing countries. The International Bank for Reconstruction and Development/The World Bank Group, Washington, DC

Hotez PJ, Brindley PJ, Bethony JM, King CH, Pearce EJ, Jacobson J (2008) Helminth infections: the great neglected tropical diseases. J Clin Invest 118:1311–1321

Hu Y, Xiao SH, Aroian RV (2009) The new anthelmintic tribendimidine is an L-type (levamisole and pyrantel) nicotinic acetylcholine receptor agonist. PLoS Negl Trop Dis 3:e499

Humphries D, Simms BT, Davey D, Otchere J, Quagraine J, Terryah S, Newton S, Berg E, Harrison LM, Boakye D, Wilson M, Cappello M (2013) Hookworm infection among school age children in Kintampo north municipality, Ghana: nutritional risk factors and response to albendazole treatment. Am J Trop Med Hyg 89(3):540–548

Igore KGG, Payne VK, Nadia NAC, Cedric Y (2020) Risk factors associated with prevalence and intensity of gastro-intestinal parasitic infections within households in Tonga sub-division, west region, Cameroon. J Infect Dis Epidemiol 6:123

Jenkins TP, Peachey LE, Ajami NJ, MacDonald AS, Hsieh MH, Brindley PJ, Cantacessi C, Rinaldi G (2018) Schistosoma mansoni infection is associated with quantitative and qualitative modifications of the mammalian intestinal microbiota. Sci Rep 8(1):12072

Jirků M, Lhotská Z, Frgelecová L, Kadlecová O, Petrželková KJ, Morien E, Jirku-Pomajbíková K (2021) Helminth interactions with bacteria in the host gut are essential for its immunomodulatory effect. Microorganisms 9:226

Kaplan RM (2004) Drug resistance in nematodes of veterinary importance. A status report. Trends Parasitol 20:477–481. https://doi.org/10.1016/j.pt.2004.08.001

Katz N, Chaves A, Pellegrino J (1972) A simple device for quantitative stool thick smear technique in Schistosomiasis mansoni. Rev Inst Med Trop Sao Paulo 14:397–400

Kay GL, Millard A, Sergeant MJ, Midzi N, Gwisai R, Mduluza T, Ivens A, Nausch N, Mutapi F, Pallen M (2015) Differences in the faecal microbiome in schistosoma haematobium infected children vs. uninfected children. PLoS Negl Trop Dis 9(6):e0003861

Keiser J, Utzinger J (2008) Efficacy of current drugs against soil-transmitted helminth infections: systematic review and meta-analysis. JAMA 299:1937–1948

Keiser J, Utzinger J (2010) The drugs we have and the drugs we need against major helminth infections. Adv Parasitol 73:197–230

Kim S (2016) Getting the most out of PubChem for virtual screening. Expert Opin Drug Discov 11:843–855

Knopp S, Mgeni AF, Khamis IS, Steinmann P, Stothard JR, Rollinson D, Marti H, Utzinger J (2008) Diagnosis of soil-transmitted helminths in the era of preventive chemotherapy: effect of multiple stool sampling and use of different diagnostic techniques. PLoS Negl Trop Dis 2:e331

Knopp S, Glinz D, Rinaldi L, Mohammed KA, N'goran EK, Stothard JR, Marti H, Cringoli G, Rollinson D, Utzinger J (2009) FLOTAC: a promising technique for detecting helminth eggs in human faeces. Trans R Soc Trop Med Hyg 103:1190–1194

Koga K, Kasuya S, Khamboonruang C, Sukhavat K, Ieda M, Takatsuka N, Kita K, Ohtomo H (1991) A modified agar plate method for detection of Strongyloides stercoralis. Am J Trop Med Hyg 45:518–521

Kvalsvig JD, Cooppan RM, Connolly KJ (1991) The effects of parasite infections on cognitive processes in children. Ann Trop Med Parasitol 85:551–568

Lahlou M (2013) The success of natural products in drug discovery. Sci Res 4(3):17–31

Lammie PJ, Fenwick A, Utzinger J (2006) A blueprint for success: integration of neglected tropical disease control programmes. Trends Parasitol 22:313–321

Lee SC, Tang MS, Lim YA, Choy SH, Kurtz ZD, Cox LM, Gundra UM, Cho I, Bonneau R, Blaser MJ, Chua KH, Loke P (2014) Helminth colonization is associated with increased diversity of the gut microbiota. PLoS Negl Trop Dis 8(5):e2880

Levecke B, Behnke JM, Ajjampur SS, Albonico M, Ame SM, Charlier J, Geiger SM, Hoa NT, Kamwa Ngassam RI, Kotze AC, Mccarthy JS, Montresor A, Periago MV, Roy S, Tchuem Tchuente LA, Thach DT, Vercruysse J (2011) A comparison of the sensitivity and fecal egg counts of the McMaster egg counting and Kato-Katz thick smear methods for soil-transmitted helminths. PLoS Negl Trop Dis 5:e1201

Li RW, Wu S, Li W, Navarro K, Couch RD, Hill D, Urban JF Jr (2012) Alterations in the porcine colon microbiota induced by the gastrointestinal nematode *Trichuris suis*. Infect Immun 80(6):2150–2157. M.D. Wilson et al

Lustigman S, Geldhof P, Grant WN, Osei-Atweneboana MY, Sripa B, Basanez MG (2012a) A research agenda for helminth diseases of humans: basic research and enabling technologies to support control and elimination of helminthiases. PLoS Negl Trop Dis 6:e1445

Lustigman S, Prichard RK, Gazzinelli A, Grant WN, Boatin BA, Mccarthy JS, Basanez MG (2012b) A research agenda for helminth diseases of humans: the problem of helminthiases. PLoS Negl Trop Dis 6:e1582

Mabaso ML, Appleton CC, Hughes JC, Gouws E (2004) Hookworm (necator americanus) transmission in inland areas of sandy soils in KwaZulu-natal, South Africa. Trop Med Int Health 9:471–476

Magnussen P, Muchiri E, Mungai P, Ndzovu M, Ouma J, Tosha S (1997) A school-based approach to the control of urinary schistosomiasis and intestinal helminth infections in children in Matuga, Kenya: impact of a two year chemotherapy programme on prevalence and intensity of infections. Trop Med Int Health 2:825–831

Maipanich W, Pubampen S, Sa-nguankiat S, Nontasut P, Waikagul J (1997) Effect of albendazole and mebendazole on soil-transmitted helminth eggs. Southeast Asian J Trop Med Public Health 28(2):321–325

Manor O, Dai CL, Kornilov SA, Smith B, Price ND, Lovejoy JC, Gibbons SM, Magis AT (2020) Health and disease markers correlate with gut microbiome composition across thousands of people. Nat Commun 11:5206

Martin I, Djuardi Y, Sartono E, Rosa BA, Supali T, Mitreva M, Houwing-Duistermaat JJ, Yazdanbakhsh M (2018) Dynamic changes in human-gut microbiome in relation to a placebo-controlled anthelminthic trial in Indonesia. PLoS Negl Trop Dis 12(8):e0006620

Mccarthy JS, Lustigman S, Yang GJ, Barakat RM, Garcia HH, Sripa B, Willingham AL, Prichard RK, Basáñez MG (2012) A research agenda for helminth diseases of humans: diagnostics for control and elimination programmes. PLoS Negl Trop Dis 6:e1601

Mehraj V, Hatcher J, Akhtar S, Rafique G, Beg MA (2008) Prevalence and factors associated with intestinal parasitic infection among children in an urban slum of Karachi. PLoS One 3(11):e3680

Miguel EA, Kremer M (2003) Worms: identifying impacts on education and health in the presence of treatment externalities. Econometrica 72:159–217

Molyneux DH, Malecela MN (2011) Neglected tropical diseases and the millennium development goals—why the "other diseases" matter: reality versus rhetoric. Parasite Vectors 4:234

Molyneux DH, Hotez PJ, Fenwick A (2005) "Rapid-impact interventions": how a policy of integrated control for Africa's neglected tropical diseases could benefit the poor. PLoS Med 2:e336

Montresor A, Crompton DWT, Hall A, Bundy DAP, Savioli L (1998) Guidelines for the evaluation of soil-transmitted helminthiasis and schistosomiasis at community level. A guide for managers of control programmes. World Health Organization, Geneva

Montresor A, Crompton DWT, Gyorkos TW, Savioli L (2002) Helminth control in school-age children: a guide for managers of control programmes. World Health Organization, Geneva

Montresor A, Gabrielli AF, Yajima A, Lethanh N, Biggs BA, Casey GJ, Tinh TT, Engels D, Savioli L (2013) Markov model to forecast the change in prevalence of soil-transmitted helminths during a control programme: a case study in Vietnam. Trans R Soc Trop Med Hyg 107:313–318

Moran M, Guzman J, Ropars AL, Mcdonald A, Jameson N, Omune B, Ryan S, Wu L (2009) Neglected disease research and development: how much are we really spending? PLoS Med 6:e1000030

Murray CJL, Vos T, Lozano R, Naghavi M, Flaxman AD, Michaud C, Ezzati M, Shibuya K, Salomon JA, Abdalla S, Aboyans V, Abraham J, Ackerman I, Aggarwal R, Ahn SY, Ali MK, Almazroa MA, Alvarado M, Anderson HR, Anderson LM, Andrews KG, Atkinson C, Baddour LM, Bahalim AN, Barker-Collo S, Barrero LH, Bartels DH, Basáñez M-G, Baxter A, Bell ML, Benjamin EJ, Bennett D, Bernabé E, Bhalla K, Bhandari B, Bikbov B, Abdulhak AB, Birbeck G, Black JA, Blencowe H, Blore JD, Blyth F, Bolliger I, Bonaventure A, Boufous S, Bourne R, Boussinesq M, Braithwaite T, Brayne C, Bridgett L, Brooker S, Brooks P, Brugha TS, Bryan-Hancock C, Bucello C, Buchbinder R, Buckle G, Budke CM, Burch M, Burney P, Burstein R, Calabria B, Campbell B, Canter CE, Carabin H, Carapetis J, Carmona L, Cella C, Charlson F,

Chen H, Cheng AT-A, Chou D, Chugh SS, Coffeng LE, Colan SD, Colquhoun S, Colson KE, Condon J, Connor MD, Cooper LT, Corriere M, Cortinovis M, De Vaccaro KC, Couser W, Cowie BC, Criqui MH, Cross M, Dabhadkar KC, Dahiya M, Dahodwala N, Damsere-Derry J, Danaei G, Davis A, Leo DD, Degenhardt L, Dellavalle R, Delossantos A, Denenberg J, Derrett S, Des Jarlais DC et al (2012) Disability-adjusted life years (DALYs) for 291 diseases and injuries in 21 regions, 1990? 2010: a systematic analysis for the global burden of disease study 2010. Lancet 380:2197–2223

Narain K, Medhi GK, Rajguru SK, Mahanta J (2004) Cure and reinfection patterns of geohelminthic infections after treatment in communities inhabiting the tropical rainforest of Assam, India. Southeast Asian J Trop Med Public Health 35:512–517

Nikolay B, Brooker SJ, Pullan RL (2014) Sensitivity of diagnostic tests for human soil-transmitted helminth infections: a meta-analysis in the absence of a true gold standard. Int J Parasitol 44:765–774

Njambi E, Magu D, Masaku J, Okoyo C, Njenga SM (2020) Prevalence of intestinal parasitic infections and associated water, sanitation, and hygiene risk factors among school children in Mwea Irrigation Scheme, Kirinyaga County, Kenya. J Trop Med 2020 Article ID 3974156

Nokes C, Bundy D (1994) Does helminth infection affect mental processing and educational achievement? Parasitol Today 10:14–18

Nokes C, Grantham-Mcgregor SM, Sawyer AW, Cooper ES, Bundy DAP (1992) Parasitic helminth infection and cognitive function in school children. Proc R Soc Lond 247:77–81

Ntie-Kang F et al (2013) AfroDb: a select highly potent and diverse natural product library from African medicinal plants. PLoS One 8(10):e78085. https://doi.org/10.1371/journal.pone.0078085

Ntie-Kang F et al (2017) NANPDB: a resource for natural products from Northern African sources. J Nat Prod 80(7):2067–2076. https://doi.org/10.1021/acs.jnatprod.7b00283

Ohkusa T, Okayasu I, Ogihara T, Morita K, Ogawa M, Sato N (2003) Induction of experimental ulcerative colitis by Fusobacterium varium isolated from colonic mucosa of patients with ulcerative colitis. Gut 52(1):79–83

Olliaro P, Seiler J, Kuesel A, Horton J, Clark JN, Don R, Keiser J (2011) Potential drug development candidates for human soil-transmitted helminthiases. PLoS Negl Trop Dis 5:e1138

Olsen A, Magnussen P, Ouma JH, Andreassen J, Friis H (1998) The contribution of hookworm and other parasitic infection to haemoglobin and iron status among children and adults in western Kenya. Trans R Soc Trop Med Hyg 92:643–649

Olsen A, Van Lieshout L, Marti H, Polderman T, Polman K, Steinmann P, Stothard R, Thybo S, Verweij JJ, Magnussen P (2009) Strongyloidiasis—the most neglected of the neglected tropical diseases? Trans R Soc Trop Med Hyg 103:967–972

Osazuwa F, Ayo OM, Imade P (2011) A significant association between intestinal helminth infection and anaemia burden in children in rural communities of Edo state, Nigeria. N Am J Med Sci 3:30–34

Osei-Atweneboana MY, Lustigman S, Prichard RK, Boatin BA, Basanez MG (2012) A research agenda for helminth diseases of humans: health research and capacity building in disease-endemic countries for helminthiases control. PLoS Negl Trop Dis 6:e1602

Osei-Fosu AK (2011) Evaluating the impact of the capitation grant and the school feeding programme on enrolment, attendance, and retention in schools: the case of Weweso circuit. J Sci Technol 31:55–64

Patridge E, Gareiss P, Kinch MS, Hoyer D (2016) An analysis of FDA-approved drugs: natural products and their derivatives. Drug Discov Today 21(2):204–207. https://doi.org/10.1016/j.drudis.2015.01.009

Prichard RK, Basanez MG, Boatin BA, Mccarthy JS, Garcia HH, Yang GJ, Sripa B, Lustigman S (2012) A research agenda for helminth diseases of humans: intervention for control and elimination. PLoS Negl Trop Dis 6:e1549

Pullan R, Smith J, Jasrasaria R, Brooker S (2014) Global numbers of infection and disease burden of soil transmitted helminth infections in 2010. Parasite Vectors 7:37

Pullan RL, Halliday KE, Oswald WE, Mcharo C, Beaumont E, Kepha S, Witek-McManus S, Gichuki PM, Allen E, Drake T, Pitt C, Matendechero SH, Gwayi-Chore MC, Anderson RM, Njenga SM, Brooker SJ, Mwandawiro CS (2019) Effects, equity, and cost of school-based and community-wide treatment strategies for soil-transmitted helminths in Kenya: a cluster-randomised controlled trial. Lancet 393(10185):2039–2050

Ramanan D, Bowcutt R, Lee SC, Tang MS, Kurtz ZD, Ding Y, Honda K, Gause WC, Blaser MJ, Bonneau RA, Lim YA, Loke P, Cadwell K (2016) Helminth infection promotes colonization resistance via type 2 immunity. Science 352(6285):608–612

Resende Co T, Hirsch CS, Toossi Z, Dietze R, Ribeiro-Rodrigues R (2007) Intestinal helminth co-infection has a negative impact on both anti-Mycobacterium tuberculosis immunity and clinical response to tuberculosis therapy. Clin Exp Immunol 147(1):45–52

Robertson LJ, Crompton DW, Sanjur D, Nesheim MC (1992) Haemoglobin concentrations and concomitant infections of hookworm and Trichuris trichiura in Panamanian primary school children. Trans R Soc Trop Med Hyg 86:654–656

Roche M, Layrisse M (1966) The nature and causes of hookworm anaemia. Am J Trop Med Hyg 15:1031–1110

Rojo-Arreola L, Long T, Asarnow D et al (2014) Chemical and genetic validation of the statin drug target to treat the helminth disease. Schistosomiasis. https://doi.org/10.1371/journal.pone.0087594

Roses AD (2008) Pharmacogenetics in drug discovery and development: a translational perspective. Nat Rev Drug Discov 7:807–817

Rubel MA, Abbas A, Taylor LJ, Connell A, Tanes C, Bittinger K, Ndze VN, Fonsah JY, Ngwang E, Essiane A, Fokunang C, Njamnshi AK, Bushman FD, Tishkoff SA (2020) Lifestyle and the presence of helminths is associated with gut microbiome composition in Cameroonians. Genome Biol 21(1):122

Sartorius B, Cano J, Simpson H, Tusting LS, Marczak LB, Miller-Petrie MK, Kinvi B, Zoure H, Mwinzi P, Hay SI, Rebollo M, Pullan RL (2021) Prevalence and intensity of soil-transmitted helminth infections of children in sub-Saharan Africa, 2000-18: a geospatial analysis. Lancet Glob Health 9(1):e52–e60

Savage DC (1977) Microbial ecology of the gastrointestinal tract. Annu Rev Microbiol. 31:107–133

Savioli L, Bundy DAP, Tomkins A (1992) Intestinal parasitic infections a soluble public health problem. Trans R Soc Trop Med Hyg 86:353–354

Schad GA, Anderson RM (1985) Predisposition to hookworm infection in humans. Science 228:1537–1540

Schaubeck M, Clavel T, Calasan J, Lagkouvardos I, Haange SB, Jehmlich N, Basic M, Dupont A, Hornef M, von Bergen M, Bleich A, Haller D (2016) Dysbiotic gut microbiota causes transmissible Crohn's disease-like ileitis independent of failure in antimicrobial defence. Gut 65:225–237

Schwenkenbecher JM, Albonico M, Bickle Q, Kaplan RM (2007) Characterization of beta-tubulin genes in hookworms and investigation of resistance associated mutations using real-time PCR. Mol Biochem Parasitol 156:167–174

Sharpton TJ, Combrink L, Arnold HK, Gaulke CA, Kent M (2020) Harnessing the gut microbiome in the fight against anthelminthic drug resistance. Curr Opin Microbiol 53:26–34

Simeon D, Callender J, Wong M, Grantham-Mcgregor S, Ramdath DD (1994) School performance, nutritional status and trichuriasis in Jamaican schoolchildren. Acta Paediatr 83:1188–1193

Smith J, Taylor EM (2016) What is next for NTDs in the era of the sustainable development goals? PLoS Negl Trop Dis 10:e0004719

Solomons NW (1993) Pathways to the impairment of human nutritional status by gastrointestinal pathogens. Parasitol Int 107:S19–S35

Steinbaum L, Mboya J, Mahoney R, Njenga SM, Null C, Pickering AJ (2019) Effect of a sanitation intervention on soil-transmitted helminth prevalence and concentration in household soil: a cluster-randomized controlled trial and risk factor analysis. PLoS Negl Trop Dis 13(2):e0007180

Steinmann P, Utzinger J, Du ZW, Jiang JY, Chen JX, Hattendorf J, Zhou H, Zhou XN (2011) Efficacy of single-dose and triple-dose albendazole and mebendazole against soil-transmitted helminths and Taenia spp.: a randomized controlled trial. PLoS One 6(9):e25003
Stephenson LS (1987) Impact of helminth infection in human nutrition. Taylor & Francis, New York
Stephenson LS, Latham MC, Kurz KM, Kinoti SN, Oduori ML, Crompton DWT (1985) Relationships of Schistosoma haematobium, hookworm and malaria infections and metrifo-nate treatment to haemoglobin levels in Kenyan school children. Am J Trop Med Hyg 34:519–528
Stephenson LS, Latham MC, Kinoti SN, Kurz KM, Brigham H (1990) Improvements in physical fitness of Kenyan schoolboys infected with hookworm, Trichuris trichiura and Ascaris lumbricoides following a single dose of albendazole. Trans R Soc Trop Med Hyg 84:277–282
Stephenson LS, Latham MC, Adam EJ, Kinoti SN, Pertet A (1993) Physical fitness, growth, and appetite of Kenyan school boys with Hookworm, Trichuris trichiura and Ascaris lumbricoideinfections are improved four months after a single dose of albendazole. J Nutr 123:1036–1046
Stephenson LS, Latham MC, Ottesen EA (2000) Malnutrition and parasitic helminth infections. Parasitology 121:S23–S28
Stoltzfus RJ, Albonico M, Chwaya HM, Savioli L, Tielsch J, Schulze K, Yip R (1996) Hemoquant determination of hookworm-related blood loss and its role in iron deficiency in African children. Am J Trop Med Hyg 55:399–404
Stoltzfus RJ, Chwaya HM, Tielsch JM, Schulze KJ, Albonico M, Savioli L (1997) Epidemiology of iron deficiency anemia in Zanzibari school children: the importance of hookworms. Am J Trop Med Hyg 65:153–159
Stromberg BE (1997) Environmental factors influencing "transmission Vet Parasitol 72(3–4):247–256
Strunz EC, Addiss DG, Stocks ME, Ogden S, Utzinger J, Freeman MC (2014) Water, sanitation, hygiene, and soil-transmitted helminth infection: a systematic review and meta-analysis. PLoS Med 11:e1001620
Tang YT, Gao X, Rosa BA, Abubucker S, Hallsworth-Pepin K, Martin J, Tyagi R, Heizer E, Zhang X, Bhonagiri-Palsikar V, Minx P, Warren WC, Wang Q, Zhan B, Hotez PJ, Sternberg PW, Dougall A, Gaze ST, Mulvenna J, Sotillo J, Ranganathan S, Rabelo EM, Wilson RW, Felgner PL, Bethony J, Hawdon JM, Gasser RB, Loukas A, Mitreva M (2014) Genome of the human hookworm *Necator americanus*. Nat Genet. 46(3):261–269. https://doi.org/10.1038/ng.2875
Taren DL, Nesheim MC, Crompton DWT, Holland CV, Barbeau I, Rivera G, Sanjur D, Tiffany J, Tucker K (1987) Contribution of ascariasis to poor nutritional status of children from Chiriqui Province, Republic of Panama. Parasitology 95:603–613
Tatala S, Svanberg U, Mduma B (1998) Low dietary iron availability is a major cause of anaemia: a nutritional survey in Lindi District of Tanzania. Am J Clin Nutr 68:171–178
Traore' M, Traore' H, Kardorff R, Diarra A, Landoure A (1998) The public health significance of urinary schistosomiasis as a cause of morbidity in two districts in Mali. Am J Trop Med Hyg 59:407–413
Uniting To Combat Ntds (2012) The London declaration on neglected tropical diseases 2012. Available: http://unitingtocombatntds.org/downloads/press/ntd_event_london_declaration_on_ntds.pdf. [Online]. Available: http://unitingtocombatntds.org/downloads/press/ntd_event_london_declaration_on_ntds.pdf. Accessed 28 Dec 2013
Ursell LK, Metcalf JL, Parfrey LW, Knight R (2012) Defining the human microbiome. Nutr Rev 70(Suppl 1):S38–S44
Utzinger J, Keiser J (2004) Schistosomiasis and soil soil-transmitted helminthiasis: common drugs for treatment and control. Expert Opin Pharmacother 5:263–285
Van Den Broek N (2003) Anaemia and micronutrient deficiencies. Br Med Bull 67:149–160
Veeresham C (2012) Natural products derived from plants as a source of drugs. J Adv Pharm Technol Res 3(4):200–201. https://doi.org/10.4103/2231-4040.104709
Vercruysse J, Behnke JM, Albonico M, Ame SM, Angebault C, Bethony JM, Engels D, Guillard B, Nguyen TV, Kang G, Kattula D, Kotze AC, Mccarthy JS, Mekonnen Z, Montresor A, Periago MV, Sumo L, Tchuente LA, Dang TC, Zeynudin A, Levecke B (2011) Assessment

of the anthelmintic efficacy of albendazole in school children in seven countries where soil-transmitted helminths are endemic. PLoS Negl Trop Dis 5:e948

Verweij JJ, Brienen EA, Ziem J, Yelifari L, Polderman AM, Van Lieshout L (2007) Simultaneous detection and quantification of *Ancylostoma duodenale*, *Necator americanus*, and *Oesophagostomum bifurcum* in fecal samples using multiplex real-time PCR. Am J Trop Med Hyg 77:685–690

Viteri FE (1994) The consequences of iron deficiency and anaemia in pregnancy on maternal health, the foetus, and the infant. SCN News 11:14–18

Warren K, Bundy D, Anderson R, Davis A, Henderson D, Jamison D, Prescott N, Senf A (1993) Helminth infection. In: Jamison D, Mosley W, Measham A, Bibadilla J (eds) Disease control priorities in developing countries. Oxford Publishing, Oxford

WHO (1997) Information series on school health – document 1. Strengthening interventions to reduce helminth infections: an entry point for the development of health-promoting schools. World Health Organization, Geneva

WHO (2002) Prevention and control of schistosomiasis and soil-transmitted helminthiasis, vol 912, WHO Technical Report Series. World Health Organization, Geneva

WHO (2006) Preventive chemotherapy in human helminthiasis: coordinated use of anthelminthic drugs in control interventions: a manual for health professionals and programme managers. World Health Organization, Geneva

WHO (2012) Soil-transmitted helminthiases: eliminating soil-transmitted helminthiases as a public health problem in children: progress report 2001–2010 and strategic plan 2011–2020. World Health Organization, Geneva

WHO (2020) Soil-transmitted helminth infections. World Health Organization, Geneva

Wishart DS, Feunang YD, Guo AC et al (2018) DrugBank 5.0: a major update to the drugbank database for 2018. Nucleic Acids Res 46:D1074–D1082. https://doi.org/10.1093/nar/gkx1037

Wu S, Li RW, Li W, Beshah E, Dawson HD, Urban JF Jr (2012) Worm burden-dependent disruption of the porcine colon microbiota by Trichuris suis infection. PLoS One. 7(4):e35470

Yazdanbakhsh M, van den Biggelaar A, Maizels RM (2001) Th2 responses without atopy: immunoregulation in chronic helminth infections and reduced allergic disease. Trends Immunol. 22(7):372–377

Young ND, Jex AR, Li B, Liu S, Yang L, Xiong Z, Li Y, Cantacessi C, Hall RS, Xu X, Chen F, Wu X, Zerlotini A, Oliveira G, Hofmann A, Zhang G, Fang X, Kang Y, Campbell BE, Loukas A, Ranganathan S, Rollinson D, Rinaldi G, Brindley PJ, Yang H, Wang J, Wang J, Gasser RB (2012) Whole-genome sequence of Schistosoma haematobium. Nat Genetics 44(2):221. https://doi.org/10.1038/ng.1065

Zerlotini A, Aguiar ERGR, Yu F et al (2013) SchistoDB: an updated genome resource for the three key schistosomes of humans. Nucleic Acids Res. https://doi.org/10.1093/nar/gks1087

Zetterstrom R (2004) Iron deficiency and iron deficiency anaemia during infancy and childhood. Acta Paediatr 93:436–439

Zimmermann MB, Hurrell RF (2007) Nutritional iron deficiency. Lancet Infect Dis 370:511–520

Trachoma

Teshome Gebre, Amir B. Kello, Esmael Habtamu, and Jeremiah M. Ngondi

Contents

T. Gebre (✉)
International Trachoma Initiative, The Task Force for Global Health, Addis Ababa, Ethiopia
e-mail: tgebre@taskforce.org

A. B. Kello
World Health Organization, Regional Office for Africa, Brazzaville, Republic of the Congo

E. Habtamu
London School of Hygiene and Tropical Medicine, London, UK

Eyu-Ethiopia, Bahir Dar, Ethiopia

J. M. Ngondi
RTI International, Washington, DC, USA

J. O. Gyapong, B. A. Boatin (eds.), *Neglected Tropical Diseases - Sub-Saharan Africa*, Neglected Tropical Diseases, https://doi.org/10.1007/978-3-031-53901-5_15

Abstract

Trachoma, the leading infectious cause of preventable blindness, is caused by ocular infection *Chlamydia trachomatis*. Blindness due to trachoma is preventable through the multipronged approach of Surgery, Antibiotics, Facial cleanliness and Environmental improvement (SAFE) strategy. Trachoma was targeted for global elimination as a public health problem by 2020; however, despite the substantial progress made in the last two decades, only 11 countries had been validated for elimination of trachoma as a public health problem by 2021. Trachoma remains endemic in 44 countries of which 26 (59%) are in sub-Saharan Africa. In addition, 85% of the global burden of active trachoma is found in sub-Saharan Africa. This chapter covers the epidemiology of trachoma in sub-Saharan Africa, underscores the continued need for SAFE strategy implementation, and outlines present-day trachoma elimination endgame challenges that affect trachoma endemic countries, mainly in sub-Saharan Africa. The recently launched NTD roadmap 2021–2030 brings renewed enthusiasm for global elimination of trachoma by 2030. A lot has been learnt in the last two decades of SAFE implementation thus building on these lessons alongside new strategies of tackling endgame challenges makes global trachoma elimination attainable in the next decade.

Introduction

Trachoma, one of the most common neglected tropical diseases (NTD) and the leading infectious cause of preventable blindness worldwide, is endemic in many of the poorest and most remote rural areas in 44 countries. The burden of trachoma on affected individuals and communities is enormous. The economic cost in terms of lost productivity from blindness and visual impairment is estimated at US$2.9–5.3 billion annually, increasing to US$8 billion when trichiasis is included (WHO 2013). Trachoma can be eliminated as a public health problem (EPHP) through the World Health Organization (WHO) endorsed strategy of surgery, antibiotics, facial cleanliness, and environmental improvements (SAFE) (WHO 1997). According to the WHO, elimination of trachoma as a public health problem is defined as the following: (1) a prevalence of trachomatous trichiasis (TT) "unknown to the health system" of <0.2% in adults aged ≥15 years (approximately 1 case per 1000 total population); (2) a prevalence of trachomatous inflammation—follicular (TF) in children aged 1–9 years of <5%, sustained for at least two years in the absence of ongoing antibiotic mass treatment, in each formerly endemic district; and (3) the existence of a system able to identify and manage incident trachomatous trichiasis cases, using defined strategies, with evidence of appropriate financial resources to implement those strategies (WHO 2016). The WHO Alliance for the global elimination of trachoma had set a target of 2020 for EPHP worldwide. However, by 2021, only 11 countries (Cambodia, China, Gambia, Ghana, Islamic Republic of Iran, Lao People's Democratic Republic, Mexico, Morocco, Myanmar, Nepal, and Oman) had been formally validated for the elimination of trachoma. The new WHO NTD Roadmap 2021–2030 has set a new goal for the global elimination trachoma by 2030 (WHO 2020).

Basic Biology and Disease Presentation

Trachoma is caused by the obligate intracellular bacterium *Chlamydia trachomatis*. Recurrent episodes of conjunctival infection and the associated chronic inflammation it causes initiate a scarring process that ultimately leads to irreversible blindness. Endemic trachoma is caused by the four ocular serovars of *C. trachomatis* (A, B, Ba, and C). Although the genital serovars (D to K) of *C. trachomatis* can infect the conjunctiva causing either ophthalmia neonatorum in infants or inclusion conjunctivitis in adults, these are usually isolated episodes for the individual, which do not lead to blinding sequelae (Burton et al. 2003).

For endemic trachoma, the average age of acquisition of the first episode of *C. trachomatis* infection is probably related to the prevailing level of infection in the community. In hyperendemic settings, infection may be acquired in early infancy, whereas in meso-endemic and hypo-endemic regions, it is probably on average in the later years of childhood. Infection is usually acquired through living in close physical proximity to an infected person, with the family as the principal unit for transmission (Barenfanger 1975). It seems universally accepted and recognized over time that, however defined, trachoma has two major phases: active or inflammatory trachoma and cicatricial or late trachoma. Active trachoma is characterized by an inflammatory response associated with the variable presence of demonstrable infection. The inflammation can vary in intensity but when severe and prolonged, it leads to tarsal scarring and corneal pannus. Cicatricial trachoma is marked by structural change in the lid with tarsal scarring and trichiasis (eyelashes scratching the eyeball). In this case, inflammation is variable and *Chlamydia* are infrequently seen (Taylor 2008).

Studies from trachoma-endemic communities show that the prevalence and duration of conjunctival chlamydial infection decline with increasing age, suggesting that there is a maturation of the immune response as individuals are repeatedly exposed to infection. However, in the early vaccine trials using whole *C. trachomatis* organisms, the acquired immunity appeared to be strain specific and relatively short lived. As a result of the relatively ineffective immune response, repeated infection of the individual by *C. trachomatis* is common within an endemic environment. This leads to a recurrent chronic inflammation, which is associated with the development of scar tissue within the conjunctiva over many years. As the scar tissue accumulates, it also contracts, causing the eyelids to roll inward toward the eye (entropion) and the eyelashes to scratch the corneal surface leading to corneal scarring and opacity. The degree to which conjunctival scarring develops probably depends on a complex interaction between the pressure of infection (load and frequency) and host-specific immunogenetic factors. It is possible that a failure of the immune response to adequately control the *Chlamydia* leads to prolonged infection episodes, which provokes more severe inflammation, tissue damage (through the release of proteases), and aberrant repair. The most serious disease sequelae from trachoma is blinding corneal opacification. The main etiological risk factor for corneal damage is the presence of trichiasis; however, a number of other factors probably contribute, such as bacterial infection and chronic conjunctival inflammation (Burton and Mabey 2009).

Taylor further describes that in an endemic setting, a well-established case of active trachoma can be very easy to diagnose using the WHO simplified grading system (Fig. 1) (Taylor et al. 1987; Solomon et al. 2020). The key sign in active trachoma is the "trachoma follicle" that is lymphoid follicles or germinal centers in the superior tarsal conjunctiva. Although the presence of superior tarsal follicles (trachomatous inflammation-follicular, TF) is used as a primary diagnostic feature in the grading of active trachoma, it is the intense inflammation shown by conjunctival inflammatory thickening and papillae (trachomatous inflammation-intense, TI) that reflects the critical process in the pathogenesis. The challenge in trachoma diagnosis is mainly in the differentiation of borderline cases from normal, or sometimes, distinguishing cases of severe inflammatory trachoma from acute bacterial or viral conjunctivitis or an eyelash pointing toward the eye but not touching in case of trichiasis. The diagnosis of

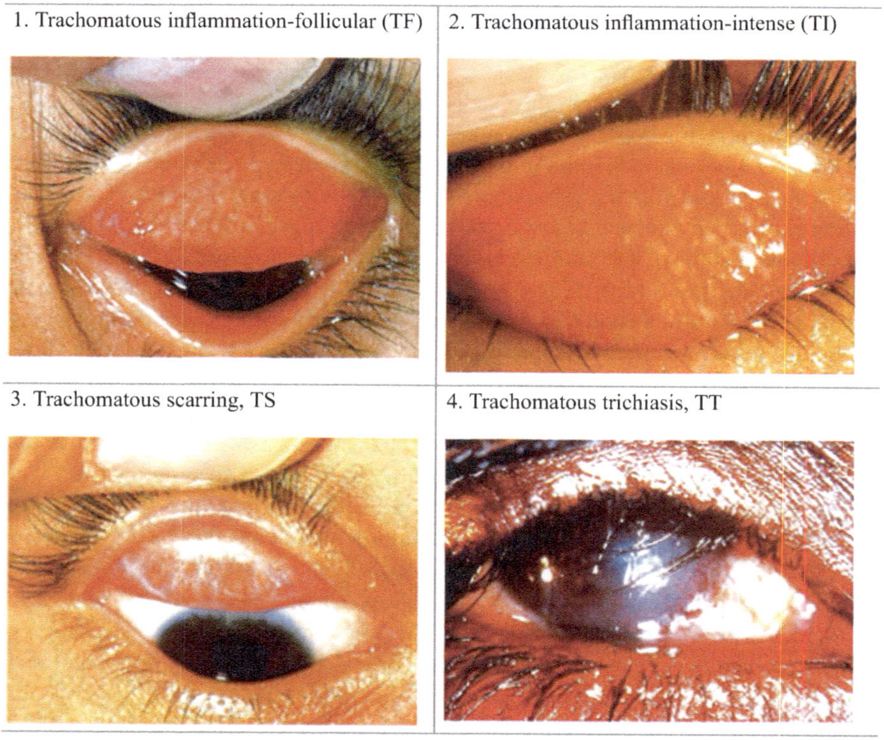

Fig. 1 The WHO simplified grading system. 1. Trachomatous inflammation—follicular (TF). 2. Trachomatous inflammation—intense (TI). 3. Trachomatous scarring (TS). 4. Trachomatous trichiasis (TT)

trachoma can be much more difficult to make in areas of low endemicity than in high endemicity ones, although severe tarsal scarring, corneal pannus with Herbert's pits, and upper lid trichiasis are almost pathognomonic in any setting.

The Epidemiology of Trachoma

Trachoma is a disease of the crèche and clusters in large or small pockets. These pockets may be countries, districts or regions, villages, or household clusters within villages, but ultimately, trachoma is a disease of individual families. Epidemiology can describe where trachoma occurs but more importantly, why some are affected, and others are not. It can also explain the patchiness of these pockets of infection and the basis of family clustering (Taylor 2008).

Trachoma is hyperendemic in many of the poorest and most remote rural areas in 44 countries in Africa, Asia, Central America and South America, Australia, and the Middle East. As of June 2021, the WHO estimated that globally, there were 136.2 million people living in districts where TF prevalence was ≥5% and the global burden of TT was estimated at 1.8 million cases unknown to the health system in 1595 districts (WHO 2021a).

One hundred years ago, trachoma was wide spread in Europe and North America but faded away during the first half of the twentieth century as living conditions improved (Mariotti et al. 2009). The clinical manifestations of trachoma change with age. As indicated in the disease biology section above, active trachoma is predominantly seen in young children, becoming less frequent and of shorter duration with increasing age (Bailey et al. 1999). A study showed that only less than 5% of the adults show signs of active trachomatous disease (West et al. 1991). The age distribution of the different signs of trachoma depends in part on the stability and endemicity of the disease in the community. In hyperendemic areas, TF and TI are commonest in preschool children, with prevalence as high as 60%–90% (Courtright et al. 1989). The onset of the blinding complications of trachoma can occur in children living in regions where the pressure of infection is high (Ngondi et al. 2007). Epidemiological surveys have generally found TT and corneal opacity to be more common in women than men. This difference has been attributed to the greater lifetime exposure of women to *C. trachomatis* infection, through closer contact with children, the main reservoirs of infection (West et al. 1991).

In addition to age and sex, environmental factors such as household access to water sources, poor hygienic conditions, presence and density of eye-seeking flies, availability of functional latrine, presence of cattle pens and cattle ownership, crowded living conditions in the family unit, and nutritional deficiencies (vitamin A) have been identified as possible risk factors for trachoma (West 2004). Several studies in sub-Saharan Africa (SSA) have reported high prevalence of active trachoma in areas where there is low access to water and sanitary conditions are poor (Berhane et al. 2008). A large risk factor analysis following a trachoma prevalence study conducted in the Amhara region of Ethiopia reported active trachoma prevalence in children to be independently associated with ocular discharge, nasal discharge, and proxy indicators of low socioeconomic status and increasing altitude, while TT was independently associated with increasing age, female gender, increasing prevalence of active trachoma in children, and increasing altitude (Ngondi et al. 2008).

Burden of Trachoma in Sub-Saharan Africa

Figure 2 show status of elimination of trachoma as a public health problem in the Africa Region in 2021. A total of 26 countries (59% of the 44 trachoma endemic countries globally) were in the African Region (WHO 2021a). Two countries (Ghana and Gambia) have been validated for the elimination of trachoma as a public health problem. Fifteen countries are thought not to be endemic for trachoma, while in three countries (Angola, Botswana, and Namibia) interventions may be required and investigations are planned to establish endemicity of trachoma. One country (Togo) has reported to have achieved the prevalence targets for elimination as a public health problem (WHO 2021a, b).

Significant progress has been made in the African Region in the control of trachoma over the last couple of years. The number of people that required treatment

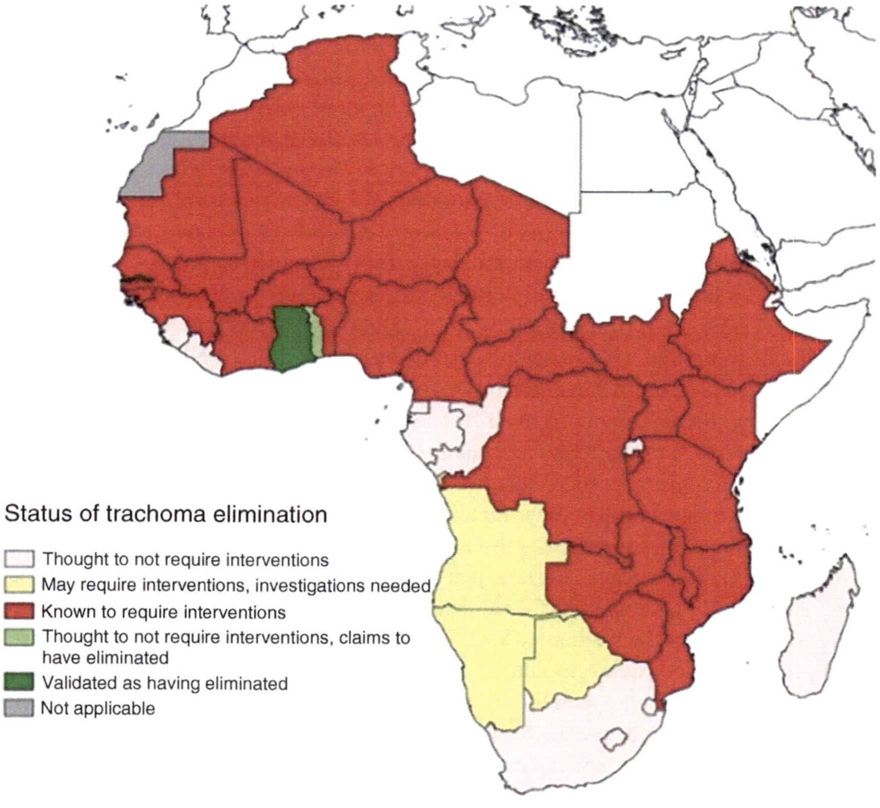

Fig. 2 Status of elimination of trachoma as a public health problem in the Africa Region, 2021. Adapted from: https://apps.who.int/neglected_diseases/ntddata/trachoma/trachoma.html

with antibiotics for trachoma in the Region reduced from 189 million in 2014 to 134 million in 2020—a reduction of 29% (WHO 2021b).

Ghana and the Gambia are the two countries in the African Region to have been validated in June 2018 and April 2021, respectively, as having eliminated trachoma as a public health problem (WHO 2021a, b).

Sub-Saharan Africa is disproportionately affected by trachoma. As of June 2021, an estimated 115.7 million (85% of global burden) people lived in districts where TF prevalence was ≥5% in 1–9-year-olds (Fig. 3), while 1442 districts (90% of global estimates) had prevalence of TT unknown to health system ≥0.2% in ≥15-year-olds (WHO 2021a). The highest prevalence of active trachoma has been reported from Ethiopia and South Sudan, where the infection often occurs in more than 50% of

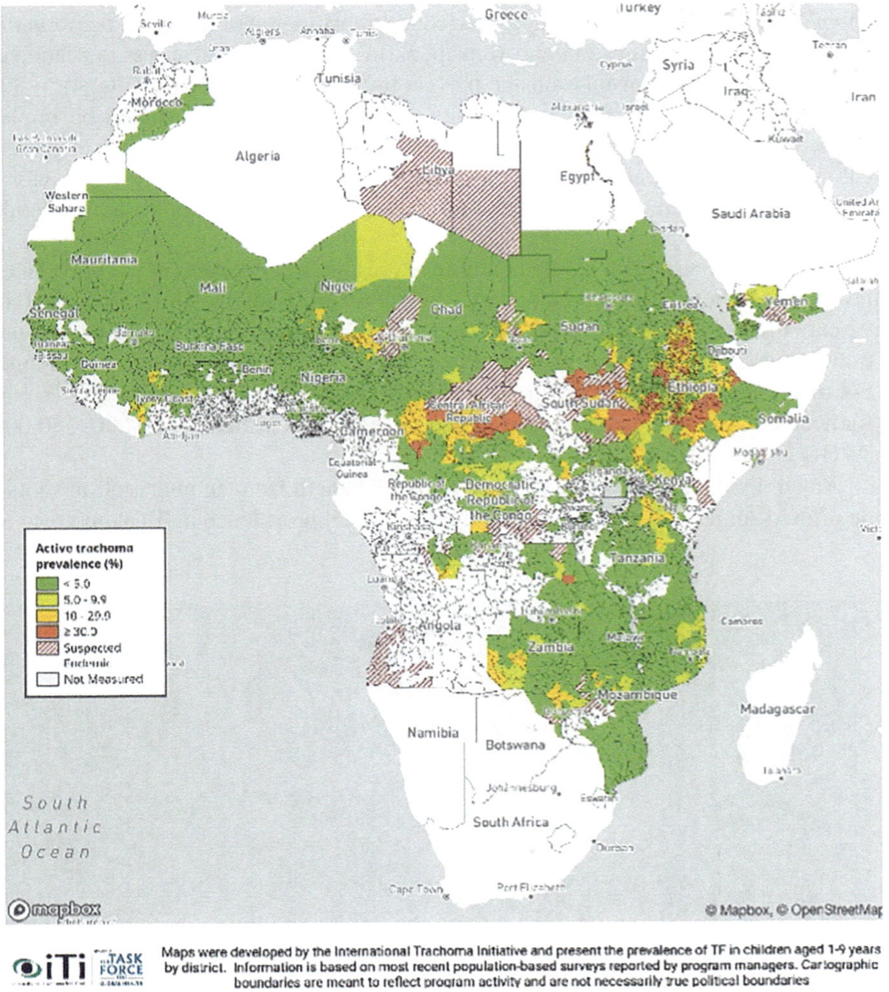

Fig. 3 Prevalence of trachomatous inflammation—follicular (TF) in children aged 1–9 years in the Africa Region. Adopted from Global Atlas of Trachoma: https://www.trachomaatlas.org/

children who are younger than 10 years and trichiasis is found in up to 19% of adults (WHO 2013). Ethiopia stands out among the endemic countries with the highest burden of trachoma: 67.2 million (49% of global burden) people living in districts where TF prevalence was ≥5% and 696 districts (43% of global estimates) with prevalence of TT unknown to health system ≥0.2% in ≥15-year-olds (WHO 2021a).

Elimination Strategies for Trachoma

The recommended strategy to eliminate trachoma as a public health problem that was endorsed by the WHO in 1993 is known as the SAFE strategy (Turner et al. 1995). SAFE is an acronym that stands for the following: Surgery for trachomatous trichiasis to prevent visual impairment and blindness from TT; antibiotics to clear ocular *C. trachomatis* infection; facial cleanliness to reduce transmission of ocular *C. trachomatis* infection; and environmental improvement, particularly improved access to water and sanitation to reduce transmission and prevent infection from recurring (Fig. 4).

Where relevant, the four components of the SAFE strategy should be implemented as an integrated approach. One should not only focus on medical and surgical interventions but also include the F and E components that address behavioral and environmental determinants for trachoma's persistence in endemic communities.

Trachoma control as a public health intervention is dependent on the prevalence levels of trachomatous inflammation—follicular (TF) among children aged 1 to 9 years old and TT among those who are aged ≥15 years, as measured by population-based prevalence surveys in districts suspected of being endemic at baseline. For trachoma control purposes, a district is defined as the normal administrative unit for health management—generally a population of approximately 100,000–250,000 (WHO 2015a, 2018).

Any individual who has TT thought to likely benefit from an intervention should be offered surgery. However, when the district-level prevalence of TT "unknown to

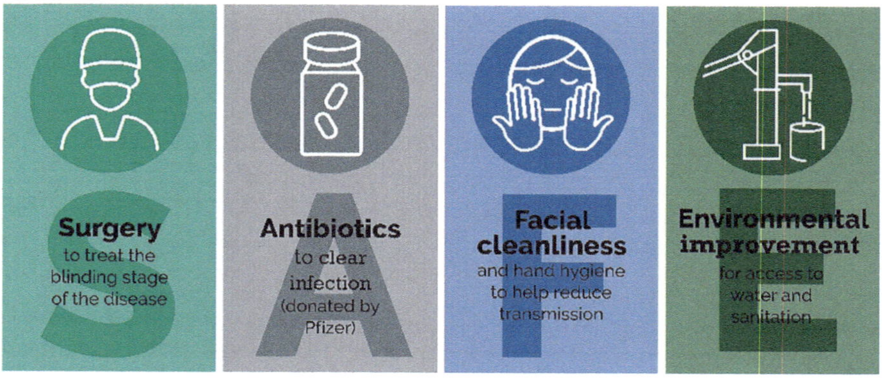

Fig. 4 Components of the SAFE Strategy for elimination of trachoma as a public health problem. Adapted from International Trachoma Initiative (https://www.trachoma.org/about-trachoma)

the health system" is ≥0.2% among people aged ≥15 years, the S component of the SAFE strategy becomes a public health-level intervention that involves active case finding and outreach surgical services provision if necessary (WHO 2021a).

The A, F, and E components of SAFE strategy are recommended for entire districts in which the TF prevalence is ≥5%. In such districts, all residents should be offered antibiotic treatment annually regardless of their individual infection status. The planned number of rounds depends on the most recent estimate of TF prevalence at the district level (Taylor et al. 2014). It is recommended that impact surveys be undertaken 6–12 months after the last planned annual round of antibiotic mass drug administration (MDA) in order to determine whether treatment should be continued or can be safely stopped. Once prevalence of TF in children aged 1–9 years is <5%, MDA is stopped but F and E activities continue. Two years after the impact survey, a district-level trachoma surveillance survey (TSS) is conducted to determine whether the TF prevalence has remained below threshold (WHO 2015a; Taylor et al. 2014).

Surgery for Trichiasis

The WHO recommends two major surgical techniques for TT: the Bilamellar Tarsal Rotation Procedure (BLTR) and the Posterior Lamellar Tarsal Rotation Procedure (PLTR). The latter is also known as the modified Trabut technique (WHO 2015b). A recently conducted randomized controlled trial has shown the PLTR technique to be better than the BLTR procedure which showed that the PLTR had 13% postoperative trichiasis at 1-year postop, compared to 22% for the BLTR (Habtamu et al. 2016). Therefore, the WHO recommends that all new surgical training be conducted using the PLTR technique. However, already practicing surgeons using BLTR procedure with good surgical outcomes are not required to shift to the PLTR technique (WHO 2015c).

The recent introduction of a mannequin-based surgical training platform known as HEAD START has contributed toward improving the quality of training of new surgeons as well as the refresher training of existing ones (Gower et al. 2014). Therefore, it is recommended that mannequin-based surgical training platform be part of any new and refresher training (WHO 2015c). The WHO further recommends that only properly trained TT surgeons that are trained and certified using the WHO Final Assessment of TT Surgeons' checklist should be allowed to operate independently without supervision. It is also recommended to provide adequate supportive supervision of TT surgeons during outreach activities to ensure that good quality surgical services are being offered to patients with TT (WHO 2015b; ICTC 2018, 2016).

Antibiotics for Trachoma

Trachoma is one of the five Preventive Chemotherapy NTDs (PC-NTDs) that are amenable to control through the administration of antibiotics to endemic communities. Drugs that are used for trachoma MDA are 1% tetracycline eye ointment (TEO) and azithromycin. TEO is applied to both eyes twice daily for 6 weeks, which makes compliance quite challenging. Azithromycin, however, can be administered orally as a single dose. Zithromax® is donated by Pfizer Inc. through the International Trachoma Initiative (ITI) for trachoma elimination purposes free of charge to trachoma-endemic countries. As of November 2021, over 958 million treatment of Zithromax® were shipped to endemic countries since the donation program started in 1999 (ITI 2021).

The need or otherwise for antibiotic MDA is guided by the prevalence of TF in children aged 1–9 years as measured by population-based prevalence surveys in districts suspected of being endemic at baseline. The recommended number of annual rounds of MDA depends on the most recent estimate of TF in children aged 1–9 years old at district level (Taylor et al. 2014; WHO 2018; ITI 2019).

All people residing in a trachoma endemic district, regardless of their individual infection status, are eligible for treatment during MDA. MDA teams should aim to achieve the highest coverage of the population as possible and certainly no lower than 80% coverage, which is defined as the number of people treated (with either azithromycin or tetracycline eye ointment) divided by the total number of residents. Children under 6 months of age are offered 1% TEO for 6 weeks (ITI 2019). Although azithromycin is said to be a safe drug even during pregnancy, some national programs prefer to treat pregnant women with 1% TEO for 6 weeks.

Children are given azithromycin in a dosage of 20 mg per kg body weight, using height as a proxy for weight (Muñoz et al. 2003). Children from 6 months to <7 years of age are offered azithromycin syrup, and those aged ≥ 7 years to ≤ 15 years are offered Zithromax® tablets based on their height using a dosing pole. Those aged ≥ 15 years are offered 4 tablets of 250 mg—a 1 g single dose. No child should be forced to take drugs, in order to reduce any risk of choking during drug administration (ICTC 2013; Addiss et al. 2019).

F&E Components

Trachoma control using the SAFE strategy involves the implementation of S and/or A, F&E at the same time in endemic districts, guided by the TT prevalence in people aged ≥ 15 years and prevalence of TF among children aged 1–9 years. Wherever A is implemented, F&E should also be implemented.

It is critical for trachoma control programs to engage with the water, sanitation, and hygiene (WASH) sector for the implementation of the F&E components for long-term sustainability in maintaining reduction of ocular *C. trachomatis* transmission. F&E interventions will also support overall improvements in people's health and well-being and achievement of the Sustainable Development Goals (SDGs). In order to achieve full F&E scale-up, the trachoma community needs to partner with other organizations already working in water and sanitation, education, and health promotion (WHO 2017; ICTC 2015).

Though essential, to achieve the desired outcomes of F&E interventions, it is not sufficient to provide the hardware, such as improving access to water and sanitation. There is a matching need to improve and better align behaviors central to transmission of infection. Theory and evidence both suggest that this "software" side of

F&E implementation should move beyond health education and information dissemination and appropriately employ a variety of behavior change techniques (Delea et al. 2018).

Challenges of Program Implementation

Persistent and Recrudescent Trachoma

There is a growing evidence that in some hyperendemic populations, the elimination threshold for active (inflammatory) trachoma (TF prevalence <5% in children aged 1–9 years) is often not met despite several years of implementation of the recommended interventions (Nash et al. 2018; Stewart et al. 2019; COR-NTD 2020). In addition, after achieving elimination threshold during impact surveys, some districts bounce back (recrudesce) with a prevalence of active disease above the elimination threshold following surveillance surveys, undermining the elimination efforts in multiple settings. Among many, one of the contributing factors for this is believed to be the weak or unsatisfactory coverage of the F&E components of the SAFE strategy. Despite the ongoing effort of trachoma elimination programs, facial cleanliness and environmental improvement practices are limited in some trachoma-endemic communities, contributing to the persistent transmission of ocular *C. trachomatis*. Failure of the currently recommended interventions to achieve elimination in hyperendemic settings is likely to challenge the success of the entire global elimination target by 2030.

TT Surgery Uptake and Quality

There has been a significant reduction in the estimated global TT backlog, from about 8.2 million in 2007 (Mariotti et al. 2009) to 1.8 million in 2020 (WHO 2021a). This is largely attributable to the unprecedented global effort to avert the risk of blindness through corrective eyelid surgery. However, a reduction in TT surgery uptake in the remaining unoperated TT cases has been reported by various programs, contributing to the relatively lower number of TT cases being operated globally. The global TT surgery productivity started to decline since 2017 (WHO 2019) and in 2019, the surgeries performed constituted only 5% of the global backlog (WHO 2021a). Unpublished data from various programs indicate that increasing number of TT cases refusing surgical services might have contributed to this decline. In Amhara region, Ethiopia, among TT cases identified from 61 districts in 2021 about 16% declined free and close to home TT surgical services. Large proportion of the remaining backlog of unoperated TT cases probably have mild disease that may not warrant surgical intervention. In a recent study conducted among 193 unoperated TT cases identified in a community-based screening in Amhara Region, Ethiopia, 47% had only a single trichiatic eyelash (Habtamu et al. 2020).

The quality of TT surgery has been a major issue for the global trachoma control program. Postoperative trichiasis rates vary considerably across settings, but 20% at 1 year is typical (Rajak et al. 2012). Other adverse outcomes, such as eyelid contour abnormalities and granuloma after surgery occur in 5–30% of TT cases that underwent surgery (Rajak et al. 2011; Habtamu et al. 2016; Gower et al. 2013). The consequences of poor surgical outcomes are huge ranging from psychological and physical consequences deterring patients from accepting surgery to adding extra burden to surgical program. In addition, there are no optimal management strategies for cases with poor surgical outcomes.

COVID-19 and Funding

The COVID-19 pandemic has resulted in a profound negative impact on every facet of life. The health care sector has been the most challenged so far with actual and anticipated needs to control the spread of infection on one hand and the delivery of regular healthcare services on the other. Trachoma control programs are no exception. Interruption of elimination activities and funding have been a major challenge. The funding cut for trachoma control came in a time where trachoma affected countries are not in position to filling the gap both due to the impact of the COVID19 on themselves and the lack of readiness from the already weak health system for NTD financing and integration. The funding cuts are anticipated to result in the potential loss of over 14 million trachoma treatments and hundreds of thousands of trichiasis surgeries to affected communities in 2022 alone, jeopardizing elimination efforts in some of the most trachoma affected countries.

Sustaining Elimination

As some countries are moving toward the endgame, establishing a system to sustaining the achieved results through appropriate surveillance and health system integrated interventions would be priority challenges that need addressing. Transitioning from the current campaign-based donor driven program to a government owned, financed, and integrated strategy with the general healthcare system is yet to be adequately implemented in most national programs. Continued efforts to ensure efficient implementation of interventions and establishing improved and effective monitoring, surveillance, and reporting tools are needed. Integrated inter- and intra-sectoral responses such as between and within eye health, WASH and education and aligned with the Sustainable Development Goals and Universal Health Coverage is not only a goal but a challenge for trachoma control programs in the next decade (WHO 2017).

Further Research

Further operational research is needed to generate evidence on the following elimination endgame challenges:

Persistent infections

- Why trachoma prevalence remains above elimination threshold in some districts despite many years of implementation of the WHO-recommended control strategy.
- Likely causes or factors contributing to persistence and recrudescence of trachoma in various settings.
- TF and its correlation with *C. trachomatis* infection and response to MDA.
- TF is considered a lagging indicator and still exists for a few years in settings where there is no *C. trachomatis* infection. Alternative indicators other than TF that can be used to measure and decide elimination threshold for active trachoma are needed.

Strategies for elimination

- Effective enhanced MDA strategies that can be used in settings that are not responding to the traditional WHO recommended annual MDA.
- Strategies that help to improve facial cleanliness, latrine access and latrine use in trachoma-endemic communities to reduce the transmission of *C. trachomatis* infection.
- Strategies that facilitate the integration of Trachoma elimination interventions into the primary healthcare system to achieve efficient and sustainable elimination.
- Strategies that help to improve TT surgical uptake in various settings.

Other areas

- Understand the phenotype of unoperated TT cases in the remaining backlog to help understand surgical refusal and devise alternative strategies for management.
- Strategies that help minimize poor postoperative surgical outcomes.
- Develop optimal management strategies for TT cases with poor surgical outcomes (mainly postoperative TT and eyelid contour abnormality).
- Evidence on the impact of cross-border movement or migration on trachoma elimination efforts.

Outlook for the Next Decade

The target of global elimination of trachoma as a public health problem was set for 2020; however, this goal was not realized. Despite these delays, substantial progress has been made; therefore, concerted action among all sectors is required to sustain and build on the progress of the past two decades. Compared to the 2002 estimates, the number of people living in trachoma endemic areas and trachomatous trichiasis cases had declined by 91% (to 139.6 million) and 74% (to 2 million), respectively, by 2020 (WHO 2019). However, national programs continue to face challenges related to persistent TF and low uptake and quality of TT surgery.

Looking at the epidemiology of trachoma, the geographic contiguity of endemic communities, and owing to the well-established traditional practices and social interactions of the populations sharing common transnational borders, establishing regional cross-border collaboration partnerships involving trachoma affected countries and implementing partners in Eastern, Southern, and Western African regions have been operational since 2015 (Gebre and Sankar 2015). These were highly instrumental in sharing experiences among neighboring districts and addressing outstanding programmatic issues. In some cases, program interventions such as MDAs and TT surgeries were jointly organized and implemented in these affected districts along the common borders. These cross-border partnerships need to be strengthened in this decade to both maximize efficiency and minimize any potential untoward impacts of cross-border movements on the trachoma elimination efforts.

The recent launch of the World Health Organization NTD roadmap 2021–2030 has renewed enthusiasm and a sense of urgency to tackle NTDs (WHO 2020). The roadmap has set new targets for the elimination of trachoma globally by 2030. The roadmap aims to improve coordination and collaboration among health partners, integrate and mainstream approaches, and promote greater ownership by governments and communities. Alongside the NTD roadmap, the WHO has also published a sustainability framework that focuses on protecting the progress achieved and, at the same time, expanding the scope to cover all 20 diseases and disease groups toward the control, elimination, and eradication of NTDs (WHO 2021c).

Trachoma elimination endgame challenges will require implementing the SAFE strategy more effectively, especially adapting the interventions to specific country settings and populations. The operational research priorities outlined above will provide much needed evidence to address the endgame challenges. In addition, planning for sustainability will help Member States, stakeholders, and partners to consolidate cross-sector collaboration for implementing the SAFE strategy and align their strategies on identifying sustainable pathways towards the road map 2030 targets.

References

Addiss D, Sarah V, Negussu N, Emerson P (2019) Safe mass drug administration and trachoma elimination. Community Eye Health 32:37

Bailey R, Duong T, Carpenter R, Whittle H, Mabey D (1999) The duration of human ocular Chlamydia trachomatis infection is age dependent. Epidemiol Infect 123:479–486. https://doi.org/10.1017/s0950268899003076

Barenfanger J (1975) Studies on the role of the family unit in the transmission of trachoma. Am J Trop Med Hyg 24:509–515. https://doi.org/10.4269/ajtmh.1975.24.509

Berhane Y, Worku A, Bejiga A, Adamu L, Alemayehu W, Bedri A, Haile Z, Ayalew A, Adamu W, Gebre T, Kebede TD, West E, West S (2008) National Survey on blindness, low vision and trachoma in Ethiopia: methods and study clusters profile. Ethiop J Health Dev 21:185–203. https://doi.org/10.4314/ejhd.v21i3.10049

Burton MJ, Mabey DCW (2009) The global burden of trachoma: a review. PLoS Negl Trop Dis 3:e460. https://doi.org/10.1371/journal.pntd.0000460

Burton MJ, Holland MJ, Faal N, Aryee EAN, Alexander NDE, Bah M, Faal H, West SK, Foster A, Johnson GJ, Mabey DCW, Bailey RL (2003) Which members of a community need antibiotics to control trachoma? Conjunctival chlamydia trachomatis infection load in Gambian villages. Invest Ophthalmol Vis Sci 44:4215–4222. https://doi.org/10.1167/iovs.03-0107

COR-NTD (2020) Tackling trachoma endgame challenges: COR-NTD 2020 virtual meeting, 12–14 November 2020

Courtright P, Sheppard J, Schachter J, Said ME, Dawson CR (1989) Trachoma and blindness in the Nile Delta: current patterns and projections for the future in the rural Egyptian population. Br J Ophthalmol 73:536–540. https://doi.org/10.1136/bjo.73.7.536

Delea MG, Solomon H, Solomon AW, Freeman MC (2018) Interventions to maximize facial cleanliness and achieve environmental improvement for trachoma elimination: a review of the grey literature. PLoS Negl Trop Dis 12:e0006178. https://doi.org/10.1371/journal.pntd.0006178

Gebre T, Sankar G (2015) The East Africa trachoma/NTD cross-border partnership. Community Eye Health 28:78

Gower EW, West SK, Harding JC, Cassard SD, Munoz BE, Othman MS, Kello AB, Merbs SL (2013) Trachomatous trichiasis clamp vs standard bilamellar tarsal rotation instrumentation for trichiasis surgery: results of a randomized clinical trial. JAMA Ophthalmol 131:294–301. https://doi.org/10.1001/jamaophthalmol.2013.910

Gower EW, Kello AB, Kollmann KM (2014) Training trichiasis surgeons: ensuring quality. Community Eye Health 27:58

Habtamu E, Wondie T, Aweke S, Tadesse Z, Zerihun M, Zewudie Z, Kello AB, Roberts CH, Emerson PM, Bailey RL, Mabey DCW, Rajak SN, Callahan K, Weiss HA, Burton MJ (2016) Posterior lamellar versus bilamellar tarsal rotation surgery for trachomatous trichiasis in Ethiopia: a randomised controlled trial. Lancet Glob Health 4:e175–e184. https://doi.org/10.1016/S2214-109X(15)00299-5

Habtamu E, Wondie T, Gobezie W, Tadesse Z, Gashaw B, Gebeyehu A, Roberts CH, Callahan EK, Macleod D, Burton MJ (2020) Effect of repeated epilation for minor trachomatous trichiasis on lash burden, phenotype and surgical management willingness: a cohort study. PLoS Negl Trop Dis 14:e0008882. https://doi.org/10.1371/journal.pntd.0008882

ICTC (2013) Preferred practices for Zithromax® mass drug administration

ICTC (2015) All you need for F&E: a practical guide to partnering and planning

ICTC (2016) Supportive supervision for trachomatous Trichiasis Programmes

ICTC (2018) Training Trichiasis surgeons for trachoma elimination programs. To be used alongside the WHO yellow manual Trichiasis Surgery for Trachoma

ITI (2019) Zithromax® Management Guide

ITI (2021) Zithromax® Shipments [WWW Document]. URL https://www.trachoma.org/. Accessed 8.15.21

Mariotti SP, Pascolini D, Rose-Nussbaumer J (2009) Trachoma: global magnitude of a preventable cause of blindness. Br J Ophthalmol 93:563–568. https://doi.org/10.1136/bjo.2008.148494

Muñoz B, Solomon AW, Zingeser J, Barwick R, Burton M, Bailey R, Mabey D, Foster A, West SK (2003) Antibiotic dosage in trachoma control programs: height as a surrogate for weight in children. Invest Ophthalmol Vis Sci 44:1464–1469. https://doi.org/10.1167/iovs.02-0234

Nash SD, Stewart AEP, Zerihun M, Sata E, Gessese D, Melak B, Endeshaw T, Chanyalew M, Chernet A, Bayissasse B, Moncada J, Lietman TM, Emerson PM, King JD, Tadesse Z, Callahan EK (2018) Ocular chlamydia trachomatis infection under the surgery, antibiotics, facial cleanliness, and environmental improvement strategy in Amhara, Ethiopia, 2011-2015. Clin Infect Dis 67:1840–1846. https://doi.org/10.1093/cid/ciy377

Ngondi J, Matthews F, Reacher M, Onsarigo A, Matende I, Baba S, Brayne C, Zingeser J, Emerson P (2007) Prevalence of risk factors and severity of active trachoma in southern Sudan: an ordinal analysis. Am J Trop Med Hyg 77:126–132
Ngondi J, Gebre T, Shargie EB, Graves PM, Ejigsemahu Y, Teferi T, Genet A, Mosher AW, Endeshaw T, Zerihun M, Messele A, Richards FO, Emerson PM (2008) Risk factors for active trachoma in children and trichiasis in adults: a household survey in Amhara Regional State, Ethiopia. Trans R Soc Trop Med Hyg 102:432–438. https://doi.org/10.1016/j.trstmh.2008.02.014
Rajak SN, Habtamu E, Weiss HA, Kello AB, Gebre T, Genet A, Bailey RL, Mabey DCW, Khaw PT, Gilbert CE, Emerson PM, Burton MJ (2011) Absorbable versus silk sutures for surgical treatment of trachomatous trichiasis in Ethiopia: a randomised controlled trial. PLoS Med 8:e1001137. https://doi.org/10.1371/journal.pmed.1001137
Rajak SN, Collin JRO, Burton MJ (2012) Trachomatous trichiasis and its management in endemic countries. Surv Ophthalmol 57:105–135. https://doi.org/10.1016/j.survophthal.2011.08.002
Solomon AW, Kello AB, Bangert M, West SK, Taylor HR, Tekeraoi R, Foster A (2020) The simplified trachoma grading system, amended. Bull World Health Organ 98:698–705. https://doi.org/10.2471/BLT.19.248708
Stewart AEP, Zerihun M, Gessese D, Melak B, Sata E, Nute AW, Astale T, Endeshaw T, Teferi T, Tadesse Z, Callahan EK, Chanyalew M, Gaudie B, Emerson PM, King JD, Nash SD (2019) Progress to eliminate trachoma as a public health problem in Amhara National Regional State, Ethiopia: results of 152 population-based surveys. Am J Trop Med Hyg 101:1286–1295. https://doi.org/10.4269/ajtmh.19-0450
Taylor HR (2008) Trachoma: a blinding scourge from the bronze age to the twenty-first century. Haddington Press, South Yarra, Vic
Taylor HR, West SK, Katala S, Foster A (1987) Trachoma: evaluation of a new grading scheme in the United Republic of Tanzania. Bull World Health Organ 65:485–488
Taylor HR, Burton MJ, Haddad D, West S, Wright H (2014) Trachoma. Lancet 384:2142–2152. https://doi.org/10.1016/S0140-6736(13)62182-0
Turner, V., Fransis, V., WHO, 1995. Achieving community support for trachoma control: a guide for district health work
West SK (2004) Trachoma: new assault on an ancient disease. Prog Retin Eye Res 23:381–401. https://doi.org/10.1016/j.preteyeres.2004.04.001
West SK, Munoz B, Turner VM, Mmbaga BB, Taylor HR (1991) The epidemiology of trachoma in Central Tanzania. Int J Epidemiol 20:1088–1092. https://doi.org/10.1093/ije/20.4.1088
WHO (1997) Future approaches to trachoma control: report of a global scientific meeting, Geneva, 17–20 June 1996. World Health Organization, Geneva
WHO (2013) Sustaining the drive to overcome the global impact of neglected tropical diseases: second WHO report on neglected diseases
WHO (2015a) Technical consultation on trachoma surveillance, September 11–12, 2014, Task Force for Global Health, Decatur, USA
WHO (2015b) Trichiasis surgery for trachoma, 2nd edn. World Health Organization, Geneva
WHO (2015c) WHO Alliance for the Global Elimination of Trachoma by 2020. Second global scientific meeting on trachomatous trichiasis, Cape Town, 4–6 November 2015. World Health Organization, Geneva
WHO (2016) Validation of elimination of trachoma as a public health problem
WHO (2017) Integrating neglected tropical diseases into global health and development: fourth WHO report on neglected tropical diseases
WHO (2018) Design parameters for population-based trachoma prevalence surveys; 2018. (No. (WHO/HTM/NTD/PCT/2018.07) Licence: CC BY-NC-SA 3.0 IGO). World Health Organization., Geneva
WHO (2019) WHO Alliance for the global elimination of trachoma by 2020: progress report on elimination of trachoma, 2018. Wkly Epidemiol Rec 20:317–328
WHO (2020) Ending the neglect to attain the Sustainable Development Goals—A road map for neglected tropical diseases 2021–2030

WHO (2021a) WHO Alliance for the global elimination of trachoma by 2020: progress report on elimination of trachoma, 2020. Wkly Epidemiol Rec 31:353–364
WHO (2021b) Status of elimination of trachoma as a public health problem 2021. Preventive Chemotherapy Data Portal
WHO (2021c) Ending the neglect to attain the Sustainable Development Goals: a sustainability framework for action against neglected tropical diseases 2021–2030

Yaws

Nsiire Agana, Cynthia Kwakye-Maclean,
and Earnest Njih Tabah

Contents

N. Agana (✉) · C. Kwakye-Maclean
Disease Control and Prevention Department, Accra, Ghana

Regional Health Directorate, Greater Accra, Accra, Ghana

National Buruli Ulcer, Leprosy, Yaws and Leishmaniasis Control Program, Ministry of Public Health, Yaoundé, Cameroon

E. N. Tabah
Disease Control and Prevention Department, Accra, Ghana

Regional Health Directorate, Greater Accra, Accra, Ghana

National Buruli Ulcer, Leprosy, Yaws and Leishmaniasis Control Program, Ministry of Public Health, Yaoundé, Cameroon

Department of Public Health, Faculty of Medicine and Pharmaceutical Sciences, University of Dschang, Dschang, Cameroon

J. O. Gyapong, B. A. Boatin (eds.), *Neglected Tropical Diseases - Sub-Saharan Africa*, Neglected Tropical Diseases, https://doi.org/10.1007/978-3-031-53901-5_16

Abstract

Yaws, a chronic skin, bone, and cartilage treponemal infection of rural tropical communities, causes painless but stigmatizing lesions, some of which may become permanent disfiguring, crippling deformities. The palmar and plantar forms of yaws are, however, painful and pose difficulties in movement and working. For this reason, yaws, which had high clinical prevalences in some communities before the first eradication efforts which ended in the mid-1960s, caused considerable rural poverty. A global eradication effort initiated by the World Health Organization (WHO) and UNICEF in 1952 reduced the prevalence of the treponemal diseases and that of yaws in particular, in endemic countries, by over 95% by 1965. But short of eradicating it, and given the poor economies and weak health systems of many of such countries then to maintain surveillance and appropriate response, yaws started to resurge in the 1970s (Asiedu et al., PLoS Negl Trop Dis 8(9):e3016, 2014). It has since been fought successfully by only Ecuador (Anselmi et al., Trop Med Int Health 8(7):634–638, 2003) and India (Narrain et al., Indian J Med Res 141(5):608–613, 2015). Some sub-Saharan African (SSA) countries, Southeast Asian countries, and the Pacific Ocean Islands remain endemic. Children under 15 years, the most affected by yaws, continue to suffer deformities and, possibly, stigma, resulting in family and community neglect and some "dropouts" of school. Country and global goals for poverty reduction and universal education cannot be achieved if diseases like yaws continue to be neglected despite the availability of effective tools to eradicate, eliminate, or, at least, control them.

Introduction

Yaws is a neglected tropical disease. It is a contagious disease of the skin, cartilage, and bones. It causes firm swellings, ulcers, and rashes on the skin and can affect any part of the body. It may cause mild pain and tenderness, with or without swelling, in the bones and joints. Children in deprived communities in the tropics are mostly affected, but no age is spared. The infective agent is a bacterium called *Treponema pallidum pertenue*. It is transmitted by persons with yaws skin lesions to susceptibles who have breaches in the skin through which the organisms enter the body. The infection stimulates some antibody production, but no lasting immunity is attained. Previously, the disease was diagnosed mostly clinically and easily treated with an appropriate dose for age single intramuscular injection of aqueous benzathine benzylpenicillin. With improved technological tools, yaws is now suspected clinically and confirmed serologically or with PCR before being treated with a single dose of oral azithromycin tablets or syrup.

Yaws is known and given local names in parts of Sub-Saharan Africa (SSA). In Ghana, it is called *gyator* or *due* (pronounced *duay*) generally in the middle and southern belt, *ekli* in the eastern border with Togo, and *jaga* in northern Ghana. It is called *buba* in Swahili and *kazuwa* in Hausa. Yaws was well-known by endemic populations before the first eradication attempts. Of late though, it is less widely known, especially among the younger generations, even though, in typically endemic communities, even children may still know yaws. A primary school child in one endemic community in Ghana was able to describe the disease accurately in the local language as "a yellow boil." The children in the class named a particular child in the village with the disease and when the team that included one of the current authors got into the village, they found the child with a papilloma on the waist that he was hiding with a very dirty pair of shorts held in place with twines and actually trying to scrape it away with a stone, it being painless. The child was clearly neglected and could not be in school like his friends. Anecdotes of local treatments of yaws in the past include rubbing the painless "bumps" with rough, loofah sponges, and using bluestone (copper sulfate) to eat away lesions. Children of the age group that get yaws feel stigmatized by their friends and tend to hide the lesions, so they are not found and treated early. This may cause the disease to continue to spread quietly and pose a risk of complications called tertiary yaws. Two young boys with gangosa deformity of the nose have been seen in an endemic area in Ghana in the not-too-distant past (picture by Dwamena, 2009).

Clinical Aspects, Diagnosis, and Treatment of Yaws

The complete description of the clinical picture of yaws is presented in *Manson's Tropical Diseases* and in the papers of C. J. Hackett, a WHO yaws expert of the 1950s and 1960s. Yaws looks like many other skin conditions which means diagnosing yaws clinically alone may overestimate or underestimate its prevalence depending on whether a wide or narrow clinical case definition is used. Variation in clinical case definitions may be necessary during the pre- and post-elimination surveillance phases of the disease. This section describes clinical yaws with focus on observations made particularly in the black skin of SSA. First, the clinical categories, stages of the disease, and some relevant epidemiological terms are defined, followed by the clinical features (lesion descriptions), diagnosis, and treatment.

Yaws Clinical Categories and Stages

The following terminologies have been associated with the clinical presentations of yaws and are defined from the onset to clarify further discussions.

Early yaws refers to lesions of the first two phases of the disease called primary and secondary yaws. Primary yaws, or mother yaws, is the first lesion, usually a single raised bump called a papilloma, that appears in 2–6 weeks at the site of entry of yaws germs into a breached skin. This papilloma may have other primary small satellite papules or papillomas near it (Fig. 1). The primary papilloma may also be two at different positions depending on the sites initially inoculated with treponemal

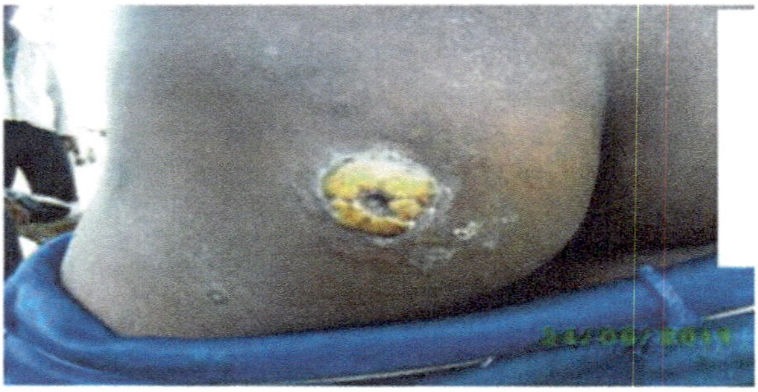

Fig. 1 Primary yaws papilloma with satellite lesions (Taken by Dr Kwakye-Maclean in 2011)

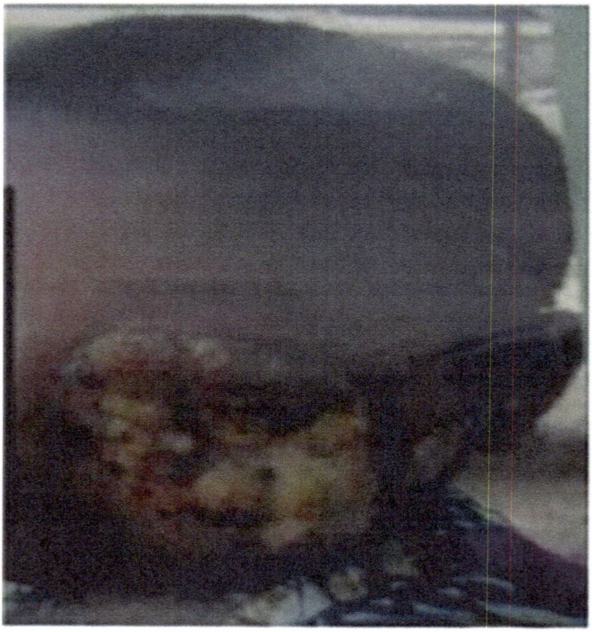

Fig. 2 Yaws primary papilloma over the left eye that healed on treatment without affecting the sight. *Agana Nsiire 2009*

organisms. One school boy presented with two large primary papillomas almost at symmetrical points on the back of the shoulders.

In one bad case of primary yaws over the left eye, the lesion had been present for about 9 months before being seen by the health system because the child was hidden away in a room from the public (Fig. 2).

Secondary yaws appears in other parts of the body, earliest 2 weeks after the primary papilloma, but may take as long as 16 weeks or longer to appear. These may show as uniform lesions or a mixture of papillomas, ulcers, macules, or papules of various forms, pustules especially in the scalp where they are likely to be infected. Other forms are anhidrosis of the skin, cracks, pitting, erosions or thickenings of palms and soles, or bone and joint pain with or without mild swelling. Secondary lesions result from organisms that enter the bloodstream from the primary

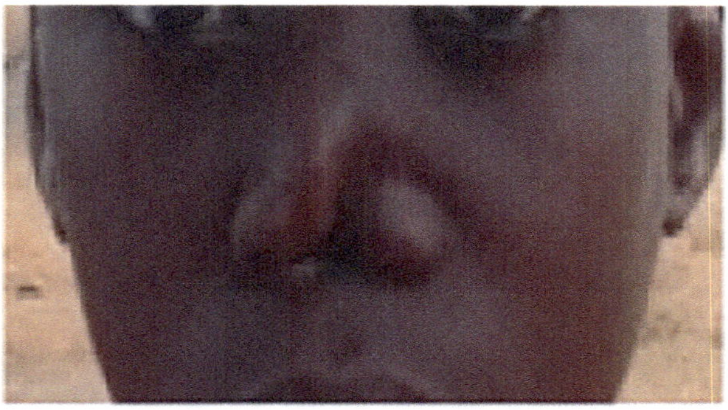

Fig. 3 Gangosa. *Dwamena 2009*

papillomas to set up skin, bone, or joint lesions in other parts of the body. Secondary yaws can therefore coexist with a primary yaws papilloma or its ulcerated stage which can make their distinction difficult except by careful history taking.

Late yaws, also called tertiary yaws, refers to a very small fraction of early yaws cases (less than 5%) who progress to late forms of yaws that destroy tissues leaving permanent scars, contractures, or other bone and cartilage deformities, typically, gangosa, gondou, sabre tibia, and chronic dactylitis (Fig. 3).

Late yaws does not normally coexist with early yaws lesions, but two 13-year-old boys from the same community reported with yaws-like lesions on the noses which progressed to permanent gangosa deformities. This is consistent with reports which suggest that gangosa and gondou can occur early in yaws infection and can be stopped with appropriate treatment or progress to permanent deformities (Perine et al. 1984). This makes it imperative to pay particular attention to lesions on the face and nose in yaws-endemic populations. Saber tibia and dactylitis may also be stopped and reversed if treated early or may progress to permanent deformities. Late yaws is noninfectious because the body's own defense mechanisms have eliminated the organism or greatly attenuated it, though no permanent specific immunity is generated.

For practical purposes of yaws eradication, primary, secondary, and tertiary yaws have been regrouped as (1) *early yaws*, which is infectious but can be treated with antibiotics and prevented and controlled with other measures like personal hygiene and avoiding close contact behavior; and (2) *late yaws*, which is noninfectious but the lesions remain permanent features of individuals unless surgically corrected where possible or functionally improved through physiotherapy. These complications are very rare nowadays, except among the highly endemic populations in the Congo Basin area.

Infectious Yaws This refers to all the skin lesions of early yaws with the potential to transmit yaws organisms to other persons. The papillomas secrete many more organisms than the other skin manifestations and are therefore most infectious. Papillomas on the palms and soles are also infectious, but the other palmar and plantar forms like cracks, erosions, and fissures which occur more on the soles than on the palms are the least infectious. Childhood behaviors like sleeping or playing together, especially close contact wrestling or other similar games promote transmission by virtue of the close contact and the higher probability of skin breaches.

Noninfectious Yaws The early bone and joint forms as well as all the late forms of yaws are noninfectious in real time as they secrete no treponemes on the surface. Technically, however, the early bone and joint forms are potentially infectious or in the infectious phase as organisms' germs circulating in the blood from the initial primary lesions or these skeletal lesions may later set up skin manifestations after latency.

Overt and Latent Yaws Overt yaws shows signs on the skin, cartilage, or bone and is confirmed serologically and treated, at the same time tracing and treating their contacts. Latent yaws is a phase during early yaws infection when all the skin, bone, and joint manifestations have disappeared without treatment or with ineffective treatment. The organisms have, however, not all died out but persist in the blood to manifest later with signs of early yaws on the body. Because of this, latent yaws is a crucial population in yaws eradication activities as not treating them allows transmission of the disease to persist in the population. Overt and latent periods may alternate every few weeks or months over 3–5 years.

Contacts Yaws contacts are those who are incubating the disease (i.e., between inoculation of germs and appearance of primary papilloma(s) which on average is 2–6 weeks but can be longer). This means that one overt case can continue infecting many people and set up a continuum of new cases and incubators that will not show overt disease for weeks and months. These incubators will be missed out when looking for cases in facilities, schools, or communities. Contacts are therefore an important population in disease eradication strategies because if they are not treated, they will later show overt disease and continue the transmission cycle.

Attenuated Yaws This term attempts to describe the current clinical and epidemiological picture of yaws due to the widespread use of antibiotics and improving sanitation which are believed to reduce the manifestation of the disease. Thus, attenuation appears in four ways: (1) reduced size and numbers of papillomas and rapid reagin concentrations in individuals; (2) longer duration of latency; (3) absence or near absence of late forms of yaws; (4) generally low prevalence due to improvement in water, sanitation, and hygiene (WASH) facilities or mass treatment in highly endemic areas (Vorst 1985). Attenuated yaws may, however, revert to aggressive yaws conditions with lowered vigilance on surveillance and response or falling WASH conditions.

Clinical Features of Yaws

Constitutional symptoms are mild in yaws, presenting as mild to moderate joint and bone pain in children at night with no obvious fever. Both children and guardians complain of these symptoms in endemic communities where the disease is known by local names. Itching is rare, occurring more in the patchy papular and macular forms.

1. *Early yaws* lesions may appear on any part of the skin from the scalp to the sole of the foot. The majority of the lesions, however, are found in the lower limbs because of the greater exposure of those parts. The bone and joint forms affect mostly the face, forearms, wrists, hands, fingers, legs, and the ankles. But nodular forms may also affect the skull. Early yaws lesions are characterized by the presence of one or more of the following:

Papillomas are pale or yellowish, painless, and firm to hard chancres usually with dark tips showing points of beginning necrosis. They vary in size from 1 to 3 cm typically with no inflammation of the skin around them. But papillomas may also have a pale halo of macular inflammation extending a centimeter or more around them. Those on the exposed parts especially the legs tend to have surrounding skin inflammation due to secondary bacterial infection. There may be pigmented skin around the lighter yaws lesions or a mixture of pale and pigmented surroundings. Because they are painless and stigmatizing, some children tend to hide them with their clothing or rub off those in accessible parts of their body with rough sponges, sand, or other objects until they become flattened, small pale, leathery blebs that feel rubbery soft rather than hard and loss of the classical yellow hue.

Ulcers: These may occur as initial lesions but are usually ulcerated papillomas. They have raised edges as papillomas that necrose from the tip. The floor is dirty and crusty or yellowish and soggy. Some papillomas may not ulcerate from the tip downward but have the skin peeled off by necrosis or use of rough objects, leaving a raw fungating surface. Both papilloma and ulcer forms attract flies because of their secretions, but there is no documented evidence of transmission by flies (Figs. 4 and 5).

Papules are of the size of rice grains or smaller, pale, painless, firm, and usually discrete. The smaller forms may be patchy and itchy.

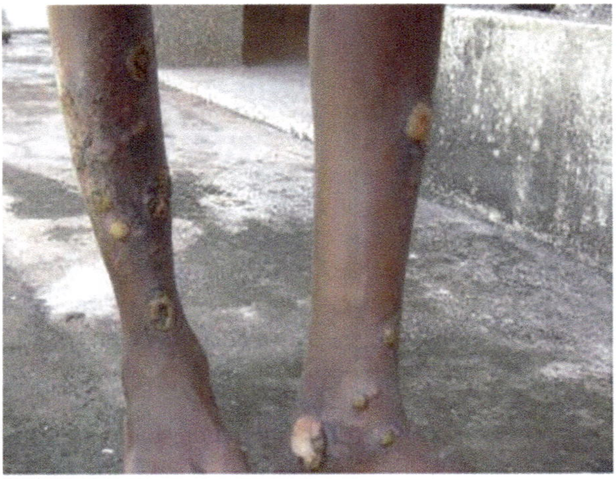

Fig. 4 Papillomata, papules and ulcers of early yaws *(Ghana Yaws Program)*

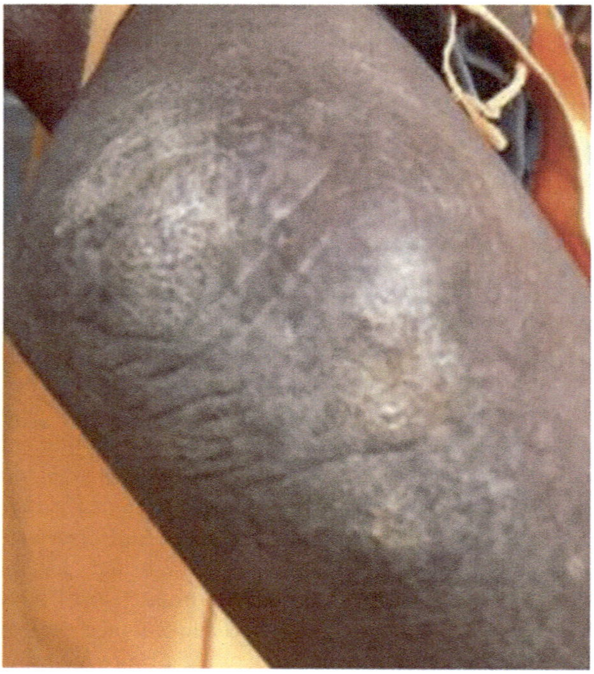

Fig. 5 Poorly developed and infected papules and macules on elbow *(Ghana Yaws Program)*

Macules are small, pale, discrete dots or patches. They may be present as mixed with other yaws lesions like papillomas and papules. The dorsum and margins of feet frequently show maculopapular rashes that may have an atypical appearance due to the exposure of those parts to hazards of the physical environment.

Palmar and plantar yaws are painful lesions of the palms and soles. The palmar lesions are less painful usually showing as hyperkeratosis. Plantar yaws especially is very painful and presents as cracks, fissures, erosions, pitting, or hyperkeratotic thickenings on the soles (Figs. 6 and 7). Papillomas may also occur on palms and soles and are similarly painful.

Bone and joint lesions: Moderate swelling and tenderness affect mostly the bones and joints of the forearm, wrist, fingers, and legs. Usually, one or two bones may show obvious swelling with other bones having subclinical osteitis. The early stages of tibia bowing may be seen with or without ulcerated lesions on the leg. It is not uncommon to see a pale, contiguous patch on the dorsum of a hand with one of the fingers of the same hand swollen and tender. The presence of bone and/or joint swelling together with suspicious skin rashes in a child in an endemic area should lead to a high suspicion of yaws (Figs. 8 and 9).

Fig. 8 Azithromycin dosages for yaws*Ganglia, nodules, and enlarged lymph nodes:* These have rarely been seen in Ghana. They can occur especially near the joints and on the head.

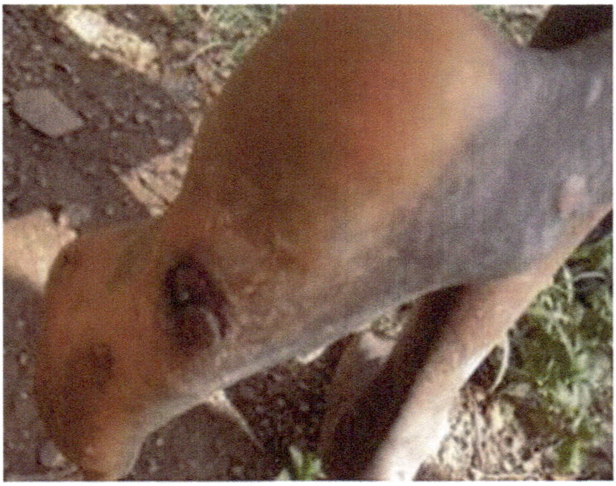

Fig. 6 Plantar yaws thickenings; usually very painful

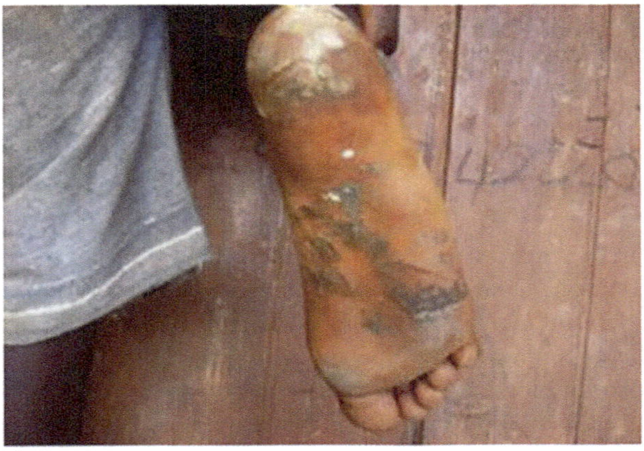

Fig. 7 Plantar yaws erosions (Dr. Kwakye-Maclean)

Yaws may present as other atypical rashes including anhidrosis with dry or pigmented markings especially on the lower limbs. Lesions on elbows and knees tend to be mixed papules, macules, and small papillomas which usually ulcerate and look soggy, but they may also be clean macules or papules or patches of them. Generally, all early yaws skin lesions are paler than the normal skin and painless except for those on the soles and palms or when secondarily infected with other bacteria. They are also frequently polymorphous—that is, many types of early lesions present together.

2. *Late yaws lesions* are rarely seen but have traditionally affected the face (gangosa and gondou), forearm and finger bones (dactylitis), leg bones (sabre tibia), and the skin as ugly and permanent scars. It should be noted that bending of long

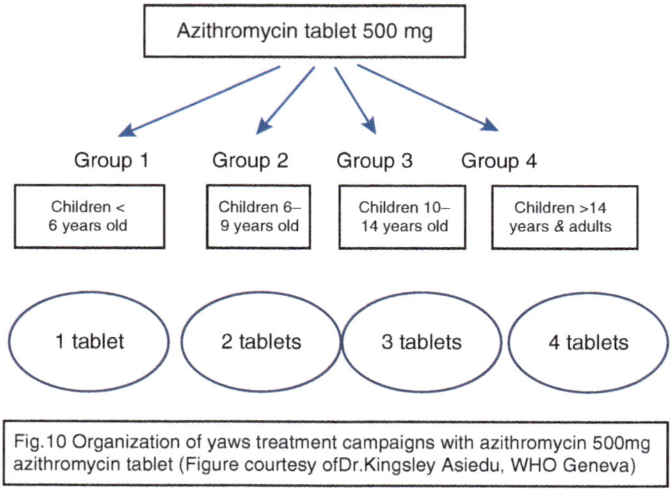

Fig. 8 Azithromycin dosages for yaws

bones occurs in early yaws but can be corrected with treatment. Sabre tibia is therefore used here to describe late yaws where the tibia deformity has become permanent.

Diagnosis of Yaws

Yaws and the treponematoses were first diagnosed clinically for treatment and decisions on mass treatment campaigns. Mitija and others, however, have indicated the availability and usefulness of new laboratory tools for yaws control activities (Mitjà et al. 2013). Yaws is now suspected using a clinical case definition and confirmed with serology before treatment. The case definition arises from the various lesions described above. For purposes of searching for overt and latent cases and contacts in eradication programs, a case definition of yaws involves anyone living in an endemic community with one or more of the following: papillomas, ulcers with raised edges, papules and macules of unknown causes, arthralgia especially in children of unknown causes, bone swellings with moderate tenderness especially in children and of unknown causes, nodules or ganglia, plantar cracks, erosions or thickenings, or a history of any of these in the last 5 years in endemic communities. Gangosa, gondou, sabre tibia, deformed finger bones and joints, and skin gummas, active or within the recent past or any skin condition not attributable to any known cause, should be suspected for yaws and serology done for confirmation of active disease. In Ghana, the use of community based surveillance volunteers (CBSVs) to facilitate the eradication process through early integrated case detection with other skin NTDs, like Buruli ulcer, leprosy, and scabies, has resulted in a modification of case definitions for this category of health worker which does not specify the

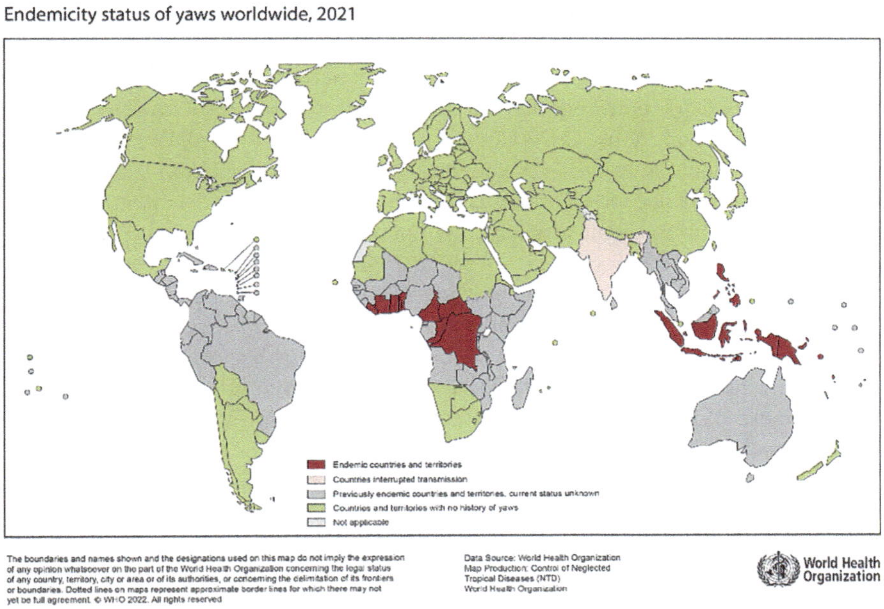

Fig. 9 Distribution of yaws worldwide, 2012

disease by name but by simple description of the visible lesion. The Ghana CBSV Community Register lists the following case definitions for identification and reporting by the volunteer: *ulcer, painless boil or swelling, change in skin color, skin itching, skin patch, thick elbow nerve, nodule, and deformed hand/foot* (Ghana skin NTD CBSV Register, 2021). These are followed on by health professionals at various levels of the district health system to confirm with more specific case definitions and by laboratory tests before treatment is initiated, for cases and sometimes for contacts, as in the case of yaws and leprosy.

The laboratory diagnosis of yaws is based on both qualitative and quantitative serology tests. The two main types of serology used are *treponemal* and *non-treponemal* tests. Treponemal tests detect treponemal antibodies which are present for life once a person is infected with any of the Treponemes. They are useful, therefore, in confirming a person's previous exposure to infection but do not confirm a current infection. Examples of treponemal tests are Treponema pallidum hemagglutination (TPHA) and Treponema pallidum passive particle agglutination (TPPA) tests. There are variants of these like *syphilis first response* used in antenatal clinics and in the field qualitatively as point-of-care tests (POCTs). Treponemal tests can also be used quantitatively to estimate the level of antibodies in the individual.

Non-treponemal tests like rapid plasma reagin (RPR) detect active disease by the presence of nonspecific antibodies that rise with infection but peter out in a month or two. They may also be used qualitatively and quantitatively but have the drawback of showing positive in other conditions like malaria, typhoid, and pregnancy.

In community surveys, they are probably more useful than treponemal tests in detecting latent cases and contacts who are harboring active infection.

Perhaps the best serological tests are the modern generation of dual POCTs being developed for both treponemal and non-treponemal tests on one platform. The main one in current use is the Chembio Dual Path Platform (DPP) that has been evaluated in Ghana (Kwakye-Maclean et al. 2017) and the Solomon Islands (Marks et al. 2014) and are accepted for use as rapid diagnostic tests for surveys, for confirmation of yaws in resistant lesions to treatment, and for surveillance in the post-elimination phase of yaws, as cost-effective tools.

DNA-based polymerase chain reaction (PCR) tests are being developed with primers specific for yaws. There is also the f-TLC or LAMP test being evaluated for use in place of classic PCR in district level laboratories because of the simplicity and faster returns of results. These will be the only tests to differentiate the various Treponemes. Old Venereal Diseases Research Laboratory (VDRL) tests and dark field microscopy are hardly used nowadays for the diagnosis of yaws.

The WHO Guide for Monitoring and Evaluation of the Global Yaws Eradication Program further provides the following definitions for the purpose:

Rumor of yaws: A rumor of yaws is a person that a community member or volunteer identified as begin potentially a suspected yaws case.

Suspected yaws case: A suspected yaws case is a person of any age who is or was living in a previously or currently endemic area, presenting with clinical signs consistent with yaws.

Probable yaws case: A probable yaws case is a suspected yaws case with a positive rapid treponemal point-of-care (PoC) test, i.e., both treponemal and control lines visible.

Serologically confirmed yaws case: A serologically confirmed yaws case is a suspected case who is confirmed through dual positive serology (either DPP-dually positive or TPHA/TPPA+RPR positive).

PCR-confirmed yaws case: A PCR-confirmed yaws case is a case that is positive to polymerase chain reaction (PCR).

Differential Diagnoses of Yaws

Common differential diagnoses are as follows:

Similar skin, bone, and cartilage lesions of the three related treponemal diseases, namely, venereal syphilis, endemic syphilis, and pinta. Pinta is however not found in SSA, and endemic syphilis affects more the mucous membranes. In venereal syphilis, there may be a chancre or history of a chancre in the genital area in a person that is sexually active, the age of onset of which is coming down and posing more differential diagnosis problems.

Scabies presents as tiny pustules that are intensely itchy, found in the finger webs and flexure areas. An itchy pustule on the penis is almost pathognomonic of scabies. Presence in other family members including adults at the same time is more frequent than with yaws.

Fungal skin infections also occur in any part of the body including the mucous membranes. Useful distinguishing features are the ring nature of fungal infections with healing in the center and active disease at the edge of the ring. Scaling is also commoner in fungal infections. The ring forms of yaws may be heaped at the edges and not show much scaling.

Eczematous lesions and urticaria of varied causes may be differentiated from yaws by careful history and the patch test in the case of allergic or contact dermatitis. The dermatophism test may also indicate if lesions are other than yaws, though a person may have both.

Leprosy has associated thickened peripheral nerves and loss of sensation on the lesions as opposed to yaws lesions which are painless but tactile sense is not lost.

Buruli ulcer may be distinguished by undermining edges of the ulcer stage as opposed to the raised edges of most yaws ulcers. The nodule, plaque, and edematous forms of Buruli ulcer may, however, be difficult to distinguish except by careful history and histology by fine-needle aspiration of contents. Buruli ulcers will, however, usually be single as opposed to the many lesions especially of secondary yaws. Corns, niduses, plantar warts, and the cracks or fissures of soles in certain occupations like farmers and in certain seasons like the Harmattan are frequently diagnosed by inexperience as plantar yaws.

Other uncommon differential diagnoses include vitiligo which is permanent, much paler than the normal skin, and on exactly the same plane with normal skin (you cannot see or feel the slightest elevation of the lesion). Cutaneous leishmaniasis with its characteristic volcano ulcers, lichen planus distinguished by its shiny surfaces and tiny white lines (Wickham's striae), psoriasis with its shiny, reddish, and itchy plaques are other differentials. Psoriasis frequent affection of the extensor aspects of the knees and elbows alone cannot be used to differentiate it from yaws which also affects those parts frequently.

Molluscum contagiosum, a chronic viral infection presenting as painless soft to firm round papules about half a centimeter and usually with the color of the normal skin, and congenital epidermolytic hyperkeratosis are other rare conditions simulating yaws. The latter, however, tends to be present from birth or at least for a long, long time and is also painless.

Sickle cell disease should always be excluded in joint and bone pain in children.

Treatment of Yaws

Yaws was first treated in the 1950s and earlier with toxic bismuth and arsenical compounds which could not be used easily for mass campaigns. These were replaced by penicillin, and fortunately, yaws, like syphilis, has remained sensitive to penicillin especially in SSA. Depot forms for single intramuscular injections to keep high blood levels for long periods have included penicillin aluminum monostearate in oil

(PAM) which was used for yaws eradication purposes in the 1950s and 1960s. It had the undesirable effect of leaving scars at injection sites due to the oil base. PAM was, therefore, replaced by aqueous benzathine benzyl penicillin which can keep blood levels high for 2–4 weeks depending on the source. The dose of depot penicillin for adult cases and children over 10 years is a single intramuscular injection of 1.2 mega-units. Half of that is given to corresponding adult and child contacts. For children 10 years and below, 0.6 megaunits is used and half of that (0.3 megaunits) for child contacts. Aqueous depot penicillin is however painful and cumbersome for mass field administration for eradication purposes. The Ghana and Papua New Guinea drug trials of 2010/2011 have, however, caused the WHO to revamp and revise the policy for yaws eradication using single oral doses of tablet or syrup azithromycin in mass drug administration at 30 mg/kg body weight. However, approximations for age to simplify program activities prescribe doses as indicated in Fig. 10. Some studies seem to indicate that yaws and trachoma MDAs could be integrated because the 20 mg/kg body weight used in trachoma have shown similar impact on yaws (Marks et al. 2015).

Epidemiology, Public Health Importance, and Eradication of Yaws

Epidemiology

The demographics of yaws have not changed over the years. Yaws affects all ages but mostly children under 15 years. Boys are more affected than girls (Touré et al. 2007). The reasons for these may lie in behavior, boys being usually more exposed and more prone to contact and bruises, which are necessary conditions for yaws transmission. Young adults are next most affected especially by plantar yaws. Poor personal hygiene is a strong contributing factor. Warm, tropical zones, especially the rural and crowded suburban communities with poor WASH facilities are more prone to yaws. But as economies grow and social facilities improve, yaws gets further and further back where the road ends.

Between 1940 and 1950, 37 SSA countries (outside South Africa) reported yaws cases to the WHO (Hackett 1953). Since 46 countries were involved in the global campaign, it means Africa formed the bulk of yaws endemic countries then. A WHO AFRO *Report of a Regional Meeting, Brazzaville, 3–6 February 1986* on "Yaws and other endemic Treponematoses" also listed additional countries that could be endemic for yaws. A 2015 review by Oriol and others estimate that about 89 million of the world's population may still be at risk of yaws, based on administrative data and the few surveys available. Today, yaws is still endemic in a number of SSA countries with wet, humid forest and/or dry, grassland areas. Generally, these countries lie within the tropics with temperatures between 20 °C and 40 °C and exceptionally exceeding these limits especially in East and South Africa with considerable highlands and which may lie within the southern temperate belt, and in the hot Sahel countries. Yaws or its close relation bejel therefore probably thrives in much of SSA (Toure 1985), but the current WHO records indicate that only nine countries have any reports or post eradication period assessments suggesting that they are yaws endemic. These are Benin, Cameroon, Central African Republic, Congo, Democratic Republic of Congo, Ghana, La Cote d'Ivoire, Liberia, and Togo (Edorh et al. 1994; Sanou 2012). A yaws sero-prevalence of 10.9 has been cited among school children in Ghana (Abdulai et al. 2018). The prevalence surveys (results unpublished) were carried out in 2011 prior to trial of mass drug administration with oral Azithromycin for yaws eradication. Similar high prevalences were reported among the autochthonous populations of the Congo Basin by Medecins Sans Frontièr piloting single-dose oral azithromycin mass drug administration in 2012 (MSF, Unpublished). Sporadic outbreaks have also been reported in Cameroon and Ghana, but these have been diagnosed only clinically. The current yaws situation on the African continent, therefore, needs a re-assessment. Guinea reported many clinically suspected cases of yaws in their administrative data in 2018 but expert surveys in 2019, though limited, revealed no confirmed cases. The highest prevalence remains among the autochthonous populations of Central African countries (Cameroon, Central African Republic, Congo, and the Democratic Republic of Congo). In the West African countries, however, the prevalence is generally low, but the distribution is wide in isolated rural areas and some suburban poor communities. Surveys in Ghana suggest that yaws may be found more in children and young adults not in school than those in school (National Yaws Eradication Program Annual Report 2008, Ghana). But this is probably becoming less so due to high school enrolments over the years.

Records compiled by C. J. Hackett for 37 SSA countries of the 1950s gave little information on the seasonality pattern to yaws, though few countries (Angola, Cameroon, Congo, Mozambique, and Sierra Leone) have indicated higher numbers of yaws during the warm, humid, rainy seasons (Hackett 1953; Harding 1949). Monthly returns to the national program in Ghana do not show clear seasonality, but this is based on clinical cases not confirmed in the laboratory. Some reports suggest, however, that yaws incidence may rise in the cold seasons as a behavioral outcome when children huddle together during sleep, as reported of the cold monsoons in parts of India. My personal experience is the finding of common typical yaws clinical lesions during the cold Harmattan season in some communities in Northern Ghana in 1994 (unpublished).

Yaws is transmitted through skin contact between infectious yaws cases and others who sustain bruises or openings of any kind during the contact, or previously, allowing the organism to enter and establish infection. Treponeme is a slow-growing oxygen-dependent intracellular organism with a doubling rate of about 30 h (Magnuson and Eagle 1948; Norris et al. 2001). This, combined with a long incubation period of up to 6 weeks or more, makes yaws a not too highly infectious disease. But the very existence of long incubating contacts and latent cases makes yaws transmission intractable unless effective interventions are put in place. Apart from age and sex, other predisposing factors for yaws transmission are poor personal hygiene, especially not bathing at least once a day, poor water supply, and overcrowding.

Public Health Importance of Yaws

The importance of yaws is derived from the direct physical effects it has on the individual and the social, cultural, and economic impact on individuals, families, and communities. For yaws, the direct physical effects are not death but the presence of ugly, discomforting, and, sometimes, painful or crippling lesions, which could be transient or permanent. Some yaws lesions, especially those on the face, leave stigmatizing effects negatively affecting social life and the education of some children. The child in Ghana mentioned above, with a large papilloma covering the left eye, was hidden away in the room until discovered by a philanthropist, treated successfully and returned to school in the care of the said philanthropist. Another young girl with early bone deformities was kept out of mingling with her colleagues until unearthed and treated by the health system before resuming active social life. Yaws can negate the sustainable development goals (SDGs) on poverty reduction, universal education, and morbidity reduction especially of the NTDs.

As a communicable disease, yaws remains a public health hazard in nine African countries, but the exact situation must be assessed in the other countries. The colonial office medical department annual reports (MDARs) of the Gold Coast Government (now Ghana) point to the high burden of yaws in the pre-campaign days, and some documentations ranked yaws as highest cause of outpatient morbidity or at same level with malaria, guinea worm, and onchocerciasis (Destombes 1999; Colonial Reports—Annual No 1919, Gold Coast 1938–1939). This clearly demonstrates the potential burden posed by the disease if allowed to spread. One survey among school children in Ghana in 2013 put yaws clinical prevalence at 1.3% and serological prevalence at 10.9% (Abdulai et al. 2018). Surveys in other SSA countries have reported prevalences ranging between 1% and 10% depending on the method and size of the study population (Coldiron et al. 2013; Touré et al. 2007; Toure 1985; Gerstl et al. 2009). Qualitative comments of the social and economic impact on rural populations are rampant in the literature, but these have never been quantified and are absent in recent research works. Anecdotal evidence and personal experience of the authors and some field workers indicate that stigma and morbidity affect children negatively in terms of their education and social life.

My personal appreciation of the current burden of yaws is that it will no longer pose concern in terms of morbidity or economic impact of the pre-eradication era. It is, however, still important at the individual, family, and community level and must be eradicated to spare resources and energy for other priorities. The growth of African economies, albeit, slow, is helping to roll this disease back to the remote areas and, with good surveillance and appropriate response by highly decentralized health systems now, yaws eradication should be realized by the SDG timeline of 2030.

Eradication of Yaws

The first global yaws eradication effort from 1952 to 1965 went alongside that of malaria eradication. Surveillance of yaws after this effort ceased in most African countries. In Cameroon, the eradication effort ended in 1970 and, with it, surveillance on yaws (Njih Tabah 2013). After outbreaks of yaws in five districts in 2007 and 2008, the Ministry of Public Health included yaws in the Buruli ulcer, Leprosy and Leishmaniasis Control Program in 2009 and in the national master plan of 2012–2014. Between 2010 and 2011, 23 out of the 58 health districts had reported clinical yaws cases through mass screening and passive notification. Cameroon has now initiated a mass drug administration (MDA) eradication effort recommended in some circumstances following the Morges Strategy guidelines of 2012. La Cote d'Ivoire similarly had no program of yaws control, though in 2010, according to routine data made available to the WHO by Professor Henri Asseh of the Buruli Ulcer program, 67 out of the 90 districts reported yaws. Yaws is now confirmed serologically in the country and some efforts are started with support from the Anesvad NGO of Spain toward the eradication goal. Ghana's first eradication effort ended in 1969, but Eastern and Central Regions were excluded in the campaign. Since then passive surveillance and treatment have been continuously carried out by facility-based medical field unit (now disease control unit) staff. Between 1980 and 1983, the WHO, the World Bank, and other partners tried a combined yaws mass treatment with immunization against yellow fever and tetanus in Ghana and neighboring countries, but it was short-lived.

Ghana revamped yaws elimination efforts in 2008. WHO and the international community have renewed efforts on the NTDs, including yaws eradication following studies on the global socioeconomic burden of the group. This resurgence of global interest, backed by the 2013 World Health Assembly resolution on the NTDs (WHA66.12 2013), propelled by the SDGs and WHO's NTDs Roadmap 2030, coupled with new single-dose oral treatment and diagnostic tools for yaws, and encouraged by successes made so far in the global eradication of guinea worm, the elimination of trachoma from many countries, the significant advances toward control and elimination of human African trypanosomiasis, lymphatic filariasis, onchocerciasis and other NTDs, especially using the WHO recommended strategies of good governance, program integration, and partner collaboration, has put yaws eradication firmly back on track. The WHO has, therefore, initiated actions to help the process including grouping countries into three categories: (1) Category C countries are traditionally non-endemic or have achieved and documented yaws elimination for which they will be so certified. (2) Category B countries are previously endemic countries with little or no knowledge of their current situation (they will carry out studies/surveys to establish their status). (3) Category A countries, like the nine in SSA know they are endemic and are piloting the new tools and trying to scale up eradication activities towards the 2030 target. The WHO has elaborated a Morges Strategy for yaws eradication actives (WHO 2012, 2013). This strategy modifies the original approach of total mass treatment (TMT for community clinical prevalence of yaws >10%), juvenile mass treatment (JMT for prevalence 5%–10%), and selective mass treatment (SMT for prevalence <5%) into *total community treatment* (TCT for proven endemic communities irrespective of prevalence) and total targeted treatment (TTT). TCT uses Azithromycin MDAs and TCT follows TCT MDAs with treatment of only cases and close contacts or more restricted MDAs.

This original Morges Strategy has seen modifications over time and also makes allowance for the TTT approach ab initio which involves active surveillance and treatment of cases and close contacts in integrated programs depending on the country's endemicity level and other peculiar circumstances including human and material resources. This is in line with the original SMT and was used successfully by India to eliminate yaws (Narain et al. 2015). MDAs theoretically target everybody, but experience in MDAs for other NTDs shows that coverages have never been 100%. The average for Ghana over the years for the integrated preventive chemotherapy NTDs targeting lymphatic filariasis, onchocerciasis and the soil transmitted helminthiasis has been 73% (below a WHO target of 75% but above a West African average of 64% as reported in the Ghana NTDs Master Plan 2021–2025. Biritwum and others have reported community MDA administrative therapeutic coverage ranges between 19 and 120% (Biritwum et al. 2019). In Cameroon, two rounds of TCT, targetting ten health districts for a total population of 642,946 ,achieved theurapeutic coverages of 92.8% and 96% respectively, between 2020 and 2023 (Tabah, 2023. unpublished report). In situations of low endemicity, therefore, TCT MDAs may not be more effective than active surveillance and response by the TTT approach.

The rest of the 20 NTDs identified by the WHO, other than guinea worm and yaws, are all targeted for elimination or control by this target date. This raises the frequently asked question: should yaws or any other disease be eradicated instead of just being controlled or eliminated?

Should Yaws be Eradicated?

This question is not rhetorical and may be considered from both the practical and philosophical viewpoints. In practical terms, disease eradication is considered to be the reduction of the disease to zero (no cases at all) and removal of the causative agent from the wild where it freely moves within the population. The causative agent may then be confined only to one laboratory. But if all live specimens are destroyed, the disease would be said to be extinct. Yaws (if the germ is the same strain) is believed to exist also among other primates like chimpanzees and gorillas (Rinaldi and King 2008), so it would seem more appropriate to talk of yaws elimination (reduction of disease to below public health importance or elimination from the human population) rather than eradication. Attempts to eradicate it in other primates, like in all zoonoses, could be extremely challenging. Conceptually though, eradication of the disease from the human population, rather than the causative organism from the world, is still possible, especially if the zoonotic strain is not proven to be transmissible to humans. That yaws has already been eliminated from a number of countries including India most recently, using the tedious tool of injections, encourages us that, with the newer and easier tools (azithromycin for oral treatment and point-of-care tests and PCR for surveillance), eradication as a disease is possible. We could, therefore, reengineer our social effort to achieve it. The key is in the reengineering of our collective effort towards this objective, goaded by the achievements of earlier efforts. But such a question always attracts more than the health professional's point of view, like those of politicians, donors, economists, and the communities with many priorities. Furthermore, even if eradication is agreed on by all, the "how-to-do-it" also arises as a philosophical and pragmatic question, whether through slow socioeconomic improvements considering the water-washed status of yaws, through accelerated drug treatment interventions or both.

Conclusion

Yaws, a communicable disease transmitted by skin contact, is a nonfatal but crippling and stigmatizing disease that affects mostly children in deprived tropical communities. It is still present in a number of SSA and other countries despite previous efforts at global eradication. Earlier campaigns, however, helped many countries, some of which, with improving socioeconomic conditions, are seeing less or no cases of yaws now. Ecuador (Anselmi et al. 2003) and India (Narain et al. 2015) have recently declared their borders free of yaws. For those countries that are still endemic with yaws, competing priorities have relegated yaws to the domain of neglected tropical diseases, though it is relatively easy to diagnose and treat. Countries that do not know their current status need surveys and other means to establish their status in the renewed global effort if necessary.

Eradication of yaws requires that contacts of yaws cases and its latent forms also be treated to remove the germ, which resides mainly in men, from the population. New and simpler treatment options and surveillance tools are available to facilitate the process as improving sanitation, water supply, and personal hygiene are scaled up. What remains, therefore, seems to be the collective will and deliberate effort in this direction. There is a resurging world community interest in the NTDs, backed by successes stories that promise that an end to the 20 WHO identified NTDs is in sight and, with it, the eradication of yaws.

Acknowledgments We would like to acknowledge the contribution and support for the following researchers in putting this chapter together:

- *K. K. Addo*, Noguchi Memorial Institute of Medical Research, University of Ghana, Legon, Accra, Ghana.
- *P. Koffi*, National Buruli ulcer Control and Yaws Eradication Program, Ministry of Health, Cote d'Ivoire.
- *Y. T. M. Barogui*, National Buruli Ulcer and Leprosy, Control Program, Ministry of Health, Cotonou, Benin,
- *N. K. Kotei*, National Buruli ulcer Control and Yaws Eradication Program, Disease Control and Prevention Department, Korle Bu, KB 493, Accra, Ghana.
- *R. Phillips*, Kumasi Center for Collaborative Research, Kwame Nkrumah University of Science and Technology, Kumasi, Ghana.
- *B. O. Quao*, National Leprosy Control Program, Disease Control and Prevention Department, Korle Bu, KB 493, Accra, Ghana.

References

Abdulai AA, Agana-Nsiire P, Biney F, Kwakye-Maclean C, Kyei-Faried S, Amponsa-Achiano K, et al. (2018) Community-based mass treatment with azithromycin for the elimination of yaws in Ghana - results of a pilot study. PLoS Negl Trop Dis 12(3):e0006303. https://doi.org/10.1371/journal.pntd.0006303

Anselmi M, Moreira J, Caicedo C, Guderian R, Tognoni G (2003) Community participation eliminates yaws in Ecuador. Trop Med Int Health 8(7):634–638. https://doi.org/10.1046/j.1365-3156.2003.01073.x

Biritwum NK, Frempong KK, Verver S, Odoom S, Alomatu B et al (2019) Progress towards lymphatic filariasis elimination in Ghana from 2000-2016: analysis of microfilaria prevalence data from 439 communities. PLoS Negl Trop Dis 13(8):e0007115

Colonial Reports—Annual, No. 1919, Annual Report on the Social and Economic Progress of the People of the GOLD COAST (1938–1939) libsysdigi.library.illinois.edu/.../5530214_1938_1939_opt.pdf

Coldiron M, Obvala D, Mouniaman-Nara I, Pena J, Blondel C, Porten K (2013) The prevalence of yaws among the Aka in The Congo. Med Sante Trop 23(2):231–232. [Article in French]

Destombes J (1999) Working Paper No. 49/99, Nutrition and economic destitution in northern Ghana, 1930–1957. A historical perspective on nutritional economics, eprints.lse.ac.uk/22388/1/wp49.pdf

Edorh AA, Siamevi EK, Adanlete FA, Aflagah EK, Kassankogno Y, Amouzou AB, Lodonou KG (1994) Resurgence of endemic yaws in Togo. Cause and eradication approach. Bull Soc Pathol Exot 87(1):17–18

Gerstl S, Kiwila G, Dhorda M, Lonlas S, Myatt M, Ilunga BK, Lemasson D, Szumilin E, Guerin PJ, Ferradini L (2009) Prevalence study of yaws in the Democratic Republic of Congo using the lot quality assurance sampling method. PLoS One 4(7) https://doi.org/10.1371/journal.pone.0006338

Ghana NTDs Master Plan 2021–2025

Ghana Health Service Skin Neglected Tropical Diseases Community-Based Surveillance Volunteer (CBSV) Register For Skin NTDs (Buruli ulcer, Leprosy, Yaws) (2021)

Hackett CJ (1953) Part III. Extent and nature of the yaws problem. Bull World Health Organ 8(1–3):129–182

Harding RD (1949) Trans R Soc Trop Med Hyg 42:348. http://www.who.int/yaws/epidemiology/en/: Yaws: epidemiological situation

Kwakye-Maclean C, Agana N, Gyapong J, Nortey P, Adu-Sarkodie Y, Aryee E, Asiedu K, Ballard CR, Binka F (2017) A single dose oral azithromycin versus intramuscular benzathine penicillin for the treatment of yaws-a randomized non inferiority trial in Ghana. PLOS Negl Trop Dis 11:e0005154. https://doi.org/10.1371/journal.pntd.0005154

Magnuson HJ, Eagle H (1948) The minimal infectious inoculum of Spirochaeta pallida (Nichols strain), and a consideration of its rate of multiplication in vivo. Am J Syph 32:1–18

Marks M, Goncalves A, Vahi V, Sokana O, Puiahi E, Zhang Z, Dalipanda T, Bottomley C, Mabey D, Solomon A (2014) Evaluation of a rapid diagnostic test for yaws infection in a community surveillance setting. PLoS Negl Trop Dis 8(9):e3156. https://doi.org/10.1371/journal.pntd.0003156

Marks M, Vahi V, Sokana O, Chi KH, Puiahi E et al (2015) Impact of community mass treatment with azithromycin for trachoma elimination on the prevalence of yaws. PLoS Negl Trop Dis 9(8):e0003988. https://doi.org/10.1371/journal.pntd.0003988

Mitjà O, Šmajs D, Bassat Q (2013) Advances in the diagnosis of endemic treponematoses: yaws, bejel, and pinta. PLoS Negl Trop Dis 7:e2283. https://doi.org/10.1371/journal.pntd.0002283 [PMC free article] [PubMed] [CrossRef] [Google Scholar]

Narain JP, Jain SK, Bora D, Venkatesh S (2015) Eradicating successfully yaws from India: the strategy & global lessons. Indian J Med Res 141(5):608–613. https://doi.org/10.4103/0971-5916.159542

National Yaws Eradication Program Annual Report (2008) Ghana
Norris SJ, Cox DL, Weinstock GM (2001) Biology of Treponema pallidum: correlation of functional activities with genome sequence data. J Mol Microbiol Biotechnol 3(1):37–62
Oriol M et al (2015) Global epidemiology of yaws: a systematic review. Lancet Global Health
Perine PL, Hopkins DR, St John RK et al (1984) Handbook of treponematoses: yaws, endemic syphilis and pinta. WHO, Geneva
Rinaldi A, King CH (eds) (2008) Yaws: a second (and maybe last?) Chance for Eradication. PLoS Negl Trop Dis Aug 2(8):e275. Published online 27 Aug 2008. https://doi.org/10.1371/journal.pntd.0000275
Sanou I (2012) Presentation captured in WHO's summary report of a consultation on the eradication of yaws, 5–7 March 2012. Morges, Geneva
Tabah EN (2013) Presentation on Yaws in Cameroon at Yaws Review Meeting in Geneva, 20–22
Toure IM (1985) Endemic Trematoses in Togo and other west African States. Rev Infect Dis 7(Suppl 2):242–244
Touré B, Koffi NM, Assi KP, Ake O, Konan DJP (2007) Yaws in Côte d'Ivoire: health problem forgotten and neglected. Bull Soc Pathol Exot 100(2):130–132
Vorst FA (1985) Clinical diagnosis and changing manifestations of treponemal infection. Rev Infectious Dis 7(Suppl 2)
WHA66.12 WHO (2013) Resolutions and decisions: WHA66.12 Neglected tropical diseases. World Health Organization, Geneva. Available: http://www.who.int/neglected_diseases/mediacentre/WHA_66.12_Eng.pdf. Accessed 22 Aug 2014
World Health Organization (2012) Eradication of yaws – the Morges Strategy. Weekly Epidemiological Record 87(20)
World Health Organization (2013) Eradication of Yaws - the Morges strategy

Drug Donation Programmes in the Control and Elimination of NTDs

Stefanie Meredith, Elizabeth Elhassan, and Boakye A. Boatin

Contents

S. Meredith (✉)
Global Health Consulting, Divonne les Bains, France
e-mail: stefanie@globalhealthconsulting.net

E. Elhassan
Kaduna, Kaduna State, Nigeria

B. A. Boatin
Institute of Parasitology, McGill University and Lymphatic Filariasis Support Centre for Africa, Accra, Ghana

J. O. Gyapong, B. A. Boatin (eds.), *Neglected Tropical Diseases - Sub-Saharan Africa*, Neglected Tropical Diseases, https://doi.org/10.1007/978-3-031-53901-5_17

Abstract

This chapter reviews the history and evolution of the medicine donation model that has been the mainstay of neglected tropical disease (NTD) control programmes since Merck & Co., Inc. made the decision to donate its medicine Mectizan® for the treatment and control of onchocerciasis. The 2012 London Declaration marked the launch of an expanded and coordinated international effort against NTDs, inspired by the World Health Organisation road map to eradicate or prevent transmission of neglected tropical diseases by the year 2020. This boosted the progress and scale-up of both the preventive chemotherapy diseases (PCDs) programmes and the innovative and intensified disease management (IDM) (WHO). We discuss the challenges, lessons learned by all the stakeholders and the unprecedented collaboration between pharmaceutical companies for effective treatment of some of the NTDs. The socio-economic impact of the donations of medicines on the NTD control programmes and what the next 10 years may hold are briefly reviewed.

Background

Since 1987, when Merck & Co., Inc. made the unprecedented decision to donate its medicine Mectizan® (ivermectin) "to all who needed it for as long as needed" for the treatment and control of onchocerciasis (Dull 1980), this drug donation model has been the mainstay of Neglected Tropical Disease (NTD) control programmes.

The London Declaration on NTDs in January 2012 marked the launch of an expanded and coordinated international effort against NTDs, inspired by the World Health Organisation road map to eradicate or prevent transmission of neglected tropical diseases by the year 2020. Pharmaceutical companies, donors, endemic countries and non-government development organisations (NGDO) committed to control, eliminate or eradicate 10 diseases by the end of the decade (WHO 2012). Currently, 12 pharmaceutical companies are involved in drug donation programmes for 13 of the NTDs recognised by the World Health Organization (2021a, b, c) as shown in Table 1.

It is not an exaggeration to say that drug donation completely changed the outlook and strategies for neglected disease control programmes and opened up the potential for pursuing disease elimination and even the prospect of eradication for some of the NTDs (Uniting to Combat NTDs 2012). Most of these donated drugs were developed several decades ago and are relatively cheap, but this does not mean that the endemic communities and ministries of health may be able to afford them

Table 1 NTD Medicine donations by diseases and pharmaceutical donors' commitment (last updated September 2021)

NTDs	Manufacturer	Product name	MOU period	Donation commitments/internal discussion	Total number of drugs donated from 2011 to 2020 (source: WHO dashboard)
Donations managed by the WHO					
Chagas	Bayer	Nifurtimox	2007–2025	Up to a total of 12,500,000 tablets for the treatment of Chagas disease	7.9 million
	Chemo Group (Mundo Sano)	Benznidazole	2020–2023	3,000 tabs (12.5mg)/105,000 tabs (100mg)	
Food-borne trematodes	Novartis	Triclabendazole	2016–2025	Up to 600,000 tablets for the treatment of fascioliasis and paragonimiasis	5.40 million (to be confirmed)
Human African trypanosomiasis (HAT)	Bayer	Nifurtimox	2009–2026	Up to 150,000 tablets for 5 years, adjustable to needs to treat human African trypanosomiasis	1.8 million
		Suramin	2002–2026	Up to 10,000 vials for 5 years, adjustable to needs to treat human African trypanosomiasis	
	Sanofi	Eflornithine, melarsoprol, pentamidine	2001–2026	Unlimited quantity for the treatment of human African trypanosomiasis	
		Fexinidazole	2019–2026		
Leprosy	Novartis	Lamprene® 100mg	2000–2025	Unlimited supply of additional bulk quantities for the treatment of severe erythema nodosum leprosum reactions	32 million
		Lamprene® 50mg			
		MDT[a] MB[b] Adult & Child		Unlimited supply for the treatment of leprosy and its complications	
		MDT PB [c]Adult & Child			
Lymphatic filariasis (LF)	Eisai	Diethylcarbamazine citrate (DEC) (100 mg)	2014 until elimination of LF, but current MOU runs till 2021	Up to 2.2 billion tablets committed for the first 7-year period for use in the preventive chemotherapy of lymphatic filariasis Extended in 2017 until elimination is achieved Amendment 1 signed for donation till end of 2021	8.2 billion
	GlaxoSmithKline (GSK)	Albendazole 400-mg tablets	Since 1997 until elimination of LF	Up to 600 million tablets annually for use in the preventive chemotherapy of lymphatic filariasis	
	Sanofi	DEC		Short-term donation to fill a gap for 2 years	120 million
Schistosomiasis	Merck KGaA	Praziquantel	Since 2007 for an unlimited period	Up to 200 million tablets annually for the treatment of schistosomiasis in school-age children (notably in Africa) Since 2017, donation scaled up to 250 million tablets annually for the treatment of schistosomiasis	1.3 billion

(continued)

Table 1 (continued)

NTDs	Manufacturer	Product name	MOU period	Donation commitments/internal discussion	Total number of drugs donated from 2011 to 2020 (source: WHO dashboard)
Soil-transmitted helminthiasis (STH)	GlaxoSmithKline	Albendazole	2012–2025	Donation expanded by 400 million tablets annually for use in the treatment of soil-transmitted helminthiases	3.24 billion
	Johnson & Johnson	Mebendazole	Until 2025	Up to 200 million tablets annually for the treatment of soil-transmitted helminthiases in school-age children	
Visceral leishmaniasis (VL)	Gilead Sciences, Inc.	Liposomal amphotericin B (lyophilised 50-mg formulation) AmBisome®	2012–2016	Up to 445,000 vials for the treatment of visceral leishmaniasis in South-East Asia and East Africa	515 ,000
			2016–2021	Up to 380 400 vials for the treatment of visceral leishmaniasis in South-East Asia and East Africa	
Yaws	EMS SA Pharma (Brazil)	Azithromycin	2019–2023	153 million (500 mg) tablets to support the global eradication of yaws	1700,00
Taeniasis and cysticercosis	Bayer	Niclosamide	2020–2024	Up to a total of 2,800,000 tablets for the treatment of taeniasis/cysticercosis	N/A
		Praziquantel		Up to a total of 1,339,000 tablets for the treatment of taeniasis/cysticercosis	
Donations managed outside the WHO					
Onchocerciasis	Merck Sharp & Dohme (MSD)	Ivermectin	Since 1987 until the elimination of onchocerciasis	Unlimited supply for the treatment of onchocerciasis	10.2 billion
Lymphatic filariasis (LF)			Since 1997 until elimination of LF	Unlimited supply for the treatment of lymphatic filariasis in countries co-endemic for onchocerciasis in Africa and Yemen	
			2017–2025	Up to 280 million tablets for use in WHO-recommended triple-therapy (IDA) MDA regimen to eliminate LF in countries not co-endemic for onchocerciasis	218 million
Trachoma	Pfizer Inc	Azithromycin	1998–2025	Unlimited quantity for the elimination of trachoma as a public health problem	1.8 Billion

Source: WHO
[a] MDT, multi-drug therapy: rifampicin, clofazimine, dapsone in blister packs
[b] Multibacillary
[c]Paucibacillary

at the scale required (Bradley et al. 2021). Even if the drugs themselves are inexpensive, there were, and still are, many other challenges ahead and issues to consider.

In this chapter, we review the history and evolution of the donation programmes and how the 2012 London Declaration and partner commitments boosted the progress and scale-up of both the preventive chemotherapy diseases (PCDs) programmes and the innovative and intensified disease management (IDM) (WHO). We will also review the challenges, lessons learned, the impact of the donations of medicines for NTD control programmes and what the next 10 years may hold.

History of the NTD Drug Donation Programmes

The ground-breaking Merck and Co. Inc. donation of Mectizan® (ivermectin, MSD) has been the subject of many case studies and has been cited as a model for successful public-private partnerships in addressing specific global health problems (Colatrella 2003, 2008; Peters and Phillips 2004; Hotez et al. 2007; Hopkins 2005; Thylefors et al. 2008). However, it is important to give a brief history of how this started, how it changed the prospect and strategy for onchocerciasis control and elimination of transmission and how it stimulated other pharmaceutical companies to make similar donations.

Until the availability of Mectizan®, through the Merck & Co., Inc. donation for the treatment and control of onchocerciasis (also known as river blindness), there was no safe, effective and affordable drug for case management and especially not on a large scale. The only strategy available to the Onchocerciasis Control Programme (OCP) that was established in 1974 was vector control using larvicides by aerial spraying. This method was used in 11 West African countries from 1974 to 2001 (Boatin 2008). Although good progress was made within the OCP area in reducing transmission of the disease, larviciding was an expensive strategy that was also limited to the more open savannah regions and not suitable for heavily forested areas in both African and Latin American onchocerciasis foci.

Twelve years after the discovery and development of ivermectin by Merck scientists, and nearly 7 years after human clinical trials in Dakar, Senegal, in 1987, Merck & Co., Inc. announced plans to donate Mectizan® for the treatment of onchocerciasis for as long as needed (Sturchio and Colatrella 2002). This decision was taken in consultation with the WHO and other stakeholders on the premise that the population who needed the medicine the most could not afford to pay for it. This decision thereby pioneered the donation of medicines for control of neglected diseases.

Mectizan® was incorporated into the OCP under strict medical conditions at first, but as large-scale community trials showed that the drug was safe, a different strategy was needed to reach the millions of people affected. At local level, a few national programmes, supported by their NGDO partners, had already begun to experiment with a community-based approach. These efforts were built on and developed by the WHO/TDR in 1994 into the innovative strategy of community distribution and mass drug administration (MDA). The stage was set for the new African Programme on Onchocerciasis Control (APOC), based on the distribution of Mectizan® alone. This resulted in the collaboration of an international, multi-sectoral coalition including the WHO, the World Bank, UNICEF, the Mectizan Expert Committee, dozens of Ministries of Health, the international donor community, more than 30 NGDOs and local community health workers (Sturchio and Colatrella 2002) and remains the largest ongoing drug donation programme.

Other Companies Followed Suit

The current list of company commitments and the NTDs supported through drug donation programmes is given in Table 1 (source: WHO) and has also been described by Bradley et al. (2021) and Engels and Zhou (2020). The success of the Mectizan® donation—both from the public health perspective and the positive corporate social responsibility impact—soon stimulated other pharmaceutical companies to look into their drugs "arsenal" for what they may have for other NTDs.

GlaxoSmithKline (GSK) committed a large-scale donation of albendazole for the preventative treatment of lymphatic filariasis in 1997 and extended this to include albendazole for treatment of soil transmitted helminths (STH) in 2012. Shortly after in 1998, *Pfizer* established the International Trachoma Initiative with the donation of azithromycin through the WHO for the global elimination of trachoma as a public health problem.

In 2000, *Novartis* made a pledge for an unlimited donation of rifampicin, clofazimine and dapsone as the multi-drug therapy for elimination of the transmission of leprosy and prevention of disabilities, followed in 2005 by the donation of triclabendazole for the treatment of fascioliasis.

Sanofi started collaboration with the WHO in 2001 to support the screening, control and treatment of sleeping sickness and in 2005 followed with an initial 5-year donation of eflornithine, melarsoprol and pentamidine through the WHO for treatment of HAT. This collaboration included support for disease management, screening of populations, disease awareness campaign, capacity building as well as the drug donation and has now been extended to donation of the new oral treatment fexinidazole, which was developed in partnership with the Drugs for Neglected Diseases Initiative (DNDi).

Bayer provided suramin for HAT and partnered with *Sanofi* in 2004 on the donation of eflornithine and donated nifurtimox as required through the WHO for Chagas disease in the Americas and in 2020 added the donation of niclosamide and praziquantel for treatment of taeniasis and cysticercosis.

Johnson & Johnson is donating mebendazole for treatment of soil-transmitted helminths (STH) since 2006, and in 2007 *Merck KGA* committed to the donation of praziquantel for schistosomiasis (SCH) treatment, extending this for the elimination of SCH as a public health problem.

The Japanese company *Eisai* added the donation of diethylcarbamazine (DEC) for lymphatic filariasis elimination in 2010. *Gilead* had partnered with the WHO since 1992 in making AmBisome available at discounted prices to "low- and middle-income countries" (LMICs), and in 2016 they announced an expansion to the drug donation and funding for access to diagnostic services and treatment for visceral leishmaniasis.

EMS Sa Pharma is now donating azithromycin to support the global eradication of yaws, and the most recent company to join in 2020 is the *Chemo Group (Mundo Sano)* donating benznidazole for Chagas disease.

The 2021 Access to Medicines (ATM) Index identified 14 donation programmes for NTDs carried out by 11 companies and operating from 1 to 74 countries. These 14 structured programmes target 11 out of 20 NTDs identified by the WHO. *Eisai, GSK, Merck KGA, MSD and Sanofi* have committed to donating their medicines (as from 2020) until elimination, eradication or control goals are achieved for the targeted disease. In 6 (out of 14) programmes, the company in question has expanded its initiatives beyond the 2020 road map goals. For example, *Merck KGA* has increased the target population supported to include all age groups needed for the treatment of schistosomiasis (ATM Index 2021).

Figures 1 and 2 show the annual total donations from 2012 to 2020 for the IDM disease and the PCT diseases. In Fig. 1, the decrease in 2018 was due to reduction in leprosy (xRef chapter on Leprosy) medicine to balance oversupply the previous year and reduction in human African trypanosomiasis (HAT) cases in the DRC (xRef chapter on HAT). The decreases in 2020 in both Figs. 1 and 2 are mainly the impact and disruption of the global pandemic (Source: WHO (2021)).

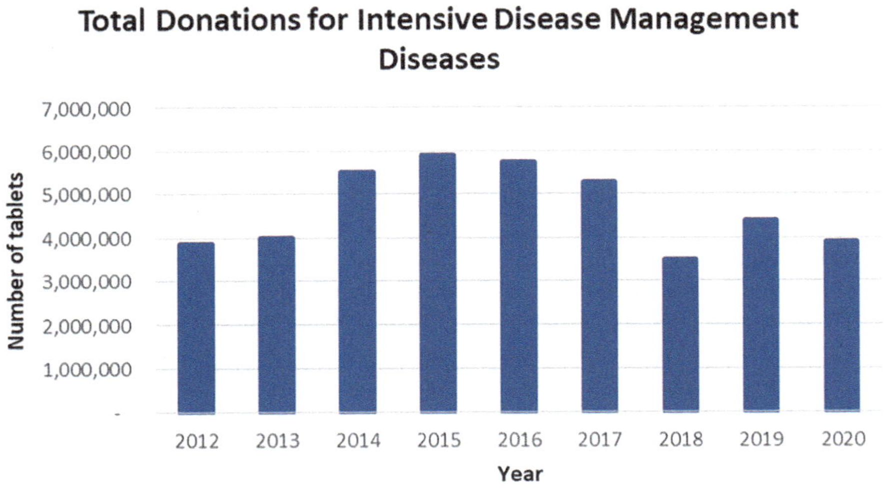

Fig. 1 Annual total donations from 2012 to 2020 for the IDM disease

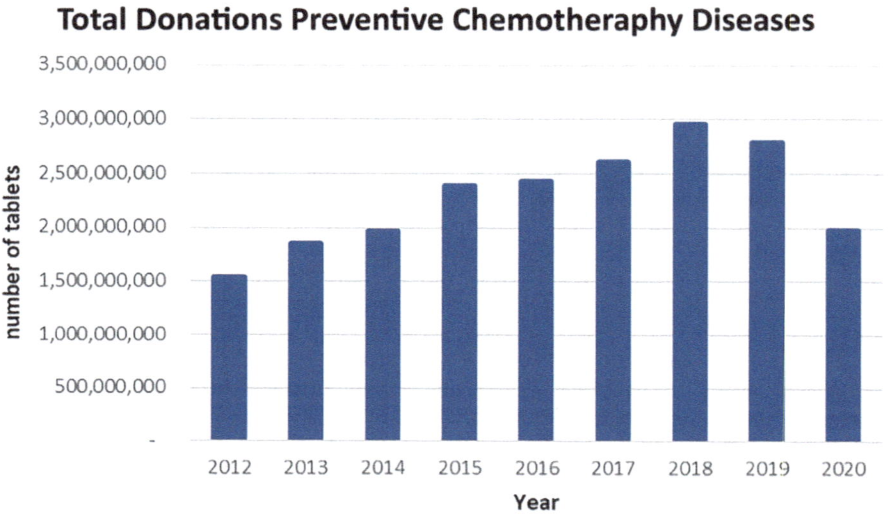

Fig. 2 Annual total donations from 2012 to 2020 for the PCT diseases

The Unprecedented Collaboration with Other Pharmaceutical Companies

One of the main reasons for the donation of albendazole for lymphatic filariasis (LF) by GSK was because research had shown the efficacy of the drugs albendazole, ivermectin and DEC in preventing the transmission of LF. Moreover, it was evident that two drug combinations, either albendazole-ivermectin or albendazole-DEC, were more effective in clearing microfilariae from the blood of infected individuals. Clinical trials confirmed the safety of co-administered albendazole-ivermectin and albendazole-DEC at the community level (Ismail et al. 1998).

In 1997, WHO, Ministries of Health, NGDOs and development partners including GSK and Merck established the Global Alliance to Eliminate Lymphatic Filariasis (GAELF) in response to a specific resolution (WHA.50.29) by the World Health Assembly (Molyneux and Zagaria 2002). Merck then extended the donation of Mectizan® to the elimination of LF in Yemen and Africa where LF and onchocerciasis are co-endemic.

The need for the co-administration and coordination of the donations of albendazole and ivermectin stimulated the "framework for co-operation in matters related to the programme to eliminate lymphatic filariasis as a public health problem" between the two companies in April 2000 (Gustavsen et al. 2009). This was an unprecedented intra-industry collaboration between two companies who are in competition commercially (Bradley et al. 2021) and, even at the time, were involved in legal suits. It is to the credit of the two companies that they were able to work through their differences and cooperate for the benefit of the endemic communities. Annual mass drug administration (MDA) with ivermectin (for onchocerciasis), albendazole (for STH and with ivermectin for LF) and praziquantel (for schistosomiasis) is the WHO-recommended treatment strategy for preventive chemotherapy. However, when community trials showed the safety of co-administering the three recommended anthelminthic drugs, this triple drug administration (TDA) (Mohammed et al. 2008; Weil et al. 2019) has also been shown to be cost-effective, as one large-scale intervention can be carried out rather than three separate MDAs. Thus, companies are now even working together on TDA therapy.

Other pharmaceutical companies involved in drug donations through the WHO for the so-called innovative and intensified disease management (IDM) are also working in partnership, e.g. Sanofi and Bayer for the treatment of human African trypanosomiasis.

Over the years since the drug donation programme first started, there have been questions posed by civil society groups and other stakeholders regarding the motivation and interest in the pharmaceutical companies' involvement in these public health initiatives. Many critics pointed to self-interest due to tax incentives for US-based companies, but although in the long term, these tax incentives may help offset some of the costs, this is not a reason for such programmes. The donation programmes can be seen as addressing the corporate social responsibility model and the presumptive obligation that these companies, as social actors, have towards all parties involved in achieving their economic activity.

However, as Gustavsen et al. (2009) point out, these programmes that go beyond the for-profit model of the companies and are making a real difference to people's health and lives are motivational for the employees. They also serve as validation for companies' research in these areas.

So although these are only part of the solution, donations of medicines by pharmaceutical companies can be a route of access to medicine for the poorest populations (ATM Index 2021).

Working in Partnership with Non-governmental Development Organisations (NGDOs)

At the beginning of the drug donation programmes, in the late 1980s and early 1990s, NGDOs had no experience of collaborating with major pharmaceutical companies, although they were already funding and supporting the affected countries on other programmes, particularly in prevention of blindness. It quickly became apparent that processes had to be developed to facilitate large-scale delivery of the drugs to millions of people annually. The NGDO Coordination Group for Onchocerciasis Control formed a coalition both at the international and national levels and, under WHO and World Bank auspices, developed a model which was subsequently adopted for other NTDs (Elhassan et al. 2018).

The benefits of this partnership were twofold at the beginning. The tie-in with NGDOs through this partnership enabled the companies to count the donation as part of their corporate social responsibility programmes, while for the NGDOs, the relationship with the drug companies enabled them to scale up their work supporting national programmes. The NGDOs would never have been able to afford to buy the drugs on the scale required. The donation programmes evolved from the NGDOs originally applying for the drugs, to a country coalition responsible for drug applications and now to the current situation where the drug applications go through the governments to the WHO and then to the companies concerned.

The role of the NGDOs has also evolved over the course of the last 20 plus years, but they have always played a key role at national, regional and international levels. At the country level, they support national programmes through providing and funding technical assistance and support for capacity building, operational research for improving programme delivery and ensuring national ownership. At the regional level, for instance, in West Africa, they collaborate with other NGDO-supported programmes particularly to address cross-border issues. At the international level, they provide support to facilitate the supply chain and are able to discuss programmatic questions with all key stakeholders and add value to the debate based on their experience on the ground.

The NGDO Coordination Group for Onchocerciasis Control has now evolved into the NTD NGO Network (NNN). It represents nearly 80 organisations and is a partner of national programmes in 100 NTD-endemic countries (Uniting to Combat NTDs 2021).

The Evolving Partnership Between the WHO, Pharmaceutical Companies and Stakeholders

The WHO has played an important role in the drug donation programmes from the early days of the Mectizan® Donation Programme, and its role has evolved with those of the programmes. From its normative role and establishing the evidence-based treatment protocols and policy guidance to both the countries and the donation programme governance committees, it now is coordinating the multiple actors towards the common goals outlined in the NTD 2021–2030 road map.

Until the mid-2000s, the drug donation programmes had mainly bilateral relationships with the WHO and the specific disease units. The WHO 2005 strategy to combat NTDs, which is based on the combination of public health measures shown in Box 1, resulted in the scaling up of medical interventions, especially the large-scale preventative treatment programmes. These necessitated closer working relationship between the pharmaceutical companies and NGDOs supporting work on the ground (Engels and Zhou 2020).

1. innovative intensified disease management
2. large-scale preventive treatment
3. integrated vector management
4. veterinary public health
5. access to water, sanitation, and hygiene

Box 1 WHO 2005 integrated strategy to combat NTDs

The WHO 2012–2020 NTD road map with its ambitious targets for prevention, control, elimination and eradication of NTDs, and the 2012 London Declaration, strengthened the relationship between the WHO, the pharmaceutical industry, NGOs and donors. The WHO has facilitating partnerships and agreements with industry and is coordinating the supply chain. As shown in Table 1, 15 of the 17 different medicines are donated through the WHO (Engels and Zhou 2020). The new WHO NTD 2021–2030 road map launched in 2021 is focussed on strengthening the response to NTDs by focusing on cross-cutting issues and disease-specific targets.

The consolidation of several of the drug donation and NTD support programmes under one roof—that of the Task Force for Global Health in Atlanta—has also led to coordination and support to the programmes as well as improved working relations with the WHO. Currently, the Task Force for Global Health houses the Mectizan Donation Programme, Children Without Worms, the International Trachoma Initiative, Zero Leprosy and the NTD Support Centre.

Lessons Learned

Despite the differences in the disease interventions and the different pharmaceutical company approaches and challenges, there are key learning points common to all the drug donation programmes that have been observed over the course of the last 25 plus years. These lessons have been well documented in publications as well as through programme evaluations and meetings (Williams et al. 2021; Bradley et al. 2021; Sturchio and Colatrella 2002; Global Health Progress 2022).

One of the key learnings that was established early on by the Mectizan Donation Programme, and validated by the other drug donation programmes since, is the role and importance of partnerships. As observed by Peters and Phillips (2004), a donation programme can only be successful through a broad coalition of organisations each with their own mandate. This needs to be underpinned by a conceptual framework on governance and management in order to work effectively.

The key components of this framework are shared vision, resources and agreements on benchmarks for performance with clear roles and responsibilities. Other essential components include accountability to stakeholders, coordination mechanisms, recognising the strengths and contributions of diverse partners, engagement and dialogue among partners, tolerance and flexibility, effective and timely reporting, monitoring and evaluation.

In addition, developing trust between different partners is critical, as is the ability to share credit.

The preventive chemotherapy drug donation programmes have learned the importance of developing non-invasive and user-friendly techniques for the definition of endemic areas, the potentials of working through and with the community and streamlining appropriate drug-management and operational research to address emerging challenges (Thylefors et al. 2008).

Table 2 Lessons learnt

Learning points	Comments
Independence from demand and supply decisions	It is important to set up independent committees and collaborate with the WHO to provide an objective operational space and to aid discussions on supply and demand
Coordination of supply chain and resources	This promotes manufacturing and supply efficiency and in-country programme functioning. It also facilitates coordination and provides a forum for the engagement of supply chain stakeholders
Operational research	Addresses questions, challenges and difficulties in the implementation of programmes and improves efficiency and effectiveness
Robust and dependable programme evaluation processes	Allow for sustainable programmes, effective post-validation surveillance systems and inform manufacturing adjustments of medicines
Programme adaptation to utilise new tools	Enables programmes to adopt new processes as evidence emerges from operational research and to respond to new strategies and tools
Monitoring and evaluation and visibility of achievements	Provides evidence for raising awareness both within the industry and with external stakeholders on the health and economic impacts of donated medicines and the importance of supporting long-term programme sustainability
Reliable financial resourcing	Donation programmes require timely release of funds and continuous investment to achieve control and elimination goals Gustavsen et al. (2009) and Williams et al. (2021)

Williams et al. (2021) describe the additional learning points concerning private-sector partnerships which resulted from the evaluation of the GSK and Merck collaboration on lymphatic filariasis. These include seven factors that benefitted the GAELF and which are summarised in the Table 2.

The IDM programmes are more complex due to the often complicated diagnostic pathways and case management. For most IDM diseases, these are still very much vertical programmes which will need to be integrated into the primary health systems of the countries for sustainability and progressing towards universal healthcare (UHC). However, the first lesson learnt by these programmes is also the importance of *partnerships* and particularly with the WHO at global, regional and national levels, recognising the facilitating/coordinating role of the WHO (but that it is not an implementing partner). A second lesson, also relevant to all donation programmes, is how building widespread involvement of different stakeholders can help create an enabling "ecosystem" at the country level.

Challenges

Early Challenges Affecting Drug Donations and Distribution

In the early days of the Merck Mectizan Donation Programme, many challenges were encountered that had to be addressed, which would subsequently inform and facilitate the other donation programmes, particularly the PCT programmes. Lawrence et al. (2015) have documented these, how they were addressed and the overall impact on onchocerciasis control. We give a brief summary of the main challenges and the ways these were managed for the donation programmes treating the PCT diseases.

- *Lack of a distribution and oversight mechanism* – initially the strategies were developed "on the hoof", but later more strategic approaches were taken. In 2005 the WHO developed the strategy to combat NTDs as a group of diseases based on five public health interventions. The WHO NTD road map 2020 focused on wide-scale implementation of these, and now the WHO road map 2030 is pivoted on impact and sustainability.
- *Insufficient implementing partners* (the NGDOs) to support scale-up and distribution across Africa so coalitions were formed at country and global levels.
- *The need for mapping endemic areas to define those which need to be treated.* Innovative mapping techniques have been developed.
- The ivermectin *dosage is based on weight/surface area.* A major challenge in the early days was how to both supply and keep the scales in any form of accuracy given that they had to be moved from place to place (operational issues). This challenge forced the need for a simplified regimen for the health workers in the field. A modification from weight-based dosing to height was established, and the now familiar measuring stick for the drug distributors was developed that is used by all the MDA programmes, as shown in Fig. 3.
- The *drug doses and packaging were inappropriate* for the context and led to operational difficulties in the field (e.g. 6-mg tablets that had to be broken in half and foil blister packs that had to be opened before the distribution). Feedback

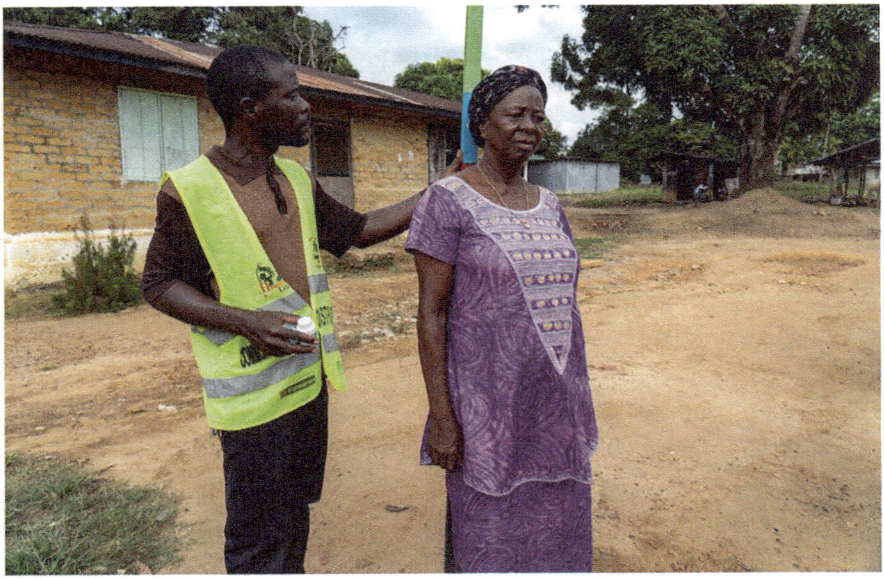

Fig. 3 Dose distribution pole used to determine dosage during MDA (credit Sightsavers/Carielle Doe)

from the field to Merck brought about the change in dosage to 3-mg tablets and repackaging in bottles. This reduced wastage losses of broken tablets and opened blister packs littering the countryside. The later drug donation programmes incorporated this learning and developed more appropriate child- and patient-friendly dosing and packaging (Figs. 4 and 5).

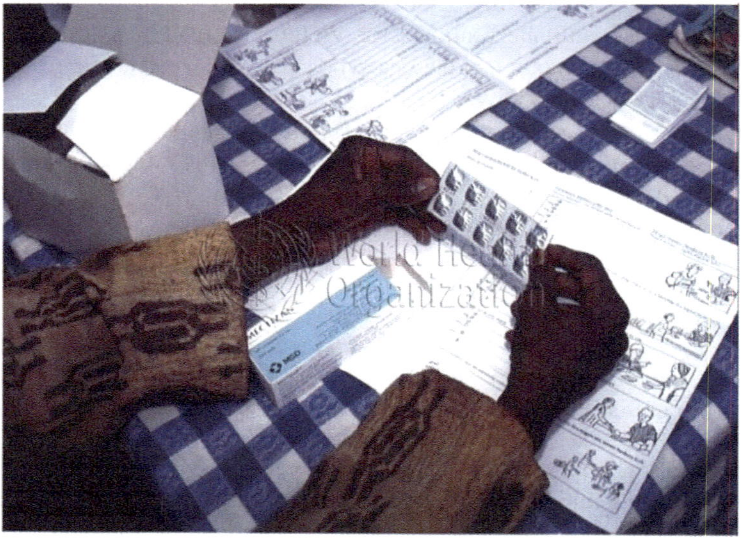

Fig. 4 6-mg Mectizan tablets in blister packs (Credit TDR/WHO)

Fig. 5 3-mg Mectizan tablets in bottles of 500 introduced in 1996 (Credit TDR/WHO)

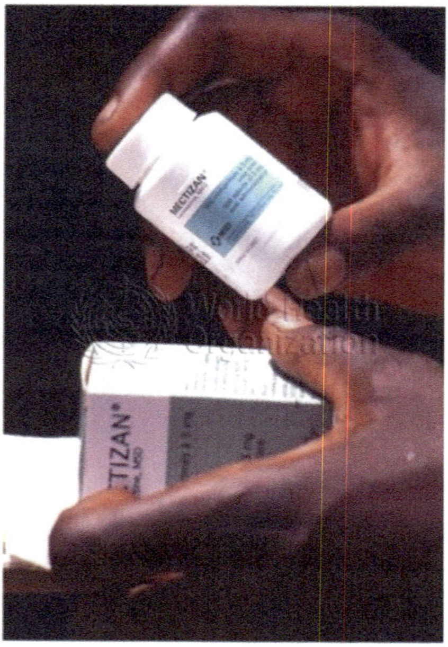

- The *challenges of introducing the new tablets to the countries and health workers* highlighted the importance of appropriate, annual training of health workers.

Continuing Challenges for Partners If the Donated Drugs Are to Reach All the People Who Need Them in Order to Achieve Elimination

- *Pharmacovigilance and serious adverse events* reporting needed strengthening and simplification for unskilled community distributors.
- *Supply chain logistics* Overall, all programmes have recognised the need for strengthening of national supply systems for in-country distribution in order to prevent supply chain ruptures and to build M&E capacity to work in normative programmes.
- After several years of essentially vertical programmes, the focus is now on *integration of all these disease control programmes into the national health systems.* This not only poses problems for the national programmes but also presents some major challenges for the drug companies in the coordination and synchronisation of delivery of different drugs so they all arrive about the same period to facilitate integrated MDA.
- Implementation resources
- Drug distributors incentives

All this impacts on health staff and community workers with limited education and many other pulls on their time—manning clinics, maternity, epidemics, shortage of transport and so forth. These factors ultimately affect the companies donating up the value chain.

- A more recent challenge, which is the result of the success of the programmes, is how to adjust to the changing disease targets and treatment with the reduction in the prevalence and incidence of the diseases.

An Example of Supply Chain Challenges for the Donation of Mectizan®—Which Were True 25 Years Ago and Are Still the Same Today

Many of the challenges in getting the drugs to the field encountered today are not so different than those faced by the Mectizan Donation Programme when it started more than 25 years ago. Although the WHO has streamlined the process of applications for drugs through the Joint Application Form (JAP) that is completed by the Ministries of Health in each country (Engels and Zhou 2020), the actual shipping process of the drugs is different for each country each time.

Application process
Ministries of Health complete the JAP and send to WHO.
Applications are sent to MDP for review by the MEC
Approved applications sent to Merck & Co. Inc.
Certificate of donation issued.
Drugs packed ready for shipment.

Supply process and challenges at country level
Country asked to assure entry of drugs ex- duty, taxes etc.
Issuing Duty Exemption certificate
Clearance in country
Supply chain management and distribution

Box 2 Example of the complexities of shipment of Mectizan®

Delays in the process illustrated in Box 2 often occur due to frequent staff turnover at the Ministries and Customs. Lack of understanding of the procedures and policy changes are not always communicated to the programmes or the companies.

Strengthening Supply Chain Management

The NTD Supply Chain Forum established in 2012 to overcome some of these critical challenges for the PCT programmes has allowed significant improvement in the importation process with a single coordinating body handling the logistics for the drugs donated through the WHO (GSK, Merck KgaA, J&J, Souza et al. 2020). This initiative is also helping harmonise data collection and analysis which is facilitating both the identification of issues and ways to address them.

The WHO monitors the data on case numbers and manages the supply chain of the medicines and the stock balances in a country which is particularly important for the drug donations for the IDM diseases. However, as described by Bradley et al. (2021), challenges still exist, and the disruption to the programmes due to the COVID-19 global pandemic, insecurity and natural disasters in certain countries has highlighted the ongoing difficulties in demand forecasting and supply chain management.

Socio-Economic Aspects of NTDs and the Impact of the Disease Control and Drug Donation Programmes

Disability-adjusted life years (DALYs) are used to give an assessment of overall disease burden and are calculated as the sum of the years of life lost due to premature mortality and the years of healthy life lost due to disability. The estimations of the DALYs for the NTDs benefiting from drug donation programmes from the 2013 Global Burden of Disease Study range from 4.24 for visceral leishmaniasis to 0.04 for leprosy as shown in Table 3.

Table 3 Disability-adjusted life years (DALYs) resulting from NTDs: from the Global Burden of Disease Study 2013 with attributing years lived with disability (YLD) and years of life lost (YLL)

Disease	DALYs (millions) 2013	YLDs (millions) 2013	YLLs (millions) 2013
Visceral leishmaniasis	4.24	0.008	4.23
Food-borne trematodiasis	3.63	3.63	0
Schistosomiasis	3.63	2.86	0.2
Hookworm	2.18	2.18	0
Lymphatic filariasis	2.02	2.02	0
Ascariasis	1.27	0.93	0.34
Rabies	1.24	0.0001	1.24
Onchocerciasis	1.18	1.18	0
Dengue	1.14	0.56	0.58
Trichuriasis	0.58	0.58	0
African trypanosomiasis	0.39	0.005	0.38
Chagas disease	0.34	0.1	0.24
Cysticercosis	0.34	0.31	0.03
Echinococcosis	0.18	0.08	0.1
Trachoma	0.17	0.17	0
Cutaneous and mucocutaneous leishmaniasis	0.04	0.04	0
Leprosy	0.04	0.04	0
Other NTDs	3.13	2.26	0.87
Total NTDS	25.17	16.95	8.21

Modified from Herricks et al. (2017)

As pointed out by Conteh and Engels (2010), DALY calculations are mainly driven by mortality rates, and many NTDs do not have high mortality rates but result in chronic suffering and long-term disabilities, trapping individuals and families in a cycle of poverty. The Global Burden of Disease 2013 estimated that NTDS account for approximately 25 DALYs (Herricks et al. 2017). However, it is generally accepted that DALYs cannot account for the actual burden of the NTDs and their socio-economic impact on the affected communities and countries—one reason being the difficulty in the collection of accurate data for morbidity and mortality for most of the NTDs. In reality, NTDs probably cost developing economies billions of dollars every year in lost revenue by interfering with labour productivity and with the wage-earning potential of individuals who are already poor and surviving on less than the US$2 a day. Moreover, infection with NTDs can limit educational opportunities for school-aged children and child development (Ochola et al. 2021).

Impact of the Donation Programmes

There has been tremendous progress in reducing the prevalence and incidence of many of the NTDs, and the overall changes since 1990 are illustrated in Fig. 6 from Qian and Zhou (2016).

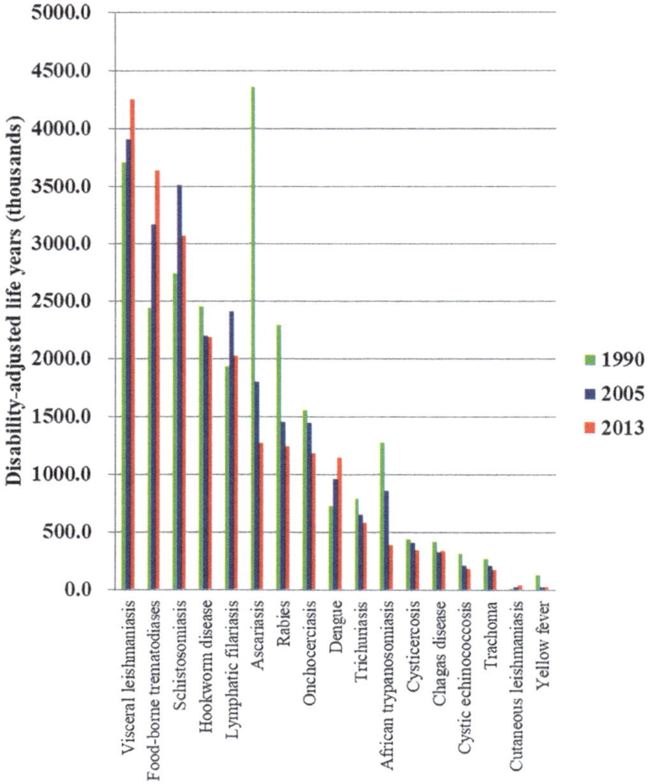

Fig. 6 Progress in reducing prevalence and incidence of NTDs. Reproduced from Qian and Zhou (2016)

The progress to date is a result of the strategic interventions outlined in the NTD road map 2012–2020, the continued commitment of the pharmaceutical industry through drug donations (and financial support) and the commitment of the various stakeholders involved in the disease control programmes.

However, the real impact of the donation of the medicines for the NTDs may never be fully quantified or determined, but as stated by Thoko Elphick-Pooley of Uniting to Combat NTDs (UTC), "Without the drug donation programmes, the progress we are seeing on NTDs today would simply not exist".

Sub-Saharan African countries have a multitude of health priorities for very limited national health budgets, and it is unlikely that the ministries of health and the endemic communities would have been able to pay for the cost of NTD treatments on the scale that has been available through the donation programmes. The 2012 London Declaration on NTDs and the WHO road map for implementation were instrumental in increasing the profile and reach of NTDs globally, regionally and nationally.

From 2015 to 2019 over one billion people have received treatment for an NTD to date. Fifteen countries in the African region have successfully eliminated at least

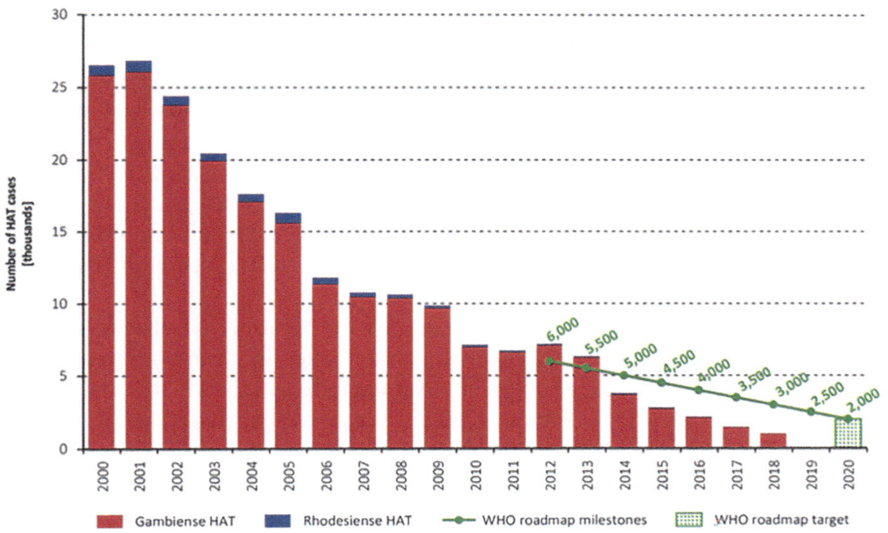

Fig. 7 Total number of reported cases of HAT (*Trypanosoma gambiense* and *Trypanosoma rhodesiense*) per year (2000–2018). Source: Monitoring the elimination of human African trypanosomiasis at continental and country level: Update to 2018. Franco et al. (2020)

one NTD since 2010, and globally, nine countries have eliminated trachoma, and 19 are free from lymphatic filariasis. With only 24 cases reported in 2020 (compared to around 10,000 in 2009), Guinea worm disease elimination progress indicates that this may be the next NTD to achieve eradication. The reported cases of HAT have reduced since 2000 with a dramatic drop since 2012 and the increased pharma commitments as shown in Fig. 7 from Franco et al. (2020) (**xRef chapter on HAT**).

The WHO NTD dashboard provides current data and overviews of the progress of the NTD programmes and donations (World Health Organization 2021a, b, c).

In summary, these NTD programmes have averted morbidity, disability and death. The socio-economic effects of the NTDs averted by the donation programmes include blindness, incessant itching, severe dermatitis, discrimination/exclusion, stigmatisation, loss of livelihood, cognitive impairment, low education and death.

Cost-Benefit and Cost-Effectiveness of the NTD Programmes

To date, much of the work on both the cost-effectiveness and the cost-benefit of the NTD programmes has been done on the PCT diseases that involve MDA. According to the WHO, each US$1 invested in MDA programs is expected to yield an estimated US$25 net benefit to the individual, averting out-of-pocket spending on health and costs in lost productivity (WHO, 2020). Other reports confirm that these MDA programmes are good value for money and cost-effective (Turner et al. 2021), although there is still limited evidence on the actual cost of these programmes (Mow et al. 2022).

As Fitzpatrick et al. (2016) and Mow et al. (2022) reported, the costing studies and unit costs are sensitive to economies of scale, whether local volunteers or health workers are involved, and if a single dose regimen such as for lymphatic filariasis and onchocerciasis is used. Where two dose regimens or small and last mile programmes are involved, the economic and financial unit costs are going to be higher.

The advocacy for the NTD programmes involving mass treatment for the elimination of the PCT NTDs typically uses the cost per person treated at less than US$0.50. Although useful for advocacy and fundraising, this is a simplification of what is actually a much more complex issue, given the wide range of different situations across the endemic regions.

For the IDM disease, the major cost drivers for control and elimination programmes are the screening and diagnostic costs that come from both passive surveillance and active screening campaigns (Sutherland and Tediosi 2019). Lenk et al. (2018) looked at the socio-economic benefits of achieving the 2020 targets for control and elimination of leprosy, HAT, visceral leishmaniasis and Chagas disease (from the 2012–2020 WHO NTD road map). Despite the limitations in data, and having to combine the limited data from various sources, the heterogeneous background and the variable quality, they conclude that "the economic benefits far exceed the necessary investments by endemic country governments and their development partners. Given the higher frequency of NTDs among the poorest households, these investments represent good value for money in the effort to improve well-being, distribute the world's prosperity more equitably and reduce inequity".

Changing Disease Epidemiology and Impact on Drug Donations

The remarkable progress in reduction of many of the NTDs since 2012—primarily through the preventive chemotherapy or MDA—has led to reviews of the epidemiology of the diseases and reassessing the objectives. For example, the goal for onchocerciasis is now elimination of transmission which will have implications in terms of duration, extent and frequency of treatment and thus also the ivermectin requirements.

For LF the current strategy is the elimination of the disease as a public health problem. Should this change to elimination of transmission, the duration of treatment and drugs required will also change. The new target date for global elimination of trachoma as a public health problem is 2030, and treatment strategies have been revised to achieve this goal.

The water, sanitation and hygiene (WASH) situation in endemic countries has implications for the many of the disease control/elimination programmes and is particularly important for schistosomiasis (SCH) and soil-transmitted helminths. Urbanisation and deteriorating WASH could result in increased transmission of soil-transmitted helminths (STHs) requiring different strategies—as evidenced by the recent strategy change for SCH programmes from treating only children to including the adult population in communities where endemicity is high. Moreover, where humans and animals share the same water point, zoonotic infections could occur, leading to infections in the human population. This is also currently evidenced by the challenge to eradication of Guinea worm by the detection of the infection in dogs in Chad in 2012. Previously it was believed that there was no animal reservoir of Guinea worm, but since then, canine infections have also been reported from Ethiopia, Mali and Cameroon (Hopkins et al. 2021).

Many of the NTDs are transmitted by vectors and intermediate hosts that may be affected by climate change. Although Tidman et al. (2021) report that at present the evidence and predictions of the impact of climate change on specific NTDs are conflicting and dependent on geography, disease, vectors and hosts, it is quite likely that there will be changes in disease transmission periods and geographic range and may well encroach on regions and populations that have been previously unaffected.

As is already the case with population growth, encroachment into the countryside can result in exposure to zoonotic or previously unknown diseases (such as the current situation with COVID-19), so there will be a need to maintain surveillance of NTDS to identify changes in epidemiology.

As shown in Table 1, the pharma partners have committed to continue their donations, either until the disease is eliminated or at least to support the goals of the 2021–2030 WHO NTD road map. However, should there be a re-emergence or emergence of NTDs, the pharmaceutical companies and partners in the drug donation programmes may need to review their current commitment and strategies.

Prospects and Conclusions: The Next 10 Years and Beyond the 2030 NTD Elimination Goal

As the requirement for mass drug administration changes with reduced disease transmission, different strategies for addressing the "last mile" and low compliance with treatment where few cases remain will need to be developed and implemented. Mechanisms for procuring or obtaining drugs for case management of those few cases are needed, as well as robust surveillance and response programming.

In addition, global and national level risk mitigation for disruptions to the NTD programmes is essential. The last few years have underscored how security issues, natural disasters and other disease outbreaks can disrupt health programmes—from drug and diagnostic supplies, service delivery to human resources. The current COVID-19 pandemic has shown us how a major health crisis can derail even the best programmes. As reported by the World Health Organization (2021a, b, c), NTD services were found to have been among the most frequently and severely affected. The COVID-19 pandemic, the resultant lockdowns, travel restrictions, changed priorities for staffing, etc. caused major delays in 2020 treatments. In addition, the global supply chain disruption also caused major problems for the pharmaceutical industry for drug manufacturing (problems with active pharmaceutical ingredients (APIs)) and logistics for shipping.

Some 10% of health workers who supervise and monitor programmes were unavailable due to the virus, and the safety of distributors, community members, health workers and partners became a concern to all. The unfortunate false perception of the use of ivermectin to treat COVID-19 infections highlighted the possibility of diversion of donated medicines and subsequent accountability.

However, there have been some valuable lessons learnt from the pandemic regarding risk mitigation, pivoting NTD programmes to support the COVID-19 needs and the guidelines WHO and partners developed for safe mass distribution of medicines during a pandemic.

So in summary, up till now, the NTD programmes have been dependent on donor financing and a lot of support from external partners. Although the support of the bilateral and philanthropic donors, NGDO stakeholders and the continued commitments of the pharmaceutical companies will be important, the goal of elimination of NTDs by 2030 will only happen when the countries take complete ownership and invest domestic resources in the NTD elimination and control programmes. Coordinated and integrated NTD programmes are, to some extent, already being implemented, although there is still a long way to go before there is sustainable integration of them into to the national health systems. An effective elimination strategy at the national level will require health system strengthening, targeted improvements in case detection, surveillance and referral services, to ensure linkage to care, monitoring and evaluation, local protocols and action plans focused on elimination. These will be necessary to ensure sustainability that enables countries to maintain elimination once targets are achieved (Meskel 2022).

Acknowledgements We are very grateful to Dr. Afework Hailemariam Tekle, World Health Organisation, for providing us with the data in Table 1 and critical reviewing and to Dr. Yao Sodahlon, Director, MDP, for constructive discussions. Thanks also go to Mr. Andy DL Wright and Ms. Brenda Colatrella for manuscript review. Sincere thanks to Mrs. Catherine Cross and Mr. John Myers for helpful comments and editing.

References

Access to Medicines Index (2021). https://accesstomedicinefoundation.org/publications/2021-access-to-medicine-index

Boatin B (2008) The onchocerciasis control programme in West Africa (OCP). Ann Trop Med Parasitol 102(Suppl 1):13–17. https://doi.org/10.1179/136485908X337427. PMID: 18718148

Bradley M, Taylor R, Jacobson J, Guex M, Hopkins A, Jensen J et al (2021) Medicine donation programmes supporting the global drive to end the burden of neglected tropical diseases. Trans R Soc Trop Med Hyg 2021(115):136–144

Colatrella B (2003) The Mectizan donation program – successful collaboration between the public and private sector. Drug Donations 33:27–28

Colatrella B (2008) The Mectizan donation program: 20 years of successful collaboration - a retrospective. Ann Trop Med Parasitol 102(Suppl 1):7–11. https://doi.org/10.1179/136485908X337418. PMID: 18718147

Conteh L, Engels T, Molyneux DH (2010) Socioeconomic aspects of neglected tropical diseases. Lancet 375:239–247

Dull B (1980) Mectizan donation and the Mectizan expert committee. Acta Leiden 59:399–404

Elhassan E, Zhang Y, Bush S, Molyneux D, Kollmann MKH, Sodahlon Y, Richards F (2018) The role of the NGDO coordination Group for the Elimination of onchocerciasis. Int. Health 10(suppl_1):i97–i101. https://doi.org/10.1093/inthealth/ihx050. PMID: 29471339

Engels D, Zhou XN (2020) Neglected tropical diseases: an effective global response to local poverty-related disease priorities. Infect Dis Poverty 9(1):10. https://doi.org/10.1186/s40249-020-0630-9

Fitzpatrick C, Fleming FM, Madin-Warburton M, Schneider T, Meheus F, Asiedu K et al (2016) Benchmarking the cost per person of mass treatment for selected neglected tropical diseases: an approach based on literature review and meta-regression with web-based software application. PLoS Negl Trop Dis 10(12):e0005037. https://doi.org/10.1371/journal.pntd.0005037

Franco JR, Cecchi G, Priotto G, Paone M, Diarra A, Grout L et al (2020) Monitoring the elimination of human African trypanosomiasis at continental and country level: update to 2018. PLoS Negl Trop Dis 14(5):e0008261. https://doi.org/10.1371/journal.pntd.0008261

Global Health Progress (2022). https://globalhealthprogress.org/collaboration/leprosy-drug-donation. Accessed January 2022

Gustavsen KM, Bradley MA, Wright AL (2009) GlaxoSmithKline and Merck: private-sector collaboration for the elimination of lymphatic filariasis. Ann Tropical Med Parasitol 103(sup1):11–15. https://doi.org/10.1179/000349809X12502035776478

Herricks JR, Hotez PJ, Wanga V, Coffeng LE, Haagsma JA, Basa Âñez M-G et al (2017) The global burden of disease study 2013: what does it mean for the NTDs? PLoS Negl Trop Dis 11(8):e0005424. https://doi.org/10.1371/journal.pntd.0005424

Hopkins A (2005) Ivermectin and onchocerciasis: is it all solved? Eye 19:1057–1066. https://doi.org/10.1038/sj.eye.6701962

Hopkins DR, Weiss AJ, Roy SL, Yerian S, Cama VA (2021) Progress toward global eradication of dracunculiasis, January 2020–June 2021. MMWR Morb Mortal Wkly Rep 70:1527–1533. https://doi.org/10.15585/mmwr.mm7044a1

Hotez PJ, Molyneux DH, Fenwick A, Kumaresan J, Sachs SE, Sachs JD, Savioli L (2007) Control of neglected tropical diseases. N Engl J Med 357(10):1018–1027. https://doi.org/10.1056/NEJMra064142

Ismail MM, Jayakody RL, Weil GJ et al (1998) Efficacy of single dose combinations of albendazole, ivermectin and diethylcarbamazine for the treatment of bancroftian filariasis. Trans R Soc Trop Med Hyg 92(1):94–97

Lawrence J, Sodahlon YK, Ogoussan KT, Hopkins AD (2015) Growth, challenges, and solutions over 25 years of Mectizan and the impact on onchocerciasis control. PLoS Negl Trop Dis 9(5):e0003507. https://doi.org/10.1371/journal.pntd.0003507

Lenk EJ, Redekop WK, Luyendijk M, Fitzpatrick C, Niessen L, Stolk WA et al (2018) Socioeconomic benefit to individuals of achieving 2020 targets for four neglected tropical diseases controlled/eliminated by innovative and intensified disease management: human African trypanosomiasis, leprosy, visceral leishmaniasis, chagas disease. PLoS Negl Trop Dis 12(3):e0006250. https://doi.org/10.1371/journal.pntd.0006250

Meskel A (2022) The road to ending visceral leishmaniasis. https://www.devex.com/news/sponsored/opinion-the-road-to-ending-visceral-leishmaniasis-

Mohammed KA, Haji HJ, Gabrielli AF, Mubila L, Biswas G, Chitsulo L, Bradley MH, Engels D, Savioli L, Molyneux DH (2008) Triple co-administration of ivermectin, albendazole and praziquantel in zanzibar: a safety study. PLoS Negl Trop Dis 2(1):e171. https://doi.org/10.1371/journal.pntd.0000171

Molyneux DH, Zagaria N (2002) Lymphatic filariasis elimination: progress in global programme development. Ann Trop Med Parasitol 96(Suppl 2):S15–S40. https://doi.org/10.1179/000349802125002374

Mow M, Thean LJ, Parnaby M, Mani J, Rafai E, Sahukhan A et al (2022) Costs of mass drug administration for scabies in Fiji. PLoS Negl Trop Dis 16(2):e0010147. https://doi.org/10.1371/journal.pntd.0010147

Ochola EA, Karanja DMS, Elliott SJ (2021) The impact of neglected tropical diseases (NTDs) on health and wellbeing in sub-Saharan Africa (SSA): a case study of Kenya. PLoS Negl Trop Dis 15(2):e0009131. https://doi.org/10.1371/journal.pntd.0009131

Peters DH, Phillips T (2004) Mectizan donation program: evaluation of a public–private partnership. Trop Med Int Health 9(4):a4–a15. suppl April

Qian M-B, Zhou X-N (2016) Global burden on neglected tropical diseases. Lancet Infect Dis 16:1113–1114

Souza AA, Holloway C, Williams T (2020) The NTD supply chain forum—strengthening the backbone of NTD programs. PLoS Negl Trop Dis 14(11):e0008818. https://doi.org/10.1371/journal.pntd.0008818

Sturchio JL, Colatrella BD (2002) Successful public-private partnerships in global health: lessons from the Mectizan donation program. In: Granville B (ed) Economics of essential medicines. Royal Institute of International Affairs, London, pp 255–274

Sutherland CS, Tediosi F (2019) Is the elimination of 'sleeping sickness' affordable? Who will pay the price? Assessing the financial burden for the elimination of human African trypanosomiasis Trypanosoma brucei gambiense in sub-Saharan Africa. BMJ Glob Health 4(2):e001173. https://doi.org/10.1136/bmjgh-2018-001173

Thylefors B, Alleman M, Twum-Danso NAY (2008) Operational lessons from 20 years of the Mectizan donation program for the control of onchocerciasis. Trop Med Int Health 13(5):689–696

Tidman R, Abela-Ridder B, de Castañeda RR (2021) The impact of climate change on neglected tropical diseases: a systematic review. Trans R Soc Trop Med Hyg 115(2):147–168. https://doi.org/10.1093/trstmh/traa192

Turner HC, Stolk WA, Solomon AW et al (2021) Are current preventive chemotherapy strategies for controlling and eliminating neglected tropical diseases cost-effective? BMJ Global Health 6:e005456. https://doi.org/10.1136/bmjgh-2021-005456

Uniting to Combat NTDs (2012) The London declaration on neglected tropical diseases. http://unitingtocombatntds.org/London-declaration-neglected-tropical-diseases/accessed 9 Nov 2021

Uniting to Combat NTDs (2021). https://unitingtocombatntds.org/reports/5th-report/ngos. Accessed 18 November 2021

Weil GJ, Bogus J, Christian M, Dubray C, Djuardi Y, Fischer PU, Goss CW, Hardy M, Jambulingam P, King CL, Kuttiat VS, Krishnamoorthy K, Laman M, Lemoine JF, O'Brian KK, Robinson LJ, Samuela J, Schechtman KB, Sircar A, Srividya A, Steer AC, Supali T, Subramanian S, DOLF IDA Safety Study Group (2019) The safety of double- and triple-drug community mass drug administration for lymphatic filariasis: a multicenter, open-label, cluster-randomized study. PLoS Med 16(6):e1002839. https://doi.org/10.1371/journal.pmed.1002839. PMID: 31233507; PMCID: PMC6590784

Williams T, Taylor R, Iwamoto M, Hida T, Gusovsky F (2021) The role of medicine donations in the global programme for the elimination of lymphatic filariasis. International health 13(Suppl 1):S39–S43. https://doi.org/10.1093/inthealth/ihaa077

World Health Organization (3 February 2012) "WHO roadmap inspires unprecedented support to defeat neglected tropical diseases". who.int. World Health Organization, Geneva. Archived from the original on April 7, 2014. Accessed January 2022

World Health Organization (2021a). https://www.who.int/teams/control-of-neglected-tropical-diseases/overview/progress-dashboard-2011-2020. Accessed January 2022

World Health Organization (2021b) Wkly Epidemiol Rec 96:461–476

World Health Organization (2021c). https://www.who.int/teams/control-of-neglected-tropical-diseases. Accessed 9 Nov 2021102631. Accessed February 2022

The Social and Economic Impact of Neglected Tropical Diseases in Sub-Saharan Africa

Margaret Gyapong, Mustapha Immurana, Alfred Manyeh, Christopher Tetteh Odopey, Laura Dean, and Alison Krentel

Contents

Abstract

NTDs are characterised by several factors, the most common of which is poverty. It is estimated that the disease costs developing countries the equivalent of a billion US dollars every year in direct costs, loss of productivity, and reduced socioeconomic and educational attainment. Since the last edition of this book several NTDs have been added to the compendium, the consequences of some NTDs such as female genital schistosomiasis (FGS) have been brought more to the

M. Gyapong (✉) · M. Immurana · A. Manyeh · C. T. Odopey
University of Health and Allied Sciences, Volta Region, Ghana
e-mail: mgyapong@uhas.edu.gh

L. Dean
Liverpool School of Tropical Medicine, Liverpool, UK

A. Krentel
The University of Ottawa, Ottawa, ON, Canada

J. O. Gyapong, B. A. Boatin (eds.), *Neglected Tropical Diseases - Sub-Saharan Africa*, Neglected Tropical Diseases, https://doi.org/10.1007/978-3-031-53901-5_18

fore, countries have eliminated some NTDs, and a new NTD road map has been developed to ensure no one is left behind in the journey towards SDG 2030 targets. Attention however still needs to be given to the socio-economic impact of NTDs and the wide-reaching effects this has on the health and well-being of affected individuals and households. This impact is not uniform as NTDs are linked to poverty and other axes of Inequity and Vulnerability, e.g. gender, disability, and ethnicity may be exacerbated. In this chapter, we address these issues while highlighting the need to focus on illness as well as disease.

Introduction

Angus Deaton's conceptualisation of well-being describes it as all the things that are good for a person that makes for a good life. Well-being includes material well-being such as income and wealth and physical and psychological well-being, represented by health and happiness, education, and the ability to participate in civil society through democracy and the rule of law (Ocholaid et al. 2021).

Neglected tropical diseases (NTDs), known as ancient diseases of poverty and their associated burdens, pose a social and economic burden and threat to the well-being of more than one billion people worldwide, predominantly in tropical and sub-tropical areas among the most vulnerable, marginalised populations (World Health Organization [WHO] 2020). This impact is not uniform as NTDs are linked to poverty and other axes of inequity such as gender, disability, and ethnicity (Gyapong et al. 2016).

The NTD burden is expressed through disability-adjusted life years (DALYs) lost. This refers to the many years of healthy life lost due to premature death or years lived with a disability. Seven of these 20 diseases, namely, lymphatic filariasis (LF), onchocerciasis, trachoma, schistosomiasis, and three soil-transmitted helminth infections (hookworm, ascariasis, and trichuriasis), are the most prevalent and are responsible for the majority of the NTD disease burden. Even though NTDs may not pose an immediate threat to mortality, their associated disability and subsequent social stigma can be extremely burdensome. They disproportionally affect poor populations of the world decreasing quality of life, worker productivity, and agricultural outputs trapping them in a cycle of poverty which is a serious detriment to economic development in many developing countries (Adelman et al. 2012; Hotez et al. 2014).

As nations aspire towards the attainment of the Sustainable Development Goals (SDGs), it is important that countries in the global south adopt well-being measures and, in that attempt, capture the lived experiences and inequities of individuals particularly as these are shaped by NTDs (Ocholaid et al. 2021).

The first WHO road map for the prevention and control of NTDs was published in 2012, and since then, a lot of progress has been made. Currently, 500 million fewer people require interventions against these NTDs than in 2010, and 40 countries, territories and areas have eliminated at least 1 NTD (WHO 2020). This progress again has made lighter the human and economic burden imposed on the world's most disadvantaged communities by NTDs. This demonstrates the effectiveness of aligning the work of WHO member states with that of its different partners which have in the past years achieved two notably important results: (1) NTD interventions are one of the "best buys" in global public health, yielding an estimated net benefit to affected individuals of about US$25 per US$1 invested in preventive chemotherapy, and (2) NTDs are important indicators of disparities in progress towards both universal health coverage and equitable access to high-quality health services, regardless of factors such as gender, age, disability, and location (WHO 2020).

The new road map for 2021–2030 sets global targets and milestones to prevent, control, eliminate, or eradicate 20 diseases and disease groups. It also sets cross-cutting targets aligned with WHO's 13th General Programme of Work and the SDGs, with strategies for achieving the targets during the next decade. It demonstrates critical gaps and the actions required to reach the 2030 targets identified through global consultation. The emergence of coronavirus disease 2019 (COVID-19) poses a threat to the achievement of the new road map for NTDs 2030. It has caused major impacts both in human health and societal activities. In particular, the economic impacts of various containment measures began to ripple across the world, and initial hopes for a swift recovery were soon dampened. As difficult as it is to assess the magnitude of the collateral damage of the COVID-19 pandemic, there is no doubt that it was and will continue to be considerable. Health systems in many low- and middle-income countries (LMICs) were already weak before the COVID-19 pandemic. They will continue to be massively challenged to the point of collapse across the world, affecting public health programmes for both infectious diseases and non-communicable diseases.

In addition, there are shortages of personal protective equipment (e.g. face masks) hitting the LMICs hardest, supply chains are strained or interrupted (e.g. donated deworming drugs might not reach the countries in need in time) due to closure of borders, and campaigns for the control and elimination of NTDs are postponed or discontinued (e.g. closure of schools will prevent deworming campaigns through the education sector) (Molyneux et al. 2020; Abdela et al. 2020). This shows clearly that the COVID-19 pandemic not only has direct but also indirect impacts on NTDs, and these are likely to revert the progress made over the past 10–15 years (Chaumont et al. 2020).

In the wake of all these happenings, the NTD community is still struggling to reach out beyond a relatively small circle of established donors even though the integration of disease control and elimination efforts has become an accepted principle (Ehrenberg et al. 2021). Fortunately for the NTD community, aside from initiatives by the more traditional donors in the NTD field, there are some major emerging economic opportunities which are quickly appearing on the horizon because of the unfolding COVID-19 pandemic. The opportunity available to include the NTD agenda within these major global efforts may not last for long, and if not seized promptly, there is a risk NTDs may be left behind. This would eventually mean billions of people who are at risk of NTDs might not benefit from these immense economic recovery and health systems strengthening efforts (Ehrenberg et al. 2021).

This chapter focuses on the social and economic impact of NTDs. Multiple models of disability exist that inform the ways in which we discuss and acknowledge the impacts of disease and illness, including how we describe and understand the concept including its determinants, consequences, and the response we promote. Within this chapter, we depart from a medical model of disability to take a broader more societal view of disability and see it as a construction of society in which an impairment is located (Mitra 2006; Mitra 2017). Social models of disability focus on the transformation of unjust systems in which the participation of people with disability is centred (Addlakha 2013; Al Ju'beh 2015; Rohwerder 2015). Impairment tends to refer to an individual's condition, whereas disability is specifically connected to the social disadvantages, discrimination, and exclusion that individuals face because of their impairment (Mitra 2017), and in this case, we focus more specifically on impairment from NTDs. We acknowledge that disability as a result of NTDs is neither purely medical nor purely social. It is rather a complex and dynamic interaction between health conditions (as a result of NTDs) and the context in which they are located that shape the interactional processes that can lead to activity limitations and participation restrictions putting the affected at risk of being in a vicious cycle of poverty and disability. Contextual factors consider individual identity-based characteristics such as age, gender, and education as well as structural factors such as physical environment, social attitudes, laws, policies, etc. which may become barriers or enablers to individual functioning. Functioning becomes the direct opposite of disability, and disability becomes an overarching term for impairment, activity limitations, and participation restrictions (WHO 2002).

Disease Versus Illness: Lay Perceptions of NTDs

The distinction between disease and illness has been well described. Whereas disease defines a pathophysiologic process, illness is defined by the complete person: physical, psychological, social, and cultural (Eisenberg 1977; Helman 1981). Illness represents an individual's unique and personal experience of being unwell. People often rationalise their illness experience through a complex web of personal encounters, experiences, and belief systems shaped by their cultural and social world. As expressed by Kleinman (1980), individuals would usually develop a personal (or adopt an existing) "explanatory model" that represents their personal conceptualisation of the cause, course, and consequences of their illness. For each of the NTDs being discussed in this volume, there is a classic medical description, research, and acceptable causes and tried and tested treatment options.

The conditions according to medicine are either bacterial infections, worm infestations, or vector transmitted (mosquito, a fly, etc.). In some instances, there is no clear-cut cause. On the other hand, many people in African societies rationalise their experience of ill health through complex personal experiences and belief systems (sociocultural and/or religious) ingrained in their cultural and social world. An explanation of ill health or physical manifestations, be it by the patient or the people in his or her society, would usually be representative of their personal conceptualisation of the cause, course, and consequences of their illness (Green et al. 2002). Patients' perceptions about a disease, its effective treatment, and the socio-economic dislocation caused by the illness and related symptoms generally have a significant impact on when and where to go for diagnosis and treatment. Beliefs about the cause and transmission of a disease vary considerably from culture to culture; however, there are some general traits that permeate many cultures. They can be dynamic in that the prolonged nature of a condition and the long period and difficulty of a healing process can lead to assumptions of mystical involvement even for people who attributed their ill-health to natural causes at the onset of symptoms. As Peeters-Grietens et al. (2012) state, the same condition can also have double causality. In other words, a particular disease may be attributed to natural/medically known causes and simultaneously to mystical causes. Therefore, while medical science describes a malfunction in the body, lay perceptions of disease causation relate to their experience with their ill health.

According to Peeters-Grietens et al. (2012) in Benin, plots of land on which crops are grown are protected by charms to prevent theft, so, in that country, Buruli can be attributed to a charm because of trespassing on another person's land thereby disrupting a social order. In Ghana, among other causes, Buruli is attributed to a curse "if your mouth is strong" or, in other words, if a person uses powerful words to curse or witchcraft because someone hates you (Renzaho et al. 2007; Stienstra et al. 2002).

In sub-Saharan Africa, there is a dearth of information on perceptions of lymphatic filariasis. Even in Ghana where studies on the subject have been conducted in the northern and southern part of the country, there are multiple differences in perceptions of aetiology. While in Southern Ghana the belief is that one gets the disease because of an imbalance in the body's constitution and other physical and sometimes spiritual causes, in the northern part of the country, the perception is that it is due to external forces like witchcraft or a curse (Ahorlu et al. 1999; Gyapong et al. 1996a, b).

In the Democratic Republic of Congo, the mode of transmission and dramatic sequelae following sleeping sickness are very well known and recognised by community members (Mpanya et al. 2012). In addition, serious prohibitions (complete rest for 6 months) are linked to those with diseases with severe consequences for individuals if they are not seen to be adhered to. Although considered shameful, having the disease signifies a shift in societal position rather than rejection by the family and community. Table 1 from the chapter on this subject in the first edition of the book now includes perceptions about genital schistosomiasis, which itself is set out in a section in the revised chapter on schis tosomiasis.

Table 1 Lay perceptions and description of disease causation

Condition	Perceived cause	Reference
Buruli	Unknown, witchcraft, drinking unclean water, unhygienic practices, playing with an infected person, and infraction against social order	Peeters-Grietens et al. (2012), Renzaho et al. (2007), Stienstra et al. (2002) and WHO (2013a)
Dracunculiasis (Guinea worm)	Poor drinking water	Sisay (2012) and WHO (2013b)
Human African trypanosomiasis (sleeping sickness)	While in some communities perceptions are in line with medical diagnosis, in others, it is attributed to witchcraft and drinking water infected by people with the disease. The fly, eating the amaranth plant, and eating pork	Kabanga et al. (2022), Mpanya et al. (2012) and WHO (2013c)
Leishmaniasis (kala-azar)	Sandflies, dirt, washing in a canal, eating, and drinking with patients	WHO (2013d)
Leprosy	Hereditary, witchcraft, and breaking of a taboo	WHO (2012)
Lymphatic filariasis	Hydrocele "fever scrotum", elephantiasis, "elephant leg"; these are because of a curse and poor body constitution and/or are hereditary	Ahorlu et al. (1999), Gyapong et al. (1996b) and WHO (2013e)
Onchocerciasis	Bite of the blackfly causing itching, several worms in the body, eating improperly cooked leaves	Aninakwah-Boahene et al. (2014) and WHO (2013f)
Schistosomiasis Genital schistosomiasis	Sign of growth, hereditary. Common among boys. For women, blood in urine happens when they menstruate	WHO (2013g) and Kukula et al. (2019)
Soil-transmitted helminths (ascaris, whipworm, and hookworm)	No information found	WHO (2015)

NTDs and Stigma: The Impact on Gender Roles and Relations

There are increasing numbers of studies that explore experience of individuals affected by NTDs; however, few explore multiple axes of oppression simultaneously. Where varying axes of oppression or social phenomenon are considered, they tend to be in relation to one principal factor such as gender or stigma, normally at preference of the researcher/s, with limited consideration of how inequalities or inequities may interact with each other and with social processes to cause both medical and social co-morbidities. Furthermore, social inequities tend to be considered in relation to their influence on health determinants and access to treatment, as opposed to their influence on illness experience (Manderson et al. 2009). Where social inequities are considered in relation to disease outcome, this tends to be from a biomedical standpoint in terms of risk, with limited exploration of the social conditions that may lead to such biomedical consequences (Allotey et al. 2005). In the following sections of this chapter, we consider how different inequities interact to shape illness experience for individuals.

Many if not all NTDs discussed in this book cause impairment, disfigurement, and sometimes permanent disability, leading to stigma and social discrimination. Many studies on stigma refer to leprosy, and this may be because rejection from leprosy dates far back to biblical times to depict a mark of disgrace or physical disorder (to such an extent that sufferers of leprosy subjected to their own mock funeral prior to banishment from their families and communities). In some cases, they endured torture and execution (Roueche 1986) and rejection either for physical or social reasons (Goffman 1963; Weiss 2008). The concept of stigma can be either enacted (when the affected person faces rejection, discrimination, or physical abuse by others) or perceived (when the affected person thinks he or she is or can be stigmatised because of their physical condition). In many cases, the perceived stigma can be worse than that enacted and may lead to known attempts of suicide, emotional stress, depression, and anxiety (Heijnders 2002; Hyland 1993; Scambler 1998; Van Brakel 2003). Stigma as an important mediator of social burden can lead to invisibility and marginalisation of infected people, emotional distress, and delayed diagnosis and treatment even to the extent that free treatment at health facilities is not sought to avoid a public appearance because of the disease (Pépin et al. 2002; Perera et al. 2007).

Discrimination and stigma heighten people's vulnerability to ill health. In all countries of the world, the burden of disease is disproportionately borne by vulnerable and marginalised groups, who often suffer from other social inequities as well as discrimination (United Nations Special Rapporteur 2003). It can also be an obstacle to prevention and treatment. This form of discrimination against people living in poverty can be reinforced by other forms of discrimination, such as on grounds of sex or race, which further increase vulnerability to neglected diseases. For example, the status of women in many countries, including their lack of ownership of resources, affects their access to prevention and treatment and means they may well be more vulnerable to stigma and the impact of stigma.

It must be noted that despite cultural or religious diversity, the expression or experience of stigma is similar worldwide. It affects mobility, interpersonal relationships, marriage, employment, leisure activities, and attendance at social and religious functions, and the depth and detail of how stigma is experienced are linked to gender, roles, relations, and expectations (Van Brakel 2003).

A recent systematic review that focused on stigma in relation to NTDs highlighted that all types of stigmas were prominent in relation to leprosy, lymphatic filariasis, podoconiosis, Buruli ulcer, onchocerciasis, and leishmaniasis. They also found evidence of all types of stigma related to Chagas disease, schistosomiasis, trachoma, and STH, although fewer studies had focused on stigmas associated with these diseases. Explorations of stigma associated with NTDs that cause less visible effects, such as schistosomiasis, are often less prioritised. This is problematic, particularly when considering internalised stigmas, and therefore presents a need for greater research into illness experience associated with NTDs that may produce less visible impairment. This is also important when considering long-term implications of stigma in association with NTDs, as even when physical impairment may be reversed or minimised, social stigma, particularly that which has caused poor mental health outcomes, can last a lifetime (Barrett 2005). Brieger et al. (1998) found that treatment with ivermectin alone had no impact on the stigma associated with onchocercal skin disease. Stigma associated with FGS was a significant obstacle for girls and women seeking care at health facilities in Ghana (Kukula et al. 2019). Health workers and community members consider boys and men to be more at risk of schistosomiasis infection than women and adolescent girls. Meanwhile, there is evidence of household gender roles (collecting water, washing, and bathing) that expose girls and women to FGS infection. Because of this misconception, workers said they had never seen a woman or girl infected with schistosomiasis; hence when younger girls reported to the health facilities with the condition, they were often stigmatised by healthcare workers. Therefore, girls with the signs and symptoms of FGS in schistosomiasis-endemic communities are often accused of sexual promiscuity and referred for STI treatment, even though they are not sexually active. Lack of awareness about FGS among health workers has created highly stigmatising experiences for women and adolescent girls. In avoiding stigmatisation, adolescent girls prefer going for home remedies and herbal medicine to treat FGS (Kukula et al. 2019).

Issues influencing variations in types of stigmas experienced have been more widely considered than in relation to gender. In relation to disease stage, Lu et al. (1988) found within India that enacted stigma experienced increased with disease progression of lymphatic filariasis and resultant lymphoedema or hydrocele. In addition, the location of hydrocele in relation to LF also caused increased stigma, which was greater when it was in the genitals as opposed to the leg (Lu et al. 1988). Aetiology of disease, as perceived by the communities, was also seen to have a bearing on enacted and internalised stigma, although this was not the case for all diseases. Schistosomiasis is often perceived to be associated with promiscuity in view of signs and symptoms in the urinary tract; however, this was not the case for diseases like LF and leprosy. Rather, more physically visible diseases were perceived to be hereditary, and therefore stigmatisation tended to be linked to moral constructions of disability that suggest impairment is the result of wrongdoing in a previous life.

Disembodiment was also often associated with internalised and anticipated stigmatisation. This was evidenced by observed covering of limbs or lack of their use. Disembodiment has been identified as particularly evident when disease or impairment was acquired in adulthood since prototypes of health and normalisation were often learned from a young age (Barrett 2005). One study noted that when individuals had low incomes and were young and female, they were likely to experience increased levels of enacted stigma (Hofstraat and van Brakel 2016). Similarities between experience in enacted, anticipated, and internalised stigma across a range of NTDs suggest there is scope for more joined-up thinking in relation to stigma management. This has been evidenced in the implementation of joint support groups for LF and leprosy, which have been shown to be a key component of rehabilitation and support to individuals living with these NTDs. The use of support groups has also been effective for individuals living with podoconiosis and Buruli ulcer (Agbo et al. 2019; Tsegay et al. 2018); however, these groups have been less integrated with other NTD support groups. All these expressions and the ways in which perceived and enacted stigma are experienced depict physical, mental, and social attributes, self-hate, self-deprivation, and lack of social acceptance and stress of illness experience as described by Goffman (1963). The complex ways in which stigma is experienced affect healthcare-seeking behaviour in many ways. Weiss (2008) suggests some key interventions to address stigma. These are (1) support, (2) enactment of laws, and (3) awareness creation to challenge sociocultural norms and behaviour.

Gender and NTDs

The specific intersection between gender and disability in many LMIC settings significantly shapes disease experience, with vulnerabilities and discrimination being centred on the ability to fulfil stereotypical or socially assigned gendered roles, such as men being economic providers and women/girls as responsible for reproductive work, including caregiving. This disproportionately affects the extent to which men and women and boys and girls are exposed to and infected by any of the NTDs being discussed and shapes whether infected patients experience enacted (a discriminatory experience because of being labelled with a disease) or felt stigma (an imagined social reaction that can change a person's identity) (Crinson 2007). It also affects their perception and healthcare-seeking behaviour in the event of the disease.

Allotey et al. (2005) argue that all functioning societies are built on the abilities of individual members to adopt specific roles and responsibilities. The roles that are assigned to males and females by culture and society, how they are played out, and how they relate to each other at the individual and broader sociopolitical levels are described by the concept of gender. That there are gender differences is not in dispute; the problem arises with the different values placed on the various roles and responsibilities and the consequences of this in creating and sustaining disadvantage and inequities. Gender roles and relations shape how individuals and societies operate, the effects of the disadvantages created, and permeate all aspects of life, from production, allocation, and distribution of resources to exposure to disease and health-promoting services.

But does all this add value to health-related outcomes, and do these need to be addressed in responses to NTDs? At the very least, retaining a gender focus on any agenda provides information on new ways of approaching health issues and identifying points for specific interventions to address specific issues. Based on the conventional health indicators provided through quantitative methodologies, current data are not strong enough to support a definitive response. Data quality is poor and often draws largely on biased samples; sex-disaggregated data are often not available and, when available, are not always used to inform responsive policy. In addition, qualitative data usually focuses on women without reference to the interaction with men within their societies or the dynamics of gender relations. There is little information to proffer explanations of gender differences from male perspectives.

In the case of sleeping sickness and leishmaniasis, men generally tend to contract the disease more than women probably because of occupational hazards and general exposure to the vector; however, women tend to be more stigmatised from the consequences of contracting the disease and can experience complications such as amenorrhea, infertility, and miscarriage (WHO 2010b).

In the DRC, women appeared to be more knowledgeable about the disease than men due to their role as caregivers which means they are more exposed to health education messages when they send sick family members to a health facility (Mpanya et al. 2012; Pépin et al. 2002). Within this context, having sexual relations when one has the condition is taboo, and gendered norms mean that women must leave their matrimonial homes to avoid sex while their male counterparts seek other women for sexual interactions (Mpanya et al. 2012).

Both men and women are susceptible and probably equally infected by onchocerciasis, schistosomiasis, soil-transmitted helminths, and sleeping sickness with women suffering more of the consequences during pregnancy because of anaemia and iron deficiency due to heavy infection (WHO 2010b).

Urogenital schistosomiasis also referred to as female or male genital schistosomiasis (FGS and MGS) is common in Africa. Adolescent girls and women with FGS can experience bleeding and unpleasant discharge from the vagina, genital lesions, and nodules in the vulva, general discomfort, and pain during sex. FGS is a cause of subfertility and miscarriage and increases vulnerability to HIV, and all of which arguably have wide-reaching social and economic consequences, although how women experience symptoms indicative of FGS, their treatment-seeking pathways and the impact on their livelihoods and well-being are poorly understood. Symptoms of MGS in adolescent boys and men include bleeding and egg deposition within semen, yellow discolouration in semen, and lumpy semen, the implications of which are poorly understood (Mbabazi et al. 2011).

FGS and MGS are sensitive, private, and possibly stigmatising conditions. Messaging, therefore, needs to be geared to the realities of females' and males' gendered experiences. This requires in-depth qualitative research to explore the context and community discourse surrounding FGS and MGS symptoms and the development of appropriate referral and treatment strategies that are accessible to all women and people and boys, regardless of where they live or how much money or resources they can access.

Although the prevalence of lymphoedema and elephantiasis is higher in women than in men when hydrocele is taken into consideration, more adult men than women appear to have severe chronic consequences of lymphatic filariasis. The consequent disfigurement results in the infected person experiencing stigma and social and psychological distress with women having a far less chance of finding a spouse than men and suffering more problems and general discrimination (WHO 2010a, b).

In relation to leprosy, a recent systematic review identified that men are more likely than women to present with multibacillary leprosy (more than five skin lesions) and present with Grade II disabilities, whereas women are more likely to present with paucibacillary disease (one to five skin lesions) than men, emphasising difference in treatment seeking and delayed disease presentation based on gender. However, the review also identified that it was common for women to not adhere to full treatment regimens, sometimes due to clinic hours conflicting with domestic or household responsibilities and resulting in physical disfigurement and impairment. Thus, long-term impact of leprosy is shaped by social factors that guide treatment adherence processes as well as stage and grading of disease at diagnosis (Dijkstra and van Elteren 2017).

Clemmons et al. (2002) report that when it comes to decision-making for MDA for onchocerciasis control, men are the key decision-makers. Women are not invited to meetings where issues are discussed and are not given feedback on the major decisions taken. They are informed about CDTI activities by word of mouth just before drug distribution. The effect of gender on NTD volunteer performance is understudied. In Kenya, Omedo et al. (2012) showed that women found it difficult to balance their daily work with being a volunteer on NTD program. Due to social hierarchies in Uganda, women's performance as NTD volunteers could be affected (Omedo et al. 2012; Parker and Allen 2011).

The perception in the Ashanti region of Ghana is that more women have Buruli ulcers than men because of sex work. However, like lymphatic filariasis, both sexes appear to be equally discriminated against since they are avoided in public places due to fear of contracting the disease, have less chance of being appointed into leadership positions, and have problematic marriages and less chance of getting a spouse (Stienstra et al. 2002). Brieger et al. (1998), in their study in Nigeria, documented some issues around stigma related to onchocerciasis from individual survey respondents. Examples of responses are as follows:

> "I am embarrassed because my whole body is covered with onchocerciasis". It is shameful for having tough skin like a frog". "I am embarrassed because it has destroyed my body". "I am ashamed when it itches me in the midst of a crowd". "It embarrasses me". "I use a long wrap whenever I go out". "It causes disrespect when one scratches in public". "People laugh at him". "People won't want to be close to someone with onchocerciasis". "Onchocerciasis leads to disrespect because it causes blindness and destroys the skin".

Experiences of disability and gender-based violence in relation to NTDs are infrequently explored outside relationships to stigma. However, a recent study that explored experiences of intimate partner violence (IPV) among Ethiopian women affected by podoconiosis (lymphoedema) found that IPV was often triggered by inability to complete household chores and influenced disease outcomes in a two-way relationship. Episodes of IPV frequently included restrictions to resource access (e.g. water and healthcare), which limited women's ability to complete self-washing or disease management, resulting in increased disease severity, limited social networks, and reduced ability to complete household activities, thus further triggering episodes of IPV (Tsegay et al. 2018). A study in Nepal also described episodes of IPV among women affected by leprosy, identifying that five out of ten leprosy-affected women had experienced sexual violence by their husbands. This was described as more likely where women were living with an alcoholic partner (van't Noordende et al. 2016). Consequently, as also suggested by Tsegay et al. (2018), within NTD programming, there is a need to improve access to IPV prevention and management services.

In relation to PCT interventions, gendered analysis has predominantly focused on attitudes and barriers towards the uptake of drugs. Challenges in drug delivery for pregnant women are a key barrier. Preventive medicines used to treat lymphatic filariasis and onchocerciasis cannot safely be administered during pregnancy (McDonald 2011). Furthermore, even though praziquantel (used to treat schistosomiasis) was deemed safe for administration in pregnancy in 2002, it is infrequently administered to pregnant women due to safety concerns (WHO 2003). Given that women living in areas endemic for NTDs may spend up to 25% of their reproductive years pregnant and another 60% lactating (Rilkoff et al. 2013), this presents them at significantly higher risk of living with long-term untreated NTD infections (United Nations Development Programme 2019).

In addition, there is consideration given to the gendered interactions between young female drug distributors with older men in communities; it has been found in some studies that older men are less likely to take drugs when they are provided by young female drug distributors than if they were delivered by men or older women (Rilkoff et al. 2013). Rilkoff et al. (2013) also present findings that when men are absent from the household, for example, when working outside the community, this can have a positive impact on the uptake of drugs for the rest of the household. When men who often function as key decision-makers in the household are absent, women have a higher level of autonomy in deciding whether to take the drug or not.

Poverty, NTDs, and Disability

The NTD-poverty relationship is mutually reinforcing as poor material circumstances also place individuals, households, and communities at increased risk of NTDs, due to poor water and sanitation infrastructure and extreme environmental conditions. NTDs are commonly referred to as diseases of poverty, and the tight interrelationship between poverty and inequity is well recognised. The interactions between poverty and NTDs should be considered at multiple levels: individual, household, community, and country. For individuals, households, and communities, the impoverishing impacts of NTDs result from direct costs linked to diagnosis and treatment and indirect from lost productivity, missed education, and caregiving responsibilities (Manderson et al. 2009). For example, in Ghana and Nigeria, for families affected by NTDs including Buruli ulcer, out-of-pocket spending for treatment has been observed to result in catastrophic health expenditure (Ackumey et al. 2012; Bangert et al. 2017). Furthermore, many impoverished at-risk populations, particularly the geographically isolated, are frequently unable to reach preventive services (Manderson et al. 2009). Countries where NTD burden is the greatest are also frequently too poor to provide infrastructure and human resources to tackle the diseases because of crippling international debt and economic disadvantage

(Manderson et al. 2009). This is exacerbated further in affected nations as associated morbidity can result in high levels of lost economic productivity. For example, in India, morbidity associated with lymphatic filariasis is thought to result in US$1 billion of economic losses (Aagaard-Hansen and Chaignat 2010).

Mental Health and NTDs

A review by Litt et al. (2012) suggests that factors associated with NTDs predispose individuals to poor mental health. These include stigma and discrimination, exclusion from participating fully in society, reduced access to health and social services, lack of educational opportunities, exclusion from income generation and employment opportunities, and restrictions in exercising civil and political rights – especially for those with physical disability and permanent disfigurement (Litt et al. 2012). For example, lower quality of life scores were associated with mental illness in patients with podoconiosis in northern Ethiopia (Mousley et al. 2015). Effects of chronic ill health and disease are frequently not limited to the affected individual, and this is particularly the case for NTDs. However, research is lacking on the impact of NTDs on the mental health of caregivers (Litt et al. 2012).

Seeking Healthcare for a Neglected Tropical Disease

Since the London Declaration in 2012, there have been important advances in the development of diagnostics and massive support from pharmaceutical companies to supply drugs for the treatment of NTDs. Available treatments are either for free or cost less than a dollar a day, and studies have shown that an integrated MDA using a combination of medicines to prevent or treat the seven most prevalent NTDs yields the best return (Hotez et al. 2007).

Except for Guinea worm, the NTDs discussed in this book can be effectively treated with preventive chemotherapy and the majority through mass drug administration (MDA), clean water, vector control, and proper sanitation. Through health education and massive campaigns, NTD control programmes in sub-Saharan Africa have succeeded in treating people over many years; however, the response to attempts at treatment and control varies from disease to disease and depends on the approach used. In the DRC, low attendance at screening sessions for human African trypanosomiasis (HAT) was attributed to, among other things, mistrust of mobile team nurses (who were perceived to leave the disease in people's blood with the use of injections), unannounced arrival of staff for screening exercises, fear of lumbar punctures, fear of unsolicited HIV/AIDS tests, other economic priorities, and the lack of confidentiality during the screening procedure itself (Mpanya et al. 2012). Fear of adverse events, the timing of MDA, the type and attitude of the distributor (community distributor or health worker), and experience with adverse events have also affected MDA for lymphatic filariasis and onchocerciasis, although the expelling of worms after drug administration has been found to be a motivating factor

(Abanobi et al. 2011; Abd Elaziz et al. 2013; Babu and Babu 2014; Gyapong et al. 2000).

Some countries in sub-Saharan Africa have been conducting MDA for many years, yet they remain unable to interrupt transmission of disease. Notably urban areas remain a challenge as populations are more mobile, are less accessible during MDA, and do not see themselves at risk for disease (Biritwum et al. 2017; Krentel et al. 2018; Njomo et al. 2014). In other countries, nomadic and pastoralist communities may not be included as part of MDA activities. With an estimated 50 million nomadic pastoralists in Africa, Gammino et al. (2020) noted in their review that factors beyond mobility alone restricted access of these communities to healthcare and included distance to health facilities, poor-quality healthcare, costs, and poor disease knowledge, among others. Complex emergencies also pose a challenge to the success of NTD elimination goals as MDA delivery may be disrupted, while case identification and healthcare provision may be interrupted. Some countries with higher NTD prevalence, South Sudan, Central African Republic, and the Democratic Republic of Congo, also experience less security (Kelly-Hope et al. 2021).

Despite the many challenges facing NTD programmes, much success has also been documented. Togo and Malawi have been certified as having eliminated LF, Ghana has eliminated trachoma, and Guinea worm has reached lowest levels in history, bringing the goal of eradication closer. In some countries, new treatment strategies for LF offer a more efficient and acceptable MDA to reach elimination goals faster (Krentel et al. 2021; Weil et al. 2019). New technologies like a diagnostic tool for loa-loa (LoaScope) have enabled the resumption of community-wide distribution in areas where MDA had ceased due to severe adverse events (Kamgno et al. 2017).

Mass drug administration uses preventive chemotherapy given to undiagnosed eligible people living in an endemic area. While taking the treatment offered during MDA may prevent the onset of chronic manifestations of the NTDs, it does not reverse the disability that people chronically affected with NTDs experience. When people experience symptoms (acute or chronic) of NTDs, their treatment-seeking behaviour will depend on where they live, their economic situation, and their understanding of their disease. People with limited resources, particularly those living in rural areas, invariably seek medical help first from the least expensive, closest, and most trusted sources. Treatment may also be sought simultaneously or sequentially before going to a hospital. The first time a person experiences a feeling of being unwell, they ask what is wrong, and then often they will deploy the wait and see what happens next approach. As soon as the condition is prolonged for longer than expected, the "why" questions begin to set in. Why is this happening to me or my family or that person at this time and what can be done about it? This leads to a quest for therapy as described by Kleinman (1980) and Helman (1981) where people irrespective of their socio-economic, religious, or cultural background seek traditional and cosmopolitan care simultaneously. Kleinman propounds that when a person is sick, there are three interlinked health sectors that are available for use: popular, folk, and professional sectors. In all societies, these sectors are available and create a pluralistic healthcare system with a wide range of options for care seekers to choose from in times of ill health. The popular sector is usually the first point of call and is usually at the home level where advice is sought from relatives and friends. Communities and parents' experience informs practice at this stage: for example, self-medication is practised with leftover drugs from the home and drugs borrowed from friends or purchased from local drug shops. Conditions like hot body, chills, general discomfort, and malaria, which are considered mild, are dealt with easily within this sector.

The folk sector has the more traditional type of healing, and treatment is sought from traditional healers, spiritualists, faith healers, herbalists, and traditional birth attendants. This sector can be visited as a first option but in most cases are contacted when the first health-seeking attempt within the household or community does not work. However, the popular and folk sectors may be used concurrently. The professional sector is the more formal sector that has people with formal professional training like doctors, nurses, and pharmacists. Others however may rush or be rushed directly to a health facility depending on what they perceived to be the cause of their condition. Due to cost and other factors, it may be the last point of call when all other options have failed, by which time the condition would have progressed severely.

In Benin, Buruli ulcer patients can access home treatments (salves, antibiotics, and analgesics), traditional medicine (herbs, divination, incantations), and hospital treatments. Over 70% of patients interviewed at home and in hospital used traditional and hospital medicine interchangeably (Peeters-Grietens et al. 2012). The availability of a pluralistic healthcare system in African settings, the cost of care, fear of stigma, and "mutilation" after treatment are some of the reasons for the delay in seeking healthcare. Interestingly the perceived cause of the disease did not contribute to a delay in care seeking for Buruli ulcer patients in Cameroon (Stienstra et al. 2002). Similarly in the Greater Accra and Ashanti Regions of Ghana, Buruli patients had multiple sources of care available to them and acted in the same way as their counterparts in Benin and Cameroon, applying salves to their nodules and "boils" at home, seeking treatment from a traditional healer, and then finally using the hospital as a last resort (Renzaho et al. 2007; Stienstra et al. 2002). For schistosomiasis, some people felt that, treatment was ineffective due to reoccurence of the disease (Onyeneho et al. 2010), hence limiting healthcare-seeking behaviour.

The Economic Impact of Having NTDs

The economic effects of NTDs go beyond those directly involved to affect families or households, the health system, and nations. These effects are in the form of direct medical (hospitalisation, laboratory, medications) and non-medical (meals, transportation) costs as well as productivity loss (due to the inability of the affected persons and their caregivers to participate in economic activities). Also, NTDs can be associated with permanent disability (Tembei et al. 2018), which would make it very difficult for such individuals to get employed. Moreover, since NTDs are regarded as one of the diseases of poverty (Hotez et al. 2007; WHO 2018), they have the potential to worsen the poverty status of affected persons via catastrophic health spending, hence reducing expenditure on other necessities of life (such as food and shelter) for the affected persons as well as their families. The poverty effects of NTDs on those who suffer permanent disability are likely to be worse.

Notwithstanding, 40% of the global burden of NTDs is found in Africa (WHO 2018). It is therefore not surprising that, in 2015, the monetary value of the number of people killed by NTDs on the African continent was Int$ 5 112,472,607 (Kirigia and Mburugu 2017). Also, in sub-Saharan Africa (SSA), the disability-adjusted life years (DALYs, healthy years of life lost due to NTDs) estimates by the Global Burden of Disease Collaborative Network (2020) in 2019 were as follows: leprosy (6108.202), leishmaniasis (226,484.4), human African trypanosomiasis (82,615.46), schistosomiasis (1,340,321), lymphatic filariasis (419,039.6), onchocerciasis (1,225,538), trachoma (75,484.43), dengue (52,270.67), ascariasis (238,643.7), trichuriasis (63,379.81), and hookworm (541,540.7). Thus, among these NTDs, schistosomiasis had the greatest healthy life years lost in 2019. Concerning gender, schistosomiasis, trachoma, dengue, trichuriasis, and hookworm had higher healthy life years lost in females than males (Table 2). The implication is that the important roles women play in terms of unpaid household work, as well as the provision of farm labour in the sub-region, will be lost.

Concerning specific NTD cost (direct and indirect cost), Mathew et al. (2020) estimated that, globally, before the MDA, on the average, the yearly economic burden of each chronic LF case was US$115, mainly due to productivity costs. Moreover, people affected by lymphatic filariasis are likely to abandon economic activities due to stigma (Norris et al. 2012). In Ghana, individuals with acute adenolymphangitis (ADL) were found to be unable to carry out their usual activities for most of the ADL period (Gyapong et al. 1996a), hence limiting their participation in economic activities. Similarly, the annual working days lost due to leprosy was estimated to be 115 days in Cameroon (Tembei et al. 2018), while in Ghana, seeking

Table 2 DALYs of common NTDs in SSA, 2019

Disease	Total burden (Years)	Burden among males (Years)	Burden among females (Years)
Leprosy	6108.202	3979.098	2129.104
Leishmaniasis	226,484.4	154,645.6	71,838.84
Human African trypanosomiasis	82,615.46	42,953.27	39,662.19
Schistosomiasis	1,340,321	650,342.6	689,978.1
Lymphatic filariasis	419,039.6	332,514.6	86,525.03
Onchocerciasis	1,225,538	654,345.6	571,192.1
Trachoma	75,484.43	34,358.34	41,126.09
Dengue	52,270.67	22,985.56	29,285.11
Ascariasis	238,643.7	121,400.2	117,243.5
Trichuriasis	63,379.81	31,640.07	31,739.74
Hookworm disease	541,540.7	207,799.7	333,740.9

Source: Global Burden of Disease Collaborative Network (2020)

care for Buruli ulcer was estimated to lead to a loss of 535 productive days among caregivers and adult patients (Amoakoh and Aikins 2013). Further, people affected by onchocerciasis in a rural farming community in Nigeria reported a reduction in concentration at work and fatigue, as well as giving up their jobs due to visual impairment (Wogu and Okaka 2021). Audibert and Etard (1998) reported 23% annual productivity loss among families affected by schistosomiasis in Mali, while between 6.6% and 32% losses in agricultural productivity have been reported among villages and households in the top 5% schistosomiasis intensity areas in Burkina Faso (Rinaldo et al. 2021).

Trachoma at the global level is estimated to cause an annual loss in productivity of US$2.9–5.3 billion (attributable to only blindness and visual impairment), which increases to US$8 billion after adding trichiasis (WHO 2021). In SSA, countries with blindness due to trachoma have low levels of economic productivity relative to those in the other Asia/islands and Middle Eastern Crescent that do not have blindness due to trachoma (Frick et al. 2003). Moreover, it has been found in Ethiopia that people affected by trachoma were more likely to be poor but less likely to engage in economic activity. Even if they engaged in economic activity, they were more likely to need assistance in performing these activities (Habtamu et al. 2015). Regarding guinea worm, it normally occurs during the planting or harvest season which significantly reduces agricultural productivity, especially given that it could affect more than half of the population in a village at a point in time (Ruiz-Tiben and Hopkins 2006).

A study by Ahmed et al. (2021) estimated the annual cost of snake bites for health facilities in three regions in Burkina Faso to be US$506,413 (US$49 per hospitalisation). Moreover, in Africa, the cost of each hookworm infection as well as its associated health outcomes has been estimated at $56 (95% UI: $52–$59). Further, the cost of each hookworm infection, employing minimum wage as an indicator of per capita productivity, has been estimated to be $24 (95% UI: $23–$25) (Bartsch et al. 2016). Similarly, Lee et al. (2019) estimated the mean cost of dengue fever among inpatients to be $26 (95% CI $23-$29) in Burkina Faso, while that of outpatients was $23 (95% CI $19-$28) and $13 (95% CI $23-$29) in Kenya and Burkina Faso, respectively. Similarly, the total cost for one episode of visceral leishmaniasis was US$450 in Sudan. Although patients are given free drugs for visceral leishmaniasis in Sudan, 53% of the total cost of each episode of visceral leishmaniasis was borne by households, and this cost constituted 40% of the yearly household income. In effect, over 75% of households experienced catastrophic out-of-pocket expenditure on visceral leishmaniasis (Meheus et al. 2013). The situation in Morocco is not different where the median total cost of an episode of paediatric visceral leishmaniasis was US$520 (Tachfouti et al. 2016), while human African trypanosomiasis had an in-direct medical cost ranging from US$60 to US$170 for people suffering from abject poverty (living on less than US$1 per day) in Kenya (Bukachi et al. 2017).

The above indicates that, while NTDs normally affect the poor in Africa, they suffer from catastrophic out-of-pocket expenditure because the cost of treatment takes all or a greater proportion of their income. Moreover, while victims of NTDs are confronted with significant losses in productive days, while others end up losing their jobs because of disability and/or stigma. These tend to perpetuate the poverty situation among victims as well as their households. The resultant effects at the macroeconomic level are declining economic growth (emanating from loss in human capital/labour productivity) as well as rising government health and social protection expenditure.

There is therefore the need to enhance efforts towards fighting NTDs because it has been estimated that $27–$42 of economic benefits are obtained by Africa with every $1 devoted to the fight against NTDs. Moreover, if the 2020 goal of eliminating NTDs in Africa had been met t, it would have resulted in productivity savings of $52 billion in SSA over the next 10 years (WHO 2018). Also, between 2021 and 2040, Zimbabwe, Rwanda, Kenya, and Ethiopia, for instance, stand to gain US$5.1 billion (at purchasing power parity (PPP)) when the WHO target on eliminating morbidity and mortality from soil-transmitted helminthiasis and schistosomiasis is achieved by 2030 (The Economist Intelligence Unit 2020).

Comments

NTDs are diseases of poverty and are themselves impoverishing, disproportionally affecting people living in poor or marginalised communities. Hotez (2009) and Person et al. (2009) suggest that women are often particularly isolated and marginalised by stigma associated NTDs. These diseases play out in multiple ways and across the generations. Children, for example, are withdrawn from school due to financial constraints on older people with disabilities. Vulnerability to different NTDs, ability to access treatment, and social and economic impact of NTDS affect different people in multiple ways, and this in turn is dependent on contextual factors. There is a real need for intersectionality (Tolhurst et al. 2012) and approaches that analyse the interplay between gender, poverty, ethnicity, disability, sexuality, religion, and age. Additionally adopting an intersectional approach is useful in terms of influencing policy as it provides an inclusiveness, which is often missing when health policies are developed.

There is clear and complex disconnect between biological cause/medical model explanation and social understanding and aetiology of disease. What is important for health promotion is that dialogue needs to be tailored to the understandings and realities of different communities. This is not straightforward but community-embedded cadres such as community health workers and community-based drug distributors have embedded knowledge that can support interventions. The focus for action needs to go beyond community messaging, to ensure that resilient health systems are developed that are responsive to the needs and realities of socially and economically disenfranchised citizens.

There is also need for qualitative research. The value of qualitative research to better understand the lived realities of affected communities as well as the ability and challenges faced by health systems to appropriately address these is often overlooked when addressing NTDs. However, increasingly those working on NTDs are aware there is a gap in data, which can only be filled by research methods which enable the depth and detail of affected communities and health providers' experiences and challenges to be better understood. For example, through qualitative and participatory approaches, the needs and priorities of frontline providers and volunteers can be documented to ensure this feed into policy and practice to support sustainable and equitable human resource strategies for integrated NTD control.

The Social and Economic Impacts of NTDs Go Beyond the Health Sector

- There is a need for intra- and inter-sectoral action, i.e. links within the health sector itself as well as between the health sector and other sectors like education, agriculture, and sanitation. For example, action to address FGS will require new partnerships within the health sector between NTDs and maternal and sexual and reproductive health and strategies at the community level to build links with relevant formal or informal providers (e.g. women who might seek help in the case of infertility).
- New partnerships beyond the health sector, for example, with Ministries of Education and gender, youth, and community services, will be essential in any discussions or action on FGS. Generating and making evidence on interventions that work need to be accessible to stakeholders who should use it in the course of their work, which requires new ways of working together within and across sectors and with stakeholders.
- Poverty alleviation actions are also necessary to address the catastrophic social and economic effects of NTDs on women and men and girls and boys.

Conclusion

In a time when maintaining the integrity of the art of medicine is of critical importance, we are instead witnessing its devaluation due to the current counterposing forces in healthcare. Disease has become the focus of the technologic and market-driven medical system, while illness and the sociocultural aspects of medicine have blurred into the background. Yet the healing tools and instruments of science are blunt and ineffective when used blindly in ignorance of the complex meaning and context of a patient's illness and the impact of this on their livelihoods and well-being. We need to foster attitudes, values, and communication skills that focus on illness, not just disease, to prepare ourselves for the challenges ahead. The new NTD road map seeks to address issues of ill health in a holistic way now more than ever to ensure that no one is left behind and the WHO definition of health being a state of complete mental, physical, and social well-being and not merely the absence of disease or infirmity becomes a reality.

Acknowledgement We wish to acknowledge the contribution of colleagues from the Liverpool School of Tropical Medicine, the Ghana Health Service, the University of Health and Allied Sciences, and the Bruyere Research Institute.

References

Aagaard-Hansen J, Chaignat CL (2010) Neglected tropical diseases: equity and social determinants. Equity Soc Determinants Public Health Programmes 135(1):303

Abanobi O, Chukwuocha U, Onwuliri C, Opara K (2011) Primary motives for demand of ivermectin drug in mass distribution programmes to control onchocerciasis. N Am J Med Sci 3(2):89

Abd Elaziz KM, El-Setouhy M, Bradley MH, Ramzy RMR, Weil GJ (2013) Knowledge and practice related to compliance with mass drug administration during the Egyptian national filariasis elimination program. Am J Tropical Med Hygiene 89(2):260

Abdela SG, van Griensven J, Seife F, Enbiale W (2020) Neglecting the effect of COVID-19 on neglected tropical diseases: the Ethiopian perspective. Trans R Soc Trop Med Hyg 114(10):730–732. https://doi.org/10.1093/trstmh/traa072

Ackumey MM, Gyapong M, Pappoe M, Maclean CK, Weiss MG (2012) Socio-cultural determinants of timely and delayed treatment of Buruli ulcer: implications for disease control. Infect Dis Poverty 1(1):1–13

Addlakha R (2013) Disability studies in india: global discourses, local realities. Routledge, New Delhi

Adelman C, Marano K, Norris J, Spantchak Y (2012) Social and economic impact review on neglected tropical diseases. (Economic Policy/Briefing Paper). In *Social and economic impact review on neglected tropical diseases* (Issue November)

Agbo IE, Johnson RC, Sopoh GE, Nichter M (2019) The gendered impact of Buruli ulcer on the household production of health and social support networks: why decentralization favours women. PLoS Negl Trop Dis 13(4):e0007317

Ahmed S, Koudou GB, Bagot M, Drabo F, Bougma WR, Pulford C, Bockarie M, Harrison RA (2021) Health and economic burden estimates of snakebite management upon health facilities in three regions of southern Burkina Faso. PLoS Negl Trop Dis 15(6):e0009464

Ahorlu CK, Dunyo SK, Koram KA, Nkrumah FK, Aagaard-Hansen J, Simonsen PE (1999) Lymphatic filariasis related perceptions and practices on the coast of Ghana: implications for prevention and control. Acta Trop 73(3):251–261

Al Ju'beh K (2015) Disability inclusive development toolkit. Bensheim, CBM

Allotey P, Gyapong M, UNICEF (2005) *The gender agenda in the control of tropical diseases: A review of current evidence*

Amoakoh HB, Aikins M (2013) Household cost of out-patient treatment of Buruli ulcer in Ghana: a case study of Obom in ga south municipality. BMC Health Serv Res 13(1):1–8

Aninakwah-Boahene D, Apraku A, Adu-Kumi M (2014) Perception of inhabitants in Bui township on causes and effects of onchocerciasis, Ghana. Int J Sci Technol Res 3(4):328–331

Audibert M, Etard J-F (1998) Impact of schistosomiasis on rice output and farm inputs in Mali. J Afr Econ 7(2):185–207

Babu BV, Babu GR (2014) Coverage of, and compliance with, mass drug administration under the programme to eliminate lymphatic filariasis in India: a systematic review. Trans R Soc Trop Med Hyg 108(9):538–549

Bangert M, Molyneux DH, Lindsay SW, Fitzpatrick C, Engels D (2017) The cross-cutting contribution of the end of neglected tropical diseases to the sustainable development goals. Infect Dis Poverty 6(1):1–20

Barrett R (2005) Self-mortification and the stigma of leprosy in northern India. Med Anthropol Q 19(2):216–230

Bartsch SM, Hotez PJ, Asti L, Zapf KM, Bottazzi ME, Diemert DJ, Lee BY (2016) The global economic and health burden of human hookworm infection. PLoS Negl Trop Dis 10(9):e0004922

Biritwum N-K, de Souza DK, Marfo B, Odoom S, Alomatu B, Asiedu O, Yeboah A, Hervie TE, Mensah EO, Yikpotey P (2017) Fifteen years of programme implementation for the elimination of lymphatic filariasis in Ghana: impact of MDA on immunoparasitological indicators. PLoS Negl Trop Dis 11(3):e0005280

Brieger WR, Oshiname FO, Ososanya OO (1998) Stigma associated with onchocercal skin disease among those affected near the Ofiki and Oyan Rivers in western Nigeria. Soc Sci Med 47(7):841–852

Bukachi SA, Wandibba S, Nyamongo IK (2017) The socio-economic burden of human African trypanosomiasis and the coping strategies of households in the southwestern Kenya foci. PLoS Negl Trop Dis 11(10):e0006002

Chaumont C, Kamara K, Baring E, Palacio K, Power A, Lancaster W (2020) The SARS-CoV-2 crisis and its impact on neglected tropical diseases: threat or opportunity? PLoS Negl Trop Dis 14(9):e0008680

Clemmons L, Amazigo UV, Bissek A-C, Noma M, Oyene U, Ekpo U, Msuya-Mpanju J, Katenga S, Seketeli A (2002) Gender issues in the community-directed treatment with ivermectin (CDTI) of the African programme for onchocerciasis control (APOC). Ann Tropical Med Parasitol 96(sup1):S59–S74

Crinson I (2007) *Concepts of health and illness section 3, labelling, and stigma. Health knowledge*

Dijkstra JIR, van Elteren M (2017) Gender and leprosy-related stigma in endemic areas: a systematic review. Lepr Rev 88(3):419–440

Ehrenberg JP, Utzinger J, Fontes G, da Rocha EMM, Ehrenberg N, Zhou XN, Steinmann P (2021) Efforts to mitigate the economic impact of the COVID-19 pandemic: potential entry points for neglected tropical diseases. Infect Dis Poverty 10(1):1–10. https://doi.org/10.1186/s40249-020-00790-4

Eisenberg L (1977) Disease and illness distinctions between professional and popular ideas of sickness. Cult Med Psychiatry 1(1):9–23

Frick KD, Hanson CL, Jacobson GA (2003) Global burden of trachoma and economics of the disease. Am J Tropical Med Hygiene 69(5_suppl_1):1–10

Gammino VM, Diaz MR, Pallas SW, Greenleaf AR, Kurnit MR (2020) Health services uptake among nomadic pastoralist populations in Africa: a systematic review of the literature. PLoS Negl Trop Dis 14(7):e0008474

Global Burden of Disease Collaborative Network (2020) Global burden of disease study 2019 (GBD 2019) results. Available at: https://vizhub.healthdata.org/gbd-results/. Accessed 10 May 2023

Goffman E (1963) Stigma Englewood cliffs. NJ: Spectrum:127–128

Green AR, Carrillo JE, Betancourt JR (2002) Why the disease-based model of medicine fails our patients. West J Med 176(2):141

Gyapong JO, Gyapong M, Evans DB, Aikins MK, Adjei S (1996a) The economic burden of lymphatic filariasis in Northern Ghana. Ann Tropical Med Parasitol 90(1):39–48

Gyapong M, Gyapong JO, Adjei S, Vlassoff C, Weiss M (1996b) Filariasis in northern Ghana: some cultural beliefs and practices and their implications for disease control. Soc Sci Med 43(2):235–242

Gyapong M, Gyapong JO, Owusu Banahene G, Remme JH, Kale OO, Adjei S (2000) Community directed treatment for filariasis control in Ghana. Am J Trop Med Hyg 62(3 supp):336

Gyapong M, Nartey A, Oti E, Page S (2016) The Social and economic impact of neglected tropical diseases in sub-saharan Africa. Springer International Publishing, Cham, Neglected Tropical Diseases - Sub-Saharan Africa, pp 349–365

Habtamu E, Wondie T, Aweke S, Tadesse Z, Zerihun M, Zewdie Z, Callahan K, Emerson PM, Kuper H, Bailey RL (2015) Trachoma and relative poverty: a case-control study. PLoS Negl Trop Dis 9(11):e0004228

Heijnders ML (2002) Understanding adherence in leprosy. *Experiencing and Making Sense of Leprosy and Its Treatment in Its Social Context. A Qualitative Study Conducted in the Eastern Terai Region of Nepal. PhD, Faculty of Health, London, South Bank University*

Helman CG (1981) Disease versus illness in general practice. J R Coll Gen Pract 31(230):548–552

Hofstraat K, van Brakel WH (2016) Social stigma towards neglected tropical diseases: a systematic review. Int Health 8(suppl_1):i53–i70

Hotez P (2009) Devastating global impact of neglected tropical diseases-health advocates are fostering a new sense of urgency to solve neglected health problems besetting people in developing countries. Microbe 4(8):366

Hotez PJ, Molyneux DH, Fenwick A, Kumaresan J, Sachs SE, Sachs JD, Savioli L (2007) Control of neglected tropical diseases. N Engl J Med 357(10):1018–1027

Hotez PJ, Alvarado M, Basáñez MG, Bolliger I, Bourne R, Boussinesq M, Brooker SJ, Brown AS, Buckle G, Budke CM, Carabin H, Coffeng LE, Fèvre EM, Fürst T, Halasa YA, Jasrasaria R, Johns NE, Keiser J, King CH, Naghavi M (2014) The global burden of disease study 2010: interpretation and implications for the neglected tropical diseases. PLoS Negl Trop Dis 8(7). https://doi.org/10.1371/journal.pntd.0002865

Hyland J (1993) A socio-cultural study of leprosy in Nepal: compliance, patient illness career patterns and health education. University of Tasmania, Tasmania

Kabanga C, Valverde Mordt O, Mbo F, Mbondo M, Olela D, Etinkum R, Nkaji D, Mukoso B, Mananasi L (2022) Communities' perception, knowledge, and practices related to human African trypanosomiasis in the Democratic Republic of Congo. Diseases 10(4):69

Kamgno J, Pion SD, Chesnais CB, Bakalar MH, D'Ambrosio MV, Mackenzie CD, Nana-Djeunga HC, Gounoue-Kamkumo R, Njitchouang G-R, Nwane P (2017) A test-and-not-treat strategy for onchocerciasis in Loa Loa–endemic areas. N Engl J Med 377(21):2044–2052

Kelly-Hope LA, Sanders AM, Harding-Esch E, Willems J, Ahmed F, Vincer F, Hill R (2021) Complex emergencies and the control and elimination of neglected tropical diseases in Africa: developing a practical approach for implementing safe and effective mapping and intervention strategies. Confl Heal 15:1–12

Kirigia JM, Mburugu GN (2017) The monetary value of human lives lost due to neglected tropical diseases in Africa. Infect Dis Poverty 6(1):1–16

Kleinman A (1980) Patients and healers in the context of culture: an exploration of the borderland between anthropology, medicine, and psychiatry, vol 3. University of California Press, California

Krentel A, Gyapong M, Ogundahunsi O, Amuyunzu-Nyamongo M, McFarland DA (2018) Ensuring no one is left behind: urgent action required to address implementation challenges for NTD control and elimination. PLoS Negl Trop Dis 12(6):e0006426

Krentel A, Gyapong M, McFarland DA, Ogundahunsi O, Titaley CR, Addiss DG (2021) Keeping communities at the Centre of efforts to eliminate lymphatic filariasis: learning from the past to reach a future free of lymphatic filariasis. Int Health 13(Supplement_1):S55–S59

Kukula VA, MacPherson EE, Tsey H, Stothard JR, Theobald S, Gyapong M (2019) A major hurdle in the elimination of urogenital schistosomiasis revealed: Identifying key gaps in knowledge and understanding of female genital schistosomiasis within communities and local health workers. PLoS Negl Trop Dis 13(3):e0007207. https://doi.org/10.1371/journal.pntd.0007207

Lee J-S, Mogasale V, Lim JK, Ly S, Lee KS, Sorn S, Andia E, Carabali M, Namkung S, Lim S-K (2019) A multi-country study of the economic burden of dengue fever based on patient-specific field surveys in Burkina Faso, Kenya, and Cambodia. PLoS Negl Trop Dis 13(2):e0007164

Litt E, Baker MC, Molyneux D (2012) Neglected tropical diseases and mental health: a perspective on comorbidity. Trends Parasitol 28(5):195–201

Lu AG, Valencia LB, Llagas L, de las Aballa L, Postrado L (1988) Filariasis: a study of knowledge, attitudes, and practices of the people of Sorsogon, final report of a project supported by the TDR social and economic research component. World Health Organization, Geneva

Onyeneho NG, Yinkore P, Egwuage J, Emukah EC (2010) Perceptions, attitudes and practices on schistosomiasis in Delta State, Nigeria. Tanzan J Health Res 12(4):278–282

Manderson L, Aagaard-Hansen J, Allotey P, Gyapong M, Sommerfeld J (2009) Social research on neglected diseases of poverty: continuing and emerging themes. PLoS Negl Trop Dis 3(2):e332

Mathew CG, Bettis AA, Chu BK, English M, Ottesen EA, Bradley MH, Turner HC (2020) The health and economic burdens of lymphatic filariasis prior to mass drug administration programs. Clin Infect Dis 70(12):2561–2567

Mbabazi PS, Andan O, Fitzgerald DW, Chitsulo L, Engels D, Downs JA (2011) Examining the relationship between urogenital schistosomiasis and HIV infection. PLoS Negl Trop Dis 5(12):e1396

McDonald M (2011) Neglected tropical and zoonotic diseases and their impact on women's and children's health. *The Causes and Impacts of Neglected Tropical and Zoonotic Diseases: Opportunities for Integrated Intervention Strategies*, 357–388

Meheus F, Abuzaid AA, Baltussen R, Younis BM, Balasegaram M, Khalil EAG, Boelaert M, Musa AM (2013) The economic burden of visceral leishmaniasis in Sudan: an assessment of provider and household costs. Am J Tropical Med Hygiene 89(6):1146

Molyneux DH, Aboe A, Isiyaku S, Bush S (2020) COVID-19 and neglected tropical diseases in Africa: impacts, interactions, consequences. Int Health 12(5):367–372. https://doi.org/10.1093/inthealth/ihaa040

Mousley E, Deribe K, Tamiru A, Tomczyk S, Hanlon C, Davey G (2015) Mental distress and podoconiosis in northern Ethiopia: a comparative cross-sectional study. Int Health 7(1):16–25

Mpanya A, Hendrickx D, Vuna M, Kanyinda A, Lumbala C, Tshilombo V, Mitashi P, Luboya O, Kande V, Boelaert M (2012) Should I get screened for sleeping sickness? A qualitative study in Kasai province, Democratic Republic of Congo. PLoS Negl Trop Dis 6(1):e1467

Mitra S (2006) The capability approach and disability. J Disabil Policy Stud 16(4):236–247. https://doi.org/10.1177/10442073060160040501

Mitra S (2017) Disability, health and human development. Palgrave Macmillan, US New York

Njomo DW, Mukoko DA, Nyamongo NK, Karanja J (2014) Increasing coverage in mass drug administration for lymphatic filariasis elimination in an urban setting: a study of Malindi town, Kenya. PLoS One 9(1):e83413

Norris J, Adelman C, Spantchak Y, Marano K (2012) Social and economic impact review on neglected tropical diseases. *Economic policy/briefing paper*. Hudson Institute, Washington DC

Ocholaid EA, Karanja DMS, Elliott SJ (2021) The impact of neglected tropical diseases (Ntds) on health and wellbeing in sub-Saharan Africa (SSA): a case study of Kenya. PLoS Negl Trop Dis 15(2):1–19. https://doi.org/10.1371/journal.pntd.0009131

Omedo MO, Matey EJ, Awiti A, Ogutu M, Alaii J, Karanja DMS, Montgomery SP, Secor WE, Mwinzi PNM (2012) Community health workers' experiences and perspectives on mass drug administration for schistosomiasis control in western Kenya: the SCORE project. Am J Tropical Med Hygiene 87(6):1065

Parker M, Allen T (2011) Does mass drug administration for the integrated treatment of neglected tropical diseases really work? Assessing evidence for the control of schistosomiasis and soil-transmitted helminths in Uganda. Health Res Policy Systems 9(1):1–20

Peeters-Grietens K, Toomer E, Um Boock A, Hausmann-Muela S, Peeters H, Kanobana K, Gryseels C, Ribera JM (2012) What role do traditional beliefs play in treatment seeking and delay for Buruli ulcer disease? –insights from a mixed methods study in Cameroon. PLoS One 7(5):e36954

Pépin J, Mpia B, Iloasebe M (2002) Trypanosoma brucei gambiense African trypanosomiasis: differences between men and women in severity of disease and response to treatment. Trans R Soc Trop Med Hyg 96(4):421–426

Perera M, Whitehead M, Molyneux D, Weerasooriya M, Gunatilleke G (2007) Neglected patients with a neglected disease? A qualitative study of lymphatic filariasis. PLoS Negl Trop Dis 1(2):e128

Person B, Bartholomew LK, Gyapong M, Addiss DG, van den Borne B (2009) Health-related stigma among women with lymphatic filariasis from The Dominican Republic and Ghana. Soc Sci Med 68(1):30–38

Renzaho AMN, Woods PV, Ackumey MM, Harvey SK, Kotin J (2007) Community-based study on knowledge, attitude, and practice on the mode of transmission, prevention, and treatment of the Buruli ulcer in ga West District, Ghana. Trop Med Int Health 12(3):445–458

Rilkoff H, Tukahebwa EM, Fleming FM, Leslie J, Cole DC (2013) Exploring gender dimensions of treatment programmes for neglected tropical diseases in Uganda. PLoS Negl Trop Dis 7(7):e2312

Rinaldo D, Perez-Saez J, Vounatsou P, Utzinger J, Arcand J-L (2021) The economic impact of schistosomiasis. Infect Dis Poverty 10(1):1–12

Rohwerder B (2015) Disability inclusion

Roueche B (1986) A lonely road (chapter 5). Med Detectives 2:68–86

Ruiz-Tiben E, Hopkins DR (2006) Dracunculiasis (Guinea worm disease) eradication. Adv Parasitol 61:275–309

Scambler G (1998) Stigma and disease: changing paradigms. Lancet 352(9133):1054–1055

Sisay GT (2012) *Perceptions and experiences of people in Gambella region of Ethiopia on Guinea worm disease eradication interventions*

Stienstra Y, van der Graaf WTA, Asamoa K, van der Werf TS (2002) Beliefs and attitudes toward Buruli ulcer in Ghana. Am J Tropical Med Hygiene 67(2):207–213

Tachfouti N, Najdi A, Alonso S, Sicuri E, Laamrani El Idrissi A, Nejjari C, Picado A (2016) Cost of paediatric visceral leishmaniasis care in Morocco. PLoS One 11(6):e0155482

Tembei AM, Kengne-Ouaffo JA, Ngoh EA, John B, Nji TM, Deribe K, Enyong P, Nkuo-Akenji T, Davey G, Wanji S (2018) A comparative analysis of economic cost of podoconiosis and leprosy on affected households in the northwest region of Cameroon. Am J Tropical Med Hygiene 98(4):1075

The Economist Intelligence Unit (2020) *Breaking the Cycle of Neglect: Reducing the Economic and Societal Burden of Parasitic Worms in Sub-Saharan Africa*

Tolhurst R, Leach B, Price J, Robinson J, Ettore E, Scott-Samuel A, Kilonzo N, Sabuni LP, Robertson S, Kapilashrami A (2012) Intersectionality and gender mainstreaming in international health: using a feminist participatory action research process to analyse voices and debates from the global south and north. Soc Sci Med 74(11):1825–1832

Tsegay G, Deribe K, Deyessa N, Addissie A, Davey G, Cooper M, Trueba ML (2018) "I should not feed such a weak woman." Intimate partner violence among women living with podoconiosis: A qualitative study in northern Ethiopia. PLoS One 13(12):e0207571

United Nations Development Programme (2019) *Discussion Paper the Gender Dimensions of Neglected Tropical Diseases*

United Nations Special Rapporteur (2003) *Para. 59*

Van Brakel WH (2003) Measuring leprosy stigma-a preliminary review of the leprosy literature. Int J Lepr Other Mycobact Dis 71(3):190–197

van't Noordende AT, van Brakel WH, Banstola N, Dhakal KP (2016) The impact of leprosy on marital relationships and sexual health among married women in eastern Nepal. J Trop Med 2016

Weil GJ, Bogus J, Christian M, Dubray C, Djuardi Y, Fischer PU, Goss CW, Hardy M, Jambulingam P, King CL (2019) The safety of double-and triple-drug community mass drug administration for lymphatic filariasis: a multicenter, open-label, cluster-randomized study. PLoS Med 16(6):e1002839

Weiss MG (2008) Stigma and the social burden of neglected tropical diseases. PLoS Negl Trop Dis 2(5):e237

WHO (2020) Ending the neglect to attain the sustainable development goals: a road map for neglected tropical diseases 2021–2030. World Health Organization. https://www.who.int/neglected_diseases/Revised-DraftNTD-Roadmap-23Apr2020.pdf, Geneva, p 196

Wogu MD, Okaka CE (2021) Prevalence and socio-economic effects of onchocerciasis in Okpuje, Owan west local government area, Edo state, Nigeria. Int J Biomed Health Sci 4(3)

World Health Organisation (2012) *WHO Fact sheet 101*

World Health Organization (2003) *Report of the WHO informal consultation on the use of praziquantel during pregnancy/lactation and albendazole/mebendazole in children under 24 months: Geneva, 8-9 April 2002*. World Health Organization

World Health Organization (2010a) *Preventive chemotherapy and transmission control databank*

World Health Organization (2010b) *Working to overcome the global impact of neglected tropical diseases: first WHO report on neglected tropical diseases* (issue WHO/HTM/NTD/2010.1). World Health Organization, Geneva

World Health Organization (2002) Towards a common language for functioning, disability, and health: ICF. The international classification of functioning, disability and health
World Health Organization (2013a) *WHO Fact sheet* 199
World Health Organization (2013b) *WHO Fact sheet 359*
World Health Organization (2013c) *WHO Fact sheet 259*
World Health Organization (2013d) *WHO Fact sheet 375*
World Health Organization (2013e) *WHO Fact sheet 102*
World Health Organization (2013f) *WHO Fact sheet 374*
World Health Organization (2013g) *WHO Fact sheet* 115
World Health Organization (2015) *WHO Fact sheet 366*
World Health Organization (2018) *Africa: The elimination of neglected tropical diseases is finally within reach*
World Health Organization (2021) *Trachoma: Key facts*

Alternative Interventions Against Neglected Tropical Diseases in SSA: Vector Control

Daniel A. Boakye, Dziedzom K. de Souza,
and Moses Bockarie

Contents

D. A. Boakye (✉) · D. K. de Souza
Noguchi Memorial Institute for Medical Research, University of Ghana, Legon, Ghana
e-mail: DBoakye@noguchi.ug.edu.gh

M. Bockarie
School of Community Health Sciences, Njala University, Bo, Sierra Leone

J. O. Gyapong, B. A. Boatin (eds.), *Neglected Tropical Diseases - Sub-Saharan Africa*, Neglected Tropical Diseases, https://doi.org/10.1007/978-3-031-53901-5_19

Abstract

Vector control is one of the strategies recommended by the World Health Organization for the control and prevention of the neglected tropical diseases (NTDs) apart from preventive chemotherapy, intensified case management, provision of safe water, sanitation and hygiene, and veterinary public health. Although an integrated approach based on a combination of strategies or one strategy targeting a group of diseases is the preferred approach, the current policy for the global elimination of most of the NTDs is based solely on chemotherapy (Molyneux, Adv Parasitol 61:1–45, 2006; *Uniting To Combat NTDs, Delivering on promises & driving progress.* Available: http://unitingtocombatntds.org/report/delivering-promises-driving-progress-second-report-uniting-combat-ntds [Online]. Accessed 16 May 2014). Dependence on preventive chemotherapy alone without measures to control vectors and intermediate hosts, for vector-borne NTDs like lymphatic filariasis (LF) and onchocerciasis, may not achieve the expected outcome (Bockarie et al., Ann Rev. Entomol 54:469–487, 2009) in the set targeted time frame. Vector control has the potential to play a very important role in the control of NTDs and is increasingly becoming a supplementary intervention strategy. However, it requires the commitment of resources, both financial and human, from disease control programs. There is the need for integration at all levels while adopting the WHO policy guidelines on IVM. Although vector control has been shown to be an effective strategy for the control of vector-borne NTDs (Bockarie et al., Ann Rev. Entomol 54:469–487, 2009; Townson et al., Bull World Health Org 83:942–947, 2005), it is faced with some challenges, major among which are insecticide resistance, multiplicity of vector species, changes in vector behavior, and cost. Further research on insecticide resistance and the effect of vector control on one disease, as, for example, malaria vector control on lymphatic filariasis, is still required.

Introduction

Vector control is one of the strategies recommended by the World Health Organization for the control and prevention of neglected tropical diseases (NTDs). The others are preventive chemotherapy, intensified case management, provision of safe water, sanitation and hygiene, and veterinary public health. An integrated approach based on a combination of strategies or one strategy targeting a group of diseases is the preferred approach. However, the main policy for the global elimination of most of the NTDs was based solely on chemotherapy especially if there were drugs that could be given to populations at risk [preventive chemotherapy] (Molyneux 2006; Uniting to Combat NTDs 2014). The generous commitments by industry have led to increased availability of drugs for onchocerciasis (Merck & Co., Inc.), lymphatic filariasis (Merck & Co., Inc., GlaxoSmithKline (GSK), and Eisai), and schistosomiasis (Merck KGaA). Dependence on preventive chemotherapy alone for vector-borne NTDs like LF and onchocerciasis may not achieve the expected outcome, looking particularly at the targeted timeframe, if measures to control vectors and intermediate hosts are not considered (Bockarie et al. 2009). Therefore, vector control is increasingly becoming a supplementary intervention strategy. The transmission efficiency of many vector-borne NTDs is poor and the vector-parasite relationship fragile for Chagas disease, human African trypanosomiasis, onchocerciasis, and lymphatic filariasis.

In its new NTD road map 2021–2030, the WHO has placed a significant emphasis on vector control and other interventions, with the aim to holistically address the control and elimination of vector-borne NTDs. Thus, one of the cross-cutting targets in the road map is to reduce by 75% the number of deaths from vector-borne neglected tropical diseases (relative to 2016) and to achieve the WHO's global vector control response goal (World Health Organization 2020).

Vector-Borne Diseases

Vector-borne diseases are those in which the causative pathogens are transmitted between hosts by vector species, mostly arthropods (mosquitoes, sandflies, blackflies, fleas, ticks, flies) but also other organisms such as snails. Vector-borne diseases include some of the most important ailments that have plagued mankind since antiquity. For example, plague (*Yersinia pestis*) transmitted by fleas decimated populations in the Middle Ages, and louse-borne fever had a major impact on armies, for example, the Spanish in 1489, the French in 1528, contributed to Napoleon's retreat from Moscow in 1812, and the death of millions of soldiers and civilians in World War I (Pierce 1974; Raoult et al. 2004). Thus, vector-borne diseases represent an important threat to human health, causing severe morbidity and mortality (Varmus et al. 2003; Hill et al. 2006). It has been estimated that vector-borne diseases are responsible for 17% of the global burden of diseases (World Health Organization 2008) and about 10% of human deaths (World Health Organization 2011; Rascalou et al. 2012) while contributing significantly to the vicious cycle of poverty by imposing an annual burden of more than 50 million disability-adjusted life years [DALYs] (Gold et al. 2002; World Health Organization 2011; Rascalou et al. 2012). The most important vector-borne disease of all times is malaria (World Health Organization 2011).

The distribution of vector-borne diseases is worldwide. For example, exposure to vector-borne helminths such as *Dirofilaria immitis* is reported to be about 45% of the total European population (Otranto et al. 2013). Other vector-borne infections like leishmaniasis, onchocerciasis, malaria, and yellow fever, among many, are found mostly in tropical and subtropical parts of the world.

Vector-borne diseases particularly those found in tropical and subtropical areas promote poverty and impair education and economic development. Some of them such as malaria cause high mortalities, others result in epidemics with high fatality rates, and many cause immense suffering and lifelong disabilities. However, many of the vector-borne diseases together with some other infections apparently do not cause high mortalities such as HIV or TB or malaria and often do not receive the attention they deserve and most often neglected.

The neglected tropical diseases (NTDs) are a group of diseases of different etiology that affects hundreds of millions of people worldwide, mostly the poor and developing economies but sometimes also found in the developed world, for example, in the United States (Hotez 2008). Although there are more than 20 diseases that would qualify as neglected in the amount of funds spent on their control, research, etc., the World Health Organization officially recognizes 17 of these (Table 1) as the major NTDs.

Table 1 List of officially recognized NTDs by the World Health Organization

Diseases	Vectors	References
Buruli ulcer	Aquatic insects?	Johnson et al. (2007), Marsollier et al. (2002) and Portaels et al. (1999)
Chagas disease	Triatomine bugs	Lent and Wygodzinsky (1979) and Ramsey and Schofield (2003)
Dengue and chikungunya	*Aedes* sp.	Higa (2011) and World Health Organization (2009)
Dracunculiasis (Guinea worm diseases)	Cyclops	Muller (1991) and Yelifari et al. (1997)
Echinococcosis	–	–
Foodborne trematodiasis	–	–
Human African trypanosomiasis (sleeping sickness)	Tsetse fly (*glossina* sp.)	Scott (1959) and Gouteux (1990)
Leishmaniasis	Sandflies (*Phlebotomus* sp., *Sergentomyia* sp.)	Claborn (2010) and Peters and Killick-Kendrick (1987)
Leprosy	–	–
Lymphatic filariasis	Mosquitoes (*Anopheles* sp., *Aedes* sp., *Culex* sp., *Mansonia* sp., *Ochlerotatus* sp.)	Zagaria and Savioli (2002) and WHO (2013)
Mycetoma, chromoblastomycosis, and other deep mycoses		
Onchocerciasis (river blindness)	Blackflies (*Simulium* sp.)	Boakye et al. (1998) and Brown (1962)
Rabies	–	–
Scabies and other ectoparasitoses	Mite (*Sarcoptes scabiei* var. *hominis*)	Arlian and Morgan (2017)
Schistosomiasis	Snails (*Biomphalaria* sp., *Bulinus* sp., *Oncemalania* sp.)	Mandahl-Barth (1965), Morgan et al. (2001) and Woodruff et al. (1999)
Soil-transmitted helminthiasis	–	–
Snake bite envenoming		
Taeniasis/cysticercosis	–	–
Trachoma	*Musca sorbens*	Emerson et al. (1999)
Yaws	–	–

The Transmission of NTDs

Of the 20 officially WHO-recognized NTDs, the transmission of the causative pathogens of 8 is directly vector borne (Table 1), and further 3 have insect involvement in their transmission. The vector-borne NTDs include some of the most debilitating infections of humans such as onchocerciasis, lymphatic filariasis, and schistosomiasis (Boakye et al. 2007; Traoré et al. 2009; Yirenya-Tawiah et al. 2011) and others like dengue which can lead to high fatalities (Carneiro da Silva et al. 2013).

The vectors are either insects or other arthropods or snails (Table 1) with diverse life histories. For example, the main blackfly species vectors of human onchocerciasis in Africa, members of the *Simulium damnosum* complex, breed in fast-flowing stretches of rivers (Le Berre 1974; Walsh 1985), while *Simulium albivirgulatum*, a vector in the Democratic Republic of Congo, breeds in much slower flowing streams (Fain et al. 1980), and yet still *Simulium neavei* group is phoretic on crabs (McMahon 1952; Van Someren and Mcmahon 1950). Another example is the different genera of mosquitoes that transmit the causative agents of lymphatic filariasis (de Souza et al. 2012).

Control of Vector-Borne NTDs

The components of a vector-borne disease are the vector, pathogen, and the host. This triad of interlinking components is all influenced by the environment. The understanding of this natural history of vector-borne diseases is critical for implementing any successful control program irrespective of whether it is an NTD or not. Thus, the control should integrate a combination of strategies targeting the pathogen and vector as well as promoting human behavior that leads to prevention or propagation of pathogen or vector.

Historical Approaches

Before the transmission of most of these infections was understood, the control of vector-borne NTDs had generally depended on treating those showing overt symptoms of the disease and/or diagnosed as having infection with a particular pathogen (Brown and Neu 1990; Ligon 2005; Haldar et al. 2011). The observation and understanding that insects, other arthropods, and snails could serve as intermediate hosts for the causative agents of diseases and parasites, bacteria, and viruses between the late 1890s to the mid-1930s (Service 1976) opened the gates for a flurry of measures to then tackle the vectors as a disease control method. These early vector control measures revolved around the view that it should be easy to eliminate the vector and therefore break the transmission of diseases once any factor in the vector life cycle is modified (Kershaw 1963). Thus, other important components of medical entomology such as studies of insect physiology were ignored or progressed slowly (Kershaw 1963). During this early phase of vector control, emphasis was therefore based on ecological science and thinking, particularly physical modification of the environment. The belief that vector-borne diseases could be eliminated easily was even more accentuated with the discovery and use of dichlorodiphenyltrichloroethane (DDT) and other insecticides in vector control (Soper et al. 1947; Chinery 1968; Duchon et al. 2009). Ecological science and thinking, the basis for earlier efforts to control pests and disease vectors, lost its prominence, and vector control became synonymous with insecticide use.

Current Approaches to the Control of Vector-Borne NTDs

The euphoria that followed the discovery and use of potent insecticides such as DDT soon led to the reality of the complexity of the situation. The drive for commercial agriculture led to its heavy use in the 1950s and 1960s (Chapin and Wasserstrom 1981). Insecticide resistance became widespread, and the negative impacts of insecticides on the environment were highlighted as eloquently expressed in the book *Silent Spring* by Rachel Carson in 1962. Subsequently, DDT was banned as an agricultural insecticide, following the Stockholm Convention in 2001. However, its use for disease vector control was never banned, but international pressure restricted its implementation in malarious countries (Stockholm Convention on Persistent Organic Pollutants 2001; Sadasivaiah et al. 2007). Vector control against NTDs was only then acceptable when there was no safe chemotherapeutic agent against a particular disease such as the control of onchocerciasis in the 1970s to early 1990s by the WHO Onchocerciasis Control Programme in West Africa (Amazigo et al. 2006).

This status quo has been maintained starting with the donation of Mectizan® (ivermectin) by Merck and Co., Inc. for as long as needed to control onchocerciasis, followed by that of albendazole by GSK against lymphatic filariasis and recent pledges during the London Declaration by other pharmaceutical companies producing drugs against vector-borne/involved NTDs such as schistosomiasis and trichiasis (Uniting to Combat NTDs 2012). This is thus the current backbone of the drive to control or eliminate NTDs such as onchocerciasis, lymphatic filariasis, trichiasis, schistosomiasis, and soil-transmitted helminthiasis (preventive chemotherapy) and the call for more research to develop drugs for the other NTDs.

In recent times new approaches and technologies to vector control have been developed and are being trialed in various settings (Wilson et al. 2020). These include the gene drive, a genetic modification method used to spread desirable traits through interbreeding populations of mosquitoes vectors of diseases (Hammond and Galizi 2017); introduction of *Wolbachia* (genus of bacteria that naturally infect some insect species) into *Aedes* vectors to reduce their longevity and transmission of dengue and other arboviruses spatial repellent chemicals used to prevent human host-vector encounters in defined spaces (Achee et al. 2012); and eave tubes with insecticide-laden electrostatic netting placed below the roof to kill any mosquitoes that are attracted to host odors emanating through the tubes (Knols et al. 2016).

Importance of Vector Control in NTDs

A factor that has mitigated against the use of vector control interventions among many others is the wrong notion and interpretation that vector control as the name implies is only against the vectors and not one of the strategies for disease control and elimination. Hence, in disease control circles, the use of bed nets is spoken of as preventing morbidity and mortality without taking into consideration the vector components. Notwithstanding this, vector control has a strategic role in the control/elimination of NTDs due to various factors. Chief among these are the fact that some diseases such as African trypanosomiasis, Chagas disease, dengue, and leishmaniasis either have imperfect drugs, vaccines are not available, and therefore, vector control remains the main strategy to prevent transmission (Townson et al. 2005). Furthermore, it is becoming clearer that other chemotherapeutic campaigns (e.g., against lymphatic filariasis) will benefit from a combination with vector control (Bockarie et al. 2009).

An efficient vector control strategy depends on a critical understanding of the biology of the vector in question. This involves being able to specifically identify the species involved in disease transmission, their distribution, life cycle, feeding behavior, resting behavior, and migratory behavior (Lambrechts et al. 2009; Pates and Curtis 2005; Radcliffe and Ragsdale 2002). Another factor in vector control is vector ecology which involves the comprehensive study of an organism and its relation to its environment (breeding sites and resting sites) and environmental factors that influence density and behavior (Ellis and Wilcox 2009).

Vector control measures can be classified into personal protection (repellents, protective clothing, insecticide vaporizers, mosquito nets, and insecticide-treated materials), environmental manipulation (closing of breeding sites, planting of trees that grow rapidly, and use lots of water and biological control), house design (ceiling and trapdoors with screens), and insecticide applications (adulticiding and larviciding). Historically, it has been shown that vector control can effectively reduce disease transmission, when done properly (Townson et al. 2005).

Methods of Vector Control for NTDs

The major vector control methods fall under insecticide use, and non-insecticide use (World Health Organization 2019), and have remained the same over the years. However, the strategy of application has evolved and also depends on the vector being targeted.

Insecticide Use

The main vector control strategy revolves around the use of insecticides. These are applied in various forms and target the immature and adult stages of the vectors.

Larviciding/Molluscicisting

The vectors of many of the NTDs (e.g., mosquitoes, blackflies, and snails) have developmental stages that are found in aquatic habitats. The control of these stages therefore is best achieved by the use of insecticides to kill the developmental stages. There are reported successes such as those of the Onchocerciasis Control Programme (OCP) in West Africa (Traoré et al. 2009), *Culex* breeding sites (Mulla 1967), and control of snail vectors (McCullough et al. 1980) by using this strategy. However, this is not easily achieved either due to the difficulties in detecting breeding sites, the dispersed nature of breeding sites (*Anopheles gambiae* s.s. breeding sites), or the sheer expanse of it (lakes for snail vectors). These apparent challenges have led to a lack of interest in the use of this strategy. Recent improvements in spatial data and analysis are now being employed to accurately determine the presence of breeding sites (Shao and Li 1985; Zhang et al. 2003; Hassan and Onsi 2004) which will resolve one of the major challenges and lead to targeted control. Furthermore, the use of bio-larvicides which are more specific to particular vectors without affecting nontarget organisms coupled with observations that this method provides a significant level of contribution toward reducing disease transmission is generating a renewed interest (Mittal 2003; Priyadarshini et al. 2012).

Adulticiding

The use of insecticides to kill the adults of most vectors is the commonest form application of insecticides. The adults of most vectors are easily observed and hence can be easily targeted. However, this is effective when the behavioral characteristics (resting, feeding, oviposition habits) of the adult stage of the vector are known and understood. Therefore, while it is useful to target most mosquito vectors of NTDs, it is not effective to target *Simulium* vectors of onchocerciasis. Methods of insecticide application against the adult stages are varied and include indoor spraying of walls and crevices where the adult are found (Pluess et al. 2010), space spraying to kill flying adults as well as outdoor resting vectors (WHO 2003), protective barriers such as insecticide-treated bed nets (Hill et al. 2006), and incorporation into traps [lethal ovitraps] (Perich et al. 2003; Ritchie et al. 2009).

Indoor residual spraying (IRS) has been used extensively against mosquito vectors of lymphatic filariasis (Kelly-Hope et al. 2013), vectors of Chagas disease (Hashimoto and Schofield 2013), and sandfly vectors of leishmaniasis (Davies et al. 2000). Space spraying is used against a variety of insects including those transmitting disease pathogens of man. The commonest form of space spraying is the use of aerosol cans for household insect control. Recently such use of aerosols is found in the airline industry where planes are sprayed with pyrethroid insecticide formulations. Control of *Aedes* mosquitoes is also amenable to the use of space spraying (Esu et al. 2010).

Prevention of Man-Vector Contact

This strategy has been used as a vector control measure since antiquity. This is done either as protective barriers, the most famous and successful is that of bed nets. The incorporation of insecticides into bed nets revolutionized the use of this personal protection method as the most advocated strategy against indoor biting mosquitoes even though in some cases, such as in Ada, Ghana, it is also used outdoors along the beach. The use of insecticide-treated bed nets is the main vector control strategy in malaria control (Hill et al. 2006). However, since the *Anopheles* mosquitoes that transmit malaria in most areas in Africa and the Pacific region are also the vectors of LF, this method is also effective against LF. Another strategy against mosquitoes to prevent man-vector contact has been improved housing and screening since the early twentieth century in Italy (Celli 1901), Greece (Ross 1913), Panama (Le Prince et al. 1916), and the United States (Boyd 1926). There is ample evidence that house screening contributed to the elimination of malaria from many parts of the world (Lindsay et al. 2002). Although screening has been shown to provide protection against indoor-biting mosquitoes similar to insecticide-treated bed nets in recent times too (Lindsay et al. 2002; Ogoma et al. 2009), this strategy has found little support in control strategies targeting mosquito-borne diseases (Smith 2004).

Challenges to Vector Control

Although vector control has been shown to be an effective strategy for the control of vector-borne NTDs (Bockarie et al. 2009; Townson et al. 2005), it is faced with some challenges, major among which are insecticide resistance, multiplicity of vector species, changes in vector behavior, and cost.

Insecticide Resistance

Insecticide usage continues to be the backbone of most vector control strategies; hence, a major challenge is the development of insecticide resistance in target vectors. This has been responsible for the failure of many campaigns against insect vectors. Currently insecticide resistance has been found in almost all genera of vectors implicated in the transmission of vector-borne diseases (Padonou et al. 2011; Flores et al. 2013; Thomas et al. 2013; Hassan et al. 2012). Although resistance to snails has not been much documented, there is still the possibility of this happening in the only chemical niclosamide (Bayluscide, Bayer, Leverkusen, Germany) recommended by the WHO against snail vectors of schistosomiasis (Rapado et al. 2013). This situation poses a threat to vector control of NTDs.

Multiplicity of Vector Species

Many of the vector-borne NTDs are transmitted by a multiplicity of vectors species of the same genus (onchocerciasis, dengue, and trypanosomiasis) or by several species of one genus or of different genera (LF). LF is vectored by species of *Aedes*, *Anopheles*, *Culex*, and *Mansonia* (Bockarie et al. 2009), and although in most geographical locations, different species of one genus are vectors, in other locations, species belonging to different genera are involved in the transmission such as species of *Anopheles* and *Mansonia* in Ghana (Ughasi et al. 2012). This multiplicity of vector species poses a challenge for vector control since these various vectors may have different breeding habitats and adult behaviors such as exophily (Service, MW 1982). Furthermore, a less important vector could take on more active role in transmission particularly when the major vector is controlled or after the introduction of MDA or the importation of different strains of the pathogen. An example was the mediation of *Aedes albopictus* as the vector for an imported variant of chikungunya fever virus in Singapore instead of the main vector *Aedes aegypti* (Yoshikawa 2010). Another challenge is that in most endemic areas, all the vectors involved in the transmission of a particular disease are not known. For example, in an endemic area of cutaneous leishmaniasis in Ghana, the vectors are still not known (Boakye et al. 2005; Kweku et al. 2011), and this makes it difficult to plan any vector control strategy.

Lack of Integrated Vector Management

Vector control options are powerful tools in the fight against vector-borne diseases. However, these should not be treated as stand-alone solutions but rather as part of well-integrated vector control programs in disease-endemic regions. For example, the use of insecticide-treated nets (ITNs) and/or IRS has been shown to reduce malaria parasite transmission by 90% or more (Nyarango et al. 2006; Protopopoff et al. 2008). Nonetheless, an important question is how best to eliminate the remaining level of transmission, the answers to which are dependent on existing challenges to integrated management. The WHO listed five reasons for the slow uptake of integrated vector control: (1) overestimation of the requirements of integrated control: "most programme managers are unaware of this possibility that an integrated approach can, in principle, be introduced without additional resources"; (2) over-sophistication of integrated control: "in short, for many mosquito species enough is already known…to base the introduction of integrated control"; (3) insufficient conviction of the advantages of integrated control; (4) misconceptions about the use of pesticides in integrated control: "project managers need to be assured that integrated control does not, nor cannot, mean exclusion of pesticides, but their more rational and correct use"; and (5) insufficient consciousness of comparative costs, effectiveness, benefits, and risks: "application of pesticides, especially via indoor residual spraying (IRS), frequently becomes a routine practice, regardless of the need or the results obtained"

(Rafatjah 1982; Beier et al. 2008). Other reasons for the slow uptake of IVM include a lack of capacity building, poorly defined roles for advocacy and legislative activities, and a general lack of intersectoral linkage within the health sector (Beier et al. 2008). The authors also discussed reasons why vector control strategies fail in many countries, and these include but are not limited to the following: the lack of intersectoral collaborations, improper use of insecticides, inability of entomologists to convince decision-makers and the failure to link disease burden with societal burden, and the lack of financial resources and logistics.

Research to Support Program Implementation

Experience with vector control methods points to the need for further research and planning. Vector control poses different challenges in different disease transmission environments or settings. In malaria vector control, for example, five research challenges have been identified:

1. The development of improved vector control for malaria elimination
2. Development of new vector control interventions, including new insecticides and formulations
3. An understanding of vector biology for the development of new control interventions
4. Development of new approaches such as the genetic modification of mosquitoes in order to reduce the high vectorial capacity in some malaria-endemic regions
5. Development of innovative cross-disciplinary technologies that are required to control outdoor biting and resting mosquito vectors, measure transmission, and educate communities about vector control (The malERA Consultative Group on Vector Control 2011).
6. These challenges address the need for broader range of insecticides with novel modes of action that can avoid the development of resistance to existing insecticides, particularly the pyrethroids, and the creation of strategies for the use of new insecticides that minimize the emergence of resistance. Furthermore, different species of *Anopheles* mosquitoes play different vectorial roles in disease transmission. Interventions such as IRS and long-lasting insecticidal nets (LLINs) only target indoor feeding and/or resting vectors (Terenius et al. 2008), and as such, there is the need to define vector species for which new tools are needed as well as creating tools that will be effective for multiple important vector species. In addition, there is the need to develop new tools that will permanently reduce the vectorial capacities of dominant disease vectors.

The research needs outlined above can be generalized for most vector-borne diseases in addition to the needs of understanding vector behavior and transmission dynamics in different areas. Environmental factors exert much influence on disease vectors, and as such the role of different vectors in different areas must further be investigated to aid vector control efforts. Knowledge of vector dispersal under the greatly changed ecologies (e.g., deforestation, irrigation) is essential for protecting rural and urban areas selected for long-term vector control operations (Zahar 1984).

Development of Personnel and Infrastructure

To most disease control programs in Africa, vector control is synonymous to bed net distribution and usage and the spraying of insecticides to kill disease vectors. Surveillance for control programs for vector-borne diseases is expensive, and the lack of the financial capabilities leads to an erosion of vector-borne diseases control infrastructure, which in turn leads to an attrition in human resource capacity (Institute Of Medicine (Us) 2008). Thus, most vector-borne disease control programs lack properly trained personnel and infrastructure. For an effective program, the type of training required and the necessity for providing adequate career structures for vector control personnel and adequate funding for the control programs must all be taken into consideration. Shawarby identified some of the key areas for the training of vector control personnel (Shawarby 1963). These include the effective application of the basic principles of vector control according to local environmental conditions, the evaluation of the results of vector control programs, the avoidance of hazards associated with the use of pesticides, the economy of materials through proper planning and organization, proper handling of transport, and equipment, and health education in the local community. While it is noted that many organizations including the WHO provide workshops aimed at specific issues related to vectors and vector-borne diseases in disease-endemic areas, such programs, however, fail to address the in-depth training needs in molecular, biological, epidemiological, and information technology techniques nor do they empower trainees to exploit these techniques effectively in real-world situations (Institute Of Medicine (Us) 2008). Efficient training strategies are necessary to address the human resource needs in all areas of vector-borne diseases, from identification and processing of vectors to gene identification and characterization and to development of GIS and other information technology-based approaches for the control of vectors and vector-borne diseases (Institute Of Medicine (Us) 2008). Building capacity and infrastructure for vector-borne diseases can only be achieved through establishing vector control programs supervised by well-trained professional staff provided with adequate budgetary support.

Sustainability and Financing

The main challenge to the implementation and sustainability of vector control activities, however, has a lot to do with the limited financial resources (Jobin 1979; Yukich et al. 2008). Another challenge is the collaboration on the part of community members and individuals. Vector control activities are usually carried out by programs without much consideration of the roles communities could play in the process. Involving communities in monitoring and environmental modification could significantly help reduce the cost associated with vector control activities in disease-endemic areas.

Conclusion

In conclusion, many infectious diseases around the world are vector borne and neglected. However, despite the availability of effective chemotherapy against these diseases, the cost of treatment remains a significant barrier in disease control. Thus, other intervention measures like vector control must supplement chemotherapy. Vector control has the potential to play a very important role in the control of NTDs. However, it requires the commitment of resources, both financial and human, from disease control programs. There is the need for integration at all levels, while adopting the WHO policy guidelines on IVM. Finally, research must play an important part in vector control programs. For example, the effect of malaria vector control on other diseases such as lymphatic filariasis must be investigated. The success in controlling onchocerciasis from the Island of Bioko (Traoré et al. 2009) as a result of larviciding by the WHO African Programme for Onchocerciasis Control as well as elimination of LF using indoor residual spraying with DDT to control malaria in the Solomon Islands (Pichon 2002; Duerr et al. 2005) underscore an important historical lesson. The campaigns for the control of vector-borne diseases have the best chance of success when multiple interventions that target different points in the disease cycle are implemented concomitantly. As the WHO has identified the importance of vector control in its 2030 road map (World Health Organization 2020), now is the time to build the capacity and gather the resources needed in order to achieve the elimination of vector-borne NTDs.

References

Achee NL, Bangs MJ, Farlow R, Killeen GF, Lindsay S, Logan JG, Moore SJ, Rowland M, Sweeney K, Torr SJ, Zwiebel LJ, Grieco JP (2012) Spatial repellents: from discovery and development to evidence-based validation. Malar J 11:164

Amazigo U, Noma M, Bump J, Benton B, Liese B, Yameogo L, Zouré H, Seketeli A (2006) Onchocerciasis. In: Jamison DT, Feachem RG, Makgoba MW, Bos ER, Baingana FK, Hofman KJ, Rogo KO (eds) Disease and mortality in sub-Saharan Africa, 2nd edn. World Bank, Washington, DC

Arlian LG, Morgan MS (2017) A review of *Sarcoptes scabiei*: past, present and future. Parasit Vectors 10:297

Beier JC, Keating J, Githure JI, Macdonald MB, Impoinvil DE, Novak RJ (2008) Integrated vector management for malaria control. Malar J 7(Suppl 1):S4

Boakye DA, Back C, Fiasorgbor GK, Sib APP, Coulibaly Y (1998) Sibling species distribution of the Simulium damnosum complex in the west African onchocerciasis control programme area during the decade 1984–1993, following intensive larviciding since 1974. Med Vet Entomol 12:345–358

Boakye DA, Wilson MD, Kweku M (2005) A review of leishmaniasis in West Africa. Ghana Med J 39:94–97

Boakye D, Baidoo H, Glah E, Brown C, Appawu M, Wilson M (2007) Monitoring lymphatic filariasis interventions: adult mosquito sampling, and improved PCR – based pool screening method for Wuchereria bancrofti infection in anopheles mosquitoes. Filaria J 6:13

Bockarie M, Pedersen E, White G, Michael E (2009) Role of vector control in the global program to eliminate lymphatic filariasis. Annu Rev Entomol 54:469–487

Boyd MF (1926) The influence of obstacles unconsciously erected against anophelines (housing and screening) upon the incidence of malaria. Am J Trop Med Hyg 6:157–160
Brown AWA (1962) A survey of simulium control in Africa. Bull World Health Organ 27:511–527
Brown KR, Neu DC (1990) Ivermectin – clinical trials and treatment schedules on onchocerciasis. Acta Leiden 59:169–175
Carneiro Da Silva MM, Gil LHVG, Marques ETA, Calzavara-Silva CE (2013) Potential biomarkers for the clinical prognosis of severe dengue. Mem Inst Oswaldo Cruz 108:755–762
Celli A (1901) The new preventative treatment of malaria in Latium. Collection of papers on malaria. Angelo Celli, 1899–1912. London School of Hygiene and Tropical Medicine, London pp. 1–12
Chapin G, Wasserstrom R (1981) Agricultural production and malaria resurgence in Central America and India. Nature 293:181–185
Chinery WA (1968) Mosquito control in an urban environment with reference to the anti-mosquito operations in Accra and Tema. Ghana Med J 7:205–209
Claborn DM (2010) The biology and control of leishmaniasis vectors. J Global Infect Dis 2:127–134
Davies CR, Llanos-Cuentas EA, Campos P, Monge J, Leon E, Canales J (2000) Spraying houses in the Peruvian Andes with lambdacyhalothrin protects residents against cutaneous leishmaniasis. Trans R Soc Trop Med Hyg 94:631–636
de Souza DK, Koudou B, Kelly-Hope L, Wilson MD, Bockarie MJ, Boakye DA (2012) Lymphatic filariasis vectors and the implications for accelerated elimination of anopheles-transmitted filariasis in West Africa. Parasites Vectors 5:259
Duchon S, Bonnet J, Marcombe S, Zaim M, Corbel V (2009) Pyrethrum: a mixture of natural pyrethrins has potential for malaria vector control. J Med Entomol 46:516–522
Duerr HP, Dietz K, Eichner M (2005) Determinants of the eradicability of filarial infections: a conceptual approach. Trends Parasitol 21:88–96
Ellis BR, Wilcox BA (2009) The ecological dimensions of vector-borne disease research and control. Cad Saude Publ 25(Suppl 1):S155–S167
Emerson PM, Lindsay S, Walraven G, Faal H, Bogh C, Lowe K, Bailey RL (1999) Effect of fly control on trachoma and diarrhoea. Lancet 353:1401–1403
Esu E, Lenhart A, Smith L, Horstick O (2010) Effectiveness of peridomestic space spraying with insecticide on dengue transmission; systematic review. Trop Med Int Health 15:619–631
Fain A, Wery M, Tilkin J (1980) Simulium albivirgulatum Wanson et Henrard, 1944, vector of Onchocerca volvulus, in the central basin region of Zaire. Preliminary note. Ann Soc Belg Med Trop 60:285–289
Flores AE, Ponce G, Silva BG, Gutierrez SM, Bobadilla C, Lopez B, Mercado R, Black WC (2013) Wide spread cross resistance to pyrethroids in Aedes aegypti (Diptera: Culicidae) from Veracruz state Mexico. J Econ Entomol 106:959–969
Gold MR, Stevenson D, Fryback DG (2002) HALYS and QALYS and DALYS, oh my: similarities and differences in summary measures of population health. Annu Rev Public Health 23:115–134
Gouteux JP (1990) Current considerations on the distribution of glossina in west and Central Africa. Acta Trop 47:185–187
Haldar AK, Sen P, Roy S (2011) Use of antimony in the treatment of leishmaniasis: current status and future directions. Mol Biol Int 2011:23
Hammond AM, Galizi R (2017) Gene drives to fight malaria: current state and future directions. Pathog Glob Health 111:412–423
Hashimoto K, Schofield CJ (2013) Elimination of Rhodnius prolixus in Central America. Parasites Vectors 22:45
Hassan AN, Onsi HM (2004) Remote sensing as a tool for mapping mosquito breeding habitats and associated health risk to assist control efforts and development plans: a case study in Wadi El Natroun, Egypt. J Egypt Soc Parasitol 34:367–382

Hassan MM, Widaa SO, Osman OM, Numiary MSM, Ibrahim MA, Abushama HM (2012) Insecticide resistance in the sand fly, Phlebotomus papatasi from Khartoum state, Sudan. Parasites Vectors 7:46

Higa Y (2011) Dengue vectors and their spatial distribution. Trop Med Health 39(Suppl 4):17–27

Hill J, Lines J, Rowland M (2006) Insecticide-treated nets. Adv Parasitol 61:77–128

Hotez PJ (2008) Neglected infections of poverty in The United States of America. PLoS Negl Trop Dis 2:e256

Institute Of Medicine (Us) (2008) Forum on microbial threats. Integrating strategies to address vector-borne disease. In: Vector-borne diseases: understanding the environmental, human health, and ecological connections, workshop summary. National Academies Press (US), Washington, DC. http://www.ncbi.nlm.nih.gov/books/NBK52950/. Accessed 10 Jul 2014

Jobin WR (1979) Cost of snail control. Am J Trop Med Hyg 28:142–154

Johnson PD, Azuolas J, Lavender CJ, Wishart E, Stinear TP, Hayman JA, Brown L, Jenkin GA, Fyfe JA (2007) Mycobacterium ulcerans in mosquitoes captured during outbreak of Buruli ulcer, southeastern Australia. Emerg Infect Dis 13:1653–1660

Kelly-Hope LA, Molyneux DH, Bockarie MJ (2013) Can malaria vector control accelerate the interruption of lymphatic fi lariasis transmission in Africa; capturing a window of opportunity? Parasites Vectors 6:39

Kershaw W (1963) Vector-borne diseases in man: a general review. Bull World Health Org 29(Suppl):13–17

Knols BG, Farenhorst M, Andriessen R, Snetselaar J, Suer RA, Osinga AJ, Knols JM, Deschietere J, Ng'habi KR, Lyimo IN, Kessy ST, Mayagaya VS, Sperling S, Cordel M, Sternberg ED, Hartmann P, Mnyone LL, Rose A, Thomas MB (2016) Eave tubes for malaria control in Africa: an introduction. Malar J 15(1):404

Kweku MA, Odoom S, Puplampu N, Desewu K, Nuako GK, Gyan B, Raczniak G, Kronmann KC, Koram K, Botero S, Boakye D, Akuffo H (2011) An outbreak of suspected cutaneous leishmaniasis in Ghana: lessons learnt and preparation for future outbreaks. Glob Health Action 4:5527

Lambrechts L, Knox TB, Wong J, Liebman KA, Albright RG, Stoddard ST (2009) Shifting priorities in vector biology to improve control of vector-borne disease. Trop Med Int Health 14:1–10

Le Berre R (1974) Simulium damnosum. In: Pal R, Wharton RH (eds) Control of arthropods of medical and veterinary importance. Springer, New York

Le Prince JA, Orenstein AJ, Howard LO (1916) Mosquito control in Panama. G.P. Putnam's Sons, New York

Lent H, Wygodzinsky PW (1979) Revision of the Triatominae (Hemiptera, Reduviidae), and their significance as vectors of chagas' disease. Bull Am Mus Nat Hist 163:3

Ligon BL (2005) Dengue fever and dengue hemorrhagic fever: a review of the history, transmission, treatment, and prevention. Semin Pediatr Infect Dis 16:60–65

Lindsay SW, Emerson PM, Charlwood JD (2002) Reducing malaria by mosquito-proofing houses. Trends Parasitol 18:510–514

Mandahl-Barth G (1965) The species of the genus Bulinus, intermediate hosts of Schistosoma. Bull World Health Org 33:33–44

Marsollier L, Robert R, Aubry J, Saint Andre JP, Kouakou H, Legras P, Manceau A-L, Mahaza C, Carbonnelle B (2002) Aquatic insects as a vector for Mycobacterium ulcerans. Appl Environ Microbiol 68:4623–4628

Mccullough FS, Gayral P, Duncan J, Christie JD (1980) Molluscicides in schistosomiasis control. Bull World Health Organ 58:681–689

Mcmahon J (1952) Phoretic association between Simuliidae and crabs. Nature 169:1018

Mittal PK (2003) Biolarvicides in vector control: challenges and prospects. J Vector Borne Dis 40:20–32

Molyneux DH (2006) Control of human parasitic diseases: context and overview. Adv Parasitol 61:1–45

Morgan JA, Dejong RJ, Snyder SD, Mkoji GM, Loker ES (2001) Schistosoma mansoni and Biomphalaria: past history and future trends. Parasitology 123(Suppl):S211–S228

Mulla MS (1967) Larvicides and larvicidal formulations for the control of Culex pipiens fatigans. Bull World Health Organ 37:311–315

Muller R (1991) Dracunculus in Africa. In: Macpherson C, Craig PS (eds) Parasitic helminths and zoonoses in Africa. Unwin Hyman, London

Nyarango PM, Gebremeskel T, Mebrahtu G, Mufunda J, Abdulmumini U, Ogbamariam A, Kosia A, Gebremichael A, Gunawardena D, Ghebrat Y, Okbaldet Y (2006) A steep decline of malaria morbidity and mortality trends in Eritrea between 2000 and 2004: the effect of combination of control methods. Malar J 5:33

Ogoma SB, Kannady K, Sikulu M, Chaki PP, Govella NJ, Mukabana WR, Gerry F, Killeen GF (2009) Window screening, ceilings and closed eaves as sustainable ways to control malaria in Dar Es Salaam, Tanzania. Malar J 8:221

Otranto D, Dantas-Torres F, Brianti E, Traversa D, Petri D, Genchi C, Capelli G (2013) Vector-borne helminths of dogs and humans in Europe. Parasites Vectors 6:16

Padonou GG, Sezonlin M, Gbedjissi GL, Ayi I, Azondekon R, Djenontin A, Bio-Bangana S, Oussou O, Yadouleton A, Boakye D, Akogbeto M (2011) Biology of Anopheles gambiae and insecticide resistance: entomological study for a large scale of indoor residual spraying in south East Benin. J Parasitol Vector Biol 3:59–68

Pates H, Curtis C (2005) Mosquito behavior and vector control. Annu Rev Entomol 50:53–70

Perich MJ, Kardec A, Braga IA, Portal IF, Burge R, Zeichner BC, Brogdon WA, Wirtz RA (2003) *Field evaluation of a lethal ovitrap against dengue vectors in Brazil*. Med Vet Entomol 17:205–210

Peters W, Killick-Kendrick R (1987) The leishmaniases in biology and medicine; volume I: biology and epidemiology. Academic, London

Pichon G (2002) Limitation and facilitation in the vectors and other aspects of the dynamics of filarial transmission: the need for vector control against anopheles transmitted filariasis. Ann Trop Med Parasitol 96:S143–S152

Pierce WD (1974) The deadly triangle: a brief history of medical and sanitary entomology. Natural History Museum of Los Angeles County, Los Angeles

Pluess B, Tanser FC, Lengeler C, Sharp BL (2010) Indoor residual spraying for preventing malaria. Cochrane Database Syst Rev 14:CD006657

Portaels F, Elsen P, Guimaraes-Peres A, Fonteyne PA, Meyers WM (1999) Insects in the transmission of mycobacterium ulcerans infection. Lancet 353:986

Priyadarshini KA, Murugan K, Panneerselvam C, Ponarulselvam S, Hwang JS, Nicoletti M (2012) Biolarvicidal and pupicidal potential of silver nanoparticles synthesized using Euphorbia hirta against Anopheles stephensi Liston (Diptera: Culicidae). Parasitol Res 111:997–1006

Protopopoff N, Van Bortel W, Marcotty T, Van Herp M, Maes P, Baza D, D'alessandro U, Coosemans M (2008) Spatial targeted vector control is able to reduce malaria prevalence in the highlands of Burundi. Am J Trop Med Hyg 79:12–18

Radcliffe EB, Ragsdale DW (2002) Aphid-transmitted potato viruses: the importance of understanding vector biology. Am J Potato Res 79:353–386

Rafatjah HA (1982) Prospects and progress on IPM in world-wide malaria control. Mosq News 42:491–498

Ramsey JM, Schofield CJ (2003) Control of chagas disease vectors. Salud Publica Mex 45:123–128

Raoult D, Woodward T, Dumler J (2004) The history of epidemic typhus. Infect Dis Clin N Am 18:127–140

Rapado LN, Pinheiro AD, Lopes PODV, Fokoue HH, Scotti MT, Marques JV, Ohlweiler FP, Borrely SI, Pereira CA, Kato MJ, Nakano E, Yamaguchi LF (2013) Schistosomiasis control using piplartine against Biomphalaria glabrata at different developmental stages. PLoS Negl Trop Dis 7:e2251

Rascalou G, Pontier D, Menu F, Gourbière S (2012) Emergence and prevalence of human vector-borne diseases in sink vector populations. PLoS One 7:e36858

Ritchie SA, Rapley LP, Williams C, Johnson PH, Larkman M, Silcock RM, Long SA, Russell RC (2009) A lethal ovitrap-based mass trapping scheme for dengue control in Australia: I. Public acceptability and performance of lethal ovitraps. Med Vet Entomol 23:295–302

Ross R (1913) Malaria prevention in Greece. Br Med J 1:1186

Sadasivaiah S, Tozan Y, Breman J (2007) Dichlorodiphenyltrichloroethane (DDT) for indoor residual spraying in Africa: how can it be used for malaria control? Am J Trop Med Hyg 77:249–263

Scott D (1959) Human trypanosomiasis in northern Ghana, 1950–56: an epidemiological review. West African Med J 8:165–184

Service, MW (1976) Mosquito ecology – field sampling methods. Wiley, New York

Service, MW (1982) Importance of vector ecology in vector disease control in Africa. Bull Soc Vector Ecol 7:1–13

Shao JJ, Li SK (1985) A preliminary study of applying Landsat TM images to identify snail habitats in Dongting Lake regions. Human Med 2:15–16

Shawarby AA (1963) Training of vector control personnel. Bull World Health Organ 29:177–182

Smith SM (2004) High- and low-tech malaria control. Science 304:1744

Soper FL, Knipe FW, Casini G, Riehl LA, Rubino A (1947) Reduction of anopheles density effected by the pre-season spraying of building interiors with DDT in kerosene, at Castel Volturno, Italy, in 1944–1945, and in the Tiber Delta in 1945. Am J Trop Med Hyg 27:177–200

Stockholm Convention on Persistent Organic Pollutants (2001) Convention text [online]. United Nations Environment Programme, Geneva. http://www.pops.int/documents/convtext/convtext_en.pdf

Terenius O, Marinotti O, Sieglaff D, James AA (2008) Molecular genetic manipulation of vector mosquitoes. Cell Host Microbe 13:417–423

The MalEra Consultative Group On Vector Control (2011) A research agenda for malaria eradication: vector control. PLoS Med 8:e1000401

Thomas TG, Prakash V, Singh S, Mandal AK, Chauhan LS (2013) Insecticide susceptibility status of Culex quinquefasciatus say, the vector of bancroftian filariasis against temephos in Delhi and National Capital Region. Jpn J Infect Dis 66:238–240

Townson H, Nathan MB, Zaim M, Guillet P, Manga L, Bos R, Kindhauser M (2005) *Exploiting the potential of vector control for disease prevention.* Bull world health. Organ 83:942–947

Traoré S, Wilson MD, Sima A, Barro T, Diallo A, Aké A, Coulibaly S, Cheke RA, Meyer RRF, Mas J, Mccall PJ, Post RJ, Zouré H, Noma M, Yaméogo L, Sékétéli AV, Amazigo UV (2009) The elimination of the onchocerciasis vector from the Island of Bioko as a result of larviciding by the WHO African programme for onchocerciasis control. Acta Trop 111:211–218

Ughasi J, Bekhard H, Coulibaly M, Adabie-Gomez D, Gyapong J, Appawu M, Wilson M, Boakye D (2012) *Mansonia africana* and *Mansonia uniformis* are vectors in the transmission of *Wuchereria bancrofti* lymphatic filariasis in Ghana. Parasites Vectors 5:89

Uniting To Combat NTDs (2012) *The London declaration on neglected tropical diseases 2012.* http://unitingtocombatntds.org/downloads/press/ntd_event_london_declaration_ on_ntds.pdf. Accessed 28 Dec 2013

Uniting To Combat NTDs (2014) *Delivering on promises & driving progress.* http://unitingtocombatntds.org/report/delivering-promises-driving-progress-second-report-uniting-combat-ntds [Online]. Accessed 16 May 2014

Van Someren VD, Mcmahon J (1950) Phoretic association between Afronurus and Simulium species, and the discovery of the early stages of Simulium neavei on freshwater crabs. Nature 166:350–351

Varmus H, Klausner R, Zerhouni E, Acharya T, Daar AS, Singer PA (2003) Public health. Grand Challenges Global Health Sci 302:398–399

Walsh JF (1985) The feeding behaviour of Simulium larvae, and the development, testing and monitoring of the use of larvicides, with special reference to the control of Simulium damnosum Theobald s.l. (Diptera: Simuliidae): a review. Bull Entomol Res 75:549–594

WHO (2003) In: WHO (ed) *Space spray application of insecticides for vector and public health pest control: a practitioner's guide.* WHO, Geneva

WHO (2013) Lymphatic filariasis: a handbook for national elimination programmes. World Health Organization, Geneva

Wilson AL, Courtenay O, Kelly-Hope LA, Scott TW, Takken W, Torr SJ, Lindsay SW (2020) The importance of vector control for the control and elimination of vector-borne diseases. PLoS Negl Trop Dis 14(1):e0007831

Woodruff DS, Carpenter MP, Upatham ES, Viyanant V (1999) Molecular phylogeography of Oncomelania lindoensis (Gastropoda: Pomatiopsidae), the intermediate host of Schistosoma Japonicum in Sulawesi. J Molluscan Stud 65:21–31

World Health Organization (2008) WHO position statement on integrated vector management. Wkly Epidemiol Rec 83:177–181

World Health Organization (2009) Dengue guidelines for diagnosis, treatment, prevention and control. WHO, Geneva

World Health Organization (2011) World malaria report. World Health Organization, Geneva

World Health Organization (2019) Vector control advisory group. Overview of intervention classes and prototype/products under vector control advisory group (VCAG) review for assessment of public health value. World Health Organization, Geneva

World Health Organization (2020) Ending the neglect to attain the sustainable development goals: a road map for neglected tropical diseases 2021–2030. World Health Organization, Geneva. Licence: CC BY-NC-SA 3.0 IGO

Yelifari L, Frempong E, Olsen A (1997) The intermediate hosts of dracunculus medinensis in northern region, Ghana. Ann Trop Med Parasitol 91:403–409

Yirenya-Tawiah DR, Rashid AA, Futagbi G, Aboagye I, Dade M (2011) Prevalence of snail vectors of schistosomiasis in the Kpong head pond, Ghana. West Afr J Appl Ecol 18:39–45

Yoshikawa MJ (2010) Dengue and chikungunya virus infection in Southeast Asia: active governmental intervention in Republic of Singapore. In: Tanaka K, Niki Y, Akatsuki Y (eds) Current topics of infectious diseases in Japan and Asia. Springer, Tokyo

Yukich J, Lengeler C, Tediosi F, Brown N, Mulligan J-A, Chavasse D, Stevens W, Justino J, Conteh L, Maharaj R, Erskine M, Mueller D, Wiseman V, Ghebremeskel T, Zerom M, Goodman C, Mcguire D, Urrutia J, Sakho F, Hanson K, Sharp B (2008) Costs and consequences of large-scale vector control for malaria. Malar J 7:258

Zagaria N, Savioli L (2002) Elimination of lymphatic filariasis: a public health challenge. Ann Trop Med Parasitol 96(Suppl 2):S3–S13

Zahar AR (1984) Vector control operations in the African context. Bull World Health Organ 62:89–100

Zhang ZY, Xu DZ, Sun ZD, Zhang B, Zhou XN, Zhou Y, Gong ZL, Liu ZJ (2003) Application of satellite image for surveillance of vegetation landscapes of Oncomelenia-snail habitats in marshland using unsupervised classification. Chin J Epidemiol 24:261–264

The Role of Health Systems in the Control of Neglected Tropical Diseases in Sub-Saharan Africa

Justice Nonvignon, Leonard Baatiema, Frances B. da Costa Vroom, Ernest Mensah, and John O. Gyapong

Contents

Abstract

Neglected tropical diseases (NTDs) mostly affect populations with low socio-economic status and limited access to healthcare. The opportunity to combat NTDs is now available facilitated by increasing partnerships and availability of effective drugs. However, weak health systems (HS) in NTD endemic areas pose

J. Nonvignon (✉) · L. Baatiema · F. B. da Costa Vroom
School of Public Health, University of Ghana, Accra, Ghana
e-mail: jnonvignon@ug.edu.gh

E. Mensah
FHI 360, Accra, Ghana

J. O. Gyapong
Centre for Neglected Tropical Diseases Research, University of Health and Allied Sciences, Ho, Ghana

J. O. Gyapong, B. A. Boatin (eds.), *Neglected Tropical Diseases - Sub-Saharan Africa*, Neglected Tropical Diseases, https://doi.org/10.1007/978-3-031-53901-5_20

a challenge to delivery of interventions at adequate levels to at-risk populations. To effectively reach target populations with NTD treatment, weak HS challenges must be addressed by fixing specific weaknesses in the HS building blocks—service delivery, health workforce, information, medical products and technologies, financing, and leadership and governance. The building blocks, however, should be tackled with a system thinking approach to have the best impact. The role of community health workers, also referred to as community volunteers or community drug distributors (CDDs), in achieving adequate treatment coverage during mass drug administration (MDA) for NTDs cannot be overemphasised. This is because, apart from mobilising communities to support in NTD preventive, treatment, and elimination activities, they also serve to extend health service beyond the reach of the formal HS and also contribute to the low cost of drug delivery to target populations. Hence, demotivating factors which often reduce their commitment to these tasks must be addressed to sustain the gains made so far. CDDs have acceptability challenges in urban areas; hence use of low cadre community health workers in urban settings must be explored. Commitment by pharmaceuticals and donors to support NTD preventive and treatment efforts must be accompanied by increased HS investment by governments in SSA to improve weak HS. Current medicines for preventive chemotherapy in NTD control are effective microfilaricides. The development of effective macrofilaricides will not only shorten the number of MDAs required to eliminate lymphatic filariasis and onchocerciasis but also significantly reduce cost. Quality and timeliness of data in NTD control programmes are affected by the literacy levels of CDDs and the largely paper-based system used to collect the data. While the CDD literacy levels will be difficult to alter in the short term, the use of electronic data collection tools and systems can be explored to address the shortfalls listed. Control of NTDs required inputs such as potable water and basic sanitation facilities that often do not fall directly under the control of ministries of health. To effectively control NTDs means building good inter-sectorial coordination mechanisms that work beyond ad hoc meetings of inter-ministerial agencies whose outcome has no policy implications.

Introduction

Neglected tropical diseases (NTDs) continue to be a major public health threat despite decades of investment, campaigns, and policy attention at the global level. NTDs affect more than one billion of the world's poorest people; households with high burden of NTDs morbidity are of the lowest socio-economic status, with often very poor access to healthcare. Previously, the NTDs were comprised of a group of 17 diseases that affect mostly impoverished populations living in underdeveloped areas, which experience a lack of the most basic social amenities (Manderson et al. 2009; World Health Organization 2010a, b). This number has been updated by the WHO to include 20 conditions (World Health Organization 2020). In fact, this list has been updated to 24 NTDs for control and elimination in the 2021–2030 road map (World Health Organization 2020). Despite this little attention has been paid to these diseases over the years, however, since 2012 there has been significant progress in efforts at eliminating many NTDs with increases in treatment coverage.

Like other diseases that affect millions of people, the NTDs have seen an increase in effective and affordable interventions and technologies devoted to their control, yet health outcomes from these interventions continue to be off target. The reality is that the health systems that are supposed to support delivery of interventions for these diseases to detect and curb a reintroduction of transmission of NTDs in many of the countries are themselves often weak. Consequently, the impact from the interventions and technologies that are introduced to those in greatest need is not optimum, neither are they comprehensive nor provided on an adequate scale. More, therefore, still needs to be done. There are two reasons why it is critical not to ignore the state of health systems in the delivery of interventions that are aimed at controlling NTDs: first, interventions that target specific diseases or a cluster of diseases—however successful they are in achieving the desired outcome—often leave unexpected and unintended consequences (negative and positive) on the broader health system (Shakarishvili et al. 2010). Evidence abounds to this effect in countries such as Malawi, Ethiopia, and Benin (Stillman and Bennett 2005). Second, much as it is important for specific diseases to have developed their own—often vertical—structures to support the delivery of interventions, such structures need to strengthen health systems and make it easy to be adapted for the control of other existing and future diseases. For these reasons, the importance of "health systems thinking" in the development and implementation of interventions targeting NTDs cannot be underestimated (Gyapong et al. 2010).

As health systems are highly context specific, there is no single set of best practices that can be put forward as a model for improved performance. But health systems that function well have certain shared characteristics. They (a) have procurement and distribution systems that deliver interventions to those in need, (b) are staffed with sufficient health workers who have the right skills and motivation, and (c) operate with financing systems that are sustainable, inclusive, and fair. The costs of healthcare should not force impoverished households even deeper into poverty. In light of this, it will therefore be impossible to achieve national and international goals—including the Sustainable Development Goals (SDGs), the Universal Health Coverage (UHC), and the London Declaration on NTDs—without greater and more effective investments in health systems and services. While more resources are needed, government ministers are also looking for ways of doing more with existing resources. In fact, the World Health Organisation road map for NTDs emphasised the need to galvanise domestic funding to complement, donor, and crowdfunding activities to make any sustainable gains and resilience within these systems towards the elimination goal by 2030 (World Health Organization 2020). They are seeking innovative ways of harnessing and focusing the energies of communities, non-governmental organisations (NGOs), and the private sector. They recognise that there is no guarantee the poor will benefit from reforms unless they are carefully designed with them in mind. Furthermore, they acknowledge that only a limited success will be realised unless the efforts of other sectors are brought to bear on achieving better health outcomes. All these are health systems issues.

The objective of this chapter is to discuss the role of health systems in the control of NTDs in SSA. It is set out as a presentation and discussion of (i) the building blocks of health systems and the health systems priorities as they relate to the NTDs and (ii) "health systems thinking" as an approach to strengthening the health systems for effectiveness of NTD interventions.

Health Systems Framework

The World Health Organisation (WHO) describes a health system as one that "consists of all organizations, people and actions whose primary intent is to promote, restore or maintain health" (de Savigny and Adam 2009). The final goals of a health system are "improving health and health equity in ways that are responsive, financially fair, and make the best, or most efficient, use of available resources" (de Savigny and Adam 2009). But intermediate goals are to improve access, coverage, safety, and quality of essential health services and products. Health systems, especially in developing countries, are generally weak, thereby compromising efforts at improving interventions to the people in the greatest need. This weakness poses a challenge to efforts towards the attainment of the health-related SDGs targets of reducing child mortality, improving maternal health, and combating HIV/AIDS, tuberculosis, and malaria. So, does the system also pose a challenge to the control of NTDs?

In order for health systems to deliver the services to those in the greatest need, the WHO has developed a framework which has six health systems building blocks: service delivery, governance and leadership, financing, human resources, information, and medical products, vaccines, and technologies (de Savigny and Adam 2009). The interactions and relationships among the building blocks serve to strengthen the system to better provide essential services. At the core of these six components or building blocks is community engagement. For a health system to be strengthened and function well to service communities in a more equitable ways towards NTDs prevention, treatment, and eventual eradication, any intervention targeting any of the six building blocks must be done with the communities involved and according to their needs. Therefore, any intervention that seeks to strengthen the health systems leads to an interaction between two or more building blocks. While the building blocks provide a useful way of clarifying essential functions, the challenges facing countries rarely manifest themselves in this way. Rather, they require a more integrated response that recognises the interdependence of each part of the health system.

In view of this and with illustrated case studies from SSA, we centre our analysis and discussion of the role of health systems in NTDs control using the WHO six building blocks framework which highlights service delivery, governance and leadership, financing, human resources, information, and medical products, vaccines, and technologies.

Health Systems and Control of NTDs: Challenge

Indicators of health status in SSA are among the worst globally. While the region has not been successful in controlling infectious diseases that have been known to the region for decades, it continues to record high incidence of pandemics like HIV/AIDS, while non-communicable diseases are on a rapid rise (Stanaway et al. 2019). The poor health indices have been attributed to less than adequate health infrastructure and inadequate numbers of skilled health staff who are often not up to date with modern healthcare delivery knowledge and who are poorly motivated. Net out-migration of scarce skilled health workforce aggravates the situation. Access to healthcare is further constrained by the inequitable financial arrangements which do not provide security from catastrophic healthcare costs to the poor. The multiple reasons for the poor health outcomes in SSA are a pointer to the weak health systems (Kruk et al. 2018). Though healthcare outcomes are generally low in SSA, the situation varies greatly from urban to rural areas as health resources are skewed in favour of the urban communities with the distribution of health facilities and skilled health work force favouring the urban areas.

Service Delivery

Governments remain the largest provider of healthcare in SSA. However, publicly owned health facilities are supported by quasi-government facilities which are mostly owned by faith-based organisations that receive public support with respect to human resources/staff, equipment, drugs, and payment of remuneration of core clinical health staff among other things. The role of private orthodox health facilities varies across the region. Private orthodox health facilities are often small. Private hospitals, pharmacies, and maternity homes including privately owned small retail drug sellers and dispensaries are mostly located in urban centres.

A three-tier healthcare delivery system is identified in most of SSA—the local government (i.e. district/county) level, regional/provincial/state level, and the national/federal level. The public healthcare delivery system is made up of a network of health facilities located at the various levels of the decentralised system. At the bottom are the community/village health posts. Health centres are found at the sub-district/sub-county levels and are basically extensions of district hospitals that are located at the district/county levels as the primary tier. The second-tier facilities are relatively large secondary hospitals that operate at the regional/provincial level. Teaching/specialised hospitals at the national level constitute the third tier (Macha et al. 2012; Mills et al. 2012). For each of these levels, health facilities at the next higher level acts as referral point for those below.

The above health system structure forms the context and state of the healthcare infrastructure used to deliver health interventions to populations in most developing countries. It is clear that the largely rural populations are constrained in access to healthcare, and, though efforts have been made in many countries to develop and implement close-to-client approaches to the delivery of healthcare services, these have been mainly on pilot basis. Thus, the age-long problem of geographical inaccessibility to healthcare services by most of the population, especially in rural communities, still exists. It is in this context that NTD control programmes operate.

Current management of NTDs include four main strategies—preventive chemotherapy (PCT), particularly mass drug administration (MDA), case management, vector control, and safe water and improved sanitation and personal hygiene (World Health Organization 2010a, b). Case management following diagnosis and PCT are the two main interventions. In the first case, an NTD is diagnosed following routine clinical methods followed by use of appropriate drugs or surgical intervention. This is seen in leprosy, yaws, leishmaniasis, Guinea worm, etc. Some NTDs such as lymphatic filariasis, schistosomiasis, onchocerciasis, soil-transmitted helminthiasis (STH), and trachoma are controlled through PCT by MDA. Preventive chemotherapy is a secondary prevention strategy where antihelminthic drugs are used as a public health tool to prevent morbidity and transmission in infected populations who may be asymptomatic (World Health Organization 2006). Populations with high prevalence of the infection are treated as a unit to reduce the prevalence of the infective agent in the population. The strategies require strong and wide coverage of health systems to make interventions available to wide sections of the population. Weak health systems in SSA means that formal health system structures are inadequate to reach large sections of the population with the needed interventions (World Health Organization 2007a). This is often worst with inadequately equipped facilities in the lower pyramids of the health systems in the NTD-endemic region in SSA where service delivery is far from optimal.

To optimise NTD control in SSA, it is imperative to invest in existing service delivery facilities, most especially at the lower pyramids of the health systems to ensure they are robust enough to provide the following: diagnostic services, treatment, with trained lower cadre staff such as the CDD, sufficient and well-equipped laboratory space, equipment, and/or inpatient wards for treatment purposes. The call to strengthen service delivery at the lower levels (PHCs) is critical because the burden of NTDs is highest at the lowest healthcare levels where majority of the NTD population and those at risk reside. For example, given the deficient capacities and less resilient nature of health facilities at the lower levels, the advent of COVID-19 disrupted and, in some cases, led to the suspension of MDA and other community-based services resulting in delays in diagnosis, treatment, morbidity management, and disability prevention in NTD-endemic regions (World Health Organization 2020).

Health Financing

In the quest to ensure that populations are not impoverished by financial constraints to healthcare, universal health coverage (UHC) has gained prominence in the past 3–5 years. The goal of UHC is to ensure that all people have access to essential healthcare services without suffering any form of hardship in their attempt to access these services. It has been suggested that the failure of countries and national governments to control and eliminate NTDs will remain a major threat to the achievement of the UHC and SDG goals. However, access to funding to support health programmes in SSA and other underserved settings of the world is a major concern. Globally, funding for NTDs is abysmally low, as low as less than 1% of development assistance for health relative to about 6% for infectious diseases (Global Burden of Disease Health Financing Collaborator Network 2020). Most governments commit an insignificant budgetary allocation to health, and this affects NTD interventions. For example, in 2001—a year after the declaration of the MDGs—leaders of SSA countries meeting in Abuja, Nigeria, committed to set aside at least 15% of their annual national budgets for the health sector. During the same meeting, the leaders called on donor partners to ensure that steps are taken to fulfil their pledge of committing 0.7% of their gross national products (GNP) to developing countries. However, in the 12 years since the pledge, seven African countries—Botswana, Burkina Faso, Malawi, Niger, Rwanda, South Africa, and Zambia—had been reported to have achieved the 15% target of the Abuja Declaration (World Health Organization 2011a, b). On a per capita basis, SSA countries spend an average of \$25–\$27 on health (purchasing power parity). Furthermore, in real terms, official development assistance to developing countries reduced between 2001 and 2009 (World Health Organization 2011a, b).

Within the context of SSA, 26 African country ministers signed the Addis Ababa Neglected Tropical Diseases Commitment in 2014 and pledged to invest more domestic resources to control NTDs. This notwithstanding, investment in NTD control efforts is still low and discouraging as evidence to date shows national NTD-related interventions from these countries manifest little gains with little or no additional investments to controlling NTDs, and out-of-pocket payments by affected individuals and communities are still telling (Cavalli et al. 2010; Fitzpatrick et al. 2017). To date, preventive chemotherapy through MDA is one of the proven interventions towards the control and eradication of NTDs. Despite major gains since the introduction of this intervention, uptake is still low, invariably resulting in low coverage levels in NTD-endemic regions owing to insufficient funding and investment by governments. This in effect has far-reaching implications on the availability and accessibility of medicines, thus undermining efforts to ensure equitable achievement of NTD elimination goals in SSA (Dean et al. 2019). Funding and investment efforts need to be boosted to ensure the UHC goals are achieved in these settings.

Funding of NTD control activities has been shared by international health partners and donors on the one hand and governments in SSA countries endemic for the NTDs on the other hand. However, partners and donors have received most credit. Partners and donors include large pharmaceutical companies such as Merck & Co. (donors of ivermectin for the control of onchocerciasis and lymphatic filariasis), GlaxoSmithKline (donating albendazole for the control of lymphatic filariasis (LF) and STH), and Novartis (donors of rifampicin, dapsone, and clofazimine for the treatment of leprosy). Drug donation schemes often include transportation of the drugs to user countries. Other partners like the USAID, DFID, Bill and Melinda Gates Foundation, and the WHO provide resources for planning, training, logistics management in country, data collection, monitoring and evaluation, and operational research. Governments' contribution to NTD control in SSA often involves infrastructure which includes the provision of office space, equipment, vehicles, human resources, and salaries for health staff. The large pool of drug distribution volunteers can be considered as local in-kind contribution to NTD control (Bockarie et al. 2013; Olds 2013).

Government financing of NTDs as described above is not expected to change significantly in the absence of deliberate change of policy direction, since they emanate from general health system units rather than new expenditures. Continued drug donations appear to have the firm assurances of the pharmaceutical companies involved until 2020 and beyond (Anderson et al. 2013). However, funding for operational activities of MDA, monitoring, surveillance, and operational research are subject to donor economic challenges. It is not uncommon to see a disruption in drug distribution when the expiration of one funding agent's commitment and the next donor funding contract is not synchronised properly leaving a financing gap which governments cannot absorb because this was not anticipated or budgeted for. In 2012, only about a third of targeted STH and schistosomiasis populations were treated due to reduced funding resulting from the economic recession in the west (Hooper et al. 2013; Olds 2013).

Nevertheless, with a change in the income status of some SSA countries—i.e. from low-income country (LIC) to lower-middle-income country (LMIC)—these countries are expected to create and tap into additional fiscal space for financing of general health services but also NTD activities. In recent years, SSA countries such as Ghana and Kenya, among others, moved to LMICs. Much as such developments are welcome, indicating that the economies of these countries are expanding, there are implications for the funding of general services, including the provision of health services. The attainment of a higher income status means that such countries become ineligible to receive funding support from key international health partners (Moss and Majerowicz 2012). This reduced funding directly affects the delivery of services for the control of NTDs which typically depend on donor support at the moment. Thus, it is important for such countries to mobilise resources from local sources to support healthcare delivery. Given the strain on public sources of financing social services—such as tax and through social health insurance—it is important for developing countries to explore new and alternative ways to fund the control of NTD, such as through public-private partnerships. One way of doing this could be for the public sector to invest in research that estimates the economic gains that private entities would derive from investing in such activities—including reduced employee absenteeism due to ill health but specifically due to such NTDs.

Importantly, the recent WHO road map for the eradication of NTDs prioritised domestic funding activities as part of measures to leapfrog efforts towards the NTD elimination goals. Against this background, and the debilitating impact of COVID-19 in most donor countries, it stands to reason that domestic funding for NTDs should increase to ensure countries are on track towards the 2030 elimination target. Countries in SSA need to earmark or shore up their funding allocations for NTDs if this is still not happening, in order to scale up mass drug administration interventions. Commitment to domestic investment in strengthening the national health system should be encouraged to ensure sustainable gains are made. As noted by researchers, even if external funding supports are available or sought, it is still important for local investments to ensure effective morbidity management, surveillance, and monitoring and evaluation to consolidate gains (Huang et al. 2021).

Further investment in research and development in NTDs to generate local and contextually relevant data is critical to devise a reliable, sustainable, robust solution for combating NTDs in SSA. Funding focus should be geared towards health systems strengthening efforts and away from the long-standing single disease-focused funding activities. Such an approach will promote a health system strengthening agenda and inspire integration among the different NTDs within the health system, from an in-country, home-based, and sustainable country-level support strategies. The private sector can also play a pivotal role in raising funds to support NTDs in the areas of therapeutics, diagnostics, and preventive measures initiated by governments. As part of efforts to strengthen the health systems, major investments in developing health system resource tracking platform are needed in order to ensure adequate and consistent allocations are made available for NTD eradication at national, continental, and global levels (Huang et al. 2021).

Medical Products, Logistics, and Technological Challenges

There are four main strategies for the control of NTDs involving interruption of transmission through the PCT, management of cases, improvement in access to improved sanitation and potable water, and health education for behaviour change. The PCT strategy is perhaps the biggest arsenal in the fight to combat NTDs. Not only does it prevent suffering from the disease but also protects sufferers from the debility and disability associated with NTDs which is the cause of the cycle of poverty. PCT hinges on the availability of potent, effective, safe, affordable, and accessible drugs. There are major challenges to ensuring that the right drugs are available in the right quantities to those who need them most. First is the affordability question considering the cost of research and development of drugs which in the case of NTDs cannot be competitively priced to recoup the investment (Oprea et al. 2009; Pollastri 2014; Stefanakis et al. 2012). This makes research and development of NTD drugs unattractive to the global pharmaceutical giants with the expertise and technology to produce the drugs. Second, even where the drugs are available, a complex and expensive supply chain network needs to be put in place to get the drugs to often remote communities at most risk of NTDs (NTDs UtC 2015). The cost is often out of the reach of developing countries that have weak health systems and invest very little in health infrastructure. While current drugs for PCT including ivermectin has proven effective against microfilaria of both *Onchocerca* and *Wuchereria*, it has little effect on the adult worm which in the case of *Onchocerca* can live and reproduce for up to 15 years in infected persons. Treatment for onchocerciasis for at least 15 years is therefore required in endemic populations to ensure effective control. The long-term treatment presents real challenges of compliance, data management, and cost. Effective and easy-to-use macrofilaricidal drugs will not only shorten the duration of PCT but will also contribute to a reduced cost of the control effort. The potential for killing the adult worm that causes LF exists. The bacteria *Wolbachia* are known to be in a symbiotic existence with the adult *Wuchereria* necessary for growth and survival of the worm. Various doses of the antibiotic doxycycline have been found to either sterilise the adult worm through inhibiting further production of microfilaria or kill it (Hoerauf et al. 2008; Makepeace et al. 2006). However, the challenge has been that at least 4 weeks of daily treatment with doxycycline is required to be effective (Bockarie et al. 2009).

In the 2021–2030 road map for NTDs to achieve the SDGs, the WHO and partners recently declared that by 2030, there should be a 90% reduction in the number of people in need of treatment against NTDs, 75% reduction in DALYs related to NTDs, 100 countries having eliminated at least 1 NTD, and at least 2 NTDs eradicated in the world. (World Health Organization 2021). To achieve these targets in SSA, finding workable solutions to existing challenges militating against NTDs control and elimination will be crucial. Encouraging efforts are ongoing to address these challenges to availability of effective and affordable drugs for the control of NTDs. The Drug for Neglected Tropical Diseases Initiative which now has about ten pharmaceutical companies has promised to make existing NTD drugs available to endemic countries at no cost through their collaboration with the WHO and international non-profit development partners (NTDs UtC 2015). In fact, global NTD drug donations have been increased significantly following pledges by the group as part of the London Declaration. In some cases drug donation has increased by more than 80% in the past 3 years (NTDs UtC 2015). This increase has occurred for both PCT and Integrated Disease Management (IDM) diseases. To address the challenge of making these donated drugs available to the affected populations, the NTD Supply Chain Forum constituted by the donating pharmaceutical companies, other global donor foundations, and logistics companies such as DHL is ensuring that the cost along the supply chain of donated NTD drugs is not the burden of affected countries.

The collaboration outlined above working around donation and distribution of currently available NTD drugs appears to be well taken care of at least to 2020 and perhaps up to 2030. However, it is no news that new more effective drugs especially microfilariae are needed to ensure that the war against NTDs can be won in good time if at all (Slingsby and Kurokawa 2013; Stefanakis et al. 2012). Developing new drugs takes up to a decade to have a promising molecule and perhaps another decade to bring it to public use through the various phases of trial coupled with the stringent licencing regimes (Olliaro et al. 2012). For drugs which are not likely to pay for the research and development invested, there is very little motivation for giant global pharmaceutical companies to invest resources in these drugs leaving investment in this area to non-profit organisations and academic institutions who may not have the technology and personnel to succeed in good time (Pollastri 2014).

The success of NTD control efforts depends on making effective, safe drugs widely available to at-risk populations and infected persons at affordable costs. The donation and long-term pledge of Merck & Co. and GlaxoSmithKline to provide ivermectin and albendazole, respectively, for the treatment of onchocerciasis, LF, and STH answer the question of drug availability. Merck is working in partnership with the WHO to donate praziquantel for control of schistosomiasis. However, this has not been adequate to meet needs of endemic countries. Merck in 2012 proposed to expand its commitment by increasing production levels and also making the drug available until elimination. Meanwhile some NTD partners are supporting schistosomiasis-endemic countries to purchase generic praziquantel to address the short fall (Bockarie et al. 2013; World Health Organization 2010a, b, 2012).

The above contexts vividly point to a situation of lack of investment in adequately supplying and equipping health facilities in SSA and other resource-poor regions with modern and evidence-based medical products and consumables towards the control and elimination of NTDs. Access to chemotherapy, diagnostics, and other medical logistics to promote NTD control and interrupt transmission of NTDs and improved diagnostics for NTDs elimination and post-elimination surveillance are unfortunately lacking. Thus, efforts to strengthen the health systems should have these on their priority list. This is consistent with an earlier proposal by Solomon et al. (2012) where he noted that country programmes for control and elimination of NTDs need "practical diagnostic tools of sufficient sensitivity and specificity to detect levels of infection that can lead to transmission" (Solomon et al. 2012).

With the current drive towards telehealth or digital health, intentional efforts should be made to invest more in the acquisition of modern point-of-care (POC) devices that can digitise data and transmit to the appropriate units to inform policy formulation and decision-making. According to the WHO, investment to integrate digital technology into diagnostics can accelerate the activities of supply chain management which can go a long way to strengthen the healthcare system. They noted that linking data from diagnostic laboratories and point-of-care (POC) test devices potentially facilitates health systems efficiency and overall improvement in patient outcomes (World Health Organization 2016).

Health Workforce

The WHO's first report on NTDS stated that "health workforce should be able to perform responsively, fairly and efficiently to achieve the best health outcomes possible, given the available resources and circumstances (e.g. there should be enough trained and competent staff evenly distributed to meet needs)" (World Health Organization 2010a, b). This suggests that health workers engaged to implement NTD activities in their locales should be adequately trained to handle the responsibility. The workforce for NTD programmes involves health workers such as nurses, disease control officers, health information officers, and community health nurses at the various health administration levels in various countries. These are salaried workers who have their core duties outlined. The other arm of the NTD workforce consists of schoolteachers for school-based MDAs and community drug distributors (CDDs) who volunteer to help their communities. Formal health sector workers act as focal persons and supervise programme activities within their locality (Amazigo et al. 2012).

Programmes in SSA have relied heavily on the volunteer CDDs to administer the drugs needed for elimination of NTDs (Amazigo et al. 2021). Their efforts have ensured the availability of treatment for communities in remote places, and, therefore, such efforts cannot be discounted. A large number of these CDDs have been with the programmes from the onset of the drug distribution in their communities. They are usually chosen or nominated by elders and other community members with the main criteria for selection being their level of literacy—which is sometimes translated as their ability to spell their own names—and trustworthiness. Apart from drug distribution, CDDs are also tasked to educate and conduct social mobilisation activities to generate interest and a desire for community members to receive the drugs.

Incentives for CDDs are an important component of NTD programmes. These can be categorised as monetary and non-monetary incentives (Baatiema et al. 2016). Like other community health volunteers, CDDs are paid some minimal allowances only when they are involved in mass campaigns and are often not remunerated for other services rendered. While CDDs understand they are volunteers, their expectation of remuneration for the services they render is quite high. The low remuneration has affected CDD enthusiasm for the programme over the years. This has led to some volunteers abandoning the programme when opportunities of earning money from other activities occur. There have been few instances of volunteers receiving drugs and registers with the intention of distributing but not making the effort to. Other incentives CDDs ask for include free medical care, first-in-line treatment at health centres, bicycles, raincoats, wellington boots, and bags to carry their medicines and registers. Compared to other public health programmes using community-directed approaches, NTD MDAs are considered the most tedious and least rewarding. Expansion of communities, increase in population, and difficult terrain all account for the apathy. There have been suggestions of paying CDDs wages, but this has been considered unsustainable and difficult to manage by donors and ministries of health for endemic countries (Bhattacharyya et al. 2001; Miller et al. 2014). While monetary incentives may be difficult to sustain, non-monetary incentives may help increase CDD enthusiasm. Various countries may have to assess their local conditions and determine appropriate incentives for their volunteers as retention of CDDs is important for programme sustainability.

The issues discussed previously contribute to high attrition rate among volunteers. Bhattacharyya et al. (2001) and Baatiema et al. (2016) indicated in their papers that attrition among volunteer community health workers is higher than that of paid community health volunteers. Emukah et al. (2008) also reported a 10.9% attrition rate among volunteers in an onchocerciasis programme in two states in Nigeria. In this study, 66% of the volunteers attributed their decision to quit to the lack of incentives by the programme. Some other reasons for attrition could also be that the substantive volunteer had travelled, others could not be reached by phone, and other CDDs had chosen other persons in their communities to replace them. These replacements were usually their children or other younger people they had worked with over the years. Others may be dropped due to poor performance (Miller et al. 2014). There appear to be fewer younger persons volunteering for MDAs. This could be due to the inadequate or lack of incentives. Several current CDDs are aging but are holding on because they are unable to find replacements in their communities. This issue will have to be addressed seriously by the health systems of SSA to ensure continued progress and attainment of the global NTD elimination targets.

Formal sector health workers play supervisory roles in ensuring the smooth implementation of the NTD programmes. A cascading structure of supervision and reporting is used in most countries from national, regional or provincial, district, and finally to community level (Hodges et al. 2011). The health workers at community health posts then train and supervise the CDDs (Dembélé et al. 2012). A number of these workers stay in their communities for years, and this is beneficial to the programme. They gain experience in NTD programme management and develop stronger bonds with the CDDS and community members. However, transfer of the health workers drains the programme of valuable experience. New supervisors may not have adequate knowledge and good grasp of the epidemiology of the diseases and the management of adverse reactions. They may also appear disconnected from the programme because NTDs are mostly managed as vertical programmes within the health service and are not part of routine health activities in the communities. It becomes an additional task "outside" of their core duties as public health nurses or disease control officers. There is also a lack of enthusiasm on the part of supervisors, who are meant to motivate the volunteers to give off their best. Since the community health officers interact with the CDDs directly, they bear the brunt of their frustration with the programme. There should be a level of integration with the health systems so NTD MDAs or activities are considered part of their core duties. Due to the vertical nature of the programme, district health management teams do not commit any of their resources to the management of the programme. Supervisory visits are sometimes made based on how much fuel allocation has been made by the programme. However, in most cases provision is just enough to collect registers from volunteers when the drug distribution is completed. Although volunteers have effectively delivered interventions in rural areas in developing countries, use of volunteers in urban areas faces challenges like willingness of people to serve as such in these areas and acceptance of volunteers by residents in urban setting (Hodges et al. 2010).

To address the problem of access to the health system, PCT is delivered in packages that do not require skilled staff at the point of delivery. Community drug distributors (CDDs) have been used to extend MDA to remote communities far removed from the formal healthcare infrastructure. Use of volunteers in MDAs has been documented to significantly reduce the cost of delivering antihelminths to at-risk populations (Conteh et al. 2010; Goldman et al. 2007). The CDDs are community-based volunteers who, with minimal orientation, distribute drugs under the supervision of peripheral health workers. While this strategy seems to have worked to some extent, its effect is less than optimum in terms of achieving and sustaining the high treatment coverage needed to achieve control objectives especially in urban populations. Factors affecting low treatment coverage are not limited to health system challenges (Dabo et al. 2013). Volunteer fatigue has been cited as a threat to achieving the goals of MDAs. Volunteers consider the opportunity cost of participation in MDAs uncompensated for by the low token rewards. As a result volunteers do a partial job leaving some eligible persons untreated or opt out of MDA leaving some communities unreached with MDA drugs (Parker and Allen 2011). Community volunteers, however, have been found to have specific motivators, key among them being cash rewards, community respect and recognition, and future hope of employment, among others (Takasugi and Lee 2012; Yasemin Dil et al. 2012). These challenges notwithstanding, the use of community-based health volunteers in delivering needed healthcare services to populations in remote communities has proven to be both effective and cost-effective in many developing countries (Lubell et al. 2010; Chanda et al. 2011; Nonvignon et al. 2012). Thus, the input of volunteers in NTD control will continue to be relevant, but health systems must be strengthened to provide effective supervision and motivation of volunteers while keeping communities continuously interested in contributing to their own health through active participation (Reich et al. 2008).

Considering the indispensable role of CDD and the different cadre of health workers in efforts towards NTDs control and elimination, there is the need for massive investment in training and retaining CDDs and other cadre of health workers, including NTD specialists. Staff need to be equipped with up-to-date knowledge on how to use modern POC diagnostics and apply the test results to effectively manage the NTD patients, especially those needing specialist care. Systems should be put in place from the national to the local health system levels to ensure CDDs are sufficiently motivated and capacity and plans built to eventually mainstream them into the general health systems well-articulated to prevent the high attrition rates.

Governance

The governance structure of NTD control programmes in many developing countries follows a similar trend as follows: at the top—central level—of the structure is a dedicated programme with a director/manager and staff. At the regional and decentralised levels, the established health system structures are used to deliver interventions. However, like general health service delivery, NTD control programmes fall within the health system, whose governance is affected by the actions of outside forces. As the Lancet Commission on Global Health Governance points out, inequities in health service delivery sometimes fall on non-state actors, and until the issue of health governance is given a regional perspective, it will be almost impossible to depend on internal and state actors alone to fix the numerous governance challenges facing healthcare delivery in developing countries; global politics and power play a key role in determining the governance structures in many countries (Ottersen et al. 2014).

Furthermore, the complexities of the health systems in many developing countries create challenges for the delivery of healthcare services. Fortunately, NTD control has not been left to country health system actors. The control of NTDs is a global endeavour with critical participation at the highest level of the global health governance—the WHO (Hotez and Pecoul 2010). The WHO has shown its leadership through the setting up of the Department of Control of Neglected Tropical Diseases which "coordinates and supports policies and strategies to enhance global access to interventions for the prevention, control, elimination and eradication of neglected tropical diseases, including some zoonotic diseases" (World Health Organization 2015a, b) and produces several policy documents to guide member countries in their effort to control NTDs while setting up implementation goals and monitoring the same (World Health Organization 2007b, 2010a, b, 2011a, b, 2012). At the international level NTD control efforts also enjoy considerable partnership with international NGDO, state development support agencies, and international collaborations including Uniting to Combat NTDs, Bill and Melinda Gates Foundation, the USAID, and DFID.

While the international community feels responsible to support the control of NTDs, it must be stated that the greatest responsibility for the control of NTDs may rest with affected country governments who must demonstrate political commitment to the control of NTDs to make the support of global partners towards NTD control have the greatest impact (Narain et al. 2010). This is important for several reasons. First is the human right dimension of health which is a fundamental responsibility of national governments (World Health Organization 1978). While nearly all governments have signed the UN convention on the right to health, few have taken steps to put in place laws to guarantee this right to their citizens (Heymann et al. 2013). Second, NTDs have gained prominence due to their impact on the socio-economic fortunes of affected populations particularly so when its greatest burden is on the poorest countries. The return on investment case for NTDs is considered very high making it imperative for governments of developing countries to pay attention to their control as a complement to poverty reduction strategies (Allen and Parker 2012; Chu et al. 2010). The benefit of NTD control is a healthy population who can contribute meaningfully to the economic development of their countries.

At the country level NTD control must leverage the advantage of the health system to multiply the impact of interventions and also for sustainability. There is the need to integrate NTDs with similar delivery strategies on the same platform to make efficient use of resources (Gyapong et al. 2010). The PCT NTDs could be managed as an integrated NTD programme. Some aspects of the IDM NTDs such as lymphoedema management in LF and management of ulcers in leprosy and Buruli ulcer can be integrated. In sub-Saharan Africa, NTDs co-exist with other conditions such as malaria, TB, and HIV/AIDS. Clear points of convergence exist either in the area of common vectors such as the *Anopheles* mosquito transmitting malaria and LF or genital schistosomiasis increasing the risk of HIV in affected women (Molyneux et al. 2009). There must be some collaboration between these programmes in the same country working in co-endemic communities to enhance their impact. Yet another collaboration that holds great potential for NTD control is that with the water, sanitation, and hygiene (WASH) programmes and ministries. Schistosomiasis, soil-transmitted helminths, trachoma, and Guinea worm are sustained through insanitary conditions and inadequate access to potable water. Sectors such as water and sanitation, though outside the health sector, play a key role in control of NTDs. Inter-sectoral approach involving these sectors to control NTDs is necessary to achieve desired results. The challenge has been how to successfully and effectively bring the health and other sectors to work together for the common good. The NTD benefits of NTD control can be mainstreamed into WASH programming to accelerate achievement of NTD targets. Currently most WASH activities use only the human rights argument for the improvement of WASH indicators (Freeman et al. 2013). Similarly, government can align policies of ministries in water and sanitation to NTD programming in endemic communities to accelerate achievement of NTD control.

Dynamic stewardship and leadership at the country level can sometimes be a challenge which requires attention. This is important because the adoption and uptake of new diagnostic technologies are sometimes slow due to leadership ineptitude or dysfunctional regulatory frameworks which often delay the introduction of new or novel diagnostic technologies or delays arising from bureaucratic red tapes in the regulatory and approval procedures of such diagnostic and technological devices. Multisectoral stakeholder engagement of policymakers and partners from both private and state actors is needed to ensure effective and timely uptake and country buy-in. Strengthening the governance and leadership systems at the national to the decentralised health units is critical to ensure the requisite infrastructure and resources such as health services, basic water supply, adequate sanitary infrastructure, and improved housing are all in place. As demonstrated in Tanzania, NTD programme governance strategies require considerable power and autonomy at the decentralised district-level agencies to ensure proper and sustainable coordination of different programmes and resources to improve governance and accountability of NTD control (Madon et al. 2018). As previously noted, NTD control requires strong leadership of the Ministry of Health to ensure better coordination (Gyapong et al. 2010), and this is still relevant and required in strengthening health systems for NTDs control.

Health Information

Successful health programmes depend on good-quality information for decision-making and planning. Health information from NTD programmes is generated at the community level by field staff who could be community health volunteers or salaried health workers and researchers. They usually generate and report disease surveillance data, treatment coverages, morbidity, and mortality data. Treatment coverage data is generated when CDDs register and record drug dosages for communities eligible for MDAs. Active case identification data, such as the recent drive to certify Ghana as being Guinea worm-free, is also carried out especially in places where specific diseases are suspected or earmarked for elimination or eradication (World Health Organization 2015a, b). Once data collection is completed, staff at the community or district are expected to summarise the data collected for onward submission to the programme. Data is reported in the same way cascading training is conducted. Each administrative level of the health system is expected to check the integrity of the data they receive to ensure good-quality data is being submitted. The information generated should be timely, accurate, and complete.

One of the biggest challenges with the NTD health information is untimely reporting. The source of information for NTD programmes makes it prone to inaccuracies and often untimely submission. This tends to affect the quality of data generated. Dembélé et al. (2012) have attested to the fact that data collection for NTD programmes can be difficult. Often, data is not reported in a timely manner, therefore affecting planning for the next years' annual programme. Untimely submission of data was also identified in a study by da-Costa Vroom et al. (2015) in two districts in Ghana. Some reasons identified for this challenge were poor numeracy skills of CDDs, poor road networks to some communities, and inadequate resources for transportation to some communities. This study also suggested that the slow reporting process starts at the sub-district level since CDDs usually kept to their deadlines for drug distribution.

CDDs generate a bulk of ongoing data collection in NTD programmes. Their literacy levels also influence the quality of data being reported by the programmes. A basic requirement for CDDs is their ability to read and write (da-Costa Vroom et al. 2015; Njomo et al. 2012). Observation of MDA registers has shown discrepancies and incomplete recording of data. A few surveys conducted to assess drug coverages of NTD data found some variance between reported treatment coverage and actual treatment coverage (Parker and Allen 2013; Worrell and Mathieu 2012). Such overestimation or underestimation can have an adverse effect on planning for logistics and drugs for the next years' MDA campaign.

Attempts have been made to introduce electronic data capture and management tools for NTD programmes. One such study by Stanton et al. (2015) tested SMS reporting tool for assessing LF morbidity burden in selected communities in Malawi and Ghana. This allowed health surveillance assistants (Malawi) and community health volunteers (Ghana) report cases of lymphoedemas and hydrocele for physician follow-up. This was a simple cost-effective tool that can easily help with disease surveillance efforts in any country. LINKS is another tool for electronic data capture for global health programmes (Pavluck et al. 2014). It has been used by the Global Trachoma Mapping Project (GTMP) to map the prevalence of trachoma in suspected endemic countries. SSA countries can employ such tools to help in disease surveillance efforts. When data from field sites reach programmes rapidly, it ensures decisions, and relevant implementation plans can be executed. More of such electronic tools can be deployed by programmes to ensure data is captured correctly and delivered in a timely fashion. Mobile phone technology has become ubiquitous and can therefore enhance community engagement through simplifying the exchange of information between staff at the community level and health programmes (Stanton et al. 2015).

National health information systems (HIS) are being touted as the best way of managing health information for countries. Data collection for health facilities and programmes should be integrated into the national HIS, for the purpose of producing timely quality information for evidence-based decisions and interventions (Bergquist and Whittaker 2012). NTD programmes are managed as vertical programmes in most health systems so data generated is donor driven rather than country needs driven. While several countries are aiming to create and use national health information systems, NTD data is left out of the HIS database. This does not allow a complete picture of the state of health in any country to be seen. Countries such as Ghana, Burkina Faso, Sierra Leone, Uganda, and Kenya have implemented the open source District Health Information System version 2 (DHIS2) to collect, collate, and manage national health data (HISP 2015). Other countries like Burundi and the Democratic Republic of Congo (DRC) have rolled it out for programme-specific purposes such as malaria and maternal health. However, none of these countries seem to be integrating NTD data with the routine national health information systems. If this integration occurs, NTDs could receive more attention and potentially benefit from national funding and not just donor funding.

Therefore, health systems strengthening efforts should incorporate interventions to address the data challenges, as without good data it will be difficult to monitor progress made or make any evidence-based decisions towards NTD elimination.

Integration of Neglected Tropical Diseases into Routine Health Services

In the past decade, there has been some effort in providing healthcare to the most disadvantaged communities through close-to-client services in an attempt to strengthen the concept of primary healthcare. Examples of such services include home-based care for malaria and fevers in children under five years in 18 sub-Saharan African countries (Nonvignon et al. 2012). Studies have shown that using community health volunteers to deliver needed services to communities is effective (Kidane and Morrow 2000; Sirima et al. 2003; Chinbuah et al. 2012) and cost-effective (Lubell et al. 2010; Chanda et al. 2011; Nonvignon et al. 2012). Scaling up coverage of these services will be crucial for integrating NTD into the general health service. Integration of healthcare services could occur vertically, horizontally, or diagonally.

The control of many diseases in developing countries has been implemented through vertical integration. Vertical integration involves the planning, direction, supervision, and implementation of health services, wholly or to a great extent, by specialised service units through workforce dedicated for that specific programme. Vertical programmes have the advantage of ensuring that people focus on the specific health programme being worked on, therefore channelling all energies and resources to that programme. However, a major challenge of this type of programme implementation is that parallel structures are created for different diseases or health programmes, each tackling a specific disease or programme. In the process, there is no coordination among programmes, and this undermines the health system, making it weak and unresponsive to any major threat that may affect the entire population. Furthermore, vertical programmes lead to inefficiency in the use of scarce health resources, providing incentives for waste of such needed resources.

Horizontal integration, on the other hand, refers to the use of existing structures or permanently created structures that tend to serve all health programmes (Gyapong et al. 2010). The use of horizontal integration minimises (if not eliminates) the problems of lack of coordination through parallel structures and inefficiency in the use of health resources. However, the extent of achieving desired goals in specific health programmes may be sacrificed as the workforce and existing structures are faced with a broad range of issues to tackle at any given time. Vertical programmes are regarded as attractive to many donors and the international community as they tend to show quick results and are easier to manage than horizontal programmes. However, for national governments and policymakers, vertical programmes are seen as diverting the critical and often limited human and financial resources, thereby being detrimental to overall health systems strengthening and responsiveness to emergencies (Schreuder and Kostermans 2001).

Thus, effective combination and coordination of both vertical and horizontal programmes (sometimes referred to as the diagonal approach) could ensure better outcomes for NTD control in developing countries. Diagonal integration means that programmes keep their specific focus but at the same time implement activities that strengthen the whole system, thus carrying the system along with it. This ensures that the entire health system would benefit from such effort as it will lead to strengthened health systems capable of meeting emergencies. According to Gyapong et al. (2010) important elements need to be working together to effectively integrate NTD control into the health system. These elements include logistics and infrastructural development, human resource development, strategies for delivery of the interventions, information systems including ICT, governance issues, and financing. For the general population, integration is essential as it enhances the accessibility and equity of services (Namadi et al. 2002). Studies have shown that successful integration requires strong leadership, good preparation and planning, and the need to address many context-specific health system hurdles (Namadi et al. 2002; Visschedijk et al. 2003).

As recently argued by Ehrenberg et al. (2020), the integration of NTD programme with existing health systems in resource-poor settings is weak and insufficiently done, and this is seen as one of the underlining factors inhibiting efforts to sustainably control NTDs. It is important for NTD activities at the PHC levels to be routine and mainstreamed into existing health systems programming and governance. In the same vein, more investment to strengthen the health systems for NTD control should adopt a diagonal approach to ensure sustainable impact of these investments in the health systems. This will avoid the usual debate of whether control of NTDs must be done in a vertical approach to achieve maximum impact or horizontally as part of the regular health system. Ultimately, there is the need for NTDs to be delivered as part of universal health coverage goals while enhancing the monitoring and evaluation aspects of their control.

Opportunities and Future Research for Policy

Despite the challenges, there are opportunities for the control of NTDs in developing countries. For example, since the London Declaration in 2012 which galvanised huge global and local support to sustain NTD control up to 2020, major strides were made although a lot of work remains to be done. In order to control and eliminate NTDs in line with the 2021–2030 road map, governments are expected to incorporate NTD control in national health plans, budgets, and development strategies including poverty reduction and to gradually take up significant financing of NTD control in SSA (World Health Organization 2020). Bolstering funding available will help support health systems strengthening efforts especially funding of NTD programmes at the PHC level and other public health systems to ensure delivery of essential and basic medicines and vaccines for NTDs.

Integration of NTD programmes has been found to increase funding for NTD control activities in some countries compared to the vertical programmes. Integration provides the platform for effective use of resources (Hooper et al. 2013; Olds 2013). For example, allowances for volunteers could be provided in such a way that they are not tied to specific programme activities, but rather individual programmes should pool resources and pay uniform amounts to volunteers.

Furthermore, global health partnerships played instrumental role in NTD control and elimination efforts, and it is important for such efforts to be reinvigorated and sustained as part of the road map for NTD elimination and control under the 2021–2030 road map. In relation to health workforce, due to the difficulty in identifying volunteers who are willing to work in urban areas, and the challenge with urban dwellers accepting these volunteers, an ideal approach will be to use health workers (e.g. community health nurses) for NTD control interventions. Furthermore, with the increasing application of ICT in health service delivery, the opportunity for the development of IT-based approaches for data collection and management could be explored.

A critical resource that remains untapped to support NTD control and elimination efforts is the optimisation of the role of communities, opinion leaders, youth groups, civil society organisations/non-government organisations, community-based organisations, and other local/grassroot actors (Madon et al. 2019). Engagement of stakeholders at the community level is critical to the functioning of health systems in respect of service delivery, provision of responsive NTD care, and resource mobilisation and governance of NTD service delivery. New efforts to control NTD should recognise the need to engage local leaders, build partnership, and work closely with them to ensure the health service provides NTD service delivery which is well tailored and responsive to the needs of the communities. Communities affected by NTDs should play a key role in designing NTD control and elimination programmes to ensure ownership and responsiveness of the interventions to their needs, culture, and traditions.

Research has previously established that the availability and uptake of drugs, vaccines, diagnostics, and vector control products play a pivotal role in the prevention and treatment of NTD. In the wake of COVID-19 and the apparently skewed interest and attention on controlling the global pandemic, there would be a tendency to overlook, and place less priority on, essential drugs and diagnostics for NTD control and elimination. It is therefore important for current scholarship to investigate the extent to which the COVID-19 pandemic has affected current and future efforts to control and prevent NTDs. An opportunity to explore health systems reforms is currently needed to leapfrog NTD control and elimination efforts in light of the fact that the COVID-19 pandemic has exacerbated the currently underinvested and fragile health systems in SSA.

Further research is needed to either refine the regime/dosage or identify subsections of endemic populations where the anti-*Wolbachia* therapy will be most appropriate and effective. As country NTD programmes approach elimination stage, there is the need to refine the tools to accurately determine end points for stopping MDA and verify elimination while minimising the risk of recrudescence. This requires further research to inform policy to guide country programmes as well as research in the areas of post-elimination surveillance to serve as early warning system to recrudescence.

Conclusion

A strong and resilient health system remains critical to the control and elimination of NTDs and the Sustainable Development Goals by 2030. Although major gains have been made in efforts to control and eliminate NTDs, critical gaps remain in the health systems which when unaddressed will undermine current gains and impede future progress towards the 2030 target for NTD control and elimination.

While support is being galvanised to improve access to NTD care at the community level, the same efforts are needed to strengthen the health systems to provide high-quality care, and this could partly be achieved through high political commitment and will on the part of national government in partnership with key development partners and other stakeholders. This is important in the light of the funding gaps and the need for more domestic sources of funding to support health systems strengthening activities for NTDs.

Another area of interest to health systems is the integration of NTD care and programmes into routine standardised care delivery processes and protocols. This should also be reflected in national-level efforts and public health governance arrangements, as a way of prioritising NTDs within and across national and local health systems.

References

Allen T, Parker M (2012) Will increased funding for neglected tropical diseases really make poverty history? Lancet 379. https://doi.org/10.1016/S0140-6736(12)60159-7

Amazigo UV, Leak SGA, Zoure HGM, Njepuome N, Lusamba-Dikassa P-S (2012) Community-driven interventions can revolutionise control of neglected tropical diseases. Trends Parasitol 28(6):231–238. https://doi.org/10.1016/j.pt.2012.03.002

Amazigo UV, Leak SG, Zoure HG, Okoronkwo C, Diop Ly M, Isiyaku S et al (2021) Community-directed distributors—The "foot soldiers" in the fight to control and eliminate neglected tropical diseases. PLoS Negl Trop Dis 15(3):e0009088

Anderson RM, Truscott JE, Pullan RL, Brooker SJ, Hollingsworth TD (2013) How effective is school-based deworming for the community-wide control of soil-transmitted helminths? PLoS Negl Trop Dis 7(2):e2027. https://doi.org/10.1371/journal.pntd.0002027

Baatiema L, Sumah AM, Tang PN, Ganle JK (2016) Community health workers in Ghana: the need for greater policy attention. BMJ Glob Health 1(4):e000141

Bergquist R, Whittaker M (2012) Control of neglected tropical diseases in Asia Pacific: implications for health information priorities. Infect Dis Poverty 1(1):3. https://doi.org/10.1186/2049-9957-1-3

Bhattacharyya K, Winch PJ, LeBan K, Tien M (October 2001) Community health worker incentives and disincentives: how they affect motivation, retention, and sustainability. Published by the Basic Support for Institutionalising Child Survival Project (BASICS II) for the United States Agency for International Development, Arlington, Virginia

Bockarie MJ, Taylor MJ, Gyapong JO (2009) Current practices in management of lymphatic filariasis. Expect Rev Anti Infect Ther 7(5):595

Bockarie MJ, Kelly-Hope LA, Rebollo M, Molyneux DH (2013) Preventive chemotherapy as a strategy for elimination of neglected tropical parasitic diseases: endgame challenges. Philos Trans R Soc Lond Ser B Biol Sci 368(1623)

Cavalli A, Bamba SI, Traore MN et al (2010) Interactions between global health initiatives and country health systems: the case of a neglected tropical diseases control program in Mali. PLoS Negl Trop Dis

Chanda P, Hamainza B, Moonga H et al (2011) Relative costs and effectiveness of treating uncomplicated malaria in two rural districts in Zambia: implications for nationwide scale-up of home-based management. Malar J 10:159. http://www.malariajournal.com/content/10/1/159

Chinbuah MA, Kager PA, Abbey M, Gyapong M, Awini E, Nonvignon J, Aduik M, Aikins M, Pagnoni F, Gyapong JO (2012) Impact of community management of fever (using antimalarials with or without antibiotics) on childhood mortality: a cluster-randomized controlled trial in Ghana. Am J Trop Med Hyg 87(suppl 5):11–20

Chu BK, Hooper PJ, Bradley MH, McFarland DA, Ottesen EA (2010) The economic benefits resulting from the first 8 years of the global programme to eliminate lymphatic filariasis (2000–2007). PLoS Negl Trop Dis 4(6). https://doi.org/10.1371/journal.pntd.0000708

Conteh L, Engels T, Molyneux DH (2010) Neglected tropical diseases 4: socioeconomic aspects of neglected tropical diseases. Lancet 375:239–247

Dabo A, Bary B, Kouriba B, Sankaré O, Doumbo O (2013) Factors associated with coverage of praziquantel for schistosomiasis control in the community direct intervention (CDI) approach in Mali (West Africa). Infect Dis Poverty 2(11)

da-Costa Vroom FB et al (2015) Data reporting constraints for the lymphatic filariasis mass drug administration activities in two districts in Ghana: a qualitative study. SAGE Open Med 3. https://doi.org/10.1177/2050312115594083

de Savigny D, Adam T (eds) (2009) Systems thinking for health systems strengthening. Alliance for Health Policy and Systems Research. WHO, Geneva

Dean L, Ozano K, Adekeye O, Dixon R, Fung EG, Gyapong M et al (2019) Neglected tropical diseases as a 'litmus test' for universal health coverage? Understanding who is left behind and why in mass drug administration: lessons from four country contexts. PLoS Negl Trop Dis 13(11):e0007847. https://doi.org/10.1371/journal.pntd.0007847

Dembélé M, Bamani S, Dembélé R, Traoré MO, Goita S, Traoré MN, Zhang Y (2012) Implementing preventive chemotherapy through an integrated National Neglected Tropical Disease Control Program in Mali. PLoS Negl Trop Dis 6(3):e1574

Dil Y, Strachan D, Cairncross S, Seidu Korkor A, Hill Z (2012) Motivations and challenges of community-based surveillance volunteers in the northern region of Ghana. J Commun Health 37. https://doi.org/10.1007/s10900-012-9569-5

Ehrenberg JP, Zhou XN, Fontes G, Rocha EM, Tanner M, Utzinger J (2020) Strategies supporting the prevention and control of neglected tropical diseases during and beyond the COVID-19 pandemic. Infect Dis Poverty 9(04):135–141

Emukah EC et al (2008) Factors affecting the attrition of community-directed distributors of ivermectin, in an onchocerciasis-control programme in the Imo and Abia states of South-Eastern Nigeria. Ann Trop Med Parasitol 102(1):45–51. https://doi.org/10.1179/136485908x252241

Fitzpatrick et al (2017) An investment case for ending neglected tropical diseases. Diseases control priority 3 (DCP3). International Bank for Reconstruction and Development/World Bank, Washington, DC

Freeman MC et al (2013) Integration of water, sanitation, and hygiene for the prevention and control of neglected tropical diseases: a rationale for inter-sectoral collaboration. PLoS Negl Trop Dis 7(9). https://doi.org/10.1371/journal.pntd.0002439.t001

Global Burden of Disease Health Financing Collaborator Network (2020) Health sector spending and spending on HIV/AIDS, tuberculosis, and malaria, and development assistance for health: progress towards sustainable development goal 3. Lancet 396(10252):693–724. https://doi.org/10.1016/S0140-6736(20)30608-5

Goldman AS et al (2007) National mass drug administration costs for lymphatic filariasis elimination. PLoS Negl Trop Dis 1(1):e67. https://doi.org/10.1371/journal.pntd.0000067

Gyapong JO, Gyapong M, Yellu N, Anakwah K, Amofah G, Bockarie M, Adjei S (2010) Integration of control of neglected tropical diseases into health-care systems: challenges and opportunities. Lancet 375:160–165

Heymann J, Cassola A, Raub A, Mishra L (2013) Constitutional rights to health, public health, and medical care: the status of health protections in 191 countries. Glob Public Health 8(6):639–653. https://doi.org/10.1080/17441692.2013.810765

HISP (2015) DHIS 2 In Action. Retrieved 04/08/2015, from https://www.dhis2.org/inaction

Hodges MH et al (2010) High coverage of mass drug administration for lymphatic filariasis in rural and non-rural settings in the Western area, Sierra Leone. Parasit Vectors 3:120. https://doi.org/10.1186/1756-3305-3-120

Hodges ME, Koroma JB, Sonnie M, Kennedy N, Cotter E, MacArthur C (2011) Neglected tropical disease control in post-war Sierra Leone using the onchocerciasis control programme as a platform. Int Health 3(2):69–74. https://doi.org/10.1016/j.inhe.2011.03.003

Hoerauf A et al (2008) Wolbachia endobacteria depletion by doxycycline as antifilarial therapy has macrofilaricidal activity in onchocerciasis: a randomized placebo-controlled study. Med Microbiol Immunol 197(3):295–311. https://doi.org/10.1007/s00430-007-0062-1

Hooper PJ et al (2013) The effects of integration on financing and coverage of neglected tropical disease programs. Am J Tropical Med Hygiene 89(3):407–410. https://doi.org/10.4269/ajtmh.13-0018

Hotez PJ, Pecoul B (2010) "Manifesto" for advancing the control and elimination of neglected tropical diseases. PLoS Negl Trop Dis 4(5):e718. https://doi.org/10.1371/journal.pntd.0000718.t001

Huang X, Toure H, Biswas G (2021) Resource tracking for neglected tropical disease programmes: the first step for developing a sustainable financing strategy. Trans R Soc Trop Med Hyg 115(2):179–181

Kidane G, Morrow RH (2000) Teaching mothers to provide home treatment of malaria in Tigray, Ethiopia: a randomised trial. Lancet 356:550–555

Kruk ME, Gage AD, Arsenault C, Jordan K, Leslie HH, Roder-DeWan S, Adeyi O, Barker P, Daelmans B, Doubova SV, English M, García-Elorrio E, Guanais F, Gureje O, Hirschhorn LR, Jiang L, Kelley E, Lemango ET, Liljestrand J, Malata A, Marchant T, Matsoso MP, Meara JG, Mohanan M, Ndiaye Y, Norheim OF, Reddy KS, Rowe AK, Salomon JA, Thapa G, Twum-Danso NAY, Pate M (2018) High-quality health systems in the sustainable development goals era: time for a revolution. Lancet Glob Health 6(11):e1196–e1252. https://doi.org/10.1016/S2214-109X(18)30386-3. Epub 2018 Sep 5

Lubell Y, Mills AJ, Whitty CJM, Staedke SG (2010) An economic evaluation of home management of malaria in Uganda: an interactive markov model. PLoS One 5:e12439

Macha J et al (2012) Factors influencing the burden of health care financing and the distribution of health care benefits in Ghana, Tanzania, and South Africa. Health Policy Plan 27(Suppl 1):i46–i54. https://doi.org/10.1093/heapol/czs024

Madon S, Malecela MN, Mashoto K, Donohue R, Mubyazi G, Michael E (2018) The role of community participation for sustainable integrated neglected tropical diseases and water, sanitation, and hygiene intervention programs: a pilot project in Tanzania. Soc Sci Med 202:28–37

Madon I, Drev D, Likar J (2019) Long-term risk assessments comparing environmental performance of different types of sanitary landfills. Waste Manage 96:96–107

Makepeace BL, Rodgers L, Trees AJ (2006) Rate of elimination of Wolbachia pipientis by doxycycline in vitro increases following drug withdrawal. Antimicrob Agents Chemother 50(3):922–927. https://doi.org/10.1128/AAC.50.3.922-927.2006

Manderson L, Aagaard-Hansen J, Allotey P, Gyapong M, Sommerfeld J (2009) Social research on neglected diseases of poverty: continuing and emerging themes. PLoS Negl Trop Dis 3(2):332. https://doi.org/10.1371/journal

Miller JS, Musominali S, Baganizi M, Paccione GA (2014) A process evaluation of performance-based incentives for village health workers in Kisoro district, Uganda. Hum Resour Health 12:19. https://doi.org/10.1186/1478-4491-12-19

Mills A, Ally M, Goudge J, Gyapong J, Mtei G (2012) Progress towards universal coverage: the health systems of Ghana, South Africa, and Tanzania. Health Policy Plan 27:i4–i12. https://doi.org/10.1093/heapol/czs002

Molyneux DH, Hotez PJ, Fenwick A, Newman RD, Greenwood B, Sachs J (2009) Neglected tropical diseases and the Global Fund. Lancet 373

Moss T, Majerowicz S (2012) No longer poor: Ghana's new income status and implications of graduation from IDA. Center for Global Development Working Paper 300. Accessed from www.cgdev.org

Namadi A, Visschedijk J, Samson K (2002) The leprosy elimination campaign in Jigawa, Nigeria: an opportunity for integration. Lepr Rev 73(2):138–146

Narain JP et al (2010) Elimination of neglected tropical diseases in the South-East Asia region of the World Health Organization. Bull World Health Organ 88(3):206–210. https://doi.org/10.2471/BLT.09.072322

Njomo DW, Amuyunzu-Nyamongo M, Magambo JK, Njenga SM (2012) The role of personal opinions and experiences in compliance with mass drug Administration for Lymphatic Filariasis Elimination in Kenya. PLoS One 7(11):e48395. https://doi.org/10.1371/journal.pone.0048395

Nonvignon J, Chinbuah MA, Gyapong M, Abbey M, Awini E, Gyapong JO, Aikins M (2012) Is home management of fevers a cost-effective way of reducing under-five mortality in Africa? The case of a rural Ghanaian district. Trop Med Int Health. https://doi.org/10.1111/j.1365-3156.2012.03018.x

NTDs UtC (2015) Country leadership and collaboration on neglected tropical diseases: Third Report of London Declaration

Olds GR (2013) Deworming the world. Trans Am Clin Climatol Assoc 124

Olliaro P, Vaillant MT, Sundar S, Balasegaram M (2012) More efficient ways of assessing treatments for neglected tropical diseases are required: innovative study designs, new endpoints, and markers of effects. PLoS Negl Trop Dis 6(5):e1545. https://doi.org/10.1371/journal.pntd.0001545.g001

Oprea L, Braunack-Mayer A, Gericke CA (2009) Ethical issues in funding Research and Development of drugs for neglected tropical diseases. J Med Ethics 35(5)

Ottersen OP, Dasgupta J, Blouin C, Buss P, Chongsuvivatwong V, Frenk J, et al (2014) The political origins of health inequity: prospects for change, Report of the Lancet-University of Oslo Commission on Global Health Governance. Accessed from https://doi.org/10.1016/S0140-6736(13)62712-9, https://doi.org/10.1016/S0140-6736(14)60050-7

Parker M, Allen T (2011) Does mass drug administration for the integrated treatment of neglected tropical diseases really work? Assessing evidence for the control of schistosomiasis and soil-transmitted helminths in Uganda. Health Res Policy Sys BioMed Central 9:3. https://doi.org/10.1186/1478-4505-9-3

Parker M, Allen TIM (2013) Will mass drug administration eliminate lymphatic filariasis? Evidence from northern coastal Tanzania. J Biosoc Sci 45(4):517–545. https://doi.org/10.1017/S0021932012000466

Pavluck A, Chu B, Mann Flueckiger R, Ottesen E (2014) Electronic data capture tools for Global Health programs: evolution of links, an android- web-based system. PLoS Negl Trop Dis 8(4):e2654. https://doi.org/10.1371/journal.pntd.0002654

Pollastri MP (2014) Finding new collaboration models for enabling neglected tropical disease drug discovery. PLoS Negl Trop Dis 8(7):e2866. https://doi.org/10.1371/journal.pntd.0002866
Reich MR, Takemi K, Roberts MJ, Hsiao WC (2008) Global action on health systems: a proposal for the Toyako G8 summit. Lancet 371(9615):865–869. https://doi.org/10.1016/s0140-6736(08)60384-0
Schreuder B, Kostermans C (2001) Global health strategies versus local primary health care priorities—a case study of national immunisation days in southern Africa. S Afr Med J 91(3):249–254
Shakarishvili G, Atun R, Berman P, Hsiao W, Burgess C, Lansang MA (2010) Converging health systems frameworks: towards a concepts-to-actions roadmap for health systems strengthening in low and middle income countries. Global Health Governance III(2)
Sirima SB, Konate A, Tiono AB et al (2003) Early treatment of childhood fevers with pre-packaged antimalarial drugs in the home reduces severe malaria morbidity in Burkina Faso. Trop Med Int Health 8:133–139
Slingsby BT, Kurokawa K (2013) The Global Health innovative technology (GHIT) fund: financing medical innovations for neglected populations. Lancet Glob Health 1(4):e184–e185. https://doi.org/10.1016/s2214-109x(13)70055-x
Solomon AW, Engels D, Bailey RL, Blake IM, Brooker S, Chen JX, Peeling RW (2012) A diagnostics platform for the integrated mapping, monitoring, and surveillance of neglected tropical diseases: rationale and target product profiles. PLoS Negl Trop Dis 6:e1746
Stanaway JD, Parisi A, Sarkar K, Blacker BF, Reiner RC, Hay SI, Crump JA (2019) The global burden of non-typhoidal salmonella invasive disease: a systematic analysis for the global burden of disease study 2017. Lancet Infect Dis 19(12):1312–1324
Stanton MC, Mkwanda SZ, Debrah AY, Batsa L, Biritwum NK, Hoerauf A, Kelly-Hope LA (2015) Developing a community-led SMS reporting tool for the rapid assessment of lymphatic filariasis morbidity burden: case studies from Malawi and Ghana. BMC Infect Dis 15:214. https://doi.org/10.1186/s12879-015-0946-4
Stefanakis R, Robertson AS, Ponder EL, Moree M (2012) Analysis of neglected tropical disease drug and vaccine development pipelines to predict issuance of FDA priority review vouchers over the next decade. PLoS Negl Trop Dis 6(10):e1803. https://doi.org/10.1371/journal.pntd.0001803.g001
Stillman K, Bennett S (2005) System-wide effects of the Global Fund: interim findings from three country studies. The Partners for Health Reformplus, Abt Associates Inc, Bethesda, MD
Takasugi T, Lee ACK (2012) Why do community health workers volunteer? A qualitative study in Kenya. Public Health 126(10)Visschedijk J, Engelhard A, Lever P et al (2003) Leprosy control strategies and the integration of health services: an international perspective. Cad Saude Publica, Rio de Jeneiro 19(6):1567–1581
World Health Organization (1978) Declaration of Alma Ata. Paper presented at the International Conference on Primary Health Care, Alma Ata, Russia
World Health Organization (2006) Preventive chemotherapy in human helminthiasis: coordinated use of anthelminthic drugs in control interventions: a manual for professionals and programme managers. World Health Organization, Geneva
World Health Organization (2007a) Everybody business: strengthening health systems to improve health outcomes: WHO's framework for action. World Health Organization, Geneva
World Health Organization (2007b) Global plan to combat neglected tropical diseases, 2008–2015. World Health Organization, Geneva
World Health Organization (2010a) First WHO report on neglected tropical diseases: working to overcome the global impact of neglected tropical diseases. World Health Organization, Geneva
World Health Organization (2010b) In: David WT, Crompton abMPP (eds) Working to overcome the global impact of neglected tropical diseases: first WHO report on neglected tropical diseases. World Health Organization, Geneva
World Health Organization (2011a) Monitoring and epidemiological assessment of mass drug administration in the global programme to eliminate lymphatic filariasis: a manual for national elimination programmes. World Health Organization, Geneva

World Health Organization (2011b) The Abuja Declaration ten years on. http://www.who.int/healthsystems/publications/Abuja10.pdf
World Health Organization (2012) Accelerating work to overcome the global impact of neglected tropical diseases—a roadmap for implementation. World Health Organization, Geneva
World Health Organization (2015a) Neglected Tropical Diseases. http://www.who.int/neglected_diseases/about/en/. Accessed 12 Aug 2015
World Health Organization (2015b) WHO certifies Ghana free of dracunculiasis. http://www.who.int/dracunculiasis/Ghana_free_of_dracunculiasis/en/. Accessed 06 Aug 2015
World Health Organization (2016) WHO Research and Development Blueprint: evaluation of ideas for potential platforms to support development and production of health technologies for priority infectious diseases with epidemic potential (No. WHO/EMP/RHT/TSN/2016.02). World Health Organization
World Health Organization (2020) Considerations for implementing mass treatment, active case-finding and population- based surveys for neglected tropical diseases in the context of the COVID-19 pandemic: interim guidance
World Health Organization (2021) Ending the neglect to attain the Sustainable Development Goals: a road map for neglected tropical diseases 2021–2030. https://www.who.int/publications/i/item/9789240010352. cited 2021 February 26
Worrell C, Mathieu E (2012) Drug coverage surveys for neglected tropical diseases: 10 years of field experience. Am J Trop Med Hyg 87(2):216–222. https://doi.org/10.4269/ajtmh.2012.12-0167

Index

J. O. Gyapong, B. A. Boatin (eds.), *Neglected Tropical Diseases - Sub-Saharan Africa*, Neglected Tropical Diseases, https://doi.org/10.1007/978-3-031-53901-5

D

GPSR Compliance

The European Union's (EU) General Product Safety Regulation (GPSR) is a set of rules that requires consumer products to be safe and our obligations to ensure this.

If you have any concerns about our products, you can contact us on ProductSafety@springernature.com

In case Publisher is established outside the EU, the EU authorized representative is:

Springer Nature Customer Service Center GmbH
Europaplatz 3
69115 Heidelberg, Germany

Batch number: 10761599

Printed by Printforce, the Netherlands